Developments in environmental biology of fishes 17

Series Editor
EUGENE K. BALON

Sturgeon biodiversity and conservation

Editors:

VADIM J. BIRSTEIN, JOHN R. WALDMAN & WILLIAM E. BEMIS

Reprinted from *Environmental biology of fishes*, Volume 48 (1–4), 1997
with addition of species and subject index

KLUWER ACADEMIC PUBLISHERS
DORDRECHT / BOSTON / LONDON

A C.I.P. Catalogue record for this book is available from the Library of Congress.

ISBN 0-7923-4517-7

Published by Kluwer Academic Publishers,
P.O. Box 17, 3300 AA Dordrecht, The Netherlands

Kluwer Academic Publishers incorporates
the publishing programmes of
D. Reidel, Martinus Nijhoff, Dr W. Junk and MTP Press.

Sold and distributed in the U.S.A. and Canada
by Kluwer Academic Publishers,
101 Philip Drive, Norwell, MA 02061, U.S.A.

In all other countries, sold and distributed
by Kluwer Academic Publishers,
P.O. Box 322, 3300 AH Dordrecht, The Netherlands

Cover design by William E. Bemis
inspired by figure of lake sturgeon, *Acipenser fulvescens*, in M.B. Trautman (1981) 'The fishes of Ohio'.

Printed on acid-free paper

Printed in the Netherlands

Contents

This volume is dedicated to the memory of Leo Semenovich Berg

Logo design by William E. Bemis inspired by the holarctic distribution of sturgeons and paddlefishes

Numae is Anishnabe for sturgeon, a painting by the Chippewas of Nawash artist Adrian Nadjiwon of Cape Croker, Lake Huron: "In the painting I wanted to depict the balance that one existed between my ancestors and the sturgeon (the sturgeon being symbolic of all the fish species that the Old People subsisted on) with the fishing line symbolising a spiritual link and balancing point between the man and the sturgeon" (1992).

Portraits of three juvenile sturgeons which originated from the Black Sea stock bred at the Propa-Gen International Aquaculture Produ tion R & D and Trading, Komadi, Hungary: the top photograph shows a 71 cm armored form of the Russian sturgeon, *Acipenser guelde staedtii*; the middle photograph shows a 77 cm naked form of *A. guldenstaedtii* (note its smooth skin and absence of scutes); and th bottom photograph is of a 54 cm *A. persicus*. Photographs by Paul Vecsei, May 1996.

Environmental Biology of Fishes **48**: 9–12, 1997.

Prelude to sturgeon biodiversity and conservation

Soaring Eagle, you have angered Great Sturgeon, and he has taken your son because you took too many fish from the lake.

Joe McLellan (1993)
in 'Nanabosho, Soaring Eagle and the Great Sturgeon', Pammican Publications, Winnipeg.

It gave me great pleasure to help with the publication of this volume. The elusive beasts it deals with sustained my enthusiasm in the early days of work on the Danube River. Already then, 40 years ago, sturgeons were so rare[1] in the middle Danube (Figure 1), that I had to shift my attention to the wild carp[2,3] which have followed since many of the sturgeons into oblivion.

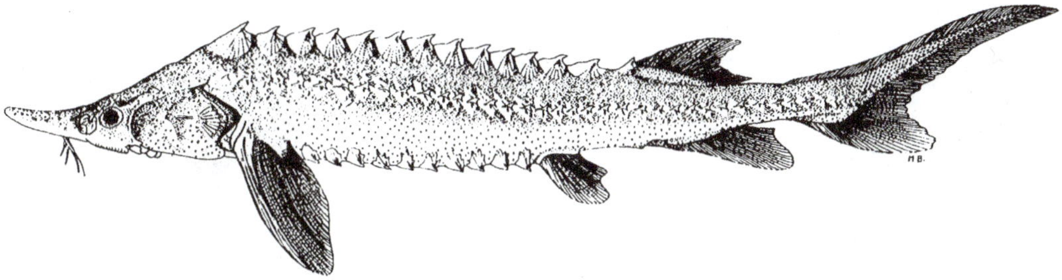

Figure 1. Specimen of *Acipenser gueldenstaedtii* 35 cm long caught on 30.7.1967 in the Danube River near Radvaň (river km 1749). Original drawing by Miriam Baradlai[1].

Often I was standing on the shore, where about 100 years ago, the Viennese court had their military fire cannon balls at giant beluga, *Huso huso*, and other sturgeon species ascending the river to spawn[4]. I was dreaming of arresting the present destruction[5,6] and of rebuilding the past (Figure 2). I envied Marsilius[7] who was able to witness here the processing of caviar (Figure 3) which I was privileged to spoon up in Russia only 30 years, no, 270 years later, and became addicted to.

---

[1] Balon, E.K. 1968. Ďalší nález mlade *Acipenser güldenstaedti colchicus* Marti, 1940 v československom úseku Dunaja (A further discovery of a juvenile Russian sturgeon in the Czechoslovak part of the Danube). Ac. Rer. Natur. Mus. Nat. Slov. (Bratislava) 14: 95–100.

[2] Balon, E.K. 1974. Domestication of the carp *Cyprinus carpio* L. Royal Ontario Mus. Life Sci. Misc. Publ., Toronto. 37 pp.

[3] Balon, E.K. 1995. The common carp, *Cyprinus carpio*: its wild origin, domestication in aquaculture, and selection as colored nishikigoi. Guelph Ichthyology Reviews 3: 1–55.

[4] Hensel, K. & J. Holčík. 1997. Past and current status of sturgeons in the upper and middle Danube River. Env. Biol. Fish. 48 (this volume).

[5] Balon, E.K. 1967. Vývoj ichtyofauny Dunaja, jej súčasný stav a pokus o prognózu d'alších zmien po výstavbe vodných diel (Evolution of the Danube ichthyofauna, its recent state and an attempt to predict further changes after the construction of the planned hydro-electric power stations and diversion schemes). Biologické práce 13: 1–121 + 24 plates.

[6] Balon, E.K. 1968. Einfluß des Fischfangs auf die Fischgemeinschaften der Donau. Arch. Hydrobiol. (Suppl. Donauforschung 3) 34: 228–249.

[7] Marsilius, A.F.C. 1726. Danubius Pannonico-Mysicus, observationibus geographicis, astronomicis, hydrographicis, historicis, physicis perlustratus et in sex Tomos digestus. Hagae Comitum, Amstelodami.

I like this volume for other reasons too. Sturgeons and paddlefishes seem to exist, like most modern fishes, in altricial and precocial forms. Becoming aware of this may inspire future workers to pay closer attention to these life history and evolutionary phenomena[8]. By doing so, the general usefulness of the theory of saltatory ontogeny[9] and of dichotomous processes in both development and evolution[10] may become more widely accepted.

Figure 2. Reconstruction of a method to catch beluga on the Danube according to description in Aelianus by Rohan-Csermak (1963) from Balon (1967)[5].

The Guest Editors of this volume agreed in advance on the order listed on the title page, although equal editorial effort should be recognized. Vadim Birstein, Robert Boyle and John Waldman convened the 1994 conference that was the source for the original drafts of most of the manuscripts included in this collection. William Bemis put the volume in final shape; he is also responsible for safeguarding the standards of all contributions as the editorial representative of *Environmental Biology of Fishes*. Alice Klingener prepared the index with suggestions and help from William Bemis and Vadim Birstein.

While this volume was being assembled a young scientific illustrator – Paul Vecsei of Montreal – became a sturgeon fanatic (Figure 4) and travelled to the Czech Republic, Hungary and Romania to obtain material for illustrations. Our second frontispiece and drawings reproduced on pp. 72, 156, 184, 208, 220, 230, 240, 290, 310, 384, 406 and 436 are only a few selected examples of his ambitious plan to illustrate the sturgeons of the world. His results on these pages may be compared to samples of earlier, historical illustrations of sturgeons on the leading pages of the three parts into which this volume is divided.

I would like to thank Paul Vecsei for his timely completion of the illustrations, and René Mijs for his exceptional understanding and patience. David Noakes, beside other help, is to be thanked for drawing my attention to the native American source of the motto used in this Prelude, Steve Crawford for bringing to my attention and for arranging permission from Adrian Nadjiwon, the Native American artist to use his painting as the first frontispiece, Tony Lelek for information on the Hungarian-Russian enterprise at Komadi, and Christine Flegler-Balon for the many corrections and administrative assistance.

Guelph, 1 August 1996 — Eugene K. Balon

[8] Bemis, W.E. & B. Kynard. 1997. Sturgeon rivers: and introduction to acipenseriform biogeography and life history. Env. Biol. Fish. 48 (this volume).

[9] Balon, E.K. 1986. Saltatory ontogeny and evolution. Rivista di Biologia/Biology Forum 79: 151–190 (in English and Italian).

[10] Balon, E.K. 1989. The epigenetic mechanisms of bifurcation and alternative life-history styles. pp. 467–501. *In*: M.N. Bruton (ed.) Alternative Life-History Styles of Animals, Perspectives in Vertebrate Science 6, Kluwer Academic Publishers, Dordrecht.

Figure 3. Sturgeon fishing, butchering and caviar processing on the Danube River at the time of Marsilius (1726)[7], from Balon (1967)[5].

Figure 4. Paul Vecsei on the Fraser River (Canada) holding a white sturgeon, *Acipenser transmontanus*. Photograph by Eugene Hoyano, August 1990.

The illustrator's note

Much time has passed since those rainy nights on the Fraser River. In the meantime my effort to illustrate charrs of the world was initiated and delivered with pride to E.K. Balon. It caught him in the middle of a new project, the present volume on sturgeons. To join in, I took off to Europe, visiting the sturgeon farm in Hungary and ultimately the Grigore Antipa Natural History Museum in Bucharest…

I consider dot stippling an unmatched medium in which important characters can be best emphasised. Most of my illustrations of sturgeons in this volume are images of live specimens, some originally from the wild now kept in ponds, others already hatchery offspring of wild caught parents. A few are from museum specimens preserved wet. The live specimens were anesthesized, laid out on a wet surface and then photographed. Care was taken to avoid parallax distortions by using long focal length lenses.

The resulting slides were projected to facilitate enlarged drawings and detailed rendering of all structures. Often light glare on mucus or wet surfaces made some structures invisible on the photographs. Theses structures had to be drawn from other frames of the same specimen. Some heads were enlarged up to 15 times so that the finished drawings contain more information than can be seen by an unaided eye. For example, the two heads of *Acipenser gueldenstaedtii* on page 436 represent over 100 hours of work. The complete views were done only about 45 cm long in order to present enough details when reproduced at 33% of their original size. My illustrations are exactly what you would see if you step far enough to avoid parallax distortions and block one eye in order to loose your stereoscopic vision.

15 August 1996

Paul Vecsei

Environmental Biology of Fishes **48**: 13–14, 1997.

Sturgeon biodiversity and conservation: an introduction

William E. Bemis[1], Vadim J. Birstein[2] & John R. Waldman[3]
[1] *Department of Biology and Graduate Program in Organismic and Evolutionary Biology, University of Massachusetts, Amherst, MA 01003, U.S.A.*
[2] *The Sturgeon Society, 331 West 57th Street, Suite 159, New York, NY 10019, U.S.A.*
[3] *Hudson River Foundation, 40 West 20th Street, Ninth Floor, New York, NY 10011, U.S.A.*

Key words: Acipenseriformes, Acipenseridae, Polyodontidae, status

This volume includes many of the papers presented at the International Conference on Sturgeon Biodiversity and Conservation which took place at The American Museum of Natural History (AMNH), New York, on 28–30 July 1994. The main goal of the Conference was to attract attention to sturgeons and paddlefishes, still the most speciose group of 'living fossil' fishes, but now fast disappearing from our planet (Birstein 1993, Bemis & Findeis 1994, Waldman 1995).

Some presentations at the conference described basic aspects of acipenseriform biology, including evolution, genetics, and life cycles. Others focused on the contemporary status of a particular species or a few species inhabiting the same basin or region; most of these contributions also addressed ongoing conservation efforts. Still other speakers examined current controversies at the interface between science and society, bringing information from a variety of sources to enrich our meeting. These three approaches are reflected by the three part organization of this volume: Part 1, Diversity and evolution; Part 2, Biology and status reports; and Part 3, Controversies, conservation and summary. We hope that the included papers offer a broad perspective about contemporary work on the phylogeny of Acipenseriformes, as well as a review of the worldwide status of almost all of the species constituting this order.

In preparing the materials for publication, we discovered several revisions in the scientific names of some species. Smith & Clugston (1997 this volume) follow Gilbert (1992), who showed that the name of the American Atlantic sturgeon has been frequently misspelled in the literature and that the original correct spelling is *Acipenser oxyrinchus* (instead of the commonly used *A. oxyrhynchus*). Ruban (1997 this volume) returns to the original spelling of the scientific name of the Siberian sturgeon, *A. baerii* (instead of *A. baeri*). We standardized the spelling of these species throughout the volume. Also, Birstein et al. (1997 this volume) presented genetic data showing that the Sakhalin sturgeon, usually considered as the same species as the American green sturgeon *A. medirostris* or as its Asian subspecies *A. medirostris mikadoi*, is in fact a distinct species, *A. mikadoi*, as it was described originally (Hilgendorf 1892). Additional treatment of these and other questions is taken up by Birstein & Bemis (1997 this volume).

Because the materials presented in different papers cover a wide geographical range, literally the whole northern hemisphere, we tried to be consistent about geographic names and to follow (insofar as possible) one resource for names. We used the New York Times Atlas (1992) as our guide for unifying geographical names throughout the volume. The biogeography of sturgeons has intrigued zoologists for more than two hundred years, and to unify comments and analyses presented by the authors of the status papers on separate species of Acipenseriformes, we wrote a new contribution overviewing the biogeography of the entire group (Bemis & Kynard 1997).

In addition to our primary affiliations, all three of us benefit from a network of institutions committed to the scientific study of fossil and recent fishes, and wish to thank our colleagues at these institutions by formally noting our courtesy appointments with them. William E. Bemis is a Research Associate in the Department of Ichthyology at the American Museum of Natural History, New York and a Research Associate in the Department of Geology, Field Museum of Natural History, Chicago. Vadim J. Birstein is a senior scientist at the Koltsov Institute of Developmental Biology, Russian Academy of Sciences, Moscow, a visiting scientist at the American Museum of Natural History, New York and Adjunct Professor of Biology at the University of Massachusetts, Amherst. John R. Waldman is a Research Associate in the Department of Ichthyology at the American Museum of Natural History, New York.

We are grateful to all persons and organizations who helped Vadim Birstein, John Waldman and Robert H. Boyle to organize the 1994 conference. Clay Hiles, Executive Director of the Hudson River Foundation for Science and Environmental Research (HRF, New York), and Robert Boyle, Co-Chairman of the Conference, Chairman of the Board of The Sturgeon Society (New York), and a member of the Board of Directors of the HRF, arranged funding for the conference through the HRF, the principal financial supporter of the conference. The President of the American Museum of Natural History (AMNH, New York), Ellen Futter, and Provost, Michael Novacek, encouraged us and provided the Kaufmann Theater of the Museum for the meetings. We also thank Thomas Lovejoy, Assistant Secretary for Environment and External Affairs of the Smithsonian Institution (Washington), and Joel Cracraft, Curator (Department of Ornithology, AMNH), for welcoming the participants at the opening ceremony. Also, we are thankful to Bill Murray, the actor, comedian, and supporter of aquatic environmental causes, for attending our opening ceremony and making a generous donation to the work of The Sturgeon Society. Stolt Sea Farms (California), provided aquacultured white sturgeon caviar (as an alternative to wild sturgeon caviar) for the conference. Pat Yazgi, President of Friends of Fishes (New York), organized two successful evening events. Finally, we thank Eugene Balon, the Editor-in-Chief of the journal Environmental Biology of Fishes, for his kind collaboration in publishing the materials of the conference as dedicated issues of the journal and a separate volume of Developments in EBF 17.

References cited

Bemis, W.E. & E.K. Findeis. 1994. The sturgeons' plight. Nature 370: 602.

Bemis, W.E. & B. Kynard. 1997. Sturgeon rivers: an introduction to acipenseriform biogeography and life history. Env. Biol. Fish. (this volume).

Birstein, V.J. 1993. Sturgeons and paddlefishes: threatened fishes in need of conservation. Cons. Biol. 7: 773–787.

Birstein, V.J. & W.E. Bemis. 1997. How many species are there in the genus *Acipenser*? Env. Biol. Fish. (this volume).

Birstein, V.J., R. Hanner & R. DeSalle. 1997. Phylogeny of the Acipenseriformes: cytogenetic and molecular approaches. Env. Biol. Fish. (this volume).

Gilbert, C.R. 1992. Atlantic sturgeon. pp. 31-39. *In:* R.A. Ashton (ed.) Rare and Endangered Biota of Florida, Vol. 2, University of Florida, Gainesville.

Hilgendorf, F. 1892. Über eine neue Stör-Art aus Nord-Japan (*Acipenser mikadoi*). Sitzungsber. Ges. naturf. Freunde, Berlin 7: 98–100.

Ruban, G.I. 1997. Species structure, contemporary distribution and status of the Siberian sturgeon, *Acipenser baerii*. Env. Biol. Fish. (this volume).

Smith, T.I.J. & J.P. Clugston. 1997. Status and management of Atlantic sturgeon, *Acipenser oxyrinchus*, in North America. Env. Biol. Fish. (this volume).

Waldman, J. 1995. Sturgeons and paddlefishes: a convergence of biology, politics, and greed. Fisheries 20: 20–21, 49.

Environmental Biology of Fishes **48**: 15–22, 1997.

Leo Semenovich Berg and the biology of Acipenseriformes: a dedication

Vadim J. Birstein[1] & William E. Bemis[2]
[1] *The Sturgeon Society, 331 West 57th Street, Suite 159, New York, NY 10019, U.S.A.*
[2] *Department of Biology and Graduate Program in Organismic and Evolutionary Biology, University of Massachusetts, Amherst, MA 01003, U.S.A.*

Received 5.3.1996 Accepted 23.5.1996

Key words: T. Dobzhansky, A. Sewertzoff, T. Lysenko, Paleonisciformes, biogeography

This volume is dedicated to the memory of Leo Semenovich Berg (1876–1950), a Russian ichthyologist and geographer. In the foreword to the English translation of Berg's remarkable treatise, 'Nomogenesis or evolution according to law', Theodosius Dobzhansky wrote: 'Berg was one of the outstanding intellects among Russian scientists. The breadth of his interests and the depth as well as the amplitude of his scholarship were remarkable. He had the reputation of being a 'walking library', because of the amount of information he could produce from his memory' (Dobzhansky 1969, p. xi). Berg was prolific, publishing 217 papers and monographs on ichthyology, 30 papers on general zoology and biology, 20 papers on paleontology, 32 papers on zoogeography, 320 papers and monographs on geography, geology, and ethnography, as well as 290 biographies, obituaries, and popular articles (Berg 1955, Sokolov 1955).

Berg was born 120 years ago, on 14 March 1876, in the town of Bendery. According to laws of the Russian Empire, Berg could not enter the university as a Jew, so he was baptized and became a Lutheran, which allowed him to study and receive his diploma in zoölogy at the Moscow University in 1898. From 1899 to 1904, he explored the fisheries and the general ecology of the Aral Sea and lakes in Turkestan and western Siberia. In 1904, Berg was appointed curator of the Ichthyology Department of the Zoological Museum (later Zoological Institute) at the Academy of Sciences in St. Petersburg. Later he held several positions in this and other institutions (Shapovalov 1951, Oliva 1951, 1952, Holčík 1976, Lindberg 1976, Oliva & Holčík 1977, 1978). As one of the most talented biologists of his time, Berg was a target of Trofim Lysenko and his followers. In January 1939, after discrediting Berg and an outstanding geneticist Nicolai Koltsov in the press, Lysenko and his accomplice, Nikolai Tsitsin, were elected in their stead as members of the Soviet Academy of Sciences. Berg was never formally recognized by the Soviet Academy for his accomplishments in biology, and only later (1946) was he elected a member of the Geography Branch of the Soviet Academy of Sciences (Figure 1).

Sturgeons and the order Acipenseriformes were a central theme in Berg's theoretical works and papers on systematics and zoogeography (Andriyashev 1955, Lindberg 1976). In December 1936, he addressed a meeting of the Biology Branch of the Soviet Academy of Sciences on 'Classification of fishes, both living and fossil'. This fundamental work was published in Russian in 1940, although some general ideas in a short form appeared earlier in English and French (Berg 1935a, 1937). The entire book was translated into English in 1947 (Berg 1947a, 1965). It was the most comprehensive study of its era on systematics and evolution of fossil and recent fishes, and it remains useful. An additional chapter, entitled 'On the position of Polypteridae in the system of fishes' appeared as a separate paper

Figure 1. Leo Semenovich Berg, ichthyologist and biogeographer.

the same year (Berg 1947b). In 1948, Berg published a second additional chapter, 'On the position of Acipenseriformes in the system of fishes'. These two chapters, as well as additional new material on fossil fishes, were included in the second Russian edition of the book which appeared only in 1955, after the author's death.

Unfortunately, the chapter on the Acipenseriformes was never translated into English. In 50 pages, Berg described the morphology, anatomy, and embryology of Acipenseriformes, comparing them to extinct Paleonisciformes and modern Elasmobranchii (Figure 2). Berg's conclusions contradicted the theory introduced by Aleksei Sewertzoff (1925, 1926, 1928), who considered acipenseriforms to be closely related to elasmobranchs. Berg wrote: 'Acipenseriformes belong to the same group of fishes as the Paleonisciformes, i.e., to the primitive Ac-

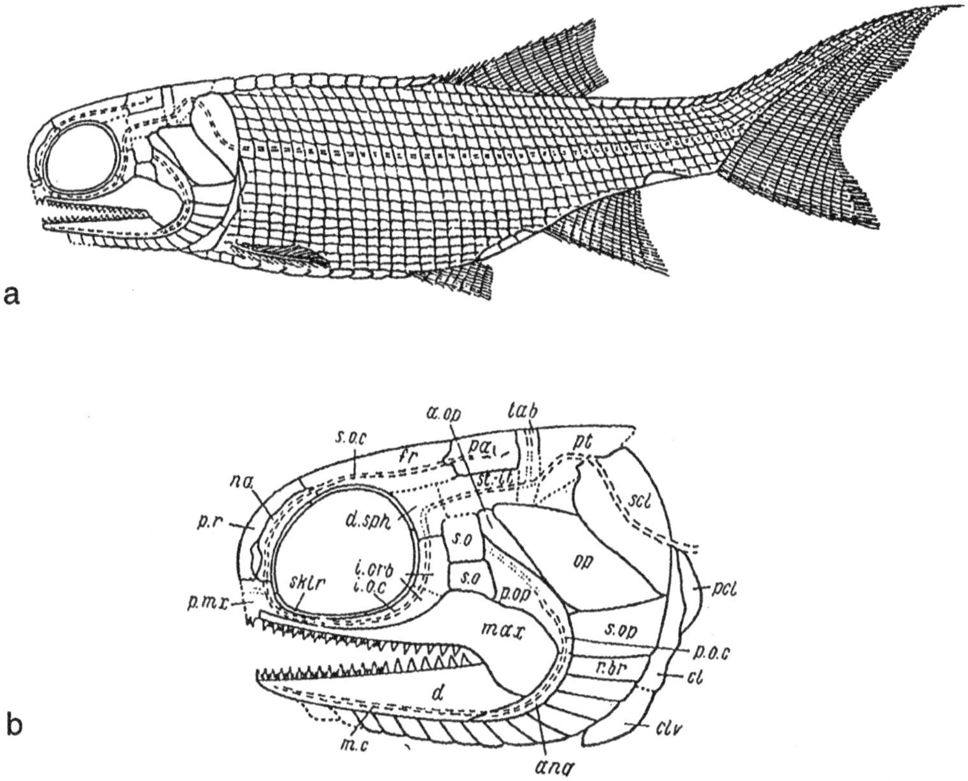

Figure 2. Berg's original reconstructions of a paleoniscid, *Ganolepis gracilis* (first published by Obruchev 1955): a – Lateral view of the entire fish, b – reconstruction of skull (ang = angular, a op = anteoperculum, cl = cleithrum, clv = clavicle, d = dentary, d sph = dermosphenotic, fr = frontal, i.o.c. = infraorbital sensory canal, i orb = infraorbital, max = maxillary, m c = Meckel's cartilage, na = nasal, op = operculum, pa = parietal, pcl - postcleithrum, p mx = premacilla, p o c = preopercular canal, p op = preoperculum, p r = postrostrale, pt = posttemporal, r br = branchiostegal rays, scl = supracleithrum, sklr = sclerotic ring, s o = suborbitalis, s o c = supraorbital canal, s op = suboperculum, st-it = supratemporal-intertemporal, tab = tabular).

tinopterygii. There is no contemporary data supporting the hypothesis on the close relationship of acipenseriforms to selachians' (Berg 1948a, p. 53). He identified three families within Acipenseriformes: 'Chondrosteidae (from the Lower Lias to the Lower Cretaceous), Acipenseridae (beginning from the Upper Cretaceous), and Polyodontidae (beginning from the Upper Cretaceous)' (Berg 1948a, p. 54). Berg's understanding of Acipenseriformes as actinopterygians is fundamental to all contemporary views (Sokolov & Berdichevskii 1989a, b, Grande & Bemis 1991, Bemis et al. 1997, this volume).

Systematics of Acipenseridae was the topic of one of Berg's early theoretical papers (Berg 1904). He included four genera in this family: *Huso* with two species, *H. huso* and *H. dauricus; Acipenser* with sixteen species; *Scaphirhynchus* with one species, *S. platorhynchus*; and *Pseudoscaphirhynchus* with three species, *P. fedtschenkoi, P. hermanni,* and *P. kaufmanni*. This division of Acipenseridae into four genera is used by most contemporary researchers (but see Jollie 1980). In his first monograph on the fishes of Russia (Berg 1911), Berg divided *Acipenser* into three subgenera: (1) *Lioniscus* Bonaparte, 1846, with one species, *A. nudiventris*; (2) *Helops* Bonaparte, 1846, also with one species, *A. stellatus*; and (3) *Acipenser* sensu stricto, which includes all other species of *Acipenser*. Later, in 1948, in the last edition of his monograph on the Russian fish fauna, Berg changed the name *Helops* Bonaparte 1846, to *Gladostomus* Holly, 1936. Historical reviews of these divisions within *Acipenser* are given by Findeis (1997) and Birstein et al. (1997) in this volume, but it is clear that we are still far from an unambiguous, synapomorphy-based diagnosis of the genus *Acipenser* (also see Birstein & Bemis 1997 this volume).

In many monographs and papers, Berg gave classic descriptions of sturgeons inhabiting Russia, eastern Europe and Asia, including their zoogeography and biology (Berg 1905a, b, 1908a, b, 1909, 1911–1913, 1916, 1923, 1932a, b, 1933, 1945, 1948b, c). His encyclopedic knowledge of the material allowed him to discuss hybrids as well as different forms within the same species. Extreme polymorphism is characteristic of many sturgeon species, which poses problems for morphological diagnoses. Berg's approach was typical for his time: recognize and name distinctive subspecies from portions of the range. Many examples are known. For instance, in the Caspian Sea, besides the typical form of the Russian sturgeon, *A. gueldenstaedtii*, Berg recognized a subspecies *A. gueldenstaedtii persicus* Borodin, 1897 or Persian sturgeon (Berg 1933, 1934a, 1948). Later this form was elevated to the rank of species, *A. persicus* (Artyukhin 1979, 1984). This species also occurred in the Black Sea (Artyukhin & Zarkua 1986, Vlasenko et al. 1989). Berg (1948b) considered the Black Sea and Sea of Azov populations of *A. gueldenstaedtii* to be a distinct subspecies, *A. gueldenstaedtii colchicus*. Within the European sterlet, *A. ruthenus,* Berg (1911, 1923, 1948a) recognized two morphs. One, with a typical long and pointed rostrum he named '*A. ruthenus* morpha *kamensis* Lovetsky, 1834', which was synonymous to *A. gmelini* Fitzinger & Heckel, 1834, and to *A. ruthenus* var. *brevirostris* Antipa, 1909. Berg described Siberian sterlet from the Ob River as *A. ruthenus* natio *marsiglii* (Berg 1949).

Berg published several well-known articles on winter and vernal (or spring) races of anadromous fishes (Berg 1934b, 1934c, 1935b). The English-speaking audience learned about these definitions only 25 years later, when Berg's article was translated into English (Berg 1959). He concluded that anadromous fishes, including sturgeons, typically consist of two main races, winter and vernal. Their characteristics are: (1) winter fish spend the coldest time of the year either in the river itself, or in the sea close to the river mouth, whereas vernal fish enter the river at higher temperatures in the spring. (2) During the coldest seasons, the winter fish are in a state of vegetative quiescence, eating little or nothing. Many 'hibernate' in holes. Vernal races have only a short period of vegetative quiescence and do not 'hibernate'. (3) The vernal races spawn in the same season in which they enter the rivers. The winter races spawn the next year. (4) The winter races usually spawn earlier than the vernal races, i.e. in a given year they mature earlier. (6) The winter race is usually larger than the vernal race. (7) The winter race is usually more fertile than the vernal race. As typical examples of the two races, Berg analyzed the behavior of the four species of sturgeons in the northern part of the Caspian Sea: *A. stellatus, A. gueldenstaedtii, H. huso,* and *A. nudiventris*. Depending on the species, one of the two races usually predominates. One of the races can disappear completely. For example, there was only a winter race of the ship sturgeon, *A. nudiventris*, in the Aral Sea (now the Aral Sea population has disappeared completely, see Zholdasova 1997 this volume). Although the sterlet, *A. ruthenus*, is a freshwater resident species, there were two races (and two morphs, as mentioned above) in the Volga, Danube, and Dnieper rivers, which migrated along the rivers to the deltas and back. Differences in races of sturgeons remain even now, despite drastic changes in the Volga, Danube, and other rivers (see Bacalbaşa-Dobrovici 1997, Hensel & Holčík 1997, Khodorevskaya et al. 1997, Kynard 1997, all this volume). Long migrations of *A. ruthenus* in major European rivers are disrupted by dams (for the situation in the Danube River see Hensel & Holčík 1996, Bacalbaşa-Dobrovici 1997, this volume). Migrating and riverine races (or populations) are discussed by: Ruban (1997 this volume) for Siberian sturgeon, *A. baerii*; Krykhtin & Svirskii (1997 this volume) for Amur River sturgeons; and Hensel & Holčík (1997 this volume) for sturgeons of the Danube River.

Profound knowledge of the distribution of Acipenseriformes played a major role in Berg's evolutionary (Berg 1922) and zoogeographic theories.

Figure 3. At the Institute of Zoology in St. Petersburg, the presence of Lev Semenovich Berg is still strong 40 years after his death. Minutes after arrival on 18.6.1990 E.A. Dorofeeva seated Eugene Balon in the chair used by L.S. Berg.

Amur River acipenserids (Berg 1909, 1911) were one of the elements of Berg's hypothesis on the relic character of the fauna of the Amur River Basin (Berg 1912, 1928). According to this hypothesis, the species constituting the Amur River fauna are remnants of the subtropical Upper Tertiary fauna that characterized the entire northern hemisphere, and which mostly disappeared as a result of cooling during the Quaternary. Berg also discussed problems of interrelationships of the Asian, European, and North American fish faunas (Berg 1950). Contemporary information about sturgeons of the Amur River is presented in two articles of this volume (Krykhtin & Svirskii 1997, Zhuang et al. 1997).

Some of Berg's other zoogeographic ideas are useful for understanding the distribution and evolution of sturgeons in the northern hemisphere. For instance, in a hypothesis explaining the similarity of elements of the Pacific and Atlantic faunas, Berg suggested two periods of exchanges between elements of the faunas of the northern parts of the two oceans (Berg 1918, 1934d, e, 1947b). Also, Berg's ideas on historic changes in the fauna of the Caspian Sea (Berg 1928c, d) are useful for understanding the

Figure 4. A portrait of L.S. Berg in his office at the Zoological Institute, St. Petersburg. Lithograph by G. Vereisky, 1950.

history and evolution of sturgeons in the Caspian and Black Sea basins.

Berg gave his first short presentation on sturgeons in 1897, when he was a student (Berg 1898). It described experiments on artificial breeding of *A. stellatus*. Much later he returned to the problem of sturgeon development, describing juveniles of *Pseudoscaphirhynchus kaufmanni* caught in the Amu Darya River (Berg 1929). In this volume detailed information on the reproductive cycle of the white sturgeon, *A. transmontanus*, is given by Doroshov et al. (1997).

Berg never stressed the ability of sturgeons to hybridize, but he described many sturgeon hybrids in detail (Berg 1911, 1932, 1948b). Some genetic aspects of acipenseriforms, including hybridization, are discussed in this volume by Birstein et al. (1997).

Berg lived and worked when the sturgeon crisis in Russia, Europe, and Asia had only started (Figure 3, 4). The desperate need for conservation measures to save sturgeons was in the future. He published only a small article describing his concern about *A. sturio* in the Baltic Sea and especially in the Neva River (Berg 1935c). He suggested that a complete ban on the catch of this species should be established for at least the next 10–15 years. Unfortunately, since then most of the species of sturgeons and paddlefishes have become threatened or endangered, a theme of many papers of this volume, and one that surely would have saddened L.S. Berg.

Acknowledgements

We are very grateful to Raissa Lvovna Berg (L.S. Berg's daughter) and Maria Berg (L.S. Berg's granddaughter) for providing an unpublished photo of L.S. Berg. Eugene Balon gave his photo of L.S. Berg's chair.

References cited

Andriyashev, A.P. 1955. L.S. Berg as a zoogeographer. pp. 116–126. *In:* E.N. Pavlovskii (ed.) To the Memory of Academician L.S. Berg, Izdatelstvo Akademii Nauk USSR, Moscow (in Russian).

Antipa, G. 1909. Fauna ichtiologica a României. Publicatiunile Fondul Vasilie Adamanchi, Academia Româna, Bucureşti 16: 1–294.

Artyukhin, E.N. 1979. Persian sturgeon in the rivers entering northern part of the Caspian Sea and perspectives of its harvest. pp. 105–115. *In:* Biological Fundamental of Sturgeon Management Development in the Water Bodies of the USSR, Nauka Press, Moscow (in Russian).

Artyukhin, E.N. 1983. Differentiation of the Persian sturgeon populations and perspectives of its artificial breeding at the Volga River hatcheries. pp. 54–61. *In:* Biological Fundamentals of Sturgeon Management, Nauka Press, Moscow (in Russian).

Artyukhin, E.N. & Z.G. Zarkua. 1986. On the problem of taxonomic rank of the sturgeon from the Rioni River (the Black Sea Basin). Voprosy Ikhtiologii 26: 61–67 (in Russian).

Bacalbaşa-Dobrovici, N. 1997. Endangered migratory sturgeons of the lower Danube River and its delta. Env. Biol. Fish (this volume).

Bemis, W.E., E.K. Findeis & L. Grande. 1997. An overview of Acipenseriformes. Env. Biol. Fish (this volume).

Berg, L.S. 1898. Experiments on artificial breeding of sevruga sturgeon in the Ural River. Izvestiya Obshchestva Lyubitelei Estestvoznaniya, Antropologii i Etnografii, Vol. 86. Dnevnik Zoologischeskogo Otdeleniya Obshchestva i Zoologichesko-go Muzeya Moscovskogo Universiteta 2: 36 (in Russian).

Berg, L.S. 1904. Zur Systematik der Acipenseriden. Zool. Anz. 27: 665–667.

Berg. L.S. 1905a. Fishes of Turkestan. Scientific results of the Aral expedition, No. 6. St. Petersburg. 261 pp. (in Russian).

Berg, L.S. 1905b. Verzeichnis der Fische von Russisch Turkestan. Ezhegodnik Zoologischeskogo Muzeya Adademii Nauk 10: 316–332 (in Russian).

Berg, L.S. 1908a. List of the Ob River basin fishes. Ezhegodnik Zoologischeskogo Muzeya Adademii Nauk 13: 69–107 (in Russian).

Berg, L.S. 1908b. List of the Kolyma River fishes. Ezhegodnik Zoologischeskogo Muzeya Akademii Nauk 13: 221–228 (in Russian).

Berg, L.S. 1909. Fishes of the Amur River basin. Zapiski Akademii Nauk 24: 1–270 (in Russian).

Berg, L.S. 1911. Fishes (Marsipobranchii and Pisces). Fauna of Russian and adjacent countries, Vol. 3, Vypusk 1, Izdatelstvo Akademii Nauk, St. Petersburg. 337 pp. (in Russian).

Berg, L.S. 1912. Über die Zusammensetzung und Herkunft der Fischfauna des Amur-Flusses mit Bezug auf die Frage von den zoogeographischen Regionen für die Süsswasserfische. Zool. Jahrb. 32: 475–520.

Berg, L.S. 1913. On the find of *Acipenser medirostris* (Ayres) in the lower reaches of the Amur River. Ezhegodnik Zoologischeskogo Muzeya Akademii Nauk 18: 16 (in Russian).

Berg, L.S. 1916. Fishes of fresh waters of Russian Empire. Moscow. 563 pp. (in Russian).

Berg, L.S. 1918. On the causes of similarity in the faunas of the northern parts of the Atlantic and Pacific oceans. Izvestiya Rossiiskoi Akademii Nauk 16: 1835–1842 (in Russian).

Berg, L.S. 1922. Nomogenesis or evolution determined by law. Moscow. 200 pp. (in Russian, English translation published by The M.I.T. Press, Cambridge, 1926. 477 pp.)

Berg, L.S. 1923. Fishes of fresh waters of Russia. 2nd ed. Gosudarstvennoe Izdatelstvo, Moscow. 365 pp. (in Russian).

Berg, L.S. 1928a. Life style and geographic morphs in sevruga. Priroda 3: 294–296 (in Russian).

Berg, L.S. 1928b. Zoogeographical divisions for Far Eastern freshwater fishes. pp. 1041–1043. *In:* Proceedings of the Third Pan-Pacific Science Congress, Vol. 1, Tokyo.

Berg, L.S. 1928c. On the origin of the northern elements in the fauna of the Caspian Sea. Doklady Akademii Nauk USSR, Seriya A 7: 107–112 (in Russian).

Berg, L.S. 1928d. Mediterranean elements in the fauna of the Caspian Sea. Priroda 7/8: 753 (in Russian).

Berg, L.S. 1929. Juvenile fishes from the lower Amu Darya River. Izvestiya Otdeleniya Prikladnoi Ikhtiologii 9: 225–230 (in Russian).

Berg, L.S. 1932a. Übersicht der Verbreitung der Susswasserfische Europas. Zoogeographica 1: 107–208.

Berg, L.S. 1932b. The freshwater fishes of the USSR and adjacent countries. 3rd ed. Part 1. Vsesoyuznyi Institute Ozernogo i Rechnogo Rybnogo Khozyaistva, Leningrad. 543 pp. (in Russian).

Berg. L.S. 1933. The freshwater fishes of the USSR and adjacent countries. 3rd ed. Part 2. Vsesoyuznyi Institut Ozernogo i Rechnogo Rybnogo Khozyaistva, Leningrad. 545–903 pp. (in Russian).

Berg. L.S. 1934a. *Acipenser güldenstädti persicus*, a sturgeon from the south Caspian Sea. Ann. Mag. Nat. Hist. Ser. 10, 13: 317–318.

Berg, L.S. 1934b. Summer and autumn races of anadromous fishes. Izvestiya Akademii Nauk USSR 5: 711–732 (in Russian).

Berg, L.S. 1934c. Summer and autumn races of anadromous fishes. Priroda 4: 36–40 (in Russian).

Berg, L.S. 1934d. On the amphiboreal (discontinuous) distribution of the marine fauna in the northern hemisphere. Izvestiya Gosudarstvennogo Geographicheskogo Obshchestva 66: 69–78 (in Russian).

Berg, L.S. 1934e. Über die amphiboreale Verbreitung des Meeresfauna in der nördlichen Hemisphäre. Zoogeographica 2: 393–409.

Berg, L.S. 1935a. Sur les unités taxonomiques chez les poissons. Bull. Mus. Nat. Hist. Nat., 2-e sér. 7: 79–84.

Berg, L.S. 1935b. Sommer- und Winterrassen bei den anadromen Fischen. Arch. Naturgesch. 4: 376–403.

Berg, L.S. 1935c. On the necessity to save sturgeon in the Neva River basin. Za Rybnuyu Industriyu Severa 7: 30–31 (in Russian).

Berg, L.S. 1937. A classification of fish-like vertebrates. Izvestiya Akademii Nauk USSR, Seriya Biologischeskaya 4: 1277–1280.

Berg, L.S. 1940a. Classification of fishes both recent and fossil. Trudy Zoologischeskogo Instituta 5: 85–517 (in Russian).

Berg, L.S. 1940b. On the position of Polypteridae in the system of fishes. Zoologischeskii Zhurnal 19: 727–740 (in Russian).

Berg, L.S. 1945. On sterlet in the White Sea basin. Priroda 6: 66–67 (in Russian).

Berg, L.S. 1947a. Classification of fishes both recent and fossil. I.W. Edwards, Ann Arbor. 437 pp.

Berg, L.S. 1947b. Some thoughts on the origin of the terrestrial, fresh-water, and marine flora and fauna. Bulletin Moskovskogo Obshchestva Ispytatelei Prirody, Otdelenye Biologischeskoe 52: 15–33 (in Russian).

Berg, L.S. 1948a. On the position of Acipenseriformes in the system of fishes. Trudy Zoologischeskogo Instituta 7: 7–57 (in Russian).

Berg, L.S. 1948b. The freshwater fishes of the USSR and adjacent countries. 4th ed., Part 1. Akademia Nauk USSR. Moscow & Leningrad. 466 pp. (in Russian, English translation published by Israel Program for Scientific Translations, Jerusalem, 1965. 505 pp.).

Berg, L.S. 1948c. Fishes of Gulf of Finland. Izvestiya VNIORKh 23: 3–24 (in Russian).

Berg, L.S. 1949. The freshwater fishes of the USSR and adjacent countries, 4th ed., Part 3. Adademia Nauk USSR, Moscow & Leningrad. 927–1382 pp. (in Russian).

Berg, L.S. 1950. On the causes of the similarity of the fish faunas in the Volga, Don, and Dnepr rivers. Trudy Kaspiiskogo Basseinovogo Filiala VNIRO 11: 5–8 (in Russian).

Berg, L.S. 1959. Vernal and hibernal races among anadromous fishes. J. Fish Res. Board Can. 16: 515–537.

Berg, L.S. 1965. Classification of fishes both recent and fossil. Thai National Documentation Center, Bangkok. 304 pp.

Berg, M.M. 1955. Systematic list of Academician L.S. Berg's works. pp. 531–560. *In:* E.N. Pavlovskii (ed.) To the Memory of Academician L.S. Berg, Izdatelstvo Akademii Nauk USSR, Moscow (in Russian).

Birstein, V.J. & W.E. Bemis. 1997. How many species are there within the genus *Acipenser?* Env. Biol. Fish (this volume).

Birstein, V.J., R. Hanner & R. DeSalle. 1997. Phylogeny of the Acipenseriformes: cytogenetic and molecular approaches. Env. Biol. Fish (this volume).

Bonaparte, C. 1846. Catalogo metodico dei pesci europei. Atti. Soc. Ital. Sci. Natur. 1–95.

Borodin, N.A. 1897. A report about a summer 1895 zoological expedition on board of the cruiser 'Uralets' in the northern part of the Caspian Sea. Vestnik Rybopromyshlennosti 1: 1–13 (in Russian).

Dobzhansky, T. 1969. Foreword to paperback edition. pp. VII–XVI. *In:* L.S. Berg. Nomogenesis or Evolution Determined by Law, The M.I.T. Press, Cambridge, 476 pp.

Doroshov, S.I., G.P. Moberg & J.P. Van Eenennaam. 1997. Observation on the reproductive cycle of cultured white sturgeon, *Acipenser transmontanus*. Env. Biol. Fish. (this volume).

Findeis, E.K. 1997. Osteology and phylogenetic interrelationships of sturgeons (Acipenseridae). Env. Biol. Fish. (this volume).

Fitzinger, L.J. & J. Heckel. 1836. Monographische Darstellung der Gattung *Acipenser*. Ann. Wiener Mus. Naturgesch. 1: 261–326.

Grande, L. & W.E. Bemis. 1991. Osteology and phylogenetic relationships of fossil and recent paddlefishes (Polyodontidae) with comments on the interrelationships of Acipenseriformes. J. Vert. Paleo. 11, supplement 1: 1–121.

Hensel, K. & J. Holčík. 1997. Past and current status of sturgeons in the upper and middle Danube River. Env. Biol. Fish. (this volume).

Holčík, J. 1976. K stému výročiu narodenia akademika Leva Semjonoviča Berga (1876–1950) (To the hundredth anniversary of the birth of Academician L.S. Berg). Biológia (Bratislava) 32: 891–892.

Holly, M. 1936. Pisces 4. Ganoidei. Das Tierreich. Lfg. 67: 1–65.

Jollie, M. 1980. Development of the head and pectoral girdle skeleton and scales in *Acipenser*. Copeia 1980: 226–249.

Khodorevskaya, R.P., G.F. Dovgopol, O.L. Zhuravleva & A.D. Vlasenko. 1997. Present status of commercial stock of sturgeons in the Caspian Sea basin. Env. Biol. Fish. (this volume).

Krykhtin, M.L. & V.G. Svirskii. 1996. Endemic sturgeons of the Amur River: kaluga, *Huso dauricus*, and Amur sturgeon, *Acipenser schrencki*. Env. Biol. Fish. (this volume).

Lindberg, G.U. 1976. Lev Semenovich Berg as ichthyologist. Voprosy Ikhtiologii 16: 721–725 (in Russian).

Lovetsky, A. 1834. Diagnosis piscium ad genus *Acipenserinum* pertinentium, praeprimis eorum qui habitat in aquis Imperii Rossici. Nouv. Mém. Soc. Natur. Moscou 3: 253–264.

Obruchev, D.V. 1955. L.S. Berg's works on fossil fishes. pp. 127–137. *In:* E.N. Pavlovskii (ed.) To the Memory of Academician L.S. Berg, Izdatelstvo Akademii Nauk USSR, Moscow (in Russian).

Oliva, O. 1951. Laureát Stalinovy ceny akademik Lev Semĕnovič Berg (The Stalin's prize laureate Academician L.S. Berg). Sovĕtská vĕda, biologie 3: 285-287.

Oliva, O. 1952. Akademik L.S. Berg, 1876–1950 (Academician L.S. Berg). Za Socialistické Zemĕdĕlství 2: 1448–1454.

Oliva, O. & J. Holčík. 1977. The hundredth anniversary of the birth of academician L.S. Berg. Folia Zoologica 26: 93–95.

Oliva, O. & J. Holčík. 1978. Dílo Lva Semĕnoviče Berga (The work of L.S. Berg). Dĕjiný Vĕd a Techniky 11: 23–28.

Sewertzoff, A.N. 1925. The place of the cartilaginous ganoids in the system and the evolution of the Osteichthyes. J. Morphol. 38: 105–155.

Sewertzoff, A.N. 1926. The development of the scales of *Acipenser ruthenus*. J. Morphol. 42: 523–560.

Sewertzoff, A.N. 1928. The head skeleton and muscles of *Acipenser ruthenus*. Acta Zool. 9: 193–319.

Shapovalov, L. 1951. Leo Semenovich Berg, 1876–1950. Copeia 1951: 184–186.

Sokolov, L.I. 1989. *Acipenser* Linnaeus, 1758. pp. 201–205. *In:* J. Holčík (ed.) The Freshwater Fishes of Europe, Vol. 1, pt. II, General Introduction to Fishes, Acipenseriformes, AULA-Verlag, Wiesbaden.

Sokolov, L.I. & L.S. Berdichevskii. 1989a. Acipenseriformes Berg, 1940. pp. 148–149. *In:* J. Holčík (ed.). The Freshwater Fishes of Europe, Vol. 1, pt. II, General Introduction of Fishes, Acipenseriformes, AULA-Verlag, Wiesbaden.

Sokolov, L.I. & L.S. Berdichevskii. 1989b. Acipenseridae Bonaparte, 1831. pp. 150–153. *In:* J. Holčík (ed.) The Freshwater Fishes of Europe, Vol. 1, pt. II, General Introduction to Fishes, Acipenseriformes, AULA-Verlag, Wiesbaden.

Sokolov, N.N. 1955. Lev Semenovich Berg. pp. 18–60. *In:* E.N. Pavlovskii (ed.) To the Memory of Academician L.S. Berg, Izdatelstvo Akademii Nauk USSR, Moscow (in Russian).

Vlasenko, A.D., A.V. Pavlov & V.P. Vasilev. 1989b. *Acipenser persicus* Borodin, 1897. pp. 345–366. *In:* J. Holčík (ed.) The Freshwater Fishes of Europe, Vol. 1, Pt. II, General Introduction to Fishes, Acipenseriformes, AULA-Verlag, Wiesbaden.

Wei, Q., F. Ke, J. Zhang, P. Zhuang, J. Luo, R. Zhou & W. Yang. 1997. Biology, fisheries, and conservation of sturgeons and paddlefish in China. Env. Biol. Fish. (this volume).

Zholdasova, I. 1997. Sturgeons and the Aral Sea ecological catastrophe. Env. Biol. Fish. (this volume).

Part 1: Diversity and evolution of sturgeons and paddlefishes

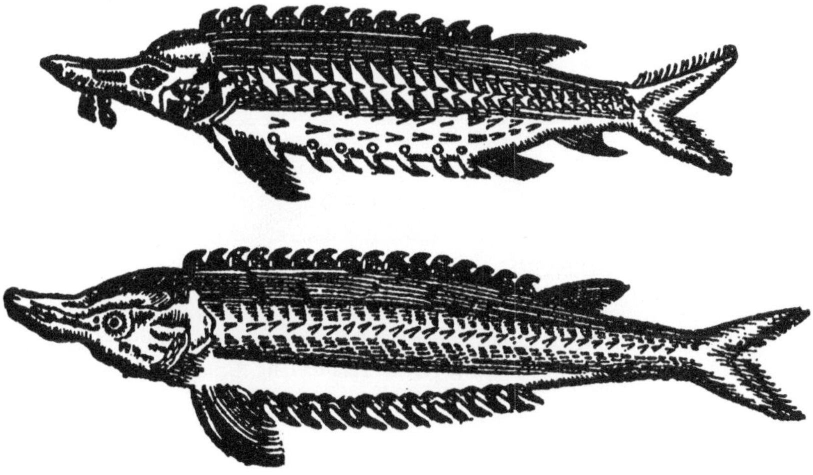

Acipenser, Aquipenser, Sturio – sturgeon woodcuts from Conrad Gesner (1558).

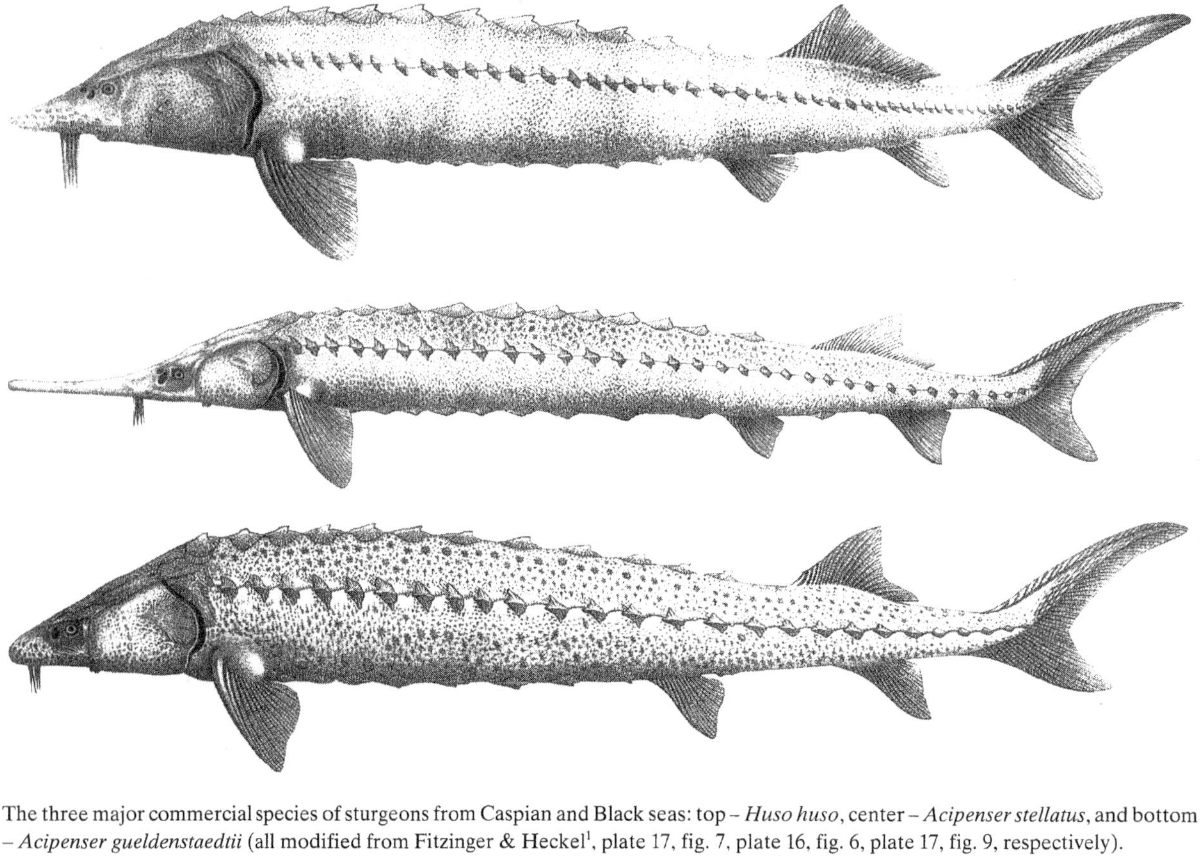

The three major commercial species of sturgeons from Caspian and Black seas: top – *Huso huso*, center – *Acipenser stellatus*, and bottom – *Acipenser gueldenstaedtii* (all modified from Fitzinger & Heckel[1], plate 17, fig. 7, plate 16, fig. 6, plate 17, fig. 9, respectively).

---

[1] Fitzinger, L.J. & J. Heckel. 1836. Monographische darstellung der Gattung *Acipenser*. Zool. Abh. Ann. Wiener Mus. Naturgesch. 1: 262–326 (note: order of authorship is reversed in some bibliographic citations).

Environmental Biology of Fishes **48**: 25–71, 1997.

An overview of Acipenseriformes

William E. Bemis[1], Eric K. Findeis[1] & Lance Grande[2]
[1] *Department of Biology and Graduate Program in Organismic and Evolutionary Biology, University of Massachusetts, Amherst, MA 01003, U.S.A.*
[2] *Department of Geology, Field Museum of Natural History, Chicago, IL 60605, U.S.A.*

Received 15.2.1995   Accepted 10.3.1996

Key words: Actinopterygii, Paleonisciformes, Acipenseridae, Polyodontidae, †Chondrosteidae, †Peipiaosteidae

Synopsis

Acipenseriformes occupy a special place in the history of ideas concerning fish evolution, but in many respects, phylogenetic studies of the group remain in their infancy. Even such basic questions as the monophyly of *Acipenser* (the largest genus) are unanswered. We define relationships based on comparative osteology, which allows us to incorporate well-preserved fossils into analyses. Acipenseriformes has existed at least since the Lower Jurassic (approximately 200 MYBP), and all fossil and recent taxa are from the Holarctic. Phylogenetic relationships among Paleozoic and Early Mesozoic actinopterygians are problematic, but most workers agree that Acipenseriformes is monophyletic and derived from some component of 'paleonisciform' fishes. ('Paleonisciformes' is a grade of primitive non-neopterygian actinopterygians, sensu Gardiner 1993.) Taxa discussed in comparison here are: †*Cheirolepis*, *Polypterus*, †*Mimia*, †*Moythomasia*, †*Birgeria*, †*Saurichthys*, *Lepisosteus* and *Amia*. We review generic diversity within the four nominal families of fossil and recent Acipenseriformes (†Chondrosteidae, †Peipiaosteidae, Polyodontidae, and Acipenseridae), and provide a cladogram summarizing osteological characters for those four groups. Monophyly of the two extant families is well-supported, but there are no comprehensive studies of all of the known species and specimens of †Chondrosteidae and †Peipiaosteidae. As a result, sister-group relationships among †Chondrosteidae, †Peipiaosteidae, and Acipenseroidei (= Polyodontidae + Acipenseridae) are unresolved. We discuss five features fundamental to the biology of acipenseriforms that benefit from the availability of our new phylogenetic hypothesis: (1) specializations of jaws and operculum relevant to jaw protrusion, feeding, and ram ventilation; (2) anadromy or potamodromy and demersal spawning; (3) paedomorphosis and evolution of the group; (4) the biogeography of Asian and North American polyodontids and scaphirhynchines; and (5) the great abundance of electroreceptive organs in the rostral and opercular regions. Finally, we summarize our nomenclatural recommendations.

Introduction and historical overview

This paper reviews the systematics of sturgeons and paddlefishes and their immediate fossil relatives in the order Acipenseriformes. We synthesize historic and current information in our effort to better understand the evolution, biogeography, and composition of the order. We emphasize generic and familial comparisons, and summarize information for all recent and well preserved fossil genera. In keeping with our objective of providing background, this paper includes several 'evolutionary scenarios' (in the sense of Gans 1986) which we hope will provoke further basic work on the group. Additional recent treatment of many of these taxa can be found in Grande & Bemis (1991, 1996a) and Findeis (1997 this volume).

Acipenseriforms are central to historical ideas about the classification and evolution of fishes. Sturgeons were often the largest freshwater animals in a fauna and quite naturally attracted attention from early naturalists and systematists. Acipenseriforms also are noteworthy because of their unusual mixture of characters, which caused early debate about their classification. Two aspects of living Acipenseriformes were especially problematic for early ichthyologists: (1) reduced ossification of the endoskeleton combined with presence of an extensive dermal skeleton; and (2) the presence of a hyostylic jaw suspension and protrusible palatoquadrate recalling the jaws of sharks. The current conventional view (developed and refined by many authors, including Müller 1846, Traquair 1877, Woodward 1891, 1895a,b, Regan 1904, Goodrich 1909, Watson 1925, 1928, Gregory 1933, Berg 1948b, Yakovlev 1977) holds that Acipenseriformes evolved from a 'paleonisciform' ancestor via paedomorphic reduction of the skeleton and specialization of the feeding system, but there is much more to the history of ideas about the systematics of this group.

Figure 1 highlights contributions to the systematics of Acipenseriformes over the last 250 years. From the time of Linnaeus through the early part of the 19th century, descriptions of most of the currently recognized species and genera were made, including *Acipenser* Linneaus 1758, *Polyodon* Lacépède 1797, and *Scaphirhynchus* Heckel 1836. Throughout this period most workers adhered to the classical idea that sturgeons must be closely related to sharks because they appeared to share a largely cartilaginous endoskeleton and similar jaw suspension. An obvious example of this was Walbaum's (1792) description of *Polyodon spathula* as '*Squalus spathula*'. By the 1830s, the first serious attempts to synthesize and revise the systematics of Acipenseriformes began, including Heckel's (1836) definition of *Scaphirhynchus* as a genus distinct from *Acipenser*. Attempts by Brandt & Ratzeberg (1833) and Fitzinger & Heckel (1836) to subdivide the genus *Acipenser* into several subgenera, however, were less successful.

It was also in the middle of the 19th century that the first important fossil acipenseriform, †*Chondrosteus,* was named by Agassiz (1844) and described by Egerton (1858). Increasingly synthetic works on higher relationships of fishes also appeared, exemplified by Müller (1846), who defined three grades of bony fishes -- Chondrostei, Holostei and Teleostei -- on the basis of increasing degrees of ossification. In doing this, Müller rejected the classical idea that sturgeons are closely related to sharks and accepted them as osteichthyans. Sewertzoff (1925, 1926b, 1928) was the only 20th century ichthyologist to seriously consider a closer link between sturgeons and chondrichthyans. Sewertzoff (1925) presented his conclusions as a phylogenetic tree, in which chondrosteans are shown as the sister group of all other bony fishes, and emphasized the presence of a protrusible palatoquadrate in both elasmobranchs and sturgeons. We now regard palatoquadrate protrusion as derived independently within chondrosteans (see additional discussion in the final section of this paper). Norris (1925) and others noted neuroanatomical similarities between sturgeons and sharks, but these are almost certainly plesiomorphic features (see Northcutt & Bemis 1993), and few workers ever accepted Sewertzoff's view (see Berg 1948b and Yakovlev 1977 for additional history and critique).

Representatives of two of the six extant genera of Acipenseriformes, *Psephurus gladius* (Martens 1862) and *Pseudoscaphirhynchus fedtschenkoi* (Kessler 1872) were discovered in the latter part of the 19th century, but apart from early papers (e.g., Handyside 1875a,b, Ivanzoff 1887), they remained poorly studied for decades. Also in the latter part of the 19th century paleontologists described and interpreted fossil taxa relevant to Acipenseriforms. Traquair (1877, 1887) considered that extant acipenseriforms were derived from 'paleonisciforms'. Traquair's (1887) ideas were the source for many subsequent interpretations of acipenseriform evolution, although we still do not sufficiently understand 'paleonisciforms' to allow us to make strong phylogenetic hypotheses about relationships within

2000	Lu (1994) - description of †*Protopsephurus*
	Zhou (1992) - redescription of †*Peipiaosteus*
	Grande & Bemis (1991) - systematics of Polyodontidae
	Gardiner & Schaeffer (1989) - reviewed paleoniscoid taxa
	Gardiner (1984a,b) - fossil chondrosteans and interrelationships of Acipenseriformes
1975	Yakovlev (1977) - acipenseriform evolution
	Schaeffer (1973) - review of chondrostean systematics
	Gardiner (1967) - classification of fossil and Recent chondrosteans
	Liu & Zhou (1965) - described †*Peipiaosteus*
	Vladykov & Greeley (1963) - reviewed Atlantic species of *Acipenser*
	Wilimovsky (1956) - described †*Protoscaphirhynchus*
1950	Neilsen (1949) - †*Birgeria* and Polyodontidae
	Berg (1948a,b) - reviewed Russian acipenserids and acipenseriform evolution
	MacAlpin (1941a) - described †*Paleopsephurus*
	Aldinger (1931, 1937) - diphyly of Polyodontidae and Acipenseridae
	Antoniu-Murgoci (1936a,b) - identified characters separating *Huso* and *Acipenser*
	Tatarko (1936) - anatomical studies of branchial arches of Acipenseridae
1925	Sewertzoff (1925) - acipenserids as sister group of Osteichthyes
1900	Nikolskii (1900) - proposed genus *Pseudoscaphirhynchus*
	Woodward (1891, 1895a,b,c) - paleontology of †*Chondrosteus*, †*Gyrosteus*
	Traquair (1877, 1887) - Acipenseriformes derived from paleoniscoids
1875	Günther (1873) - proposed genus *Psephurus*
	Kessler (1872) - described *Acipenser* (=*Pseudoscaphirhynchus*) *fedtschenkoi*
	Duméril (1870) - extensive splitting of *Acipenser*, not accepted
	Brandt (1869) - elevation of subgenus *Husones* to genus *Huso*
	Egerton (1858) - anatomical description of †*Chondrosteus*
1850	Müller (1846) - proposed Chondrostei, Holostei, Teleostei: sturgeons are Osteichthyes
	Bonaparte (1838) - proposed family name Polyodontidae
	Heckel (1836) - proposed genus *Scaphirhynchus*
	Fitzinger & Heckel (1836) - proposed subdivision of *Acipenser*
	Brandt & Ratzeberg (1833) - proposed subdivision of *Acipenser*
1825	
	Rafinesque (1820) - described *Acipenser* (=*Scaphirhynchus*) *platorynchus*
1800	
	Lacepède (1797) - proposed *Polyodon*
	Walbaum (1792) - described *Squalus* (= *Polyodon*) *spathula*: i.e., *Polyodon* originally considered to be a shark
1775	Georgi (1775) - described *Acipenser* (=*Huso*) *huso*
1750	Linnaeus (1758) - described *Acipenser sturio*: sturgeons regarded as related to sharks

Figure 1. Selected events in the history of acipenseriform systematics since Linnaeus.

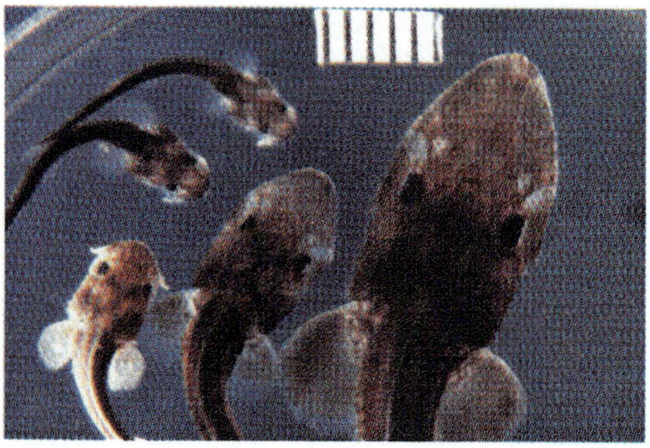

Figure 2. A partial growth series of *Scaphirhynchus platorynchus* larvae showing allometric lengthening of the rostrum. Like other living acipenseriforms, growth of the rostral region is positively allometric during early life. Scale marks are millimeters.

the group. Cope (1883) described the first fossil paddlefish, †*Crossopholis magnicaudatus*, and Woodward (1891, 1895a,b,c, 1909) reviewed the fossil history of sturgeons in papers which remain useful to this day.

With the exception of Berg's remarkable synthetic works (e.g., Berg 1940, also see Birstein & Bemis 1997 this volume), 20th century ichthyologists rarely incorporated paleontological data into their ideas about acipenseriform systematics. Thus, the ichthyological tradition of this century emphasized regional faunas and keys for sturgeons and paddlefishes, such as for territories of the former Soviet Union (Berg 1911, 1933, 1948a), the western North Atlantic (Bigelow & Schroeder 1953, Vladykov & Greeley 1963), eastern Atlantic and Mediterranean (Svetovidov 1984), and European freshwaters (Holčík 1989). Other collected works, such as Binkowski & Doroshov (1985), Williot (1991), and Gershanovich & Smith (1995) summarized much basic biological and distribution data, but did not attempt to examine acipenseriform interrelationships.

During the 1930s and 1940s, a period in which diphyletic origins were proposed for several groups (e.g., tetrapods; Holmgren 1933, Jarvik 1942), Aldinger (1937) proposed that paddlefishes and sturgeons were derived from separate early Mesozoic ancestors. In a detailed study of a Triassic species of †*Birgeria* from east Greenland, Nielsen (1949) examined and supported Aldinger's hypothesis that this genus is closely related to living paddlefishes. Few workers ever accepted the Aldinger-Niel hypothesis (see Yakovlev 1977 for history and a tailed critique), and it was rendered even more likely by the cladistic definition of Acipensero (Grande & Bemis 1991).

Several additional extinct genera of Acipens formes based on relatively complete skeletons w described in the 20th century, and more are be found at the time of writing of this paper. New penseriforms include a Jurassic paddlefish, †*I topsephurus* Lu 1994; a Cretaceous paddlefish, † *leopsephurus* MacAlpin 1941a (also see MacAl 1941b, 1947); a Cretaceous sturgeon, †*Protoscap hynchus* Willimovsky 1956; two Jurassic (or earl Cretaceous) genera, †*Stichopterus* Reis 1910 (see Yakovlev 1977, 1986) and †*Peipiaosteus* Li Zhou 1965 (also see Bai 1983, Zhou 1992, Grand Bemis 1996; see Jin 1995 and Jin et al. 1995 for m taxa described since this paper was accepted). M authors of the type descriptions of these fossil t included comparisons with recent acipenserifor but the descriptions of all four genera pre-date widespread application of cladistics to frame ph genetic questions and organize character infor tion. Another problem with some of these pape that reconstruction of the fossils was based on tant sturgeons and paddlefishes, which makes it ficult to separate observation from interpretati Comments on relationships were provided by & Zhou (1965), Nelson (1969) and Jollie (1980), none of these treatments explicitly traced relati ships of living and fossil forms, nor did they incl synapomorphy schemes. It was not until later, w Gardiner (1984b) published the first generic le cladogram including fossil and recent Acipens forms, that interest in their phylogenetic interre tionships began to grow. Gardiner's (1984b) ana sis was controversial because he suggested that p dlefishes were diphyletic, a conclusion rejected Grande & Bemis (1991). More recently, Zh (1992) provided a different tree, which we critic in our analysis below.

Molecular and karyological approaches to s tematics of Acipenseriformes are still at the leve initial surveys (e.g., Fontana & Colombo 1974, D gerkus & Howell 1976, Birstein & Vasiliev 1987), though increasingly comprehensive (e.g., Birst et al. 1997 this volume). Published molecular phylogenetic research including Acipenseriformes is limited to questions concerning higher relationships among Actinopterygii (e.g., Normark et al. 1991), and no study has yet included all living species of sturgeons and paddlefishes. Nothing approaching the comprehensive morphological-molecular-karyological-data sets now available for many groups of tetrapods (e.g., plethodontid salamanders, Wake & Larson 1987) has been attempted for Acipenseriformes or indeed for actinopterygians generally.

From this brief history, it is clear that phylogenetic studies of Acipenseriformes are still in their infancy. Some barriers to phylogenetic study seem 'intrinsic' to these fishes. In particular, acipenseriforms often exhibit great individual and ontogenetic variation. It is critical to better understand and distinguish between these types of variation in any comprehensive phylogenetic review, and this in itself is a daunting task. Extensive variation confused systematists such as Duméril (1870), who proposed more than 40 new species of *Acipenser* that were rejected by later workers. Variation is frequently noted in other contexts. For example, in a large population study of shortnose sturgeon, *Acipenser brevirostrum*, Dadswell et al. ([1], p. 2) noted that specimens ranged '. . . from sharp-plated, rough-skinned individuals to flat-plated, smooth-skinned' in the St. John Estuary in New Brunswick. There is also much individual variation in the pattern of skull roofing bones, as illustrated for *A. fulvescens* by Jollie (1980, perhaps even more extreme than variation in skull roofing bones reported for *Amia* by Jain 1985 and Grande & Bemis 1997). Although it has not been the subject of formal study, rostral shape in *Scaphirhynchus* is positively allometric during early life, as shown by the photograph of a growth series in Figure 2. The rostrum provides other well known examples of variation. For example, rostral length and width of the North American species of *Acipenser* varies ontogenetically, geographically and interspecifically (Vladykov & Greeley 1963). Ontogenetic allometry of paddle growth in *Polyodon spathula* was concisely described by Thompson (1934, also see Grande & Bemis 1991). At the start of the feeding larva period, North American paddlefishes have a barely detectable paddle. But soon afterwards, the paddle grows with positive allometry to make up more than half of the total body length. Later in life, paddle growth shows negative allometry with respect to total length. Even after *Polyodon spathula* achieves reproductive maturity, there can be significant qualitative morphological changes, such as the appearance of new ossification centers in the neurocranium (Grande & Bemis 1991). Many acipenseriforms achieve very large sizes at maturity, and may continue to grow for many years thereafter, but most systematic studies and collections are disproportionately weighted towards more easily studied (and stored) juvenile and 'sub-adult' specimens. We have already pointed out the necessity of collecting and including large adult specimens in phylogenetic studies (Grande & Bemis 1991, 1997). In studying acipenseriforms, this goal is often impractical, if not impossible, due to depletion or extinction of many populations. In particular, members of depleted populations of acipenseriforms rarely achieve the historically reported maximum sizes of individuals prior to exploitation (e.g., *Acipenser transmontanus*, Galbreath 1985). Another example of an intrinsic barrier to phylogenetic study is the potentially large but unknown role of natural hybridization (see Birstein et al. 1997 for review), and varying anthropogenic impacts on hybridization ranging from creation and release of new hybrids to selective overfishing of one species to large scale alterations in river systems. For example, some workers suggest that the hybridization frequency of shovelnose and pallid sturgeons (*Scaphirhynchus platorynchus* and *S. albus*) increased as a result of dredging, damming, and channelizing big-river habitats (Carlson et al. 1985, Phelps & Allendorf 1983).

The main reason, however, why phylogenetic studies of Acipenseriformes are still in their infancy is that few people have ever concentrated on the systematics of the group. This is unfortunate, because systematics offers the only mechanism for comprehensive comparative studies, and such stud-

[1] Dadswell, M.J., B.D. Taubert, T.S. Squires, D. Marchette & J. Buckleye. 1984. Synopsis of biological data on shortnose sturgeon, *Acipenser brevirostrum* LeSueur 1818. NOAA Technical Report NMFS 14.

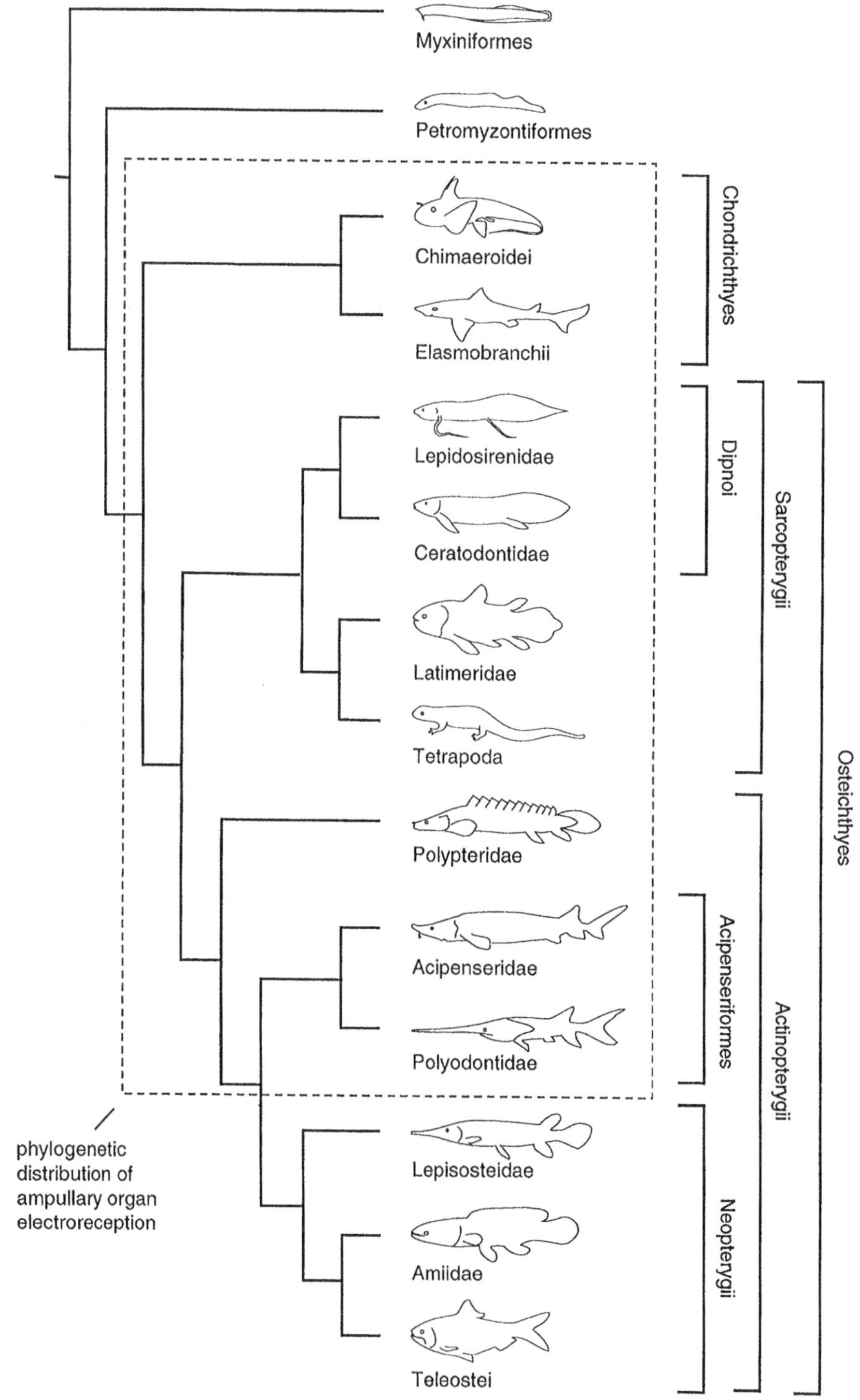

Figure 3. Tree of living craniates showing generally accepted interpretation of relationships for stem Actinopterygii. This tree is based on cladograms summarized by Patterson (1982), Lauder & Liem (1983), Maisey (1986), Schultze (1987), and Northcutt & Bemis (1993). Taxa enclosed in dotted outline are those craniates possessing ampullary electroreception; see Northcutt (1986) for discussion and analysis.

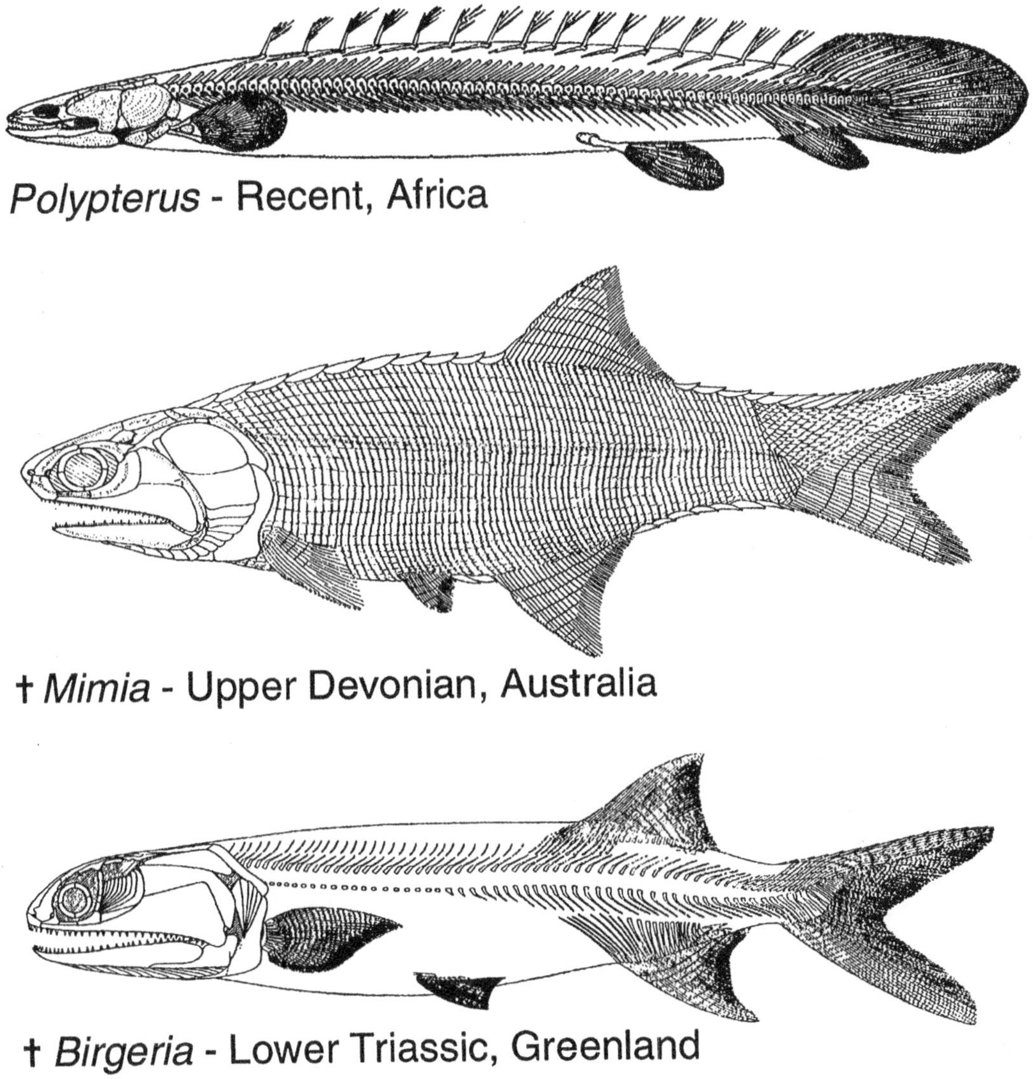

Figure 4. Some living and fossil outgroups of Acipenseriformes: a – *Polypterus* represents a clade generally considered to be the living sister group of all other living Actinopterygii . The rhombic ganoid scales are omitted in this diagram (from Dean 1895). b – †*Mimia* is known from many beautifully preserved specimens (from Gardiner 1984a). c – †*Birgeria* from the Triassic of east Greenland (from Nielsen 1949). †*Birgeria* shares three synapomorphies with Acipenseriformes discussed in the text and tables.

ies are critical to promoting awareness of a group. Acipenseriforms are increasingly threatened in their native ranges (e.g., Birstein 1993, Bemis & Findeis 1994), yet only recently has this translated into more rigorous systematic inquiry (Rochard et al. 1991). There are many outstanding systematic problems which could influence global conservation efforts for the group. For example, we cannot answer here such basic questions as: 'is the genus *Acipenser* monophyletic?' or 'how many valid species of *Acipenser* should we recognize?' These questions will necessarily absorb a great deal of future research because of the broad geographic range occupied by the species of *Acipenser* as well

Family †Chondrosteidae

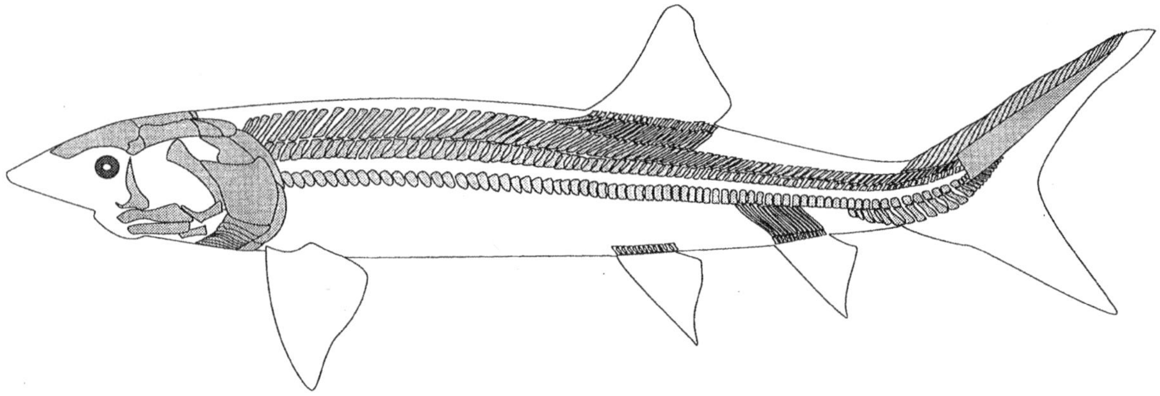

†*Chondrosteus* - Jurassic, Europe

Figure 5. Reconstruction of †*Chondrosteus* from Woodward (1895c). †*Chondrosteus* lacks body scales and has a projectile jaw system.

as the several 'intrinsic' barriers to study described in the preceding paragraph.

Our research program on Acipenseriformes emphasizes generic level relationships using comparative osteology and developmental studies of the skeleton and other tissues. The skeleton provides an excellent source of phylogenetic data which can be reliably recovered from specimens prepared in many different ways. It also allows us to incorporate well-preserved fossils, which give other insights into the evolutionary history of actinopterygians. Acipenseriformes is an old group, known from as far back as the Lower Jurassic of Europe. Certain well-known fossil taxa show that the basic body plans of living sturgeons and paddlefishes were well established by the end of the Cretaceous, and earlier fossils belonging to both of the extant families are being found. For example, Lu (1994) recently described †*Protopsephurus*, an Upper Jurassic paddlefish from China, so that Polyodontidae is as old as the middle Mesozoic. All fossil Acipenseriformes come from the northern hemisphere, which is consistent with the Holarctic range of living species. Finally, although sturgeons and paddlefishes are often loosely called 'living fossils', this does not mean that features present in living sturgeons and paddlefishes are necessarily primitive. Such hy-

Table 1. Selected references for some outgroup taxa.

Sources of osteological data	
†*Cheirolepis*	Pearson & Westoll 1979
Polypterus	Allis 1922 and pers. obs.
†*Mimia*	Gardiner 1984a
†*Moythomasia*	Gardiner 1984a
†*Birgeria*	Nielsen 1949
†*Saurichthys*	Rieppel 1992, Stensiö 1925
Lepisosteus	Wiley 1976 and pers. obs.
Amia	Grande & Bemis 1997

Table 2. Species and biogeographic ranges of †Chondrosteidae Egerton 1858.

†***Chondrosteus*** **Agassiz 1844**	
†*C. acipenseroides* Agassiz 1844	Lower Jurassic – England
†***Strongylosteus*** **Egerton 1858**	
†*S. hindenburgi* Pompeckj 1914	Lower Jurassic – Germany
†***Gyrosteus*** **Agassiz 1844**	
†*G. mirabilis* Agassiz 1844	Jurassic – England

Family †Peipiaosteidae

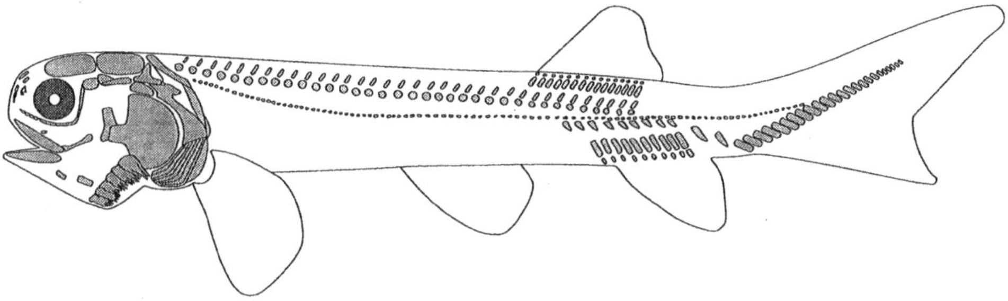

†*Peipiaosteus* - Jurassic, China

igure 6. Reconstruction of †*Peipiaosteus* from Zhou (1992). See further comments and revised interpretations and drawings in Grande & emis (1996).

otheses must be tested by outgroup comparisons o other actinopterygians.

election of taxa for outgroup comparison

igure 3 shows the relationships of living Polypteriae, Acipenseriformes, Lepisosteidae, Amiidae, nd Teleostei as currently understood (Lauder & iem 1983). There is now widespread acceptance hat, among living fishes, Polypteridae is the sister roup of all other Actinopterygii (Goodrich 1928, atterson 1982) and that sturgeons and paddlefishs together form the next extant group on the claogram. Teleostei includes more than 20000 living pecies, whereas Polypteridae, Acipenseriformes, episosteidae and Amiidae together only contain bout 45 living species. Of these 45 living species, 27 re Acipenseriformes, and this order also shows the argest total biogeographic range of any living clade of non-teleostean actinopterygians. Because of their diversity and phylogenetic position as a basal group within Actinopterygii, Acipenseriformes is essential for comparative studies within extant and fossil Actinopterygii.

able 3. Species and biogeographic ranges of †Peipaiosteidae iu & Zhou 1965 (also see Grande & Bemis 1996a).

***Peipaiosteus* Liu & Zhou 1965**	
P. pani Liu & Zhou 1965	Upper Jurassic/Lower Cretaceous – China
P. fengningensis Bai 1983	Upper Jurassic/Lower Cretaceous – China
***Stichopterus* Reis 1910**	
S. woodwardi Reis 1910	Lower Cretaceous – Trans-Baikal
S. popovi Yakovlev 1986	Lower Cretaceous – Mongolia

The earliest known complete skeletons of actinopterygians are from the Devonian (see Long 1995), but isolated scales are reported from the Upper Silurian. By the Carboniferous their diversification had produced a great variety of fishes, commonly known as 'paleonisciforms' (a grade, see Gardiner 1993). 'Typical' Paleozoic and Mesozoic paleonisciforms have heavy, rhombic scales armoring the body, a heterocercal tail, a well ossified skull with solid bony cheeks, and large eyes. More than 200 genera of paleonisciforms are known, but many are poorly preserved or inadequately studied. Increasing knowledge of the anatomy of certain Paleozoic genera such as †*Cheirolepis* (Pearson & Westoll 1979, Pearson 1982), †*Mimia* and †*Moythomasia* (Gardiner 1984a) allows their placement with greater certainty within the phylogenetic scheme for recent actinopterygians, and we follow Gardiner & Schaeffer (1989) in placing these genera near the base of Actinopterygii and including them as outgroups in our analysis of Acipenseriformes.

Several living and fossil genera are relevant outgroups for analyzing relationships among Acipenseriformes (Figure 4). Two Mesozoic genera often

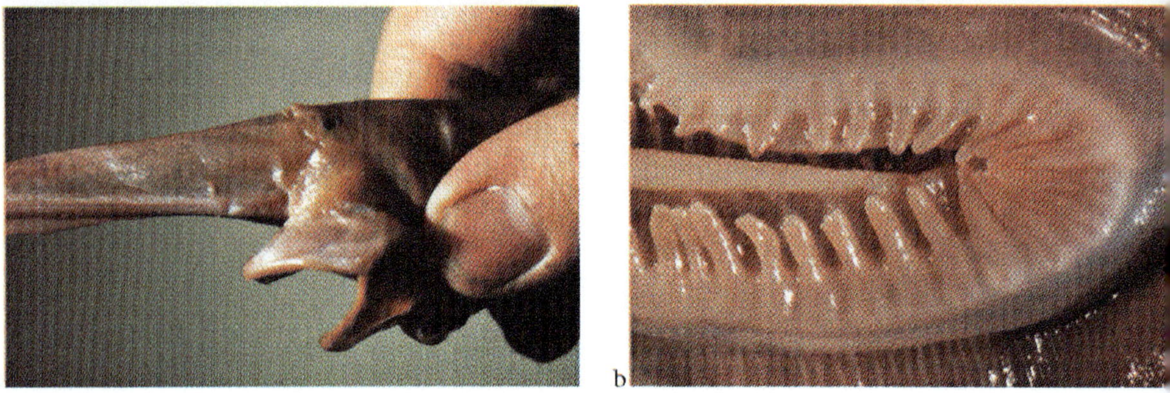

a b

Figure 7. Aspects of the feeding system of the Chinese paddlefish, *Psephurus gladius*: a – A preserved specimen with its jaws i projected position. This projection system is shared by all living and fossil Acipenseriformes except for North American paddl (*Polyodon*). b – View of the first typical gill arch to show gill rakers. The gill rakers of *Psephurus* are short, stubby and unsuited for feeding. This is the plesiomorphic condition for Polyodontidae (see Grande & Bemis 1991, fig. 26).

linked with Acipenseriformes are †*Saurichthys* (a widespread and speciose genus from the Triassic and Early Jurassic, see Rieppel 1992 for review) and †*Birgeria* (particularly the species from the Triassic of east Greenland; see Nielsen 1949 and Yakovlev 1977). In their summary phylogeny, Gardiner & Schaeffer (1989, their fig. 12; their 'chondrostean group' is equivalent to Acipenseriformes here) show a group containing †*Saurichthys* as the immediate sister group of Acipenseriformes. This position, however, is only one of several equally parsimonious possibilities from their cladistic analysis, and so must be regarded as uncertain. Rieppel (1992), in a review of the genus †*Saurichthys*, concluded that Acipenseriformes, †*Saurichthys*, and †*Birgeria* form an unresolved trichotomy.

In selecting the outgroup taxa listed in Table 1, we were guided by their putative phylogenetic positions, the availability of detailed osteological scriptions, and the general desirability of incluc a spectrum of taxa. Better understanding of the lationships of Acipenseriformes to other group Actinopterygii can be achieved by a detailed sp men-based review of these and other taxa, inc ing preparation of many of the known fossils Grande & Bemis 1996 for example of †*Peipi teus*), but this is far beyond our present purpos

Diversity of fossil and recent Acipenseriformes specification of ingroup taxa

In this section, we briefly review the known and their geographic and geological ranges. We identify which taxa were the sources of chara information for our analysis. The text is sup

Table 4. Species and biogeographic ranges of fossil and extant Polyodontidae Bonaparte 1838.

†***Protopsephurus* Lu 1994 – China**	
†*P. liui* Lu 1994	Upper Jurassic – China
†***Paleopsephurus* MacAlpin 1941a – North America**	
†*P. wilsoni* MacAlpin 1941a	Upper Cretaceous – Montana
***Psephurus* Günther 1873 – China**	
P. gladius (Martens 1862)	Yangtze River drainage, China
†***Crossopholis* Cope 1883 – North America west of Rocky Mountains**	
†*C. magnicaudatus* Cope 1883	Lower Eocene – Wyoming
***Polyodon* Lacépède 1797 – North America east of Rocky Mountains**	
P. spathula (Walbaum 1792)	Mississippi River drainage
†*P. tuberculata* Grande & Bemis 1991	Lower Paleocene – Montana

igure 8. Green River paddlefish, †*Crossopholis magnicaudatus*. More than twelve complete specimens of this taxon are known, al- ough it remains one of the rarest fishes in the Green River Formation (Grande & Bemis 1991).

nented by tabular summaries (Tables 2-5) and il- ustrations (Figure 5-16).

. Family †Chondrosteidae Egerton 1858

Chondrosteus acipenseroides Agassiz 1844 from he Lower Jurassic of England is based on multiple, omplete specimens. Some authors consider the ge- us †*Srongylosteus* (as represented by †*S. hinden- ergi* Pompeckj 1914 from the Lower Jurassic of Germany) to be synonymous with †*Chondrosteus*; he few reported differences between these genera eed new study. In the absence of modern studies, apers by Egerton (1858), Traquair (1887) and Hen- ig (1925) remain useful. Traquair (1887) empha- ized morphological similarities between †*Chon- lrosteus* and acipenserids, including jaws free from he cheek, reduced scales, and reduced ossification, nd his work provided the basis for the classic in- erpretation that †*Chondrosteus* and Acipenseri- lae are sister groups, an interpretation rejected by Grande & Bemis (1991, also see discussion of node Acipenseroidei below).

†*Gyrosteus mirabilis* is known from several in- omplete specimens from the Jurassic of England summarized by Woodward 1891, 1895a,b,c) which are not suited for detailed study. These were spectacularly large fish: the ossified portion of one hyomandibula exceeds 50 cm, compared to an entire hyomandibula only 12 cm long from a two meter *Acipenser oxyrinchus*. As far as we can tell, †*Gyrosteus* appears to be undiagnosable as a genus distinct from †*Chondrosteus*.

We used published data on †*Chondrosteus acipenseroides* and †*Chondrosteus* (= †*Strongylosteus*) *hindenbergi* from Egerton (1858), Traquair (1887) and Hennig (1925) in our phylogenetic analysis.

2. Family †Peipiaosteidae Liu & Zhou 1965

Yakovlev (1977, 1986) does not consider that †Peipiaosteidae is distinct from †Chondrosteidae. We disagree, and find that the two families probably do not even form a monophyletic group.

Two genera and four nominal species are recognized in †Peipiaosteidae (Table 3). †*Peipiaosteus* is known from two very similar species from the Upper Jurassic of Northern China. †*Peipiaosteus pani* was described from about 40 specimens (Liu & Zhou 1965). More recently, †*P. fengningensis* was described by Bai (1983). In a recent review, Zhou

a

Figure 9. The living North American paddlefish, *Polyodon spathula*: a – A pond-reared specimen from a commercial fish farm in Missouri. b – Gill rakers and entrapped plankton in an adult *Polyodon*.

(1992) provided a new reconstruction (Figure 6). †*Stichopterus woodwardi* Reis 1910 occurs in the Lower Cretaceous of Trans Baikal; this species was treated in more detail by Yakovlev (1977). A second species, †*Stichopterus popovi* Yakovlev 1986 comes from the Lower Cretaceous of Mongolia. Yakovlev (1977) questioned whether †*Stichopterus* and †*Peipiaosteus* warrant separate generic status. In answering this question, Zhou (1992) summarized some striking differences between the two genera, such as the presence of an endopterygoid and palate in †*Stichopterus* (not found in †*Peipiaosteus*) and the presence of rhombic scales in the upper lobe of the caudal fin in †*Stichopterus* (not found in †*Peipiaosteus*). All of this material warrants new additional specimen-based study.

We used published data on †*Peipiaosteus pani* from Liu & Zhou (1965) and Zhou (1992) for our analysis (see Grande & Bemis 1996 for data on †*Peipiaosteus* collected after this paper was prepared; two additional genera, †*Yanosteus* and †*Spherosteus* are treated there).

3. Family Polyodontidae Bonaparte 1838

Unlike sturgeons, paddlefishes are usually regarded as primary freshwater fishes (e.g., Swift et al. 1986), and although individuals may occasionally stray into coastal marine environments (Vladykov & Greeley 1963) all fossil paddlefishes are from freshwater deposits. This makes the family partic-

ularly useful for biogeographic interpretati Grande & Bemis (1991) studied osteology and r tionships among four genera of paddlefishes († *leopsephurus*, *Psephurus*, †*Crossopholis* and *Po don*; Table 4). Recently, a new Mesozoic ge †*Protopsephurus* was described (Lu 1994) and pu tively assigned to Polyodontidae. †*Protopsephu*

Figure 10. A cleared and double stained juvenile specimen o luga, *Huso huso*. Bone is stained red, cartilage stained blue. N the endochondral rostrum sheathed by dorsal and ventral ro bones, the projectile jaws, reduction of the lateral line cana simple tubular bones, and the series of scutes on the trunk.

does have a paddle-shaped rostrum and the large jaws characteristic of other paddlefishes and Lu (1994) reports (but does not illustrate) the presence of stellate bones in this material. This placement seems reasonable, but more study of this material is needed.

†*Paleopsephurus wilsoni* MacAlpin 1941a from the late Cretaceous Hell Creek formation of Montana is the only species of the genus. It is known from a single partial skull and caudal fin (MacAlpin 1941b, 1947). Further preparation of the remaining material revealed important new details, such as the presence of stellate bones in the paddle, which had been overlooked or misinterpreted by MacAlpin (see Grande & Bemis 1991). This species has short, triangular-shaped gill rakers, so it clearly was not specialized for filter feeding.

The large, piscivorous Chinese paddlefish, *Psephurus gladius*, is restricted to the Yangtze River (Figure 7). Described as *Polyodon gladius* by Martens (1862), it was transferred to the new genus *Psephurus* by Günther (1873) because of its moderate number of comparatively shorter gill rakers and its smaller number of large caudal fulcra. Relatively few papers focus on *Psephurus* (Handyside 1875a,b, Nichols 1928, 1943, Tatarko 1936, 1939, MacAlpin 1947, Vasetskiy 1971, Liu & Zeng 1988, Yu et al. 1986, Grande & Bemis 1991, Liu et al. 1995). Although it probably never reached the 7 meters total length commonly cited for this species, *Psephurus* reached at least 4 meters (Grande & Bemis 1991).

Table 5. Species and geographic ranges of some fossil and all recent Acipenseridae, Bonaparte 1831. Additional fossil species are listed in Wilimovsky (1956).

***Huso* Brandt 1869 – Eurasia**	
H. huso (Linnaeus 1758)	Black, Caspian, Mediterranean seas
H. dauricus (Georgi 1775)	Amur River drainage
***Acipenser* Linnaeus 1758 – Holarctic**	
A. oxyrinchus Mitchill 1815	North America – Atlantic coast
A. brevirostrum Le Sueur 1818	North America – Atlantic coast
A. fulvescens Rafinesque 1817	North America – central United States
A. sturio Linnaeus 1758	Europe – Atlantic coast, Mediterranean Sea
A. naccarii Bonaparte 1836	Adriatic Sea
A. stellatus Pallas 1771	Black, Caspian, Mediterranean seas
A. gueldenstaedtii Brandt & Ratzeberg 1833	Black, Caspian seas
A. persicus Borodin 1897	Black, Caspian seas
A. nudiventris Lovetzky 1828	Black, Caspian, Aral seas
A. ruthenus Linnaeus 1758	Rivers of east-central Europe
A. baerii Brandt 1869	Rivers of north coast of Russia
A. schrenckii Brandt 1869	Amur River drainage, Sea of Okhotsk
A. dabryanus Duméril 1868	China
A. sinensis Gray 1834	China, south Japan
A. medirostris Ayres 1854	North America, Asia – Pacific coast
A. transmontanus Richardson 1836	North America – Pacific coast
†*A. albertensis* Lambe 1902	Upper Cretaceous – Alberta
†*A. toliapicus* Agassiz 1844	Lower Eocene – England
†*A. ornatus* Leidy 1873	Miocene – Virginia
***Scaphirhynchus* Heckel 1836 – North America**	
S. platorynchus (Rafinesque 1820)	Mississippi River drainage
S. suttkusi Williams & Clemmer 1991	Mobile Bay drainage
S. albus (Forbes & Richardson 1905)	Mississippi River drainage
†*Protoscaphirhynchus* Wilimovsky 1956 – Montana	
†*P. squamosus* Wilimovsky 1956	Upper Cretaceous – Montana
***Pseudoscaphirhynchus* Nikolskii 1900 – Aral Sea drainages**	
P. kaufmanni (Bogdanow 1874)	Syr-Darya River
P. hermanni (Severtzoff in Kessler 1877)	Syr-Darya River
P. fedtshenkoi (Kessler 1872)	Amu-Darya River

Figure 11. A large specimen of kaluga, *Huso dauricus*, from the Amur River near Khabarovsk, Siberia. Photograph courtesy of Viktor Svirskii, TINRO, Vladivostok.

Large specimens are especially poorly represented in systematic collections, and this species is now severely threatened owing to construction of dams and overfishing (Wei et al. 1997 this volume).

Although †*Crossopholis magnicaudatus* (Figure 8) is comparatively rare in the extensively collected fauna of Fossil Lake (Lower Eocene of Wyoming) it is known from an excellent series of complete specimens (Grande & Bemis 1991). Originally described by Cope (1883) and later redescribed in detail by

b

Figure 12. Two North American species of Acipenser: a – A shortnose sturgeon, *Acipenser brevirostrum*, being stripped by B. Kynard and a student for captive production of eggs; b – lake sturgeon, *Acipenser fulvescens* juvenile.

ande & Bemis (1991), †*Crossopholis* is a modest-sized polyodontid that was clearly piscivorous as idenced by the presence of fish in the body cavity several specimens. Interestingly, traces of ampuly organs can be seen between stellate bones in : paddle of some specimens.

The extant North American paddlefish, *Polyon spathula* (Figure 9) is an intensively studied ecies (see bibliographies in Grande & Bemis 1991, llard et al. 1986 and Graham 1997 this volume). *lyodon* is perhaps best known for its filter-feed; habit based on numerous thin, elongate gill rak; unique to this genus among Acipenseriformes. *lyodon spathula* occurred as far north as Lake ie (Trautman 1981), but the species typically inbits large river systems, and, prior to commercial ploitation, was common in the Mississippi River d its tributaries (Gengerke 1986, Russell 1986, Graham 1997). A Lower Paleocene species from Montana, †*Polyodon tuberculata* Grande & Bemis 1991, shares the elongate gill rakers.

†*Pholidurus disjectus*, from the Jurassic of England, historically regarded as a polyodontid (Woodward 1895) was removed from Polyodontidae by Grande & Bemis (1991). Thus, the family Polyodontidae as presently known is restricted to North America and Asia.

We used representatives of the four better known genera of paddlefishes (†*Paleopsephurus wilsoni*, *Psephurus gladius*, †*Crossopholis magnicaudatus* and *Polyodon spathula*) in our phylogenetic analysis. Because only a few details are available concerning †*Protopsephurus* Lu 1994, we leave it as an unresolved multichotomy with other paddlefishes.

4. Family Acipenseridae Bonaparte 1831

As summarized in Table 5, the family Acipenseridae includes four extant genera (*Huso, Acipenser, Scaphirhynchus* and *Pseudoscaphirhychus*) and one putative fossil genus (†*Protoscaphirhynchus*). Other fossil and subfossil material has been described, but most of it is fragmentary. Many populations of sturgeons are severely depleted (e.g., Holčík et al. 1989, Birstein 1993 and many papers in this volume) or extinct (e.g., Aral Sea ship sturgeon, *A. nudiventris*, see Zholdasova 1997 this volume). A thorough anatomical description of one species, the sterlet, *A. ruthenus*, is available (Marinelli & Stren-

Figure 13. Map of eastern Europe and Asia showing a region of high diversity for Acipenseridae. Three of the four extant genera of sturgeons occur in the area of eastern Europe and central Asia shown in the enlargement at the bottom of the page.

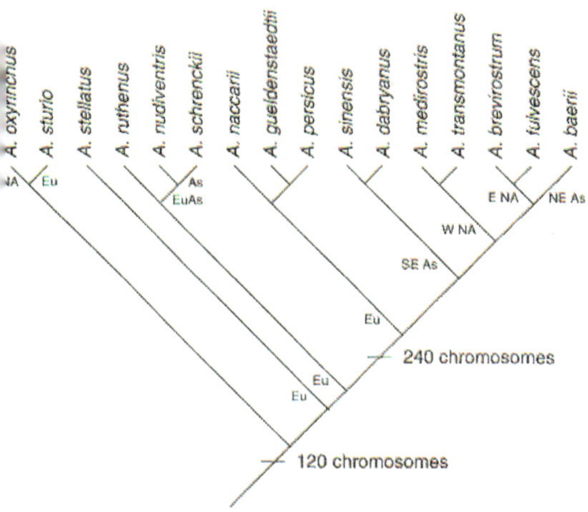

ıre 14. An interpretation of phylogenetic relationships within *ɒenser* based on Artyukhin (1995a and pers. comm.). This : is based on karyological data and biogeographic interpreta-ı; this genus is in need of much additional study.

· 1973). Sturgeons are also the subject of many nparative anatomical treatments (e.g., Jessen '2, 1973) and much developmental research (e.g., nsburg & Dettlaff 1991, Dettlaff et al. 1993). Find- (1997) described and reviewed skeletal anatomy extant sturgeons to make a cladistic analysis of ·ir interrelationships.

The genus *Huso* Brandt 1869 is known from two ant species from Eurasia (Figure 10, 11). Com-nly known as beluga, *H. huso* is the largest fish to ter freshwater, historically reaching lengths of 6 10 meters (Balon 1967, 1968, Pirogovskii et al. 39, Baruš & Oliva 1995). The kaluga, *H. dauricus* abits the Amur River system, where it is a target an aggressive fishery (Krykhtin & Svirskii 1997 s volume). Even as juveniles, these preferentially civorous sturgeons target fish as prey. The status *Huso* as a genus separate from *Acipenser* was un-ar to many 19th century workers, who considered subgenus of *Acipenser* (e.g., Fitzinger & Heckel 36). Recently, Jollie (1980) perpetuated this inter-:tation, although since Brandt (1869), Berg 04), and Antoniu-Murgoci (1936a,b), it has been sy to distinguish this genus. Findeis (1997) provid-additional new characters separating *Huso* from other extant acipenserids. Unexpectedly, mole-lar data given by Birstein et al. (1997) consistently st *Huso* within *Acipenser*.

With 17 extant species, *Acipenser* is the most spe-

Figure 15. Color photograph of an aquarium specimen of the large Amu-Darya shovelnose sturgeon, *Pseudoscaphirhynchus kaufmanni*. These sturgeons are threatened due to environmental degradation in the region of the Aral Sea. Photograph courtesy of Boris Goncharov, Koltsov Institute of Developmental Biology, Moscow.

ciose and problematic taxon within Acipenseriformes (Figure 12). The fossil record of *Acipenser* has not helped phylogeneticists. Most described fossil species are known only by fragmentary material insufficient for differential diagnosis. There has never been a comprehensive phylogenetic study of the genus, no doubt due to its vast geographic range and difficulties obtaining certain species for study. The validity of several recent species, such as *Acipenser mikadoi* is generally questioned, although other evidence suggests that this species is distinct from *A. medirostris* (see Birstein et al. 1993, 1997). Many species of *Acipenser* are endemic to eastern Europe and Asia (Figure 13) and are poorly known outside of this region. Traditional ideas about relationships within *Acipenser* relied on biogeography

Figure 16. Color photograph of a captive reared specimen of the common shovelnose sturgeon, *Scaphirhynchus platorynchus* from the central United States.

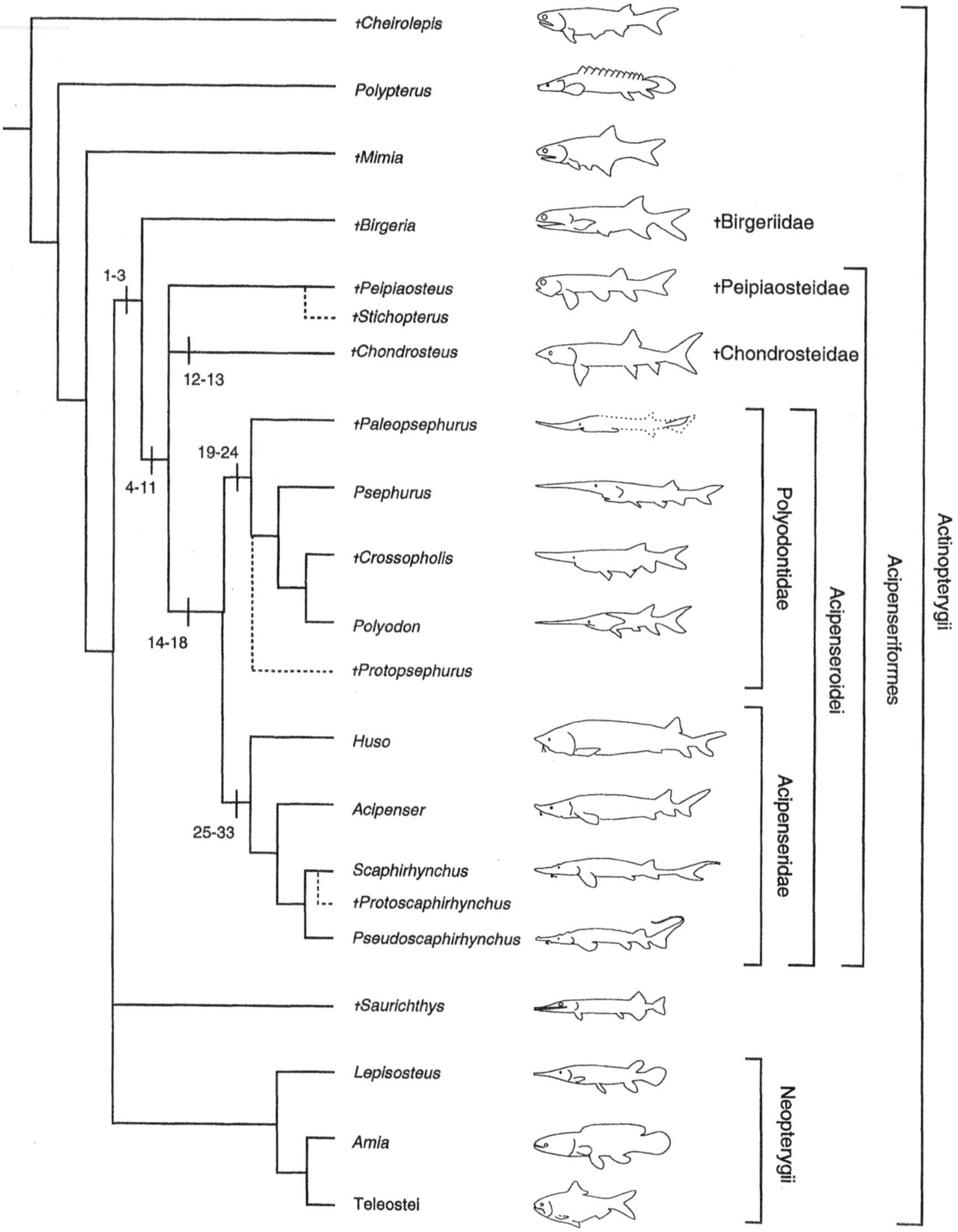

Figure 17. Preliminary cladogram of acipenseriform interrelationships. Character numbers correspond to Tables 6 and 7. Arrangement *Cheirolepis, Polypterus* and † *Mimia* follows Gardiner & Schaeffer (1989), and placement of † *Saurichthys* follows Rieppel (1992). F characters used in interpreting relationships of outgroups, see Rieppel (1992). For characters used in interpreting intrafamilial relatio ships of Polyodontidae, see Grande & Bemis (1991). For characters used in interpreting intrafamilial relationships of Acipenseridae, s Findeis (1997). Acipenseriform interrelationships are further resolved in Grand & Bemis (1996, see note added in proofs on page 71). V follow current convention for arrangement of neopterygians (see Grande & Bemis 1997).

as a first approximation (Figure 14, courtesy of E. Artyukhin, see Artyukhin 1994, 1995a,b; also see Birstein et al. 1997). Meaningful biological generalizations are difficult, for species of *Acipenser* range in size from the tiny sterlet (*A. ruthenus*) and shortnose sturgeon (*A. brevirostrum*) to the immense white sturgeon (*A. transmontanus*), Atlantic sturgeon (*A. oxyrinchus*) and common sturgeon (*A. sturio*). Several species have some exclusively freshwater populations (e.g., *A. fulvescens* Vladykov & Greeley 1963, *A. baerii* in Lake Baikal, Ruban 1997 this volume), while stocks of other species spend a great percentage of time at sea (e.g., *A. medirostris*). Findeis (1997) studied seven species of *Acipenser* (*A. brevirostrum*, *A. oxyrinchus*, *A. transmontanus*, *A. fulvescens*, *A. medirostris* and *A. ruthenus*), but failed to find any osteological synapomorphies of the genus. *Acipenser* may be paraphyletic, and it

Table 6. Character key. Characters for acipenseriform outgroups taken from sources listed in Table 1.

Synapomorphies of †*Birgeria* + Acipenseriformes
1. Reduction of the opercle
2. Elongate posterior extension of parasphenoid
3. Body scaling reduced to tiny isolated elements or absent

Synapomorphies of Acipenseriformes
4. Palatoquadrates with an anterior symphysis
5. Palatoquadrate with broad autopalatine portion, palatoquadrate bridge, and quadrate flange
6. Presence of a triradiate quadratojugal bone
7. Gill-arch dentition confined to first two hypobranchials and upper part of first arch
8. Subopercle possesses an anterior process
9. Preopercular canal in a series of ossicles, mandibular canal short or absent
10. Infraorbital sensory canal in a series of ossicles
11. Loss of premaxillary and maxillary bones

Synapomorphies of †Peipiaosteidae

None here; see Grande & Bemis (1996, written after this paper was accepted)

Synapomorphies of †Chondrosteidae
12. Anterior part of palatopterygoid club-shaped
13. Complete loss of trunk scalation

Synapomorphies of Acipenseroidei
14. Loss of opercle
15. Reduction in number of branchiostegals supporting gill cover
16. Endocranium with extensive rostrum
17. Dorsal and ventral rostral bones
18. Ventral process of posttemporal bone

Synapomorphies of Polyodontidae
19. Many small stellate bones make up lateral supports for the paddle
20. Series of very elongate dorsal and ventral medial rostral bones, with cylindrical cross-sections
21. Unique shape of subopercle
22. Elongate anterior and posterior divisions of the fenestra longitudinalis in the skull roof
23. Posttemporal with elongate anterior arm suturing into the dermosphenotic
24. Single branchiostegal with branched posterior edge

Synapomorphies of Acipenseridae
25. Five scute rows along trunk
26. Pectoral fin spine
27. Antorbital bone
28. Commissure of occipital canals in median extrascapular bone (see Grande & Bemis 1996 for correction)
29. Rostral canals curve lateral to barbels
30. Supracleithrum tightly joined to dermal skull roof
31. Opercular wall formed by cleithrum and clavicle
32. Cardiac shield formed by clavicle
33. Cleithral process limits mobility of pectoral fin spine

certainly warrants much additional systematic research. One important point, however, is that all seven of the species surveyed here share twelve osteological synapomorphies with the tribe Scaphirhynchini (Findeis 1997). This conclusion is reflected in our classification below.

The systematic and conservation status of the three nominal species of central Asian *Pseudoscaphirhynchus* is unclear. These are the smallest extant sturgeons (Figure 15), and can be readily distinguished from *Scaphirhynchus* by differences in gill raker anatomy and the scalation of the caudal peduncle (Findeis 1997). Two of the three species (*P. fedtschenkoi* and *P. hermanni*) appear to have become extinct in the recent decades, and the third species (*P. kaufmanni*) is threatened (see Birstein 1993 and Zholdasova 1997 for review). They were endemic to the Amu-Darya and Syr-Darya rivers, tributaries of the Aral Sea in Central Asia (Kazakhstan, Uzbekistan and Turkmenistan; see Berg 1948a, Tleuov & Sagitov 1973). This region has been subjected to extreme environmental degradation, including pesticide pollution and water diversion projects (Ellis 1990). Some classical anatomical descriptions of *Pseudoscaphirhynchus* were made (e.g., Ivanzoff 1887, Sewertzoff 1925), as well as recent embryological studies (Goncharov et al. 1991, Schmalhausen 1991), but very few specimens of

Table 7. Preliminary survey of characters for Figure 17, mostly based on literature (Table 1) (? = unknown, N = inapplicable, P = polymorphic). See Grande & Bemis (1966) for analysis made after this paper was written (* this character should be scored as a '1').

Char	Pol	Bir	Cho	Pei	Pal	Psep	Cro	Pol	Hus	Aci	Psu	Sca	Lep	Ami
1.	0	1	1	1	1	1	1	1	1	1	1	1	0	0
2.	0	1	1	1	1	1	1	1	1	1	1	1	0	0
3.	0	1	1	1	1	1	1	1	1	1	1	1	0	0
4.	N	0	1	1	?	1	1	1	1	1	1	1	N	N
5.	N	0	1	1	?	1	1	1	1	1	1	1	N	N
6.	0	?	1	1	1	1	1	1	1	1	1	1	0	0
7.	0	?	?	1	?	1	1	1	1	1	1	1	0	0
8.	0	0	1	1	1	1	1	1	1	1	1	1	0	0
9.	0	0	1	1	1	1	1	1	1	1	1	1	0	0
10.	0	0	1	1	1	1	1	1	1	1	1	1	0	0
11.	0	0	1	1	1	1	1	1	1	1	1	1	?	0
12.	N	0	1	0	0	0	0	0	0	0	0	0	N	N
13.	0	0	1	0	?	0	0	0	0	0	0	0	0	0
14.	0	0	0	0	1	1	1	1	1	1	1	1	0	0
15.	1	0	0	0	1	1	1	1	1	1	1	1	0	0
16.	0	0	0	0	1	1	1	1	1	1	1	1	0	0
17.	0	0	0	0	1	1	1	1	1	1	1	1	0	0
18.	0	?	0	0	1	1	1	1	1	1	1	1	0	0
19.	0	0	0	0	1	1	1	1	0	0	0	0	0	0
20.	0	0	0	0	1	1	1	1	0	0	0	0	0	0
21.	0	0	0	0	1	1	1	1	0	0	0	0	0	0
22.	0	0	0	0	1	1	1	1	0	0	0	0	0	0
23.	0	0	0	0	1	1	1	1	0	0	0	0	0	0
24.	0	0	0	0	1	1	1	1	0	0	0	0	0	0
25.	0	0	0	0	0	0	0	0	1	1	1	1	0	0
26.	0	0	0	0	0	0	0	0	1	1	1	1	0	0
27.	0	0	0	0	?	0	0	0	1	1	1	1	0	0
28.	P	0	?	0	0*	0	0	0	1	1	1	1	P	0
29.	N	0	0	0	?	0	0	0	1	1	1	1	N	N
30.	0	0	0	0	0	0	0	0	1	1	1	1	0	0
31.	0	0	0	0	0	0	0	0	1	1	1	1	0	0
32.	0	0	0	0	0	0	0	0	1	1	1	1	0	0
33.	0	0	0	0	0	0	0	0	1	1	1	1	0	0

Pseudoscaphirhynchus are available in systematic collections outside Russia and Uzbekistan, and there has never been a modern specimen based systematic review of all three species. For example, *P. kaufmanni* has been described as having two 'morphs' in the Amu-Darya river, with only the diminutive morph persisting as of the last available information, but whether these were two separate species is unclear (see Berg 1948a).

The genus *Scaphirhynchus* (American shovelnose sturgeons) is represented by three nominal species from North America (Figure 16). They are restricted to freshwater rivers and prefer open, free-flowing reaches with soft, silty bottoms. *Scaphirhynchus* distinctively possesses an elongate, armored caudal peduncle and several osteological characters defined by Findeis (1997). Common shovelnose sturgeon (*Scaphirhynchus platorynchus*) rarely exceed one meter total length, but pallid sturgeon (*S. albus*) can be much larger (Bailey & Cross 1954). Pallid sturgeons are endangered, and apparently readily hybridize with common shovelnose sturgeons. The rare Alabama sturgeon (*S. suttkussi*) closely resembles common shovelnose sturgeon; its status as a distinct species is debated.

A partially intact fossil from the Upper Cretaceous of Montana was described as †*Protoscaphirhynchus squamosus* Wilimovsky 1956. Although Gardiner (1984b) questioned its placement within Acipenseridae, the specimen does possess an elongate caudal peduncle with armoring scales extremely similar to those of *Scaphirhynchus*. No features clearly distinguish it from *Scaphirhynchus*, so †*Protoscaphirhynchus* may not warrant separate generic status. It appears to belong in Scaphirhynchini (as used here) indicating that the tribe was definitely present in North America at least 65 million years before present.

Material described as an acipenserid from the Selma Formation of Alabama (†*Propenser* Applegate 1970) is now interpreted as portions of a coelacanth and a pycnodont.

We used *Huso huso*, *Acipenser brevirostrum*, *Scaphirhynchus platorynchus*, and *Pseudoscaphirhychus kaufmanni* for our phylogenetic analysis (Table 7, Figure 17).

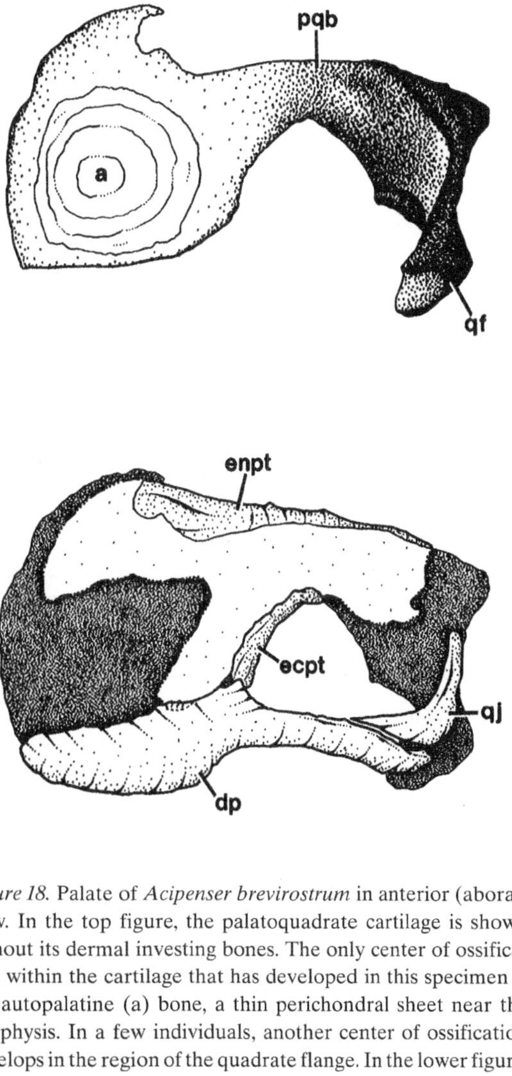

Figure 18. Palate of *Acipenser brevirostrum* in anterior (aboral) view. In the top figure, the palatoquadrate cartilage is shown without its dermal investing bones. The only center of ossification within the cartilage that has developed in this specimen is the autopalatine (a) bone, a thin perichondral sheet near the symphysis. In a few individuals, another center of ossification develops in the region of the quadrate flange. In the lower figure, the palatoquadrate cartilage is indicated with heavy stipple. It has four investing dermal bones: dp, dermopalatine; ecpt, ectopterygoid, qj, quadratojugal, and enpt, entopterygoid. The entopterygoid bone extends on both the oral and aboral surfaces of the palate; very light stippling demarcates the oral portion of the entopterygoid as seen through the palatoquadrate cartilage. Total length of specimen: 870 mm. UMA 24-116-2-47.

Phylogenetic interpretation

This section surveys characters and interpretations for the generic-level phylogenetic analysis summarized in Figure 17. The characters are listed or described in the text and Table 6 and 7. All characters

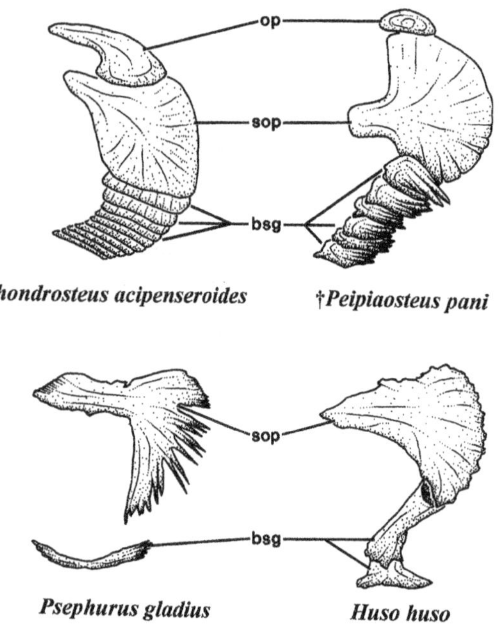

Figure 19. Lateral views of the opercular series of four taxa of Acipenseriformes to show loss of the opercle in Acipenseroidei (Polyodontidae + Acipenseridae), hypertrophy of the subopercle, and reduction in the number of branchiostegal bones: †*Chondrosteus acipenseroides* (from Traquair 1887), †*Peipiaosteus pani* (from Zhou 1992), *Psephurus gladius* (from Grande & Bemis 1991), and *Huso huso* (original, CAS 37541).

could be reliably scored in at least 16 of the 18 taxa listed in Table 7, but most soft-tissue and cartilaginous characters cannot be scored in the fossil taxa (see Grande & Bemis 1997 for additional explanations of this approach). Incorporated into this diagram are published interpretations of relationships for the genera of Polyodontidae (from Grande & Bemis 1991) and Acipenseridae (from Findeis 1993, 1997). Consult those references for characters within the two families.

The starting points for our analysis are Patterson (1982), Gardiner & Schaeffer (1989), Grande & Bemis (1991), Rieppel (1992) and Findeis (1997). For those characters that are well-known, and explained in detail elsewhere, we provide abbreviated descriptions. For other characters, new information warrants presentation, and in a few cases, new illustrations to clarify definitions. Each subsection concludes with notations on other potential characters, comments on alternative interpretations, and a statement concerning the robustness of the clade.

*Characters of †*Birgeria *+ Acipenseriformes*

We specify three characters as synapomorphic for this group.

Character 1. Reduction in size of the opercle
This character was noted by Gardiner & Schaeffer (1989). The largest element of the opercular series of Acipenseriformes is the subopercle bone (Figure 19). It is the most dorsal element of the opercular series in acipenserids and polyodontids, but a smaller, more dorsal opercle bone is present in †*Chondrosteus* and †*Peipiaosteus*. The opercle of †*Birgeria* is also small (Nielsen 1949, Gardiner & Schaeffer 1989). In *Polypterus*, †*Mimia*, †*Moythomasia*, *Lepisosteus*, *Amia* and most other actinopterygians (McCallister 1968), the opercle is the largest element of the opercular series, and we interpret its reduction to a small bone (as in †*Chondrosteus* and †*Peipiaosteus*) as synapomorphic for Acipenseriformes. The opercle is eventually lost in Acipenseroidei (character 14, below).

Character 2. Elongate posterior extension of parasphenoid
The parasphenoid of all Acipenseriformes has an elongate posterior extension. The extension underlies a series of vertebral segments partially or completely fused into the occipital region of the neurocranium. The parasphenoid bones of †*Chondrosteus* and †*Peipiaosteus* are shorter than those of living Acipenseriformes, but still much longer than in any of the outgroups, including †*Birgeria*, which is noted for having a more elongate parasphenoid than other paleonisciforms (Nielsen 1949). In the outgroup taxa, a vertebral joint close to the neurocranium is probably necessary if head-lift is to play any role in raising the upper jaw (discussed in final section of this paper). Because acipenseriforms project their jaws and do not utilize head-lift when feeding, the elongate parasphenoid and extension of the neurocranium present no interference.

Character 3. Body scaling reduced to tiny isolated elements or absent
This character was defined by Patterson (1982), noted by Gardiner & Schaeffer (1989) and accepted by Grande & Bemis (1991, character 5). The scales of †*Cheirolepis* are not typical for stem actinopterygians in that they are small tubercles, but they are not isolated from each other as in our ingroup taxa. Acipenseriformes differ from all of the outgroups we considered except †*Birgeria*, which has reduced scalation that Nielsen (1949) regarded as very similar to that of *Polyodon*.

*Discussion of clade †*Birgeria *+ Acipenseriformes*

Like Berg (1948b) and Yakovlev (1977) and in contrast to Aldinger (1937) and Nielsen (1949), our analysis does not support a sister group relationship between †*Birgeria* and paddlefishes. However, †*Birgeria* shares at least three features with Acipenseriformes. These three characters warrant additional specimen based study in †*Birgeria* and other outgroups not considered here.

Characters of Acipenseriformes

We report eight osteological synapomorphies of Acipenseriformes keyed to the cladogram in Figure 17.

Character 4. Palatoquadrates with an anterior symphysis
Woodward (1891) and many others have noted this feature, but it was more explicitly defined by Patterson (1982). Gardiner & Schaeffer (1989) and Grande & Bemis (1991, character 1) regarded this feature as an acipenseriform synapomorphy. The symphysis is a flexible cartilaginous connection in all recent Acipenseriformes, and even though the cartilage is not preserved, the configuration of the upper jaw suggests its presence in †*Chondrosteus* and †*Peipiaosteus*. (More study of this character in †*Peipiaosteus* is needed, however.) The palatoquadrate symphysis appears to be absent in †*Cheirolepis*, *Polypterus*, †*Mimia*, †*Birgeria*, *Amia*, and other actinopterygians. Gardiner & Schaeffer (1989) included as part of their character definition that the palatoquadrates of Acipenseriformes 'do not articulate with the neurocranium', but we note that the palatoquadrate is guided by the cartilaginous basitrabecular processes during projection, so their additional qualification is unclear (see discussion under 'Putative and problematic characters of Acipenseriformes', below).

Character 5. Palatoquadrate with broad autopalatine portion, palatoquadrate bridge, and quadrate flange
Schaeffer (1973) and Patterson (1982) considered palatoquadrate shape to be an acipenseriform character, but did not specify which aspects are synapomorphic. An illustration of the palatoquadrate of a large adult *Acipenser brevirostrum* is shown in Figure 18. All acipenseriform palatoquadrates have a broad autopalatine plate (Figure 18a), a narrow palatoquadrate bridge (pqb), and a quadrate flange (qf). In comparison, †*Cheirolepis*, *Polypterus*, †*Mimia*, and *Amia* possess roughly triangular palatoquadrates with thin anterior tips that extend to contact the ethmoid region of the neurocranium.

Character 6. Presence of a triradiate quadratojugal bone
This character was defined by Patterson (1982) and accepted by Grande & Bemis (1991, character 6). The condition of the quadratojugal is unknown in †*Birgeria* (Nielsen 1949). Figure 18 shows the quadratojugal of *Acipenser brevirostrum*. An example of the triradiate quadratojugal in a polyodontid is illustrated in Grande & Bemis (1991, fig. 35A,C).

Character 7. Gill-arch dentition confined to first two hypobranchials and upper part of first arch
This character was discussed by Nelson (1969, p. 257) defined by Patterson (1982), noted by Gardiner & Schaeffer (1989) and accepted by Grande & Bemis (1991, character 2). It is illustrated in Grande & Bemis (1991, fig. 17D). More information on this character in †*Birgeria* is needed.

Character 8. Subopercle possesses an anterior process
The subopercle possesses an anterior process that

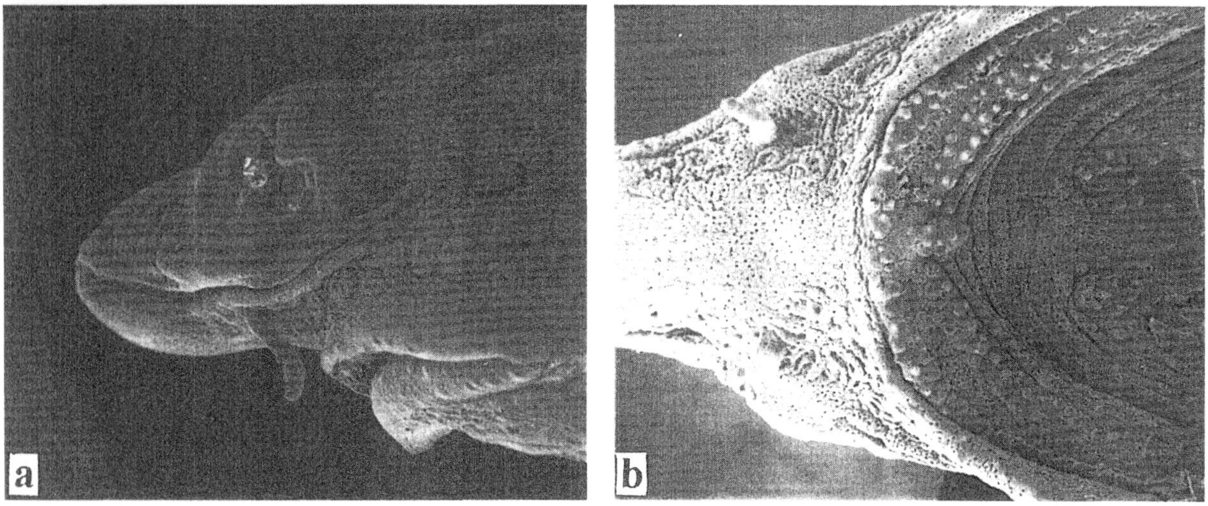

Figure 20. Scanning electron micrographs of *Polyodon spathula* larva: a – Lateral view of the head. The infraorbital lateral line canal is still a groove at this point in development and can be seen continuing onto the rostral region. The olfactory pit has not yet completely subdivided into anterior and posterior nares. Many clusters of ampullary electroreceptors are visible on the cheek region dorsal to the upper jaw (see Bemis & Grande 1992). b – Oral view of the upper jaw of a similar specimen. The teeth of the upper jaw are erupting in two series. Additional erupting teeth can be seen at the leading edge of infrapharyngobranchial 2 (see Grande & Bemis 1991 fig. 17B).

overlaps the hyomandibula and attaches the subopercle to the hyoid arch (Figure 19). The process is present, but smaller in †*Chondrosteus* and †*Peipiaosteus* than in acipenserids and polyodontids. It was included as part of a polyodontid character by Grande & Bemis (1991, character 17) but redefined by Findeis (1993) as an acipenseriform character. Nielsen (1949, p. 302, #13) considered the unusual subopercle of †*Birgeria* 'lobate' and similar to that of *Polyodon*. This appears to be a superficial similarity, and in any event †*Birgeria* lacks the anterior process of the subopercle. †*Cheirolepis*, *Polypterus*, †*Mimia*, *Lepisosteus*, and *Amia* possess oblong or circular subopercles lacking the anterior processes.

Character 9. Preopercular canal in a series of ossicles, mandibular canal short or absent

This character was discussed by Jollie (1980), defined by Patterson (1982) and redefined in modified form by Gardiner & Schaeffer (1989) and Grande & Bemis (1991, character 4). It may be linked to the next character.

Character 10. Infraorbital sensory canal in a series of ossicles

In recent acipenseriforms, †*Chondrosteus* and †*Peipiaosteus*, the infraorbital sensory canal is carried by a series of small canal bones. This contrasts with the condition in †*Cheirolepis*, *Polypterus*, †*Mimia*, †*Birgeria*, *Amia* and other outgroups in which large circumorbital bones carry the infraorbital sensory canal.

Character 11. Loss of premaxillary and maxillary bones

This interpretation derives from observations of a developmental series of *Scaphirhynchus* made by Findeis (1991). In contrast to the solid bony cheek found in outgroups such as †*Cheirolepis*, *Polypterus*, †*Mimia*, †*Birgeria* and *Amia*, the cheek of recent sturgeons and paddlefishes is composed largely of skin, and none of the tooth bearing bones of the upper jaw are exposed on the surface of the cheek. There is no bone in the upper jaw of acipenseriforms homologous to the premaxilla of the outgroup osteichthyans surveyed. Most historic and contemporary authors (e.g., Traquair 1887, Parker 1882, Woodward 1891, Sewertzoff 1928, Jollie 1980,

Grande & Bemis 1991) have regarded the bone on the leading edge of the upper jaw of Acipenseriformes as homologous to the maxilla of other osteichthyans. Instead, Findeis (1991) interpreted this bone as a dermopalatine, one of a series of investing bones of the palatoquadrate present in many osteicthyans (e.g., dipnoans, Rosen et al. 1981). A new illustration labeled with this interpretation is shown in Figure 18. If the upper jaw of Acipenseriformes is a composite of only the palatoquadrate and its investing bones, then both the maxilla and premaxilla must have been lost.

Putative and problematic characters of Acipenseriformes

This section describes some additional features which either cannot be surveyed in all acipenseriform taxa or which we cannot easily define. For example, sensory barbels were probably present at this node, but we cannot evaluate their condition in fossils nor are they present in living outgroups, so we do not know whether 2 barbels (as in living Polyodontidae) or 4 barbels (as in living Acipenseridae) is the plesiomorphic condition. Similarly, alone among living actinopterygians, embryonic paddlefishes and sturgeons share a hairpin-loop shaped pronephros (Ballard & Needham 1964, Bemis & Grande 1992).

Three cranial character complexes provide most of the characters of Acipenseriformes: the cheek region, the jaws and the opercular series. It is possible that some of the characters within these complexes are linked. For example, paedomorphosis is thought to cause 'global' changes in morphology (Bemis 1984), so that several seemingly distinct skeletal characters actually change as a unit. In Acipenseriformes, the apparent loss of cheek bones and presence instead of small ossicles to carry the sensory canals might be linked to such a common underlying process. Evaluation of this putative character complex, however, requires far more information about the development of the skeleton than is currently available.

The cartilaginous basitrabecular processes on the midventral surface of the neurocranium of recent Acipenseriformes serve as 'pivot points' guiding projection of the palatoquadrates. These processes are present in all Acipenseridae and *Psephurus*. The loss of these processes in *Polyodon* is a secondary condition associated with immobility of the upper jaw. Sewertzoff (1928) described ligaments connecting the basitrabecular processes and the palatoquadrate, but Findeis (1993) found no discrete ligaments in this position in any of the nine species of acipenserids he surveyed for this character. Because they also possess hyostylic jaw suspensions, we infer (but did not feel certain about scoring) that basitrabecular processes were present in †*Chondrosteus* and †*Peipiaosteus*. †*Cheirolepis*, *Polypterus*, †*Mimia*, *Lepisosteus*, and *Amia* lack basitrabecular processes.

Gardiner & Schaeffer (1989) regarded a large, blade shaped hyomandibula as a synapomorphy of Acipenseriformes. This may be true, but is difficult to define in an unambiguous way. It is difficult to decide, for example, whether the hyomandibula of †*Mimia* (which is integrated into a more typical actinopterygian suspensorium) is 'blade shaped'. Also, in all recent Acipenseriformes, the interhyal is hypertrophied. Interhyal hypertrophy was regarded by Gardiner & Schaeffer (1989) and Grande & Bemis (1991, character 3) as a synapomorphy of Acipenseriformes. Although this is probably correct, it is impossible to evaluate this character based only on published descriptions of the fossil taxa surveyed here because the interhyal is cartilaginous.

Gardiner & Schaeffer (1989) noted that the basisphenoid of Acipenseriformes lacks a parabasal canal; as we understand their description, this appears to be plesiomorphic for the group. Similarly, the polarity of change in myodome characters noted at this level by Gardiner & Schaeffer (1989) is not obvious to us. The myodomes present in Acipenseridae are small and separated, but similar to those of †*Mimia*.

We do not understand a putative acipenseriform character noted by Gardiner & Schaeffer (1989, p. 176): 'rostral bones reduced, numerous', for rostral bones are not particularly large or stable features in any of the stem actinopterygians known to us.

Many workers assert that absence of jaw teeth characterizes Acipenseriformes or some clade within the order (e.g., Gardiner 1984b, character 12). However, as stated, this character is problematic because acipenseriform larvae and juveniles possess jaw teeth. Figure 20 shows the oral region of a *Polyodon* larva, with two rows of conical teeth erupting in the upper jaw and a single row in the lower jaw (see also Bemis & Grande 1992). In large adult *Polyodon*, however, (>25 kg) teeth are not visible from the surface of the bone, and sections show that they are completely embedded in the jaw. The pattern of ontogenetic tooth loss is different in sturgeons: larvae have teeth, but the teeth and their attachment bones are absent in adults, and are generally considered to be shed during growth. (The biting surfaces of adult sturgeons are composed of thick collagenous pads, Nelson 1969.) Thus, this character is more complicated than usually stated, and available data for †*Peipiaosteus* and †*Chondrosteus* are inconclusive concerning the mode of ontogenetic tooth loss.

Gardiner & Schaeffer (1989) also noted that acipenseriform scales are devoid of normal peg and socket articulations, except on the caudal lobe of †*Chondrosteus*. By analogy with phylogenetic changes in scales of dipnoans (Bemis 1984), we suspect that the loss of peg and socket articulations is related to the general phenomenon of loss of scalation in acipenseriforms (character 3 above), and that both might be interpreted as paedomorphic. Similarly, Gardiner & Schaeffer (1989) regarded the absence of ganoin from fin rays as an acipenseriform character; this also could result from paedomorphosis (see Bemis 1984 for discussion of phylogenetic loss of ornament in dipnoans). There are typically a few ganoid tubercles along the leading edge of the pectoral fin in, for example, *Acipenser oxyrhynchus*, but these do not extend into the area of the fin membrane proper. Both of these putative characters deserve study.

Discussion of clade Acipenseriformes

Based on the taxa and characters surveyed, there is little doubt concerning monophyly of Acipenseriformes. Questions concerning sister-group relationships of the two Mesozoic families †Chondrosteidae and †Peipiaosteidae are discussed below, but on the basis of available evidence we cannot reach any conclusion and thus Figure 17 shows †Peipiaosteidae, †Chondrosteidae and Acipenseroidei as an unresolved trichotomy (see Grande & Bemis 1996 for new information on this trichotomy).

Characters of †Peipiaosteidae

Characters of this group are the subject of another study (Grande & Bemis 1996), and at present, we do not accept any of the characters proposed by earlier workers.

Putative and problematic characters of †Peipiaosteidae

A character often cited as evidence of monophyly of †*Peipiaosteus* is paired lateral line scales (Liu & Zhou 1965, Zhou 1992). We think this is probably a misinterpretation, for in several of the figured specimens, the right and left trunk canals are preserved as small tubular ossicles in close proximity to each other as we have found for taxa such as †*Crossopholis* (Grande & Bemis 1991).

Zhou (1992) noted several characters may be synapomorphies of †*Peipiaosteus* and †*Stichopterus*. For example, the parasphenoid of †*Peipiaosteus* has a broad anterior process, but this character has not been fully assessed in †*Stichopterus*. In view of the reported differences between these two genera (Zhou 1992), it is important to make additional specimen based study and comparison before accepting them.

†*Peipiaosteus* shares with Polyodontidae and Acipenseridae the presence of small, tripartite denticles on the shoulder girdle (Liu & Zhou 1965). Although this character is presently unknown in †*Stichopterus* and †*Chondrosteus*, it may be a synapomorphy of Polyodontidae, Acipenseridae and †Peipiaosteidae.

Liu & Zhou (1965) also noted that in †Peipiaosteidae, Acipenseridae and Polyodontidae, the pelvic fins originate anterior to the position of origin of the dorsal fin, whereas in †*Chondrosteus*, the pelvic fins originate opposite the dorsal fin. Again, this may be a synapomorphy of Polyodontidae, Acipenseridae and †Peipiaosteidae.

Finally, †*Peipiaosteus* shares with Polyodontids the presence of branchiostegals (there is only a single branchiostegal in paddlefishes) with branched posterior tips. This condition also occurs in †*Stichopterus*. Specimen based study is needed before proposing it as a synapomorphy of †Peipiaosteidae and Polyodontidae.

Discussion of †Peipiaosteidae

Too little information is currently available to demonstrate the monophyly of this group. Indeed, only with additional specimen study will we be able to determine whether †*Stichopterus* and †*Peipiaosteus* are congeneric, as proposed by Yakovlev (1977) and rebutted by Zhou (1992). Also, Zhou (1992) considered †Peipiaosteidae as the sister group of Acipenseridae. As discussed under clade Acipenseroidei (below), we disagree with Zhou's (1992) analysis (see Grande & Bemis 1996 for new information on †Peipiaosteidae).

Characters of †Chondrosteidae

Osteological apomorphies of †Chondrosteidae are listed in Table 6, coded in Table 7, and keyed to the cladogram in Figure 17.

Character 12. Anterior part of palatopterygoid club-shaped

In contrast to all other Acipenseriformes, the palate of †*Chondrosteus* has a complete sheathing of bone on its oral surface. This is because the anterior end of the palatopterygoid is club-shaped, rather than deeply notched and having a distinct ectopterygoid process (see Figure 18 for palatoquadrate of *Acipenser*; additional illustration in Grande & Bemis 1991 fig. 79). We do not interpret this condition as a fusion of an autopalatine ossification with the palatopterygoid, although the effect on the shape of the bone would be similar.

Character 13. Complete loss of trunk scalation

Liu & Zhou (1965) noted this as a feature distinguishing †*Chondrosteus* from all other Acipenseriformes. This character warrants additional specimen based study of well preserved †*Chondrosteus*, especially because the scales present in †*Peipiaosteus*, Polyodontidae and Acipenseridae are very small and potentially difficult to detect in anything but the best fossils.

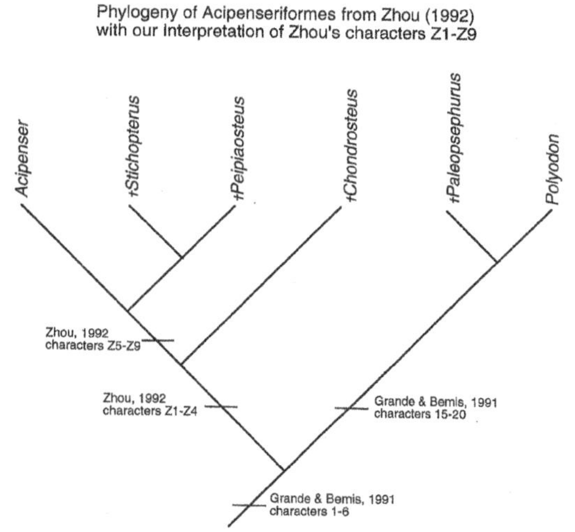

	Anc	Aci	Sti	Pei	Cho	Pal	Pol
Z1. Quadratojugal sickle shaped	?	1	1	1	1	0	0
Z2. Infraorbital canal "L" shaped	?	1	1	1	1	0	0
Z3. Mandibular canal lost	0	1	1	1	1	1	1
Z4. Palatopterygoid short and broad	1	1	1	1	1	0	0
Z5. Suprangular absent	0	1	1	1	0	?	?
Z6. <7 bones in opercular series	0	1	1	1	0	1	1
Z7. Opercular bones branched	0	1	1	1	0	1	1
Z8. Broad opercular - pectoral gap	?	?	?	?	?	?	?
Z9. Juveniles have hook-like rostral bone	?	?	?	?	?	?	?

Figure 21. Tree proposed by Zhou (1992). A table showing our interpretation of his nine relevant characters (characters Z1-Z9) is shown. This tree is two steps longer than the one we propose in Figure 17 (treelength 9 versus treelength 7 for the same 6 taxa). In text, see Discussion of clade Acipenseroidei. Anc = 'ancestor'.

Discussion of clade †Chondrosteidae

A modern specimen based study of all species of this group, including further preparation of previously described materials, is needed.

Characters of Acipenseroidei

Grande & Bemis (1991) concluded that sturgeons and paddlefishes are sister groups, which they united in the suborder Acipenseroidei. This opinion contrasts with the interpretation of Schaeffer (1973) and others, who regarded †Chondrosteidae as the sister group of Acipenseridae. It also differs from ideas of Aldinger (1937) and Nielsen (1949). Grande & Bemis (1991) discovered four characters defining Acipenseroidei, which we modify and supplement here.

Character 14. Loss of opercle
This character was defined by Grande & Bemis (1991, character 7) as a synapomorphy of Acipenseroidei. As defined in character 1 above, the subopercle replaces the opercle as the largest bone of the opercular series in Acipenseriformes. However, †*Chondrosteus* and †*Peipiaosteus* retain small opercles, which are entirely missing in Polyodontidae and Acipenseridae.

Character 15. Reduction in number of branchiostegals supporting gill cover
This character was defined as an acipenseroid synapomorphy by Grande & Bemis (1991, character 8). †*Chondrosteus* possesses at least eight branchiostegals in a continuous series along the lateral and ventral faces of the operculum. †*Peipiaosteus* possesses at least seven branchiostegals. Acipenserids typically possess two (occasionally three branchiostegals are present; see Findeis 1997 this volume). The branchiostegals are elongate in *Huso* and some species of *Acipenser* (e.g., *A. oxyrinchus*), shorter in other species of *Acipenser* (e.g., *A. brevirostrum*), and much shorter in scaphirhynchines. With the possible exception of †*Protopsephurus*, polyodontids possess only a single branchiostegal separate from the subopercle (character 24, below).

Character 16. Endocranium with extensive rostrum
Grande & Bemis (1991, character 9) defined this character. This extension of the ethmoid region of the endocranium ranges from an elongate, thin sword or paddle shaped structure in polyodontids to a broad, flattened shovel shape in scaphirhynchines. It is absent in †*Peipiaosteus*, †*Chondrosteus*, †*Birgeria*, †*Cheirolepis*, †*Mimia, Polypterus* and *Amia*. †*Saurichthys* is sometimes regarded as a potential sister group of Acipenseriformes (e.g., Gardiner & Schaeffer 1989, Rieppel 1992), partly because of its rostral shape. However, the rostrum of †*Saurichthys* is based upon extended jaws, superficially similar to those of *Lepisosteus*, so its rostrum is not homologous with that of acipenseroids.

Character 17. Dorsal and ventral rostral bones
Grande & Bemis (1991, character 10) defined this character. In our initial formulation, only the ventral keel of rostral bones was emphasized, but the presence of dorsal rostral bones should be included as well. Dorsal and ventral rostral series are always present in Polyodontidae and Acipenseridae. Rostral bones of polyodontids are so constant that homologues are readily recognizable throughout the family, but acipenserids lack this constancy. Some variation in the rostral bones of sturgeons correlates with the underlying shape of the rostrum, but the number and arrangement also vary between individuals with comparably shaped rostra (e.g., Jollie 1980). Neither †*Chondrosteus* nor †*Peipiaosteus* possesses extensive series of bones anterior to the frontals and parasphenoid, much less entire series of bones, and, based on outgroup comparison with †*Cheirolepis*, *Polypterus*, and other basal actinopterygians, the absence of rostral bones is plesiomorphic.

Character 18. Ventral process of posttemporal bone
The posttemporal bones bear prominent ventral processes that closely articulate with the neurocranium. Among sturgeons and paddlefishes, these ventral processes of the posttemporal bone are always present, although most strikingly developed in *Polyodon* (e.g., Grande & Bemis 1991, fig. 10B). The extreme condition in *Polyodon* was noted as a putative synapomorphy of this genus (Grande & Bemis 1991, character 39). The posttemporal bones of †*Chondrosteus* lack a ventral process and are separated from the dermal skull roof by the lateral extrascapular bones. Similarly, the posttemporal bone of †*Peipiaosteus* is not integrated into the skull roof, so that there was not a close association between its dermal skull roof and neurocranium in this region. No similar feature of the posttemporal bone occurs in *Polypterus*, †*Mimia*, or other actinopterygians.

Putative and problematic characters of Acipenseroidei
Some skeletal features at the level of Acipenseroidei are parts of character complexes. For example, the endochondral rostrum and rostral bones (characters 16 and 17) are presumably coordinated neo-

morphic features because the bones more or less closely ensheath the underlying endochondral rostrum. For now we accept these as separate characters.

Discussion of clade Acipenseroidei
The monophyly of Acipenseroidei sensu Grande & Bemis (1991) was questioned by Zhou (1992). Zhou's tree is shown in Figure 21, as well as a table summarizing the nine characters he identified to challenge the monophyly of Acipenseroidei.

Zhou's (1992) analysis is problematic for three basic reasons. First, the polarity of change in the proposed characters was not explained and cannot be assessed without explicit outgroup comparisons to non-acipenseriform taxa not included in Zhou's paper. This problem applies in particular to characters Z1 (quadratojugal sickle shaped) and Z2 (infraorbital canal 'L' shaped). Second, apparently only some of the taxa included in Zhou's phylogeny were examined for all characters. For example, Zhou interpreted characters Z6 (the presence of fewer than 7 opercular bones) and Z7 (branched opercular bones) as synapomorphies of a clade including †Peipiaosteidae + *Acipenser*. But both of these characters are also present in *Polyodon* and †*Paleopsephurus*, so Zhou's tree requires these two characters to be derived at least twice. Third, Zhou's scoring or definitions for at least three characters are problematic. Character Z3 (loss of mandibular canal) applies to *Polyodon* and †*Paleopsephurus* as well as to all of the other taxa in Zhou's analysis (see our character 9 above). The definition of character Z8 (broad opercular-pectoral gap) is vague, and in any event problematic to score in fossils. Finally, character Z9 (hook-like rostral bones in juveniles) is an ontogenetic feature with unknown development in extant taxa. Zhou's rostral bones seem to be canal bones bearing rostral portions of the infraorbital canal, and we doubt that this shape character is actually sufficiently similar in *Acipenser* and †*Peipiaosteus* to warrant recognition as a synapomorphy.

We scored characters Z1-Z9 as well as we could determine (Figure 21), and found that Zhou's proposed tree is two steps longer than the tree we would propose for these taxa (CI: 1.00 versus CI: 0.78). In view of this and our five well investigated synapomorphies (characters 14-18 above), we retain Acipenseroidei sensu Grande & Bemis (1991).

Characters of Polyodontidae

In this section, we simply list the characters described by Grande & Bemis (1991) with minor modifications. Figure 17 shows the same branching pattern within Polyodontidae as reported by Grande & Bemis (1991). The recent description of a new fossil polyodontid, †*Protopsephurus* Lu 1994, raises some questions that can only be answered by a side-by-side analysis of specimens. This discovery pushes back to the Mesozoic the presence of polyodontids in Asia. Based on current information, however, we cannot specify the phylogenetic placement of †*Protopsephurus* relative to the other members of the family, and so indicate it as a basal multichotomy within Polyodontidae.

Character 19. Many small stellate bones make up lateral supports for the paddle
This character was defined by Grande & Bemis (1991, character 15) as a synapomorphy of Polyodontidae. These bones may be present in †*Protopsephurus* (Lu 1994), but additional clarification and illustration is needed.

Character 20. Series of very elongate dorsal and ventral medial rostral bones, with cylindrical cross-sections
This character was defined by Grande & Bemis (1991, character 16) as a synapomorphy of Polyodontidae. Findeis (1993) proposed that the cylindrical cross sectional shape should also be noted to help differentiate these bones from the elongate rostral bones present in some sturgeons. Elongate rostral bones are present in †*Protopsephurus* (Lu 1994).

Character 21. Unique shape of subopercle
This character was defined by Grande & Bemis (1991, character 17) as a synapomorphy of Polyodontidae. The anterior arm of the subopercle, although more elongate in Polyodontidae than in any

other Acipenseriformes, is present in all Acipenseriformes (character 8, above), but the fan of rod-like ossifications in combination with it are unique to Polyodontidae. This feature is not well understood in †*Protopsephurus* (Lu 1994).

Character 22. Elongate anterior and posterior divisions of the fenestra longitudinalis in the skull roof
This character was defined by Grande & Bemis (1991, character 18) as a synapomorphy of Polyodontidae. This feature is not well understood in †*Protopsephurus* (Lu 1994).

Character 23. Posttemporal with elongate anterior arm suturing into the dermosphenotic
This character was defined by Grande & Bemis (1991, character 19) as a synapomorphy of Polyodontidae. It appears to be present in †*Protopsephurus* (Lu 1994).

Character 24. Single branchiostegal with branched posterior edge
This character was defined by Grande & Bemis (1991, character 20) as a synapomorphy of Polyodontidae (also see Nielsen 1949, p. 302). Lu (1994) noted that several branchiostegals occur in †*Protopsephurus*. It is important to confirm this in additional specimens, because if true, then it would help to place †*Protopsephurus* within Polyodontidae.

Putative and problematic characters of Polyodontidae
Findeis (1993) noted that in *Psephurus* and *Polyodon*, the incurrent and excurrent nares, as well as the bulk of the olfactory capsule, lie dorsal to a line drawn through the longitudinal axis of the eye. In sturgeons, the nares and olfactory capsule lie ventral to a line drawn through the eye. It is difficult to polarize this character, because of variation within outgroup genera such as *Polypterus*, and impossible to score it reliably in fossils. Nevertheless, the dorsal location of the olfactory system in paddlefishes may be synapomorphic for Polyodontidae.

In living and fossil paddlefishes, the supracleithrum attaches by ligaments to the cleithrum, unlike the condition in living sturgeons and †*Chondrosteus*, which have an immobile sutural connection between these two bones. Based on our outgroup comparisons, the ligamentous attachment is probably a synapomorphy of Polyodontidae, although we do not yet know the condition in †*Peipiaosteus*.

In the pectoral fin skeleton of living polyodontids, a single radial caudal to the propterygium articulates with the scapulocoracoid (Grande & Bemis 1991). This differs from the condition in sturgeons, and most of our outgroup taxa, which have three independent radials articulating with the scapulocoracoid (Findeis 1993). This cartilaginous character cannot be scored in fossil acipenseriforms because it is not preserved.

Discussion of clade Polyodontidae
There is little doubt of the monophyly of Polyodontidae. With the exception of the new genus †*Protopsephurus*, intergeneric relationships within the family are well understood (Grande & Bemis 1991). It seems probable that †*Protopsephurus* will emerge as the sister group of all other Polyodontidae, for it appears to share most of the polyodontid characters except for the single branchiostegal (character 24, above).

Characters of Acipenseridae

Characters of Acipenseridae listed here derive from Grande & Bemis (1991) and Findeis (1997 this volume). Osteological synapomorphies of Acipenseridae are listed in Table 6, coded in Table 7, and keyed to the cladogram in Figure 17. Figure 17 reports the same branching arrangement within Acipenseridae proposed by Findeis (1997).

Character 25. Five scute rows along trunk
This character of sturgeons was originally mentioned by Linnaeus (1758) and has been noted by many workers since. Grande & Bemis (1991, character 11) and Findeis (1997) regard this as a synapomorphy of Acipenseridae. Gardiner & Schaeffer (1989), however, suggested that up to six rows of scutes may be present in acipenserids, but we are unaware of any examples of six scute rows in fossil or extant sturgeons.

Character 26. Pectoral fin spine
The presence of a pectoral fin spine was accepted by Grande & Bemis (1991, character 13) as a synapomorphy of Acipenseridae. A redefinition of this character was made by Findeis (1997), who found that independent fin rays do not actually fuse to form the spine (see Jollie 1980) but are instead encased in a sheath of dermal bone.

Character 27. Antorbital bone
One of the last bones to form in the dermal skull roof of all acipenserids surveyed is the discrete antorbital bone lying on the postnasal wall between the orbit and olfactory bulb (Findeis 1997). It is absent in polyodontids, †*Chondrosteus* and †*Peipiaosteus*. This bone does not carry any portion of the supraorbital lateral line canal, and so is not homologous to the nasal bone in our actinopterygian outgroups.

Character 28. Commissure of occipital canals in median extrascapular bone
In all acipenserids, the median extrascapular bone is present and prominent in the posterior part of the skull roof because it bears the commissure of the occipital canals. The bone develops late in ontogeny, but unlike the surrounding anamestic bones, has an unusually constant shape (triangular) throughout Acipenseridae (Findeis 1997). No other acipenseriforms possess a median triangular bone in this region of the skull (see Grande & Bemis 1991 on the nuchal bone of †*Paleopsephurus*, which does not carry lateral line canals). This character is defined by Findeis (1997 this volume).

Character 29. Rostral canals curve lateral to barbels
The sensory canals extending into the rostrum of paddlefishes and sturgeons are continuous with the infraorbital canal. In sturgeons, the canals curve laterally around the outer pair of barbels, and then converge toward the ventral midline of the rostrum. In paddlefishes, the canal leading into the rostrum is straight. No other acipenseriforms or outgroup taxa surveyed exhibit the laterally curved condition. This character is defined by Findeis (1997).

Character 30. Supracleithrum tightly joined to dermal skull roof
An elongate dorsal process of the supracleithrum is tightly bound by connective tissue to the inner surface of the posttemporal bone. This strong connection of the pectoral girdle to the skull roof is unique to acipenserids among acipenseriforms and basal actinopterygians generally. This character is defined by Findeis (1997 this volume).

Character 31. Opercular wall formed by cleithrum and clavicle
In all sturgeons, a vertical wall of bone defines the posterior wall of the opercular chamber. The opercular wall is composed of thin medial laminae of the cleithrum and clavicle. It is unique to acipenserids among all taxa surveyed. Also see Findeis (1997).

Character 32. Cardiac shield formed by clavicle
The cardiac shield is formed by laminar extensions of the clavicle, which extend toward the ventral midline to form a dermal 'shield' overlying the pericardial cavity. No other actinopterygians surveyed possess this feature. It is illustrated by Findeis (1997).

Character 33. Cleithral process limits mobility of pectoral fin spine
A process of the cleithrum extends laterally to brace the fin spine. It is present in all acipenserids, but absent in all other taxa surveyed. It is probably part of a character complex including the presence of the fin spine (character 26, above). See Findeis (1997) for illustration and further discussion.

Putative and problematic characters of Acipenseridae
Synapomorphies of Acipenseridae derive principally from the skull roof, the pectoral girdle, and dermal ossifications such as scutes. Many other characters which could be included at this level of analysis are cartilaginous, and thus difficult or impossible to score in fossils. They are, however, helpful in studying relationships among recent Acipenseridae (Findeis 1997).

We noted the reduction in the number of branchiostegals as a character of Acipenseroidei (char-

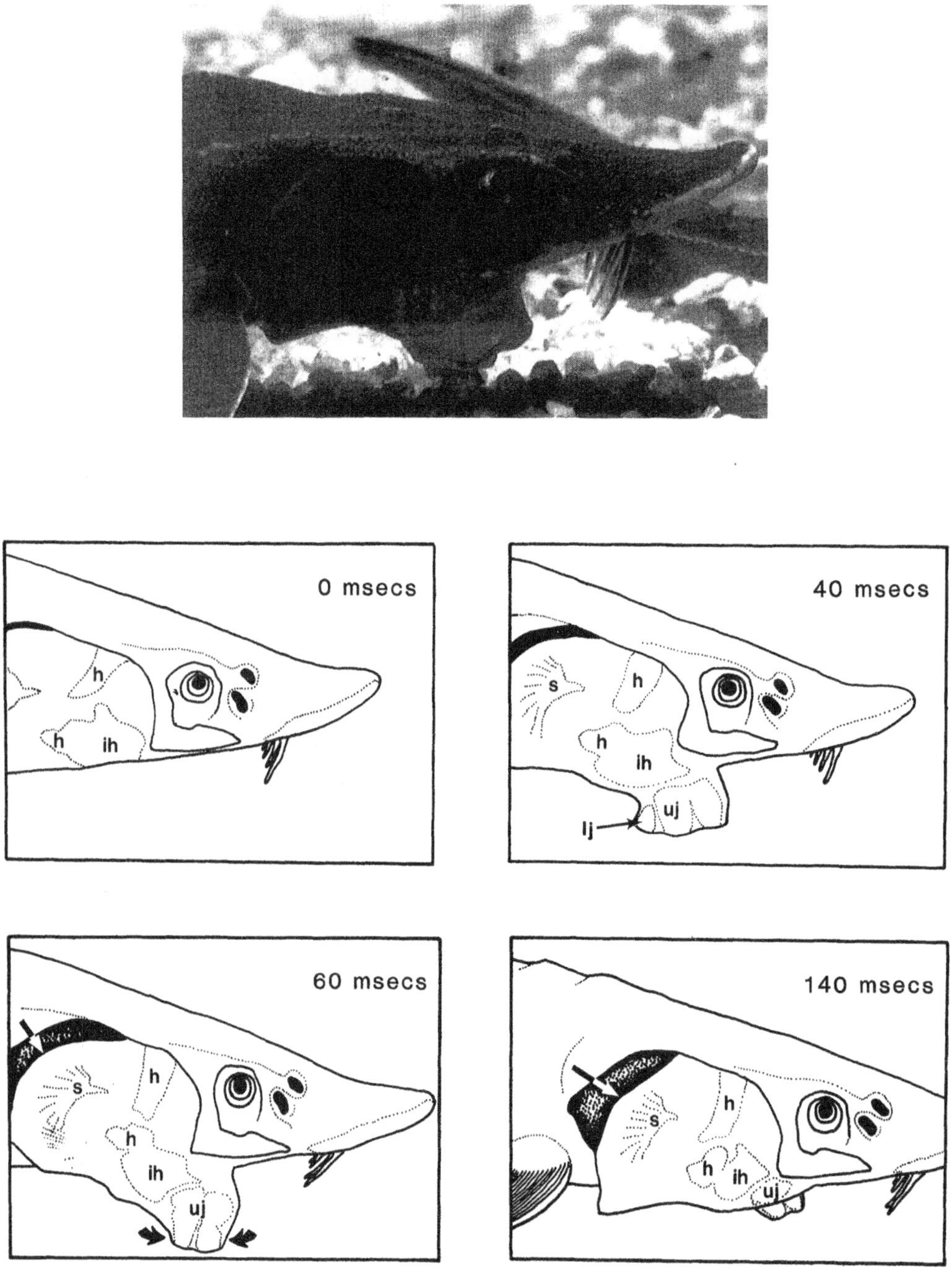

Figure 22. Functional morphology of jaw projection in a juvenile white sturgeon, *Acipenser transmontanus*. In the photograph, an individual is shown with the jaws near maximum projection and gape. The bottom four line drawings, traced from high speed film, show kinematics of jaw projection and the skeletal elements involved in jaw projection (based on 'dimpling' of the overlying skin and manipulated in cleared and double stained specimens). Jaw projection can be very rapid, and it involves at least three discrete stages: projection itself (40 msecs); adduction of the jaws (60 msecs); and recovery of the jaws to the resting position (140 msecs and beyond). Also note that opercular abduction occurs during the jaw recovery phase (arrows). h = hyomandibula; ih = interhyal; lj = lower jaw; s = subopercle; uj = upper jaw.

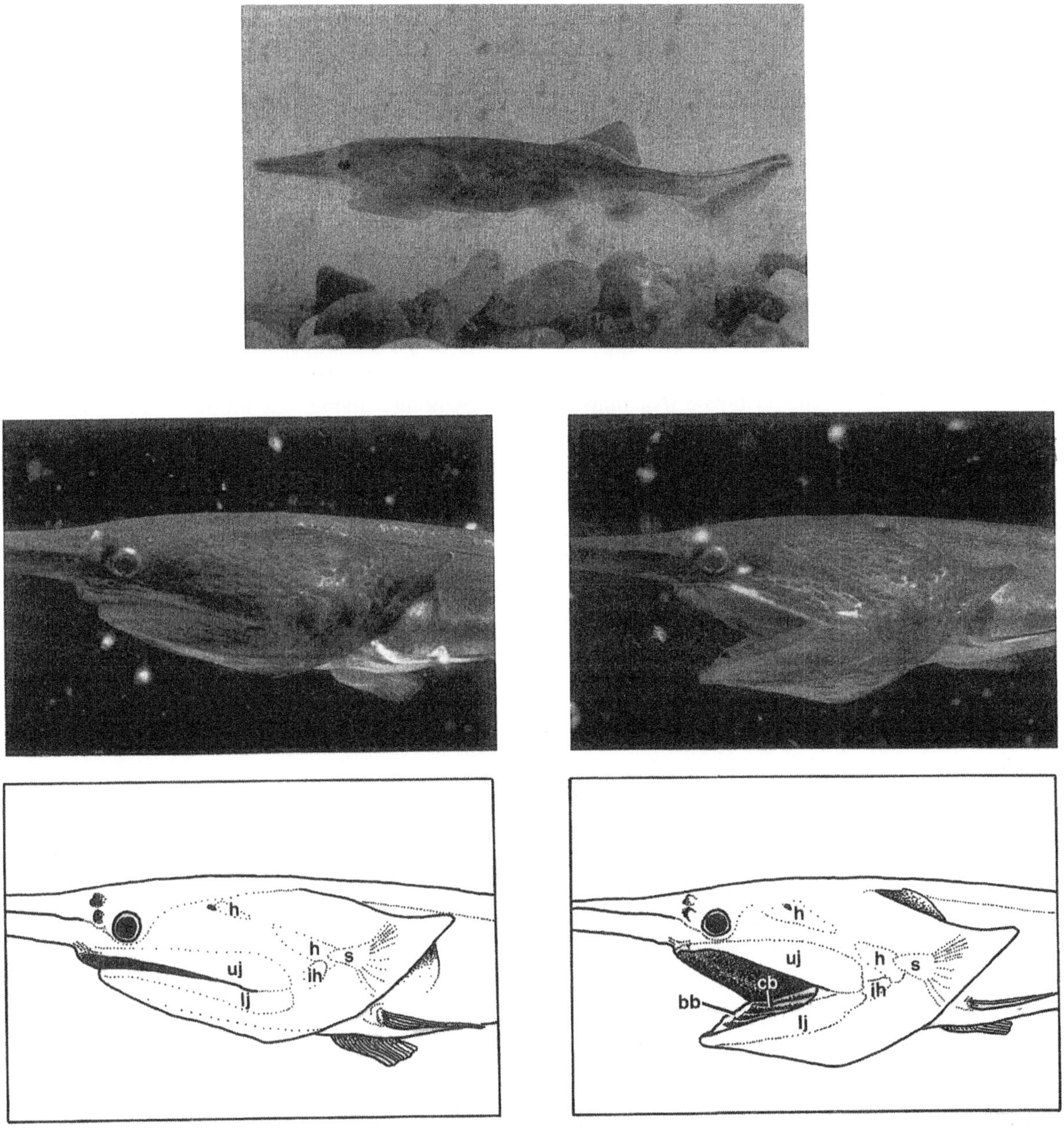

Figure 23. Functional morphology of feeding in North American paddlefish, *Polyodon spathula*. The top photograph shows a juvenile too small to filter feed; at this size, the gill rakers are not yet developed and feeding is raptorial. Such individuals can eat surprisingly large food items relative to their size including other juvenile paddlefishes (Yeager & Wallus 1982). The bottom figures show a filter feeding juvenile with the jaws closed (left column) and open (right column). The bright white objects in the photographs are individual *Daphnia*; one can be seen directly in front of the open mouth in the upper right photograph. The tracings of the photographs show in dotted outlines the positions of the major visible bony elements. Note that unlike the Chinese paddlefish *Psephurus* (shown in Fig. 7), the upper jaw of *Polyodon* is fixed at its anterior contact with the neurocranium. This fixed condition is clearly derived within Polyodontidae (see Grande & Bemis 1991) (bb = basibranchial series, cb = ceratobranchial; see Figure 7 for other abbreviations).

acter 15). One to three branchiostegals are present in Acipenseridae, whereas there is consistently a single branchiostegal in Polyodontidae (with the possible exception of †*Protopsephurus*; see Lu 1994).

Findeis (1993) noted that the jugal bone has a prominent anterior process that extends beneath the orbit to contact the rostrum, uniting the dermal skull with the rostrum. Although living paddlefishes and outgroup actinopterygians lack this anterior process of the jugal, it is impossible to score it in †*Chondrosteus* and †*Peipiaosteus* based on currently available descriptions.

In all living sturgeons, the postero-dorsal margin of the operculum is emarginate. This allows a respiratory current of water to flow into the dorsal portion of the opercular chamber, over the gills, and then exit the opercular chamber ventrally (Burggren 1978). This character complex includes specializations in the shapes of the gill arches and filaments (Findeis 1989). Living paddlefishes lack the special respiratory water-flow and the associated branchial specializations, but, depending on how it is defined, paddlefishes can be considered to have a dorsally emarginate operculum. Moreover, few of the features comprising this system can be scored reliably in fossils because they are soft tissues. Thus we consider that the opercular water flow system is probably synapomorphic for sturgeons, but cannot assess this with certainty.

Supraneurals are absent in the caudal peduncle of sturgeons (Findeis 1993). In contrast, polyodontids and †*Chondrosteus* have a complete row of supraneurals in this region of the axial skeleton (Grande & Bemis 1991). The condition in †*Peipiaosteus* is unclear, and in outgroup taxa such as *Polypterus* and †*Mimia*, it is variable, so we defer analysis of this character until it is better understood.

Cartilaginous components of the pectoral girdle may provide characters useful for understanding relationships within Acipenseridae, but these features cannot be scored in †*Peipiaosteus* and †*Chondrosteus* based on available descriptions. For example, a coracoid shelf, formed by the flattening of the coracoid wall of the scapulocoracoid along the cardiac shield, is probably synapomorphic for Acipenseridae because its presence is correlated with that of the cardiac shield (character 32; Findeis 1997 uses aspects of this character in intrafamilial phylogenetic analysis of Acipenseridae). Similarly, the supracleithral cartilage on the ventral surface of the supracleithrum appears to be unique to acipenserids.

A feature unique to Acipenseridae among living basal actinopterygians is a basipterygial process extending ventrally from the antero-ventral edge of the basipterygium. Unfortunately, because the process is cartilaginous, the condition is unknown in †*Chondrosteus* and †*Peipiaosteus*.

Among living actinopterygians, sturgeons uniquely possess a palatal complex composed of several plates of cartilage (Findeis 1997). This probably plays an important role in feeding because it forms one of the two opposing surfaces used in crushing food (Findeis 1993). Unfortunately, this cartilaginous feature cannot be studied in †*Peipiaosteus* and †*Chondrosteus*.

For the present analysis, we also disregard several interesting cartilaginous features of the hyoid arch and branchial skeleton described by Findeis (1997).

Discussion of clade Acipenseridae
Many characters confirm that this clade is monophyletic. Additional characters relevant to recovering a generic-level phylogeny of living Acipenseridae are given by Findeis (1997).

Evolutionary questions and scenarios

Some interesting aspects of acipenseriform evolution concern the distribution of functional characters such as feeding systems or reproductive features. These features cannot be confirmed in fossils and have incongruent distributions among living Osteichthyes. We also are intrigued by the current biogeographic distribution of paddlefishes and shovelnosed sturgeons, as well as such issues as the origin of the rostrum and its extreme hypertrophy in Polyodontidae. In each of the five cases we discuss, the cladogram presented in Figure 17 serves to organize our discussion.

Feeding systems, jaw protrusion and ram ventilation

The feeding system derived for all Acipenseriformes is based on a jaw projection system very different from that of late Paleozoic and early Mesozoic paleonisciforms. As exemplified by outgroup taxa such as †*Cheirolepis*, †*Mimia*, *Polypterus* and *Lepisosteus*, the premaxilla and maxilla were primitively sutured into the bones of the cheek, so that forward projection of the jaws was impossible (Schaeffer & Rosen 1961). Although the maxilla is not sutured to the cheek in bowfins, jaw protrusion in *Amia calva* is still very limited (Lauder 1980). Head lift is also an essential aspect of jaw opening and feeding in outgroup actinopterygians (Lauder 1980) and sarcopterygians (Bemis & Lauder 1986, Bemis 1987a). The importance of head lift and limitations to jaw protrusion are most parsimoniously interpreted as plesiomorphic conditions in Actinopterygii.

Living sturgeons and primitive paddlefishes (i.e., polyodontids other than *Polyodon*) stand in sharp contrast to all of these outgroup taxa. Their jaws are highly mobile, so that the upper jaw can be 'projected' far out to capture prey, and head lift plays no role in feeding at all. This is illustrated in the case of a white sturgeon, *Acipenser transmontanus*, in Figure 22. Within acipenserids, progressive modifications of the jaw projection system allow even greater specialization on benthic prey (Findeis 1997). *Psephurus gladius* can also project its jaws, as shown in Figure 7 (also see Grande & Bemis 1991, figures 28C and 33B). This ability allows sturgeons and *Psephurus* dietary variety, including both fast moving (fish) and benthic (mollusk and other invertebrate) prey. Interestingly, within Polyodontidae, *Polyodon* exhibits secondary reduction in mobility of its upper jaw, a specialization associated with filter feeding in this genus. In the top of Figure 23, a young juvenile *Polyodon* is shown; at this size, it is incapable of filter feeding both because of hydrodynamic considerations and because its gill rakers are not yet long enough to function as a filtering mechanism (Rosen & Hales 1981). Note that the upper jaw is firmly attached to the head. In the bottom four photographs of Figure 23, the maximum extent of mouth opening during filter feeding is shown. Again, the upper jaw remains firmly attached to the neurocranium. A photograph of filled gill rakers is shown in Figure 9b. Thus, sturgeons and *Polyodon* specialize the feeding apparatus in different directions, one primarily towards benthic prey and the other toward pelagic prey (see Yakovlev 1977, for much the same idea).

Another functional anatomical aspect is the protractor hyomandibularis muscle, which is not preserved in fossils, but which is probably a synapomorphy of all Acipenseriformes. In *Acipenser*, this muscle develops as a slip of mandibular arch musculature which migrates to its new attachment sites on the neurocranium and hyomandibula (Sewertzoff 1928). The protractor hyomandibularis muscle is large in all acipenseriforms (e.g., Danforth 1913, Luther 1913). It plays an important role in protruding the jaw, and, via a simple mechanical linkage, also facilitates the opening of the jaws. Amongst all vertebrates, it is very unusual for a mandibular arch muscle to function in jaw opening (Bemis 1987b).

Many unresolved issues concern the morphological transformation from a typical paleonisciform feeding system to the acipenseriform system. The seeming absence of clearly intermediate morphotypes and questions about the functional adequacy of such an intermediate form has provoked several speculations, perhaps most comprehensively by Yakovlev (1977). Yakovlev (1977) interpreted that the transformation in jaw morphology and function must have occurred very rapidly, and that it probably took place in the Triassic in northern Asia. Yakovlev (1977) further suggested that transformation of the feeding system was initially driven by paedomorphosis, and that subsequent stabilization of the new morphology occurred because it offered a feeding specialization unique among early Mesozoic fishes -- benthophagy.

Our approach to studying such transformations is different: instead of focusing on the absence of morphotypes, or any other type of negative data, we stress the importance of producing a phylogenetic hypothesis or cladogram on which to map and construct putative functional systems. By turning our focus away from ancestor speculation to the comparative analysis of outgroups, we think headway will be made in understanding many aspects of cra-

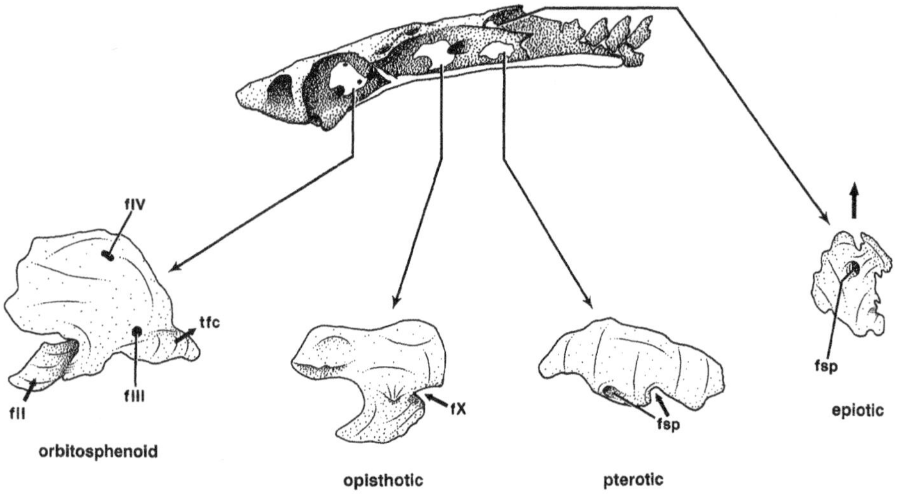

Figure 24. Direct evidence for paedomorphosis in the neurocranium of *Acipenser brevirostrum*. In a few cases we can directly observe post-reproductive changes in morphology, such as the neurocranium of this large shortnose sturgeon, which exhibits four ossification centers within the neurocranium. Only the very largest (=oldest) individuals exhibit these bones. These neurocranial ossifications serve no obvious function, and they are highly variable from one individual to the next. These four neurocranial bones are readily homologized to those of other basal osteichthyans, and we regard their variable presence as direct evidence of paedomorphosis. fII = foramen for optic nerve; fIII = foramen for oculomotor nerve; fIV = foramen for trochlear nerve; fX = foramen for vagal nerve; fsp = foramen for spinal nerve; tfc = trigeminofacialis chamber. Total length of specimen: 870 mm. UMA 24-116-2-47.

nial specialization in acipenseriforms. Related issues we consider essential are fresh re-examinations of traditional homologies using extensive developmental material. This may provide alternative interpretation for certain structures, such as Findeis' (1991) conclusion that the so-called maxilla of Acipenseriformes is actually a dermopalatine (see discussion of character 11 above). This reinterpretation is pivotal to understanding the loss of many bones from the cheek region of Acipenseriformes.

Ram ventilation and the origin of filter feeding

A characteristic observation about *Polyodon spathula* in aquaria or rearing ponds is that they never stop swimming, and if confined in a container too small to allow swimming, then they sink to the bottom. Sinking occurs because the volume of air in the swim bladder is not sufficient to achieve neutral buoyancy. Constant swimming by paddlefishes also generates a constant flow of water over the gills which allows for ram ventilation. *Polyodon* depends on ram ventilation, as evidenced by the absence of a buccal valve and the inability to completely close either the opercular chamber or the mouth. As further evidence of the critical role of ram ventilation, Burggren & Bemis (1992) found that juvenile paddlefishes are nearly as aerobic as cruising bluefin tuna, *Thunnus thunnus*. Speculating on the connection between ram ventilation and the evolution of filter feeding within Polyodontidae, they linked the potential to evolve filter feeding to the prior evolution of ram ventilation. According to this reasoning, many of the specializations associated with filter feeding in *Polyodon* (flattened gill arches, elongate gill rakers, and secondary fixation of the upper jaw to the neurocranium) were possible only because buccal-opercular pumping and branchial arch mediated respiratory movements were unnecessary.

Unlike *Polyodon*, sturgeons such as *Acipenser* and *Scaphirhynchus* can buccal pump to generate respiratory flow (e.g., Burggren 1978), which allows them to be more sedentary than paddlefishes. What is missing, however, are basic data on the possible coupling of respiration with locomotion in *Huso* and *Psephurus*, which our current phylogenetic analysis indicates as critical for understanding the evolution of respiration in Acipenseroidei. Living stem actinopterygians and sarcopterygians are not likely to help in a broader analysis of the evolution of ram ventilation because none of them exhibits constant, sustained swimming.

Potamodromy, anadromy and demersal spawning

All living species of sturgeons and paddlefishes migrate upstream to spawning sites, and many species of Acipenseridae are anadromous. Spawning runs of some species of sturgeons are legendary, exceeding 2500 km for *Huso huso* in the Danube and Volga basins (Balon 1967, Hensel & Holčík 1997 this volume, Khodorevskaya et al. 1997 this volume). Typically, living acipenseriforms spawn very large numbers of eggs onto shallow, gravely sites (e.g., Ryder 1888, Disler 1949, Soin 1951, Dragomirov 1953, 1957, Buckley & Kynard 1985). Adults depart the site soon after egg deposition, and do not provide any parental care.

The phylogenetic distribution of spawning modes in the other living non-teleostean actinopterygians offers little information about the evolution of these reproductive features (see Balon 1975, 1985). None of the other living species of non-teleostean actinopterygians typically inhabits salt water, and none exhibits such extensive upstream spawning migrations (for general review, see Breder & Rosen 1966). Spawning *Polypterus* deposit small numbers of individual eggs on leaves, rocks or other substrates. At least some species of *Lepisosteus* migrate upstream to spawn, but spawn far fewer eggs than do sturgeons and paddlefishes and typically deposit them on structures in the water, such as leaves. *Amia* spawn in nests defended by the male (Ballard 1986).

Extant stem sarcopterygians are similarly uninformative on the question: coelacanths are combined livebearers (Wourms et al. 1991, Balon 1985, 1990, 1991), and lungfishes either spawn individual eggs onto aquatic vegetation (as in *Neoceratodus*, Kemp 1987) or the males guard the eggs in a nest (as in *Protopterus* and *Lepidosiren*, Greenwood 1987, Kerr 1919).

Evidence for anadromy or spawning migrations in fossil Acipenseriformes or outgroups is obviously uncertain, but this has not prevented speculation.

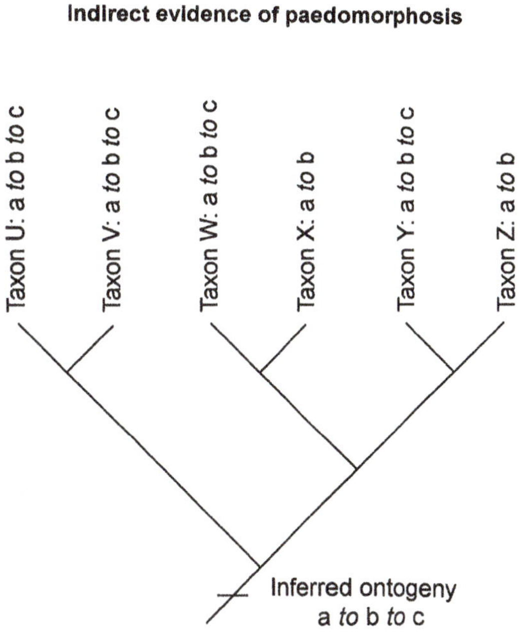

Figure 25. Method for analyzing 'less direct' evidence of paedomorphosis. The cladogram for taxa U through Z is based on character information other than the developmental characters shown. Development is studied in taxa U through Z. It is found that state 'a' always gives rise to state 'b' during ontogeny, and that in most clades, state 'b' goes on to form state 'c'. This information is then 'laid out' on the cladogram; in taxa X and Z, development has been truncated at state 'b', providing indirect evidence that paedomorphosis has occurred. Modified from Bemis (1984).

Zhou (1992) suggested that †*Peipiaosteus* may have been anadromous based on the geographic distribution of fossils. Nielsen (1949) noted that specimens of †*Birgeria groenlandica* occurred in small and large sizes only, and regarded the absence of intermediate sizes as evidence of migration. Yakovlev (1977) also speculated on the evolution of reproductive mode in acipenseriforms. At present, however, we cannot provide a compelling explanation for the evolution of characteristic spawning patterns of Acipenseriformes.

Paedomorphosis

Paedomorphosis has long been invoked in explaining many anatomical features of living Acipenseriformes, chiefly the largely cartilaginous endoskeleton (Traquair 1887, Woodward 1891, 1895a,b,c). Many workers equated these changes with 'degeneracy' of the skeleton (Goodrich 1909), which is merely another way of emphasizing that this is a secondary condition. Some workers speculate that paedomorphosis was the driving force behind many changes in acipenseriform anatomy. Paraphrasing Yakovlev (1977, p. 141), paedomorphosis ('fetalization') played a decisive role in destroying the archaic structure of the paleonisciform type of jaw, a critical step leading to formation of the acipenseriform jaw (see also Tsessarsky 1992).

Paedomorphosis is of both theoretical and practical interest in systematics (Nelson & Platnick 1981, Fink 1982), and many groups of fishes, such as lungfishes (Dipnoi), provide clear examples of paedomorphic loss of skeletal elements and degree of ossification (Bemis 1984). There are only two types of evidence available to support hypotheses of paedomorphic change. The first is 'relatively direct evidence', such as that provided by Grande & Bemis (1991) for paddlefishes. Large adult paddlefishes continue to ossify new endochondral bones after onset of reproductive maturity. We note here that this phenomenon also occurs in sturgeons, such as a series of adult *Acipenser brevirostrum* that demonstrates progressive ossification of at least four neurocranial bones (Figure 24). Additional observations of large *A. brevirostrum* confirm delayed ossification of several hyobranchial elements. 'Less direct evidence', such as that provided for dipnoans by Bemis (1984) derives from a combination of phylogenetic analysis and developmental data from other taxa. For example, if the ontogeny of a structure is studied in an outgroup and found to proceed from state A to state B to state C during development, whereas the ontogeny of the same structure in the ingroup always stops at state B, then paedomorphosis can be hypothesized to have occurred (Figure 25). Thus, recognition of paedomorphosis by less direct means relies solely on the relative congruence (or lack thereof) of developmental patterns within a phylogenetic framework.

Apart from the direct evidence of paedomorphosis presented by Grande & Bemis (1991), no study to date has provided adequate analysis of the role of paedomorphosis in acipenseriform evolution. This is because without a cladogram to specify the arrangement of taxa, it is not possible to organize the developmental data within a phylogenetic context. Moreover, the analysis shown in Figure 17 confirms that many of the most relevant outgroups are fossil taxa, which are unlikely to provide the necessary developmental information for a formal analysis. Finally, we note that only recently have ontogenetic data become available for the skeleton in taxa such as *Polypterus* (e.g., Bartsch & Gemballa 1992) which are necessary for any study of paedomorphosis in Acipenseriformes.

Pacific biogeography of Polyodontidae and Scaphirhynchinae

An interesting aspect of acipenseriform biology concerns links between Asian and North American freshwater taxa (Grande & Bemis 1991). Yakovlev (1977) asserted that Acipenseriformes originated in freshwaters of northeastern Asia in the Triassic and subsequently dispersed throughout the Holarctic. His interpretation is based on the occurrence of taxa such as †*Stichopterus* in the Jurassic of Northeastern Asia. A problem with his interpretation is that the location of the earliest fossils is not a reliable clue to identifying a place of origin for a group (Grande 1985). Also, Yakovlev (1977) did not consider sister group relationships of Acipenseriformes (see Bemis & Kynard 1997 for such analysis, which suggests Europe as the place of origin for the group).

A basic question about the biogeography of Acipenseriformes is to explain the presence in North America of †*Crossopholis*, *Polyodon*, †*Protoscaphirhynchus*, and *Scaphirhynchus*, which have Asian sister taxa, †*Protopsephurus*, *Psephurus* and *Pseudoscaphirhynchus*. Patterson (1981) clearly outlined the criteria for analyzing the biogeography of primary freshwater fishes. His central point is the need to resolve three taxon statements for endemic taxa to identify vicariant distributions. Our phylogenetic analyses allow some progress on this both for Polyodontidae and Scaphirhynchini, although the placement of the two taxa of greatest interest (†*Protopsephurus* and †*Protoscaphirhynchus*) needs additional study.

As noted above, recent and fossil paddlefishes are restricted to freshwater. Although we cannot conclude this with certainty, †*Protopsephurus* appears to be the sister taxon of all other paddlefishes (see discussion of clade Polyodontidae above). If we add †*Protopsephurus* to our earlier analysis (Grande & Bemis 1991), and still assume strict vicariance, then we conclude that North America and China shared a link at least as early as the Upper Jurassic. This trans-Pacific affinity is intriguing because many taxa of fishes from the western United States share trans-Pacific ties (Grande 1985, 1994). Similar conclusions were reached by Jin (1995) in an article received after this manuscript was accepted; see Grande & Bemis (1996) for discussion of Jin (1995).

We regard †*Protoscaphirhynchus* as a member of tribe Scaphirhynchini, and note that no characters clearly distinguish it from *Scaphirhynchus*. All recent members of the tribe Scaphirhynchini are confined to large rivers, and generally prefer a strong current and soft, silty bottoms. Thus their known distribution is more likely to have resulted from vicariance rather than trans-oceanic dispersal. In this case, the trans-Pacific link between central Asia and North America east of the Rocky Mountains must be at least as early as the late Cretaceous because of the presence of †*Protoscaphirhynchus* in eastern Montana. We predict that scaphirhynchine fossils will eventually be recovered from late Mesozoic deposits of eastern Asia.

Electroreception as the dominant sensory system of paddlefishes

Chondrichthyans, sarcopterygians, amphibians, and non-neopterygian actinopterygians share a common organ system for electroreception (groups enclosed by dotted outline in Figure 3). The sense organs of this system are known variously as ampullae of Lorenzini (chondrichthyans), rostral organ (coelacanths), or ampullary organs (aquatic amphibians, lungfishes, polypterids, sturgeons and paddlefishes; see Jorgenson et al. 1972 and Northcutt 1986 for review). Among all of these taxa, paddlefishes have many more individual ampullary organs than any other groups, with estimates of 70 000 organs in an adult *Polyodon* (Nachtrieb 1912). The chief location of these organs is the rostrum, where the organs are surrounded by stellate bones in the case of paddlefishes or contained in pockets between the rostral bones of sturgeons. Although different workers use different terminologies (e.g., Norris 1925, Nikolskaya 1983), the ampullary organs of *Polyodon* are distributed on the paddle, cheek, and opercular flap in a characteristic way that is constant from individual to individual (Bemis & Northcutt personal observation). Ampullary organs of *Polyodon* increase in number throughout life by subdivision of existing organs. There is one important unanswered question concerning *Polyodon*: does it use electroreception to detect swarms of zooplankton, and if so, how? This will be a challenging behavior to study, for it is difficult to measure or to mimic the electrical field of a plankton swarm, and it is probably impossible to completely denervate electroreceptive input (see Kalmijn 1974 for discussion of plankton and electroreception).

Interesting differences in two other sensory systems of sturgeons and paddlefishes say much about their different sensory worlds. First, polyodontids have two small barbels on the ventral surface of the rostrum in contrast to the four, large, often fimbriated barbels on the rostrum of acipenserids. In both families, the surfaces of the barbels are covered with chemoreceptive organs (taste buds), although sturgeons have many more chemoreceptive organs on their barbels than do paddlefishes. This may be linked to the evolution of benthic habits in Acipenseridae in contrast to the mid-water habits of paddlefish. The second basic difference in the sensory systems of paddlefishes and sturgeons concerns the relative sizes of their eyes. Paddlefishes have absolutely smaller eyes than do comparably sized sturgeons. The behavioral meaning of this has never been rigorously evaluated, but it suggests that visual information may be less important for adult paddlefishes than for sturgeons. The relative de-emphasis on chemosensory and visual systems in paddlefishes is possibly countered by their extraordinarily high number of ampullary organs. These ideas might be tested by comparing the size of target areas in the brain for each sensory system (e.g.,

Northcutt 1978, New & Bodznick 1985) or by a combination of physiological and behavioral approaches.

As in all of the groups enclosed by dotted outline in Figure 3, the ampullary organs of Acipenseriformes are restricted to the head. There are two types of exceptions which prove this rule. First, in some taxa, such as skates of the genus *Raja*, very long ampullary organs radiate onto the broad pectoral fins from sensory capsules located on the head (Raschi 1986). In this case, the ampullary organ tube has been lengthened so that the organ may be sampling a different environment, but the sensory epithelium is still based in the head region. Second, in taxa such as the Australian lungfish, *Neoceratodus forsteri*, the presence of ampullary organs on the trunk is linked to a recurrent branch of the anterior lateral line nerve (Northcutt 1986). The functional significance of locating the sampling portion of an electroreceptor caudal to head in each of the two cases (taxa) is untested, but perhaps it allows better localization of point electrical fields by triangulation (Kalmijn 1974). If this idea is correct, then it may help explain the unusual body shape of paddlefishes. We propose that the long rostrum and trailing opercular tip of paddlefishes reflect extreme specialization for electroreception. In a juvenile paddlefish, the rostrum and operculum together extend for more than one half the body length, so that the long trailing tip of the operculum functionally places electroreceptors far more caudal than might otherwise be possible given that electroreceptors are plesiomorphically restricted to the head. We think too little attention has been paid to the dominance of this sensory system in Polyodontidae.

Nomenclatural recommendations

Our current classification of the genera of Acipenseriformes follows:

Order Acipenseriformes Berg 1940
incertae sedis
  Family †Peipiaosteidae Liu & Zhou 1965
    Genus †*Stichopterus* Reis 1910
    Genus †*Peipiaosteus* Liu & Zhou 1965
Suborder †Chondrosteoidei *sensu* Grande & Bemis 1991
  Family †Chondrosteidae Egerton 1858
    Genus †*Chondrosteus* Agassiz 1844
    Genus †*Gyrosteus* Agassiz 1844
Suborder Acipenseroidei *sensu* Grande & Bemis 1991
  Family Polyodontidae Bonaparte 1838
  incertae sedis
      Genus †*Protopsephurus* Lu 1994
    Subfamily †Paleopsephurinae Grande & Bemis 1991
      Tribe †Paleopsephurini Grande & Bemis 1991
        Genus †*Paleopsephurus* MacAlpin 1941a
    Subfamily Polyodontinae sensu Grande & Bemis 1991
      Tribe Psephurini Grande & Bemis 1991
        Genus *Psephurus* Günther 1873
      Tribe Polyodontini sensu Grande & Bemis 1991
        Genus †*Crossopholis* Cope 1883
        Genus *Polyodon* Cope 1883
  Family Acipenseridae Linnaeus 1758
    Subfamily Husinae sensu Findeis 1993
        Genus *Huso* Brandt 1869
    Subfamily Acipenserinae *sensu* Findeis 1993
      Tribe Acipenserini sensu Findeis 1993
        Genus *Acipenser* Linnaeus 1758
      Tribe Scaphirhychini Bonaparte 1846
        Genus *Scaphirhynchus* Heckel 1836
        Genus *Pseudoscaphirhynchus* Nikolskii 1900
        Genus †*Protoscaphirhynchus* Wilimovsky 1956

Conclusions

We have four basic conclusions:

(1) The phylogeny for Acipenseridae is still in need of additional testing. A more detailed analysis of acipenserid phylogeny is needed for planning for the conservation of sturgeons and for understanding their evolutionary history and biogeography. Therefore, it is very important to conduct broad, comparative morphological developmental and molecular studies of the species of Acipenseridae. Ideally, a future comparative morphological investigation should emphasize completeness by including all well-preserved fossil and extant species of Acipenseridae. Fossil taxa are, in most cases, known only from the type descriptions and a number of newly discovered nearly complete skeletons remain undescribed. In addition to reexamining the patterns of generic interrelationships proposed by Findeis (1997 this volume) and reviewed here, future work should specifically target the question: Is *Acipenser* monophyletic? There should be a serious effort to examine all existing specimens, particular-

ly type series, held in systematic collections in eastern Europe and Asia. It will be vital to undertake at least some new field work to collect the specimens needed for comprehensive study, but because of current threats to many species of Acipenseriformes, any sampling must be done in such a way that it does not further threaten already diminished populations. The effort to better understand acipenserid biodiversity should not be postponed for even a short period, or we may not know in time which taxa are in most need of conservation.

(2) Our information about living and fossil Polyodontidae is more complete than for any other clade within Acipenseriformes. Still, there is much to do, including incorporation of the newly described †*Protopsephurus* into a phylogenetic framework, and attention to the functional morphology of feeding and, in particular, electroreception. Especially needed are new anatomical studies of *Psephurus*, including its development.

(3) Future paleontological work on fossil Acipenseriformes should focus on redescription and analysis of †Chondrosteidae and †Peipiaosteidae. In particular, it is important to resolve whether †*Peipiaosteus* and †*Stichopterus* are congeneric as suggested by Yakovlev (1977). Our currently available osteological characters for paddlefish and sturgeons will also need to be reassessed as more complete and detailed descriptions of †Chondrosteidae and †Peipiaosteidae become available. Finally, it will also be important to restudy Mesozoic and Paleozoic actinopterygians, particularly †*Birgeria*, †*Saurichthys*, and †*Phanerorhynchus*, to better assess the sister group of Acipenseriformes.

(4) In any future studies of acipenseriform systematics, it is vital to include a control for ontogenetic and individual variation. As noted above, the documented variations in morphology, including post-reproductive changes in ossification patterns, mean that growth series of individuals need to be studied.

Acknowledgements

We thank Vadim Birstein, John Waldman, Robert Boyle, and the Hudson River Foundation for organizing the International Conference on Sturgeon Biodiversity and Conservation. Penny Jaques, Nancy Ruddock and Judy Shardo helped in manuscript and figure preparation. Vadim Birstein provided advice and translation help for Russian papers. Wen Jun Li and Dan Hui assisted with Chinese papers. John Waldman, Vadim Birstein and two anonymous reviewers improved our manuscript. Our work has been supported by grants from NSF (BSR 8806539 and BSR 9220938), the Whitehall Foundation, and the Tontogany Creek Fund.

References cited

Agassiz, L. 1844. Recherches sur les poisson fossiles. 2 (2, p. 280). Petitpierre Imprimiere, Neuchatel.

Aldinger, H. 1931. Über Reste von *Birgeria* (Pisces, Palaeoniscidae) aus der alpinen Trias. N. Jahrb. Geol. Paläont. Stuttgart 66: 167–180.

Aldinger, H. 1937. Permische Ganoidfische aus Ostgronland. Medd. Grønland 102: 1–392.

Allis, E.P. 1922. The cranial anatomy of *Polypterus*, with special reference to *Polypterus bichir*. J. Anat. 56: 189–294.

Antoniu-Murgoci, A. 1936a. Particularités anatomiques caractérisant les especes des genres *Huso* et *Acipenser* des eaux roumaines. Compt. rend. Acad. Sci. Roum. 1: 1–3.

Antoniu-Murgoci, A. 1936b. Particularités anatomiques qui différencient le genre *Huso* du genre *Acipenser* des eaux roumaines. Ann. Sci. Univ. Jassy 23:94–103.

Applegate, S. 1970. The vertebrate fauna of the Selma formation of Alabama, part VIII. The fishes. Fieldiana Geol. Mem. 3: 381–433.

Artyukhin, E.N. 1994. On the relationships of the Amur sturgeon, *Acipenser schrencki*. Sturgeon Quart. 2(3): 7.

Artyukhin, E.N. 1995a. On biogeography and relationships within the genus *Acipenser*. Sturgeon Quart. 3(2): 6–8.

Artyukhin, E.N. 1995b. An ecological approach to zoogeography of Acipenseridae. pp. 7–12. *In*: A.D. Gershanovich & T.I.J. Smith (ed.) Proceedings of International Symposium on Sturgeons, 6–11 September 1993, VNIRO Publishing, Moscow.

Ayres, W.E. 1854. Descriptions of new species of California fishes. Proc. Calif. Acad. Sci. 1: 1–22.

Bai, Y. 1983. A new *Peipiaosteus* from Fengning of Hebei, China. Vert. Palas. 21:341–346 (in Chinese).

Bailey, R.M. & F.B. Cross. 1954. River sturgeons of the American genus *Scaphirhynchus*: characters, distribution, and synonymy. Pap. Mich. Acad. Sci. 39: 169–208.

Ballard, W.W. 1986. Stages and rates of normal development in the holostean fish, *Amia calva*. J. Exp. Zool. 238: 337–354.

Ballard, W.W. & R.G. Needham. 1964. Normal embryonic stages of *Polyodon spathula* (Walbaum). J. Morphol. 114: 465–478.

Balon, E.K. 1967. Vývoj ichtyofauny Dunaja, jej súčasný stav a pokus o prognózu d'alších zmien po výstavbe vodných diel (Evolution of the Danube ichtyofauna, its recent state and an attempt to predict further changes after the construction of the planned hydro-electric power stations and diversion schemes). Biologické práce 13: 1–121.

Balon, E.K. 1968. Einfluß des Fischfangs auf die Fischgemeinschaften der Donau. Arch. Hydrobiol. (Suppl. Donauforschung 3) 34: 228–249.

Balon, E.K. 1975. Reproductive guilds of fishes: a proposal and definition. J. Fish. Res. Board. Can. 32: 821–864.

Balon, E.K. 1985. Early life histories of fishes: new developmental, ecological and evolutionary perspectives. Developments in Environmental Biology of Fishes 5, Dr W. Junk Publishers, Dordrecht. 280 pp.

Balon, E.K. 1990. Epigenesis of an epigeneticist: the development of some alternative concepts on the early ontogeny and evolution of fishes. Guelph Ichthyol. Rev. 1: 1–42.

Balon, E.K. 1991. Probable evolution of the coelacanth's reproductive style: lecithotrophy and orally feeding embryos in cichlid fishes and in *Latimeria chalumnae*. Env. Biol. Fish. 32: 249–265.

Baruš, V. & O. Oliva (ed.) 1995. Mihulovci *Petromyzontes* a ryby *Osteichhyes* (Lampreys and fishes). Fauna ČR and SR volume 28/1, Academia, Praha. 623 pp.

Bartsch, P. & S. Gemballa. 1992. On the anatomy and development of the vertebral column and pterygiophores in *Polypterus senegalus* Cuvier 1829 ('Pisces', Polypteriformes). Zool. Jahrb. Anat. 22: 497–529.

Bemis, W.E. 1984. Paedomorphosis and the evolution of the Dipnoi. Paleobiology 10:293–307.

Bemis, W.E. 1987a. Feeding systems of living Dipnoi: anatomy and function. pp. 249–275. *In*: W.E. Bemis, W.W. Burggren & N.E. Kemp (ed.) The Biology and Evolution of Lungfishes, Alan R. Liss, New York.

Bemis, W.E. 1987b. Functional mechanics of jaw-opening and jaw-closing in the paddlefish, *Polyodon spathula*. Amer. Zool. 27: 87A (abstract).

Bemis, W.E. & E.K. Findeis. 1994. The sturgeons' plight. Nature 370: 602.

Bemis, W.E. & L. Grande. 1992. Early development of the actinopterygian head. I. External development and staging of the paddlefish *Polyodon spathula*. J. Morphol. 213: 47–83.

Bemis, W.E. & B. Kynard. 1997. Sturgeon rivers: an introduction to acipenseriform biogeography and life history. Env. Biol. Fish. (this volume).

Bemis, W.E. & G.V. Lauder. 1986. Morphology function of the feeding apparatus of the lungfish, *Lepidosiren paradoxa* (Dipnoi). J. Morphol. 187: 81–108.

Berg, L.S. 1904. Zur Systematic der Acipenseriden. Zool. Anz. 27: 665–667.

Berg, L.S. 1911. Fishes (Marsipobranchii and Pisces). Fauna of Russia and adjacent countries. Vol. 3, Vypusk 1. Izdatelstvo Akademii Nauk, St. Petersburg. 337 pp. (in Russian).

Berg, L.S. 1933. Freshwater fishes of the USSR and adjacent countries, 3rd ed. Part 2. Vsesoyuznyi Institute Ozernogo i Rechnogo Rybnogo Khozyaistva, Leningrad. 545–903 pp. (in Russian).

Berg, L.S. 1940. Classification of fishes both recent and fossil. Trudy Zoologicheskogo Instituta 5: 85–517 (in Russian). Thai National Documentation Center, Bangkok. pp.1–304 (English translation of Rusian original, 1965).

Berg, L.S. 1948a. Freshwater fishes of the USSR and adjacent countries, 4th edition. Part 1. Akademia Nauk USSR, Moscow & Leningrad. 466 pp. (in Russian, English translation published by Israel Program for Scientific Translations, Jerusalem. 505 pp.)

Berg, L.S. 1948b. On the position of the Acipenseriformes in the system of fishes. Trudy Zool. Inst. 7: 7–57 (in Russian).

Bigelow, H.B. & W.C. Schroeder. 1953. Fishes of the Gulf of Maine. U.S. Fish. Bull. 53: 1–577.

Binkowski, F.P. & S.I. Doroshov (ed.) 1985. North American sturgeons: biology and aquaculture potential. Developments in Env. Biol. Fish. 6, Dr W. Junk Publishers, Dordrecht. 163 pp.

Birstein, V.J. 1993. Sturgeons and paddlefishes: threatened fishes in need of conservation. Cons. Biol. 7: 773–787.

Birstein, V.J. & W.E. Bemis. 1997. Leo Semenovich Berg and the biology of Acipenseriformes: a dedication. Env. Biol. Fish. (this volume).

Birstein, V.J., R. Hanner & R. DeSalle. 1997. Phylogeny of the Acipenseriformes: cytogenetic and molecular approaches. Env. Biol. Fish. (this volume).

Birstein, V.J., A.I. Poletaev & B.F. Goncharov. 1993. The DNA content in Eurasian sturgeon species determined by flow cytometry. Cytometry 14: 377–383.

Birstein, V.J. & V.P. Vasiliev. 1987. Tetraploid-octaploid relationships and karyological evolution in the order Acipenseriformes (Pisces). Karyotypes, nucleoli, and nucleolus organizer regions in four acipenserid species. Genetica 72: 3–12.

Bogdanov, M. 1874. A report on a newly discovered acipenserid fish at the meeting of Zoological Section. Trudy Sankt-Peterburgskogo Obshchestva Ispytatelei Prirody 5: 48 (in Russian).

Bonaparte, C.L. 1831. Saggio di una distribuzione metodico degli animali vertebrati. Giorn. Arc. Sci, Lett. Arti 49: 1–77.

Bonaparte, C.L. 1836. Iconografia della fauna italica per le quatro classi degli animali vertebrati. Tomo III. Pesci. Roma. Fasc. 15–18: 80–93.

Bonaparte, C.L. 1838. Selachorum tabula analytica. Nouv. Ann. Sci. Nat. 2: 195–214.

Bonaparte, C.L. 1846. Catalogo metodico dei pesci Europi. Napoli.

Borodin, N. 1897. A report about the summer 1895 zoological expedition on board of the cruiser 'Uralets' in the northern part of the Caspian Sea. Vestnik Rybopromyshlennosti 1 :1–31 (in Russian).

Brandt, J.F. 1869. Einige worte über die europaich-asiatischen Störarten (Sturionides). Bull. Acad. Imper. Sci. St.-Petersb. 14: 171–175.

Brandt, J.F. & J.C.T. Ratzeberg. 1833. Medizinische Zoologie, oder getreue Darstellung und Beschreibung der Thiere, die in der Arzneimittellehre in betracht kommen, in systematischer folge herausgegeben. Berlin. (2 vol. 1829–1833. 63 fig.)

Breder, C.M., Jr. & D. Rosen. 1966. Modes of reproduction in fishes. Natural History Press, Garden City. 941 pp.

Buckley, J. & B. Kynard. 1985. Habitat use and behavior of prespawning shortnose sturgeon, *Acipenser brevirostrum*, in the Connecticut River. pp. 111–117. *In*: F. P. Binkowski & S. I. Doroshov (ed.) North American Sturgeons: Biology and Aquaculture Potential, Developments in Env. Biol. Fish. 6, Dr W. Junk Publishers, Dordrecht.

Burggren, W.W. 1978. Gill ventilation in the sturgeon, *Acipenser transmontanus*: unusual adaptations for bottom dwelling. Respir. Physiol. 34: 153–170.

Burggren, W.W. & W.E. Bemis. 1992. Metabolism and ram gill ventilation in juvenile paddlefishes, *Polyodon spathula* (Chondrostei: Polyodontidae) Physiol. Zool. 65: 515–539.

Carlson, D.M., W.L. Pflieger, L. Trial & P.S. Haverland. 1985. Distribution, biology, and hybridization of *Scaphirhynchus albus* and *Scaphirhynchus platorynchus* in the Missouri and Mississippi rivers. pp. 51–59. *In*: F.P. Binkowski & S.I. Doroshov (ed.) North American Sturgeons: Biology and Aquaculture Potential, Developments in Env. Biol. Fish. 6, Dr W. Junk Publishers, Dordrecht.

Cope, E.D. 1883. A new chondrostean from the Eocene. Amer. Nat. 17: 1152–1153.

Danforth, C.H. 1913. The myology of *Polyodon*. J. Morphol. 24: 107–146.

Dean, B. 1895. Fishes living and fossil. MacMillan, New York. 300 pp.

Dettlaff, T.A., A.S. Ginsburg & O.I. Schmalhausen. 1993. Sturgeon fishes: developmental biology and aquaculture. Springer Verlag, Berlin. 300 pp.

Dillard, J.G., L.K. Graham & T.R. Russell. 1986. The paddlefish: status, management, and propagation. North Central Division, American Fisheries Society, Special Publication 7. 159 pp.

Dingerkus, G. & W.M. Howell. 1976. Karyotypic analysis and evidence of tetraploidy in the North American paddlefish, *Polyodon spathula*. Science 194: 842–844.

Disler, N.N. 1949. Development of the skin sensory organs of the lateral line of *Acipenser stellatus*. Trudy Inst. Morph. Zhiv. im. A.N. Severtsova 1: 333–362 (in Russian).

Dragomirov, N.I. 1953. Development of the sevryuga larvae during yolk nutrition. Trudy Inst. Morph. Zhiv. im. A.N. Severtsova 10: 244–263 (in Russian).

Dragomirov, N.I. 1957. Larval development of the Volga-Caspian sturgeon *Acipenser güldenstädti* Brandt. Trudy Inst. Morph. Zhiv. im. A.N. Severtsova 20: 187–231 (in Russian).

Duméril, A.H.A. 1867. Prodrome d'une monographie des esturgeons et description des espèces de l'Amérique du Nord qui appartiennent au sous-genre *Antaceus*. Nouv. Arch. Mus. Hist. Nat. Paris 3: 131–188.

Duméril, A.H.A. 1868. Note sur trois poissons: *Acipenser dabryanus*, *Polyodon gladius* et *Peristethidion prionocephalum*, accompagnée de quelques considérations générales sur les groupes auxquels ces espèces appartiennent. Nouv. Arch. Mus. Nat. Hist. Nat. Paris 4: 93–116.

Duméril, A.H.A. 1870. Histoire naturelle des poissons, ou ichthyologie générale, Vol. 2, Paris. 624 pp.

Egerton, P. de M.G. 1858. On *Chondrosteus*, an extinct genus of the Sturionidae, found in the Lias Formation at Lyme Regis. Phil. Trans. Roy. Soc. Lond. 148: 871–885.

Ellis, W.S. 1990. The Aral: a Soviet sea lies dying. Nat. Geog. 177: 71–93.

Findeis, E.K. 1989. Morphological specializations of the branchial arches of sturgeons (Acipenseridae). Amer. Zool. 29:123A (abstract).

Findeis, E.K. 1991. Ontogeny and homology of the 'maxilla' of sturgeons: evolution of mobile jaw elements in early actinopterygian fishes. Amer. Zool. 31: 8A (abstract).

Findeis, E.K. 1993. Skeletal anatomy of the North American shovelnose sturgeon *Scaphirhynchus platorynchus* (Rafinesque 1820) with comparisons to other Acipenseriformes. Ph.D. Dissertation, University of Massachusetts, Amherst. 444 pp.

Findeis, E.K. 1997. Osteology and phylogenetic interrelationships of sturgeons (Acipenseridae). Env. Biol. Fish. (this volume).

Fink, W.L. 1982. The conceptual relationship between ontogeny and phylogeny. Paleobiology 8: 254–264.

Fitzinger, L.J. & J. Heckel. 1836. Monographische dorstellung der Gattung *Acipenser*. Zool. Abh. Ann. Wiener Mus. Naturgesch. 1: 262–326. (Note: order of authorship is reversed in some bibliographic citations).

Fontana, F. & G. Colombo. 1974. The chromosomes of Italian sturgeons. Experientia 30: 739–741.

Forbes, S.A. & R.E. Richardson. 1905. On a new shovelnose sturgeon from the Mississippi River. Bull. Ill. St. Lab. Nat. Hist. 7:37–44.

Galbreath, J.C. 1985. Status, life history, and management of Columbia River white sturgeon, *Acipenser transmontanus*. pp. 119–125. *In*: F.P.Binkowski & S.I. Doroshov (ed.) North American Sturgeons: Biology and Aquaculture Potential, Developments in Env. Biol. Fish. 6, Dr W. Junk Publishers, Dordrecht.

Gans, C. 1986. Scenarios: Why? pp. 1–10. *In*: R. E. Foreman, A. Gorbman, J.M. Dodd & R. Olsson (ed.) Evolutionary Biology of Primitive Fishes, Plenum Press, New York.

Gardiner, B.G. 1967. Further notes on paleoniscids with a classification of the Chondrostei. Bull. Brit. Mus. Nat. Hist., Geol. Ser. 8: 255–325.

Gardiner, B.G. 1984a. The relationships of the paleoniscid fishes, a review based on new specimens of *Mimia* and *Moythomasia* from the upper Devonian of Western Australia. Bull. Brit. Mus. Nat. Hist., Geol. Ser. 37: 173–248.

Gardiner, B.G. 1984b. Sturgeons as living fossils. pp. 148–152. *In*: N. Eldredge & S.M. Stanley (ed.) Living Fossils, Springer Verlag, New York.

Gardiner, B.G. 1993. Basal actinopterygians. pp. 611–619. *In*: M. J. Benton (ed.) The Fossil Record 2, Chapman and Hall, London. 845 pp.

Gardiner, B.G. & B. Schaeffer. 1989. Interrelationships of lower actinopterygian fishes. Zool. J. Linn. Soc. 97: 135–187.

Gengerke, T.W. 1986. Distribution and abundance of paddlefish in the United States. pp. 22–35. *In*: J.G. Dillard, L.K. Graham & T.R. Russell (ed.) The Paddlefish: Status, Management and Propagation, North Central Division, American Fisheries Society, Special Publication 7, Columbia.

Georgi, J.G. 1775. Bemerkungen einer Reise im russischen Reich. Volume 1. Saint Petersburg. 352 pp.

Gershanovich, A.D. & T.I.J. Smith (ed.) 1995. Proceedings of International Symposium on Sturgeons, September 6–11, 1993, VNIRO Publishing, Moscow. 360 pp.

Ginsburg, A.S. & T.A. Dettlaff. 1991. The Russian sturgeon *Acipenser gueldenstaedti*. I. Gametes and early development up to time of hatching. pp. 15–65. *In*: T.A. Dettlaff & S.G. Vassetzky (ed.) Animal Species for Developmental Studies, Volume 2, Vertebrates, Plenum Press, New York.

Goncharov, B.F., O.I. Shubravy & V.K. Uteshev. 1991. Reproduction and early development of *Pseudoscaphirhynchus kauf-*

manni Bogdanov under artificial environment conditions. Sov. J. Dev. Biol. 22: 286–296.

Goodrich, E.S. 1909. Vertebrata craniata: cyclostomes and fishes. 518 pp. *In*: R. Lankester (ed.) A Treatise on Zoology, Part 9, Adam & Charles Black, London.

Goodrich, E.S. 1928. *Polypterus* a paleoniscid? Paleobiologica 1: 87–92.

Graham, K. 1997. Contemporary status of the North American paddlefish, *Polyodon spathula*. Env. Biol Fish. (this volume).

Grande, L. 1985. The use of paleontology in systematics and biogeography, and a time control refinement for historical biogeography. Paleobiology 11: 1–11.

Grande, L. 1994. Repeating patterns in nature, predictability, and 'impact' in science. pp. 61–84. *In*: L. Grande & O. Rieppel (ed.) Interpreting the Hierarchy of Nature, Academic Press, New York.

Grande, L. & W.E. Bemis. 1991. Osteology and phylogenetic relationships of fossil and recent paddlefishes (Polyodontidae) with comments on the interrelationships of Acipenseriformes. J. Vert. Paleo. 11, supplement 1: 1–121.

Grande, L. & W.E. Bemis. 1996. Interrelationships of Acipenseriformes, with comments on "Chondrostei". pp. 85–115. *In*: M.L.J. Stiassny, L.R. Parenti & G.D. Johnson (ed.) Interrelationships of Fishes, Academic Press, San Diego.

Grande, L. & W.E. Bemis. 1997. A comprehensive phylogenetic study of amiid fishes (Amiidae) based on comparative skeletal anatomy. J. Vert. Paleo. Supplement. (in press).

Gray, J.E. 1834. Characters of two new species of sturgeons (*Acipenser*, Linn.). Proc. Zool. Soc. Lond. 1834: 122–123.

Greenwood, P.H. 1987. The natural history of African lungfishes. pp.163–179. *In*: W.E. Bemis, W.W. Burggren & N.E. Kemp (ed.) The Biology and Evolution of Lungfishes, Alan Liss, New York.

Gregory, W.K. 1933. Fish skulls: a study of the evolution of the natural mechanism. Trans. Amer. Phil. Soc. 23: 75–481.

Günther, A. 1873. Report on a collection of fishes from China. Ann. Mag. Nat. Hist. (4)12: 239–250.

Handyside, P.D. 1875a. On the anatomy of a new species of *Polyodon*, the *Polyodon gladius* of Martens, taken from the river Yang-Tse-Kiang, 450 miles above Woosung. Part I. Its external characters and structure. Proc. Roy. Soc. Edinb. 8: 50–51.

Handyside, P.D. 1875b. On the anatomy of a new species of *Polyodon*, the *Polyodon gladius* of Martens, taken from the river Yang-Tse-Kiang, 450 miles above Woosung. Part II. Its nervous and muscular systems. Proc. Roy. Soc. Edinb. 8: 136–137.

Heckel, J.J. 1836. *Scaphirhynchus*, eine neue fischgattung aus der ordnung der Chondropterygier mit freien kiemen. Ann. Wien. Mus. 1: 68–78.

Hennig, E. 1925. *Chondrosteus hindenburgi* Pomp., ein 'Stör' des württemberg. Oelschiefers (Lias). Paleontographica Stuttgart 67: 115–133.

Hensel, K. & J. Holčík. 1997. Past and current status of sturgeons in the upper and middle Danube. Env. Biol. Fish. (this volume).

Holčík, J. (ed.) 1989. The freshwater fishes of Europe, Volume 1, Part II, General introduction to fishes, Acipenseriformes. AULA-Verlag, Wiesbaden. 469 pp.

Holčík, J., R. Kinzelbach, L.I. Sokolov & V.P. Vasilev. 1989. *Acipenser sturio* Linnaeus, 1758. pp. 367–394. *In*: J. Holčík (ed.) The Freshwater Fishes of Europe, Volume 1, Part II, General Introduction to Fishes, Acipenseriformes, AULA-Verlag, Wiesbaden.

Holmgren, N. 1933. On the origin of the tetrapod limb. Acta Zool. 14: 186–195.

Ivanzoff, N. 1887. Der *Scaphirhynchus*. Vergleichend-anatomische Beschreibung. Bull. Soc. Mosc. 1887: 1–41.

Jain, S.L. 1985. Variability of dermal bones and other parameters in the skull of *Amia calva*. Zool. J. Linn. Soc. 84: 385–395.

Jarvik, E. 1942. On the structure of the snout of crossopterygians and lower gnathostomes in general. Zool. Bidr. Uppsala 27: 235–675.

Jarvik, E. 1980. Basic structure and evolution of vertebrates, Volume 1. Academic Press, New York. 575 pp.

Jessen, H.L. 1972. Schultergürtel und Pectoralflosse bei Actinopterygiern. Fossils & Strata 1: 89–107.

Jessen, H.L. 1973. Interrelationships of actinopterygians and brachiopterygians: evidence from pectoral anatomy. pp. 227–232. *In*: P.H. Greenwood, R.S. Miles & C. Patterson (ed.) Interrelationships of Fishes, Academic Press, London.

Jin, F. 1995. Late Mesozoic Acipenseriformes (Osteichthyes: Actinopterygii) in Central Asia and their biogeographic implications. pp. 15–21 *In*: A. Sun & Y. Wang (ed.) Sixth Symposium on Mesozoic Terrestrial Ecosystems and Biota, Short Papers, China Ocean Press, Beijing.

Jin, F. , Y.-P. Tian, Y.-S. Yang & S.Y. Deng. 1995. An early fossil sturgeon (Acipenseriformes, Peipiaosteidae) from Fenging of Hebei, China. Vert. Palasiatica 33:1–18.

Jollie, M. 1980. Development of the head and pectoral girdle skeleton and scales in *Acipenser*. Copeia 1980: 226–249.

Jorgenson, J., Å. Flock & J. Wersäll. 1972. The Lorenzian ampullae of *Polyodon spathula*. Z. Zellforsch. Mikros. Anat. 130: 362–377.

Kalmijn, A.J. 1974. The detection of electric fields from inanimate and animate sources other than electric organs. pp. 147–200. *In*: A. Fessard (ed.) Handbook of Sensory Physiology: Electroreceptors and Other Specialized Receptors in Lower Vertebrates, Springer-Verlag, Berlin.

Kemp, A. 1987. The biology of the Australian lungfish, *Neoceratodus forsteri* (Krefft 1870). pp. 181–198. *In*: W.E. Bemis, W.W. Burggren & N.E. Kemp (ed.) The Biology and Evolution of Lungfishes, Alan Liss, New York.

Kerr, J.G. 1919. Text-book of embryology, Volume 2, Vertebrata with the exception of Mammalia, MacMillan, London. 591 pp.

Kessler, K.F. 1872. On a remarkable fish of the family of sturgeons discovered by M.A.P. Fedchenko in the Syr-Darya River. Izvestiya Obshchestva Lyubitelei Estestvoznaniya, Antropologii, i Etnografii 10: 70–76 (in Russian).

Kessler, K.F. 1873. On a remarkable fish of the family of sturgeons discovered by M.A.P. Fedchenko in the River Suir-dar. Ann. Mag. Nat. Hist. 4 Ser. 12: 269–276.

Kessler, K.F. 1877. The Aralo-Caspian expedition. IV. Fishes of the Aralo-Caspian-Pontine region. St-Petersburg.

Khodorevskaya, R.P., G.F. Dovgopol, O.L. Zhuravleva & A.D. Vlasenko. 1997. Present status of commercial stock of sturgeons in the Caspian Sea basin. Env. Biol. Fish. (this volume).

Krykhtin, M.L. & V.G. Svirskii. 1997. Endemic sturgeons of the Amur River: kaluga, *Huso dauricus*, and Amur sturgeon, *Acipenser schrencki*. Env. Biol. Fish. (this volume).

Lacepède, B.G.E. 1797. Sur la *Polyodon* Fueille (*Spatularia folium*). 1. Bulletin der Sciences de la Société Philomatique de Paris 2: 49.

Lambe, L.M. 1902. New genera and species from the Belly River series (Mid-Cretaceous). Contrib. Can. Paleo. 3: 25–81.

Lauder, G.V. 1980. Evolution of the feeding mechanism in primitive actinopterygian fishes: a functional anatomical analysis of *Polypterus*, *Lepisosteus*, and *Amia*. J. Morphol. 13: 283–317.

Lauder, G.V. & K. Liem. 1983. The evolution and interrelationships of the actinopterygian fishes. Bull. Mus. Comp. Zool. 150: 95–197.

Le Sueur, C.A. 1818. Description of several species of chondropterygious fishes of North America, with their varieties. Trans. Amer. Phil. Soc. 1: 383–394.

Leidy, J. 1873. Notice of fossil vertebrates from the Miocene of Virginia. Proc. Nat. Acad. Sci. Phila. 1873: 15.

Linnaeus, C. 1758. Systema naturae per regna tria naturae, secundum classes, ordines, genera, species, cum characteribus, differentiis, synonymis, locis, Tomus I. Editio decima, reformata. Holmiae. 824 pp.

Liu, C., T.A. Georgi, X. Diao & J. Liu. 1995. Biology of the Chinese paddlefish, *Psephurus gladius* (Martens). pp. 13–22. *In*: A.D. Gershanovich & T.I.J. Smith (ed.) Proceedings of International Symposium on Sturgeons, 6–11 September 1993, VNIRO Publishing, Moscow.

Liu, C.-H. & Y.-J. Zeng. 1988. Notes on the Chinese paddlefish, *Psephurus gladius* (Martens). Copeia 1988: 483–484.

Liu, H.T. & J.J. Zhou. 1965. A new sturgeon from the upper Jurassic of Liaoning, north China. Vert. Palasiatica 9: 237–248.

Long, J.A. 1995. The rise of fishes. Johns Hopkins, Baltimore. 223 pp.

Lovetzky, A. 1831. Sur les espèces du genre esturgeon, appartenant aux eaux de l'empire Russe. Bull. Sci. Nat. 23: 131–134 (French translation of Lovetzky 1828)

Lu, L. 1994. A new paddlefish from the Upper Jurassic of northeast China. Vert. Palasiatica 32: 134–142.

Luther, A. 1913. Muskulatur der Ganoiden und Dipneusten. Acta Soc. Sci. Fennicae 41: 1–72.

MacAlpin, A. 1941a. *Paleopsephurus wilsoni*, a new polyodontid fish from the Upper Cretaceous of Montana, with a discussion of allied fish, living and fossil. Bull. Geol. Soc. Amer. 52: 1989 (abstract).

MacAlpin, A. 1941b. *Paleopsephurus wilsoni*, a new polyodontid fish from the Upper Cretaceous of Montana, with a discussion of allied fish, living and fossil. Ph.D. Thesis, University of Michigan, Ann Arbor.

MacAlpin, A. 1947. *Paleopsephurus wilsoni*, a new polyodontid fish from the upper Cretaceous of Montana, with a discussion of allied fishes, living and fossil. Contrib. Mus. Paleo. Univ. Mich. 6: 167–234.

Maisey, J. 1986. Heads and tails: a chordate phylogeny. Cladistics 2: 201–256.

Marinelli, W. & A. Strenger. 1973. Vergleichende Anatomie und Morphologie der Wirbeltiere, IV Lieferung. Franz Deuticke, Wien.

Martens, E. 1862. Über einen neuen *Polyodon* (*P. gladius*) aus dem Yangtsekiang und über die sogenannten Glaspolypen. Monatsberichte Deutsch. Akad. Wissensch. Berlin 1861: 476–480.

McCallister, D.E. 1968. Evolution of branchiostegals and classification of teleostome fishes. Nat. Mus. Can. Bull. 221: 1–239.

Mitchill, S.L. 1815. The fishes of New York described and arranged. Trans. Lit. Phil. Soc. New York 1: 355–492.

Müller, J. 1846. On the structure and characters of the Ganoidei, and on the natural classification of fish. Scient. Mem. 4: 499–542 (English translation of Müller 1844).

Nachtrieb, H.F. 1912. The lateral line system of *Polyodon*. Proc. 7th Int. Zool. Congr. 1912: 215–220.

Nelson, G.J. 1969. Gill arches and the phylogeny of fishes, with notes on the classification of vertebrates. Bull. Amer. Mus. Nat. Hist. 141: 475–552.

Nelson, G.J. & N. Platnick. 1981. Systematics and biogeography. Cladistics and vicariance. Columbia University Press, New York. 567 pp.

New, J.G. & D. Bodznick.1985. Segregation of electroreceptive and mechanoreceptive lateral line afferents in the hindbrain of chondrostean fishes. Brain Res. 336: 89–98.

Nichols, J.T. 1928. Chinese fresh-water fishes in the American Museum of Natural History's collections. A provisional check-list of the fresh-water fishes of China. Bull. Amer. Mus. Nat. Hist. 58: 1–62.

Nichols, J.T. 1943. The fresh-water fishes of China. American Museum of Natural History, New York, Natural History of Central Asia 9: 1–322.

Nielsen, E. 1949. Studies on Triassic fishes from East Greenland. 2. *Australosomus* and *Birgeria*. Meddr. Grønland. 146: 1–309.

Nikolskaya, M.P. 1983. Development of the ampullar electroreceptor system in the ontogeny of *Polyodon* and Acipenseridae. Doklady Akademii Nauk USSR 268: 19–22 (English translation of Russian original).

Nikolskii, A.M. 1900. *Pseudoscaphirhynchus rossikowi*, n. gen. et spec. Ann. Mus. Imp. Sci. St. Petersburg 4:2 57–260 (in Russian).

Normark, B.B., A.R. McCune & R.G. Harrison. 1991. Phylogenetic relationships of neopterygian fishes, inferred from mitochondrial DNA sequences. Mol. Biol. Evol. 8:819–834.

Norris, H.W. 1925. Observations upon the peripheral distribution of the cranial nerves of certain ganoid fishes (*Amia*, *Lepidosteus*, *Polyodon*, *Scaphirhynchus*, and *Acipenser*). J. Comp. Neurol. 39: 345–416.

Northcutt, R.G. 1978. Brain organization in the cartilaginous fishes. pp. 117–193. *In*: E.S. Hodgson & R.F. Mathewson (ed.) Sensory Biology of Sharks, Skates, and Rays, Office of Naval Research, Arlington.

Northcutt, R.G. 1986. Electroreception in nonteleost bony fishes. pp. 257–285. *In*: T.H. Bullock & W. Heiligenberg (ed.) Electroreception, Wiley-Interscience, New York.

Northcutt, R.G. & W.E. Bemis. 1993. Cranial nerves of the coelacanth, *Latimeria chalumnae* (Osteichthyes, Sarcopterygii, Actinistia) with comparisons to other craniates. Brain, Behav. Evol. 42, suppl.1: 1–76.

Pallas, P.S. 1771. Zoographia Rossa-Asiatica, sistens omnium animalium in extenso imperio Rossico et adjacentibus maribus observatorum recensionem, domicilia, mores et descriptiones anatomen atque icones plurimorem, III. Petropoli.

Parker, W.K. 1882. On the structure and development of the skull in sturgeons (*Acipenser ruthenus* and *A. sturio*). Phil. Trans. Roy. Soc. Lond. 173: 139–185.

Patterson, C. 1975. The braincase of pholidophorid and leptolepid fishes, with a review of the actinopterygian braincase. Phil. Trans. Roy. Soc. Lond. 269: 275–579.

Patterson, C. 1981. The development of the North American fish fauna- a problem of historical biogeography. pp. 265–281. *In*: P.L. Forey (ed.) Chance, Change and Challenge, The Evolving Biosphere, British Museum (Natural History), London.

Patterson, C. 1982. Morphology and interrelationships of primitive actinopterygian fishes. Amer. Zool. 22: 241–259.

Patterson, C. & D.E. Rosen. 1977. A review of the ichthyodectiform and other Mesozoic teleost fishes, and the theory and practice of classifying fossils. Bull. Amer. Mus. Nat. Hist. 158: 81–172.

Pearson, D.M. 1982. Primitive bony fishes, with especial reference to *Cheirolepis* and palaeonisciform actinopterygians. Zool. J. Linn. Soc. 74: 35–67.

Pearson, D.M. & T.S. Westoll. 1979. The Devonian actinopterygian *Cheirolepis* Agassiz. Trans. Roy. Soc. Edinb. 70: 337–399.

Phelps, S.R. & F. Allendorf. 1983. Genetic identity of pallid and shovelnose sturgeon (*Scaphirhynchus albus* and *S. platorynchus*). Copeia 1983: 696–700.

Pirogovskii, M.I., L.I. Sokolov & V.P. Vasilev. 1989. *Huso huso* (Linneaus, 1758). pp. 156–200. *In*: J. Holčík (ed.) The Freshwater Fishes of Europe, Volume 1, Part II, General Introduction to Fishes, Acipenseriformes, AULA-Verlag, Wiesbaden.

Pompeckj, J.F. 1914. Fische: Paläontologia. pp. 1107–1147. *In*: Handwörtebuch der Natur, III, Berlin.

Rafinesque, C.S. 1817. Additions to the observations on the sturgeons of North America. Amer. Monthly Mag. Crit. Rev. 1: 288.

Rafinesque, C.S. 1820. Ichthyologia Ohiensis, or natural history of the fishes inhabiting the river Ohio and its tributary streams, preceded by a physical description of the Ohio and its branches. Lexington.

Raschi, W.A. 1986. Morphological analysis of the ampullae of Lorenzini in selected skates (Pisces, Rajoidei). J. Morphol 189: 225–247.

Regan, C.T. 1904. The phylogeny of the Teleostomi. Ann. Mag. Nat. Hist. 7 ser. 13: 329–349.

Reis, O.M. 1910. Die Binnenfauna der Fischschiefer in Transbaikalien. Recherch. Géol. Minièr. Sibérie. 29: 1–68.

Richardson, J. 1836. The fish, Volume 3, fauna Boreali-Americana; or the zoology of the northern parts of British America: containing descriptions of the objects of natural history collected on the late northern land expeditions, under the command of Sir John Franklin, R.N. Richard Bentley, London. 327 pp.

Rieppel, O. 1992. A new species of the genus *Saurichthys* (Pisces: Actinipterygii) from the Middle Triassic of Monte San Giorgio (Switzerland), with comments on the phylogenetic interrelationships of the genus. Palaeontographica Abt. A. 221: 63–94.

Rochard, E., P. Williot, G. Castlenaud & M. Lepage. 1991. Elements de systematique et de biologie des populations sauvage d'esturgeons. pp. 475–507. *In*: P. Williot (ed.) Acipenser, CEMAGREF Publ., Bordeaux.

Rosen, D.E., P.L. Forey, B.G. Gardiner & C. Patterson. 1981. Lungfishes, tetrapods, paleontology, and plesiomorphy. Bull. Amer. Mus. Nat. Hist. 169: 159–276.

Rosen, R.A. & D.C. Hales. 1981. Feeding of the paddlefish, *Polyodon spathula*. Copeia 1981: 441–455.

Ruban, G.I. 1997. Species structure, contemporary distribution and status of the Siberian sturgeon, *Acipenser baerii*. Env. Biol. Fish. (this volume).

Russell, T.R. 1986. Biology and life history of the paddlefish – a review. pp. 2–20. *In*: J.G. Dillard, L.K. Graham & T.R. Russell (ed.) The Paddlefish: Status, Management and Propagation, North Central Division, American Fisheries Society, Special Publication 7, Columbia.

Ryder, J.A. 1888. The sturgeons and sturgeon industries of the eastern coast of the United States, with an account of experiments bearing upon sturgeon culture. Bull. U.S. Fish Comm. 8: 231–328.

Schaeffer, B. 1973. Interrelationships of chondrosteans. pp. 207–226. *In*: P.H. Greenwood, R.S. Miles & C. Patterson (ed.) Interrelationships of Fishes, Academic Press, London.

Schaeffer, B. & D.E. Rosen. 1961. Major adaptive levels in the evolution of the actinopterygian feeding mechanism. Amer. Zool. 1: 187–204.

Schmalhausen, O.I. 1991. Prelarval development of *Pseudoscaphirynchus kaufmanni*. Sov. J. Dev. Biol. 22: 296–302.

Schultze, H.-P. 1987. Dipnoans as sarcopterygians. pp. 39–74. *In*: W.E. Bemis, W.W. Burggren & N.E. Kemp (ed.) The Biology and Evolution of Lungfishes, Alan Liss, New York.

Sewertzoff, A.N. 1925. The place of the cartilaginous ganoids in the system and the evolution of the Osteichthyes. J. Morphol. 38: 105–155.

Sewertzoff, A.N. 1926a. Studies on the bony skull of fishes. I. Structure and development of the bony skull of *Acipenser ruthenus*. Q. J. Micros. Sci. 70: 451–540.

Sewertzoff, A.N. 1926b. The development of the scales of *Acipenser ruthenus*. J. Morphol. 42: 523–560.

Sewertzoff, A.N. 1928. The head skeleton and muscles of *Acipenser ruthenus*. Acta Zool. 9: 193–319.

Soin, S.G. 1951. Materials on the development of sturgeons of the Amur River. Tr. Amurskoy Ichtiol. Eksped. 1945–1949, Izd. Moskovs. Obshtchestva Ispytat. Prirody, Moskva, volume 2: 223–232 (in Russian).

Stensiö. E.A. 1925. Triassic fishes from Spitzbergen. Kungl. Svenska Vetenskaps Akademiens Handligar. 3 series 2: 1–261.

Svetovidov, A.N. 1984. Acipenseridae. *In*: P.J.P. Whitehead, M.-L. Bauchot, J.-C. Hureau, J. Nielsen & E. Tortonese (ed.) Fishes of the North-eastern Atlantic and the Mediterranean, Vol. 1, UNESCO, Paris.

Swift, C.C., C.R. Gilbert, S.A. Bortone, G.H. Burgess & R.W. Yerger. 1986. Zoogeography of the freshwater fishes of the southeastern United States: Savannah River to Lake Ponchartrain. pp. 213–265. *In*: C.H. Hocutt & E.O. Wiley (ed.) The Zoogeography of North American Freshwater Fishes, Wiley, New York.

Tanaka, S. 1908. Notes on some Japanese fishes, with descriptions of fourteen new species. J. Coll. Sci., Imper. Univ. Tokyo. 23: 1–54.

Tatarko, K. 1936. Der Kiemendeckelapparat und seine Verbindung mit dem Hyoid- und Kieferbogen bei den Acipenseridae. Trav. Inst. Zool. Biol. 3: 5–67.

Tatarko, K. 1939. The mandibular and hyoid arches and gill cover in Polypterini, Lepidosteoidei, and Amiodei. Pap. Anim. Morph. Kiev. 5: 17–71 (in Russian).

Thompson, D.H. 1934. Relative growth in *Polyodon*. Nat. Hist. Surv. Ill. Biol. Notes 2: 1–6.

Tleuov, R.T. & N.I. Sagitov. 1973. Acipenserids of the Amu-Darya River. FAN Publ., Tashkent. 155 pp. (in Russian).

Traquair, R.H. 1877. The ganoid fishes of the British Carboniferous formations, Part 1, Paleoniscidae. Monogr. Paleont. Soc. 31: 1–60.

Traquair, R.H. 1887. Notes on *Chondrosteus acipenseroides*. Geol. Mag. 4: 248–257.

Trautman, M.B. 1981. The fishes of Ohio. Ohio State University Press, Columbus. 782 pp.

Tsessarsky, A.A. 1992. Evolution of the jaw apparatus of sturgeons and the origin of paddlefishes. Doklady Akad. Nauk 323: 957–960.

Vasetskiy, S.G. 1971. Fishes of the family Polyodontidae. J. Ichthyol. 11: 18–31.

Vladykov, V. & J.R. Greeley. 1963. Order Acipenseroidei. pp. 24–60. *In*: H.B. Bigelow & W.C. Schroeder (ed.) Fishes of the Western North Atlantic, Sears Foundation for Marine Research, Yale University, New Haven.

Wake, D.B. & A. Larson. 1987. Multidimensional analysis of an evolving lineage. Science 238: 42–48.

Walbaum, J. 1792. Petri Artedi renovati, pars i et ii (et iii–v) i.e., Bibliotheca et philosophia ichthyologica, cura Iohannis Iulii Walbaumii edidit. 5 parts in 3 volumes. Grypeswaldiae, (1788) 1789 (-1793) 4 pl. Part 3 is subtitled 'Petri Artedi Sueci Genera piscium in quibis systema totum ichthyologia proponitur cum classibus, ordinibus, generum characteribus, specierum differentiis, observationibus plurimis. Redactis speciebus 242 ad genera 52. Ichthyologiae, pars iii. Emendata et aucta a Iohanne Iulio Walbaum.' 723 pp.

Watson, D.M.S. 1925. The structure of certain paleoniscids and the relationships of that group with other bony fishes. Proc. Zool. Soc. Lond. 1925: 815–870.

Watson, D.M.S. 1928. On some points in the structure of paleoniscid and allied fish. Proc. Zool. Soc. Lond. 1928: 49–70.

Wei, Q., F. Ke, J. Zhang, P. Zhuang, J. Luo, R. Zhou & W. Yang. 1997. Biology, fisheries, and conservation of sturgeons and paddlefish in China. Env. Biol. Fish. (this volume).

Wiley, E.O. 1976. The phylogeny and biogeography of fossil and recent gars (Actinopterygii: Lepisosteidae). Univ. Kansas Mus. Nat. Hist. Misc. Pub. 64: 1–111.

Wilimovsky, N.J. 1956. *Protoscaphirhynchus squamosus*, a new sturgeon from the upper Cretaceous of Montana. J. Paleo. 30: 1205–1208.

Williams, J.D. & G.H. Clemmer. 1991. *Scaphirhynchus sutkussi*, a new sturgeon (Pisces: Acipenseridae) from the Mobile Basin of Alabama and Mississippi. Bull. Mus. Nat. Hist. 10: 17–31.

Williot, P. (ed.) 1991. Acipenser. CEMAGREF Publ., Bordeaux. 519 pp.

Woodward, A.S. 1891. On the paleontology of sturgeons. Proc. Geol. Assoc. 11: 24–44.

Woodward, A.S. 1895a. The fossil fishes of the upper Lias of Whitby, Part 1. Proc. Yorkshire Geol. Polytech. Soc. 24–42.

Woodward, A.S. 1895b. The fossil fishes of the upper Lias of Whitby, Part 2. Proc. Yorkshire Geol. Polytech. Soc. 155–170.

Woodward, A.S. 1895c. Catalogue of the fossil fishes in the British Museum (Natural History), Part III. British Museum (National History), London. 544 pp.

Woodward, A.S. 1909. The fossil fishes of the English chalk, Part 5. Monogr. Palaeontogr. Soc. 43: 153–184.

Wourms, J.P., J.W. Atz & M.D. Stribling. 1991. Viviparity and the maternal-embryonic relationship in the coelacanth *Latimeria chalumnae*. pp. 225–248. *In*: J.A. Musick, M.N. Bruton & E.K. Balon (ed.) The Biology of *Latimeria chalumnae* and Evolution of Coelacanths, Kluwer Academic Publishers, Dordrecht.

Yakovlev, V.N. 1977. Phylogenesis of Acipenseriforms. pp. 116–143. *In*: V.V. Menner (ed.) Essays on Phylogeny and Systematics of Fossil Fishes and Agnathans, USSR Akademy of Sciences, Moscow (in Russian).

Yakovlev, V.N. 1986. Fishes. pp. 178–179. *In*: A.P. Rasnitsyn (ed.) Insects in the Early Cretaceous Ecosystems of the West Mongolia, Nauka, Moscow (in Russian).

Yeager, B. & R. Wallus. 1982. Development of larval *Polyodon spathula* (Walbaum) from the Cumberland River in Tennessee. pp. 73–77. *In*: Fifth Annual Larval Fish Conference, Louisiana Cooperative Fisheries Research Unit.

Yu, Z., Z. Deng, Y. Zhao & X. Huang. 1986. Observations on the goandal development of *Psephurus gladius* (Martens) in the Changjiang River below the Gezhouba Dam Acta Hydrobiol. Sinica 10: 295–296 (in Chinese).

Zholdasova, I. 1997. Sturgeons and the Aral Sea ecological catastrophe. Env. Biol. Fish. (this volume).

Zhou, Z. 1992. Review on *Peipiaosteus* based on new material of *P. pani*. Vert. Palas. 30:85–101 (in Chinese).

Note added in proofs: Although this paper was published in 1997, its completion and submission preceded that of Grande & Bemis (1996). Thus, Grande & Bemis (1996) provide some additional information on phylogenetic interrelationships of acipenseriforms (particularly †Chondrosteidae and †Peipiaosteidae) that is not included here.

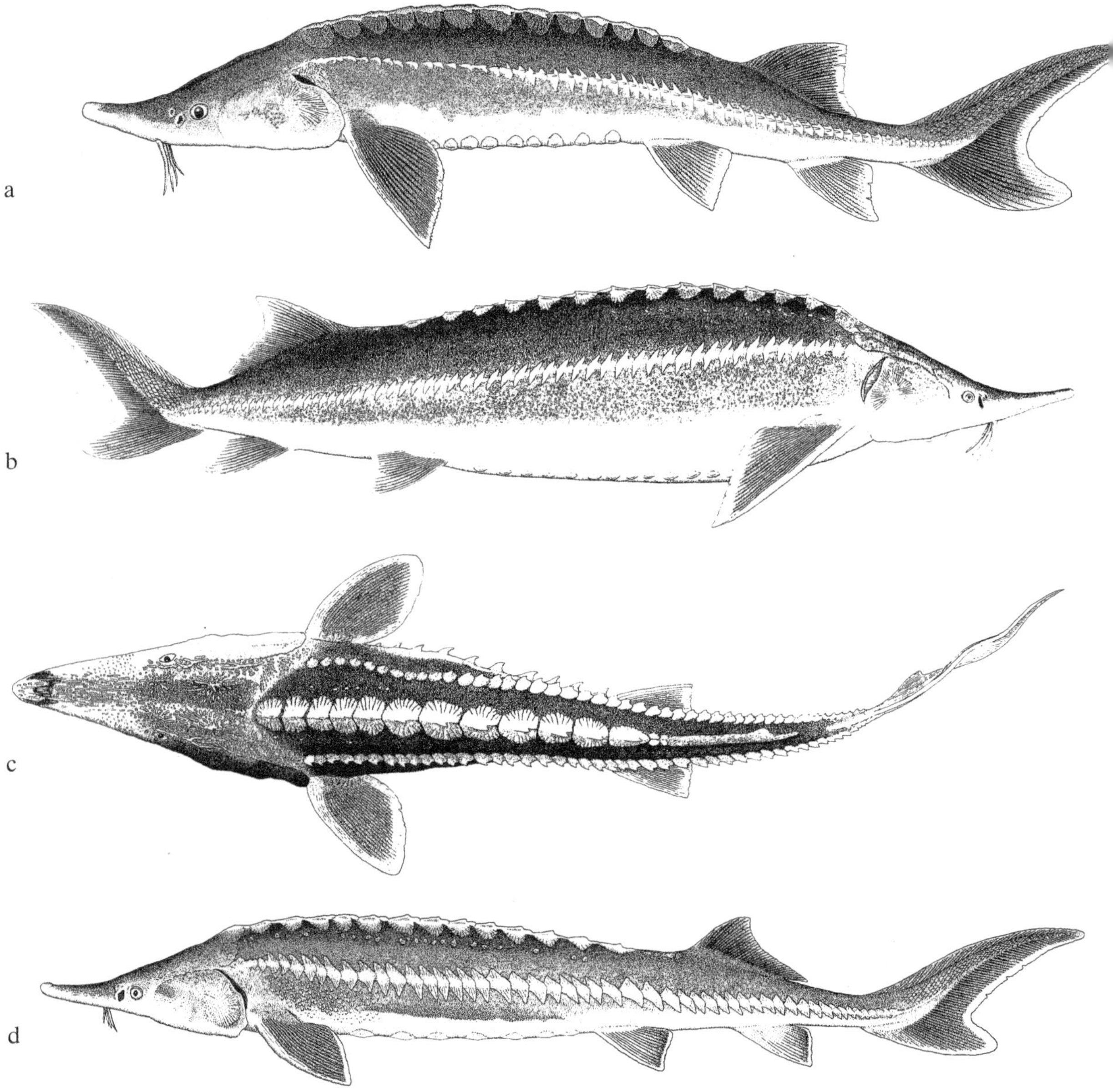

Sturgeons of the Danube River and one of their hybrids: a – *Acipenser ruthenus* 43 cm TL from the Fish Culture Research Institute, Szarvas, Hungary; b – a large *A. ruthenus* 76 cm TL from an old collection at the Grigore Antipa Natural History Museum, Bucharest, Romania; c – dorsal view of an early juvenile *Huso huso* 23 cm TL from Royal Ontario Museum collection; d – a F_2 bester (= beluga × sterlet) 69 cm TL, a cross of F_1 *A. ruthenus* × *H. huso* from the captive breeding at Propa-Gen International (Aquaculture Production R & D and Trading)', Komadi, Hungary. Originals by Paul Vecsei, 1996.

Environmental Biology of Fishes **48**: 73–126, 1997.

Osteology and phylogenetic interrelationships of sturgeons (Acipenseridae)

Eric K. Findeis
Department of Biology, University of Massachusetts, Amherst, MA 01003-0027, U.S.A.
Current address: Department of Biology, Wake Forest University, Winston-Salem, NC 27109, U.S.A.

Received 18.8.1995 Accepted 22.3.1996

Key words: Actinopterygii, cladistics, evolutionary morphology, benthic fishes, peramorphosis, *Huso, Acipenser, Pseudoscaphirhynchus, Scaphirhynchus*

Synopsis

Sturgeons (Acipenseridae) are an ancient and unique assemblage of fishes historically important to discussions of actinopterygian evolution. Despite their basal position within Actinopterygii, rigorous comparative morphological studies of acipenserids have never been made, and most ideas about acipenserid evolution hinge on an untested impression that shovelnose sturgeons (Scaphirhynchini) are phylogenetically primitive. This impression promoted ideas that: (1) the earliest acipenserids were highly benthic and evolved secondarily into pelagic predators, and (2) paedomorphosis has dominated mechanisms affecting their morphological change. Using cladistic methods, this study examines generic level interrelationships within Acipenseridae. Representatives of the four acipenserid genera *Huso, Acipenser, Pseudoscaphirhynchus*, and *Scaphirhynchus*, as well as their acipenseriform outgroups Polyodontidae, †Peipiaosteidae, and †Chondrosteidae, were surveyed for skeletal characters. Sixty-nine characters are identified and described to support the first generic level cladogram of Acipenseridae. *Huso* is phylogenetically primitive within Acipenseridae and the sister group to a redefined subfamily Acipenserinae. *Acipenser* is not supported by any characters identified in this study, but the tribe Scaphirhynchini comprising *Scaphirhynchus* and *Pseudoscaphirhynchus* is found to be monophyletic. The cladogram contradicts historical ideas about acipenserid evolution because *Huso* defines an outgroup morphology and life history founded on pelagic habitats and piscivory. In contrast, acipenserines, and more markedly scaphirhynchines, are benthic predators possessing character complexes for benthic feeding, respiration, locomotion, and protection. Also, the pattern of character acquisition within Acipenseridae suggests that peramorphosis played a central role in acipenserid evolution. Peramorphic addition and enlargement of the skeleton and scalation defines most characters at all nodes within Acipenseridae, and repudiates paedomorphosis as a major trend in evolution within the family Acipenseridae.

Introduction

The family Acipenseridae includes 25 species in four currently recognized genera in a Holarctic distribution. This is an ancient assemblage with recognizable acipenserid fossils known from the Upper Cretaceous and fossil relatives extending the origin of Acipenseriformes into the Lower Jurassic (†*Chondrosteus* Egerton, 1858). Acipenserids are distinctive in morphology and behavior, with numerous features such as armoring trunk scutes, a ventral mouth, rostral chemosensory barbels, and a flattened body contributing to their benthic habitats and behaviors. Their commercial importance, uniqueness, and almost universal endangered status (see Birstein 1993) has promoted a modern ground swell of interest in sturgeon biology (e.g., Binkowski & Doroshov 1985, Williot 1991, Gershanovich & Smith 1995), including their systematics and evolution (Bemis et al. 1997 this volume).

A better understanding of acipenserid interrelationships is central not only to organizing modern studies in sturgeon biology, but to examining evolutionary and functional attributes of the lineage. Out of only four extant clades of non-teleostean actinopterygians (Lauder & Liem 1983), acipenserids are the most speciose and have the largest biogeographic range. As such, they are pivotal to evolutionary studies of Actinopterygii and more precise understanding of their morphology and phylogenetic position is crucial for cladistic comparison to other taxa. Further, given their age and unique adaptations, extant acipenseriforms differ markedly from other actinopterygians, with distinct, parallel evolutionary solutions to the challenges of aquatic life faced by all fishes.

The history of acipenserid systematics is long, with major studies from the 19th century addressing species and generic recognition (Rafinesque 1820, Brandt & Ratzeberg 1833, Fitzinger & Heckel 1836, Brandt 1869, Dúmeril 1867, 1870). Unfortunately, there has never been a consensus about recognition of genera or subgenera with current usage accepting the genera *Huso, Acipenser, Scaphirhynchus*, and *Pseudoscaphirhynchus*. As many as six subgenera (or genera) have been proposed to subdivide *Acipenser*, and most attempts subsumed *Huso* into *Acipenser* (Brandt & Ratzeberg 1833, Fitzinger & Heckel 1836, Dúmeril 1870). While *Huso* is now recognized as an independent genus (defined by Brandt 1869), its diagnostic characters (Tatarko 1936, Antoniu-Murgoci 1936a, b) remain untested. At a higher level, *Huso* has remained paired with *Acipenser* within the subfamily Acipenserinae (sensu Bonaparte 1838), and the shovelnose sturgeon genera *Scaphirhynchus* and *Pseudoscaphirhynchus* are universally accepted individually and composing the subfamily Scaphirhynchinae (Scaphirhynchini of Bonaparte 1846). However, anatomical characters accepted as defining genera or subfamilies have never been examined rigorously.

In this study, I use comparative osteology and cladistics to survey for and test skeletal characters relevant to acipenserid phylogeny. Previous studies of acipenserid systematics focused on external features that are significantly variable and pose problems for a phylogenetic analysis. The only studies emphasizing the skeleton were made by Antoniu-Murgoci (1936a, b, 1942), but her analysis only included Rumanian species and only examined portions of the skeleton. Comparative osteology can develop numerous phylogenetic characters and notably allows use of fossil taxa as outgroups (e.g., Nielsen 1949, Patterson 1973, 1982, Gardiner 1984). Cladistic methods focus on defining characters at specific phylogenetic nodes to determine interrelationships (Hennig 1966, Wiley 1981), but also arrays character transformations as focal changes defining taxa. This provides insight into evolutionary processes underlying phylogeny and allowing evaluation of performance and behavior. Cladistics and comparative osteology have dramatically increased our understanding of actinopterygian evolution, but acipenserids have been neglected. Only recently have interrelationships within Acipenseriformes been convincingly defined (Grande & Bemis 1991, Findeis 1993, Bemis et al. 1997) recognizing outgroups necessary for a new examination of acipenserid characters (Figure 1). These studies accept the Polyodontidae as the sister group to Acipenseridae (together composing the Acipenseroidei), with an unresolved trichotomy among the Acipenseroidei, †Chondrosteidae, and †Peipiaosteidae with Acipenseriformes (but see Grande & Bemis 1996).

Morphological and systematic studies of acipenserids pose certain problems making comprehensive phylogenetic studies difficult. Specimens of many species are not readily available, so taxonomic coverage is difficult. Previous morphological descriptions are not detailed (e.g., Marinelli & Strenger 1973) or comprehensive enough (e.g., Tatarko 1936, Antoniu-Murgoci 1936a, b, 1942) to support cladistic analyses and new descriptions are needed (Findeis 1993). Acipenseriforms are largely cartilaginous and acipenseriform fossils do not show cartilage preservation, leaving incomplete data for comparison. Because of these problems, this study focuses on generic interrelationships within Acipenseridae.

Although little definitive phylogenetic work has

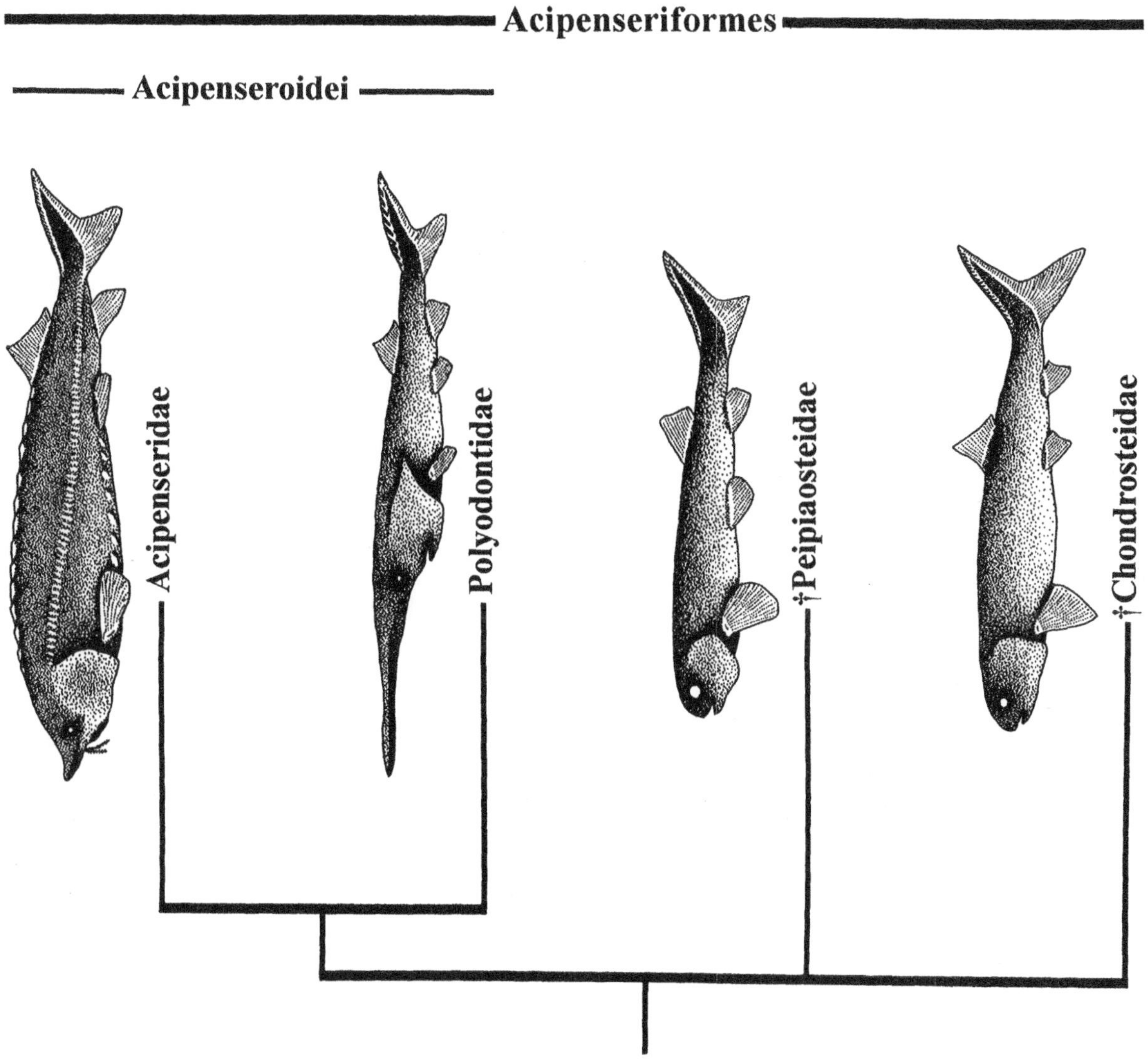

Figure 1. Proposed phylogenetic relationships within Acipenseriformes: Characters defining this cladogram are from Grande & Bemis (1991), Findeis (1993), and Bemis et al. (1997 this volume). These characters support a monophyletic Acipenseroidei comprising Acipenseridae and Polyodontidae. The Acipenseroidei forms an unresolved trichotomy in Acipenseriformes with the fossil taxa †Peipiaosteidae and †Chondrosteidae. For outgroup comparison with Acipenseridae, all three families are used when possible. Illustrated taxa are †*Chondrosteus acipenseroides* Agassiz 1844, †*Peipiaosteus pani* Liu & Zou 1965, *Psephurus gladius* (Martens 1862), and *Huso huso* (Linnaeus 1758).

been made of the Acipenseridae at any level, two themes dominate evolutionary studies of the family. The first suggests that shovelnose sturgeons (Scaphirhynchini) represent the primitive condition among acipenserids (Schmalhausen 1991, Birstein 1993). In this scenario, sturgeons evolved initially as benthic predators, with *Huso* and the more predaceous species of *Acipenser* rising subsequently into midwater environments. Second, classic ideas about acipenserids focus on paedomorphosis as the dominant mechanism in their evolution (Woodward 1889, Goodrich 1909, Gregory 1933). These persisting hypotheses have never been examined rigorously and the results from this phylogenetic study contradict them.

Materials and methods

Phylogenetic methods

Interrelationships among genera of Acipenseridae were examined using cladistic methods as generally espoused by Hennig (1966) and Wiley (1981). Skeletal features identified as potentially relevant characters for phylogenetic reconstruction were compared with outgroup taxa to: (1) determine which subgroups possess or lack the character; and (2) define the polarity of change in character states. Polarity assessments were made through comparison with all genera of Acipenseridae, Polyodontidae (personal observations; Grande & Bemis 1991), †*Chondrosteus* (Traquair 1887, Woodward 1889, Watson 1925, 1928, Hennig 1925), †*Peipiaosteus* (Liu & Zhou 1965, Bai 1983, Zhou 1992), and multiple non-acipenseriform taxa such as †*Mimia* and †*Moythomasia* (Gardiner 1984a), †*Cheirolepis* (Pearson & Westoll 1979), †*Birgeria* (Nielsen 1949), †*Saurichthys* (Rieppel 1992), *Polypterus, Lepisosteus*, and *Amia* (personal observations). Characters were accepted only when skeletal features were consistent within ingroups compared to outgroups (for respective polarities). Rather than provide a matrix with proposed character states, putative synapomorphies are defined and directed to specific phylogenetic nodes. Character descriptions provide information on character morphology of ingroups and outgroups to unambiguously define apomorphic and plesiomorphic character states.

Specimens and preparation

Specimens examined included representatives of all four acipenserid genera. Skeletal characters generally focus on bone since most taxa possess ossified endoskeletons (and fossils are predicated on preservation of bone), but cartilage is dominant in the acipenserid endoskeleton. Conventional techniques of skeletal preparation can damage or destroy cartilage, so several techniques were used to assess morphology of specimens examined in this study. Large, newly collected specimens were hand cleaned to preserve cartilages. Several adult specimens were prepared with dermestid beetles resulting in near total loss of cartilage, but leaving intact dermal skeletons. Many small specimens were prepared by clearing and double staining (Dingerkus & Uhler 1977) to examine intact skeletons. Specific osteological structures were examined in preserved specimens of multiple taxa to confirm or deny putative characters identified from skeletons. Developmental series were examined in cleared and double stained preparations to clarify structural interpretations amidst morphological variation and to interpret homology recognition.

Scaphirhynchus

Scaphirhynchus platorynchus was accepted as generally typical for the genus in this study. Nine newly collected adult specimens were hand cleaned and multiple preserved specimens were examined externally and through dissection. Two large adult specimens from the Shedd Aquarium (FMNH 98285, 98286) were prepared with dermestid beetles. An extensive developmental series ranging from prehatching embryos to small juveniles was cleared and double stained with addition of a larger juvenile (AMNH 4485) with slightly more advanced skeletal development. *Scaphirhynchus albus* was examined externally and non-invasively to confirm skeletal characters of the genus (see Bailey & Cross 1954, Carlson 1985). No specimens of *S. suttkusi* were available for examination, but this species is very similar to *S. platorynchus* (Williams & Clemmer 1991).

Pseudoscaphirhynchus

Only two specimens of the dwarf morphotype of *P. kaufmanni* were available for examination of the genus in this study. One was cleared and double stained (MCZ 27653) and one was examined externally with additional description from Ivanzoff (1887) and Sewertzoff (1926a, 1928). This species is accepted as representative for the genus in this study, although *P. hermanni* differs markedly from *P. kaufmanni* (Berg 1948a) and is assumed to be specialized. *Pseudoscaphirhynchus fedtschenkoi* has been reported to be morphologically variable (Berg 1948a) with distinct morphotypes that differ in varying degrees compared to *P. kaufmanni*.

Acipenser

I examined several species of *Acipenser* to assess morphological variation within the genus to allow comparisons with other genera. Four adult *A. brevirostrum*, two large juvenile *A. medirostris*, two large juvenile *A. oxyrinchus*, and the head of an adult *A. oxyrinchus* (2.4 m total length) were hand cleaned. I cleared and double stained several small specimens of *A. brevirostrum, A. transmontanus*, and heads of *A. ruthenus*. Abbreviated developmental series of *A. brevirostrum* and *A. transmontanus* were cleared and double stained. Marinelli & Strenger (1973) described *A. ruthenus* and several European species of *Acipenser* are described in Sewertzoff (1926a, b, 1928), Tatarko (1936), and Antoniu-Murgoci (1936a, b, 1942).

Huso

Huso huso is the only species of the genus examined in this study, but is similar in morphology to *H. dauricus* by all accounts (Berg 1948a). Of three small juveniles, one was cleared and double stained, one sectioned, and one examined intact (all CAS 37541). Two moderate sized juveniles (FMNH 96852, 96853; each approximately one meter total length) were examined externally and by dissection to confirm character states between small and larger sizes.

Terminology of acipenserid taxa

In the description of characters, characters are grouped according to the levels of acipenserid taxa that they define. Accordingly, character descriptions often refer to higher level taxa defined in this study that are not consistent with historic usage. Taxa recognized within Acipenseridae include the four genera *Huso, Acipenser, Scaphirhynchus*, and *Pseudoscaphirhynchus*. At higher levels, the taxa recognized in this study are the Husinae (comprising only *Huso* and usually referred to here with only the generic name), Acipenserinae (comprising *Acipenser, Pseudoscaphirhynchus*, and *Scaphirhynchus*), and Scaphirhynchini (comprising *Pseudoscaphirhynchus* and *Scaphirhynchus*).

Characters and phylogenetic comparisons

I define 69 characters used to analyze interrelationships of genera within Acipenseridae. All characters described and analyzed here are skeletal to allow outgroup comparison with the fossil acipenseriform taxa †*Chondrosteus* (Traquair 1887, Hennig 1925, Watson 1925, 1928), †*Peipiaosteus* (Liu & Zhou 1965, Bai 1983, Zhou 1992), and the fossil polyodontids †*Paleopsephurus* and †*Crossopholis* (Grande & Bemis 1991), as well as extant polyodontids. Several soft tissue characters such as presence of four barbels, the spiracle, and the ligaments of the jaws and hyoid arches may characterize specific nodes within Acipenseridae, but are not included here. Soft tissue characters are problematic in a cladistic analysis of Acipenseridae because Polyodontidae is the only relevant extant outgroup (Bemis et al. 1997), leaving any putative familial level characters as two taxon statements (Polyodontidae vs. Acipenseridae).

Character 1. Trunk bracketed with five rows of scutes – Acipenseridae

Five scute rows are present along the trunk in acipenserids (shown in *Huso* in Figure 1). The five rows are distinguishable as three groups: (1) the dorsal scute row extending from the dermal skull to the predorsal scale of the dorsal fin; (2) paired flank scute rows bearing the trunk canal from the supracleithrum into the base of the caudal fin; and (3) paired ventral scute rows spanning the pectoral and pelvic fins. Shape, number, and size of individual scutes varies dramatically, but their position and presence as complete groups is consistent within

Acipenseridae. Some species of *Acipenser* regress the ventral scutes (e.g., *A. fulvescens*) almost completely (Dúmeril 1867), but the row is always present in juveniles and most adults. Dorsal scute morphology has been used as a phylogenetic character by some workers (Antoniu-Murgoci 1936a, 1942), but no clear evidence of cladistically useful character states based on scute shape has been demonstrated.

In addition to scutes, acipenserids possess three novel and discrete series of other scales: (1) median predorsal and preanal scales at the anterior bases of these fins, (2) a variable assemblage of scales anterior to the anal fin, and (3) scales on the caudal peduncle (Findeis 1993). Rhombic caudal scales supporting the caudal fin are plesiomorphically present in all acipenseriforms except †*Peipiaosteus* (Liu & Zhou 1965) and are typical for palaeoniscids.

Scutes and other large scales develop as bony plates with their approximate adult morphology distinct from smaller scales of the skin. Sewertzoff (1926b) suggested that scutes develop as composites of these trunk scales, but I find no scale assemblages coalescing in early ontogeny of scutes or other named scales. Scutes are the earliest scales to appear, and develop as series distinct from other scales. Other named scales or scale groups appear later in ontogeny, but also as discrete units.

No large trunk scales are present in any other acipenseriform taxa. *Psephurus* bears small scales studding the skin of the trunk and *Polyodon* possesses denticular scales restricted to the anterior base of the median fins (Grande & Bemis 1991). †*Crossopholis* and †*Paleopsephurus* possess fringed scales infiltrating the skin of the trunk (Grande & Bemis 1991), but they are still small, scattered scales. Scales in polyodontids are comparable in size and distribution to the isolated trunk scales present in all acipenserids examined, but not scutes or other named scales. †*Chondrosteus* (Traquair 1887) possesses no known trunk scales. †*Peipiaosteus* (Liu & Zhou 1965) has thin, paired scales bracketing the putative trunk canal, but no other trunk scales.

Character 2. Dermal bone forms a pectoral fin spine – Acipenseridae

The pectoral fin of all acipenserids is supported along its anterior edge by a thick pectoral fin spine. The fin spine is composed of dermal bone extending from the leading edge of the propterygium to typically sheath two fin rays (**lpt1, lpt2**, Figure 2a). One ray is included in the weak spine of *Scaphirhynchus platorynchus* and rarely three are included in taxa with strong fin spines (occasionally in *Acipenser oxyrinchus*).

The base of the pectoral fin spine is an expanded propterygium (**ptg**) articulating with the scapulocoracoid (Findeis 1993). This enlarged propterygium is also diagnostic of acipenserids and included within this character as an associated feature of the fin spine. Dermal bone molds to the propterygium and anterior fin rays to unify the spine with its propterygial base (Figure 2a). Fusion of dermal bone with the fin rays is extensive and segmentation of the lepidotrichia is lost as they merge into the spine in adults (compare **lpt1** and **lpt2** in Figure 2a). In large (= old) specimens, the dermal spine fuses with a perichondral ossification of the propterygium (Figure 2b).

Dermal bone and their incorporated fin rays appear separately in ontogeny. Most authors have assumed that fin rays fuse into the fin spine (e.g., Vladykov & Greeley 1963, Jollie 1980, Grande & Bemis 1991), but my observations suggest that lepidotrichia are simply covered by the expanding dermal bone (Figure 2b). Recognition of a dermal component in the pectoral fin spine defines this character.

Polyodontids (Grande & Bemis 1991), †*Chondrosteus* (Traquair 1887), and †*Peipiaosteus* (Liu & Zhou 1965) lack a pectoral fin spine. No accessory dermal bone and no enlarged propterygium is present in *Polyodon* or *Psephurus*. Individual fin rays at the leading edge of the pectoral fin are recognizable in †*Chondrosteus* (Traquair 1887, Hennig 1925) and †*Peipiaosteus* (Liu & Zhou 1965, Zhou 1992), so no dermal sheath could be present in these outgroups.

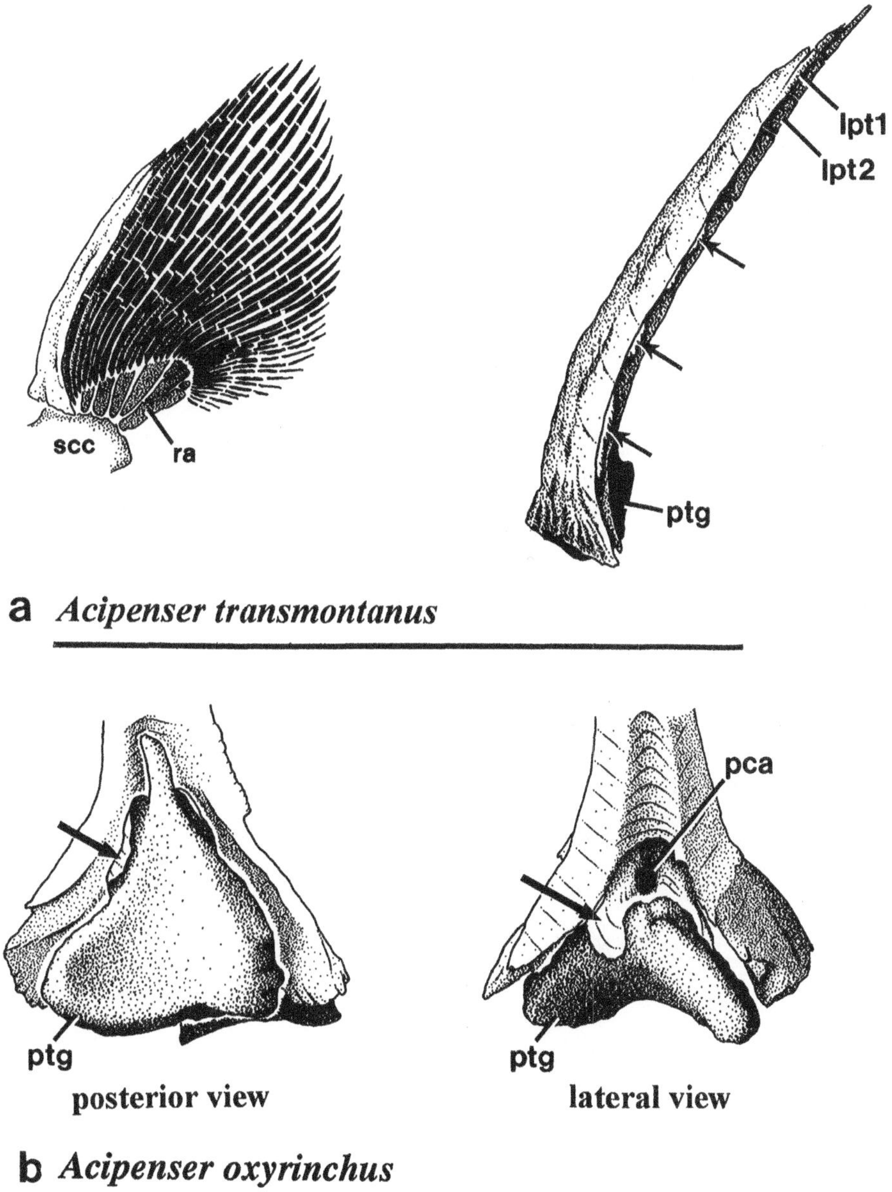

Figure 2. Pectoral fin spine of Acipenseridae (Character 2): a – Illustrations of the pectoral fin spine of *Acipenser transmontanus* in an intact fin (left) and isolated (right). The fin spine supports the leading edge of the pectoral fin. At higher magnification, dermal bone contacts the propterygium (**ptg**) at the base and encompasses one fin ray completely (**lpt1** is exposed distally) and shows bone encroaching onto the second ray (arrows onto **lpt2**). This juvenile specimen (cleared and double stained) shows only partial development of the spine, but in adults the first two fin rays are fully encompassed by dermal bone. b – Illustrations of the base of the fin spine in a large *A. oxyrinchus* (2.4 m total length) in posterior (left) and lateral (right) views. The propterygium is braced within the dermal spine and fused to the dermal bone via its perichondral ossification (arrows). The canal bearing blood vessels and nerves into the fin (**pca**) is a portion of the propterygium and shows the extent of fusion. Two fin rays are fused into this spine, but they cannot be distinguished from the dermal spine at the base where the segmented lepidotrichia merge fully into the new bone (scc=scapulocoracoid, ra=pectoral radials and metapterygium).

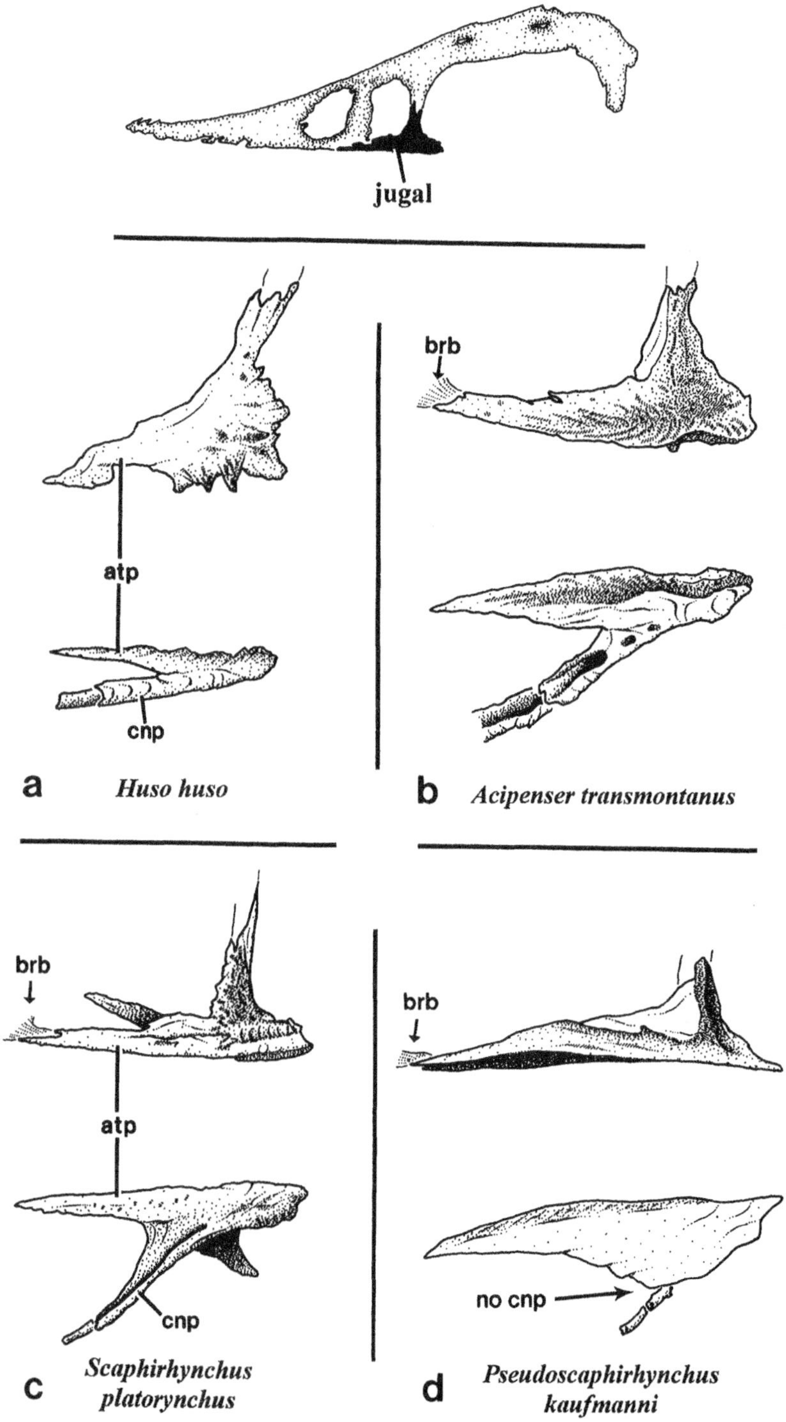

Figure 3. Jugal of representatives of all genera of Acipenseridae: The jugal defines the posteroventral margin of the orbit (top in black). Illustrations include lateral and ventral views for each genus. The anterior process of the jugal (**atp**, Character 3) extends anterolaterally beneath the orbit and lacks any portion of the infraorbital canal. The canal extends anteromedially within the canal process of the jugal (**cnp**) before entering a series of tube bones in all acipenserids except *Pseudoscaphirhynchus* (Character 57) (brb = border rostral bones).

Character 3. Jugal possesses an anterior process – Acipenseridae

The jugal of all acipenserids examined extends a prominent anterior process undercutting the orbit (**atp**, Figure 3). It is small in *Huso* (Figure 3a), but large in all species of *Acipenser* examined and *Scaphirhynchus*, and massive in *Pseudoscaphirhynchus* (Character 57). The anterior process extends anterolaterally to contact the neurocranium.

The anterior process does not bear the infraorbital canal that curls anteromedially within a separate, smaller canal process of the jugal (**cnp**, Figure 3). Thus, passage of the infraorbital canal, typical of the circumorbital series, does not apply to the anterior process of the jugal of acipenserids. The extended jugal braces the dermal skull roof against the neurocranium anterior to the orbit (Findeis 1993) and conforms the dermal skull to the expanded neurocranium of acipenserids. Ethmoid expansion of the neurocranium coincides with a broadened postnasal wall as a probable character of the Acipenseridae not scorable in fossil taxa, but coincident jugal expansion restricted to acipenserids suggests that ethmoid expansion is an acipenserid character.

No polyodontid possesses homologizable circumorbital bones as the infraorbital canal is carried

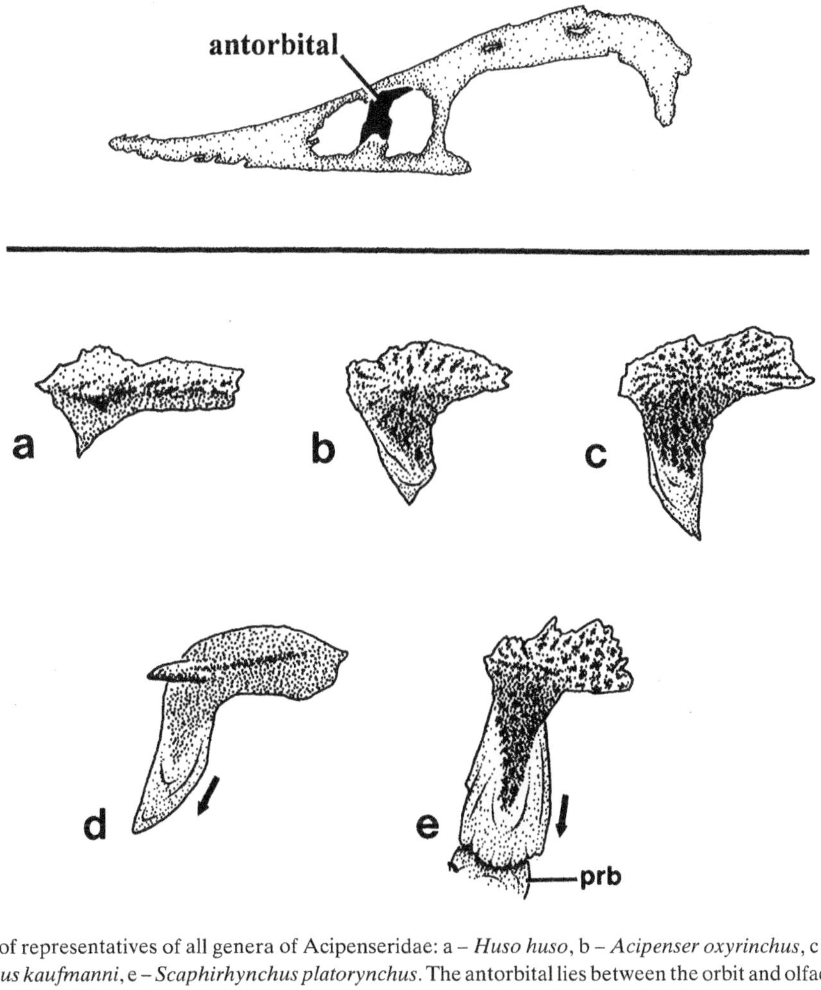

Figure 4. Antorbital of representatives of all genera of Acipenseridae: a – *Huso huso*, b – *Acipenser oxyrinchus*, c – *A. brevirostrum*, d – *Pseudoscaphirhynchus kaufmanni*, e – *Scaphirhynchus platorynchus*. The antorbital lies between the orbit and olfactory opening dorsally in the dermal skull (top in black; Character 4). It lacks any portion of the supraorbital canal and extends ventrally as a variably sized wedge-shaped process. The ventral wedge is small in *Huso* (a), variably sized in *Acipenser* (b, c), but elongate in scaphirhynchines (arrows in d, e; Character 46). It contacts an enlarged postrostral bone (**prb**) in *Scaphirhynchus* (Character 61).

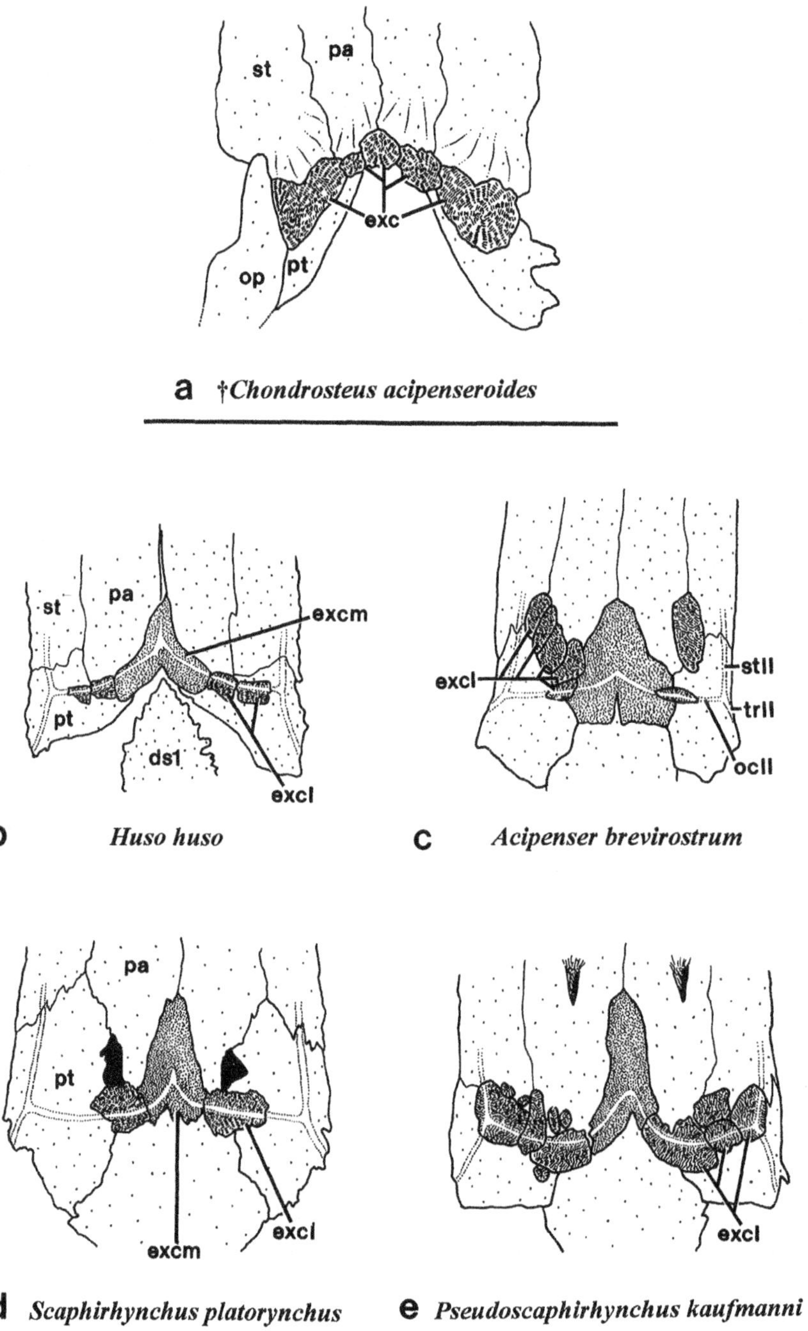

Figure 5. Extrascapular bones of the posterior skull roof in †*Chondrosteus* and representatives of all genera of Acipenseridae: The median extrascapular (**excm**; Character 5) is a midline bone that bears the commissure of the occipital canal (**ocll**; Character 24) in all acipenserids. The lateral extrascapulars (**excl**) carry the occipital canal from the posttemporal (**pt**) to the median extrascapular in all acipenserids except *Pseudoscaphirhynchus*, where the lateral extrascapular series includes the origin of the occipital canal (Character 56). Lateral extrascapulars typically form a series of canal bones or are variable in position, but are clustered in *Scaphirhynchus* (Character 60). †*Chondrosteus* (a – redrawn from Traquair, 1887) possesses a series of extrascapulars (**exc**), but no distinct median and lateral extrascapulars. **pa** = parietal, **st** = supratemporal, **op** = opercle, **stll** = supratemporal canal, **trll** = trunk canal, **ds1** = first dorsal scute.

by tube bones independent of the skull roof (Grande & Bemis 1991). The infraorbital canal of †*Peipiaosteus* (Zhou 1992) is similarly borne by tube bones rather than a dominant series of elements and is not comparable with acipenserids. An anterior process of the jugal has been reconstructed for †*Chondrosteus* (e.g., Traquair 1887, Gardiner & Schaeffer 1989), but it apparently bears the infraorbital canal unlike acipenserids.

Character 4. Antorbital bone present – Acipenseridae

The antorbital defines the anterodorsal corner of the orbit (Findeis 1993) and is present in all acipenserids examined. It lies dorsally between the orbit and olfactory capsule as a small plate that angles ventrolaterally onto the postnasal wall as ventral points (Figure 4).

The antorbital varies in size and shape, but not position or presence, in Acipenseridae. Its ventral points are short or absent in *Huso* (Figure 4a) and variably extended in most species of *Acipenser* (Figure 4b, c), but antorbitals of scaphirhynchines (Figure 4d, e) possess elongate ventral processes (see Character 46). Antorbitals rest alongside the nasal, but never bear the supraorbital canal. As with the anterior process of the jugal (Character 3), presence and ventrolateral extension of the antorbital is consistent with the expanded neurocranium as the skull is covered locally by appearance of a new bone.

The antorbital is not present in polyodontids (Grande & Bemis 1991), and has not been described in †*Chondrosteus* (Traquair 1887, Hennig 1925). Zhou (1992) names an antorbital in †*Peipiaosteus*, but it is a tube bone carrying the supraorbital canal not comparable to acipenserids. The only positionally comparable bone in †*Mimia* (Gardiner 1984a) and other palaeoniscids is the nasal, but the nasal ubiquitously bears the supraorbital canal.

Character 5. Median extrascapular bone present – Acipenseridae

The median extrascapular (**excm**) is a triangular plate present between the posterior ends of the parietals (**pa**) and anterior to the first dorsal scute (**ds1**, Figure 5). It is present in all acipenserids examined, and bears the commissure of the occipital canals (see Character 24 below).

The median extrascapular develops extremely late, first appearing as splints bracketing the occipital canal before expanding to cover the posterior skull. Several bones may compose the median extrascapular initially, but the ossification is single in all adult specimens examined. Expansion beyond the occipital canals and consistent shape suggests that the median extrascapular has developed phylogenetically from an anamestic bone into a permanent bone of the skull.

No similar median element in the posterior skull roof is known from any other acipenseriform. Grande & Bemis (1991) provisionally identified a

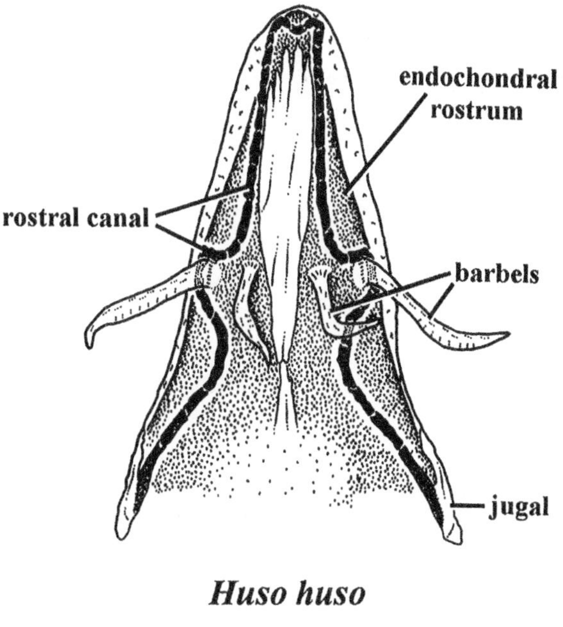

Figure 6. The rostral canal and rostrum of *Huso huso*: The rostral canal extends onto the ventral surface of the rostrum from the jugal (see Character 3) to converge medially before arching laterally around the outer barbels (Character 6). After curling around the outer barbels, the rostrum canals extend anteriorly in parallel before converging at the rostral tip.

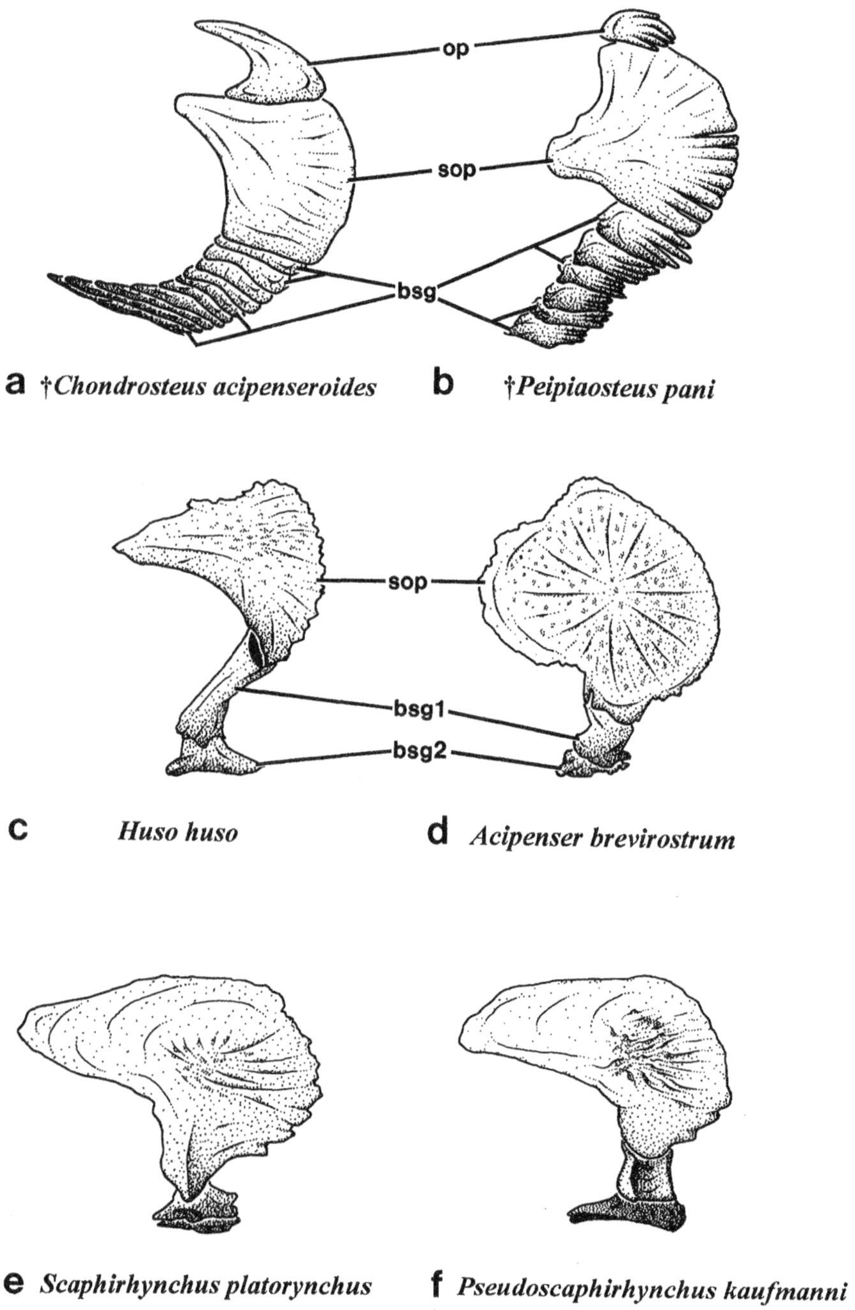

Figure 7. The opercular series of †*Chondrosteus*, †*Peipiaosteus*, and representatives of all genera of Acipenseridae: †*Chondrosteus* (a – redrawn from Woodward 1889) and †*Peipiaosteus* (b – redrawn from Zhou 1992) possess serial branchiostegals (**bsg**) supporting the operculum beneath the subopercle (**sop**). Acipenserids (c, d, f) typically possess two branchiostegals (Character 7), with a vertical, rectangular first branchiostegal (**bsg1**) and small second branchiostegal (**bsg2**). *Scaphirhynchus* (e) possesses a short, triangular first branchiostegal (Character 62). The subopercle of scaphirhynchines is dorsoventrally compressed with an anteroposteriorly elongate shape (Character 47).

'median nuchal' in †*Paleopsephurus*, but the specimen is too poorly preserved to identify an autonomous element or note whether posttemporal extensions typical of polyodontids expand over this area. All other polyodontids conclusively lack a median extrascapular. †*Chondrosteus* (Traquair 1887) possesses multiple small bones that cross the posterior midline of the skull roof (Figure 5a), but they are small bones more comparable to lateral extrascapulars. †*Peipiaosteus* (Liu & Zhou 1965, Zhou 1992) is not well described in this area, but no bone separates the parietals.

Character 6. The rostral canal arches laterally – Acipenseridae

Rostral sensory canals extend ventromedially under the rostrum, but then curl laterally around the outer barbels before recurving to parallel the rostral midline (Figure 6). The rostral canal develops with this lateral arch originally and is not displaced by the developing barbels.

In extant polyodontids the rostral canals converge smoothly to parallel the midline to the anterior tip of the rostrum without disruption. Incomplete remains of rostral tube bones in †*Crossopholis* are consistent with this condition (Grande & Bemis 1991), but are missing in †*Paleopsephurus*. Rostral canals of †*Chondrosteus* (Hennig 1925, and

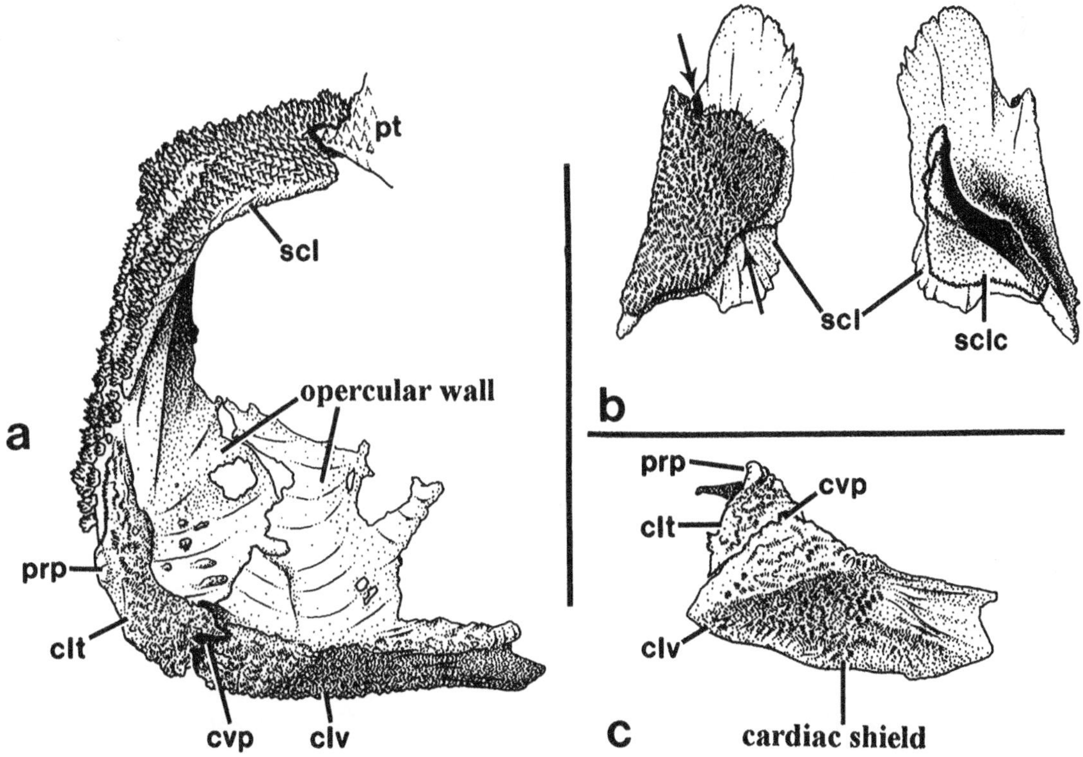

Figure 8. Dermal pectoral girdle of *Acipenser brevirostrum*: a – Anterolateral view of an intact pectoral girdle showing the medial opercular wall (Character 10) rising vertically from the cleithrum (**clt**) and clavicle (**clv**), the clavicle process (**cvp**) binding the cleithrum and clavicle (Character 13), the propterygial restraining process (**prp**; Character 12), and the close contact between the supracleithrum (**scl**) and posttemporal (**pt**). b – Dorsal (left) and ventral (right) views of the supracleithrum. The anterior shelf of the supracleithrum (Character 9) extends as a flat, unornamented area anterior to the trunk canal (arrows). The supracleithral cartilage (**sclc**; Character 15) is bound to the ventral surface of the supracleithrum. c – Ventral view of the pectoral girdle showing the expanded clavicle and cleithrum forming the cardiac shield (Character 11) and the extension of the propterygial restraining process (Character 12) laterally from the cleithrum.

indications from Gardiner & Schaeffer 1989) are apparently also straight. †*Peipiaosteus* possesses a rostral canal that extends anteriorly onto the snout and bends dorsally to terminate in a cluster of bones (Zhou 1992). In †*Mimia* (Gardiner 1984a) and *Polypterus*, the rostral canal is straight and no other outgroups possess an abrupt lateral arch.

Character 7. The opercular series includes a serial pair of branchiostegals – Acipenseridae

The dominant opercular bone of acipenseroids is the subopercle, but acipenserids typically possess two branchiostegals that extend ventrally and then medially in support of the operculum (Figure 7). Branchiostegal one (**bsg1**) forms an elongate vertical support before branchiostegal two (**bsg2**) angles medially to undercut the head. Branchiostegal one is rectangular and variably elongate in *Huso* and all species of *Acipenser* examined (Figure 7c, d), and *Pseudoscaphirhynchus* (Figure 7f), but is stubby and triangular in *Scaphirhynchus* (Figure 7e; Character 46). Occasionally three branchiostegals are present, but the third was diminutive and did not significantly enlarge the opercular series when present.

Polyodontids possess a single branchiostegal separate from the subopercle (Findeis 1993, Bemis et al. 1997). The opercular series of †*Chondrosteus* (Traquair 1887, Hennig 1925) and †*Peipiaosteus* (Liu & Zhou 1965, Zhou 1992) include multiple, serial branchiostegals not vertically elongate (Figure 7a, b). This condition is plesiomorphically similar to most palaeoniscids and suggests that reduction to two branchiostegals is diagnostic of Acipenseridae.

Character 8. Loss of supraneurals in the caudal peduncle – Acipenseridae

Supraneurals of all acipenserids examined are serially present above neural arches in the thoracic trunk between the neurocranium and dorsal fin, but absent beneath the dorsal fin and in the caudal peduncle. Supraneural homologs (Grande & Bemis 1991) support the caudal fin, leaving a gap in the peduncle.

Polyodontids (Grande & Bemis 1991) and †*Chondrosteus* (reconstruction from Woodward 1889) possess supraneurals throughout the axial skeleton. †*Peipiaosteus* apparently lacks supraneurals in the peduncle (Zhou 1992), but †*Peipiaosteus* is small and supraneurals are difficult to identify and may have never ossified. Among other outgroups, distribution of supraneurals is variable. *Polypterus* and †*Pteronisculus* (and several palaeoniscids; Gardiner 1984a) possess a complete series of supraneurals, but †*Mimia* (Gardiner 1984a), *Lepisosteus, Amia*, and neopterygians typically possess incomplete series. Inconsistency among outgroups clouds analysis of this character, but a complete supraneural series in polyodontids suggests that a peduncle gap is characteristic of Acipenseridae. This correlates with the flattened peduncle of acipenserids that presumably puts spatial constraints on the vertically oriented supraneurals unlike other acipenseriform taxa.

Character 9. Supracleithral shelf undercuts the posttemporal – Acipenseridae

The supracleithrum (**scl**) expands anteriorly as a broad shelf undercutting the posttemporal (**pt**) in all acipenserids examined (Figure 8a, b). This shelf broadly contacts the underside of the posttemporal and the bones are bound tightly to unify the pectoral girdle and skull roof. This shelf does not bear the trunk canal that passes from the posttemporal to the supracleithrum at their surface suture (arrows in Figure 8b).

Extant polyodontids possess a mobile supracleithral-posttemporal articulation (Grande & Bemis 1991). *Polyodon, Psephurus*, and †*Crossopholis* all show slight supracleithral undercutting of the posttemporal with a short portion of the supracleithrum extending anterior to the exit of the trunk canal. However, length of this process is not equivalent to acipenserids. The supracleithrum of †*Chondrosteus* (Traquair 1887) broadly contacts the posttemporal, but without an anterior process.

†*Peipiaosteus* lacks tight contact between these bones (Zhou 1992).

Character 10. Presence of a medial opercular wall – Acipenseridae

The medial opercular wall is a vertical shelf defining the internal surface of the opercular chamber (Findeis 1993). It comprises expanded laminae curving dorsally from the cleithrum and clavicle (**clt, clv**, Figure 8a). The medial opercular wall is present in all acipenserids examined. It is not as expansive in *Huso* as in acipenserines, but the corresponding opercular chamber is more slender in *Huso*.

Polyodontids possess slender pectoral girdle bones with laterally exposed faces that never angle internally or form expanded laminae (Grande & Bemis 1991). Clavicles and cleithra of †*Chondrosteus* (Traquair 1887) and †*Peipiaosteus* (Liu & Zhou 1965) are also slender and lack internal expansion. No expanded opercular wall is present in *Polypterus, Lepisosteus, Amia*, or †*Mimia* (Gardiner 1984a) and other palaeoniscids.

Character 11. Presence of a cardiac shield – Acipenseridae

The cardiac shield is a ventral shelf formed by expansion of the clavicle and cleithrum. The shield expands posteromedially as a flat, exposed plate that meets its contralateral partner at the midline (Figure 8c). The cardiac shield covers the pericardial cavity and is found in all acipenserids examined.

The cardiac shield is almost completely flat in *Scaphirhynchus* (Findeis 1993), but angles centrally through the anteroposterior axis in other genera, with a frontally flat medial region and dorsally angled lateral face (Figure 8c). This central angle is acute (up to 20°) in species with cylindrical body shapes (all species of *Acipenser* examined and *Huso*), but shallow in the flatter *Pseudoscaphirhynchus*. The cardiac shield and opercular wall of acipenserids meet in a sharply angled anteroventral edge (approximately 160°). This angle brackets the scapulocoracoid with the cardiac shield supporting the expanded coracoid shelf (see Character 14).

The clavicle and cleithrum of polyodontids (Grande & Bemis 1991), †*Chondrosteus* (Traquair 1887), and †*Peipiaosteus* (Liu & Zhou 1965) are slender bones lacking shelves. The clavicles of *Psephurus* and †*Crossopholis* possess short posterior wedges smaller than the cardiac shield, but they do not flare posteromedially or meet medially (Grande & Bemis 1991). No similar cardiac shields are present in *Polypterus, Amia*, or other distant outgroups.

Character 12. Presence of a propterygium restraining process – Acipenseridae

The propterygium restraining process extends from the lateral edge of the cleithrum (**prp**, Figure 8a, c) to wrap anterolateral to the propterygial fossa of the scapulocoracoid (see Character 16) and its articulating propterygium. It brackets the pectoral fin spine anteriorly to support and limit movement of the fin in all acipenserids examined.

The propterygium restraining process is a prominent, posteriorly curved process in *Scaphirhynchus* and extended ridge in *Pseudoscaphirhynchus*, but is less prominent in *Huso* and *Acipenser* (Figure 8a). It is well developed in juveniles, but is allometrically reduced in adults as the cleithrum thickens with dermal ornament obscuring the shallow ridge. When a distinct process is absent, this region is enlarged as a robust cleithral notch enfolding the base of the fin spine. This character tacitly includes such a cleithral notch that acts as a pectoral fin spine constraint when the process is less obvious.

No similar process is present in any other acipenseriform group. Outgroups possess no pectoral fin spine (see Character 2) or, among extant polyodontids, a propterygial fossa (Character 16) allowing frontal pivoting of the propterygium. †*Crossopholis* possesses a shallow notch in the cleithrum aligned with the fin (Grande & Bemis 1991), but it is an isolated condition within Polyodontidae.

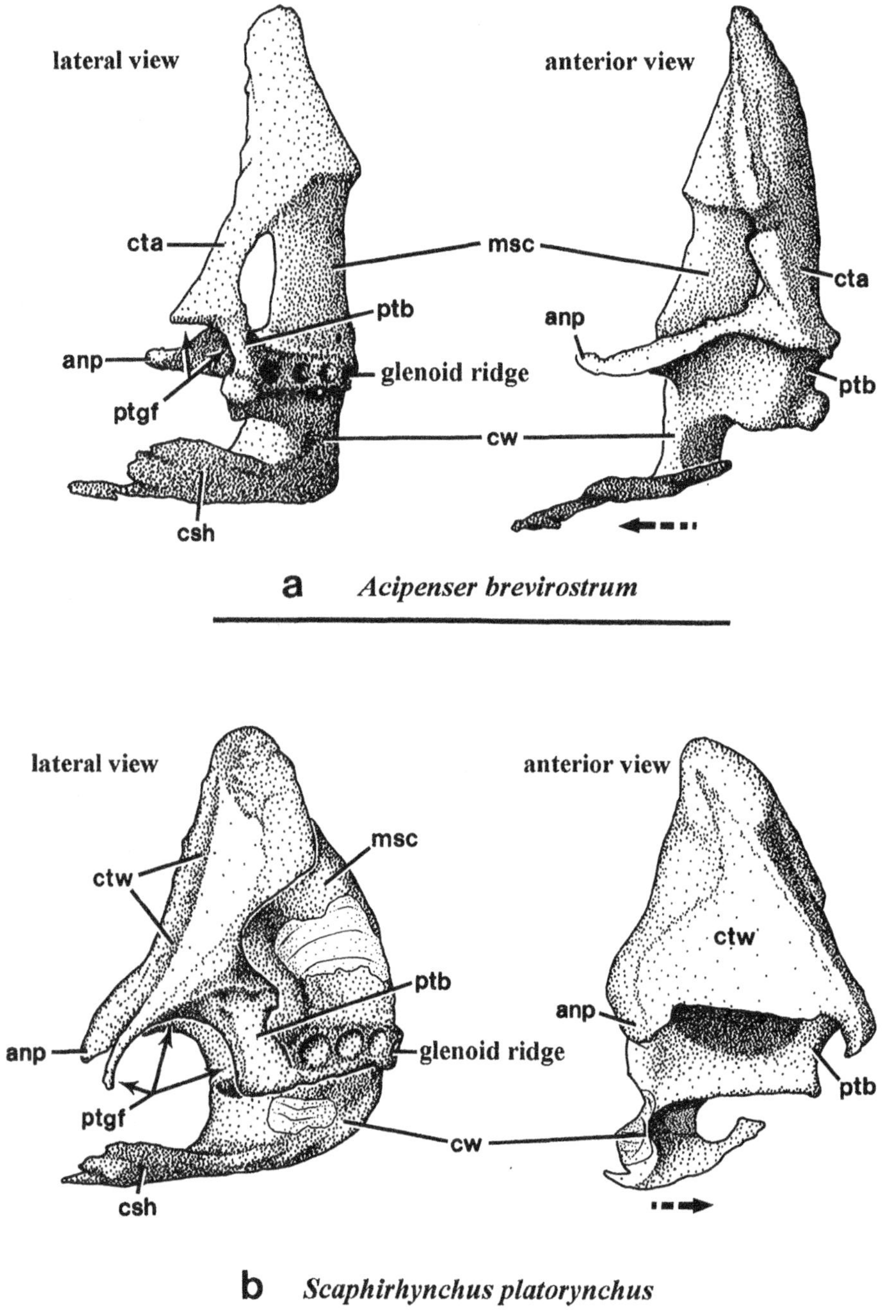

Figure 9. Left scapulocoracoid of *Acipenser brevirostrum* (a) and *Scaphirhynchus platorynchus* (b) in lateral and anterior views: The ventral scapulocoracoid extends from the coracoid wall (**cw**) as a flat shelf of cartilage (**csh**; Character 14) medially in all acipenserids (arrows pointing left) except *Scaphirhynchus*, where it forms a small footprint that flares laterally (arrows pointing right; Character 68). Acipenserids also possess a broad propterygial fossa (**ptgf**) at the anterior end of the glenoid ridge where the propterygium (see Character 2) pivots on the propterygial bridge (**ptb**; Character 16). The propterygial fossa is open anteriorly in all acipenserids except *Scaphirhynchus*, where a thin ridge encircles the fossa anteriorly (Character 67). The anterior face of the scapulocoracoid is L-shaped, spanning the anterior process of the middle region (**anp**) to the propterygial bridge, then rises dorsally as the cleithral arch (**cta**) in most acipenserids (a), but is expanded as a broad cleithral wall (**ctw**) in *Scaphirhynchus* (b; Character 66). **msc** = mesocoracoid arch.

Character 13. Clavicle process interdigitates with the cleithrum – Acipenseridae

The clavicle process is a small wedge that interdigitates with the cleithrum in all acipenserids examined (**cvp**, Figure 8a, c). The clavicle process extends along the anteroventral edge of the pectoral girdle in *Scaphirhynchus* (Findeis 1993), but typically undercuts the clavicle slightly in other acipenserids (Figure 8 for *Acipenser*). The process interlocks with a opposing groove in the cleithrum to solidify the pectoral girdle.

No clavicle process is present in extant polyodontids or †*Crossopholis* (Grande & Bemis 1991). This character is a small feature not easily noted from illustrations or descriptions of †*Chondrosteus* (Traquair 1887, Hennig 1925) or †*Peipiaosteus* (Liu & Zhou 1965), but their slender pectoral girdles are not suggestive of such an interlocking process.

Character 14. Coracoid shelf spreads over the cardiac shield – Acipenseridae

The coracoid wall (**cw**) extends ventrally from the middle region of the scapulocoracoid to spread onto the cardiac shield as a flat coracoid shelf (**csh**, Figure 9; Findeis 1993). The coracoid shelf makes immediate contact with the cleithrum and spreads anteromedially almost to the tip of the clavicle in all acipenserids (Figure 9a) except *Scaphirhynchus*. In *Scaphirhynchus*, the coracoid shelf spreads laterally and is restricted to the cleithrum (Figure 9b; see Character 68).

In extant polyodontids, the coracoid wall is cylindrical and extends anteromedially separate from the cleithrum before curving anteroventrally onto the clavicle. Shape of this coracoid process corresponds to the slender dermal girdle. The scapulocoracoid is largely cartilaginous and not scorable in fossil taxa, but shape of the dermal girdle of †*Crossopholis* and †*Paleopsephurus* (Grande & Bemis 1991), †*Chondrosteus* (Traquair 1887), and †*Peipiaosteus* (Liu & Zhou 1965) is similar to extant polyodontids, suggesting that a correspondingly cylindrical coracoid process would be consistent in outgroup acipenseriforms.

Character 15. Supracleithral cartilage present – Acipenseridae

The supracleithral cartilage lies under the supracleithrum (Figure 8b). It varies in shape among acipenserids from an elliptical, tall cartilage in some species of *Acipenser* to an elongate, shallow element in *Huso* and *Scaphirhynchus*.

The supracleithral cartilage is absent in extant polyodontids, polypterids, lepisosteids, and amiids. It never ossifies and is unknown from any fossil acipenseriform. Lacking †*Chondrosteus* or †*Peipiaosteus* as outgroups clouds cladistic analysis of this character within Acipenseriformes, but its absence in extant polyodontids suggests it is restricted to Acipenseridae.

Character 16. Propterygial fossa present in scapulocoracoid – Acipenseridae

The propterygial fossa is a broad, semicircular opening in the scapulocoracoid between the glenoid ridge and the propterygium restraining spine of the cleithrum (**ptgf**, Figure 9; see Character 12). It is bounded posteriorly by the propterygial bridge (**ptb**) and dorsally by the vertical cleithral arch (**cta**, Figure 9a) or cleithral wall (**ctw**, Figure 9b; see Character 66). The propterygial fossa opens a space for the rotating propterygium that articulates with the propterygial bridge (see Character 2) and is filled with pectoral musculature that pulls the pectoral fin spine forward. The propterygial fossa is a rounded notch (Figure 9a) in all acipenserids except *Scaphirhynchus*, where it forms a half circle (Figure 9b; see Character 67).

The glenoid ridge extends completely to the cleithrum in extant polyodontids. The propterygium of these taxa is small and can articulate with the glenoid ridge without need for an anterior fossa. Although this character cannot be confirmed in fossil taxa, lack of a cleithral notch (see Character 12) in †*Chondrosteus* (Traquair 1887, Hennig 1925) or †*Peipiaosteus* (Liu & Zhou 1965) suggests that there would be no opposing opening in the scapulocoracoid. No propterygial fossa is present in †*Mimia* (Gardiner 1984a) and other palaeoniscids.

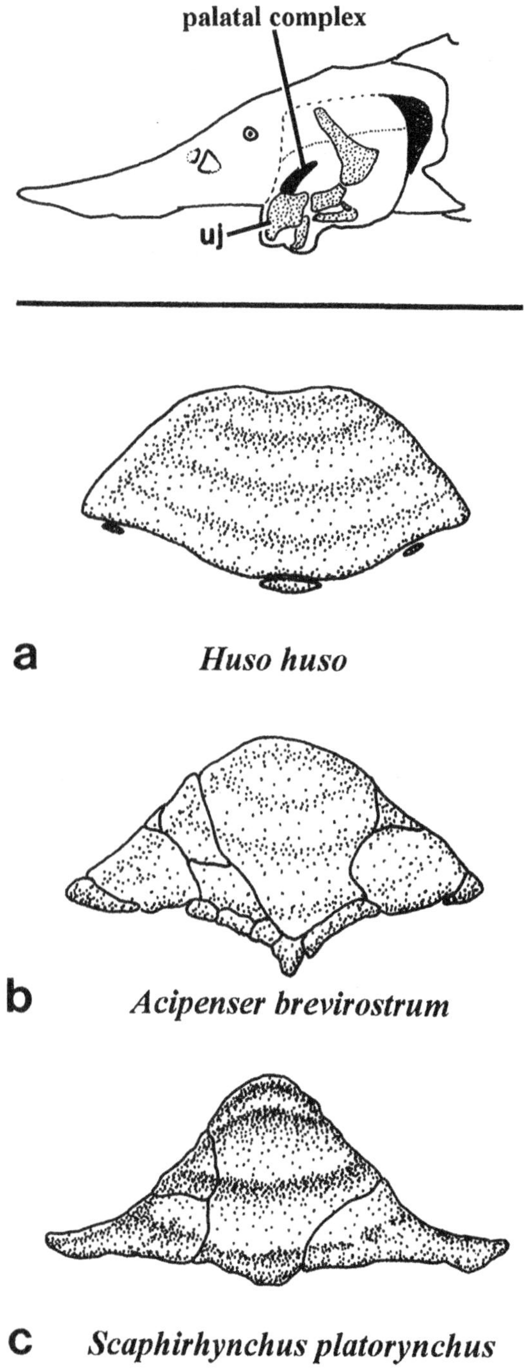

Figure 10. Palatal complex of representatives of three genera of Acipenseridae: The palatal complex is a cartilaginous plate bound to the palatoquadrate to expand the functional surface of the upper jaw (top in black; Character 18). It is formed by a single dominant cartilage in *Huso*, but comprises multiple large and small cartilages in other acipenserids.

Character 17. Basipterygial process present – Acipenseridae

This basipterygial process extends from the ventral surface of the basipterygium, flaring around the anterior edge of the pelvic fin musculature (Findeis 1993). It is present in all acipenserids examined.

Extant polyodontids possess a segmented basipterygium different from acipenserids, but also lacking a ventral process. The basipterygium is cartilaginous in adults and unknown from fossil acipenseriforms. Lacking †*Chondrosteus* and †*Peipiaosteus* as outgroups clouds cladistic analysis of this character, but the pelvic plate of †*Moythomasia* (Gardiner 1984a) is complete and lacks a ventral process. *Polypterus, Lepisosteus*, and *Amia* possess differing basipterygial morphologies, but all lack a ventral process.

Character 18. Palatal complex present – Acipenseridae

The palatal complex comprises an integrated group of flat cartilages embedded within loose skin posterior to the upper jaw and opposite the ventral tongue pad (Figure 10). This complex is attached to and functionally linked with the palatoquadrate. Number of cartilages forming the complex and their individual shapes vary, but it is present in all acipenserids examined.

The palatal complex is composed of several large plates of cartilage unified as a wedge with thin lateral arms in *Scaphirhynchus* and *Pseudoscaphirhynchus* (Figure 10c). The palatal complex of all species of *Acipenser* examined is deeper and more rounded than in scaphirhynchines, and typically includes numerous small cartilages arrayed around a large central and paired lateral plates (Figure 10b; see Antoniu-Murgoci 1942). *Huso* possesses a broad, shallow palatal complex formed by one dominant plate with occasional thin cartilages along the anterior edge (Figure 10a). Number of cartilages forming the palatal complex has been used as a character within Acipenseridae (Antoniu-Murgoci 1936b, Sokolov 1989), but this variation makes it difficult to polarize character states and no interpretation on shape or composition is made here.

No palatal complex is present in other extant acipenseriforms or actinopterygians, but as a cartilaginous character its presence or absence is unknown in †*Chondrosteus* or †*Peipiaosteus*.

Character 19. Posteroventral edge of hyomandibula forms a wedge – Acipenseridae

The hyomandibulae of all acipenserids examined are expanded posteriorly to form a posterior wedge in the ventral cartilaginous head (Figure 11). Position of the wedge varies from a dorsal position that tapers ventrally in *Huso* and all species of *Acipenser* examined (Figure 11b, c) to a ventral position forming a flat edge in scaphirhynchines (Figure 11d; Character 50). The ventral end is cartilaginous, but the ossified core reflects the wedge by posterior flaring and thinning of the perichondral center.

The hyomandibula of *Psephurus* is slender, with a ventral head tapering with a continuously curved posterior edge (Figure 11a). The ossified center of *Psephurus* and †*Crossopholis* expands ventrally (Grande & Bemis 1991), but not as sharply as in acipenserids and with a rounded posterior edge. The hyomandibula of †*Paleopsephurus* is incomplete and not well known (Grande & Bemis 1991, MacAlpin 1947). The ossified cores of †*Chondrosteus* (Tra-

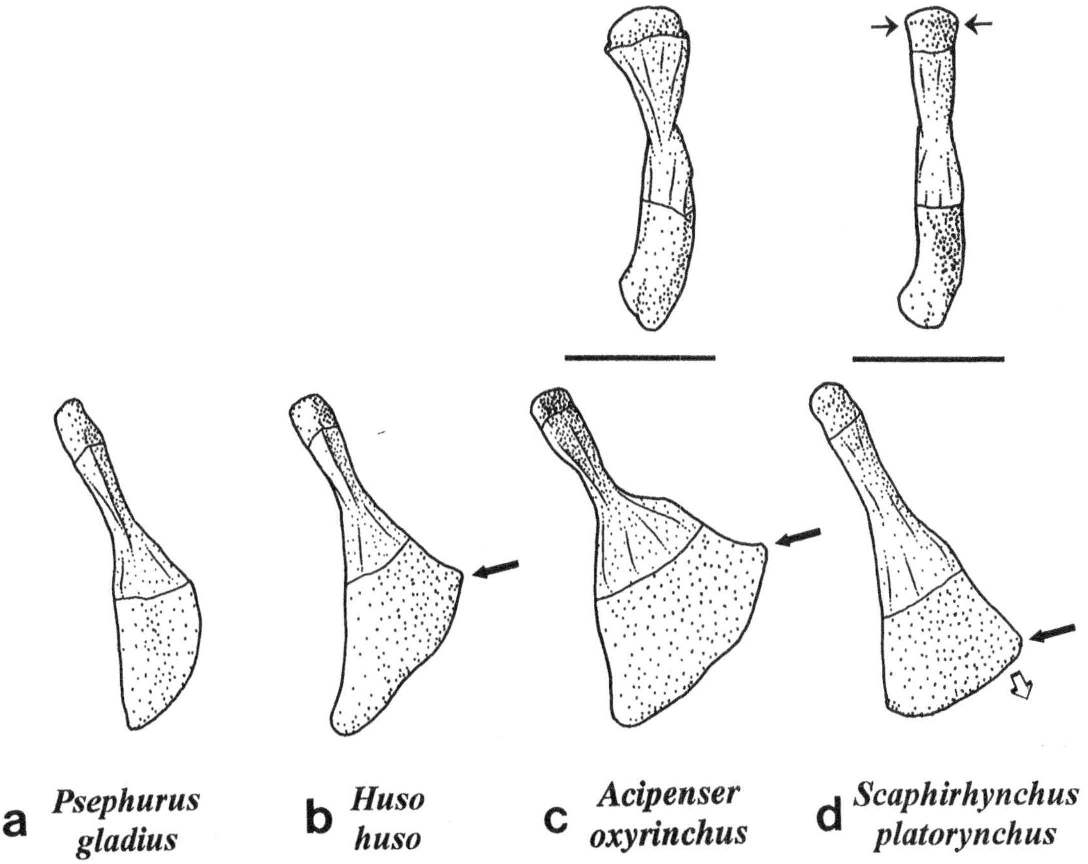

Figure 11. Hyomandibula of *Psephurus* and representatives of three genera of Acipenseridae: The hyomandibula is shown in lateral view in all taxa and also in anterior view in *Acipenser* (c) and *Scaphirhynchus* (d). The hyomandibula of *Psephurus* (a) possesses a convex posterior edge, but is expanded posteriorly in all acipenserids examined (heavy arrow in b, c, d; Character 19). The posterior wedge is dorsal in *Huso* (b) and *Acipenser* (c), but ventral in scaphirhynchines (hollow arrow; Character 50). Also, the dorsal tip is broad in *Huso* and *Acipenser*, but thin in scaphirhynchines (thin arrows; Character 49).

quair 1887), †*Gyrosteus* (Woodward 1889), and †*Peipiaosteus* (Liu & Zhou 1965) all expand ventrally as in all other acipenseriforms, but since this cartilaginous character is unpreserved in fossils they do not assist in assigning polarity. More distant outgroups lack the hyostylic jaw suspension of acipenseriforms and are useless in assessing this character. I accept *Psephurus* as representing the plesiomorphic condition.

Character 20. Anterior ceratohyal is flat and rectangular, with a mandibulo-hyoid process and asymmetric hypohyal joint – Acipenseridae

Anterior ceratohyals of all acipenserids examined are dorsoventrally flattened and rectangular in oral view (Figure 12). The lateral head is centrally grooved with a rising posterior ridge articulating with the posterior ceratohyal and anterolateral process that is the origin of the mandibulo-hyoid ligament (arrows, Figure 12c, d, e). The medial end is blunt, with an asymmetric, anteriorly displaced facet for the hypohyal (**hh**, displaced arrows of Figure 12c, d, e).

Anterior ceratohyals of *Psephurus* are cylindrical, hourglass-shaped elements lacking distinct processes and possessing a central hypohyal facet on the medial head (Figure 12a). The lateral end lacks a distinct mandibulo-hyoid ligament process or central groove and the posterolateral head is a blunt tip articulating with the posterior ceratohyal. The anterior ceratohyal of †*Paleopsephurus* is unknown, but that of †*Crossopholis* is also cylindrical based on shape of its ossified center (Grande & Bemis 1991). Anterior ceratohyals of *Polyodon* are flattened and not comparable to other taxa (Grande & Bemis 1991). Cylindrical morphologies are inferred for †*Chondrosteus* (Traquair 1887) and †*Gyrosteus* (Figure 12b; Woodward 1889) based on their cylindrical core. The hyobranchial skeleton of †*Peipiaosteus* is unpreserved (Liu & Zhou 1965).

Character 21. Anterior shelf protrudes from hypobranchial one – Acipenseridae

An anterodorsal shelf expands from hypobranchial one (**hb1**) as a wedge overhanging the anterior ceratohyal and hypohyal (**hh**) in all acipenserids examined (arrow in Figure 13b, c). The shelf is concave medially, opening a gap where the hypohyal extends dorsally to articulate with basibranchial one (Figure 13b, c). The anterior shelf, in concert with the exposed hypohyal and anterior basibranchial, forms a flat anterior edge of the ventral branchial skeleton. This edge bears connective tissue ridges

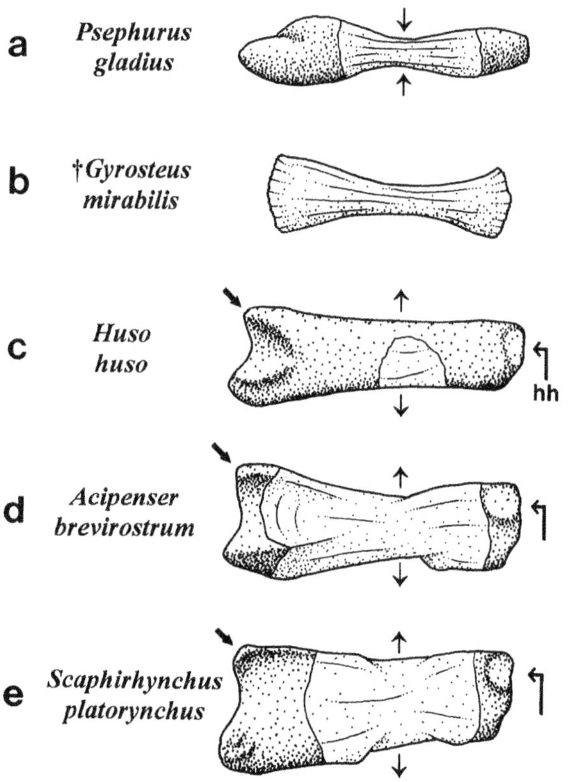

Figure 12. Anterior ceratohyal of *Psephurus*, †*Gyrosteus*, and representatives of three genera of Acipenseridae: The anterior ceratohyal is narrow and cylindrical in acipenserid outgroups (a, b; †*Gyrosteus* redrawn from Woodward 1889), but expanded centrally as a flattened rectangular element (thin arrows in c, d, e; Character 20). The acipenserid anterior ceratohyal also possesses an anterolateral process supporting the mandibulo-hyoid ligament (heavy arrows in c, d, e) and an asymmetric facet for the hypohyal on the medial tip (**hh** and bent arrows in c, d, e).

used for prey processing opposite to the palatal complex (Character 18).

Hypobranchial one of *Psephurus* and †*Crossopholis* is cylindrical with a flattened dorsal surface bearing teeth (Figure 13a), but with no expanded shelf (Grande & Bemis 1991). *Polyodon* possesses a small hypobranchial one not comparable to other acipenseriform taxa.

Hypobranchials are typically cartilaginous (although a small ossification occurs in large acipenserids and †*Crossopholis*) and not known from †*Chondrosteus* or †*Peipiaosteus*. Anterior shelves are not present in †*Mimia* (Gardiner 1984a), *Polypterus, Lepisosteus*, or *Amia*.

Character 22. Hypobranchial three makes a bicontact joint with basibranchial one – Acipenseridae

Hypobranchial three articulates with basibranchial one at two sites in all acipenserids examined. The anterior end is crescentic, with a short dorsal process articulating with the posterior tip of basibranchial one and a ventral prong that curls anteroventrally under basibranchial one (Figure 13c, inset).

Hypobranchial three of *Psephurus* and *Polyodon* is slender and makes single contact with the basibranchial (Grande & Bemis 1991). Hypobranchial three of †*Crossopholis* (when preserved) is apparently similar (Grande & Bemis 1991), with a narrow medial end suggesting a slender tip. It is cartilaginous in adults and not known from †*Chondrosteus* or †*Peipiaosteus*. Hypobranchial three of †*Mimia* (Gardiner 1984a) and *Polypterus* possess single joints with the basibranchial.

Character 23. Pectoral scales are elongate with multiple toothlike extensions – Acipenseridae

Pectoral scales studding the inner surface of the opercular chamber are elongate elements with multiple recurved tips (Figure 14c, d). Typically, three to five recurved tips overhang opposing depressions within the scale. Morphology of these scales varies in *Acipenser* and *Huso*, although the basic pattern is consistent. Scaphirhynchines lack these scales (Character 54).

Pectoral scales of fossil and extant polyodontids (Grande & Bemis 1991) are small, round elements with a single, recurved process (Figure 14a). †*Peipiaosteus* possesses pectoral scales with three processes extending from a narrow tip (Liu & Zhou 1965), but processes of these scales fan directly from the base and are not elevated, recurved tips that overhang the scale (Figure 14b). Such scales have not been described in †*Chondrosteus*. Individual sections of acipenserid pectoral scales (with a single tip) are more robust than, but similar to, the pectoral scales of polyodontids.

Character 24. Commissure of the occipital canals – Acipenseridae

In all acipenserids examined, the contralateral occipital canals enter the median extrascapular and merge as a short common canal extending anteriorly (Figure 5).

Occipital canals of *Polyodon, Psephurus*, and †*Crossopholis* are borne by tube bones that expand in adults but do not meet at the midline (Grande & Bemis 1991). Extrascapulars, and passage of lateral line canals, of †*Paleopsephurus* are unknown. A complete row of extrascapulars is present in †*Chondrosteus acipenseroides* (Traquair 1887), but not in †*C. hindenburgi* (Hennig 1925). These extrascapulars are small (Figure 5a), and the commissure might be variable, but a persistent canal has not been described in these bones. †*Peipiaosteus* lacks extrascapulars in described specimens (Liu & Zhou 1965, Zhou 1992).

Character 25. Basitrabecular processes form flattened shelves – Huso

Basitrabecular processes of *Huso* are elongate shelves that flare laterally under the orbit (Figure 15a). They are flat ventrally, extending flush from the base of the neurocranium with little or no ventrolateral curvature. The groove carrying the palatine ramus of the facial nerve and segregating the

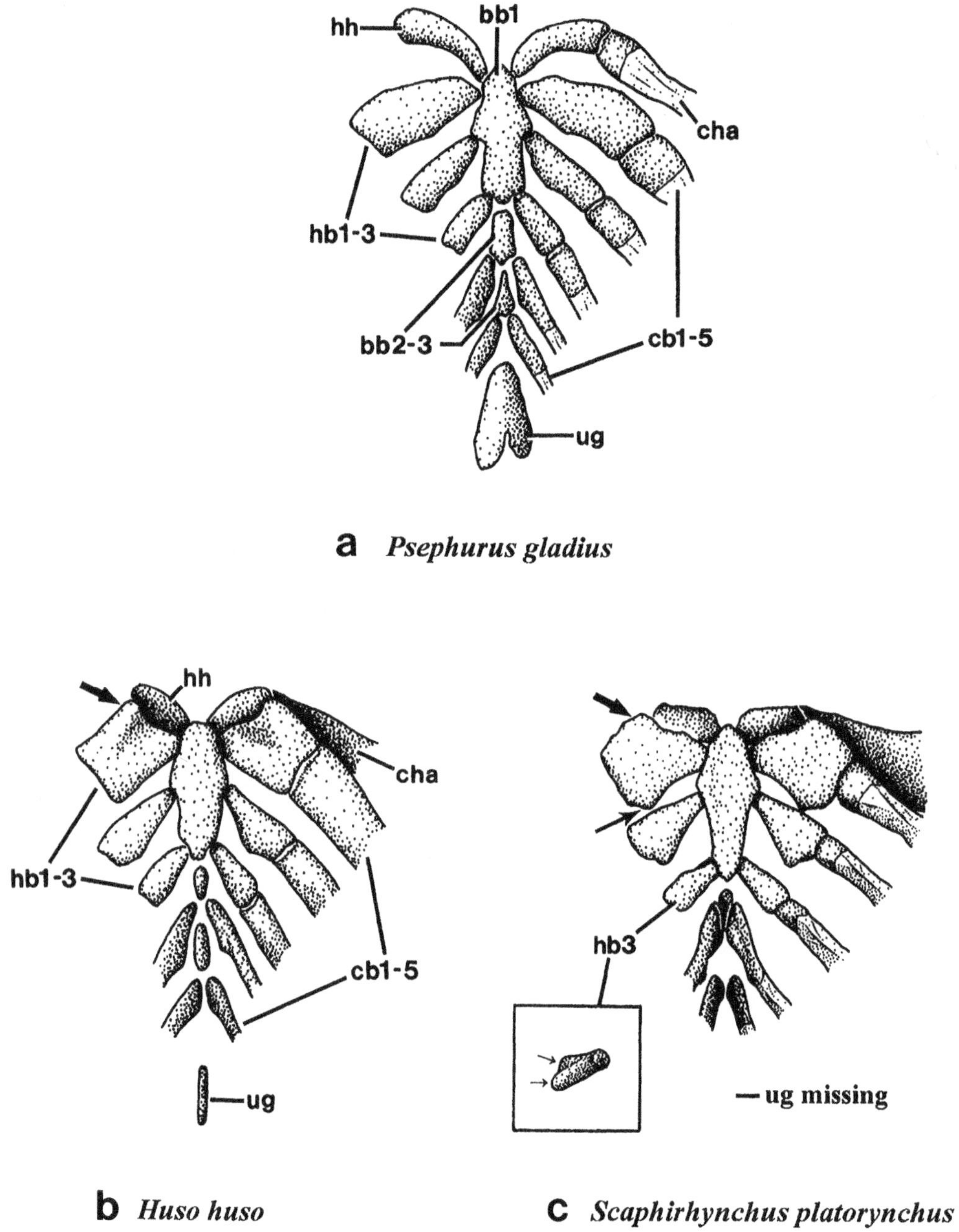

Figure 13. Ventral hyobranchial skeleton in *Psephurus* and representatives of *Huso* and *Scaphirhynchus*: The skeleton is shown in oral view. Hypobranchial three (**hb3**) of *Scaphirhynchus* is also shown in lateral view in the inset of (c). The first hypobranchial (**hb1**) is expanded with an anterior shelf in acipenserids (heavy arrow in b, c; Character 21). It also possesses a posterior wedge contacting hypobranchial two (**hb2**) in scaphirhynchines (thin arrow; Character 51). Hypobranchial three (**hb3**) makes double contact with basibranchial one (**bb1**) through a ventral process undercutting the basibranchial (inset in c) in all acipenserids (Character 22). The unidentified median cartilage (**ug**) of *Psephurus* and *Huso* is missing in acipenserines (Character 37). **hh** = hypohyal, **bb** = basibranchial, **cb** = ceratobranchial, **hb** = hypobranchial.

basitrabecular process medially (Findeis 1993) is thin and shallow.

This morphology contrasts with the inferred plesiomorphic condition of *Psephurus* and several species of *Acipenser* (e.g., *A. ruthenus, A. transmontanus*) with basitrabecular processes that protrude ventrally (Figure 15b) and are isolated by deep grooves medially. Basitrabecular morphology varies widely among other taxa. Some species of *Acipenser* (e.g., *A. oxyrinchus*) possess basitrabecular processes that curl medially over the palatine groove to occasionally complete a nerve canal. *Acipenser brevirostrum* possesses stubbier processes that flare laterally unlike other species of *Acipenser* examined (Figure 15c). Scaphirhynchines possess small processes and a basitrabecular cartilage (Figure 15d, e; Character 43). Variation within *Acipenser* makes basitrabecular morphology a difficult character to analyze, but the morphology of *Huso* is unique.

Character 26. No palatoquadrate-interhyal joint – Huso

The interhyal (**ih**) of *Huso* does not contact the palatoquadrate, articulating only with Meckel's cartilage (**mk**, Figure 16c). It is unexpanded anterodorsally and the palatoquadrate lacks a corresponding interhyal process (Figure 16c). This character was described by Tatarko (1936) and has been used as a *Huso* character by other authors (e.g., Berg 1948a, Antoniu-Murgoci 1936b).

All other extant acipenseriforms possess jaw-interhyal joints with mutual contact of the interhyal, Meckel's cartilage, and palatoquadrate (Figure 16a, b, d, e). Osteologically, contacting extensions of the interhyal (**ihpa**) and palatoquadrate (**pq**) are found in extant polyodontids (Figure 16a, b), all species of *Acipenser* examined (Figure 16d; Tatarko 1936, Antoniu-Murgoci 1936b), and scaphirhynchines (Figure 16e), but ratio of palatoquadrate/Meckel's cartilage contact with the interhyal varies. Meckel's

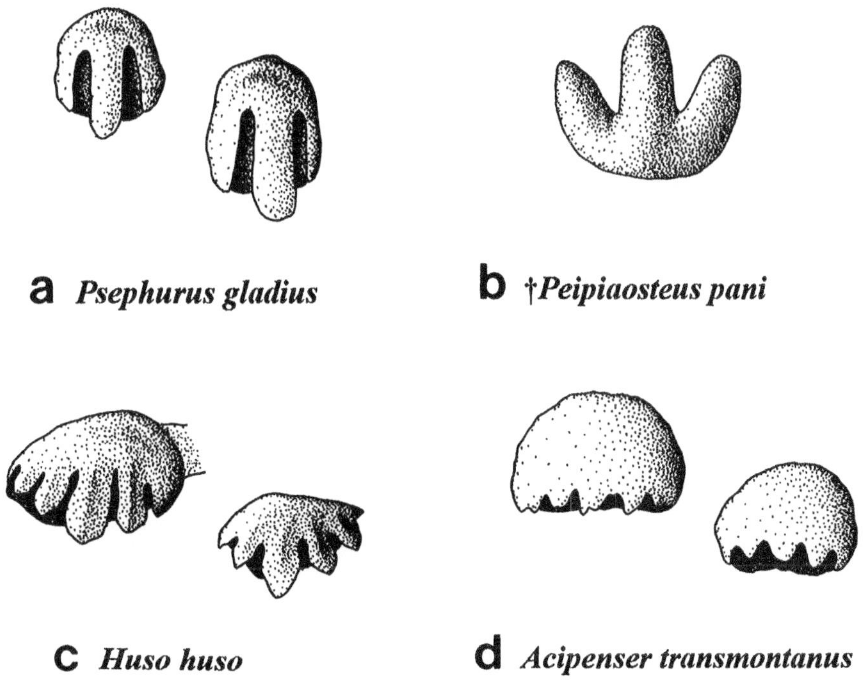

Figure 14. Scales of the pectoral girdle in *Psephurus*, †*Peipiaosteus*, and representatives of *Huso* and *Acipenser*: The pectoral scales of acipenserids (c, d) possess multiple recurved tips overhanging an elliptical base (Character 23). Polyodontids (a – redrawn from Grande & Bemis 1991) possess pectoral scales with a single process and the scales of †*Peipiaosteus* (b – redrawn from Liu & Zhou 1965) possess three processes fanning from a flat base.

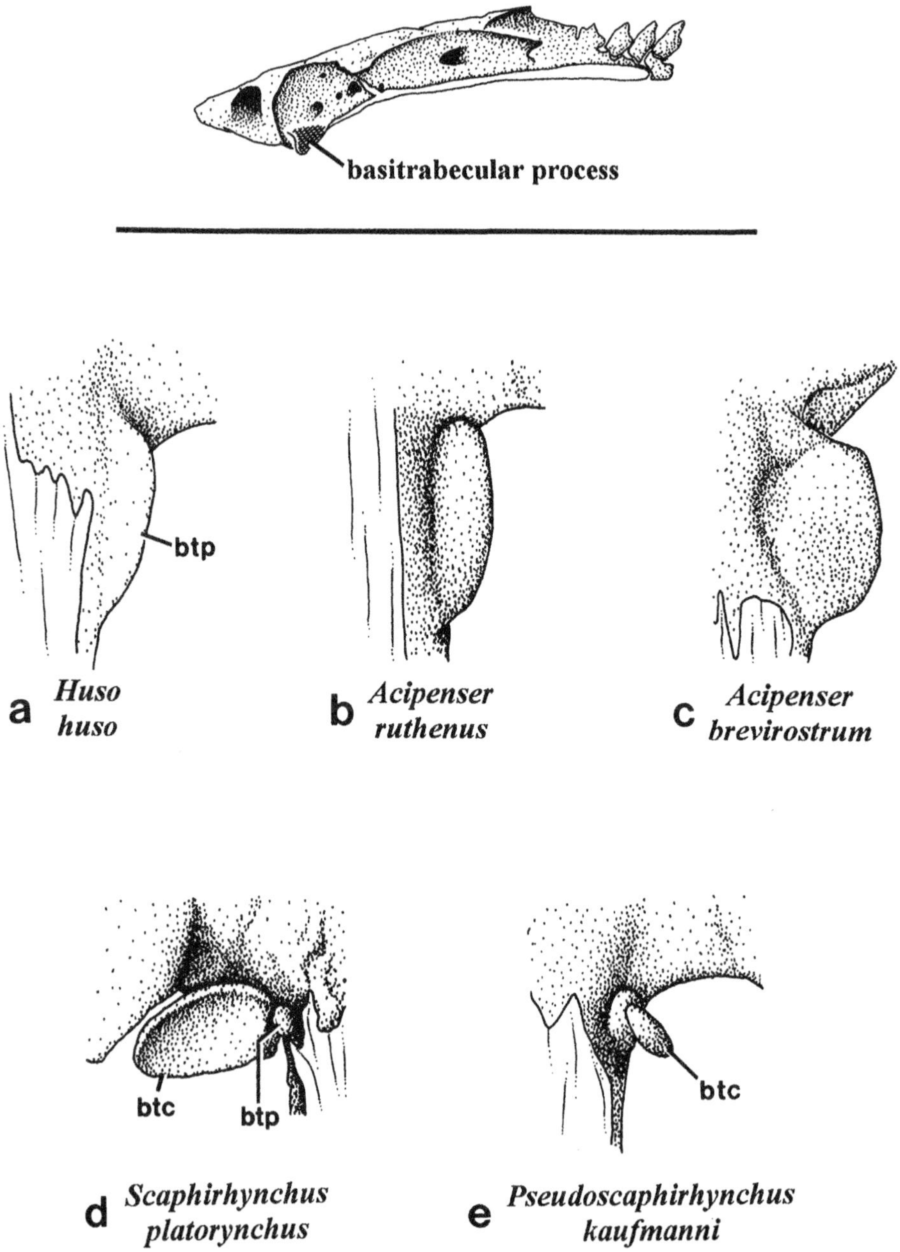

Figure 15. Basitrabecular processes of representatives of all genera of Acipenseridae: Basitrabecular processes (**btp**) extend ventrally beneath the orbit (top in cross-hatch). Morphology varies within *Acipenser* (b, c), but they are flattened shelves in *Huso* (a; Character 25). Scaphirhynchines (d, e) possess indepent basitrabecular cartilages (**btc**) bound to small basitrabecular processes (Character 43).

cartilage dominates the jaw joint in polyodontids (Figure 16a, b) with its posterior end expanded as a bulb displacing the palatoquadrate (also in *Huso*, Figure 16c). Scaphirhynchines and all species of *Acipenser* examined possess jaw joints with approximately equal contact among elements (Figure 16d, e) or jaw joints with dominant palatoquadrate-interhyal contact (i.e. *A. oxyrinchus*).

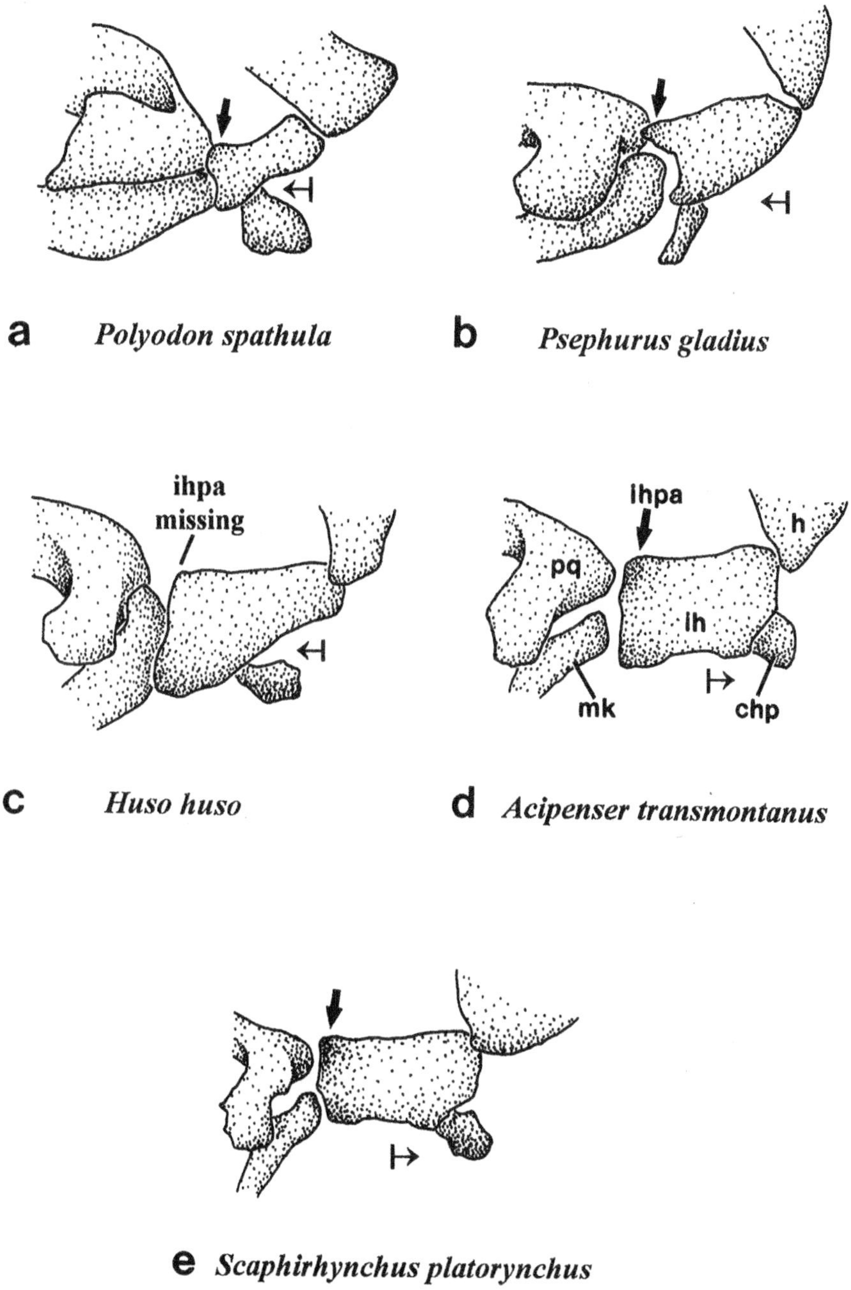

Figure 16. Jaw joint of polyodontids and representatives of three genera of Acipenseridae: The jaw joint is shown with anterior facing left. The interhyal (**ih**) contacts the palatoquadrate (**pq**) in all extant acipenseriformes except *Huso* (c; Character 26) where contacting extensions of the interhyal (**ihpa**, heavy arrows) and palatoquadrate are not present. The joint with the posterior ceratohyal (**chp**) is anterior in *Huso* and polyodontids (arrows to left in a, b, c), but displaced posteriorly in acipenserines (arrows pointing right in d, e; Character 36). **h** = hyomandibula, **mk** = Meckel's cartilage.

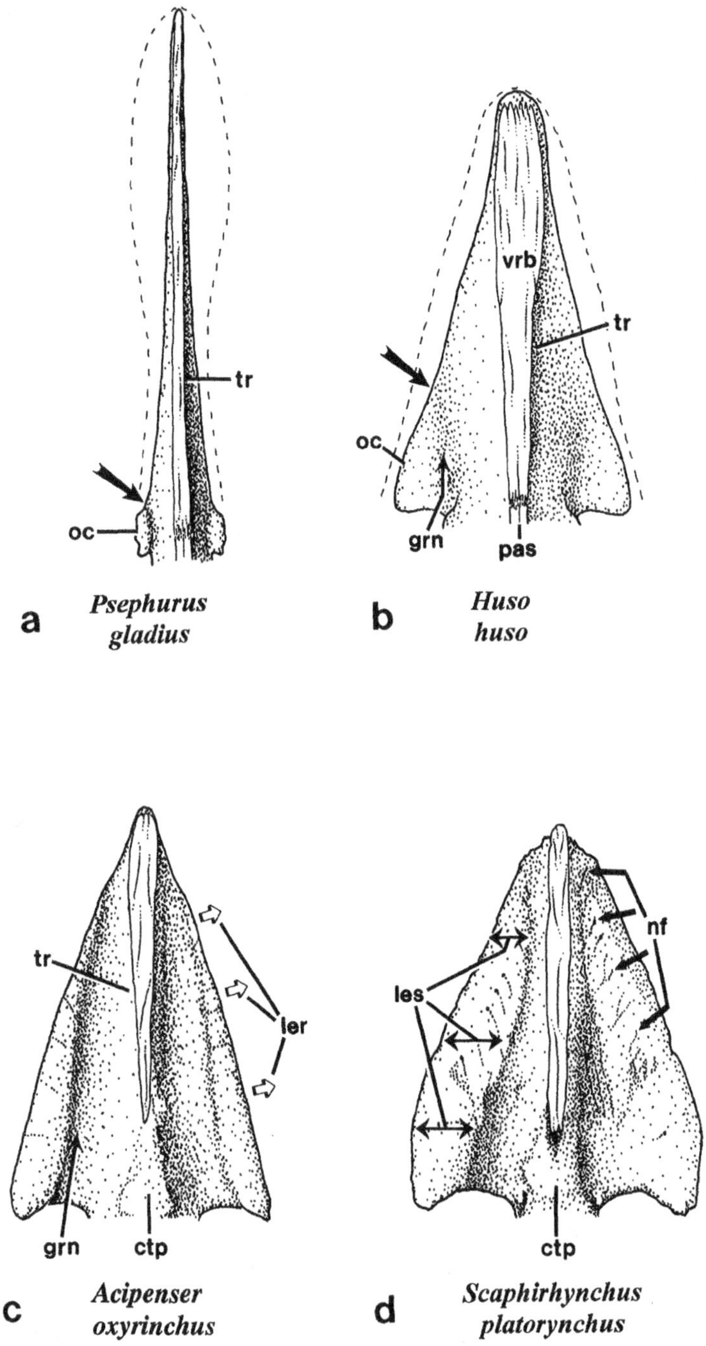

Figure 17. Ventral rostrum of *Psephurus* and representatives of three genera of Acipenseridae: The rostrum is thin and tapers in *Psephurus* (a) and *Huso* (b) with the groove for the rostral nerves (**grn**) opening laterally alongside the endochondral rostrum (large arrow in a, b). Acipenserines (c, d) possess lateral ethmoid ridges (**ler**, open arrows) that bracket the groove to the tip of the rostrum (Character 27). They are expanded as the lateral ethmoid shelves of scaphirhynchines (**les**; Character 39). Foramina for the nerves that innervate the dorsal ampullary organs (**nf** and arrows) indicate expansion of the lateral ethmoid shelves. The parasphenoid (**pas**) is exposed as it contacts the ventral rostral bones in *Huso* and polyodontids (a, b), but is covered by the central trabecular process in acipenserines (**ctp** in c, d; Character 31). **tr** = trabecular ridge, **vrb** = ventral rostral bones.

Character 27. Presence of lateral ethmoid ridges – Acipenserinae

Lateral ethmoid ridges form the lateral edges of the endochondral rostrum in scaphirhynchines (**les**, Figure 17d) and all species of *Acipenser* examined (**ler**, Figure 17c). They are ventrolaterally expanded bands of cartilage lateral to the grooves housing the rostral nerves (**grn**; Findeis 1993). These ridges extend seamlessly from the median rostrum and ethmoid region of the neurocranium without obvious boundaries. In acipenserines, the lateral ethmoid ridges support a 'hard' rostrum of cartilage and dermal bone. Lateral ethmoid ridges of *Acipenser* curve ventrally as thin ridges parallel to the median trabecular ridge (**tr**). These ridges expand into broad lateral ethmoid shelves in scaphirhynchines (see Character 39).

Huso and extant polyodontids possess endochondral rostra that indent anterior to the olfactory bulbs (Figure 17a, b). In *Huso*, the ethmoid region possesses expanded olfactory bulbs contributing to the postnasal wall as in acipenserines (see Characters 3, 4), but no lateral ethmoid ridges extend anteriorly (Figure 17b) as the rostral grooves (**grn**) pass under the olfactory capsule (**oc**) and open anteriorly alongside the trabecular ridge (**tr**, Figure 17b) without cartilage bracketing them laterally. The rostrum correspondingly tapers rather than remaining expanded as a shield as in acipenserines. *Psephurus* and *Polyodon* lack a broad ethmoid region, with a bare olfactory capsule exposed along the thin endochondral rostrum (Figure 17a). In *Huso* and polyodontids, the rostral surface expands laterally beyond the endochondral rostrum as a 'soft' rostrum supported by skin. The rostrum converges anteriorly in *Huso* and the rostral skin is not supported by any bones. The rostrum expands as a paddle in polyodontids supported peripherally by an extensive series of stellate bones (Grande & Bemis 1991).

Character 28. Dorsal rostral series forms an expanded shield – Acipenserinae

Dorsal rostral bones in all acipenserines examined form a broad shield expanding anterior to the olfactory cavity (**drb**, Figure 18c, d). The dorsal rostral shield broadens in concert with expansion of the endochondral rostrum of acipenserines (see Character 27).

Huso and polyodontids (Grande & Bemis 1991) possess a narrow dorsal rostral series with consistent breadth along the rostrum (Figure 18a, b). Lacking rostral expansion anterior to the olfactory bulb, there is no space for an expanded rostral shield in the outgroups. The expanded dorsal rostral series is likely linked with the lateral ethmoid ridges as the bones occupy the expanded rostral surface in acipenserines. Initially developing bones of the dorsal rostral series form a median row similar to *Huso* and polyodontids that ontogenetically and phylogenetically expands by addition of novel bones (Findeis 1993).

Character 29. Presence of border rostral bones – Acipenserinae

Border rostral bones (**brb**) are a thin series (one or two bones wide) of small bones along the posterolateral edge of the rostrum anterior to the jugal (**j**, Figure 18c, d). They are positionally separate and ontogenetically distinct from the dorsal rostral series as a novel group of rostral bones in scaphirhynchines (Findeis 1993) and all species of *Acipenser* examined (Figure 18c, d). Although variable in number, border rostral bones form a consistent assemblage.

Spatial separation of dorsal and border rostral series is typical in *Acipenser oxyrinchus, A. transmontanus*, and *A. ruthenus*, with an open corridor housing ampullary organs of the dorsal rostrum (Figure 18c). Only in large adults of *A. brevirostrum* (among species of *Acipenser* examined) did this gap partially close to make identification of the groups difficult. In scaphirhynchines, bones entirely cover the rostrum (Figure 15, 16), but by addition of ampullary bones between these series (Character 40). The multiplicity of rostral bone groups confuses character recognition in the dermal rostrum, but each series is ontogenetically discrete.

No bones lateral to the dorsal rostral series are

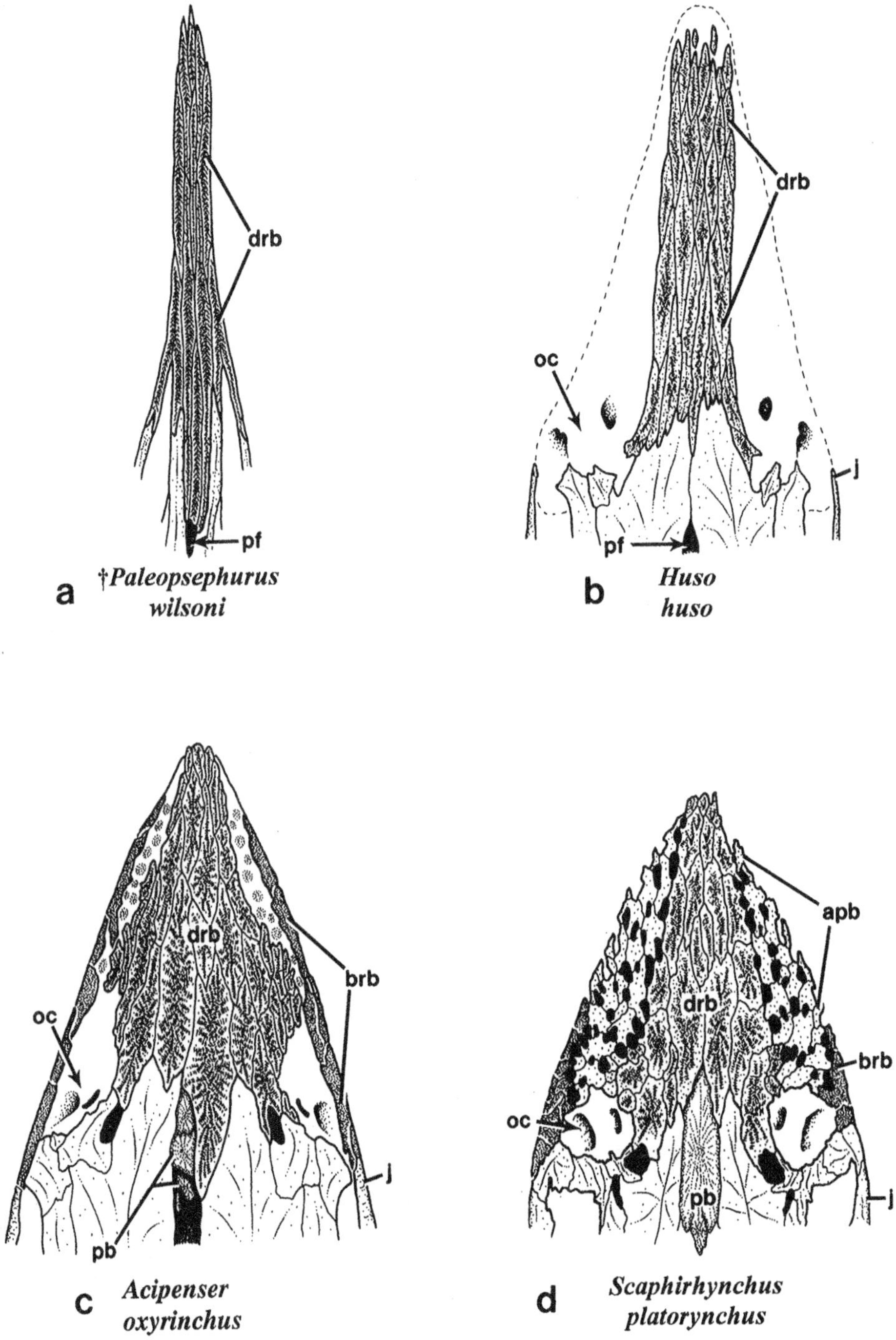

Figure 18. Dorsal bones of the rostrum of †*Paleopsephurus* and representatives of three genera of Acipenseridae: Dorsal rostral bones (**drb**) form a rectangular cover in polyodontids and *Huso*, but are expanded as a shield (Character 28) and accompanied by border rostral bones (**brb**) extending from the jugal (**j**) in acipenserines (Character 29). The dorsal rostral and border rostral series are separated by an ampullary field (stippled circles in c) in *Acipenser*, but this area is filled with ampullary bones (**apb**) in scaphirhynchines (d; Character 40). The pineal fontanelle (**pf**) is open between the frontal bones in polyodontids and *Huso* (a, b), but filled by pineal bones (**pb**) in acipenserids (Character 32). The pineal fontanelle is partially filled in juveniles of *Acipenser oxyrinchus* (b), but covered in other acipenserines (as in d). **oc** = olfactory capsule opening in dermal skull.

present in *Huso* (Figure 18b). Stellate bones supporting the lateral rostrum in polyodontids are present lateral to the endochondral rostrum only within the skin of the paddle (Grande & Bemis 1991). Border rostral bones lie directly on the rostrum posterolaterally, while polyodontid stellate bones are isolated anterolaterally.

Character 30. First ventral rostral bone in single – Acipenserinae

The first ventral rostral bone of acipenserines (**vrb1**, Figure 19c, d, e) is a single bone that contacts the parasphenoid as the most posterior bone of the ventral rostral series (Findeis 1993). It is only modestly larger than other ventral rostral bones in all species of *Acipenser* examined, but is elongate in scaphirhynchines (Character 42).

In *Huso* and all polyodontids (Figure 19a, b), the ventral rostral series possesses paired bones contacting the parasphenoid. These bones were named vomers by Grande & Bemis (1991), but they develop on the rostrum well anterior to the parasphenoid and are not present in †*Chondrosteus* (Traquair 1887, Hennig 1925) or †*Peipiaosteus* (Liu & Zhou 1965), suggesting that they are novel bones of the rostrum and rostral series of acipenseroids.

Character 31. Central trabecular process present – Acipenserinae

The central trabecular process is a ventral cartilaginous wedge between the ethmoid and orbital regions of the neurocranium in all acipenserines examined (**ctp**, Figure 20c, d). This process protrudes posterior to the trabecular ridge (**tr**, Figure 17) before rising sharply under the orbit as the ethmoid-orbital angle (Findeis 1993) to open a ventral space for the jaws (Figure 20c, d). The central trabecular process covers the posterior end of the first ventral rostral bone and anterior extension of the parasphenoid (shown in Figure 17c, d).

In *Huso* and extant polyodontids (Grande & Bemis 1991), the paired first ventral rostral bones (see Character 30) contact the parasphenoid on the surface of the neurocranium without a cartilage cover (Figure 17a, b). These taxa also lack an ethmoid-orbital angle as the ventral neurocranium curves smoothly into the orbit in *Huso* (Figure 20b), *Psephurus*, and †*Paleopsephurus* (Grande & Bemis 1991), or is straight in the derived polyodontids *Polyodon* (Figure 20a) and †*Crossopholis*. The central trabecular process correlates phylogenetically with ventral displacement of the jaws of acipenserines and possibly deflects them ventrally during projection.

Character 32. Pineal bones present – Acipenserinae

Pineal bones develop within the pineal fontanelle between the frontals (**pf**, Figure 18b, c) in juveniles and typically expand to fill the space in adults (**pb**, Figure 18c, d). Number of pineal bones varies, but they are consistently present in all acipenserines examined.

Pineal bones are anamestic bones that are developmentally constrained by bones surrounding the pineal fontanelle. Pineal bones fill the pineal fontanelle completely in *Scaphirhynchus* and most species of *Acipenser*, but *A. oxyrinchus* and *A. sturio* are often recognized by an open fontanelle (Figure 18c; Vladykov & Greeley 1963, Magnin 1963, Magnin & Beaulieau 1963). However, pineal bones appear even in these species, but do not enlarge to fill the fontanelle except in very large adults.

The pineal fontanelle occasionally closes by expansion of the frontals. *Pseudoscaphirhynchus kaufmanni* shows both morphologies, with the frontals separated by pineal bones (illustrated by Sewertzoff 1926a) or also exhibiting frontal closure posteriorly. Specimens of *Acipenser medirostris* and *A. oxyrinchus* also partially close the pineal fontanelle by frontal expansion. While pineal bones appear universally in small juveniles, frontal expansion occurs only in adults and all specimens showing frontal expansion also possess pineal bones.

No pineal bones are present in *Huso* or any polyodontid (Grande & Bemis 1991). The pineal fontanelle of these taxa (when present) is narrow between frontals (Figure 18a, b). The fontanelle is reduced or absent in large juveniles of *Huso*, but no

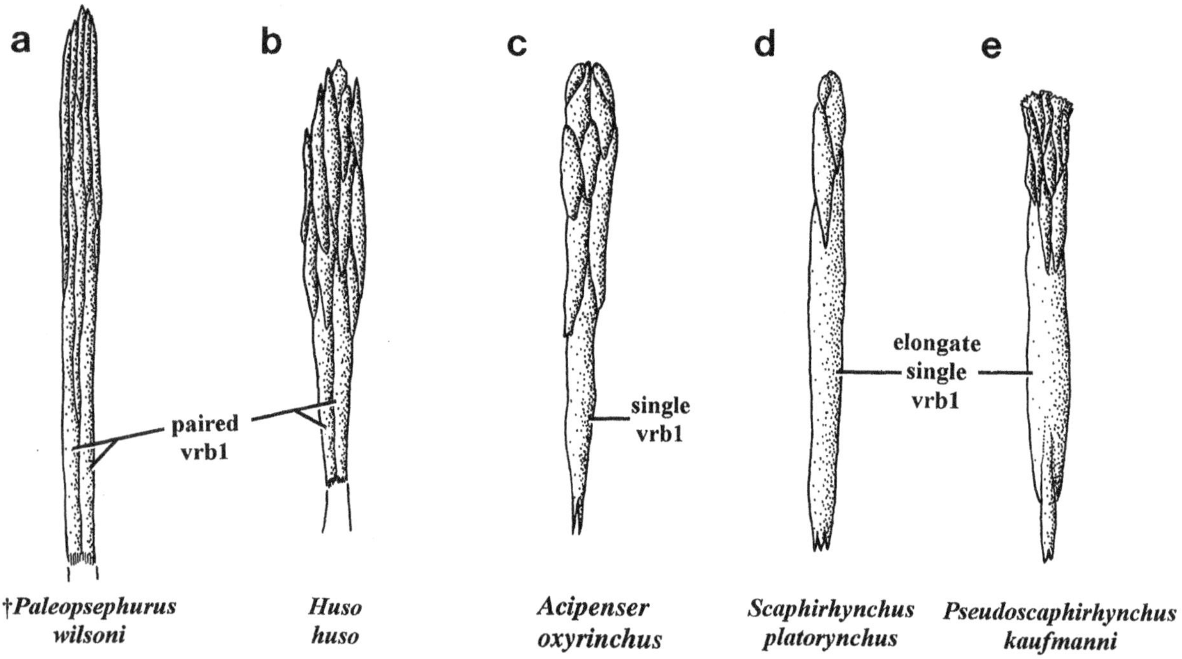

Figure 19. Ventral rostral bones in †*Paleopsephurus* and representatives of all genera of Acipenseridae: The series is shown in ventral view with anterior to top. The first ventral rostral bone (most posterior) is paired in polyodontids and *Huso* (a, b), but single in acipenserines (c, d, e; Character 30). This bone is elongate in scaphirhynchines (d, e; Character 42), with the remaining bones clustered at the anterior tip of the rostrum.

pineal bones are present and the frontal lacks dermal ornament medially, suggesting that closure occurs by frontal expansion. There are often two openings in polyodontids (Grande & Bemis 1991), but only one in *Huso*. These outgroups possess frontals making medial contact, and the frontals of †*Chondrosteus* (Traquair 1887, Hennig 1925) and †*Peipiaosteus* (Liu & Zhou 1965) uniformly close in the skull roof.

Character 33. Lower jaw is straight – Acipenserinae

The lower jaw of all acipenserines examined is short and transverse under the head. Meckel's cartilage curves slightly opposite the upper jaw to define the mouth near the symphysis, but the lateral end of Meckel's cartilage and the dentary is almost entirely linear throughout its length (Figure 21c).

The posterior end of Meckel's cartilage and the dentary of *Huso* and *Psephurus* (Grande & Bemis 1991) angles dorsally (Figure 21a, b) toward the jaw joint. *Polyodon* is similar, but less strikingly since its elongate jaws coincidentally flatten all jaw features. The dentary of all fossil polyodontids (Grande & Bemis 1991) and †*Chondrosteus* (Traquair 1887) is angled, suggesting that an angled lower jaw is plesiomorphic within Acipenseriformes. Zhou (1992) reconstructs the dentary of †*Peipiaosteus* with a slight angle and I accept an angled lower jaw as plesiomorphic for Acipenseridae.

Character 34. Dermopalatine shelf present – Acipenserinae

The medial dermopalatine broadens as a ventral shelf in all acipenserines examined (Figure 22c). This dermopalatine shelf deepens the upper jaw as an anteromedial surface often opposing a protruding dentary shelf (Figure 21c). The dermopalatine shelf is consistently present, but the dentary shelf is not universal and not used as a character here. The dermopalatine shelf occurs where the dermopala-

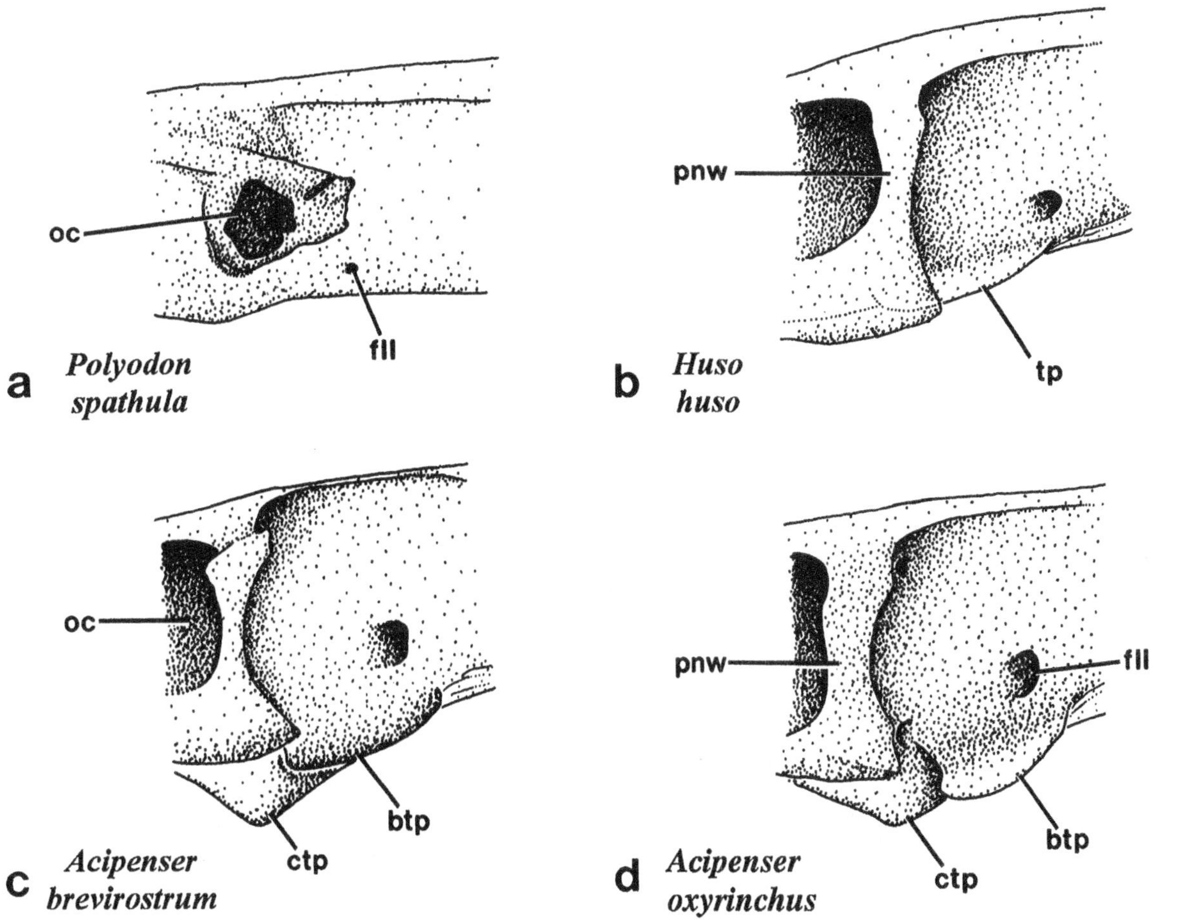

Figure 20. Lateral view of the orbit and ethmoid regions of the neurocranium in *Polyodon* and representatives of three genera of Acipenseridae: The ventral surface of the neurocranium is flat in *Polyodon* (a) and rises smoothly in *Psephurus* and *Huso* (b), but angled by the central trabecular process (**ctp**) in acipenserines (Character 31). **oc** = olfactory capsule, **pnw** = postnasal wall, **btp** = basitrabecular process, **fII** = foramen of the optic nerve.

tine contacts the palatoquadrate medially and bears teeth in juveniles.

Dermopalatines of *Huso* (Figure 22b), all polyodontids (Figure 22a; Grande & Bemis 1991), †*Chondrosteus* (Traquair 1887), and †*Peipiaosteus* (Liu & Zhou 1965) possess flat anteroventral edges. These taxa also bear teeth along the full edge of the dermopalatine, not in a restricted portion. Similarly, the lower jaws of outgroups lack a dentary shelf (Figure 21a, b).

Character 35. Prearticular bones are small and thin – Acipenserinae

The prearticular is a small, splinter-like bone on the inner surface of Meckel's cartilage (Figure 21c). Its size and shape conforms to the thin, straight jaws of acipenserines (Character 33). The prearticular is absent in scaphirhynchines (Character 48).

The prearticular of *Psephurus* (Grande & Bemis 1991) and *Huso* (Figure 21a, b) is a flat, triangular bone covering the posterodorsally angled inner surface of Meckel's cartilage (see Character 33). This bone is elongate and thin in †*Crossopholis* and *Polyodon* (Grande & Bemis 1991), but these taxa possess thin jaws. †*Paleopsephurus* is not preserved correctly to see the prearticular. †*Chondrosteus* possesses a large, flat prearticular (Watson 1928), but it is apparently not present in †*Peipiaosteus* (Liu & Zhou 1965, Zhou 1992). Despite the multiplicity of outgroup morphologies, the acipenserine prearticular is characteristically reduced.

Character 36. Interhyal-posterior ceratohyal joint displaced posteriorly – Acipenserinae

The interhyal-posterior ceratohyal joint occurs on the posterior half of the interhyal in all acipenserines examined (Figure 16d, e). Precise position of the joint ranges from the posterior edge (**chpf**, Figure 16d; e.g., *A. oxyrinchus, A. transmontanus*) to a more central position within the interhyal (Figure 16e; e.g., scaphirhynchines, *A. brevirostrum*).

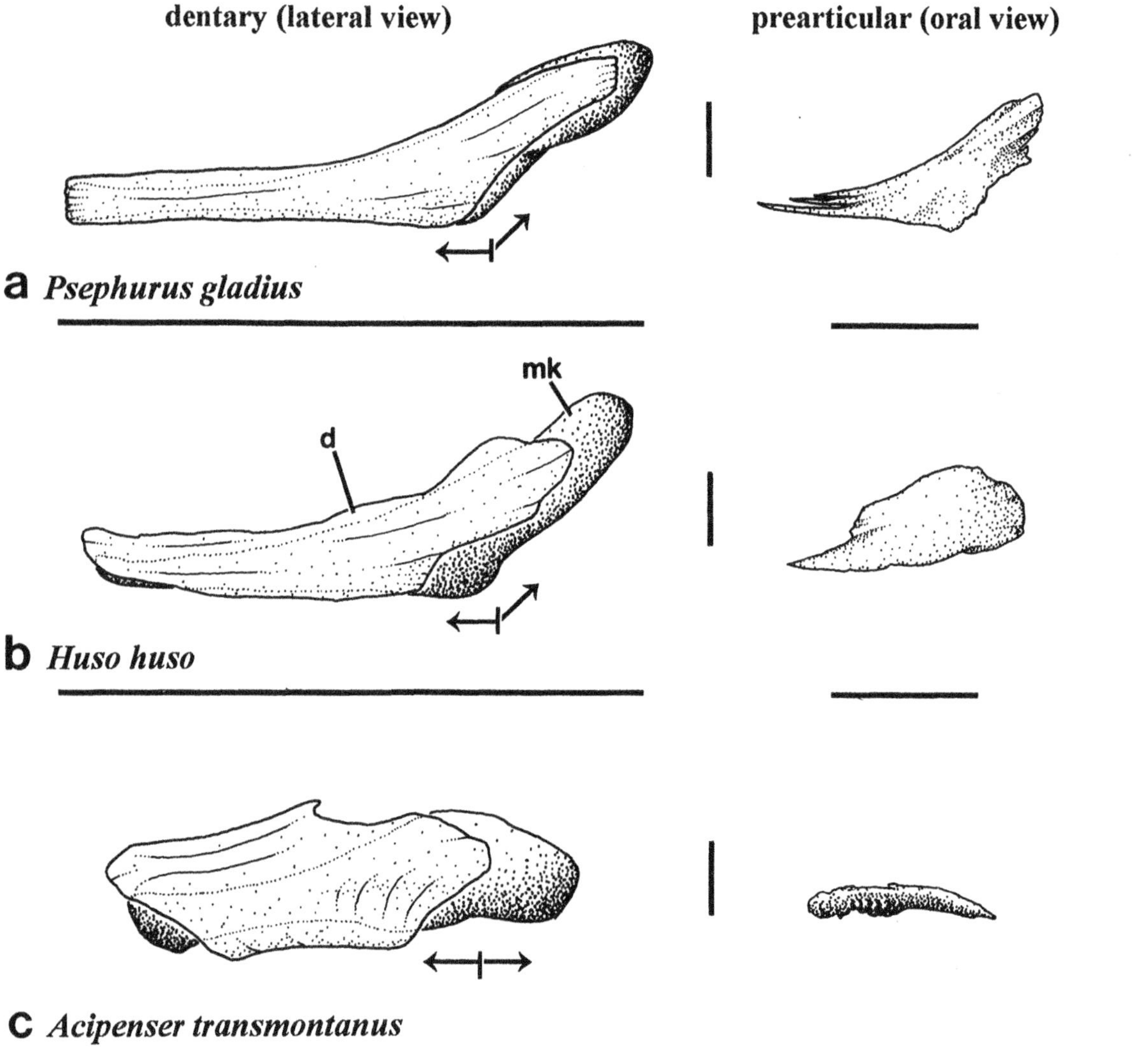

Figure 21. Lower jaw and prearticular of *Psephurus* and representatives of *Huso* and *Acipenser*: The lower jaw is shown in lateral view to the left, with the prearticular shown in oral view to the right. Meckel's cartilage angles posterodorsally to the jaw joint in *Psephurus* and *Huso* (arrows in a, b), but is straight in acipenserines (c; Character 33). Coincidentally, the flat prearticular of the outgroups (a, b) is reduced to a thin bone in acipenserines (Character 35) and absent in scaphirhynchines (Character 48).

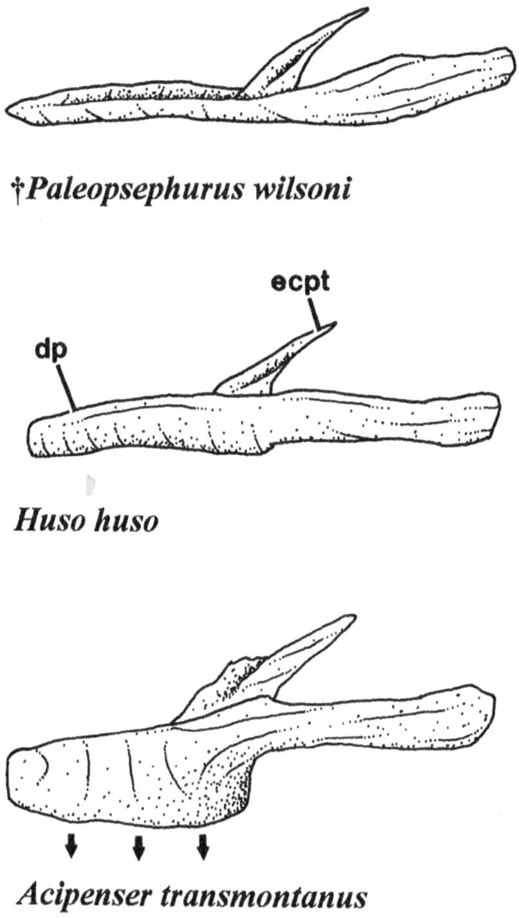

Figure 22. Dermopalatine of †*Paleopsephurus* and representatives of *Huso* and *Acipenser*: The dermopalatine is flat along the anteroventral edge in †*Paleopsephurus* and *Huso* (a, b), but expanded medially in acipenserines (arrows in c; Character 34). The dermopalatine (**dp**) and ectopterygoid (**ecpt**) fuse in *Scaphirhynchus* (Character 69), but are separate elements in all other acipenseriforms (as shown in a, b, c).

The interhyal-posterior ceratohyal joint of *Huso* and extant polyodontids (Grande & Bemis 1991) occurs anteriorly on the interhyal (Figure 16a, b, c). This is a cartilaginous character not scorable in fossil acipenseriforms, but the anterior joint is consistent in all extant outgroups.

Character 37. Loss of median cartilage posterior to branchial arches – Acipenserinae

The ventral midline of the branchial arch skeleton of all acipenserines examined includes a large basibranchial one and one or two small posterior basibranchials (Figure 13c), but no other elements. The posterior basibrachials are paired serially with ceratobranchial four and five (e.g., **bb2** associated with **cb4** in Figure 13).

The branchial arch skeleton of *Huso* and extant polyodontids possesses a median cartilage posterior to the serial basibranchials (**ug**). The median cartilage of *Huso* is thin and short (Figure 13b), but forms a plate in *Polyodon* and *Psephurus* (Figure 13a). Grande & Bemis (1991) named it an unidentified median cartilage and it has no postional affinity with any branchial arch. This element is cartilaginous and not scorable in fossils, but is present in all extant outgroups.

Character 38. Loss of haemal spines anterior to the caudal fin – Acipenserinae

Elongate basiventrals with haemal spines are present in a short span of the vertebral column entering the caudal fin, but are missing in the caudal peduncle. In all acipenserines examined, three to four basiventrals with haemal spines (**bvh**) are typically found anterior to the hypurals (**hyp**, Figure 23b) supporting the anterior fin rays of the caudal fin as an anterior shelf characteristic of Acipenseriformes (Findeis 1993, Bemis et al. 1997). More anterior basiventrals never possess haemal spines (**bv**, Figure 23b).

Polyodontids (Grande & Bemis 1991) and *Huso* (Figure 23a) possess haemal spines throughout the caudal peduncle and extending anteriorly over the anal fin. Elongate haemal spines may be dependent on a deep peduncle, with the flattened peduncle of acipenserines too spatially restricted to accommodate them.

Character 39. Lateral ethmoid shelves present – Scaphirhynchini

The lateral ethmoid ridges of scaphirhynchines are broad and ventrally flat, forming lateral ethmoid shelves (**les**, Figure 17d) lateral to the groove carrying the rostral nerves (**grn**). The lateral ethmoid shelves are penetrated several millimeters from the lateral edge by foramina carrying nerves to ampullary organs of the dorsal rostrum (**nf** and arrows, Figure 17d). Size and shape of the rostrum varies, but expanded lateral ethmoid shelves are present in all scaphirhynchines.

In all species of *Acipenser* examined, the lateral ethmoid ridges end in a thin ventral edge (Figure 17c). No foramina penetrate these lateral ethmoid ridges, suggesting that cartilage is added lateral to the foramina in scaphirhynchines. Since *Huso* and polyodontids do not possess lateral ethmoid ridges, this is a two taxon statement at the generic level, but I accept the flattened morphology of scaphirhynchines as derived. The cylindrical head of *Huso* sug-

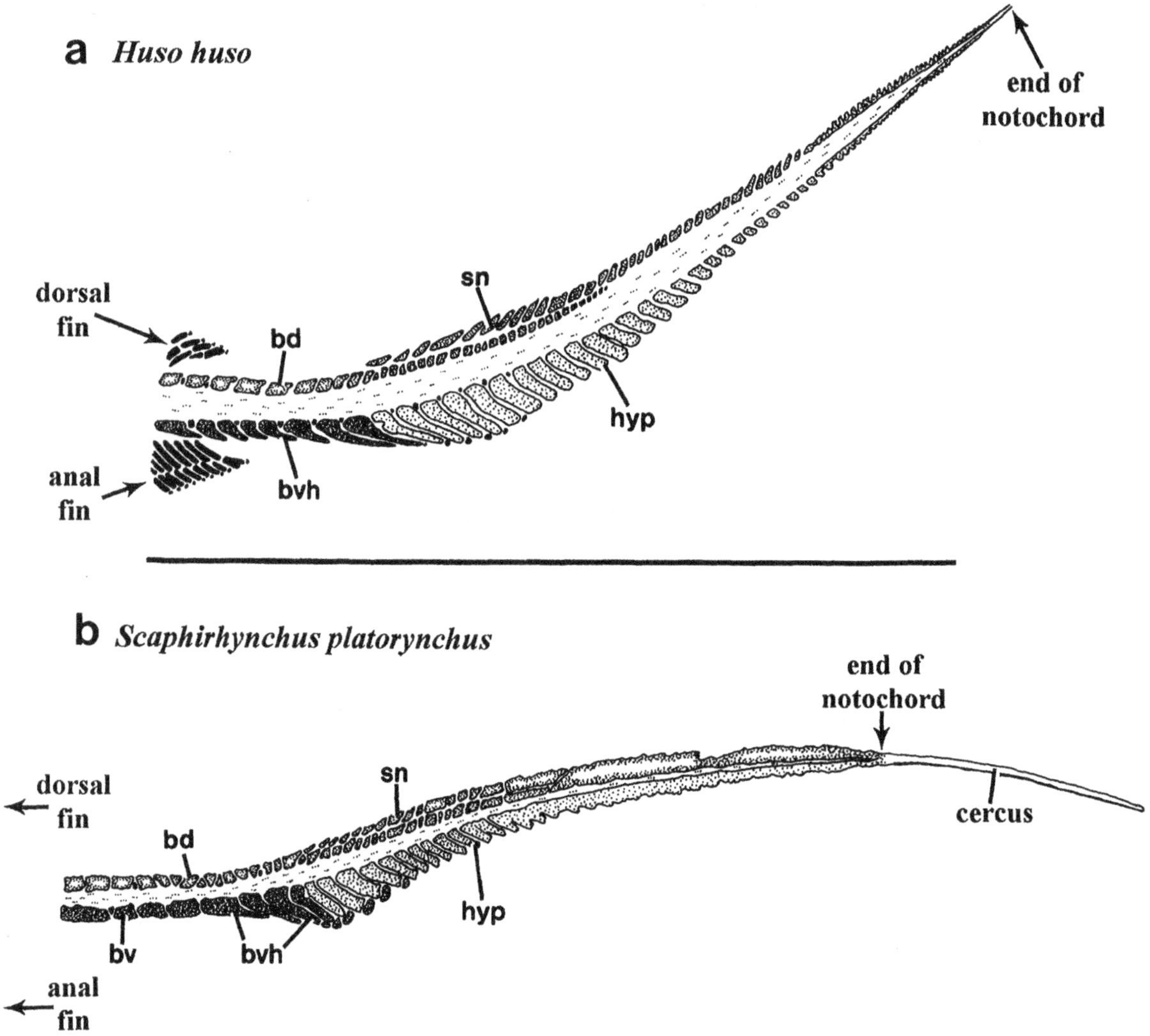

Figure 23. Caudal fin endoskeleton of *Huso huso* and *Scaphirhynchus platorynchus*: Anterior faces to left. Hypurals (**hyp**) are stippled lightly while the more anterior basiventrals (**bv**) and basiventrals with haemal spines (**bvh**) are dark. Basiventrals with haemal spines overlap the anal fin in *Huso*, but are restricted to the caudal fin in acipenserines (Character 38). The elongate caudal peduncle of *Scaphirhynchus* (b; Character 64) displaces the dorsal and anal fins anteriorly, but possessing only three-four basiventrals with haemal spines is typical for acipenserines. The notochord and bracketing cartilages end in unison in *Huso* (a) and *Acipenser*, but a cartilaginous core supporting the cercus continues in scaphirhynchines (Character 53) beyond the terminus of the notochord. **bd** = basidorsals, **sn** = supraneural homologs.

gests that the generally cylindrical skulls and rostra typical of most species of *Acipenser* are plesiomorphic.

Character 40. Ampullary bones present – Scaphirhynchini

Ampullary bones compose an ossification group covering the endochondral rostrum between the dorsal rostral series and peripheral border rostral bones (Figure 18d). They are small, rounded (*Scaphirhynchus*) or larger, rectangular (*Pseudoscaphirhynchus*) plates that interweave among the dorsal rostral ampullary field. Ampullary bones develop as a group well after appearance of the dorsal rostral series (Findeis 1993) and anterior to the simultaneously developing border rostral bones (**brb**, Figure 18c, d) to cover the broad scaphirhynchine rostrum (see Character 39).

The border rostral series is a putative synapomorphy of the Acipenserinae (Character 29) distinct from the ampullary series. Border rostral bones of *Acipenser* are positionally separate from dorsal rostral bones in early ontogeny and usually in adults (Figure 18c). No interior rostral bones fill this gap in small juveniles of any species of *Acipenser* examined, suggesting that the ampullary series is missing in *Acipenser*. Broadening of the rostrum in scaphirhynchines (see Character 39) opens its dorsal surface unlike any outgroup and this area is covered by a novel ampullary series.

Character 41. Central spines present on the parietal, posttemporal, supracleithrum, and anterior dorsal rostral bones – Scaphirhynchini

Central spines are present in *Scaphirhynchus platorynchus*, *S. albus*, and *Pseudoscaphirhynchus kaufmanni*, variably present in *P. fedtschenkoi* (Berg 1948a), but absent in *S. suttkusi* and *P. hermanni* (Berg 1948a). Within *Pseudoscaphirhynchus*, *P. kaufmanni* is most similar to *Scaphirhynchus* and seems to be representative for the basal morphotype of the genus, with presence of spines putatively plesiomorphic. *Scaphirhynchus suttkusi* may have lost discrete spines, but a ridge is present in photographs in Williams & Clemmer (1991) and probably represents a residual spine. When present, spines are present on the parietal, posttemporal, supracleithrum, and multiple anterior dorsal rostral bones.

Spines of *Scaphirhynchus* are shallow and ridge-like, but elevate into sharp posterior points. They are prominent in juveniles, but become allometrically smaller in adults as dermal ornament thickens and obscures them. Spines of the posttemporal and supracleithrum are the largest in *Scaphirhynchus*, with posteriorly curved tips identical to and in series with the flank scutes. Spines of *P. kaufmanni* are large, with posttemporal and supracleithrum spines similar to *Scaphirhynchus* accompanied by huge parietal and frontal spikes (see Character 55). Weak spines occur occasionally on the supratemporal of *Scaphirhynchus*, but they are rarely visible and not present in *P. kaufmanni*.

Some skull bones of *Acipenser* possess central ridges, but never manifest spines and anterior dorsal rostral bones are plates without raised ridges or spines. *Huso* never possesses spines or strong ridges in any skull bones and polyodontids possess little dermal ornament on the skull roof and no discrete ridges (Grande & Bemis 1991). †*Chondrosteus* (Hennig 1925) and †*Peipiaosteus* (Liu & Zhou 1965) possess obvious dermal ornament, but no spines.

Character 42. First ventral rostral bone is elongate – Scaphirhynchini

The first ventral rostral bone of scaphirhynchines is elongate, accounting for over two-thirds the length of the ventral rostral series (**vrb1**, Figure 19d, e). It is a wide, flat bone slightly keeled centrally with additional, smaller bones of the ventral rostral series clustered anteriorly. *Scaphirhynchus* typically possesses four or five additional ventral rostral bones (Findeis 1993), while *Pseudoscaphirhynchus kaufmanni* can possess twice as many (nine in Figure 19e).

This is a qualitative character since first ventral rostral bones of all species of *Acipenser* examined are also the longest of the series (Figure 19c). However, in *Acipenser* they are not as dominantly elongate and are thin. Similarly, *Huso* and all polyodontids (Grande & Bemis 1991) possess paired first ventral bones only modestly longer than other bones of the series (see Character 30).

Character 43. Basitrabecular cartilages present – Scaphirhynchini

Basitrabecular cartilages are present in all scaphirhynchines lateral to the small basitrabecular processes (Figure 15d, e). The basitrabecular cartilage of *Scaphirhynchus* is large and curved (Figure 15d) along the postnasal wall (Findeis 1993). It is much smaller in *Pseudoscaphirhynchus*, being barely larger than the diminutive basitrabecular process itself (Figure 15d). In *Scaphirhynchus*, basitrabecular cartilages develop as outgrowths of the basitrabecular processes that separate from the neurocranium in small juveniles. Sewertzoff (1928) identified basitrabecular cartilages as outgrowths of the palatoquadrate and homologized them as pharyngomandibulars, but they develop distinct from the mandibular arch.

All species of *Acipenser* examined, *Huso*, and *Psephurus* possess large basitrabecular processes (Figure 15a, b, c), but no independent, accessory cartilage.

Character 44. Orbital notch present – Scaphirhynchini

The orbital notch indents into the neurocranium from the dorsal edge of the orbit (Findeis 1993). It contains passage of the profundal ramus that scatters onto the dorsal surface of the neurocranium and the superficial ophthalmic ramus of the anterodorsal lateral line nerve. The orbital notch lies beneath the supraorbital canal and an ampullary field anterodorsal to the orbit.

The dorsal orbit of *Acipenser* and *Huso* is semicircular, without indentation. *Huso* and all species of *Acipenser* examined possess shallow anterodorsal depressions in the postnasal wall homologous to the orbital notch, but they are restricted within the orbit and do not open dorsally. Extant polyodontids possess open orbits without boundaries other than the bare olfactory bulbs (Figure 20a). They do possess a broad channel dorsomedial to the olfactory bulb, but this is open for their protractor hyomandibularis muscle and is not comparable to the orbital notch.

Character 45. Posttemporal processes present – Scaphirhynchini

Posttemporal shelves are cartilaginous processes of the neurocranium that interlock with the posttemporal. In scaphirhynchines, they flare posteromedially as thick processes with short distal tips that angle posteriorly. They possess medially angled anteroventral faces that converge to the base of the neurocranium. The dorsal surface is rounded and indented from the surface of the neurocranium. Because of their cylindrical shape, I refer to them as posttemporal processes (Findeis 1993).

In all species of *Acipenser* and *Huso* examined, the posttemporal shelves have flat lateral faces that converge without curvature to the base of the neurocranium. The dorsal edges are also flat, extending smoothly from the dorsal surface of the neurocranium and the posterolateral tips are short processes that extend linearly from the shelves. Polyodontids possess posttemporal shelves similar to *Acipenser* and *Huso*, but they are broader, flatter, and not as elongate.

Character 46. Ventral process of the antorbital is elongate – Scaphirhynchini

The antorbital possesses elongate ventral processes along the postnasal wall in scaphirhynchines (see Character 4). In *Scaphirhynchus*, it is a straight, rectangular process contacting the postrostral bone (Figure 4e; see Character 61). In *Pseudoscaphirhynchus*, the ventral process tapers with an anteriorly curving point (Figure 4d). The elongate ventral process correlates with the expanded rostrum (see Character 39) to cover the postnasal wall.

The antorbital of *Huso* possesses a short wedge

(Figure 4a). All species of *Acipenser* examined possess a ventral point of varying size (Figure 4b, c; see Character 4), but none are as elongate as in scaphirhynchines.

Character 47. Subopercle is distinctively shaped – Scaphirhynchini

The subopercle is distinctive in scaphirhynchines by being elongate with a prominent anterior process, flattened dorsal edge, and abrupt anteroventral notch that angles ventrally (Figure 7e, f). This shape is typical of *Scaphirhynchus, Pseudoscaphirhynchus* and also †*Protoscaphirhynchus* (Wilimovsky 1956).

Subopercle shape varies within Acipenseridae. *Huso* and several species of *Acipenser* (e.g., *A. oxyrinchus, A. ruthenus*) possess a triangular subopercle (Figure 7c), while other species (e.g., *A. brevirostrum, A. gueldenstaedtii*) possess circular plates with blunt anterior and ventral processes (Figure 7d). No species of *Acipenser* examined possesses a subopercle as dorsoventrally compressed as scaphirhynchines. The scaphirhynchine subopercle correlates with the flattened head limiting space for the operculum. As a shape character, it is difficult to segregate subopercle areas as characters, but overall bone shape is unique.

Character 48. Loss of prearticular bones – Scaphirhynchini

Prearticular bones are generally present on the oral surface of Meckel's cartilage (see Character 35), but are absent in scaphirhynchines.

Prearticular bones are present in *Psephurus* (Figure 21a) and †*Crossopholis* (Grande & Bemis 1991), †*Chondrosteus* (Traquair 1887, Watson 1925), *Huso* (Figure 21b), and all species of *Acipenser* examined (Figure 21c). Prearticulars are unknown in †*Paleopsephurus* and †*Peipiaosteus* (Liu & Zhou 1965), but otherwise ubiquitous distribution of the bone in outgroups suggests it is plesiomorphically present.

Character 49. Dorsal head of the hyomandibula is cylindrical – Scaphirhynchini

The dorsal head of the hyomandibula of scaphirhynchines is cylindrical and ends in a small elliptical tip (Figure 11d) that articulates with the neurocranium as a loose joint.

Psephurus (Grande & Bemis 1991), *Huso*, and all species of *Acipenser* examined possess hyomandibulae with the dorsal head expanded frontally (Figure 11c). In extant polyodontids, the dorsal head is rectangular. In *Huso* and *Acipenser*, the head is elliptical with an expanded bony shaft at its base (Figure 11c). The frontally expanded hyomandibula forms a restrictive joint allowing pivoting in only one plane.

Character 50. Ventral edge of the hyomandibula is flat – Scaphirhynchini

The ventral head of the hyomandibula of scaphirhynchines ends in a flat edge roughly perpendicular to the long axis of the element (Figure 11d). The interhyal joint opens on the anteroventral face, but is not isolated from the ventral edge.

The hyomandibula of *Psephurus* (Grande & Bemis 1991), *Huso*, and all species of *Acipenser* examined taper ventrally into narrow anteroventral tips (Figure 11a, b, c) that articulate with the interhyal. Ventral flattening of the hyomandibula of scaphirhynchines correlates with overall flattening of the head and consequent restricted dorsoventral space for the hyoid skeleton.

Character 51. Hypobranchial one contacts hypobranchial two – Scaphirhynchini

In scaphirhynchines, the posterior edge of hypobranchial one forms a wedge contacting hypobranchial two (Figure 13). Contact is slight, but ligaments consolidate the elements as a unit.

In *Psephurus* (Grande & Bemis 1991), hypobranchials are roughly cylindrical, without expansion of the oral surface (Figure 13a). *Huso* and all species of *Acipenser* examined possess rectangular first hypo-

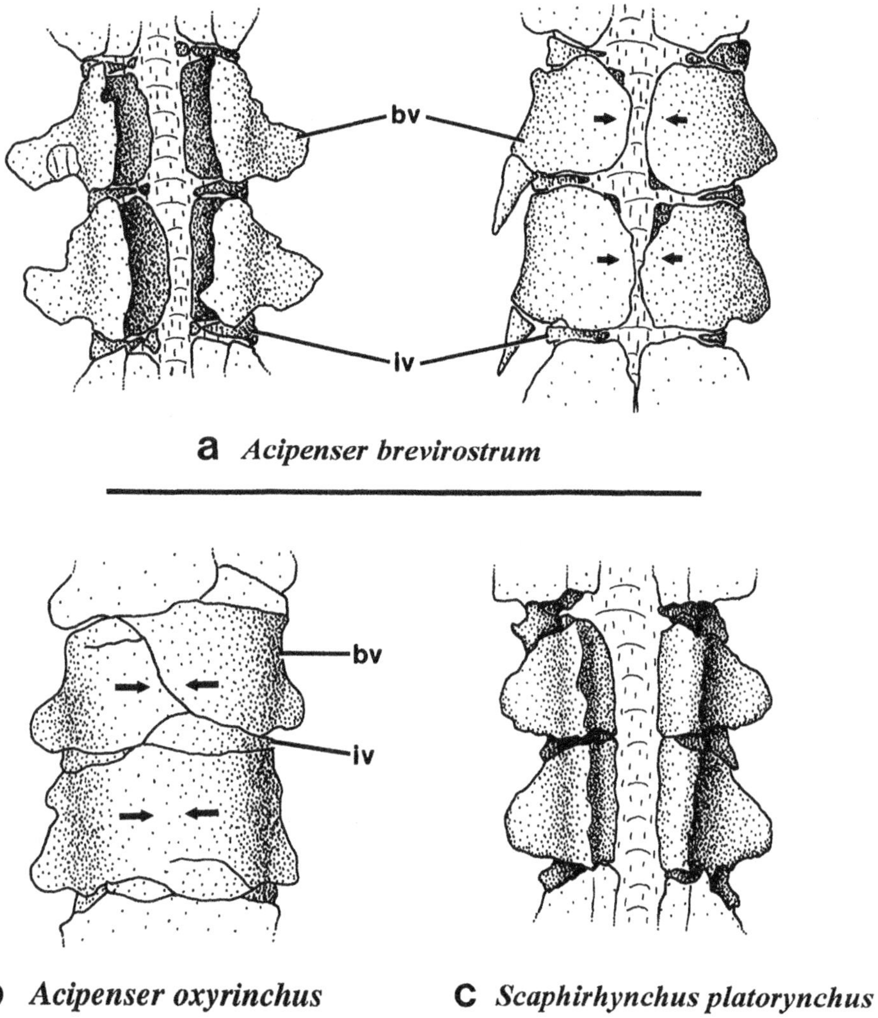

Figure 24. Ventral axial skeleton of representatives of *Acipenser* and *Scaphirhynchus*: Ontogeny of the haemal arch is shown in *A. brevirostrum* (a) with a small juvenile (left) possessing only slight extension of the basiventrals (**bv**) compared to the nearly complete haemal arch of an adult (right). This adult specimen possessed a complete haemal arch more anteriorly similar to that of *A. oxyrinchus* (b) with contralateral sealing of the basiventrals and interventrals (**iv**). Scaphirhynchines never possess a haemal arch and lack ventral expansion of the basiventrals even in large adults (c; Character 52).

branchials in the tongue pad (see Character 21), but they lack a posterior wedge contacting hypobranchial two (Figure 13b). Connective tissue ridges are restricted to hypobranchial one in *Huso* and *Acipenser*, but overlap hypobranchial two in scaphirhynchines. Juvenile *Pseudoscaphirhynchus* possess a tooth plate on hypobranchial two (Berg 1948a), suggesting that the functional palate is expanded generally in scaphirhynchines.

Character 52. Loss of a haemal arch – Scaphirhynchini

Ventral edges of basiventrals of scaphirhynchines end flatly with a slight ventral lip (Figure 24c). The basiventrals are open ventrally, lacking a haemal canal until reaching the caudal fin supported by haemal spines (see Character 38). In the peduncle, the basiventrals extend more ventrally to form a ventral groove, but never converge toward the midline.

In adults of *Polyodon, Huso*, and *Acipenser*, ventrolateral edges of the basiventrals curl to the midline to complete a haemal arch (Figure 24a, b). This is a problematic character because it develops late, only in adults by progressive elongation and merging of the basiventrals and interventrals (**bv, iv**, Figure 24a). In the thoracic region, independent ventral plates also integrate into the haemal arch. Juvenile specimens of *Acipenser* are similar to *Scaphirhynchus*, but with a slightly more pronounced basiventral edge (Figure 24a, c). However, while all species of *Acipenser* examined close a haemal arch, even large *Scaphirhynchus* lack any expansion of the basiventrals.

Character 53. Caudal fin filament present – Scaphirhynchini

The caudal fin filament (or cercus) is found in *Scaphirhynchus* and *Pseudoscaphirhynchus kaufmanni* extending from the tip of the caudal fin (Figure 23). It is formed by a thin core of cartilage sheathed by small scales. It originates posterior to the notochord, developing distinctly later than the caudal endoskeleton in small juveniles.

All species of *Scaphirhynchus, P. kaufmanni*, and certain morphotypes of *P. fedtschenkoi* (Berg 1948a) possess a cercus. It is not present in *P. hermanni*, but presumably due to secondary loss. The fin filament is short in *Scaphirhynchus*, rarely exceeding five centimeters in length, but can reach 13 centimeters or more in *P. kaufmanni*.

No other extant acipenseriforms possess a caudal fin filament and its scalation would be visible in fossils if present. The caudal endoskeleton of outgroup acipenseriforms possesses dorsal and ventral sheets of cartilage that merge and terminate with the notochord (Figure 23a).

Character 54. Loss of pectoral scales – Scaphirhynchini

Pectoral scales with elevated, recurved tips typically found on the opercular wall of acipenseriforms (see Character 23) are absent in scaphirhynchines. Universal presence of these scales in all other acipenserids and their outgroups suggests that this loss defines Scaphirhynchini.

Character 55. Spikes present on the frontals – Pseudoscaphirhynchus

Spines of the frontals, parietals, and anterior dorsal rostral bones of *Pseudoscaphirhynchus kaufmanni* are large spikes. Spikes are always present in *P. kaufmanni*, but are lacking in *P. hermanni* and *P. fedtschenkoi* (Berg 1948a). *Pseudoscaphirhynchus hermanni* occasionally has weak spines on anterior dorsal rostral bones (Berg 1948a), perhaps suggesting that a morphotype possessing spines is plesiomorphic for the genus. I accept *P. kaufmanni* as representative for the genus, with spines as a scaphirhynchine character (Character 41) secondarily lost in other species of the genus.

Spines of *Scaphirhynchus* are raised ridges similar to, but more shallow than spikes of *P. kaufmanni*. This variation in morphology and presence among scaphirhynchines makes it difficult to define distinct types of spines or spikes. While spikes of *Pseudoscaphirhynchus* are distinct in size from the spines of *Scaphirhynchus*, they are similar in shape and not accepted as discrete shape characters here. Nevertheless, no spines are present on the frontal in *Scaphirhynchus*, and presence of frontal spikes is distinctive of *Pseudoscaphirhynchus*.

Character 56. Lateral extrascapulars enclose the tri-radiation of the trunk, occipital, and supratemporal canals – Pseudoscaphirhynchus

The occipital canal (**ocll**) is carried by a lateral extrascapular series (**excl**, Figure 5e) in *Pseudoscaphirhynchus* that includes its origin separating from the trunk (**trll**) and supratemporal canals (**stll**). The lateralmost lateral extrascapular encompasses the tri-radiation of these canals as this series of bones (usually three) intrudes centrally over the posttemporal (Figure 5e).

In all other acipenserids examined (Figure 5b, c, d), splitting of these canals occurs within the post-

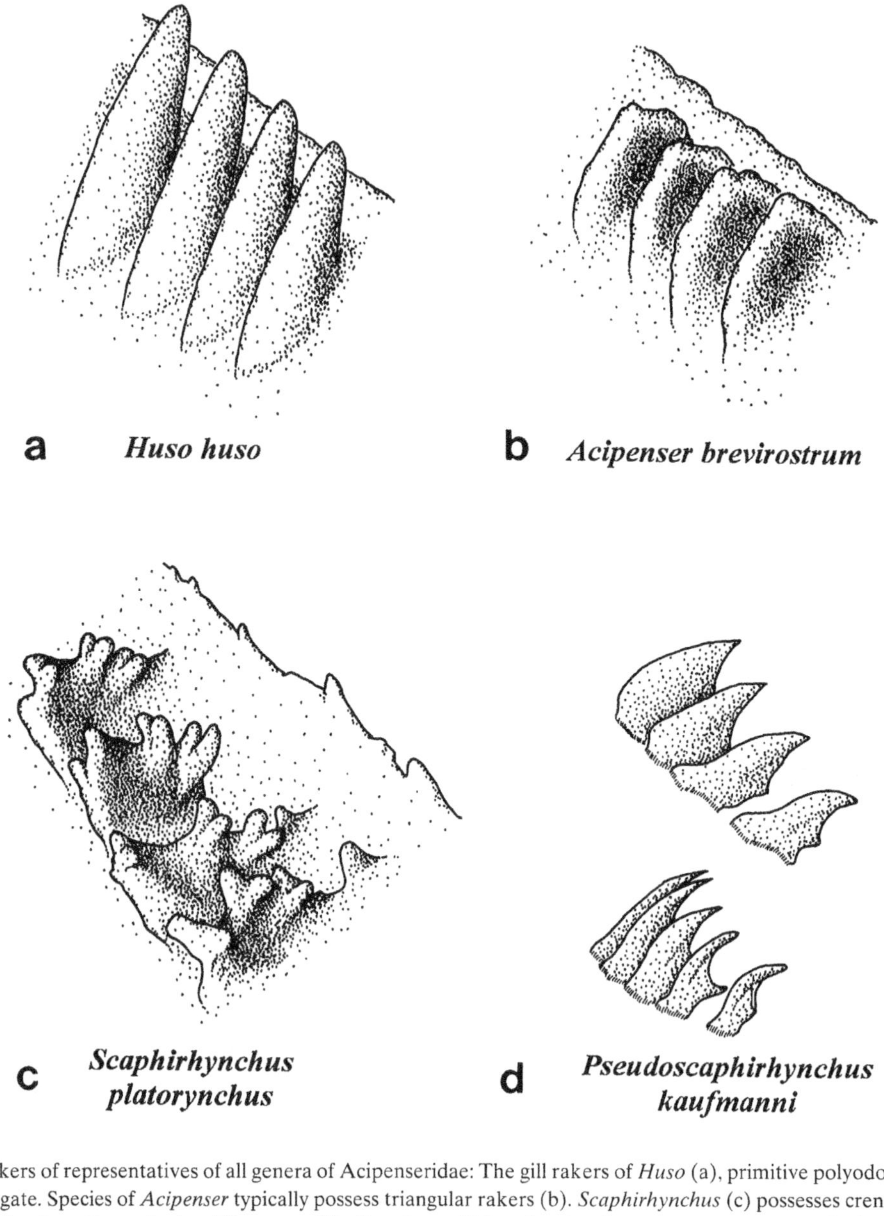

Figure 25. Gill rakers of representatives of all genera of Acipenseridae: The gill rakers of *Huso* (a), primitive polyodontids, and †*Chondrosteus* are elongate. Species of *Acipenser* typically possess triangular rakers (b). *Scaphirhynchus* (c) possesses crenelated rakers with several paired nubs along the oral edge (Character 63). *Pseudoscaphirhynchus kaufmanni* possesses pronged rakers with two, orally pointed tips (Character 58).

temporal as the occipital canal is carried initially by the posttemporal before entering the lateral extrascapulars. Pattern of the canals themselves is invariable in acipenseriforms, suggesting that the dermal patterning changed in *Pseudoscaphirhynchus*.

Character 57. Jugal is large and lacks a canal process – Pseudoscaphirhynchus

The jugal of *Pseudoscaphirhynchus* is huge, extending anteriorly to completely undercut the orbit and most of the olfactory opening, but lacks a canal process (see Character 3) enclosing the infraorbital canal (Figure 3d). The posteromedial jugal is expand-

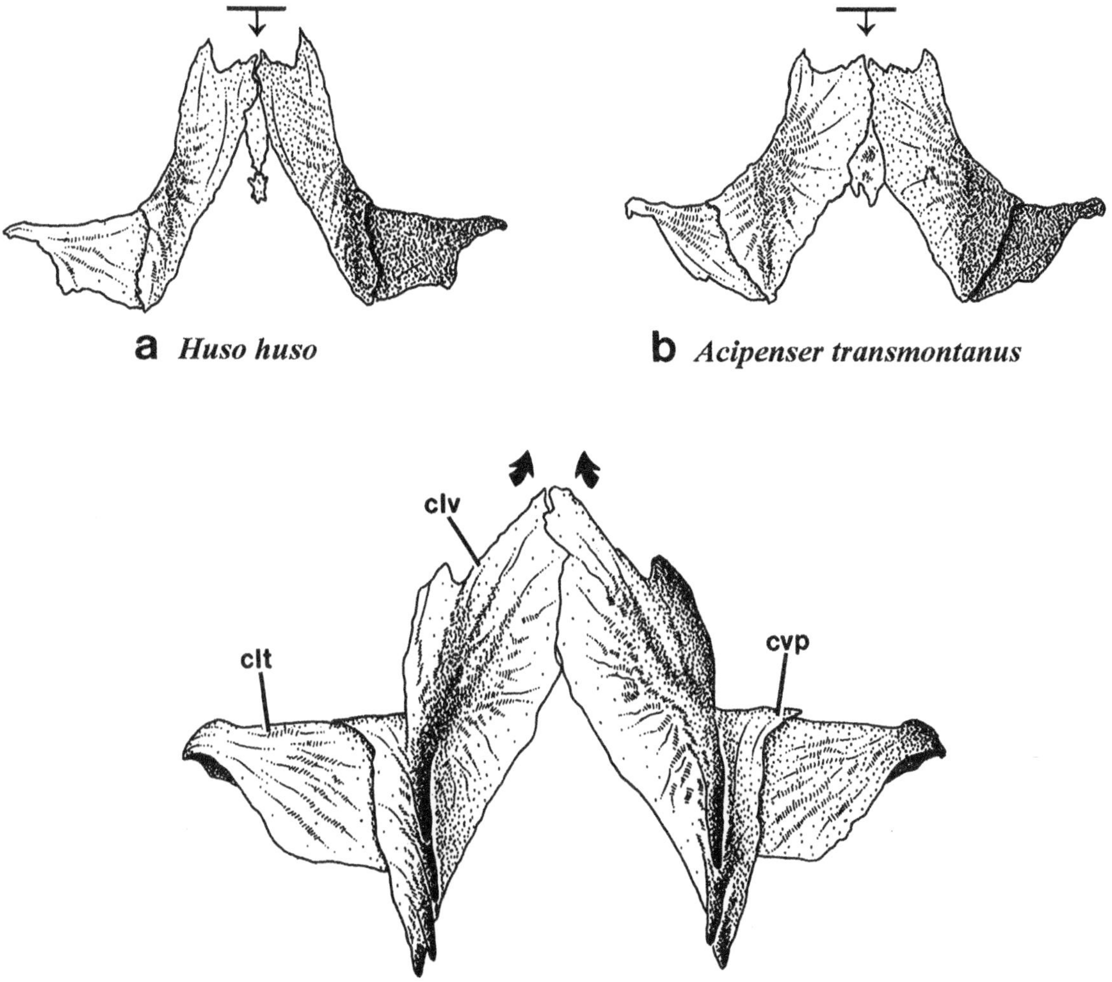

a *Huso huso* **b** *Acipenser transmontanus*

c *Pseudoscaphirhynchus kaufmanni*

Figure 26. Ventral views of the dermal pectoral girdle of representatives of *Huso, Acipenser*, and *Pseudoscaphirhynchus*: The anteromedial edge of the clavicles (**clv**) meet roughly linearly (thin arrows in a, b) in all acipenserids except *Pseudoscaphirhynchus* where they angle anteriorly to form a discrete wedge (curved arrows in c; Character 59). **clt** = cleithrum, **cvp** = clavicle process.

ed, but as an integral portion of the bone, not a process. This character is predicated on *P. kaufmanni* as representative for *Pseudoscaphirhynchus* due to lack of specimens of other species.

The jugals of *Scaphirhynchus* and all species of *Acipenser* examined undercut the orbit, but not to the posterior edge of the olfactory opening (Figure 3a, b, c). The anterior process of *Huso* is small and barely contacts the neurocranium. Relative size is a qualitative character, but jugals of all other acipenserids examined possess canal processes (Figure 3) missing in *P. kaufmanni*.

Character 58. Gill rakers are pronged – Pseudoscaphirhynchus

Gill rakers of *Pseudoscaphirhynchus kaufmanni* are split distally into a crescentic edge with paired pronged tips. These rakers split from a thin base into a curved outer (relative to the branchial arch) tip more elongate than a small, inner tip (Figure 25d). Larger rakers may possess only an outer tip, but the base remains expanded where the inner tip would protrude. Gill rakers of *P. fedtschenkoi* are described as lanceolate (Berg 1948a) and undescribed in *P. hermanni*, but lacking clear description, I accept *P. kaufmanni* as diagnostic of the genus.

Raker shape varies greatly among acipenserids. Rakers of *Huso, Psephurus*, and several species of *Acipenser* (e.g., *A. oxyrinchus*) are elongate and lanceolate, tapering from a narrow base to a blunt tip (Figure 25a). Most other species of *Acipenser* (e.g., *A. brevirostrum*) possess triangular rakers (Figure 25b), but none possess rakers with pronged tips. The crenelated rakers of *Scaphirhynchus* are distinct (Character 63) and not comparable to *Pseudoscaphirhynchus*.

Character 59. Clavicles tips meet as an anteromedial wedge – Pseudoscaphirhynchus

The clavicles of *Pseudoscaphirhynchus* converge medially as an anteromedial wedge (Figure 26c). This anterior wedge is externally visible and confirmable from photographs in *P. hermanni* (Berg 1948a). Clavicles of *Pseudoscaphirhynchus* also distinctively possess strong central ridges running through the anteroposterior axis that possess three or four spikes in *P. kaufmanni* (Figure 26c), but is a plain ridge in *P. hermanni* (Berg 1948a).

In *Scaphirhynchus, Huso*, and all species of *Acipenser* examined, anteromedial edges of the clavicles converge more linearly to the midline (Figure 26a, b). The clavicles typically meet in a slight anterior point, but this wedge is not prominent as in *Pseudoscaphirhynchus*. Also, outgroups do not possess a robust central ridge, much less any spikes along the clavicle.

Character 60. Lateral extrascapulars are clustered – Scaphirhynchus

Lateral extrascapulars form a positionally restricted cluster alongside the median extrascapular in *Scaphirhynchus* (**excl**, Figure 5d). Number of bones varies from two to seven and they may fuse in ontogeny (Findeis 1993), but they collectively occupy an consistent area (Figure 5d). Localization of the lateral extrascapulars is consistent in all species of *Scaphirhynchus*.

There is a significant variation in the dermal skull of acipenserids (Parker 1882, Jollie 1980) and position, size, and number of the lateral extrascapular bones is variable. In *Pseudoscaphirhynchus* (Figure 5e; see Character 56) and *Huso* (Figure 5b), lateral extrascapulars are typically canal bones. Whereas *Huso* typically possesses only canal-bearing lateral extrascapulars (Figure 5b), *Pseudoscaphirhynchus* possesses multiple anamestic bones (Figure 5e). All species of *Acipenser* examined show extensive variation in number and position of lateral extrascapulars (Figure 5c).

Character 61. Complete circumorbital series present – Scaphirhynchus

The dermal skull roof of *Scaphirhynchus* is unique within Acipenseridae in possessing a complete circumorbital series with contact between the antorbital (see Character 46) and an enlarged postrostral bone of the border rostral series (**prb**, Figure 4e). With postrostral contact, the antorbital is widened ventrally with a flat edge. Antorbital shape and postrostral presence are also diagnostic of *Scaphirhynchus*, but correspond to completion of the circumorbital series and are subsumed into this character.

Huso, Pseudoscaphirhynchus, and all species of *Acipenser* examined lack a complete circumorbital series, with the antorbital isolated dorsally. No postrostral is identifiable in these taxa as the border rostral series is thin without any expanded bones (see Figure 18c, d).

Character 62. Branchiostegal one is short and triangular – Scaphirhynchus

Branchiostegal one is a thin, triangular bone in *Scaphirhynchus* with a flat ventral edge that converges dorsally (Figure 7e) to contact the subopercle within a broad, vertical groove.

In *Acipenser* and *Huso*, branchiostegal one is typically rectangular and hourglass-shaped (Figure 7c, d) as it drops vertically before angling ventromedially under the operculum. All scaphirhynchines possess a flattened head with restricted vertical space in the operculum, but address these restrictions differently. *Pseudoscaphirhynchus* has a rectangular branchiostegal, but it is short and indents medially immediately from the subopercle. Branchiostegal one of *Scaphirhynchus* is vertical, but dorsoventrally compressed in shape to define this character. The independent branchiostegal of polyodontids (Grande & Bemis 1991, Findeis 1993) and branchiostegal series of †*Chondrosteus* (Traquair 1887, Hennig 1925) and †*Peipiaosteus* (Liu & Zhou 1965, Zhou 1992) are not comparable (see Character 7).

Character 63. Gill rakers are crenelated – Scaphirhynchus

Gill rakers of *Scaphirhynchus* possess short double-branched nubs at regular and close intervals along the oral edge (Figure 25c). Basic raker shape is a right triangle, with a vertical posterior edge and gradually descending anterior edge bearing the nubs. *Scaphirhynchus albus* possesses slender rakers with a single pair of nubs on the first arch (Bailey & Cross 1954), but rakers of other arches are similar to *S. platorynchus*.

These rakers are distinctive compared to most acipenserids, but apparently similar to the branched rakers of *Acipenser baerii* (Sokolov & Vasil'ev 1989). They differ from the pronged rakers of *Pseudoscaphirhynchus* (see Character 58) in possessing blunt nubs and being generally triangular. The lanceolate rakers of *Huso* and several species of *Acipenser* are elongate (Figure 25a) and the triangular rakers of other species of *Acipenser* (Figure 25b) are symmetric and lack nubs.

Character 64. Caudal peduncle is flattened and elongate – Scaphirhynchus

The caudal peduncle of *Scaphirhynchus* is elongate and dorsoventrally flattened with a rectangular cross-section. Elongation occurs suddenly and rapidly in small juveniles to achieve adult dimensions by elongation of vertebral cartilages, with individual vertebral segments over twice the length of typical vertebral segments (Findeis 1993). Total number of segments composing the peduncle is consistent with other acipenserids. †*Protoscaphirhynchus squamosus* (Wilimovsky 1956) possesses an elongate peduncle linking this fossil species to *Scaphirhynchus*.

The caudal peduncle of all other acipenserids is short, with a cylindrical cross-section. Vertebral cartilages within the peduncle of the outgroups are unlengthened and retain a recognizable segmental organization.

Character 65. Caudal peduncle and preanal area armored – Scaphirhynchus

Six groups of peduncle scales make mutual contact with the flank scutes and anal scale series to completely armor the posterior trunk from the vent and dorsal fin to the caudal scales of the caudal fin (Findeis 1993). †*Protoscaphirhynchus* possesses arrays of peduncle scales similar to *Scaphirhynchus*, but the specimen is too weathered to make detailed comparisons.

Similar peduncle and preanal scale rows are found in *Pseudoscaphirhynchus, Huso*, and many species of *Acipenser*, but they are not organized into plates as numerous, large, or with precise interdigitation as *Scaphirhynchus. Acipenser oxyrinchus* is the best scaled species of *Acipenser* examined, possessing paired rows of scales preanally and on the peduncle, but they remain separate rows without mutual contact. *Pseudoscaphirhynchus kaufmanni* possesses robust preanal scales and paired scales on the dorsal peduncle, but ventral peduncle scales are not organized. More typically in other acipenserids, peduncle and anal scales form single rows of variable size and separation.

Character 66. Cleithral wall present – Scaphirhynchus

The cleithral wall of the scapulocoracoid is a triangular sheet that spreads over the cleithrum in *Scaphirhynchus* (**ctw**, Figure 9b). It spans the anterior process of the middle region (**anp**), the dorsal tip of the mesocoracoid arch (**msc**), and the anterior edge of the propterygial bridge (**ptb**, Figure 9b).

The scapulocoracoid of all other acipenserids examined possesses a thin cleithral arch (**cta**) extending dorsally, but not expanded medially (Figure 9a). The anterior edge of the middle region is isolated as a horizontal spar extending medially from base of the cleithral arch (**cta**) to the anterior process (**anp**, Figure 9a). Thus, the anterior face of the scapulocoracoid of outgroups is L-shaped, with a vertical cleithral arm and horizontally extended anterior process.

Character 67. A thin process encircles the propterygial fossa – Scaphirhynchus

The scapulocoracoid of *Scaphirhynchus* possesses a thin process extending under the propterygium restraining process of the cleithrum (**prp**, Figure 8a). This process curls anteroventrally as a half-circle around the propterygial fossa (**ptgp**, Figure 9b).

This process is not present in any other acipenserid (Figure 9a) and the propterygial fossa is open primitively anteriorly against the cleithrum.

Character 68. Coracoid shelf restricted to the cleithrum – Scaphirhynchus

The coracoid shelf (**csh**) of the scapulocoracoid in *Scaphirhynchus* spreads from the coracoid wall (**cw**) medially and laterally onto the cleithrum (arrows in Figure 9b). The largest portion spreads anterolaterally into the angle between the opercular wall and cardiac shields (Findeis 1993). The medial side slightly overlaps the clavicle, but ends in a small footprint. In ventral view, the coracoid shelf is roughly triangular, with the broad anterior edge converging posteriorly and symmetrically on the cardiac shield.

The coracoid shelf of *Pseudoscaphirhynchus, Acipenser*, and *Huso* spreads anteromedially (Figure 9a) onto the clavicle as an elongate sheet, but without lateral expansion (see Character 14). The scapulocoracoid of polyodontids extends similarly as a thick cylindrical process. This matches the anteriorly extended pectoral girdles of the outgroups compared to the compact pectoral girdle of *Scaphirhynchus*. The coracoid shelf corresponds to the dermal girdle and implicitly includes a pyramidal scapulocoracoid shape as a character of *Scaphirhynchus*.

Character 69. Dermopalatine and ectopterygoid bones fuse – Scaphirhynchus

The dermopalatine and ectopterygoid develop as separate bones that fuse in adult *Scaphirhynchus* (Findeis 1991, 1993). All adult specimens of *S. platorynchus* examined exhibited complete fusion, while all small juveniles retain separate elements, suggesting that fusion occurs in moderately sized juveniles.

In †*Paleopsephurus* (Grande & Bemis 1991), an elongate process positionally similar to the acipenserid ectopterygoid is probably an independent bone (Figure 22a), but the short process of *Psephurus* and *Polyodon* suggests that the ectopterygoid is lost in these genera. In *Huso, Acipenser*, and *Pseudoscaphirhynchus*, the ectopterygoid is independent and closely contacts the dermopalatine (Figure 22b, c), but even in large specimens of *A. brevirostrum* and *A. oxyrinchus* there was never fusion between these bones. A small ectopterygoid process is present in †*Peipiaosteus* (Zhou 1992), but it is unclear if it is a separate bone or fused. †*Chondrosteus* has an independent ectopterygoid (= pterygoid of Traquair 1887).

Generic level phylogeny of Acipenseridae

The character distribution from this analysis supports a new cladogram of genera of Acipenseridae

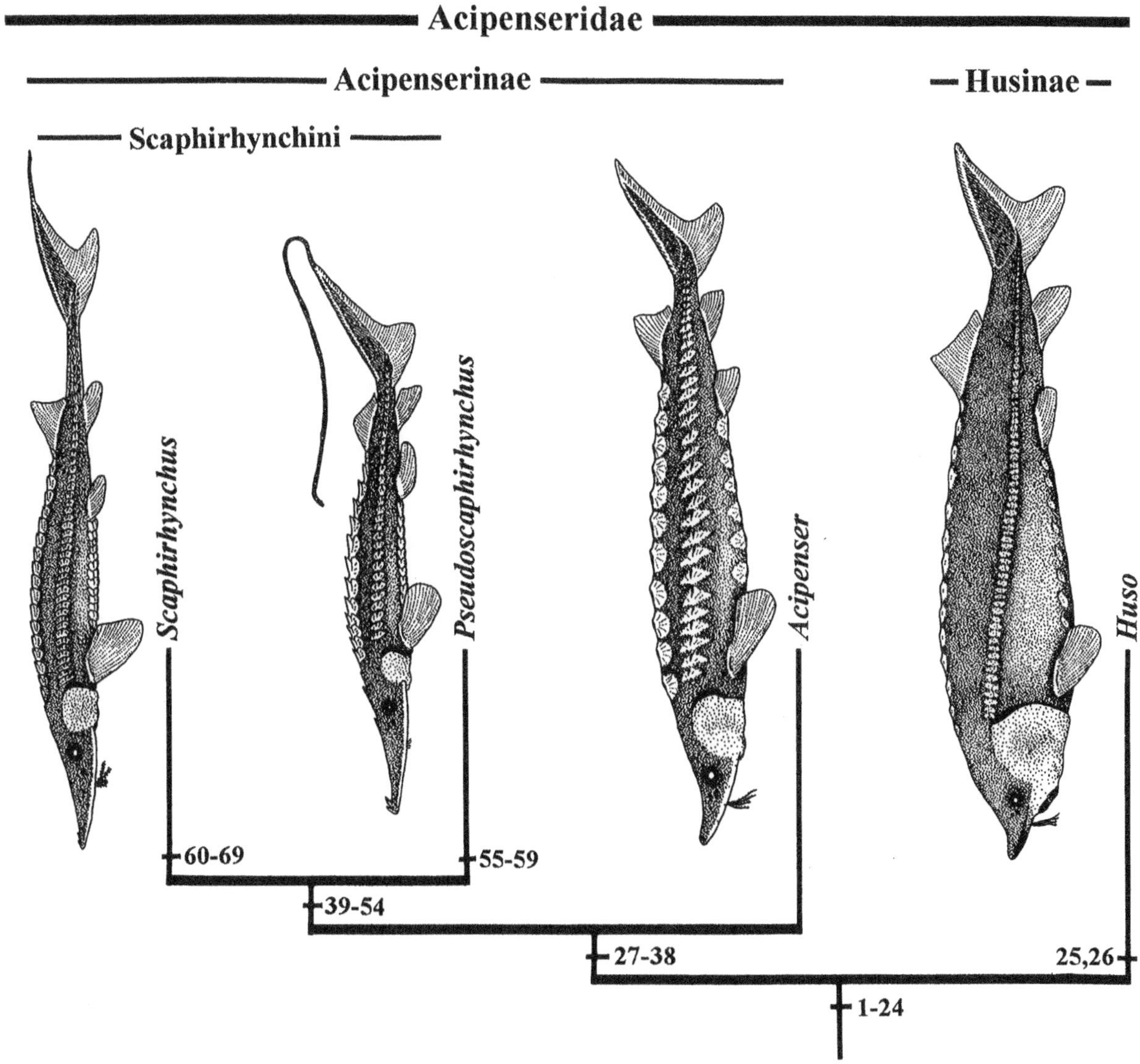

Figure 27. Proposed interrelationships of the genera of Acipenseridae: *Huso* is defined by two synapomorphies forming the subfamily Husinae recognized as the sister group to the Acipenserinae. *Acipenser* is not recognized by any skeletal synapomorphies. Shovelnose sturgeons of *Scaphirhynchus* and *Pseudoscaphirhynchus* compose Scaphirhynchini.

(Figure 27). The family Acipenseridae is supported by 24 skeletal synapomorphies (Characters 1–24). The genus *Huso* is defined by two synapomorphies (Characters 25, 26) and is the sister group to a redefined subfamily Acipenserinae defined by 12 synapomorphies (Characters 27–38) comprising all other genera. *Scaphirhynchus* and *Pseudoscaphirhynchus* compose a monophyletic tribe Scaphirhynchini based on 16 synapomorphies (Characters 39–54). *Pseudoscaphirhynchus* is defined by five synapomorphies (Characters 55–59) and *Scaphirhynchus* is defined by 10 synapomorphies (Characters 60–69). Each node is supported by characters from different skeletal complexes suggestive of multiple evolutionary events occurring at one node rather than a one event affecting multiple characters.

A remaining phylogenetic problem is the lack of identified synapomorphies defining the genus *Acipenser*, leaving this largest genus (17 species) as potentially untenable. Characters typically ascribed to *Acipenser* in the literature are plesiomorphic for Acipenserinae. Interspecific variability in morphological features within *Acipenser* is rampant, clouding the applicability of most characters used in phylogenetic reconstruction. It is possible that *Acipenser* is paraphyletic and closer examination of the genus is warranted.

Taxonomic recommendations

Order Acipenseriformes Berg 1940
- suborder †Chondrosteodei sensu Grande & Bemis 1991
- suborder Acipenseroidei sensu Grande & Bemis 1991
  - family Polyodontidae Bonaparte 1838
  - family Acipenseridae Linnaeus 1758
    - subfamily **Husinae** new name
      - genus *Huso* Brandt 1869
    - subfamily **Acipenserinae** new usage
      - tribe **Acipenserini** undefined taxon
        - genus *Acipenser* Linnaeus 1758
      - tribe **Scaphirhynchini** Bonaparte 1846
        - genus *Pseudoscaphirhynchus* Nikolskii 1900
        - genus *Scaphirhynchus* Heckel 1836

Evolutionary patterns within Acipenseridae

Based on this cladogram, scaphirhynchines are derived sturgeons, with *Huso* the most phylogenetically primitive genus within Acipenseridae (Figure 27). This counters widely held interpretations that implicate scaphirhynchines as representing the primitive condition within Acipenseridae (Schmalhausen 1991, Birstein 1993). Scaphirhynchines are generally accepted as phylogenetically primitive for two reasons: (1) the premise that acipenserids originated in freshwater basins of Triassic northern Asia (Berg 1948b, Yakovlev 1977) with *Pseudoscaphirhynchus* a local remnant of the original radiation; and (2) the observation that scaphirhynchines are heavily scaled, ancient looking sturgeons easily interpreted as primitive. That the earliest acipenserid fossils are from the upper Cretaceous of North America (not Europe or Asia), and that putatively primitive aspects of scaphirhynchine morphology have never been examined phylogenetically have not impeded such hypotheses. Typically, the interpretation that scaphirhynchines are plesiomorphic within Acipenseridae is accepted a priori, and has driven evolutionary discussions of the family.

A second dominant tenet to historical perspectives on acipenserid evolution is paedomorphosis. In recognizing chondrosteans with cartilaginous endoskeletons and a reduced dermal skeleton as secondarily de-ossified compared to their palaeoniscid ancestors (e.g., Traquair 1887, Woodward 1891, Goodrich 1909, Gregory 1933), early authors initiated a persistent theme presenting sturgeons as degenerate and primitive among extant actinopterygians. Although acipenserids possess extensive scalation, they remain perceived as paedomorphs.

The phylogeny and characters presented in this study allow for new interpretations about the evolution of the Acipenseridae. The cladogram supported here reverses classic evolutionary scenarios for Acipenseridae, suggesting instead that acipenserids show progressive layering of peramorphic characters in phylogeny and have evolved into increasingly benthic fishes.

Benthic cruising as a scenario in acipenserid evolution

Based on this phylogeny of Acipenseridae, a major trend in acipenserid evolution was to become increasingly benthic. However, acipenserids never achieved the extreme benthic specializations exhibited by some fishes exemplified by such features as flattened bodies, extensive camouflage, and stationary behaviors. Instead, acipenserids remain cylindrical in cross-section and are relatively active fishes that interact with the substrate focally for predation, while developing locomotory abilities allowing mobile benthic foraging. This proposed behavioral style I term benthic cruising is unique to Acipenseridae and incorporates a medley of benthic and non-benthic features. Benthic cruising will be addressed according to features that dominate acipenserid evolution including: (1) feeding, (2) respiration, (3) rostral expansion and head flattening, (4) scalation, and (5) locomotion. Characters are examined at several phylogenetic levels to show increasing specialization within Acipenseridae.

The linchpin to this phylogeny is defining *Huso* as the sister group to a redefined subfamily Acipenserinae. *Huso* is distinct from all other acipenserids morphologically by lacking 12 acipenserine synapomorphies and ecologically as both species of *Huso* maintain a life history style distinct compared to other acipenserids, but comparable to the polyodontid *Psephurus*. Additionally, although *Huso* shares 24 synapomorphies with acipenserids, many skeletal features plesiomorphically resemble those of *Psephurus* in shape of the rostrum and associated bones, lack of several bone groups in the dermal skull, and possessing anterior facing jaws compared to acipenserines. While these character states are plesiomorphic, similarities in life history among *Huso* and *Psephurus* provide a backdrop to interpreting morphological and behavioral changes in acipenserid evolution.

Living *Huso* (e.g., Antipa 1933, Berg 1948a) and *Psephurus* (Liu & Zeng 1988) prey dominantly on fishes, and fossil †*Crossopholis* occur with fishes enclosed in its remains (Grande & Bemis 1991), suggesting that acipenseroids were originally piscivorous predators. Piscivory in itself does not contradict a benthic feeding pattern, but midwater species included in the diet of adult *Huso* (Berg 1948a, Pirogovskii et al. 1989), and fishes enclosed in †*Crossopholis*, confirm that they are pelagic predators. In contrast, most species of *Acipenser* and all scaphirhynchines focus on benthic prey such as molluscs, crustaceans, and substrate oriented fishes. Secondary characteristics such as the cylindrical body and head shape of *Huso*, polyodontids, †*Chondrosteus*, and †*Peipiaosteus* further bolster suggestions that nonacipenserine acipenseriforms are neither benthic in morphology nor in life history.

Feeding

As benthic cruisers, acipenserines forage by feeding focally from the substrate. Hyostylic jaw suspension and resulting jaw projection is characteristic of the Acipenseriformes, but nonacipenserid acipenseriforms universally possess jaws that face anteriorly. *Huso* also possesses large, anterior jaws that typically extend onto the lateral surface of the head. These jaws are positioned to open anteriorly and are not obviously linked to benthic feeding.

Nevertheless, acipenserids, including *Huso*, have modified the jaws and skull to exploit benthic prey. Acipenserids possess 'internal' jaws based on the novel palatal complex which extends the oral surface of the upper jaw posterodorsally (Character 18) and the sharply defined tongue pad with biting ridges on the expanded first hypobranchials (Chracter 21). Double articulation of hypobranchial three with basibranchial one (Character 22) further consolidates the ventral hyobranchial skeleton within the tongue pad. Functionally, the dorsal palatal complex shears across the tongue pad as the upper jaw is projected and retracted to hold prey. This mechanism acts to retain prey during winnowing of ingested substrate and is a putative benthic specialization.

Within Acipenserinae, acipenserines possess ventral jaws restricted beneath the orbit. The jaws are sequestered behind the central trabecular process (Character 31) and expanded postnasal wall (see Character 3) that form a barrier anterior to the orbit. At rest, the jaws are held entirely within the confines of the head. The jaws are short and transverse to fit ventrally, and the lower jaw is straightened (see Characters 33, 35) opposite the upper jaw. During feeding, these jaws open obligately ventrally in an ideal position for benthic feeding. Posterior displacement of the interhyal-posterior ceratohyal joint (Character 36) facilitates projection by spatially separating the jaws from the non-projectile anterior ceratohyal.

Scaphirhynchines have essentially identical jaws, but the tongue pad expands onto hypobranchial two as it is included into the tongue pad by posterior expansion of hypobranchial one (Character 51). Although their rakers differ (Characters 58, 63), both scaphirhynchine genera possess complex rakers ideal for retaining small prey (fishes and/or insect larvae) that they prefer, as well as to winnow detritus inevitably ingested while feeding in their preferred habitats.

These characters organize a feeding morphotype that initially develops prey processing surfaces supportive of jaw projection in acipenserids and then restricts the jaws into their benthic specific ventral orientation in acipenserines.

Respiration

Acipenserids possess an accessory respiratory shunt allowing them to pull water into the buccal cavity from the opercular chamber whenever the mouth is occluded (Burggren 1978). This shunt allows acipenserids to respire when the buccal cavity is blocked by substrate ingested during benthic feeding or large prey that acipenserids preferentially feed on fill the mouth during ingestion. Juvenile sturgeons ingest remarkably large prey requiring extensive processing before swallowing. The subopercle is bound to the hyomandibula and automatically flares laterally during jaw movements to open the shunt during ingestion or winnowing.

The respiratory shunt is predicated on a dorsal opening in the operculum (not a character here), solid pectoral girdle (Characters 9, 13), and curved gill filaments that direct water ventrally out of the buccal cavity (Findeis 1989). The branchiostegals of acipenserids (Character 7) arch ventrally and then medially to support an opercular cavity open to dorsal inflow and ventral outflow. The medial opercular wall (Character 10) stabilizes the opercular chamber as an open, concave structure promoting unimpeded water outflow. Such skeletal stabilization of the girdle is central to opening the dorsal incurrent channel and then directing outflow ventrally.

This basic morphology of the pectoral skeleton is typical of acipenserids, and dimensional changes among genera are not recognized as discrete characters here. The medial opercular wall is larger in acipenserines than in *Huso*, and the girdle is more compact in *Scaphirhynchus* than other genera. These changes do not reflect obvious functional changes so much as changes in overall body flattening (see below).

Rostrum and flattening

Many benthic fishes are flattened to allow direct, stable interaction with the substrate. Outgroup acipenseriforms and *Huso* possess cylindrical heads and trunks, but varying degrees of flattening occur within Acipenserinae. This is most apparent with expansion and flattening of the head, rostrum, and trunk within Acipenseridae that develops a platform parallel to the substrate.

Appearance of the jugal (Character 3), antorbital (Character 4), and median extrascapular (Character 5) all suggest that skull expansion occurs in Acipenseridae. The neurocranium is broader in general than in polyodontids and these bones reflect this expansion dermally.

The rostrum of *Huso* is thin and cylindrical, but the rostrum of all acipenserines is expanded by addition of the lateral ethmoid ridges (Character 27), broadening of the dorsal rostral series (Character 28), and appearance of border rostral bones (Character 29). While the acipenserine rostrum appears flatter than in *Huso*, this is due more to addition of the lateral ethmoid ridges that broaden the ventral face of the rostrum than actual dorsoventral compression of the head in *Acipenser*. Rostrum expansion is likely associated with expansion of electrosensory ampullary fields and hydrodynamics of a head with ventral jaws. An indicator of skull expansion in acipenserines is addition of pineal bones in the widely opened pineal fontanelle (Character 32) and ventral extension of the antorbital (see Character 4).

Scaphirhynchines possess flattened rostra enlarged by expansion of the lateral ethmoid shelves (Character 39) and dorsal coverage with the ampullary bones (Character 40). The skull itself is wider and flatter than in *Acipenser* as indicated by additional extension of the antorbital (Character 46). Head flattening is not indicated by skull characters that simply form a shallower cover of the head, but by dorsoventral compression of the hyomandibula (Character 50) and operculum (Characters 47, 62) beneath the skull.

Further, scaphirhynchines are the only acipenserids to extensively flatten their trunk. In contrast to the cylindrical trunk of *Huso* and *Acipenser*, the ventral surface of scaphirhynchines is nearly flat between the ventral scute rows. Among acipenserids, only scaphirhynchines make extensive contact with the substrate, and their highly flattened belly serves as an ideal resting platform as *Scaphirhynchus* is especially benthic and mobile on the substrate (see below).

Initially, flattening of the head and trunk seems to support an expanded suite of ampullary organs as well as conform the skull to ventral displacement of the jaws. Trunk compression only occurs in advanced sturgeons, as *Acipenser* generally remains a genus of mobile predators while scaphirhynchines become obligate benthivores.

Scalation

Extensive scalation correlates with benthic habits in many groups of fishes, and may serve for protection against the substrate or to counter predation. While acipenserid outgroups lack any extensive scalation, the most distinctive hallmark of acipenserids is their scutes (Character 1). A precise role for scutes is unknown, but they are prominent and sharp in juveniles as a tight assemblage covering well over 50% of the trunk, suggesting that those ages are most reliant on scalation. In fact, juvenile acipenserids prefer benthic habitats (Richmond & Kynard 1995) and even juvenile *Huso huso* focuses on benthic prey (Pirogovskii et al. 1989) when scutes are most developed. Scutes become allometrically reduced in adults in size, roughness, and cohesiveness (they become well separated within the rows) and actually regress significantly and are progressively covered by skin in some species (e.g., *H. huso, Acipenser fulvescens, A. nudiventris*). As large fishes, adult sturgeons do not face serious predation and tend to be more active above the substrate, and we might regard scutes as an adaptation for juveniles rather than adults in Acipenseridae.

Scaphirhynchines retain well developed scutes in adults that are broader and more tightly overlapping than other acipenserids. Also, scales on the peduncle are large and organized as multiple, discrete groups in *Scaphirhynchus* (Character 65) with the posterior trunk completely armored by overlapping scales. Peduncle scales are present in all other acipenserids, but vary in number and size and never form an extensive cover (see Character 1). Purpose of extensive adult scalation and peduncle scalation in *Scaphirhynchus* is unclear, but correlates well with their benthic life history and persistent contact with the substrate.

Locomotion

Benthic cruising implies mobile interaction with the substrate. Acipenserids are strong swimmers with several features facilitating this behavior. Morphology of the caudal fin is not a character here due to variation within Acipenseridae and among acipenseriforms, but the hypochordal lobe is often reduced to allow sweeping of the tail while close to the substrate. More pelagic taxa (e.g., *Huso, Acipenser oxyrinchus*) possess deep caudal fins similar to those of †*Chondrosteus* and polyodontids, while more benthic species possess either abbreviated (e.g., *A. brevirostrum*) or obliquely folded hypochordal lobes (*Scaphirhynchus*).

Pectoral fin spines of acipenserids (see Characters 2, 12, 16) are neither sharp, nor possess serrated edges exposed for protection. Instead, they are permanent lateral processes erected during swimming as diving planes. Acipenserids can excellently maintain position over variable bottoms by precisely regulating inclination of the spines to modulate depth during swimming. Pectoral fin spines of *Scaphirhynchus* are uniquely curved and are used as 'legs' to shuffle along the substrate during feeding and exploration, something not done by other acipenserids. In this scenario, fin spines are locomotory stabilizers for depth control. Polyodontids lack fin spines, but are midwater fishes not requiring the same precision during locomotion.

In all cases, acipenserids show phylogenetic trends suggesting increasing abilities as benthic fishes. *Huso* is morphologically and behaviorally pelagic as adults, but possess underlying characters indicating an original shift to benthic behaviors in Acipenseridae. This shift is likely associated with evolutionary success of juvenile interval in *Huso* when benthic prey (notable other acipenserids) and predator avoidance is most important. *Huso* then rises into pelagic environments when large enough to exploit larger prey. Regardless of the initial reason to exploit benthic habitats in primitive acipenserids, original opportunities afforded by the rich benthic prey fauna continued with evolution of preferential benthic cruising acipenserines.

Many species of *Acipenser* depend on benthic abilities. Ventral displacement of the mouth allows for benthic foraging, but restricts their options to capture midwater prey. Unlike the piscivorous *Huso* and *Psephurus*, all species of *Acipenser* and Scaphirhynchini feed wholly or partially on benthic prey. Several species become piscivorous as adults (e.g., *A. transmontanus, A. oxyrinchus*), but these large species still never prey consistently on pelagic prey to such an extent as *Huso*. Far more commonly, species of *Acipenser* are molluscivores, insectivores, or generalized benthic predators. While the focus here is on feeding, the other suites of characters are critical to allowing acipenserines to interact with the substrate and together compose a successful benthic cruiser.

Scaphirhynchines are more obligate benthivores with the flattened trunk and head promoting stability and mobility along the substrate. *Scaphirhynchus* is the most derived acipenserid genus in each category and exemplifies character acquisitions pointing to sequential entry into benthic habitats during evolution of the Acipenseridae.

Peramorphy as a mechanism in acipenserid evolution

A persistent theme overshadowing studies of the Acipenseridae is the putative role of paedomorphosis dominating their evolutionary change. Paedomorphosis has been invoked to generally explain loss of ossification of the endoskeleton and loss of dermal elements in Acipenseridae compared to other fossil and extant actinopterygians. Paedomorphic themes are so pervasive that they have possibly promoted or capitalized on the idea that scaphirhynchines are plesiomorphic sturgeons, with the more lightly armored species of *Acipenser* and *Huso* being natural evolutionary destinations within the family.

However, paedomorphy has focused on the well known Acipenseridae, not the larger assemblage Acipenseriformes. Our current understanding of polyodontids and fossil acipenseriforms now indicate that loss of endochondral ossification, dermal cheek bones, and trunk scalation (the most obvious paedomorphic characters) are synapomorphies of Acipenseriformes and plesiomorphic to Acipenseridae. Thus, while reduction of the skeleton exemplifying paedomorphic events are valid phylogenetic characters and evolutionary events, they characterize the origin of Acipenseriformes and define a morphotype underlying Acipenseridae.

Paedomorphosis is also accepted as the primary process driving evolution in Acipenseriformes and Acipenseridae. However, ideas that paedomorphy dictates morphological change in acipenseriforms conflicts with the suggestion that selective forces for performance might drive evolutionary patterns themselves and only result in paedomorphy secondarily. For instance, evolution of projectile jaws in acipenseriforms requires loss of the dermal cheek to free the mobile palatoquadrate. Invoking paedomorphosis as a preliminary route removing the cheek originally ignores the possibility of precedent selection for jaw projection.

In contrast to loss and reduction characters consistent with paedomorphosis, numerous characters defined in this study entail addition or enlargement of skeletal elements at all levels with Acipenseridae and suggest a progressive role for peramorphosis (gerontomorphosis, Balon 1983, 1985, 1989) in acipenserid evolution. For the purposes of this discussion, peramorphic characters are simply defined as transformations that result in addition of new structures or enlargement of preexisting structures compared to outgroups. A peramorphic pattern of character acquisition contrasts sharply with ideas of paedomorphy in evolution of Acipenseridae. In fact, basal to Acipenseridae, the Acipenseroidei (Polyodontidae + Acipenseridae) is defined by a character complex including both paedomorphic (e.g., loss of the opercle, reduced branchiostegal number) and peramorphic (e.g., presence of a rostrum) characters (see Findeis 1993, Bemis et al. 1997). This suggests that a shift away from dominant paedomorphic mechanisms toward peramorphy begins prior to evolution of Acipenseridae. However, the most dominant peramorphic characters in number and morphological results appear within Acipenseridae.

Acipenseridae is defined by a pervasive array of peramorphic characters. Scutes (Character 1), the pectoral fin spine (Character 2), the antorbital and median extrascapular bones (Characters 4, 5), the supracleithral cartilage (Character 15), and palatal complex (Character 18) all appear without precursor in Acipenseridae. Many other characters are peramorphic enlargements of features such as the skull (Character 3), pectoral girdle (Characters 9–14), pelvic girdle (Character 17), hyobranchial skeleton (Characters 19–22), pectoral scales (Character 23), and even the occipital canal (Character 24).

Peramorphic characters defining Acipenserinae include the expanded rostrum (Characters 27–29), central trabecular process (Character 31), and dermopalatine shelf (Character 34) that are all additions to the plesiomorphic (e.g., *Huso*) condition. Acipenserines are also the first level within Acipenseridae with numerous anamestic bones in the skull roof including pineal bones (Character 32) and anamestic lateral extrascapulars. Anamestic bones generally lack consistent organization, filling areas of the skull with variable numbers and sizes of bones between individuals and species. While they can define peramorphic characters, they may also represent general peramorphic processes beyond cladistic characterization.

Scaphirhynchines continue this peramorphic trend with extreme expansion of the rostrum (Characters 39, 40), appearance of basitrabecular cartilages (Character 43), ventral expansion of the antorbital (Character 46), posterior expansion of hypobranchial one (Character 51), and appearance of the caudal fin filament (Character 53).

While *Pseudoscaphirhynchus* is generally defined by characters that are peramorphically neutral, several peramorphic transformations occur in *Scaphirhynchus*. The caudal peduncle is elongate (Character 64), and armored (Character 65). This genus possesses a circumorbital series based on expanded antorbital and postrostal bones (Character 61), an expanded cleithral wall (Character 66), and fusion of the dermopalatine and ectopterygoid (Character 69).

While putative paedomorphic characters occur at several nodes of the cladogram (Character 8 for Acipenseridae; Characters 35, 37, 38 for Acipenserinae; Characters 48, 52, 54 for Scaphirhynchini), they are generally associated with shape changes that require reduction or loss of elements to facilitate the morphological shift to benthic cruising and are significantly outnumbered by peramorphic characters. Many of these peramorphic characters are not linked and occur at all major nodes within Acipenseridae, supporting peramorphy, not paedomorphy, as a dominant influence in acipenserid evolution.

Conclusions

(1) *Huso* exemplifies a precedent primitive condition within Acipenseridae as a pelagic sturgeon compared to other acipenserids. However, characters defining Acipenseridae (including *Huso*) appear to be linked to benthic behaviors and may indicate importance for benthic orientation for juvenile acipenserids. Thus, many features diagnostic of acipenserids may have evolved in support of the juvenile period of life history, not for adults.

(2) Acipenserines are clearly more benthically oriented than *Huso*. Acipenserines possess ventral mouths and a broadened, somewhat flattened rostrum and head. These features define a morphology allowing more precise benthic foraging.

(3) *Scaphirhynchus* represents an ultimate benthic morphology within Acipenseridae. Their increased dermal armor, highly flattened head and trunk, and pectoral 'legs' allow for stable benthic exploration. *Pseudoscaphirhynchus* is similarly flattened, but this genus lacks features that facilitate the level of benthic behaviors characteristic of *Scaphirhynchus*.

(4) In the scenario of benthic cruising, initial opportunities for *Huso* to exploit substrate habitats were expanded upon by phylogenetically successive acipenserid taxa that focused on substrate habitats and prey.

(5) Peramorphy is a dominant mechanism in acipenserid evolution. Appearance of new elements and enlargements of preexisting features are typical for characters defined in this study.

Future research on Acipenseridae

The most immediate problem in acipenserid phylogeny is alpha-level systematics of *Acipenser*. Morphological variation within and among species of *Acipenser* is a stumbling block for comparative work, but most species do possess a basic morphology definable for phylogenetic analysis. Morphological characters such as shape of the trunk scales, scutes and even morphometric characters might be fruitful for species comparisons. For instance, *Acipenser brevirostrum* and *A. fulvescens* are similar to each other in these characters, but not to *A. oxyrinchus*. Similar biogeographically localized groups of species within *Acipenser* are recognizable, but too many species are currently undescribed and all are necessary for a phylogenetic analysis. Recent work on karyological and molecular systematics suggests interesting patterns among the species of *Acipenser* that would be ideal starting points for a complete analysis of potential characters.

At the evolutionary level, many of the hypotheses presented here require additional ecological information and functional investigation. Detailed ontogenetic studies on behavior and morphology, external and internal, will certainly define characters for phylogenetic analysis within *Acipenser*. The cladogram presented here is a first step toward better understanding acipenserids, but future studies on several fronts are necessary to make progress in phylogenetic and evolutionary studies of sturgeons.

Acknowledgements

This project is a portion of a doctoral dissertation completed at the University of Massachusetts. I wish to thank Willy Bemis and Lance Grande for their assistance and dialogue throughout the work. Also, I thank Vadim Birstein, John Waldman, Willy Bemis and Eugene Balon for reviewing this manuscript specifically and for their energies in editing this volume on acipenseriform biology. I am grateful for all the help of the Kahrs family (especially Peter and Steven Kahrs) and Gary Pinson of Osage Catfisheries Inc. who assisted in collection of adult and developmental specimens of *Scaphirhynchus* with facilities, finances, and help that made field work rewarding and beneficial. This project was supported at different times by the Lerner-Grey and Donn E. Rosen Funds of the American Museum of Natural History and a National Science Foundation Grant BSR-9220938. Finally, I thank Carolanne Milligan for reading the manuscript and supporting me during the writing and final preparation of this study.

References cited

Agassiz, L. 1844. Recherches sur les poissons fossiles. 2 (2, p. 280). Petitpierre Imprimière, Neuchâtel.

Antipa, G. 1933. Les sturions de la Mer Noire leau biologie et les mesures nécessaires pour leur protection. Bull. Sec. Sci. Acad. Roum. 16: 1–17.

Antoniu-Murgoci, A. 1936a. Particularités anatomiques caractérisant les espèces des genres *Huso* et *Acipenser* des eaux roumaines. Compt. Rend. Séanc, Acad. Sci. Roum. 1: 1–3.

Antoniu-Murgoci, A. 1936b. Particularités anatomiques qui différencient le genre *Huso* du genre *Acipenser* des eaux roumaines. Ann. Sci. Univ. Jassy 23: 94–103.

Antoniu-Murgoci, A. 1942. Contributions à l'étude des Acipenserides de Roumanie. Ann. Sci. Univ. Jassy, 2, Sci. Natur. 28: 289–385.

Bai, Y. 1983. A new †*Peipiaosteus* from Fengning of Hebei, China. Vertebrata Palasiatica 21: 341–346.

Bailey, R.M. & F.B. Cross. 1954. River sturgeons of the american genus *Scaphirhynchus*: characters, distribution and synonymy. Pap. Mich. Acad. Sci. 39: 169–208.

Balon, E.K. 1983. Epigenetic mechanisms: reflections on evolutionary processes. Can. J. Fish. Aquat. Sci. 40: 2045–2058.

Balon, E.K. 1985. Early life histories of fishes: new developmental, ecological and evolutionary perspectives. Dev. Env. Biol. Fish. 5, Dr W. Junk Publishers, Dordrecht. 280 pp.

Balon, E.K. 1989. The epigenetic mechanisms of bifurcation and alternative life-history styles. pp. 467–501. *In:* M.N. Bruton (ed.) Alternative Life-History Styles of Animals, Perspectives in Vertebrate Science 6, Kluwer Academic Publishers, Dordrecht.

Bemis, W.E., E.K. Findeis & L. Grande. 1997. An overview of Acipenseriformes. Env. Biol. Fish. (this volume).

Berg, L.S. 1940. Classification of fishes both recent and fossil. Reprint edition with English translation (1965). Thai National Document Center, pp. 85–517.

Berg, L.S. 1948a. The freshwater fishes of the USSR and adjacent countries, Vol. 1, Part 1. Akademia Nauk USSR, Moscow & Leningrad (in Russian, English translation published by Israel Program for Scientific Translations, Jerusalem. 505 pp.).

Berg, L.S. 1948b. On the position of the Acipenseriformes in the system of fishes. Trudy Zoologicheskogo Instituta 7: 7–57 (in Russian).

Binkowski, F.P. & S.I. Doroshov (ed.) 1985. North american sturgeons: biology and aquaculture potential. Dev. Env. Biol. Fish. 6, Dr W. Junk Publishers, Dordrecht. 163 pp.

Birstein, V.J. 1993. Sturgeons and paddlefishes: threatened fishes in need of conservation. Conservation Biology 7: 773–787.

Bonaparte, C.L. 1838. Selachorum tabula analytica. Nouv. Ann. Sci. Natur. 2: 195–214.

Bonaparte, C.L. 1846. Catalogo metodico dei pesci europi. Atti. Soc. Ital. Sci. Natur. 1–95.

Brandt, J.F. 1869/1870. Einige Worte über die europäischen-asiatischen Störarten (Sturionides). Bull. Acad. Imper. Sci. St.-Petersb. 14: 171–175.

Brandt, J.F. & J.C.T. Ratzeberg. 1833. Medizinische Zoologie, oder getreue Darstellung und Beschreibung der Tiere, die in der Arzneimittellehre in Betracht kommen, in systematischer Folge herausgegeben, Vol. 2. Berlin. 364 pp.

Burggren, W.W. 1978. Gill ventilation in the sturgeon, *Acipenser transmontanus*: unusual adaptations for bottom dwelling. Respir. Physiol. 34: 153–170.

Dingerkus, G. & L.D. Uhler. 1977. Enzyme clearing of alcian blue stained whole small vertebrates for demonstration of cartilage. Stain Technol. 52: 229–232.

Duméril, A.H.A. 1867. Prodrome d'une monographie des esturgeons et description des espèces de l'Amérique du Nord qui appartiennent au sous-genre *Antaceus*. Nouv. Arch. Mus. Hist. Nat. Paris 3: 131–188.

Duméril, A.H.A. 1870. Histoire naturelle des poissons, ou ichthyologie générale, Vol. 2, Ganoïdes, Dipnés, Lophobranches. Paris. 624 pp.

Egerton P. de M.G. 1858. On †*Chondrosteus*, an extinct genus of the Sturionidae, found in the Lias Formation at Lyme Regis. Phil. Trans. Roy. Soc. London 148: 871–885.

Findeis, E.K. 1989. Morphological specializations of the branchial arches of sturgeons (Acipenseridae). Amer. Zool. 29: 123A.

Findeis, E.K. 1993. Skeletal anatomy of the North American shovelnose sturgeon *Scaphirhynchus platorynchus* (Rafinesque, 1820) with comparisons to other Acipenseriformes. Ph.D. Thesis, University of Massachusetts, Amherst. 444 pp.

Fitzinger, L.J. & J. Heckel. 1836. Monographische Darstellung der Gattung *Acipenser*. Ann. Wiener Mus. Naturgesch. 1: 261–326.

Gardiner B.G. 1984a. The relationships of the palaeoniscid fishes, a review based on new specimens of *Mimia* and *Moythomasia* from the upper Devonian of Western Australia. Bull. British Mus. Natur. Hist., Geol. Ser. 37: 173–248.

Gardiner B.G. 1984b. Sturgeons as living fossils. pp. 148–152. *In:* N. Eldredge & S.M. Stanley (ed.) Living Fossils, Springer Verlag, New York.

Gardiner B.G. & B. Schaeffer. 1989. Interrelationships of lower actinopterygian fishes. Zool. J. Linn. Soc. 97: 135–187.

Gershanovich, A.D. & T.I.J. Smith (ed.) 1995. Proceedings of International Symposium on Sturgeons, 6–11 September 1993, VNIRO Publishing, Moscow. 360 pp.

Goodrich, E.S. 1909. Vertebrata craniata. First fascicle: cyclostomes and fishes. pp. 1–518. *In:* R. Lankester (ed.) A Treatise on Zoology, Part IX, Adam and Charles Black, London.

Grande, L. & W.E. Bemis. 1991. Osteology and phylogenetic relationships of fossil and recent paddlefishes (Polyodontidae) with comments on the interrelationships of Acipenseriformes. J. Vert. Paleo. 11, supplement 1: 1–121.

Grande, L. & W.E. Bemis. 1996. Interrelationships of Acipenseriformes with comments on "Chondrostei". pp. 85–115. *In:* M.L.J. Stiassny, L.R. Parenti & G.D. Johnson (ed.) Interrelationships of Fishes, Academic Press, New York.

Gregory, W.K. 1933. Fish skulls: a study of the evolution of natural mechanisms. Trans. Amer. Philisoph. Soc. 23: 75–481.

Heckel, J.J. 1836. *Scaphirhynchus*, eine neue Fischgattung aus der Ordnung der Chondropterygier mit freien Kiemen. Ann. Wiender Mus. Naturgesch. 1: 68–78.

Hennig, E. 1925. †*Chondrosteus hindenburgi* Pomp., ein 'Stör' des Württemberg. Oelschiefers (Lias). Paleontographica Stuttgart 67: 115–133.

Hennig, W. 1966. Phylogenetic systematics. University of Illinois Press, Urbana. 263 pp.

Ivanzoff, N. 1887. Der *Scaphirhynchus*. Vergleichend-anatomische Beschreibung. Bull. Soc. Imp. Natur. Moscou 1887: 1–41.

Jollie, M. 1980. Development of the head and pectoral girdle skeleton and scales in *Acipenser*. Copeia 1980: 226–249.

Lauder, G.V. 1980. Evolution of the feeding mechanism in primitive actinopterygian fishes: a functional anatomical analysis of *Polypterus, Lepisosteus*, and *Amia*. J. Morph. 13: 283–317.

Lauder, G.V. & K. Liem. 1983. The evolution and interrelationships of the actinopterygian fishes. Bull. Mus. Comp. Zool. 150: 95–197.

Linnaeus, C. 1758. Systema naturae, Editio X [Systema naturae per regna tria naturae, secundum classes, ordines, genera, species, cum characteribus, differentiis, synonymis, locis. Tomus I. Editio decima, reformata.] Holmiae. 824 pp.

Liu, C.H. & Y.J. Zeng. 1988. Notes on the Chinese paddlefish, *Psephurus gladius* (Martens). Copeia 1988: 483–484.

Liu, H.T. & J.J. Zhou. 1965. A new sturgeon from the upper Jurassic of Liaoning, north China. Vertebrata Palasiatica 9: 237–248.

MacAlpin, A. 1947. *Paleopsephurus wilsoni*, a new polyodontid fish from the upper Cretaceous of Montana, with a discussion of allied fish, living and fossil. Contrib. Mus. Paleontol., Univ. Michigan 6: 167–234.

Magnin, E. 1963. Recherches sur la systématique et la biologie des Acipenseridés *Acipenser sturio* L., *Acipenser oxyrhynchus* Mitch. et *Acipenser fulvescens* Raf. Ann. Stat. Cent. Hydrobiol. Appliq. 9: 8–242.

Magnin, E. & G. Beaulieau. 1963. Etude morphométrique comparée de l'*Acipenser oxyrhynchus* Mitchill du Saint-Laurent et l'*Acipenser sturio* Linné de la Gironde. Natur. Can. 40: 4–38.

Marinelli, W. & A. Strenger. 1973. Vergleichende Anatomie und Morphologie der Wirbeltiere, IV Lieferung. Franz Deuticke, Wien. pp. 309–460.

Martens, E. 1862. Über einen neuen *Polyodon* (*P. gladius*) aus dem Yangtsekiang und über die sogenannten Glaspolypen. Monats. Deutsch. Akad. Wissen. Berlin 1861: 476–480.

Nielsen, E. 1949. Studies on Triassic fishes from East Greenland. II. *Australosomus* and *Birgeria*. Meddelelser om Grønland 146:L 1–309.

Nikolskii, A.M. 1900. *Pseudoscaphirhynchus rossikowi*, new genus et new species. Ann. Mus. Zool. Acad. Imp. St.-Petersb. 4: 257–260.

Patterson, C. 1973. Interrelationships of holosteans. pp. 233–305. *In:* P.H. Greenwood, R.S. Miles & C. Patterson (ed.) Interrelationships of Fishes, Academic Press, London.

Patterson, C. 1982. Morphology and interrelationships of primitive actinopterygian fishes. Amer. Zool. 22: 241–259.

Pearson, D.M. & T.S. Westoll. 1979. The Devonian actinopterygian †*Cheirolepis* Agassiz. Trans. Roy. Soc. Edinburgh 70: 337–399.

Pirogovskii, M.I., L.I. Sokolov & V.P. Vasil'ev. 1989. *Huso huso* (Linnaeus, 1758). pp. 156–200. *In:* J. Holčík (ed.) The Freshwater Fishes of Europe, Vol. 1, Pt. II, General Introduction to Fishes, Acipenseriformes, AULA-Verlag, Wiesbaden.

Rafinesque, C.S. 1820. Ichthyologia Ohiensis, or natural history of the fishes inhabiting the river Ohio and its tributary streams, preceded by a physical description of the Ohio and its branches. Lexington. 90 pp.

Richmond, A.M. & B. Kynard. 1995. Ontogenetic behavior of shortnose sturgeon, *Acipenser brevirostrum*. Copeia 1995: 172–182.

Rieppel, O. 1992. A new species of the genus *Saurichthys* (Pisces: Actrinopterygii) from the middle Triassic of Monte San Giorgio (Switzerland), with comments on the phylogenetic interrelationships of the genus. Paleontographica Abt. A 221: 63–94.

Schmalhausen, O.I. 1991. Development of the large Amu-Dar shovelnose larvae. Ontogenez 22: 493–513 (in Russian, English translation Soviet J. Devel. Biol. 22: 302–315).

Sewertzoff, A.N. 1926a. Studies on the bony skull of fishes. I. Structure and development of the bony skull of *Acipenser ruthenus*. Quart. J. Microscop. Sci. 70: 451–540.

Sewertzoff, A.N. 1926b. The development of the scales of *Acipenser ruthenus*. J. Morph. 42: 523–560.

Sewertzoff, A.N. 1928. The head skeleton and muscles of *Acipenser ruthenus*. Acta Zool. 9: 193–319.

Sokolov, L.I. 1989. *Acipenser* Linnaeus, 1758. pp. 201–205. *In:* J. Holčík (ed.) The Freshwater Fishes of Europe, Volume 1, Part II, General Introduction to Fishes, Acipenseriformes, AULA-Verlag, Wiesbaden.

Sokolov, L.I. & V.P. Vasilev. 1989. *Acipenser baeri* Brandt, 1869. pp. 263–284. *In:* J. Holčík (ed.) The Freshwater Fishes of Europe, Volume 1, Part II, General Introduction to Fishes, Acipenseriformes, AULA-Verlag, Wiesbaden.

Tatarko, K. 1936. Der Kiemendeckelapparat und seine Verbindung mit dem Hyoid- und Kieferbogen bei den Acipenseridae. Trav. Inst. Biol. Acad. Sci. Ukraine 3: 5–67.

Traquair, R.H. 1887. Notes on *Chondrosteus acipenseroides*. Geol. Mag. 4: 248–257.

Vladykov, V.D. & J.R. Greeley. 1963. Order Acipenseroidei. pp. 24–60. *In:* H.B. Biglow, C.M. Breder, D.M. Cohen, G.W. Mead, D. Merriman, Y.H. Olsen, W.C. Schroeder, L.P. Schultz & J. Tee-Van (ed.) Fishes of the Western North Atlantic, Mem. Sears Found. Mar. Res. 1.

Watson, D.M.S. 1925. The structure of certain paleoniscids and the relationships of that group with other bony fishes. Proc. Zool. Soc. London 1925: 815–870.

Watson, D.M.S. 1928. On some points in the structure of paleoniscids and allied fish. Proc. Zool. Soc. London 1928: 49–70.

Weisel, G.F. 1978. The integument and caudal filament of the shovel nose sturgeon, *Scaphirhynchus platorynchus*. Amer. Mid. Natur. 100: 179–189.

Wiley, E.O. 1981. Phylogenetics: the theory and practice of phylogenetic systematics. John Wiley & Sons, New York. 439 pp.

Wilimovsky, N.J. 1956. †*Protoscaphirhynchus squamosus*, a new sturgeon from the upper Cretaceous of Montana. J. Paleo. 30: 1205–1208.

Williams, J.D. & G.H. Clemmer. 1991. *Scaphirhynchus suttkusi*, a new sturgeon (Pisces: Acipenseridae) from the Mobile Basin of Alabama and Mississippi. Bull. Alabama Mus. Nat. Hist. 10: 17–31.

Williot, P. (ed.) 1991. Acipenser. CEMAGREF Publ., Bordeaux. 519 pp.

Woodward, A.S. 1889. On the paleontology of sturgeons. Proc. Geol. Assoc. 11: 24–44.

Yakovlev, V.N. 1977. Phylogenesis of Acipenseriformes. pp. 116–143. *In:* V.V. Menner (ed.) Essays on Phylogeny and Systematics of Fossil Fishes and Agnathans, USSR Academy of Sciences, Moscow (in Russian).

Zhou, Z. 1992. Review on †*Peipiaosteus* based on new materials of †*P. pani*. Vertebrata Palasiatica 30: 85–101.

Environmental Biology of Fishes **48**: 127–155, 1997.

Phylogeny of the Acipenseriformes: cytogenetic and molecular approaches

Vadim J. Birstein[1], Robert Hanner[2] & Rob DeSalle[3]
[1] *The Sturgeon Society, 331 West 57th Street, Suite 159, New York, NY 10019, U.S.A.*
[2] *Department of Biology, University of Oregon, Eugene, OR 97405–1210, U.S.A.*
[3] *Department of Entomology, American Museum of Natural History, New York, NY 10024, U.S.A.*

Received 20.3.1995 Accepted 21.3.1996

Key words: sturgeon, paddlefish, *Huso, Acipenser, Scaphirhynchus, Pseudoscaphirhynchus, Polyodon, Psephurus*, karyotype, chromosome, macrochromosome, microchromosome, genome, DNA content, 18S rRNA gene, cytochrome *b*, 12S mtrRNA gene, 16s mtrRNA gene, rate of molecular evolution, phylogeny, evolution

Synopsis

The review of the data on karyology and DNA content in Acipenseriformes shows that both extant families, the Polyodontidae and Acipenseridae, originated from a tetraploid ancestor which probably had a karyotype consisting of 120 macro- and microchromosomes and DNA content of about 3.2–3.8 pg per nucleus. The tetraploidization of the presumed 60-chromosome ancestor seems to have occurred at an early time of evolution of the group. The divergence of the Acipenseridae into Scaphirhyninae and Acipenserinae occurred without polyploidization. Within the genus *Acipenser*, polyploidization was one of the main genetic mechanisms of speciation by which 8n and 16n-ploid species were formed. Individual gene trees constructed for sequenced partial fragments of the 18S rRNA (230 base pairs, bp), 12S rRNA (185 bp), 16S rRNA (316 bp), and cytochrome *b* (270 bp) genes of two Eurasian (*A. baerii* and *A. ruthenus*) and two American (*A. transmontanus* and *A. medirostris*) species of *Acipenser*, *Huso dauricus*, *Pseudoscaphirhynchus kaufmanni*, *Scaphirhynchus albus*, and *Polyodon spathula* showed a low level of resolution; the analysis of a combined set of data for the four genes, however, gave better resolution. Our phylogeny based on molecular analysis had two major departures from existing morphological hypotheses: *Huso dauricus* is a sister-species to *Acipenser* instead of being basal to all acipenseriforms, and *Scaphirhynchus* and *Pseudoscaphirhynchus* do not form a monophyletic group. The phylogenetic tree constructed for the cytochrome *b* gene fragments (with inclusion of 7 additional species of *Acipenser*) supported the conclusion that octoploid species appeared at least three times within *Acipenser*.

Introduction

Although a few species of *Acipenser* require revision, usually 24–25 extant sturgeon and two paddlefish species, *Polyodon spathula* and *Psephurus gladius*, are included in the Acipenseriformes (Rochard et al. 1991, Birstein 1993a). The extant members of this order form the monophyletic sister-group of all extant Neopterygii (e.g., Lepisosteidae, Amiidae, and Teleostei; Bemis et al. 1997 this volume). Most ichthyologists regard Polypteridae as the sister group of Acipenseriformes + Neopterygii (Patterson 1982). A comparison of partial sequences of 28S rRNAs supported this relationship (Le et al. 1993). In contrast to earlier works, Grande & Bemis (1991) concluded that paddlefishes and sturgeons are sister taxa, and that extinct Mesozoic genera such as *Chondrosteus* lie outside this clade.

Within Acipenseriformes, all workers agree that the Acipenseridae and Polyodontidae diverged prior the Late Cretaceous (Berg 1948a, Yakovlev 1977, 1986, Grande & Bemis 1991, Jin 1995, Grande & Bemis 1996, Bemis et al. 1997). Within the Acipenseridae, the subfamily Scaphirhynchinae (Central Asian species of *Pseudoscaphirhynchus* plus North American species of *Scaphirhynchus*) was traditionally considered the sister group of all other sturgeons (Berg 1905, Vasiliev 1985). Another interpretation, based on osteology, was proposed by Findeis (1993, 1997 this volume), who found that *Huso* lacks many characters found in *Acipenser*, *Scaphirhynchus* and *Pseudoscaphirhynchus*. On this basis, he considered Scaphirhynchinae as a derived group within Acipenseridae (Findeis 1997). This point of view was disputed since the description of the *Pseudoscaphirynchus* species 120 years ago (Kessler 1877, Berg 1905), while other researchers considered Scaphirhyninae as the oldest group within Acipenseridae (Bogdanov 1887). The examination of generic relationships within Acipenseriformes using molecular phylogenetic methods was one of the main goals of this paper.

According to Nesov & Kaznyshkin (1977), extant species of *Acipenser* belong to different evolutionary lineages which diverged long time ago. Artyukhin & Andronov (1990) and Artyukhin (1995) proposed that there were at least four main regions in which the speciation and spread of sturgeons took place: (1) the Ponto-Caspian area; (2) China-western America; (3) the Atlantic area; and (4) eastern North America. The group of Ponto-Caspian species includes most of the Eurasian species (*Acipenser gueldenstaedtii*, *A. persicus*, *A. stellatus*, *A. ruthenus*, *A. nudiventris*, *Huso huso*, *A. baerii*; Berg 1948b, Holčík et al. 1989, Pirogovsky et al. 1989, Shubina et al. 1989, Vlasenko et al. 1989a, b, Sokolov & Vasilev 1989a–c, Ruban 1997 this volume), as well as Amur River endemics (*A. schrenckii* and *Huso dauricus*; Artyukhin 1994, Krykhtin & Svirskii 1997 this volume), and, possibly, an Adriatic species *A. naccarii* (Tortonese 1989, Rossi et al. 1991). The origin of Ponto-Caspian species might have been associated with brackish-water derivatives of the Tethys Sea. The oldest extinct forms of the Ponto-Caspian group from the Upper Cretaceous occur in Central Asia (Nesov & Kaznyshkin 1983).

The China-western American group consists of *Acipenser sinensis*, *A. dabryanus*, *A. medirostris*, and *A. transmontanus*; the external morphology of the last two species is very similar (Findeis 1993). All of these species seem to have common Tertiary roots (Artyukhin & Andronov 1990). The two extant paddlefish species also indicate a trans-Pacific pattern that is strengthened by the inclusion of fossil taxa, such as Eocene *Crossopholis* and Paleocene *Paleopsephurus* (Grande & Bemis 1991, Jin 1995, Bemis et al. 1996, Grande & Bemis 1996).

The third group includes European and American Atlantic sturgeons (Vladykov & Greeley 1963, Kinzelbach 1987, Holčík et al. 1989). Probably, the European *A. sturio* has many primitive characters of the genus (Nesov & Kaznyshkin 1977). Once considered a subspecies of *A. sturio*, the American Atlantic sturgeons were subsequently split off as a separate species, *A. oxyrinchus* (Vladykov & Greeley 1963). Then, two subspecies, *A. o. oxyrinchus* and *A. o. desotoi*, were described within *A. oxyrinchus* (Vladykov 1955, Vladykov & Greeley 1963). The origin and spread of the Atlantic sturgeon probably reflect close Tertiary links between Europe and North America (Artyukhin & Andronov 1990). The freshwater sturgeons of eastern North America, belonging to the fourth group, the lake sturgeon, *A. fulvescens*, and shortnose sturgeon, *A. brevirostrum*, are possibly closely related (Vladykov & Greeley 1963) and may have originated on the Eastern coast of North America (Artyukhin & Andronov 1990).

Evidence for monophyly of these four groups remains uncertain, and the relationships among them are unknown. Recently, Artyukhin (1995) published a first phylogenetic tree based on general data of morphology, biogeography and karyology (but not the DNA content) of *Acipenser*. Atryukhin's scheme is the first modern attempt to reconstruct relationships within the genus *Acipenser* (see Bemis et al. 1997, for a history of the 19th century attempts to subdivide *Acipenser* into subgenera).

Artyukhin & Andropov (1990) wrote: 'It is quite evident that methods of biochemical genetics and karyology, as well as paleontological data and cur-

Table 1. Chromosome numbers and DNA content in the Acipenseriformes, Lepisosteiformes, and Amiiformes[1].

Species	Chromosome numbers	DNA content in pg	Ploidy, n (according to the DNA content)	Reference
Order Acipenseriformes				
family **Polyodontidae**				
North America				
Polyodon spathula				
Tennesse River, Alabama	120	–	–	Dingerkus & Howell (1976)
Kentucky	–	3.9[2]	4	Tiersch et al. (1989)
Missouri	–	4.89[2]	4	Blacklidge & Bidwell (1993)
Moscow Aquarium, Russia	–	3.17[2]	4	Birstein et al. (1993)
family **Acipenseridae**				
subfamily **Acipenserinae**				
Europe				
Acipenser gueldenstaedtii				
Volga River	250 ± 8	–	–	Birstein & Vasiliev (1987)
		7.87[2]	8	Birstein et al. (1993)
Caspian Sea	247 ± 8	–	–	Vasiliev (1985)
Sea of Azov	250 ± 8	–	–	Vasiliev (1985), Arefjev (1989a)
Italy, cell culture	256 ± 8	–	–	Fontana et al. (1995)
A. naccarii				
Italy	239 ± 7	–	–	Fontana & Colombo (1974)
	246 ± 8	–	–	Fontana (1994)
Italy, cell culture	246 ± 8	–	–	Fontana et al. (1995)
A. nudiventris				
Black Sea	118 ± 2	–	–	Arefjev (1983), Vasiliev (1985)
Aral Sea, Uzbekistan (Central Asia)	–	3.90[2]	4	Birstein et al. (1993)
A. persicus				
Caspian Sea	> 200	–	–	Fashkhami (pers. comm.)
A. ruthenus				
Volga River	118 ± 2	–	–	Vasiliev (1985), Birstein & Vasiliev (1987)
	–	3.74[2]	4	Birstein et al. (1993)
Don River	118 ± 3	–	–	Arefjev (1989b)
Danube, Yugoslavia	116 ± 4	–	–	Fontana et al. (1977)
Danube, Slovakia	118 ± 3	–	–	Rab (1986)
Italy (aquaculture)	118 ± 4	–	–	Fontana (1994)
Italy, cell culture	118 ± 9	–	–	Fontana et al. (1995)
A. stellatus				
Volga River	118 ± 2	–	–	Birstein & Vasiliev (1987)
	–	3.74[2]	4	Birstein et al. (1993)
A. sturio				
Italy	116 ± 4	–	–	Fontana & Colombo (1974)
	–	3.6[4]	4	Fontana (1976)
	–	3.2[6]	4	Mirsky & Ris (1951)
Huso huso				
Don River	118 ± 2	–	–	Serebryakova et al. (1983), Arefjev (1989b)
Volga River	118 ± 2	–	–	Birstein & Vasiliev (1987)
	–	3.17[2]	4	Birstein et al. (1993)
Italy	116 ± 4	–	–	Fontana & Colombo (1974)
	–	3.6[4]	4	Fontana (1976)
Asia				
Siberia				
Acipenser baerii				
Lena River	248 ± 5	–	–	Vasiliev et al. (1980)
Italy (aquaculture)	246 ± 8			Fontana (1994)

Table 1. (Continued).

Species	Chromosome numbers	DNA content in pg	Ploidy, n (according to the DNA content)	Reference
Far East and China				
A. mikadoi (*A. medirostris mikadoi*)				
Tumnin (Datta) River	[500?][3]	14.20[2]	16	Birstein et al. (1993)
A. schrenckii				
Amur River	[240?][5]	–	–	Sereberyakova (1970)
A. sinensis				
Yangtze River	264 ±	–	–	Yu et al. (1987)
Huso dauricus				
Amur River	[120?][5]	–	–	Serebryakova (1969)
	–	3.78[2]	4	Birstein et al. (1993)
North America				
A. brevirostrum				
Florida and South Carolina, USA	[360? or 500?][3]	13.08[2]	12 (?) or 16	Blacklidge & Bidwell (1993)
A. fulvescens				
Wisconsin	[250?][3]	8.90[2]	8	Blacklidge & Bidwell (1993)
A. medirostris				
Washington	[250?][3]	8.82[2]	8	Blacklidge & Bidwell (1993)
A. oxyrinchus desotoi				
Florida	[120?][3]	4.55[2]	4	Blacklidge & Bidwell (1993)
A.o. oxyrinchus				
Halifax, cell culture	(99–112)[7]	–	–	Li et al. (1985)
A. transmontanus				
San Francisco Bay, California, cell culture	(230 ±)[8]	–	–	Hedrick et al. (1991)
	–	9.55[2]	8	Blacklidge & Bidwell (1993)
Snake River, Idaho	–	9.12[2]	8	Blacklidge & Bidwell (1993)
Columbia River, Washington	–	9.59[2]	8	Blacklidge & Bidwell (1993)
Italy (aquaculture)	248 ± 8	–	–	Fontana (1994)
subfamily **Scaphirhynchinae**				
Central Asia				
Pseudoscaphirhynchus kaufmanni				
Amu Darya River, Uzbekistan	[120?][3]	3.47[2]	4	Birstein et al. (1993)
North America				
Scaphirhynchus platorynchus				
Illinois	112 ±	3.6[2]	4	Ohno et al. (1969)
Order Lepisosteiformes				
family **Lepisosteidae**				
North America				
Lepisosteus oculatus	68 ±	2.8[4]	2	Ohno et al. (1969)
L. osseus	56	–	–	Ojima & Yamano (1980)
L. platostomus	54	–	–	Ueno (1985), cit. in Suzuki & Hirata (1991)
Order Amiiformes				
family **Amiidae**				
North America				
Amia calva	46 ±	2.4[4]	2	Ohno et al. (1969)
		2.3[6]	2	Mirsky & Ris (1951)
	46	2.0[4]	2	Suzuki & Hirata (1991)

[1] All species investigated karyologically so far are listed; the DNA content values for all species studied are given.
[2] Determined by flow cytometry.
[3] Chromosome number is assumed on the basis of the DNA content.
[4] Determined by microdensitometry of Feulgen-stained nuclei.
[5] Only macrochromosomes were counted precisely.
[6] Determined by the biochemical Schmidt-Thankauser method.
[7] Chromosome number was determined in cardiac tissue cell culture.
[8] Chromosome number was determined in cell cultures: the modal 2n in a spleen cell line was 219, and in a heart cell line, 237–243 (Hedrick et al. 1991).

Table 2. Characteristics of karyotypes of several acipenseriform species[1].

Species	Chromosome number	Number of large metacentrics plus medium meta/submetacentrics, M + m/sm	Number of large telocentrics	Number of microchromosomes	Approximate arm number, NF	Reference
1. Tetraploid species						
Family **Polyodontidae**						
Polyodon spathula	120	8 + 36	4	72	164	Dingerkus & Howell (1976)
Family **Acipenseridae**						
Acipenser nudiventris						
Black Sea, Russia	118 ± 3	8 + 46	4	60 ± 3	172	Arefjev (1983)
A. ruthenus						
Sea of Azov, Russia	118 ± 3	8 + 50	4	56 ± 3	176	Arefjev (1989a)
Volga River, Russia	118 ± 2	8 + 50	4	56 ±23	176	Birstein & Vasiliev (1987)
Danube, Slovakia	118 ± 4	8 + 50	4	56 ± 4	176	Rab (1986)
Danube, Yugoslavia	116 ± 2	8 + 48	4	56 + 4	170	Fontana et al. (1977)
A. stellatus						
Volga River, Russia	118 ± 2	8 + 48	4	58 ± 2	174	Birstein & Vasiliev (1987)
A. sturio						
Italy	116 ± 4	8 + 48	4	56 ± 4	172	Fontana & Colombo (1974)
Huso huso						
Volga River, Russia	118 ± 2	8 + 54	4	52 ± 2	180	Birstein & Vasiliev (1987), Arefiev & Nikolaev (1991)
Sea of Azov, Russia	118 ± 3	8 + 54	4	52 ± 3	180	Serebryakova et al. (1983), Arefiev (1989b)
Italy	116 ± 4	8 + 52	4	52 ± 4	176	Fontana & Colombo (1974)
Scaphirhynchus platorynchus						
USA	112 ± 2	8 + 52	4	48 ±	172	Ohno et al. (1969)
2. Octoploid species						
Family **Acipenseridae**						
Acipenser baerii						
Lena River, Siberia, Russia	248 ± 5	16 + 42		190	308	Vasiliev et al. (1980)
A. gueldenstaedtii						
Volga River, Russia	250 ± 8	16 + 76		155	339	Birstein & Vasiliev (1987)
Sea of Arzov, Russia	250 ± 8	16 + 82		152	348	Arefjev (1989a)
A. naccarii						
Italy (wild)	239 ± 7	16 + 76		147	331	Fontana & Colombo (1974)
Italy, aquaculture	241 ± 3	16 + 74		151	331	Arlati et al. (1995)
A. sinensis						
China	264 ± 3	16 (?) + 82		166	362	Yu et al. (1987)

[1] A revision of data from papers mentioned as References. The numbers of m/sm, microchromosomes, and NF are given approximately, since it is impossible to discriminate the form and exact number of small chromosomes and microchromosomes. For octoploid species, the number of telocentrics and microchromosomes is given together because there is no clear size difference between these two kinds of chromosomes. Usually there are 2–5 middle and/or small-sized telocentric pairs in these karyotypes.

rent views on paleogeography will provide useful tools for resolving complex relationships and phylogeny of sturgeons'. In the first part of this paper we review all cytogenetic data available on Acipenseriformes and make some new conclusions relevant to the four groups within *Acipenser* mentioned above. In the second part we describe experimental data on the molecular phylogeny of Acipenseriformes. Because multiple gene regions have been useful in other groups of fishes (reviews in Normark et al. 1991, Stock et al. 1991a, Meyer 1993, Patterson et al. 1993), we believed that they might also provide reasonable character state information for acipenseriforms. Consequently we amplified and sequenced partial fragments of 18S rRNA, 12S rRNA, 16S rRNA, and cytochrome *b* genes of a few representatives of all lineages of this order. We included the phylogenetic analysis of the combined molecular and morphological data for all the species we studied. Also, we examined relationships among representatives of four species groups of the genus *Acipenser* recognized by Artyukhin (1995), using data for a partial sequence of the cytochrome *b* gene.

Acipenseriform cytogenetics: an overview

Main karyotypic characteristics, DNA content, and polyploidy

Karyotypes of about half of all sturgeon species have been described and the DNA content in most sturgeon species and American paddlefish has been measured (Table 1). Acipenseriform karyotypes investigated so far have two particular characteristics: (1) they are large; and (2) they consist of macro- and microchromosomes. According to the number of chromosomes (2n), the species can be divided into two groups: those with about 120 chromosomes (e.g., *Huso huso*, *H. dauricus*, *Acipenser ruthenus*, *A. stellatus*, *A. nudiventris*, *A. sturio*, and *Polyodon spathula*), and those with 240 chromosomes (e.g., *A. gueldenstaedtii*, *A. naccarii*, *A. baerii*, *A. schrenckii*, and *A. transmontanus*). By comparison to the 120-chromosome species, the 240-chromosome species are tetraploids.

The size of macrochromosomes in both groups is between 2–5 μm, and the majority of macrohromosomes are the meta- and submetacentrics (Table 2). One third to one half of the chromosome number in both groups is comprised of microchromosomes of a very small size (about 1 μm). Karyotypes of the 120-chromosome species typically consist of 4 pairs of large metacentrics (no. 1–4), 5 pairs of large but somewhat smaller metacentrics (no. 5–9), about 20 pairs of medium-sized metacentrics and/or submetacentrics of gradually decreasing size (no. 10–30), one pair of comparatively large telocentrics (no. 30), one pair of small telocentrics, and approximately 56 ± 4 microchromosomes of different form (Table 2). The difference between karyotypes of representatives of two lineages of the extant acipenseriforms, the Polyodontidae (*Polyodon spathula*) and Acipenseridae (all other 120-chromosome species in Table 1), as well as of the lineages within the Acipenseridae (*Huso huso*, the 120-chromosome species of the genus *Acipenser*, and *Scaphirhynchus platorynchus*), seems to be small. Evidently, the ancestral acipenseriform karyotype was preserved in these fishes without dramatic changes during diverisification of the group.

In general, few karyotypic changes are noticeable among the species of *Acipenser* with 120-chromosomes (Table 2). The karyotype of *Huso huso* is more symmetric than those of the 120-chromosome species of *Acipenser*, i.e., it contains more biarmed chromosomes and fewer microchromosomes (see Morescalchi 1973). Also, there is a small difference among the 120-chromosome species in the size of a pair of large telocentrics (no. 30): it is small in *A. sturio* (Fontana & Colombo 1974) and *A. nudiventris* (Arefjev 1983) and it is as large as pair no. 8 or 9 in *Huso huso* (Fontana & Colombo 1974, Birstein & Vasiliev 1987, Arefjev 1989b) and *A. ruthenus* (Rab 1986, Birstein & Vasiliev 1987).

The similarity of karyotypes of the 120-chromosome acipenseriforms points to a generally slow rate of karyological evolution. This correlates with a slow rate of nuclear DNA evolution: practically all genome fractions (both the repeated and unique sequences) are homologous in *Acipenser ruthenus*, *A. stellatus*, *A. gueldenstaedtii*, and *H. huso*, and the number of nucleotide substitutions in the first fraction is 0–2.65, and 1.5–2.7% in the second (Kedrova et al. 1980). These data were supported by our results from sequencing the 18S genes, which are almost invariable among acipenseriforms (see below).

A low degree of protein evolution is usually characteristic of the acipenseriforms, especially the 120-chromosome species. A mean heterozygosity for three freshwater species, *Polyodon spathula*, *Scaphirhynchus platorynchus* and *S. albus*, is between 0.010 and 0.017 (Carlson et al. 1982, Phelps & Allendorf 1983) and is the highest (of all species investigated) in the anadromous *Acipenser stellatus*, 0.093 (Ryabova & Kutergina 1990). All of these species are 120-chromosome forms. The mean heterozygosity for other osteichthyans is 0.051 (Ward et al. 1992). The complete lack of genetic divergence at the protein level between the two species of *Scaphirhynchus* is very unusual for fishes, especially because freshwater fishes typically exhibit subpopulational differentiation significantly higher than that of anadromous and especially marine species (Ward et al. 1994).

The DNA content (2C) of all of the 120-chromosome species is about 3.7–3.9 pg (in *Polyodon spathula* it is a little lower, 3.2 pg), whereas in the 240-chromosome species it is twice as high, 7.9–8.3 pg (Birstein et al. 1993, Table 1). Moreover, the same tendency in DNA content is present in American sturgeons whose karyotypes have not been investigated: *A. oxyrinchus desotoi* (2C = 4.6 pg) is evidently a 120-chromosome subspecies, whereas *A. fulvescens* and *A. medirostris* (American form) are 240-chromosome species, 2C = 8.8–8.9 pg (Blacklidge & Bidwell 1993; the slightly higher DNA content for all American species as compared with the Eurasian species in Table 1 is due to the different methods used by Birstein et al. 1993, and Blacklidge & Bidwell 1993). DNA content in *Pseudoscaphirhynchus kaufmanni* (3.2 pg) is slightly lower than in the sterlet, *A. ruthenus* (3.7 pg), which was used by Birstein et al. (1993) as a standard species for comparative measurments due to its invariable DNA content. However, this difference is so small that we predict *P. kaufmanni* is a 120-chromosome species.

The sterlet, *Acipenser ruthenus*, has other properties which attest to its genetic stability. The basic chromosome number in three generations of the 'bester', the fertile hybrid between beluga, *Huso huso*, and sterlet, *A. ruthenus*, does not differ from that of parental species, and a gradual displacement of karyotypic parameters (numbers of bi- and uniarmed chromosomes) towards those of the sterlet occurs (Arefjev 1989b). The sterlet has evidently not only invariable DNA content, but also a dominant karyotype. The situation is very unusual, because, as a rule, the karyotypes of fish interspecies hybrids are more variable than karyotypes of the parental species (Arefjev 1991, Arefjev & Filippova 1993).

The Asian green (Sakhalin) sturgeon, *A. mikadoi* (or *A. medirostris mikadoi* as explained below), and the American shortnose sturgeon, *A. brevirostrum*, have even higher DNA contents than the 240-chromosome forms. The DNA content of *A. mikadoi* is 14.2 pg per nucleus, four times higher than in the 120-chromosome species or roughly twice that of the 240-chromosome forms (Birstein et al. 1993). In American green sturgeon, *A. medirostris*, it is about half of this value (8.8 pg, Blacklidge & Bidwell 1993). Therefore, the American green sturgeon seems to be a typical 240-chromosome form, while the Sakhalin sturgeon might be predicted to have twice the number of chromosomes, or around 500. If so, the Sakhalin sturgeon would have the highest diploid number reported in vertebrates. But polyploidization can occur without an increase in chromosomes number, as in several species of sharks (see below), and only direct karyotypic study can address the chromosome number and morphology of the Sakhalin sturgeon. DNA content data support the old point of view that the Asian form of *A. medirostris* is a separate species, *A. mikadoi* (Hilgendorf, 1892), or a subspecies, *A. medirostris mikadoi* (Shmidt 1950, Lindberg & Legeza 1965; in Table 1 it is mentioned as a species, see Birstein 1993b). Although Blacklige & Bidwell (1993) consider *A. brevirostrum* to be an allopolyploid (12n = 360), a descendant of ancestral spontaneous triploids, allopolyploidy is unknown in other acipenserids, and it is more logical to propose that this is a 16n-ploid.

An investigation of active nucleoli gave additional information about ploidy in the acipenseriforms. There are different modal numbers of nucleoli per nucleus in species investigated (Table 3): 2–4 (maximum 6) in the 120-chromosome species, and 6–8 (maximum 11–12) in the 240-chromosome species of *Acipenser* (Birstein & Vasiliev 1987, Fontana 1994). In *A. ruthenus*, 4n, the NORs are located in two pairs of small chromosomes, a pair of metacentrics and a pair of telocentrics (possibly no. 30; Rab 1986, Birstein & Vasiliev 1987, Fontana 1994). According to Arefjev (1993), in octoploid *A. baerii* NORs are located also on two pairs of chromosomes, while Fontana (1994) found two quadruplets bearing NORs in *A. baerii*, *A. naccarii*, and *A. transmontanus*. The average number of nucleoli in *Polyodon spathula* is 4 (Dingerkus & Howell 1976), as many as in the 120-chromosome species of *Acipenser* (Table 3). Usually the number of nucleoli in diploid teleosts is half that of the 120-chromosome acipenserforms and equals 1–2 nucleoli per nucleus (review in Birstein 1987).

The tendency seen in 120-chromosome sturgeons to have more nucleoli than in diploid teleosts is apparently caused by their high ploidy. Ohno et al. (1969) arranged the first 64 chromosomes of *Scaphirhynchus platorynchus*, 4n = 116± into 16 groups of four homologs, and Dingerkus & Howell (1976) divided the karyotype of 120-chromosome *Polyodon spathula* into 30 groups consisting of four chromosomes of similar morphology. If it is taken into consideration that this species has 4 active nucleoli per nucleus (see above), it is clear that *P. spathula* is a tetraploid. The same procedure of chromosome grouping (but with greater uncertainity) can be done for *A. ruthenus* and *H. huso* (Birstein & Vasiliev 1987). From this comparison, one can conclude that the 120-chromosome species are in reality tetraploids, while the 240-chromosome species are really octoploids, and the ploidy of *A. mikadoi* and *A. brevirostrum* possibly is 16n.

Further evidence that the 120-chromosome species are tetraploids comes from the existence of duplicated loci, a common characteristic of polyploids. A high level of duplicated loci (31%) was found in *A. stellatus* (Nikanorov et al. 1985, Ryabova & Kutergina 1990); duplicated loci were also found in *Huso huso* (Slynko 1976). In *A. stellatus*, duplicated loci Ldh3 and Ldh4 are located at two different chromosomes (Kutergina & Ryabova 1990). In *Polyodon spathula* the expression of dupli-

Table 3. Number and location of nucleolar organizer regions (NORs) in Acipenseriformes (data on Ag-staining).

Species	Chromosome number	Number of NORs per nuclei[1]	Number of NORs-bearing chromosomes and NORs' location[2]	Reference
Family **Polyodontidae**				
Polyodon spathula	120	4		Dingerkus & Howell (1976)
Family **Acipenseridae**				
1. Tetraploid species				
Acipenser ruthenus	118	2–3 (1–6)	Two pairs (T & m), telomeric	Birstein & Vasiliev (1987)
	118	–	Two pairs, telomeric	Fontana (1994)
A. stellatus	118	2–3 (1–6)	Two pairs (both (?) m), telomeric	Birstein & Vasiliev (1987)
Huso huso	118	2–3 (1–6)	Two pairs (both (?) m), telomeric	Birstein & Vasiliev (1987)
2. Octoploid species				
Acipenser baerii	250	4 (2–6)	Two pairs (T & m)	Arefjev (1993)
	250	–	Two quadruplets	Fontana (1994)
A. gueldenstaedtii	250	6–8 (2–12)	–	Birstein & Vasiliev (1987)
A. naccarii	246	–	Two quadruplets	Fontana (1994)
A. transmontanus	248	–	Two quadruplets	Fontana (1994)

[1] The modal number; a variation in the number is given in the parenthesis.

[2] T means medium-sized telocentric, and m, microchromosome; telomeric means telomeric location of NORs.

Table 4. Natural hybridization of sturgeon species and their ploidy.

Interspecies hybrids	Intergenera hybrids
1. Caspian Sea basin[1]	
(a) Volga River	
A. ruthenus (4n) × *A. stellatus* (4n)	*H. huso* (4n) × *A. gueldenstaedtii* (8n)
A. stellatus (4n) × *A. ruthenus* (4n)	*H. huso* (4n) × *A. ruthenus* (4n)
A. nudiventris (4n) × *A. gueldenstaedtii* (8n)	
A. gueldenstaedtii (8n) × *A. ruthenus* (4n)[2]	
A. gueldenstaedtii (8n) × *A. stellatus* (4n)	
A. gueldenstaedtii (8n) × *A. persicus* (8n)[3]	
(b) Kama River	*Huso huso* (4n) × *A. nudiventris* (4n)
	H. huso (4n) × *A. gueldenstaedtii* (8n)
	H. huso (4n) × *A. stellatus* (4n)
	A. ruthenus (4n) × *H. huso* (4n)
(c) Ural River	
A. nudiventris (4n) × *A. stellatus* (4n)	*H. huso* (4n) × *A. stellatus* (4n)
A. stellatus (4n) × *A. nudiventris* (4n)	
(d) Kura River	
A. nudiventris (4n) × *A. stellatus* (4n)	*H. huso* (4n) × *A. nudiventris* (4n)
A. stellatus (4n) × *A. nudiventris* (4n)	
A. nudiventris (4n) × *A. gueldenstaedtii* (8n)	
(e) Sefir-Rud River	
A. nudiventris (4n) × *A. stellatus* (4n)	
A. nudiventris (4n) × *A. gueldenstaedtii* (8n)	
(f) Caspian Sea	*H. huso* (4n) × *A. persicus* (8n)
2. Sea of Azov basin[4]	
Don River	
A. ruthenus (4n) × *A. stellatus* (4n)	
3. Black Sea basin[5]	
(a) Danube	
A. ruthenus (4n) × *A. stellatus* (4n)	*A. gueldenstaedtii* (8n) *H. huso* (4n)
A. ruthenus (4n) × *A. nudiventris* (4n)	*A. stellatus* (4n) × *H. huso* (4n)
A. stellatus (4n) × *A. ruthenus* (4n)	*A. nudiventris* (4n) × *H. huso* (4n)
A. ruthenus (4n) × *A. gueldenstaedtii* (8n)	*H. huso* (4n) × *A. stellatus* (4n)
A. stellatus (4n) × *A. gueldenstaedtii* (8n)	
A. nudiventris (4n) × *A. gueldenstaedtii* (8n)	
A. sturio (4n) × *A. gueldenstaedtii* (8n)	
(b) Black Sea	
A. gueldenstaedtii (8n) × *A. sturio* (4n)	
A. gueldenstaedtii (8n) × *A. nudiventris* (4n)	
4. Siberian rivers	
Main rivers (Yenisey, Lena, Ob, Kolyma)[6]	Amur River[7]
A. baerii (8n) × *A. ruthenus* (4n)	*Huso dauricus* (4n) × *A. schrenckii* (8n?)
5. Central Asia[8]	
Amu-Darya River	
Pseudoscaphirhyncus kaufmanni (4n) × *P. hermanni*	
6. North America[9]	
Missouri and Mississippi Rivers	
Scaphirhynchus albus (?) × *S. platorynchus* (4n)	

[1] Data from Berg (1911, 1948b), Kozhin (1964), Kozlov (1970), Legeza (1971), and Keyvanfar (1988).
[2] In the early 1950s, this hybrid was the most numerous (46% of all hybrids caught; Konstantinov et al. 1952).
[3] Data from Vlasenko et al. (1989b).
[4] Data from Berg (1948b) and Kozhin (1964).
[5] Data from Antipa (1909), Antoni-Murgoci (1946), Banarescu (1964), and Berg (1948b).
[6] Data from Berg (1948b).
[7] Data from Berg (1948b), Wei et al. (1996), and Krykhtin & Svirskii (1996).
[8] Data from Nikolskii (1938) and Berg (1948b).
[9] Data from Carlson et al. (1985).

cated loci is much lower, only 6% (Carlson et al. 1982). Why this is so is unknown. In this species duplicated loci for insulin, glucagon and glucagon-like peptide were found (Nguyen et al. 1994).

As for the 240-chromosome octoploid species, it is evident that two forms of vitellogenin monomers in American *A. transmontanus* (Bidwell et al. 1992) and two forms of growth hormones in Russian *A. gueldenstaedtii* (Yasuda et al. 1992) are a result of polyploidization. A higher ploidy level seems to be a reason for a higher heterozygosity in *A. gueldenstaedtii* compared to the 120-chromosome species of *Acipenser* from the same geographic area (Slynko 1976, Keyvanfar 1988, Kuzmin 1991). Possibly, a high level of ploidy causes a high variation in the mean heterozygosity of American octoploid *A. transmontanus* (0.014–0.069, Bartley et al. 1985). Hemoglobin, the only protein examined in the 16n-ploid *A. mikadoi*, is more heterogeneous (11 electrophoretic fractions) than in tetra- (8–9 fractions) or octoploid (7–8 fractions) species of *Acipenser* (Lukyanenko & Lukyanenko 1994).

Polyploidization is a relatively uncommon genetic mechanism in vertebrates, occurring only in lampreys, elasmobranchs, acipenseriforms, some groups of teleosts (salmonids, cyprinids, and catostomids), amphibians (anurans), and lizards (review in Birstein 1987). To date, polyploidization is unknown in birds or mammals. It seems that the polyploid state, and karyotypic and genomic similarity of different acipenseriforms contribute to easy interspecific and even intergeneric hybridization within the Acipenseridae.

The Acipenseridae is the only group among vertebrates all members of which can hybridize with each other in the wild if their spawning grounds overlap (Table 4). The unique easiness of hybridization of acipenserids was described by Russian ichthyologists more than 100 years ago (Ovsyannikov 1870, Zograf 1887). Some hybrids (such as the artificially obtained 'bester', *H. huso* × *A. ruthenus*, and its reciprocal hybrid, Nikolyukin 1970), have the desirable characteristics of fast growth and high viability, are fertile (which is also unusual for vertebrate hybrids) and are widely used in aquaculture (Williot et al. 1993). Besters inherit a phenotype intermediate between the parental species. Of 29 characters studied, 9 deviated toward the maternal species (beluga), and 18 deviated toward the paternal species (sterlet) (Krylova 1981). The meristic characters (the number of dorsal, lateral, and ventral scutes) deviated toward the maternal species (beluga). The ease of hybridization of sturgeon species allowed to show the maternal inheritance of some behavior characters of sturgeons (Marshin et al. 1969).

Problem of the ancestral karyotype

Data presented above support the hypothesis of the tetraploid origin of 120-chromosome acipenseriforms from a 60-chromosome ancestor before the radiation of this order (Dingerkus & Howell 1976, Carlson et al. 1982). The karyotypes of Lepisosteidae (gars) and Amiidae (bowfins) are relevant to understanding the proposed acipenseriform ancestral karyotype. Gars have approximately 60 chromosomes (Table 1), but it seems that many karyological changes have occurred during their evolution. The karyotype of *Lepisosteus oculatus*, 2n = 68, consists of many meta- and acrocentric macrochromosomes, as well as many microchromosomes (Ohno et al. 1969), while the karyotype of *L. osseus*, 2n = 56, lacks microchromosomes (Ojima & Yamano 1980). The karyotype of *Amia calva* is even more reduced, 2n = 46, but it still includes microchromosomes (Ohno et al. 1969, Suzuki & Hirata 1991). The cellular DNA content of gars and *Amia* ranges from 2.0 to 2.8 pg per nucleus, which is approximately half that of the 120-chromosome acipenseriforms. Therefore, it is quite possible that the common ancestor of acipenseriforms and neopterygians had a karyotype of about 60 chromosomes consisting of micro- and macrochromosomes, with a DNA content about 2.0 pg per nucleus.

Extant polypterids, which are the living members of the basal actinopterygian group Cladistia, have 36 bi-armed chromosomes (except *Polypterus weekesii*, 2n = 38; reviews in Vervoort 1980, Suzuki et al. 1988, 1989). The DNA content in polypterids is considerably higher than in the acipenseriforms, 2C = 12–13 pg per nucleus (Vervoort 1980). It is evident that the polypterids are a cytogenetically advanced group. Molecular data also support this conclusion: the 18S rRNA sequences in the species of the genera *Polypterus* and *Erpetoichthys* are very similar to each other, but both are highly divergent from those of other gnathostomes (Stock et al. 1991a).

The karyotypes of chondrichthyans are more informative. Chondrichthyans are mostly tetraploids, 4n = 90–104 (reviews in Schwartz & Maddock 1986, Asahida et al. 1988, 1993, Asahida & Ida 1989, 1990, Stingo et al. 1989, Stingo & Rocco 1991), and karyotypes contain macro- and microchromosomes. According to cellular DNA content and DNA re-association kinetics data (Olmo et al. 1982, Ida et al. 1985), the ploidy level of a few species of sharks is higher, and in these cases polyploidization occurred without a change in the chromosome number (phenomenon known as cryptopolyploidy; Wagner et al. 1993). The only extant chondrichthyan which possibly retains a diploid karyotype is the spotted ratfish, *Hydrolagus colliei*: it has the lowest chromosome number among elasmobranchs, 2n = 58, and the lowest DNA content, 2C = 3.0 pg (Ohno et al. 1969). But this could be a derived condition, for insufficient taxa have been studied to draw any conclusion. Moreover, changes in DNA content occur even during ontogenesis of this species: about 10% of the genome is eliminated in somatic tissues as compared with the germ cells (Stanely et al. 1984).

In contrast to Acipenseriformes, the reduction of chromosome number through fusion of micro- and small chromosomes into macrochromosomes was the main evolutionary karyotypic trend in chondrichthyans (Schwartz & Maddock 1986, Stingo et al. 1989). As a result, the karyotypes of advanced chondrichthyans consist of a smaller number of mainly bi-armed chromosomes, 2n = 50–70. Another considerable difference is that acipenseriforms have numerous bi-armed macrochromosomes, whereas the most generalized, plesiomorphic elasmobranchs have only 2–3 pairs of bi-armed macrochromosomes, and up to 50 pairs represented by small telocentrics or by microchromosomes (Schwartz & Maddock 1986, Stingo & Rocco 1991). Moreover, the average DNA content in elasmobranchs is much higher than in acipenseriforms (reviews in Schwartz & Maddock 1986, Birstein 1987, Asahida et al. 1988, 1993). Based on general characteristics of the karyotypes of spotted ratfish, gars, and sturgeons, Ohno (1970) and later Dingerkus (1979) proposed that the ancestral karyotype of gnathostomes consisted of approximately 50–60 macro- and microchromosomes.

Recently a karyotype consisting of macro- and microchromosomes was described in another living fossil fish, the coelacanth *Latimeria chalumnae* (Bogart et al. 1994). It appears to include 16 pairs of macro- and eight pairs of microchromosomes. This karyotype resembles to a high extent that of one of the most primitive living frogs, *Ascaphus truei*, but this resemblance could be coincidental.

Karyotypes of acipenseriforms generally resemble karyotypes of primitive amphibians, not anurans as in *L. chalumnae*, but, instead, urodeles belonging to the family Hynobiidae. With several exclusions, the karyotypes of hynobiids, 2n = 56–62, consist of a few large and middle-sized pairs of bi-armed macrochromosomes, a few pairs of telocentric macrochromosomes, and 15–20 pairs of microchromosomes (Morescalchi et al. 1979, King 1990, Kohno et al. 1991). It seems that the ancestral karyotype of these amphibians consisted of 60 macro- and microchromosomes. Because the karyotypes of hynobiids, as well as those of acipenseriforms, include a large number of bi-armed macrochromosomes, they should be considered derived (as compared, for instance, with those of the most ancient forms of chondrichthyans).

Cytogenetic data and phylogeny of the Acipenseriformes

It is impossible to infer generic interrelationships within the Acipenseridae from cytogenetic data. Divergence of the three lines within the family occurred without polyploidization, and the ancestors of all three lineages within the acipenserids seem to have been tetraploids, 4n = 120. If the genus *Huso* originated as the first outshoot within the Acipenseridae (as proposed by Findeis 1993, 1997), then this event was not accompanied by a substantial karyotypic change. The ancestral karyotype seems to have been retained without significant modification, since the karyotype of *Huso huso* is only slight-

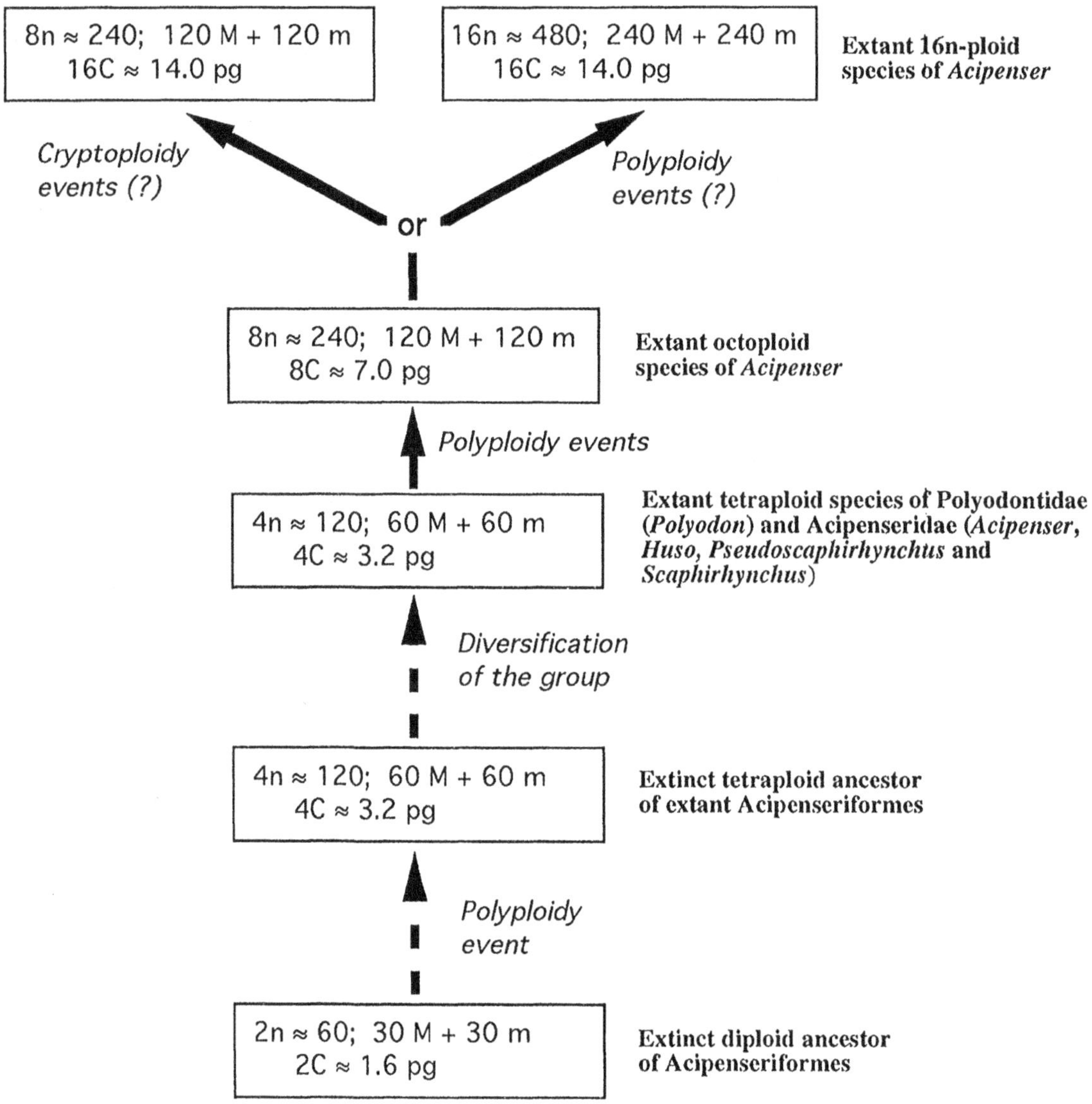

Figure 1. A schematic representation of changes in ploidy in Acipenseriformes. M = macro-, m = microchromosomes.

ly more symmetric than karyotypes of other acipenserids. A schematic course of possible cytogenetic evolution within the Acipenseriformes is presented in Figure 1.

Cytogenetic data are helpful for understanding some relationships within the genus *Acipenser*, where polyploidization was one of the main genetic mechanisms of speciation. The diversification of this genus was accompanied by an appearance of octoploid (according to the karyotypic and DNA content data) and 16n-ploid (according to DNA content data) species. The octoploid 240-chromosome sturgeon species seem to have originated independently in different regions. The closely related *A. gueldenstaedtii* and *A. persicus* may have a common origin and a common octoploid ancestor. Molecular data point to the close relatedness of *A. gueldenstaedtii* to *A. baerii* (see below).

Acipenser medirostris and *A. transmontanus* have many similar characteristics in morphology, biology, and ecology, as well as overlapping ranges in western North America (Hart 1973, Scott & Grossman 1973). Moreover, according to Artyukhin & Andronov (1990), some ecological and biological characteristics are common to these two species and the Chinese sturgeon, *A. sinensis*. All three species are octoploid (Table 1). Possibly, tetraploidization occurred in the ancestor of all three species. If the difference between the karyotype of *A. sinensis* and *A. transmontanus* (8n = 264 ± 3 and 230 ± or 248 ± 8, respectively is real, then it means that at least two polyploidization events took place in the ancestral 120-chromosome form. The next step of polyploidization which occurred in the 240-chromosome ancestor of *A. medirostris* and *A. mikadoi*, resulted in the formation of the genome of *A. mikadoi* (which is, therefore, the youngest in this group of species). The Amur River sturgeon, *A. schrenckii*, also seems to be an octoploid (but these data are preliminary, see Table 1), as the species of the trans-Pacific *A. sinensis-A. medirostris-A. transmontanus* group, and lives in a close geographic area. According to Artyukhin (1994, 1995), *A. schrenckii* is closely related to the Ponto-Caspian species *A. nudiventris* and *A. ruthenus*.

It is difficult to draw any conclusions concerning the ancestor of *A. brevirostrum*, the species with the second highest level of DNA content. The other species of the Eastern Coast of North America, *A. oxyrinchus*, is closely related to the European 120-chromosome Atlantic sturgeon, *A. sturio*, and, according to the DNA content, has the same ploidy (Table 1). Possibly, *A. brevirostrum* is also related to some European sturgeons (*A. nudiventris*; see below), and originated from these European-related ancestors through polyploidization. Because of its high ploidy level, *A. brevirostrum* might be a young species among East American representatives of the genus *Acipenser*. As for another American species, the freshwater lake sturgeon, *A. fulvescens*, which is also an octoploid like *A. oxyrinchus* (according to its DNA content), its origin and relationships with the other species are still unclear, although it is morhologically similar to *A. brevirostrum* (Vladykov & Greeley 1963, Findeis 1993). Therefore, the cytogenetic data provide additional information (ploidy level can be considered as a character in cladistic terms) to the hypothetical grouping of species within the genus *Acipenser* based on the species morphology, ecology, biogeography and possible area of origin (Artyukhin & Andronov 1990, Artyukhin 1994, 1995).

Among fishes, an analogous mode of speciation through multiple independent tetraploidization events is characteristic of only one group of teleosts, the family Cyprinidae (review in Buth et al. 1991). A tetraploid ancestor of two species, *Cyprinus carpio* and *Carassius auratus*, appeared through tetraploidization about 16–20 million years ago (Risinger & Larhammar 1993, Larhammar & Risinger 1994). Hexaploids in Eurasia and Africa (reviews in Vasiliev 1985, Buth et al. 1991, Golubtsov & Krysanov 1993), as well as a 16/20n-ploid Asian species *Diptychus dipogon* (Yu & Yu 1990) also were formed within Cyprinidae. Moreover, as a result of a tetraploidization event which occurred approximately 50 million years ago an ancestor of another family, Catostomidae, appeared within this group (Uyeno & Smith 1972). Catostomids are considered to have been tetraploid since then (Ferris & Whitt 1979, Ueno et al. 1988, Tsoi et al. 1989). Tetraploids were also formed in another family closely related to Cyprinidae, the Cobitidae (review in Vasiliev 1985). Only one other group of teleosts, the Salmonidae (which includes three subfamilies, Coregoninae, Thymallinae, and Salmoninae, sensu Nelson 1994) also originated from a tetraploid ancestor (Gold 1979, Allendorf & Thorgaard 1984). By contrast with the acipenseriforms, cyprinids, and catostomids, a decrease in chromosome number through centromeric fusion was characteristic for different lineages of salmonids (Vasiliev 1985, Buth et al. 1991). Therefore, a gradual increase in chromosome number through polyploidization has occurred a few times during the history of actinopterygians.

Molecular phylogeny of the Acipenseriformes

Our first objective was to search for molecular synapomorphies of Acipenseriformes and its major included clades using representatives of all extant genera (except the Chinese paddlefish, *Psephurus gladius*, due to our inability to obtain suitable tissue). Our data were polarized using other Actinopterygii, *Polypterus* and *Amia*. We also attempted a synthesis of morphological, karyological and molecular characters as related to relationships among Acipenseriformes. Finally, we examined interrelationships among representatives of Artyukhins species groups proposed for *Acipenser* (see above).

Whereas our examination of acipenseriform taxa concentrated on comparisons of species belonging to different genera or species within the genus *Acipenser*, most previous workers have concentrated on intraspecies structure using the control region (D-loop) of the mtDNA. Buroker et al. (1990) showed that in the American white sturgeon, *Acipenser transmontanus*, mtDNA size varies between 16.1 and 16.7 kb depending on the number of tandemly repeated 82 nucleotide sequences in the control region of the mtDNA. Nearly 50% of the individuals studied by Brown et al. (1992a) were heteroplasmic (i.e., had multiple copies of different mtDNA types within an individual) for length variation, with six different mtDNA length variants found (Brown et al. 1992b, 1993). Fifty percent of *A. medirostris* studied were also heteroplasmic; D-loops of these individuals included from one to four repeats (Brown et al. 1996). The average size of mtDNA of the lake stugeon, *A. fulvescens*, is approximately the same as that of white sturgeon, 16.6–16.9 kb (Guénette et al. 1993, Ferguson et al. 1993) or 16.1–16.5 kb (Brown et al. 1996). No heteroplasmy was detected in *A. fulvescens* and *A. oxyrinchus* (Brown et al. 1996). All individuals of *A. fulvescens* studied had one of five possible mtDNA size variants which closely corresponded to *A. transmontanus* with one to five repeat units. In *A. oxyrinchus*, nearly every individual was fixed for mtDNA roughly equivalent in size to the smallest repeat found in the other species. Restriction analysis of mtDNA (Bowen & Avise 1990, Avise 1992) and partial sequencing of the control region (Miracle & Campton 1995, Ong et al. 1996, Wirgin et al. 1997 this volume) were used for inferring relationships between subspecies and populations of *A. oxy-*

Table 5. List of sturgeon species and blood samples studied.

Species number	Species	Geographical region	Number of blood (or tissue) samples	Collector
1.	*Acipenser baerii*[1]	Lena River, Siberia, Russia (Moscow Aquarium)	2	V. Birstein
2.	*A. brevirostrum*	Connecticut River, MA, USA	(eggs)	B. Kynard
3.	*A. gueldenstaedtii*[1]	Volga River, Russia (Moscow Aquarium)	2	V. Birstein
4.	*A. medirostris*	Columbia River, OR, USA	1	J. North
5.	*A. mikadoi*	Tumnin River, Russia	2 (fragments of muscles)	E. Artuykhin
6.	*A. naccarii*	Ferrara, Italy (Aquarium)	2	F. Fontana
7.	*A. nudiventris*[1]	Aral Sea, Uzbekistan (Moscow Aquarium)	2	V. Birstein
8.	*A. oxyrinchus oxyrinchus*	Hudson River	2 (fragments of muscles)	J. Waldman
9.	*A. ruthenus*[1]	Volga River, Russia (Moscow Aquarium)	2	V. Birstein
10.	*A. stellatus*[1]	Volga River, Russia (Moscow Aquarium)	2	V. Birstein
11.	*A. transmontanus*	Columbia River, OR, USA	2	J. North
12.	*Huso dauricus*[1]	Amur River, Siberia, Russia (Moscow Aquarium)	2	V. Birstein
13.	*Pseudoscaphirhynchus kaufmanni*[1]	Amu-Darya River, Uzbekistan (Moscow Aquarium)	2	V. Birstein
14.	*Scaphirhynchus albus*	Yellowstone River, MT, USA	2	H. Bollig
15.	*Polyodon spathula*[1]	Moscow Aquarium	1	V. Birstein

[1] These samples were used for the DNA content measurements in Birstein et al. (1993); see Table 1 above.

rinchus. Both subspecies, *A. o. oxyrinchus* and *A. oxyrinchus desotoi*, exhibited low mtDNA diversity.

The order and transcriptional polarity of three mitochondrial genes in *A. transmontanus* (genes for cytochrome *b*, threonine and proline tRNAs) are identical to those of other vertebrates (Gilbert et al. 1988, Brown et al. 1989, Buroker et al. 1990). The whole sequence of the cytochrome *b* gene for *A. transmontanus*, as well as partial sequences of the same gene for *Scaphirhynchus platorynchus* and *Polyodon spathula* were published recentely (Brown et al. 1989, Normark et al. 1991). Partial sequences of the same gene for *A. brevirostrum*, *A. oxyrinchus*, *Scaphirhynchus albus*, and *S. suttkusi* were submitted by W. Schill into GenBank under numbers Z22822, L35111, L35110, and L35112, respectively. Although acipenseriforms were included in a higher level phylogenetic analysis based on cytochrome *b* sequence (Normark et al. 1991), no direct evidence on the utility of other gene regions for phylogenetic analysis is available.

Because phylogenetic divergence within the Acipenseriformes is potentially broad, based on the fossil record (e.g., Grande & Bemis 1991, Bemis et al. 1997), no single gene region can be assumed to be adequately broadly informative on all phylogenetic levels as a source of characters. Consequently we chose to examine several gene regions as potential sources of characters, including well characterized gene regions from mitochondrial DNA (16S rDNA, 12S rDNA, and cytochrome *b*) and one nuclear gene region (18S rDNA). Below we discuss each of these four gene regions.

Materials and methods

Specimens

Species used in this study and location of the fishes are listed in Table 5. With three exceptions (*Acipenser brevirostrum*, *A. mikadoi*, and *A. oxyrinchus*), blood samples were taken, mixed with buffer (100 mM Tris, 100 mM EDTA, and 2% SDS; 0.5 ml of blood and 5 ml of buffer), and the blood cells lysed in this solution were kept in a freezer at –70°C. Most Russian specimens examined were the same individuals used for DNA content measurements by Birstein et al. (1993). Also, we isolated DNA from alcohol-fixed samples of muscles of *Amia calva* and *Polypterus senegalus* provided by Paul Vrana (American Museum of Natural History, New York).

DNA extraction, amplification, and sequencing

DNA was isolated from each sample using a standard phenol preparation (Hillis et al. 1990, DeSalle et al. 1993). We examined partial sequences of three ribosomal genes (two mitochondrial and one nuclear) and a partial sequence of cytochrome *b*. PCR products were prepared for DNA sequencing in several ways. In all cases the nuclear 18S rDNA fragments were GeneCleaned (BIO 101; Palumbi et al. 1991) and directly sequenced. PCR products of the mitochondrial genes (12S, 16S, and cytochrome *b*) were either GeneCleaned and directly sequenced or cloned into the TA vector (INVITROGEN) and sequenced (in such cases, at least two clones for each taxon were used to establish the sequence). We used the following primers: in the 18S gene region, 18sai0.7 (5′-ATTAAAGTTGTTGCGGTTT-3′) and 18sai0.79 (5′-TTAGAGTGCTYAAAGC-3′) (Wheeler et al. 1993), in the 12S gene region, 12SA (5′-GGTGGCATTTTATTTTATTAGAGG-3′) and 12SB (5′-CCGGTCTGAACTCAGATCACGT-3′) (Kocher et al. 1989, Hedges et al. 1993b), in the 16S gene region, 16SA (5′-CGCCTGTTTACCAAAAACAT-3′) and 16SB (5′-CCGGTCTGAACTCAGATCACGT-3′) (Palumbi et al. 1991), and in the cytochrome *b* region, H15149 (5′-AAACTGCAGCCCCTCAGAATGATATTTGTCCTCA-3′) (Kocher et al. 1989) and L14724 (5′-CGAAGCTTGATATGAAAAACCATCGTTG-3′) (Meyer et al. 1990). All sequencing was performed using the Sequenase system (U.S. Biochemicals) and double stranded templates. The sequences reported in this paper have been deposited in the EMBL Nucleotide Sequence Database (accession no. X95003–X95061).

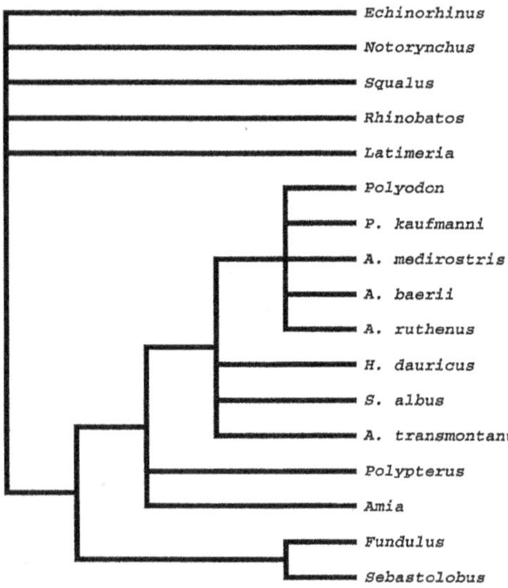

Figure 2. A phylogenetic tree for a combined (18sai0.7 plus 18sai0.79; 229 bp) region of fishes: eight acipenseriform species studied, *Polypterus senegalus*, *Amia calva*; four chondrichthyans, *Notorynchus cepedianus* (Hexanchiformes, Hexanchidae; Bernardi & Powers 1992), *Echinorhinus cookei* (Squalifomes, Squalidae; Bernardi & Powers 1992, Bernardi et al. 1992, M91179, GenBank), *Squalus acanthoides* (Squalifomes, Squalidae; Bernardi & Powers 1992, Bernardi et al. 1992, M91181, GenBank), and *Rhinobatos lentiginosus* (Stock & Whitt 1992, M97576, GenBank); *Latimeria chalumnae* (Stock et al. 1991b, L11288, GenBank); and two teleosts, *Fundulus heteroclitus* (Cyprinodontiformes, Fundulidae; Bernardi et al. 1992, M91180, GenBank), and *Sebastolobus altivelis* (Scorpaeniformes, Scorpaenidae; M91182, GenBank). *Squalus acanthias* and *Rhinobatos lentiginosus* were used as outgroups.

Outgroup choice

We chose *Polypterus senegalus*, a representative of a lineage often considered to be the sister group of Acipenseriformes plus Neopterygii (Patterson 1982), as our outgroup. Because the use of multiple outgroups is recommended in phylogenetic analyses (Watrous & Wheeler 1981), we also used *Amia calva* (Amiidae), generally regarded as the living sister species of teleosts (see Patterson 1973), as an outgroup. In the analysis of the partial sequence of the 18S gene we used two chondrichthyan species, *Squalus acanthias* and *Rhinobatos lentiginosus* (Bernardi et al. 1992, M91179; Stock & Whitt 1992, M97576), as outgroups.

DNA sequence alignment and phylogenetic analysis

We used equal weights for all nucleotide positions in all analyses. When multiple parsimonious trees were obtained for a particular analysis, successive weighting based on the retention index was used to choose among these multiple parsimonious trees (Carpenter 1988). DNA sequences for the mitochondrial 16S and 12S rRNA regions and the nuclear 18S rRNA regions were aligned using the program MALIGN (Wheeler & Gladstein 1993). Gap costs were varied in order to explore the effects of alignment parameters on phylogenetic inference (the results of varying alignment parameters on phylogenetic inference are discussed in detail in Fitch & Smith 1983, Gatesy et al. 1993, Hillis et al. 1994, Wheeler 1995, Wheeler et al. 1995). In most cases our alignments were extremely stable (i.e., alignment columns did not change by altering gap costs) and this stability suggests a low level of 'alignment ambiguity' (Gatesy et al. 1993). Consequently, the methods of 'culling' (Gatesy et al. 1993) and 'eliding' (Wheeler et al. 1995) were not applied to infer alignment. It was trivial to align cytochrome *b* sequences, which were also performed using MALIGN with a gap cost of 8. Parsimony trees for each of the four individual gene regions were generated separately using the PAUP 3.1 program (Swofford 1993) to examine the signal inherent in each gene region. Sequence alignments using a gap cost of 8 were arbitrarily chosen and were combined (Kluge 1989, Ernisse & Kluge 1993) into a single data matrix. Phylogenetic hypotheses were generated from this combined data matrix using PAUP. The degree of support for particular nodes in these trees was examined using the Bremer support index (Bremer 1988, Donoghue et al. 1993, Kallersjo et al. 1993).

Results and discussion

Gene regions

18S rDNA

The 18S rRNA gene is relatively slowly evolving in vertebrates and has been useful at higher taxonomic levels (e.g., Stock et al. 1991a). We used two sequencing primers in our analyses, 18sai0.7 and 18sai0.79. These primers are in a region of the 18S gene that varies in insects and other organisms (Wheeler et al. 1993). The 18sai0.7 sequences were invariant in all acipenseriform species investigated, whereas the 18sai0.79 fragment was variable at several positions. A low degree of 18S rRNA sequence divergence between *Scaphirhynchus* and *Polyodon* was reported previously (Stock et al. 1991a).

The phylogenetic tree for combined data sets for both fragments of the 18S gene for all fish species investigated so far is presented in Figure 2. Species used and origin of the sequences are explained in the legend to this figure. *Echinorhinus cookei* and *Rhinobatos lentiginosus* were chosen as outgroups. The tree statistics are shown in Table 6.

There is a high degree of similarity between the gene fragments under consideration in all four chondrichthyans and *Latimeria chalumnae*, and they differ from these fragments in actinopterigyans. There are two putative synapomorphies of acipenseriformes in the 18sai0.7 region: (1) all acipenserids and *Polyodon spathula* had an insertion of A between 684 and 685 comparatively to the homologous sequence of *L. chalumnae*, and (2) a T between positions 773 and 774 relative to *L. chalumnae*. Also, a change of A to T in position 771 (*L. chalumnae*) seems to be synapomorphic for all acipenseriform species. The most variable region of 18sai0.79 region in all groups of fishes examined so far lies between T and G in position 851 and 855 (*L. chalumnae*). In chondrichthyans and *L. chalumnae* there are three nucleotides, TCG, whereas in teleosts there are 7 nucleotides, TTCTCCT or TCTTTCT, and in species studied by us, only a part of the latter sequence, CCT.

16S mitochondrial rDNA

We obtained sequences for two parts of this gene: (1) a 147 nucleotide sequence using the 16SA primer (146 nucleotides for *Amia calva* and *Polypterus senegalus*), and (2) a 169 nucleotide sequence with the 16SB primer (164 nucleotides in *Amia calva*). Both regions were highly conserved, although there were differences in the 3′-part of the 16Sb fragment.

12S mitochondrial rDNA

Two overlapping short stretches of 12S mtrDNA, 12SA and 12SB, were sequenced. The final contiguous sequence consisted of 183 nucleotides in *Huso dauricus* and *Pseudoscaphirhynchus kaufmanni*, 184 nucleotides in *Polyodon spathula, A. medirostris, A. baerii, Scaphirhynchus albus*, and *Amia calva*, and 185 nucleotides in *Polypterus senegalus*. The 12S region is more variable than the 16S regions and yields several phylogenetically informative characters for examining higher level relationships (see below). Extremely low levels of variability, however, exist within the genera *Acipenser* and *Huso*.

Cytochrome b

The amplified region consisted of 270 base pairs, from the 7th to 97th codons according to Normark et al. (1991). Most variation occurs at the third posi-

Table 6. Tree statistics.

Gene	Number of characters	Apomorphies, total number	Number of phylogenetically informative characters	Number of trees	Number of steps	Consistency index	Retention index
(18sai0.7+ 18sai0.79) fragment	241	39	24	6	47	0.94	0.98
12S gene fragment	189	51	10	6	61	0.71	0.67
16S gene fragment	318	74	23	2	86	0.82	0.81
18S gene fragment	229	12	7	3	14	0.88	0.91
Cytochrome b gene fragment	270	163	56	3	187	0.62	0.54
Combined molecular characters	1006	300	96	1	354	0.65	0.59
Cytochrome *b* with additional species of *Acipenser*	270	163	56	2	216	0.54	0.52

tions of codons. For comparable lengths of the 16S and 12S sequences, the cytochrome *b* gene for the species examined by us has from three to four times more nucleotide changes.

Generic relationships within Acipenseriformes

Alignment and phylogenetic inference
First, we examined phylogenetic signal in the individual gene character sets by constructing separate cladograms for each of the four genes studied. Second, due to the small number of apomorphic characters for each gene, we used a combined approach (Miyamoto 1985, Kluge 1989, Ernisse & Kluge 1993) to infer phylogenetic relationships.

We used aggressive alignment parameters for the program MALIGN (build; treeswapping; random sequence addition; Wheeler & Gladstein 1993) in our computer searches. We avoided rearranging computer generated alignments based on eye judgment because of the arbitrary and inherently non-repeatable nature of this approach and because the choice of gap:change cost ratios in DNA sequence alignment can greatly affect final alignment (Fitch & Smith 1983, Waterman et al. 1992, Gatesy et al. 1993). We performed several alignments with varying gap:change ratios, and found alignments for all three rRNA were very stable. There were few alignment ambiguities as judged by comparing alignments generated at gap:change costs of 2, 4, 8 or 16. The phylogenetic hypotheses generated by these various alignments were congruent, further supporting our notion that the alignments are very stable.

Individual gene trees
We report three consensus gene trees from sequences aligned using a gap:change ratio of 4 for the three rRNA genes (Figure 3). As noted above, cytochrome *b* sequences were aligned with a gap cost of 8 and the resulting gene tree is also shown in Figure 3. The tree statistics for each of the gene trees, including the number of apomorphies and phylogenetically important characters, is given in Table 6. The phylogenetically informative characters and

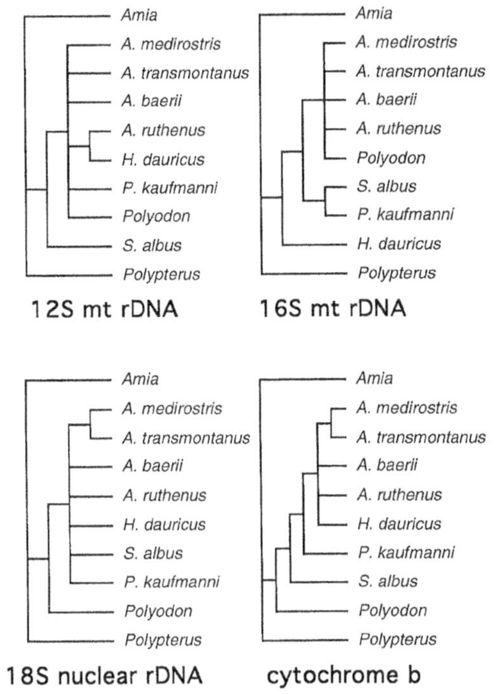

Figure 3. Consensus trees for individual gene regions examined for the 12S mtrDNA (189 bp), 16S mtrDNA (318 bp), 18S rDNA (229 bp), and cytochrome *b* genes (270 bp). *Amia calva* and *Polypterus senegalus* are outgroups. The tree statistics are in Table 6.

their positions in the sequences for each gene are shown in Figure 4.

In general, each gene tree alone showed low levels of resolution (Figure 3). Successive weighting of the individual character sets did not result in the choice of a single or fewer trees. Although character congruence was high as indicated by the relatively high consistency and retention indices, the number of informative characters for each character set was so low (Figure 4) that resolution of only a few nodes in each single gene tree was demonstrated.

One general observation, however, is that *Acipenser* was not found to be monophyletic in any of the four gene trees (Figure 3). For 12S mtDNA, *Huso dauricus* clustered with *A. ruthenus*, and no characters were found that hypothesized the remaining species of *Acipenser* as a group. For 16S mtDNA, *Polyodon spathula* clustered with the four species of *Acipenser* surveyed, and no character was found

```
                            12S              16S                 18S

                         11111111         1111111122222233      11112
                       3644555667 35677890222348839999901 3422260
Node                   5808567564 00338710689746991236704 3003553
-------------------------------- ----------------------- -------
A. medirostris         TCTAATGCAC ACGCTCACTAAGTTGAGTATGTA ATTTATT
A. transmontanus       TCTAATGCAC ACGCTGACTAAGTTGAGTATGTA ATTTATT
A. baerii              TCTAATGCAC ACGCTGACTAAGTTGAGTATGTA ATTTATC
A. ruthenus            TCTAATGC-T ACGCTGACTAACTTGAGTATOTA ATTTATC
Huso dauricus          TCTAATGC-T ACGCTGACCTAGTT-A-TATGTA ATTTACC
Scaph. albus           TCCGTGGCAC TCGCTGACCTAGTGGATGGTGTA ATTTACC
Pseudosc. kaufm.       TCCAATGC-C ACGCTGACCTAGTGGAGTGTGTA ATTTATC
Polyodon spathula      TCCAATGC-C ATGCTGAATAAGTTGAGGGTGAA ATATTTC
Amia calva             ATTAAGATAC CTATCAGCCTT-CT-GGTACTTG -CA-TCC
Polypterus seneg.      ATCGTAAT-G CTATCAGACTT-CT-G---CTAG -CA-TCC
```

```
                                        cytochrome b

                                      11111111111111111111211112222222222222222
                        1233344666677911122233445666778888899001111224445555566
Node                   69101908034625014703728143258570367928170369020790235814
-------------------------------------------------------------------------------
A. medirostris         GaCCCACTCAGTTATTGTCACTTTTGCCCTaCTCGGATATCACCCACACCCCGTCC
A. transmontanus       GaCCCATTCAGTTATTGTCACTCTTGCCCTGTTCGGATAcCACCCACACCCTGTCC
A. baerii              AGCCCACTCAOTTATCATCACCTTTGCCTTGCTCGAATGTTACCCACACCCCGTCC
A. ruthenus            AGCCCACTCAGTTAcCGTCACCTTCGCCCTaCTCGAATGTTACCCACACCCCGTCC
Huso dauricus          GGCCCACCCAGTTATTATCCTCTTTGCCTTGTCCGGACGcTCTCCACACCCCGcCC
Scaph. albus           AatCtACTCAGTTAcCACCGCCCCCCCCTTGCTCGCATAcCCCCCACGCCCTAcCC
Pseudosc. kaufm.       AatCtACTCAGTTAcCATtACCTTTGCCTTGCTCGAATGcTACtCACACCCCGctC
Polyodon spathula      TaCTCATCCAGTCAcTTCCACTCTTATCCCCCTCGACCATCACCTCCGTCCCAcAT
Amia calva             TatCGCTATCTACCACTCtCTATCCAATTCaTCTATCCAcCCTATCTGGTATTctC
Polypterus seneg.      GatTACCATCTATCACTTCACCTTCCATTATCTTACAAAcACCATCTATTACCTtT
```

Figure 4. Phylogenetically informative characters (nucleotide composition at variable sites) for four gene regions studied. Numbers above the sequences refer to base pair position from the first base in the amplified fragments. The first position for the amplified fragment for 12S (No. 35) corresponds to No. 570 in the published 12S sequence for *Latimeria chalumnae* (Hedges et al. 1992), in 16S (No. 30) corresponds to No. 176 for *L. chalumnae* (Hedges et al. 1992), in 18S (No. 33) corresponds to No. 682 for *L. chalumnae* (Stock et al. 1991b), and in cytochrome *b* gene (No. 6) corresponds to No. 25 for *L. chalumnae* (Normark et al. 1991).

uniting *Acipenser* as monophyletic. In the 18S rDNA tree, there was again no evidence of monophyly of *Acipenser*. Finally, the cytochrome *b* tree grouped *A. medirostris*, *A. transmontanus*, *A. ruthenus* and *A. baerii* with *Huso dauricus*. *Acipenser medirostris* and *A. transmontanus* shared apomorphies in the 18S rDNA and cytochrome *b* trees, and such a grouping was not ruled out by either of the other two trees. *Pseudoscaphirhynchus* and *Scaphirhynchus* emerged as monophyletic in only one tree (16S, Figure 3).

Combined molecular tree and comparison to existing morphological hypothesis

The DNA sequence characters for the four gene regions were combined into a data matrix and each character was given equal weight. One parsimony tree was obtained using this combined molecular data matrix (Figure 5, Table 6).

Findeis (1993, 1997) used osteological characters for constructing a morphological phylogenetic hypothesis which focused on generic relationships among Acipenseriformes. Only a single parsimony tree resulted from his data set. He examined six species of acipenseriforms not included in our molecular survey (*Scaphirhynchus platorynchus*, *A. brevi-*

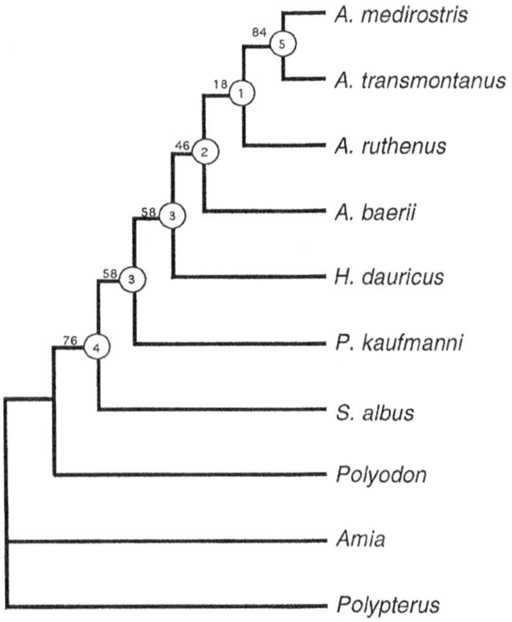

Figure 5. The single parsimony tree obtained for the combined set of molecular characters. *Amia calva* and *Polypterus senegalus* are outgroups. Decay indices (Bremer 1988, Donoghue et al. 1993) are shown at each node. Numbers above the branches represent bootstrap values computed by 1000 replications. The tree statistics are given in Table 6.

rostrum, *A. fulvescens*, *A. oxyrinchus*, and *Huso huso*). He did not examine the osteology of three species studied by us (*A. baerii*, *H. dauricus*, and *S. albus*). Thus, there is incomplete overlap of taxa surveyed by the two phylogenetic approaches, morphological and molecular. A further complication is that Findeis (1993, 1997) does not provide evidence that *Acipenser* is monophyletic, whereas our total molecular data set does (Figure 5).

Comparison of the osteological and molecular trees shows two major differences: (1) the placement of *Huso dauricus* and (2) the sister relationship of *Scaphirhynchus* and *Pseudoscaphirhynchus*. Findeis (1993, 1997) found that *Huso* is basal to the other Acipenseridae, and that the clade including all other sturgeons was well supported by osteological characters. Our combined molecular data, however, suggest that *Huso dauricus* is a sister-species to the genus *Acipenser* (Figure 5). Perhaps, this conflict between the molecular and morphological information results from the small number of molecular characters that are pertinent to the *Huso-Acipenser* sister group hypothesis. It is interesting, however, that many 19th century systematic studies placed *Huso* within *Acipenser* and that *Huso* was not elevated to a separate generic status until Brandt (1869). Later the generic status of *Huso* was still debated (for instance, Nikolukin 1970, Artyukhin 1995). Evidently, *Huso* warrants new attention from systematists.

The sister relationship of *Scaphirhynchus* and *Pseudoscaphirhynchus* is strongly supported by the morphological data (Findeis 1993, 1997), but it is not supported by our molecular characters. In Figure 5 *Scaphirhynchus* emerged as the sister taxon of all other sturgeons, and *Pseudoscaphirhynchus* emerged as the sister taxon of *Huso* and *Acipenser*. This is an interesting difference between the two phylogenies because conventional pre-cladistic ideas about relationships within Acipenseridae suggest that *Scaphirhynchus* plus *Pseudoscaphirhynchus* are basal members of the family (e.g., Zograf 1887). The fact that *Pseudoscaphirhynchus* and *Scaphirhynchus* did not group together is supported in our combined tree by relatively high decay indices (4 and 3 at the pertinent nodes in Figure 5) and bootstraps. We suspect that the traditional idea of this monophyly may be incorrect.

Relationships within the genus Acipenser

For this investigation we used partial sequences of the cytochrome *b* genes of eight Eurasian and four American species of the genus *Acipenser*. This data set includes 7 species absent in the previous molecular analysis because we did not sequence the ribosomal genes for these taxa. Taxa chosen represent all four species groups proposed by Artyukhin (1995, see above). The result of phylogenetic analysis is presented in Figure 6, and tree statistics, in Table 6.

Two parsimony trees were obtained. In the cytochrome *b* analysis, *Acipenser* is not monophyletic, and two main clades of species are seen in both trees. Two western American species, *A. medirostris* and *A. transmontanus*, group together and are a

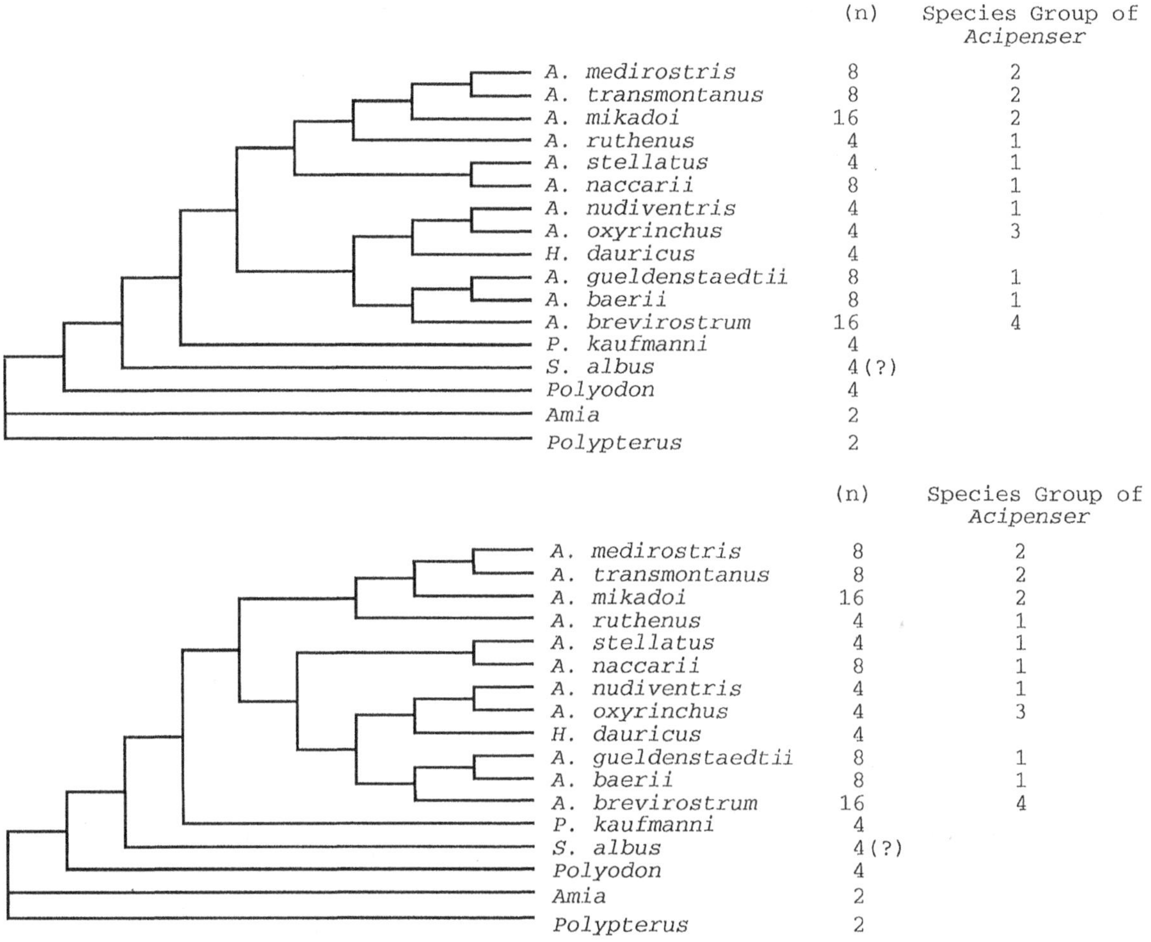

Figure 6. The two parsimony trees based on partial sequences (270 bp) of the cytochrome *b* gene regions from 11 species of *Acipenser* (representatives of all groups of species in Table 5), *Huso dauricus*, *Pseudoscaphirhynchus kaufmanni*, *Scaphirhynchus albus*, and *Polyodon spathula*. As in the previous analyses, *Amia calva* and *Polypterus senegalus* were used as outgroups. The ploidy of fishes and the group number for species of *Acipenser* are given to the right of species names. The tree statistics is shown in Table 6.

sister-group to *A. mikadoi*, whereas *A. ruthenus* is basal to all these species. Except for the freshwater *A. ruthenus*, all other species of the clade are typically anadromous sturgeons. The difference in the cytochrome *b* gene sequences between *A. mikadoi* and *A. medirostris* supports the previous assumption based on DNA ploidy that *A. mikadoi* is a separate species in spite of the fact that it is morphologically indistinguishable from *A. medirostris* (Birstein et al. 1993, Birstein 1994b). Unexpectedly, the ancestral form of this group seems to be related to *A. ruthenus*.

The second group is also unexpected. It consists of two clades. The first clade includes the European *A. nudiventris* grouped with the eastern American *A. oxyrinchus* (and *Huso dauricus*, see discussion above). Their sister-group consists of the European *A. gueldenstaedtii*, Siberian *A. baerii*, and an eastern American species *A. brevirostrum*. These relationships suggest that: (1) the Atlantic group of species is related to the Ponto-Caspian *A. nudiventris*; (2) a Ponto-Caspian European *A. gueldenstaedtii* is closely related to a Ponto-Caspian Siberian *A. baerii*; (3) both Ponto-Caspian *A. gueldenstaedtii* and *A. baerii* are related to the eastern North American

A. brevirostrum; and (4) both clades show transatlantic relationships for the species.

The position of a small clade consisting of *A. stellatus* and *A. naccarii*, is unresolved: in the first tree it is grouped with the first main clade, whereas in the second tree it clusters with the second main clade (Figure 6). Traditionally, the octoploid *A. naccarii* was considered to be closely related to the octoploid *A. gueldenstaedtii* (e. g., Tortonese 1989, Rossi et al. 1991, Artyukhin 1995), but not to the tetraploid *A. stellatus*.

According to the tree in Figure 6, ploidization occurred at least three times within the *Acipenser*: two octoploid ancestral forms were formed independently in *A. mikadoi-A. medirostris-A. transmontanus*, *A. gueldenstaedtii-A. baerii-A. brevirostrum*, and in *A. stellatus-A. naccarii*; two polyploidization events followed resulting in the appearance of *A. mikadoi* and *A. brevirostrum*. These data support our assumption (see above) that ploidization played a significant role in speciation within the *Acipenser*, but contradict a simple scheme of hypothetical relationships of the species of *Acipenser* published by Artyukhin (1995, see also Bemis et al. 1997). Except the close relatedness of *A. transmontanus* and *A. medirostris*, which is supported by other molecular data (Brown et al. 1996), the other relationships in Artyukhin's tree (Artyukhin 1995) are not supported by our molecular data.

We caution that the trees in Figure 6 are preliminary ones. Evidently, additional data should be obtained for better resolution of relationships among the species. We have already sequenced longer regions of the cytochrome *b* gene, as well as other genes (Birstein & DeSalle 1997). Our data do show, however, that phylogenetic relationships within the *Acipenser* can be reconstructed using even a partial sequence of the cytochrome *b* gene.

General lack of molecular variability among the genera of Acipenseriformes

We initially chose the gene regions described above because of the high degree of variability shown in other taxa of comparable divergence times (Meyer & Wilson 1990, Normark et al. 1991, Stock et al. 1991a, Hedges et al. 1993a). One surprising result of our study is the general lack of variability in the gene regions studied as compared to other animal groups, including teleosts (reviews in Meyer 1993, Meyer et al. 1993, Patarnello et al. 1994), some amphibians, most mammals and insects (for instance, Irwin et al. 1991, Hedges et al. 1993b, Wheeler et al. 1993). The slow rate of molecular evolution in acipenseriforms may be correlated with slow karyotypic evolution in these fishes (see above).

The other group of fishes with a low rate of evolution in 18S and mitochondrial genes is Chondrichthyes (Bernardi & Powers 1992, Martin et al. 1992, Martin & Palumbi 1993). Slow evolution of the 18S genes, as in acipenseriforms (see Figure 2), could be related to polyploidy (cryptoploidy) in these fishes (reviews in Schwartz & Maddock 1986, Birstein 1987, Stingo & Rocco 1991). But the nucleotide substitution rate in the cytochrome *b* gene in sharks is one sixth that of primates (Martin et al. 1992, Martin & Palumbi 1993), and this characteristic cannot be attributed to ploidy differences. Perhaps, it is caused by differences in the rate of accumulation of silent transversions (Martin & Palumbi 1993), but this remains a peculiar problem. It is interesting that the sequence of the region of the 18S gene under discussion in sharks is more similar to that in the coelacanth than to that in acipenseriforms (Figure 2).

It is evident that the molecular data (at least those presented here), as well as the cytogenetic data (see above), have restrictions in application to the phylogeny of Acipenseriformes at the generic level. Low levels of variability of the genes commonly used as phylogenetic tools (12S, 16S, 18S, and cytochrome *b*) suggests that some other, rapidly evolving gene regions such as the mitochondrial control region (D-loop, Shedlock et al. 1992) or larger portions of the cytochrome *b* or other mitochondrial structural genes (Normark et al. 1991) might be helpful for examining relationships among the genera of Acipenseridae.

In the meantime, our results suggest that the cytochrome *b* gene could be used for investigation of relationships among species of *Acipenser*. The cytochrome *b* data suggest that *Acipenser* is not monophyletic (due to the insertion of *Huso dauricus* into this group), as considered before. We hope that future analyses involving longer regions of the cytochrome *b* gene and, possibly, some other protein coding genes will help to establish mono- or polyphyly of this genus.

Conclusions

(1) Little cytogenetic change has occured during the evolution of Acipenseriformes. Polyodontidae and Acipenseridae presumably originated from a tetraploid ancestor whose karyotype consisted of 120 macro- and microchromosomes with a DNA content about 3.2–3.8 pg per nucleus. Tetraploidization of the 60-chromosome ancestor possibly occurred at the early times of evolution of the Acipenseriformes, probably, during the origin of this group in the Mesozoic.

(2) No conclusions regarding interrelationships within Acipenseridae among *Huso*, *Acipenser*, *Scaphirhynchus*, and *Pseudoscaphirhynchus* can be made based on cytogenetic data. Divergence of these lineages of sturgeons occurred without polyploidization.

(3) Diversification within *Acipenser* was accompanied by appearence of octoploids (according to karyotypic and DNA content data) and 16n-ploids (according to DNA content data). The octoploid 240-chromosome sturgeon species have about 240 chromosomes and may have originated independently in different geographic areas. The two 16n-ploid species, *A. mikadoi* and *A. brevirostrum*, may be the youngest species within the genus.

(4) A study of partial sequences of genes from mitochondrial DNA (16S rDNA, 315 bp; 12S rDNA, 189 bp, and cytochrome *b*, 270 bp) and of one nuclear gene region (18S rDNA, 230 bp) demonstrated very low levels of variability in the eight acipenseriform species surveyed (*Polyodon spathula*, *Huso dauricus*, four species of *Acipenser*, *Scaphirhynchus albus*, and *Pseudoscaphirhynchus kaufmanni*). This low variability is unusual for these genes, which are commonly used as phylogenetic tools.

(5) The molecular tree based on combined data from all four genes had two major departures from the existing morphological hypothesis (Findeis, 1993, 1997): *Huso dauricus* was a sister-species to the genus *Acipenser* instead of being basal to all acipenseriforms, and *Scaphirhynchus* and *Pseudoscaphirhynchus* did not form a monophyletic group.

(6) A partial sequence of the cytochrome *b* gene (270 bp) was used to examine relationships within the genus *Acipenser*. Seven additional species of Acipenser were included in this part of the study (*A. brevirostrum*, *A. gueldenstaedtii*, *A. mikadoi*, *A. naccarii*, *A. nudiventris*, *A. oxyrinchus*, and *A. stellatus*). The data support the hypothesis that octoploid species appeared at least three times within the *Acipenser*. Also they show close relationships between the Eurasian *A. ruthenus* and the Pacific *A. mikadoi*-*A. medirostris*-*A. transmontanus*, between the European *A. gueldenstaedtii*, Siberian *A. baerii*, and American *A. brevirostrum*, between two European species, *A. stellatus* and *A. naccarii*, as well as a possible trans-Atlantic relationship between the Eurasian *A. nudiventris* and American *A. oxyrinchus* suggesting limited utility of geographic locality as an indicator of relationship.

Acknowledgements

We are very grateful to all colleagues who helped us to collect blood and tissue samples: Serge Gamalei (Moscow Aquarium, Moscow), Boris Goncharov (Institute of Developmental Biology, Russian Academy of Sciences, Moscow), Evgenii Artyukhin (Central Laboratory of Regeneration of Fish Resources, St. Petersburg), Francesco Fontana (University of Ferrara), Herb Bollig (US Fish and Wildlife Service, South Dakota), Boyd Kynard (National Biological Survey, Massachusetts), John North (Department of Fish and Wildlife, Oregon), and John Waldman (The Hudson River Foundation, New York). We are also indebted to Paul Vrana for his kind permission to use his collection of tissue samples. We are extremely thankful to Willy Bemis for his patient reading of the manuscript and his important notes which allowed us to improve the manuscript immensely.

References cited

Allendorf, F.W. & G.H. Thorgaard. 1984. Tetraploidy and the evolution of salmonid fishes. pp. 1–53. *In:* B.J. Turner (ed.) Evolutionary Genetics of Fishes, Plenum Press, New York.

Antipa, G. 1909. Fauna ichtiologica a României. Publicatiunile Fondul Vasilie Adamanchi, Academia Romāna, Bucureşti 16: 1–294.

Antoniu-Murgoci, A. 1946. Sur l'hybridation chez les estrogeons et description de deux formes nouvelles. Acad. Roum. Bul. Sec. Sci. 29: 308–313.

Arefjev, V.A. 1983. Polykaryogrammic analysis of the ship *Acipenser nudiventris* Lovetsky (Acipenseridae, Chondrostei). Voprosy Ikhtiologii 23: 209–218 (in Russian, English translation J. Ichthyol. 23: 26–35).

Arefjev, V.A. 1989a. Study of karyotype of the sturgeon *Acipenser gueldenstaedti* Brandt (Chondrostei). Tsitologia i Genetika 23: 7–10 (in Russian).

Arefjev, V.A. 1989b. Karyotype variability in successive generations after hybridization between the great sturgeon, *Huso huso* (L.), and the sterlet, *Acipenser ruthenus* (L.). J. Fish Biol. 35: 819–828.

Arefjev, V.A. 1991. Cytogenetic aspects of differences in the quality of spawners of reciprocal hybrids between bester and beluga sturgeon. pp. 134–150. *In:* A.D. Gershanovich (ed.) Biological Principles of Commercial Aquaculture of Sturgeons, VNIRO Publishing, Moscow (in Russian).

Arefjev, V.A. 1993. NOR-banding studies of *Acipenser baeri* karyotype. pp. 30–31. *In:* International Symposium on Sturgeons, Abstract Bulletin, September 6–11, 1993, VNIRO, Moscow.

Arefjev, V.A. & O.P. Filippova. 1993. Karyotypic analysis of the hybrid between Russian and Siberian sturgeon in relation to its supposed fertility and to cytogenetic aspects of hybridization in Acipenseridae. pp. 80–81. *In:* International Symposium on Sturgeons, Abstract Bulletin, September 6–11, VNIRO, Moscow.

Arefjev, V.A. & A.I. Nikolaev. 1991. Cytological analysis of the reciprocal hybrids between low- and high-chromosome acipenserids, the great sturgeon, *Huso huso* (L.), and the Russian sturgeon, *Acipenser gueldenstaedti* Brandt. Cytologia 56: 495–502.

Arlati, G., L.A. Belysheva & T.I. Kaydanova. 1994. Karyologic analysis of *Acipenser naccarii* (Bonaparte). pp. 119–123. *In:* A.D. Gershanovich & T.I.J. Smith (ed.) Proceedings of the International Symposium on Sturgeons, 6–11 September 1993, VNIRO Publishing, Moscow.

Artyukhin, E.N. 1994. On the relationship of the Amur sturgeon, *Acipenser schrencki*. Sturgeon Quart. 2(3): 7.

Artyukhin, E.N. 1995. On biogeography and relationships within the genus *Acipenser*. Sturgeon Quart. 3(2): 6–8.

Artyukhin, E.N. & A.E. Andronov. 1990. A morphological study of the green sturgeon, *Acipenser medirostris* (Chondrostei, Acipenseridae), from the Tumnin (Datta) River and some aspects of the ecology and zoogeography of Acipenseridae. Zoologicheskii Zhurnal 69: 81–91 (in Russian, English translation J. Ichthyol. 30: 11–21).

Asahida, T. & H. Ida. 1989. Karyological notes on four sharks in the order Carcharhiniformes. Japan. J. Ichthyol. 36: 275–280.

Asahida, T. & H. Ida. 1990. Karyotypes of two rays, *Torpedo tokionis* and *Dasyatis matsubarai*, and their systematic relationships. Japan. J. Ichthyol. 37: 71–75.

Asahida, T., H. Ida & T. Inoue. 1988. Karyotypes and cellular contents of two sharks in the family Scyliorhinidae. Japan. J. Ichthyol. 35: 215–219.

Asahida, T., H. Ida, H. Terashima & H.Y. Chang. 1993. The karyotype and cellular DNA content of a ray, *Mobula japonica*. Japan. J. Ichthyol. 40: 317–322.

Avise, J.C. 1992. Molecular population structure and the biogeographic history of a regional fauna: a case history with lessons for conservation biology. Oikos 63: 62–76.

Bănărescu, P. 1964. Pisces-Osteichthyes. Fauna Republicii Populare Romîne 13, Editura Academiei Republicii Populare Romîne, Bucureşti. 962 pp.

Bartley, D.M., G.A.E. Gall & B. Bentley. 1985. Preliminary description of the genetic structure of white sturgeon, *Acipenser transmontanus*, in the Pacific North-West. pp. 105–109. *In:* F.P. Binkowski & S.I. Doroshov (ed.) North American Sturgeons: Biology and Aquaculture Potential, Dr W. J. Junk Publishers, Dordrecht.

Bemis, W.E., E. Findeis & L. Grande. 1997. An overview of Acipenseriformes. Env. Biol. Fish. (this volume).

Berg, L.S. 1905. Fishes of Turkestan. Scientific results of the Aral expedition, No. 6. St. Petersburg. 261 pp. (in Russian).

Berg, L.S. 1948a. On the position of the Acipenseriformes in the system of fishes. Trudy Zoologicheskogo Instituta 7: 7–57 (in Russian).

Berg, L.S. 1948b. The freshwater fishes of the USSR and adjacent countries, Vol. 1, Part 1. Akademia Nauk USSR, Moscow & Leningrad (in Russian, English translation published by Israel Program for Scientific Translations, Jerusalem. 505 pp.)

Bernardi, G., P. Sordino & D.A. Powers. 1992. Nucleotide sequence of the 18S ribosomal ribonucleic acid gene from two teleosts and two sharks and their molecular phylogeny. Mol. Mar. Biol. Biotechnol. 1: 187–194.

Bernardi, G. & D.A. Powers. 1992. Molecular phylogeny of the prickly shark, *Echinorhinus cookei*, based on a nuclear (18S rRNA) and a mitochondrial (cytochrome *b*) gene. Mol. Phylogenet. Evol. 1: 161–167.

Bidwell, C.A., K.J. Kroll, E. Severud, S.I. Doroshov & D.M. Carlson. 1991. Identification and preliminary characterization of white sturgeon (*Acipenser transmontanus*) vitellogenin mRNA. Gen. Comp. Endocrinol. 83: 415–424.

Birstein, V.J. 1987. Cytogenetic and molecular aspects of vertebrate evolution. Nauka Press, Moscow. 284 pp. (in Russian).

Birstein, V.J. 1993a. Sturgeons and paddlefishes: threatened fishes in need of conservation. Cons. Biol. 7: 773–787.

Birstein, V.J. 1993b. Is *Acipenser medirostris* one or two species? Sturgeon Quart. 1(2): 8.

Birstein, V.J. & R. DeSalle. 1997. Molecular phylogeny of Acipenserinae. Mol. Phylogenet. Evol. 7 (in press).

Birstein, V.J., A.I. Poletaev & B.F. Goncharov. 1993. The DNA content in Eurasian sturgeon species determined by flow cytometry. Cytometry 14: 377–383.

Birstein, V.J. & V.P. Vasiliev. 1987. Tetraploid-octaploid relationships and karyological evolution in the order Acipenseriformes (Pisces). Karyotypes, nucleoli, and nucleolus-organizer region in four acipenserid species. Genetica 73: 3–12.

Blacklidge, K.H. & C.A. Bidwell. 1993. Three ploidy levels indicated by genome quantification in Acipenseriformes of North America. J. Hered. 84: 427–430.

Bogart, J.P., E.K. Balon & M.N. Bruton. 1994. The chromosomes of the living coelacanth and their remarkable similarity to those of one of the most ancient frogs. J. Hered. 85: 322–325.

Bowen, B.W. & J.C. Avise. 1990. Genetic structure of Atlantic and Gulf of Mexico populations of sea bass, methaden, and sturgeon: influence of zoogeographic factors and life-history patterns. Mar. Biol. 107: 371–381.

Brandt, J.F. 1869/1870. Einige Worte über die europaischen-asiatischen Störarten (Sturionides). Bull. Acad. Imper. Sci. St.-Petersb. 14: 171–175.

Bremer, K. 1988. The limits of amino-acid sequence data in angiosperm phylogenetic reconstruction. Evolution 42: 795–803.

Brown, J.R., A.T. Beckenbach & M.J. Smith. 1992a. Influence of Pleistocene glaciations and human intervention upon mitochondrial DNA diversity in white sturgeon (*Acipenser transmontanus*) populations. Can. J. Fish. Aqua. Sci. 49: 358–367.

Brown, J.R., A.T. Beckenbach & M.J. Smith. 1992b. Mitochondrial DNA length variation and heteroplasmy in populations of white sturgeon (*Acipenser transmontanus*). Genetics 132: 221–228.

Brown, J.R., A.T. Beckenbach & M.J. Smith. 1993. Intraspecific DNA sequence variation of the mitochondrial control region of white sturgeon (*Acipenser transmontanus*). Mol. Biol. Evol. 10: 326–341.

Brown, J.R., K. Beckenbach, A.T. Beckenbach & M.J. Smith. 1996. Length variation, heteroplasmy and sequence divergence in the mitochondrial DNA of four species of sturgeon (*Acipenser*). Genetics 142: 525–535.

Brown, J.R., T.L. Gilbert, D.J. Kowbel, P.J. O'Hara, N.E. Buroker, A.T. Beckenbach & M.J. Smith. 1989. Nucleotide sequence of the apocytochrome B gene in white sturgeon mitochondrial DNA. Nucl. Acids Res. 17: 4389.

Buroker, N.E., J.R. Brown, T.A. Gilbert, P.J. O'Hara, A.T. Beckenbach, W.K. Thomas & M.J. Smith. 1990. Length heteroplasmy of sturgeon mitochondrial DNA: an illegitimate elongation model. Genetics 124: 157–163.

Buth, D.G., T.E. Dowling & J.R. Gold. 1991. Molecular and cytological investigations. pp. 83–126. *In:* I.J. Winfield & J.S. Nelson (ed.) Cyprinid Fishes, Systematics, Biology and Exploitation, Chapman & Hall, London.

Carlson, D.M., M.K. Kettler, S.E. Fisher & G.S. Whitt. 1982. Low genetic variability in paddlefish populations. Copeia 1982: 721–725.

Carpenter, J.M. 1988. Choosing among multiple equally parsimonious cladograms. Cladistics 4: 291–296.

DeSalle, R., A.K. Williams & M. George. 1993. Isolation and characterization of animal mitochondrial DNA. pp. 176–204. *In:* E.A. Zimmer, T.J. White, R.L. Cann & A.C. Wilson (ed.) Methods in Enzymology, Vol. 224, Molecular Evolution: Producing the Biochemical Data, Academic Press, San Diego.

Dingerkus, G. 1979. Chordate cytogenetic studies: an analysis of their phylogenetic implications with particular reference to fishes and the living coelacanth. Occ. Pap. Calif. Acad. Sci. 134: 111–127.

Dingerkus, G. & W.M. Howell. 1976. Karyotypic analysis and evidence of tetraploidy in the North American paddlefish, *Polyodon spathula*. Science 177: 664–669.

Donoghue, M.J., R.G. Olmstead, J.F. Smith & J.D. Palmer. 1992. Phylogenetic relationships of Dipsacales based on *rbcl* sequences. Ann. Missouri Bot. Gard. 79: 333–345.

Ernisse, D.E. & A.G. Kluge. 1993. Taxonomic congruence versus total evidence, and amniote phylogeny inferred from fossils, molecules, and morphology. Mol. Biol. Evol. 10: 1170–1195.

Ferguson, M.M., L. Bernatchez, M. Gatt, B.R. Konkle, S. Lee, M.L. Malott & R.S. McKinley. 1993. Distribution of mitochondrial DNA variation in lake sturgeon (*Acipenser fulvescens*) from the Moose River basin, Ontario, Canada. J. Fish Biol. 43 (Suppl. A): 91–101.

Ferris, S.D. & G.S. Whitt. 1979. Evolution of the differential regulation of duplicate genes after polyploidization. J. Mol. Evol. 12: 267–317.

Findeis, E.K. 1993. Osteology of the North American shovelnose sturgeon *Scaphirhynchus platorynchus* Rafinesque 1820, with comparisons to other Acipenseridae and Acipenseriformes. Ph.D. Thesis, University of Massachusetts, Amherst. 449 pp.

Findeis, E.K. 1997. Osteology and phylogenetic interrelationships of sturgeons (Acipenseridae). Env. Biol. Fish. (this volume).

Fitch, W. & T. Smith. 1983. Optimal sequence alignment. Proc. Nat. Acad. Sci. USA 80: 1382–1386.

Fontana, F. 1976. Nuclear DNA content and cytometry of erythrocytes of *Huso huso* L., *Acipenser sturio* L., and *Acipenser naccarii* Bonaparte. Caryologia 29: 127–138.

Fontana, F. 1994. Chromosomal nucleolar organizer regions in four sturgeon species as markers of karyotype evolution in Acipenseriformes (Pisces). Genome 37: 888–892.

Fontana, F. & G. Colombo. 1974. The chromosomes of Italian sturgeons. Experientia 30: 739–742.

Fontana, F., D. Jankovič & S. Živkovič. 1977. Somatic chromosomes of *Acipernser ruthenus* L. Arh. biol. Nauka 27: 33–35.

Fontana, F., M. Lanfredi, R. Rossi, P. Bronzi & G. Arlati. 1995. Established cell lines from three sturgeon species. Sturgeon Quart. 3 (4): 6–7.

Gatesy, J., R. DeSalle & W. Wheeler. 1993. Alignment-ambiguous nucleotide sites and the exclusion of systematic data. Mol. Phylogenet. Evol. 2: 152–157.

Gilbert, T.L., J.R. Brown, P.J. O'Hara, N.E. Buroker, A.T. Beckenbach & M.J. Smith. 1988. Sequence of $tRNA^{The}$ and $tRNA^{Pro}$ from white sturgeon (*Acipenser transmontanus*) mitochondria. Nucl. Acids Res. 16: 11825.

Gold, J.R. 1979. Cytogenetics. pp. 353–405. *In:* W.S. Hoar, D.J. Randall & J.R. Brett (ed.) Fish Physiology, Vol. 8, Academic Press, New York.

Golubtsov, A.S. & E. Yu. Krysanov. 1993. Karyological study of some cyprinid species from Ethiopia. The ploidy differences between large and small *Barbus* of Africa. J. Fish Biol. 42: 445–455.

Grande, L. & W.E. Bemis. 1991. Osteology and phylogenetic relationships of fossil and recent paddlefishes (Polyodontidae) with comments on the interrelationships of Acipenseriformes. J. Verteb. Paleontol. 11, supplement 1: 1–121.

Grande, L. & W.E. Bemis. 1996. Interrelationships of Acipenseriformes, with comments on "Chondrostei". pp. 85–115. *In:* M.L.J. Stiassny, L.R. Parenti & G.D. Johnson (ed.) Interrelationships of Fishes, Academic Press, New York.

Guénette, S., R. Fortin & E. Rassart. 1993. Mitochondrial DNA variation in lake sturgeon (*Acipenser fulvescens*) from the St. Lawrence River and James Bay drainage basins in Quebec, Canada. Can. J. Fish. Aquat. Sci. 50: 659–664.

Hart, J.L. 1973. Pacific fishes of Canada. Fish. Res. Board Can. Bull. 180, Ottawa. 740 pp.

Hedges, S., C.A. Hass & L.R. Maxson. 1993a. Relations of fish and tetrapods. Nature 363: 501–502.

Hedges, S.B., R.A. Nussbaum & L.R. Maxson. 1993b. Caecilian phylogeny and biogeography inferred from mitochondrial DNA sequences of the 12S rRNA and 16S rRNA genes (Amphibia: Gymnophiona). Herpetological Monographs 7: 64–76.

Hedrick, R.P., T.S. McDowell, R. Rosemark, D. Aronstein & C.N. Lannan. 1991. Two cell lines from white sturgeon. Trans. Amer. Fish. Soc. 120: 528–534.

Hilgendorf, F. 1892. Über eine neue Stör-Art aus Nord-Japan (*Acipenser mikadoi*). Sitzungsber. Ges. naturf. Freunde, Berlin 7: 98–100.

Hillis, D.M., M.W. Allard & M.M. Miyamoto. 1994. Analysis of DNA sequence data: phylogenetic inference. pp. 456–487. *In:* E.A. Zimmer, T.J. White, R.L. Cann & A.C. Wilson (ed.) Methods in Enzymology, Vol. 224, Molecular Evolution: Producing the Biochemical Data, Academic Press, San Diego.

Hillis, D.M., A. Larson, S.K. Davis & E.A. Zimmer. 1990. Nucleic acids III: sequencing. pp. 318–370. *In:* D.M. Hillis & C. Moritz (ed.) Molecular Systematics, Sinauer Associates, Sunderland.

Hinegardner, R. & D.E. Rosen. 1972. Cellular DNA content and the evolution of teleostean fishes. Amer. Nat. 106: 621–644.

Holčík, J., R. Kinzelbach, L.I. Sokolov & V.P. Vasil'ev. 1989. *Acipenser sturio* Linnaeus, 1758. pp. 366–394. *In:* J. Holčík (ed.) The Freshwater Fishes of Europe, Vol. 1, Pt. II, General Introduction to Fishes, Acipenseriformes, AULA-Verlag, Wiesbaden.

Ida, H., I. Sato & N. Miyawaki. 1985. Karyotypes of two species in the order Torpediniformes. Japan. J. Ichthyol. 32: 107–111.

Jin, F. 1995. Late Mesozoic acipenseriforms (Osteichthyes: Actinopterygii) in Central Asia and their biogeographical implications. pp. 15–21. *In:* A. Sun & Y. Wang (ed.) Sixth Symposium on Mesozoic Terresrial Ecosystems and Biota, Short Papers, China Ocean Press, Beijing.

Irwin, D.M., T.D. Kocher & A.C. Wilson. 1991. Evolution of the cytochrome *b* gene of mammals. J. Mol. Evol. 32: 128–144.

Kallersjo, M., J.S. Farris, A.G. Kluge & C. Bult. 1992. Skewness and permutation. Cladistics 8: 275–287.

Kedrova, O.S., N.S. Vladytchenskaya & A.S. Antonov. 1980. Single and repeated sequence divergency in fish genomes. Molekulyarnaya Biologiya 14: 1001–1012 (in Russian, English translation Molec. Biol. 14: 787–797).

Kessler, K.F. 1877. Fishes of the Aralo-Caspian-Pontine region. Trudy Aralo-Kaspiiskoi ekspeditsii 4: 190–196 (in Russian).

Keyvanfar, A. 1988. Étude comparative des protéines sériques et cellulaires de quatre espèces d'esturgeons anadromes de la Mer Caspienne. Ann. Inst. océonagr. (Paris) 64: 25–64.

King, M. 1990. Animal cytogenetics, Vol. 4: Chordata 2. Amphibia. Gebruder Borntraeger, Berlin. 241 pp.

Kinzelbach, R. 1987. Das ehemalige Vorkommen des Störs, *Acipenser sturio* (Linnaeus, 1758), im Einzugsgebiet des Rheins (Chondrostei: Acipenseridae). Z. Angew. Zool. 74: 167–200.

Kluge, A.G. 1989. A concern for evidence and a phylogenetic hypothesis of relationships among *Epicrates* (Boidae, Serpentes). Syst. Zool. 38: 7–25.

Kocher, T.D., W.K. Thomas. A. Meyer, S.V. Edwards, S. Pääbo, F.X. Villablanca & A.C. Wilson. 1989. Dynamics of mitochondrial DNA evolution in animals: amplification and sequencing with conserved primers. Proc. Nat. Acad. Sci. USA 86: 6196–6200.

Kohno, S., M. Kuro-o & C. Ikebe. 1991. Cytogenetics and evolution of hynobiid salamanders. pp. 67–88. *In:* D.M. Green & S.K. Sessions (ed.). Amphibian Cytogenetics and Evolution, Academic Press, San Diego.

Konstantinov, K.G., N.I. Nikolyukin & N.A. Timofeeva. 1952. On the biology of sturgeon hybrids. Doklady Akademii Nauk SSSR 86: 417–420 (in Russian).

Kozhin, N.I. 1964. Acipenserids of the USSR and their reproduction. Trudy VNIRO 52: 21–59 (in Russian).

Kozlov, V.I. 1970. Natural hybrid between ship sturgeon, *Acipenser nudiventris derjavini* Borzenko, and Kura River stellate sturgeon, *A. stellatus* Pallas. Voprosy Ikhtiologii 10: 631–636 (in Russian).

Krykhtin, M.L. & V.G. Svirskii. 1997. Endemic sturgeons of the Amur River: kaluga, *Huso dauricus*, and Amur sturgeon, *Acipenser schrenkii*. Env. Biol. Fish. (this volume).

Krylova, V.D. 1980. Morphometric characteristics of a hybrid between the beluga, *Huso huso*, and the sevruga, *Acipenser stellatus*. Voprosy Ikhtiologii 20: 875–882 (in Russian, English translation J. Ichthyol. 20: 89–97).

Kutergina, I.G. & G.D. Ryabova. 1990. Genetic analysis of duplicated loci for lactate dehydrogenase Ldh3 and Ldh4 ingeritance in stellate sturgeon. Genetika 26: 952–955 (in Russian).

Kuzmin, Ye. V. 1991. Comparative study of isozymes of muscle malate dehydrogenase of the Ob population of the Siberian sturgeon, *Acipenser baeri*, and the Don and Kama sterlet, *Acipenser ruthenus*. Voprosy Ikhtiologii 31: 342–346 (in Russian, English translation J. Ichthyol. 31: 139–144).

Larhammar, D. & C. Risinger. 1994. Molecular genetic aspects of tetraploidy in the common carp, *Cyprinus carpio*. Mol. Phylogenet. Evol. 3: 59–68.

Le, H.L.V., G. Lecointre & R. Perasso. 1993. A 28S rRNA based phylogeny of the gnathostomes: first steps in the analysis of conflict and congruence with morphologically based cladograms. Mol. Phylog. Evol. 2: 31–51.

Legeza, M.I. 1971. On the hybrids of the Caspian Sea acipenserids. Trudy TSNIORKH 3: 196–206 (in Russian).

Li, M.F., V. Marrayatt, C. Annand & P. Odense. 1985. Fish cell culture: two newly developed cell lines from Atlantic sturgeon (*Acipenser oxyrhynchus*) and guppy (*Poecilia reticulata*). Can. J. Zool. 63: 2867–2874.

Lindberg, G.U. & M.I. Legeza. 1965. Fishes of the Sea of Japan and adjacent areas of the Okhotsk and Yellow seas, Pt. 2. Nauka Press, Moscow. 391 pp. (in Russian).

Lukyanenko, V.V. & V.I. Lukyanenko. 1994. Ecological-physiological heterogeneity of sturgeons. Uspekhi Sovremennoi Biologii 114: 428–440 (in Russian).

Madsden, C.S., J.E. Brooks, E. de Kloet & S. R. de Kloet. 1994. Sequence conservation of an avian centromeric repeated DNA component. Genome 37: 351–355.

Marshin, V.G., V.V. Ponomarenko & G.P. Smirnova. 1969. Inheritance of some behavior characters in interspecies hybridization of acipenserids. pp. 192–208. *In:* Genetics, Selection, and Hybridization of Fishes, Nauka Press, Moscow (in Russian).

Martin, A.P., G.J.P. Naylor & S.R. Palumbi. 1992. Rates of mitochondrial DNA evolution in sharks are slow compared with mammals. Nature 357: 153–155.

Martin, A.P. & S.R. Palumbi. 1993. Protein evolution in different cellular environments: cytochrome *b* in sharks and mammals. Mol. Biol. Evol. 10: 873–891.

Meyer, A. 1993. Molecular phylogenetic studies of fishes. pp. 3–41. *In:* A. R. Beamumont (ed.) Evolution and Genetics of Aquatic Organisms, Chapman and Hall, London.

Meyer, A., C.H. Biermann & G. Orti. 1993. The phylogenetic position of the zebrafish (*Danio rerio*), a model system in developmental biology: an invitation to the comparative method. Proc. R. Soc. Lond. B 252: 231–236.

Meyer, A. & A.C. Wilson. 1990. Origin of tetrapods inferred from their mitochondrial DNA affiliation to lungfish. J. Mol. Evol. 31: 359–364.

Meyer, A., T.D. Kocher, P. Basasibwaki & A. Wilson. 1990. Monophyletic origin of Lake Victoria cichlid fishes suggested by mitochondrial DNA sequences. Nature 347: 550–553.

Miracle, A.L. & D.E. Campton. 1995. Tandem repeat sequence variation and length heteroplasmy in the mitochondrial DNA D-loop of the threatened Gulf of Mexico sturgeon, *Acipenser oxyrhynchus desotoi*. J. Hered. 86: 22–27.

Mirsky, A.E. & H. Ris. 1951. The DNA content of animal cells and its evolutionary significance. J. Gen. Physiol. 34: 451–462.

Miyamoto, M.M. 1985. Consensus cladograms and general classification. Cladistics 1: 186–189.

Morescalchi, A. 1973. Amphibia. pp. 233–348. *In:* A.B. Chiarelli & E. Capanna (ed.) Cytotaxonomy and Vertebrate Evolution, Academic Press, London.

Morescalchi, A., G. Odierna & E. Olmo. 1979. Karyology of the primitive salamanders, family Hynobiidae. Experientia 35: 1434–1435.

Nelson, J.S. 1994. Fishes of the World, 3rd edition. John Wiley & Sons, Toronto. 600 pp.

Nesov, L.A. & M.N. Kaznyshkin. 1983. New sturgeons from the Cretacious and Paleogene of the USSR. pp. 68–76. *In:* V.V. Menner (ed.) Contemporary Problems of Paleoichthyology, Nauka Press, Moscow (in Russian).

Nguyen, T.M., T.P. Mommsen, S.M. Mims & J.M. Conlon. 1994. Characterization of insulins and proglucagon-derived peptides from a phylogenetically ancient fish, the padlefish (*Polyodon spathula*). Biochem. J. 300: 339–345.

Nikolyukin, N.I. 1970. Hybridization within the family Acipenseridae and perspectives of its use in sturgeon breeding. Trudy VNIRO 76: 56–69 (in Russian).

Nikonorov, S.I., G.D. Ryabova, I.G. Kutergina & M.V. Ofitserov. 1985. Electrophoretic analysis of genetic variability of the starred sturgeon *Acipenser stellatus*. Doklady Akademii Nauk SSSR. Seriya Biologiya 284: 209–211 (in Russian; English translation Doklady Biological Sciences 284: 570–572).

Normark, B.B., A.R. McCune & R.G. Harrison. 1991. Phylogenetic relationships of neopterygian fishes, inferred from mitochondrial DNA sequences. Mol. Biol. Evol. 8: 819–834.

Ohno, S. 1970. Evolution by gene duplication. Springer-Verlag, Heidelberg. 160 pp.

Ohno, S., J. Muramoto, C. Stenius, L. Christian, W.A. Kittrel & N.B. Atkin. 1969. Microchromosomes in holocephalian, chondrostean, and holostean fishes. Chromosoma 26: 35–40.

Ojima Y. & T. Yamano. 1980. A chromosome study of the holostean long nose gar *Lepisosteus osseus*. Chrom. Inform. Serv. 28: 7–8.

Olmo, E., V. Stingo, O. Cobror, T. Capriglione & G. Odierna. 1982. Repetitive DNA and polyploidy in selachians. Comp. Biochem. Physiol. 73B: 739–745.

Ong, T.-L., J. Stabile, I. Wirgin & J.R. Waldman. 1996. Genetic divergence between *Acipenser oxyrinchus oxyrinchus* and *A. o. desotoi* as assessed by mitochondrial DNA sequencing analysis. Copeia 1996: 464–469.

Ovsyannikov, F.V. 1870. On the artificial breeding of the sterlet. pp. 191–200. *In:* Trudy II Syezda Russkikh Estestvoispytatelei, Pt. 2, Moscow (in Russian).

Palumbi, S.R., A.P. Martin, W.O. McMillan, S.R. Romano, G. Grabowsky & L. Stice. 1991. The simple fools guide to PCR, version 2. Department of Zoology, University of Hawaii, Honolulu. 15 pp.

Patarnello, T., L. Bargelloni, F. Caldara & L. Colombo. 1994. Cytochrome *b* and 16S rRNA sequence variation in the *Salmo trutta* (Salmonidae, Teleostei) species complex. Molec. Phylogenet. Evol. 3: 69–74.

Patterson, C. 1973. Interrelationships of holosteans. pp. 233–305. *In:* P.H. Greenwood, R.S. Miles & C. Patterson (ed.) Interrelationships of Fishes, Academic Press, London.

Patterson, C. 1982. Morphology and interrelationships of primitive actinopterygian fishes. Amer. Zool. 22: 241–259.

Patterson, C., D.M. Williams & C.J. Humphries. 1993. Congruence between molecular and morphological phylogenies. Annu. Rev. Ecol. Syst. 24: 153–188.

Phelps, S.R. & F. Allendorf. 1983. Genetic identity of pallid and shovelnose sturgeon (*Scaphirhynchus albus* and *S. platorynchus*). Copeia 1983: 696–700.

Pirogovskii, M.I., L.I. Sokolov & V.P. Vasil'ev. 1989. *Huso huso* (Linnaeus, 1758). pp. 156–200. *In:*: J. Holčík (ed.) The Freshwater Fishes of Europe, Vol. 1, Pt. II, General Introduction to Fishes, Acipenseriformes, AULA-Verlag, Wiesbaden.

Rab, P. 1986. A note on the karyotype of the sterlet, *Acipenser ruthenus* (Pisces, Acipenseridae). Folia Zool. 35: 73–78.

Risinger, C. & D. Larhammar. 1993. Multiple loci for synapse protein SNAP-25 in the tetraploid goldfish. Proc. Nat. Acad. Sci. USA 90: 10598–10602.

Rochard, E., P. Williot, G. Castelnaud & M. Lepage. 1991. Éléments de systèmatique et de biologie des populations sauvages d'esturgeons. pp. 475–507. *In:* P. Williot (ed.) Acipenser, CEMAGREF Publ., Bordeaux.

Rossi, R., G. Grandi, R. Trisolini, P. Franzoni, A. Carrieri, B.S. Dezfuli & E. Vecchietti. 1991. Osservazioni sulla biologia e la pesca della storione cobice *Acipenser naccarii* Bonaparte nella parte terminale del fiume Po. Atti Soc. Ital. Sci. Nat. Mus. Civ. Stir. Natur. Milano 132: 121–142.

Ruban, G.I. 1997. Species structure, contemporary distribution and status of the Siberian sturgeon, *Acipenser baerii*. Env. Biol. Fish. (this volume).

Ryabova, G.D. & I.G. Kutergina. 1990. Analysis of allozyme variability in the stellate sturgeon, *Acipenser stellatus* (Pallas), from the northern Caspian Sea. Genetika 26: 902–911 (in Russian).

Schwartz, F.J. & M.B. Maddock. 1986. Comparison of karyotypes and cellular DNA contents within and between major lines of elasmobranchs. pp. 148–157. *In:* T. Uyeno, R. Arai, T. Tanuichi & K. Matsuda (ed.) Indo-Pacific Fish Biology: Proceedings of the Second International Conference on Indo-Pacific Fishes, Ichthyol. Soc. Japan, Tokyo.

Scott, W.B. & E.J. Grossman. 1973. Freshwater fishes of Canada. Fish Res. Board Can. Bull. 184, Ottawa. 966 pp.

Serebryakova, E.V. 1969. Some data on chromosome complements of the acipenserids. pp. 105–113. *In:* Genetics, Selection, and Hybridization of Fishes, Nauka Press, Moscow (in Russian).

Serebryakova, E.V. 1970. Chromosome complements of hybrids between acipenserids with different karyotypes. pp. 413–419. *In:* Distant Hybridization of Plants and Animals, Vol. 2, Kolos, Moscow (in Russian).

Serebryakova, E.V., V.A. Arefiev, V.P. Vasiliev & L.I. Sokolov. 1983. The study of the karyotype of giant sturgeon, *Huso huso* (L.) (Acipenseridae, Chondrostei) with reference to their systematic position. pp. 63–69. *In:* Genetics of Commercial Fishes and Objects of Aquaculture, Legkaya i Pishchevaya Promyshlennost, Moscow (in Russian).

Shedlock, A.M., J.D. Parker, D.A. Crispin, T.W. Pietsch & G.C. Burmer. 1992. Evolution of the salmonid mitochondrial control region. Mol. Phylogenet. Evol. 1: 179–192.

Shmidt, P.Yu. 1950. Fishes of the Sea of Okhotsk. Izdatelstvo Akademii Nauk, Moscow. 370 pp. (in Russian).

Shubina, T.N., A.A. Popova & V.P. Vasilev. 1989. *Acipenser stellatus* Pallas, 1771. pp. 395–443. *In:* J. Holčík (ed.) The Freshwater Fishes of Europe, Vol. 1, Pt. II, General Introduction to Fishes, Acipenseriformes, AULA-Verlag, Wiesbaden.

Slynko, V.I. 1976. Multiple molecular forms of malate dehydrogenase and lactate dehydrogenase in Russian sturgeon (*Acipenser gueldenstaedti* Br.) and beluga (*Huso huso* L.). Doklady Akademii Nauk SSSR, Seriya Biologiya 228: 470–472 (in Russian; English translation Doklady Biological Sciences 228: 201–204).

Sokolov, L.I. & V.P. Vasilev. 1989a. *Acipenser nudiventris* Lovetsky, 1828. pp. 206–226. *In:* J. Holčík (ed.) The Freshwater Fishes of Europe, Vol. 1, Pt. II, General Introduction to Fishes, Acipenseriformes, AULA-Verlag, Wiesbaden.

Sokolov, L.I. & V.P. Vasilev. 1989b. *Acipenser ruthenus* Linnaeus, 1758. pp. 227–262. *In:* J. Holčík (ed.) The Freshwater Fishes of Europe, Vol. 1, Pt. II, General Introduction to Fishes, Acipenseriformes, AULA-Verlag, Wiesbaden.

Sokolov, L.I. & V.P. Vasilev. 1989c. *Acipenser baeri* Brandt, 1869. pp. 263–284. *In:* J. Holčík (ed.) The Freshwater Fishes of Europe, Vol. 1, Pt. II, General Introduction to Fishes, Acipenseriformes, AULA-Verlag, Wiesbaden.

Stanley, H.P., H.E. Kasinsky & N.C. Bols. 1984. Meiotic chromatin diminution in a vertebrate, the holocephalian fish *Hydrolagus colliei* (Chondrichthyes, Holocephali). Tissue and Cell 16: 203–215.

Stingo, V. & L. Rocco. 1991. Chondrichthyan cytogenetics: a comparison with teleostean. J. Mol. Evol. 33: 76–82.

Stock, D.W., J.K. Gibbons & G.S. Whitt. 1991a. Strength and limitations of molecular sequence comparisons for inferring the phylogeny of the major groups of fishes. J. Fish Biol. 39(Suppl. A): 225–236.

Stock, D.W., K.D. Moberg, L.R. Maxson & G.S. Whitt. 1991b. A phylogenetic analysis of the 18S ribosomal RNA sequence of the coelacanth *Latimeria chalumnae*. Env. Biol. Fish. 32: 99–117.

Stock, D.W. & G.S. Whitt. 1992. Evidence from 18S ribosomal RNA sequences that lampreys and hagfishes form a natural group. Science 257: 787–789.

Suzuki, A. & J. Hirata. 1991. Chromosomes and DNA content of *Amia calva*. Chromos. Inform. Serv. 50: 34–37.

Suzuki, A., K. Yamanaka, T. Urushido & N. Kondo. 1988. A banding chromosome study of African fish family Polypteridae (Pisces: Polypteriformes). Proc. Japan. Acad. 64B: 299–302.

Suzuki, A., K. Yamanaka, T. Urushido & N. Kondo. 1989. N-banding chromosomes of two polypterids, *Polypterus senegalus* and *Calamoichthys calabaricus* (Polypteriformes, Pisces). Chromos. Inform. Serv. 46: 23–25.

Swofford, D.L. 1993. PAUP: Phylogenetc Analysis Using Parsimony: Version 3.1. Computer program distributed by the Illinois Natural History Survey, Champaign. 154 pp.

Tiersch, T.R., R.W. Chandler, S.S. Wachtel & S. Elias. 1989. Reference standards for flow cytometry and application in comparative studies of nuclear DNA content. Cytometry 10: 706–710.

Tortonese, E. 1989. *Acipenser naccarii* Bonaparte, 1836. pp. 284–293. *In:* J. Holčík (ed.) The Freshwater Fishes of Europe, Vol.

1, Pt. II, General Introduction to Fishes, Acipenseriformes, AULA-Verlag, Wiesbaden.

Tsoi, S.C.M., S.-C. Lee & W.-C. Chao. 1989. Duplicate gene expression and diploidization in an Asian tetraploid catostomid, *Myxocyprinus asiaticus* (Cypriniformes, Catostomidae). Comp. Biochem. Physiol. 93B: 27–32.

Ueno, K., A. Nagase & Y.-J. Ye. 1988. Tetraploid origin of the karyotype of the Asian sucker, *Myxocyprinus asiaticus*. Japan. J. Ichthyol. 34: 512–514.

Uyeno, T. & G.R. Smith. 1972. Tetraploid origin of the karyotype of catostomid fishes. Science 175: 644–646.

Vasiliev, V.P. 1985. Evolutionary karyology of fishes. Nauka Press, Moscow. 298 pp. (in Russian).

Vasiliev, V.P., L.L. Sokolov & E.V. Serebryakova. 1980. Karyotypes of the Siberian sturgeon, *Acipenser baeri*, of the Lena River and some aspects of karyotype evolution in Acipenseriformes. Voprosy Ikhtiologii 20: 814–822 (in Russian; English translation J. Ichthiol. 20: 37–45).

Vervoort, A. 1980a. Karyotypes of Polypteridae (Osteichthyes). Experientia 36: 646–647.

Vervoort, A. 1980b. Karyotypes and nuclear DNA contents of Polypteridae (Osteichthyes). Experientia 36: 646–647.

Vladykov, V.D. 1955. A comparison of Atlantic sea sturgeon with a new subspecies from the Gulf of Mexico (*Acipenser oxyrhynchus desotoi*). J. Fish. Res. Board Can. 12: 754–761.

Vladykov, V.D. & J.R. Greeley. 1963. Order Acipenseroidei. pp. 24–60. *In:* H.B. Bigelow, C.M. Breder, D.M. Cohen, G.W. Mead, D. Merriman, Y.H. Olsen, W.C. Schroeder, L.P. Schultz & J. Tee-Van (ed.) Fishes of the Western North Atlantic, Mem. Sears Found. Mar. Res. 1.

Vlasenko, A.D., A.V. Pavlov, L.I. Sokolov & V.P. Vasilev. 1989a. *Acipenser gueldenstaedti* Brandt, 1833. pp. 294–344. *In:* J. Holčík (ed.) The Freshwater Fishes of Europe, Vol. 1, Pt. II, General Introduction to Fishes, Acipenseriformes, AULA-Verlag, Wiesbaden.

Vlasenko, A.D., A.V. Pavlov & V.P. Vasilev. 1989b. *Acipenser persicus* Borodin, 1897. pp. 345–366. *In:* J. Holčík (ed.) The Freshwater Fishes of Europe, Vol. 1, Pt. II, General Introduction to Fishes, Acipenseriformes, AULA-Verlag, Wiesbaden.

Wagner, R.P., M.P. Maguire & R.L. Stallings. 1993. Chromosomes. A Synthesis. Wiley-Liss, New York. 523 pp.

Ward, R.D., D.O. Skibinski & M. Woodwark. 1992. Protein heterozygosity, protein structure, and taxonomic differentiation. Evol. Biol. 26: 23–159.

Ward, R.D., M. Woodwark & D.O.F. Skibinski. 1994. A comparison of genetic diversity levels in marine, freshwater, and anadromous fishes. J. Fish Biol. 44: 213–232.

Waterman M.S., M. Eggert & E. Lander. 1992. Parametric sequence comparisons. Proc. Nat. Acad. Sci. USA 89: 6090–6093.

Watrous, L. & Q. Wheeler. 1981. The out-group comparison method of character analysis. Syst. Zool. 30: 1–11.

Wei, Q., F. Ke, J. Zhang, P. Zhuang, J. Luo, R. Zhou & W. Yang. 1997. Biology, fisheries, and conservation of sturgeons and paddlefish in China. Env. Biol. Fish. (this volume).

Wheeler, W.C. 1995. Sequence alignment, parameter sensitivity, and the phylogenetic analysis of molecular data. Syst. Biol. 44: 321–331.

Wheeler, W.C., P. Cartwright & C.Y. Hayashi. 1993. Arthropod phylogeny: a combined approach. Cladistics 9: 1–39.

Wheeler, W.C., J. Gatesy & R. DeSalle. 1995. Elision: a method for accomodating multiple molecular sequence alignments with alignment-ambiguous sites. Molec. Phylogenet. Evol. 4: 1–9.

Wheeler, W.C. & D.L. Gladstein. 1993. MALIGN, Version 1.85. Program and documentation. AMNH, New York. 27 pp.

Williot, P., P. Bronzi & G. Arlati. 1993. A very brief survey of status and prospects of freshwater sturgeon farming in Europe (EEC). pp. 32–36. *In:* P. Kestmont & R. Billard (ed.) Workshop on Aquaculture of Freshwater Species (except Salmonids), Spec. Publ. No. 20, Europ. Aquacult. Soc., Ghent.

Wirgin, I.I., J.E. Stabile & J.R. Waldman. 1997. Molecular analysis in the conservation of sturgeons and paddlefish. Env. Biol. Fish. (this volume).

Yakovlev, V.N. 1977. Phylogenesis of acipenseriforms. pp. 116–146. *In:* V. V. Menner (ed.) Essays on Phylogeny and Systematics of Fossil Fishes and Agnathans, Akademiya Nauk SSSR, Moscow (in Russian).

Yakovlev, V.N. 1986. Fishes. pp. 178–179. *In:* L.P. Tatarinov (ed.) Insects in the Early Cretaceous Ecosytems of the West Mongolia, The Soviet-Mongolian Palaeontological Expedition Transactions, Vol. 28, Nauka Press, Moscow (in Russian).

Yasuda, A., K. Yamaguchi, T. Noso, H. Papkoff, A.L. Polenov, C.S. Nicoll & H. Kawaguchi. 1992. The complete sequence of growth hormone from sturgeon (*Acipenser guldenstaedti*). Biochim. Biophys. Acta 1120: 297–304.

Yu, X., T. Zhou, K. Li, Y. Li. & M. Zhou. 1987. On the karyosystematics of cyprinid fishes and a summary of fish chromosome studies in China. Genetica 72: 225–236.

Yu, X.-Y. & X.-J. Yu. 1990. A schizothoracine fish species, *Diptychus dipogon*, with a very high number of chromosomes. Chrom. Inform. Serv. 48: 17–18.

Zograf, N.Yu. 1887. Materials to understanding of the organization of the sterlet. Izvestiya Obschestva Lyubitelei Estestoznaniya, Antropologii i Etnographii 52 (3). 72 pp. (in Russian).

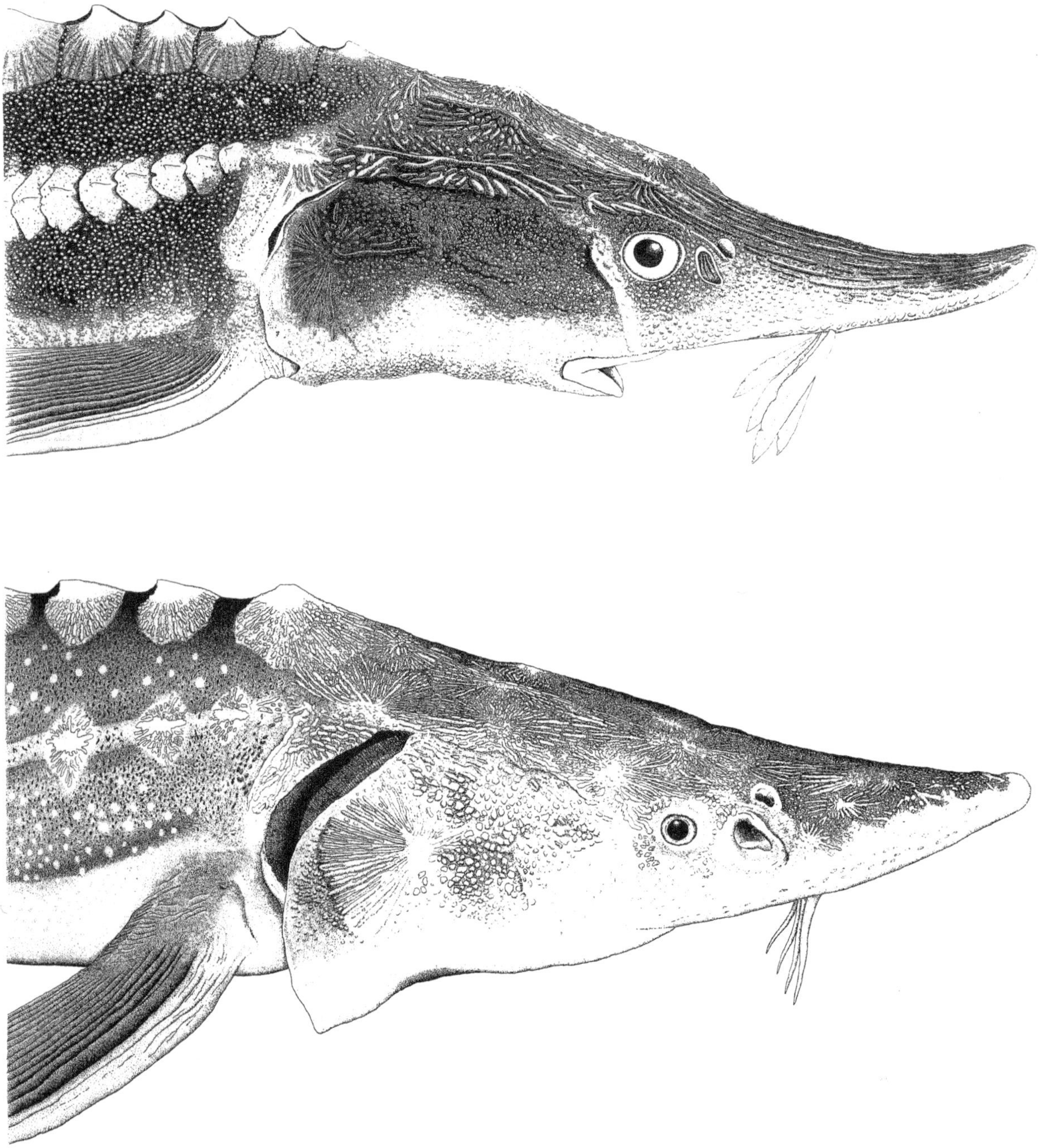

Portraits of a juvenile *Huso huso* 23 cm TL from the Ryal Ontario Museum collection (given originally as Caspian Sea fish to Montreal Expo 1967) above the head of *Acipenser schrenckii* 81 cm TL from the Amur River stock held at the Propa-Gen International, Komadi, Hungary. Originals by Paul Vecsei, 1996.

Environmental Biology of Fishes **48**: 157–163, 1997.

How many species are there within the genus *Acipenser*?

Vadim J. Birstein[1] & William E. Bemis[2]
[1] *The Sturgeon Society, 331 West 57th Street, Suite 159, New York, NY 10019, U.S.A.*
[2] *Department of Biology and Graduate Program in Organismic and Evolutionary Biology, University of Massachusetts, Amherst, MA 01003, U.S.A.*

Received 23.4.1996 Accepted 17.5.1996

Key words: Acipenser baerii, A. brevirostrum, A. dabryanus, A. fulvescens, A. gueldenstaedtii, A. medirostris, A. mikadoi, A. naccarii, A. nudiventris, A. oxyrinchus, A. persicus, A. ruthenus, A. schrenckii, A. sinensis, A. stellatus, A. sturio, A. transmontanus, Huso huso, H. dauricus

In their paper in this volume Bemis et al. (1997) ask: 'How many valid species of *Acipenser* should we recognize?' Although a partial answer to this question is presented in their Table 5, we discovered in the course of preparing this volume that some additional commentary is needed. In fact, there are two questions: (1) how many species should be recognized? and (2) what scientific names should be used for some of the species? The sympatric distributions of most species of sturgeons set the stage for much confusion about species boundaries, but the situation is actually much more complicated. Confusion about the number of species of sturgeons living within the same basin can result from the often close morphological and meristic similarities of certain species of *Acipenser*, particularly during juvenile period. Moreover, we still have remarkably inadequate knowledge of the comparative anatomy of the species of *Acipenser*: no modern study has ever attempted a comprehensive examination of all species, and it is impossible to rely on literature for the sorts of comparisons that must be made (for more on this general problem, see Grande & Bemis 1991, 1997). Most classical descriptions and comparative anatomical studies relied upon small sample sizes. Voucher specimens of large sturgeons are especially rare in most historical collections, and type specimens (if available at all) are seldom prepared in ways that are suitable for making detailed anatomical comparisons (e.g., many skins are simply overstuffed with straw, so that all internal structures are lost). Intraspecific morphological and meristic polymorphisms occur in all species of acipenserids, and in most cases we have very poor knowledge of differences that develop during ontogeny, particularly changes in such features as the shape of the rostrum (Bemis et al. 1997). Another problem is the ease of hybridization between different species of sturgeons (reviewed in Birstein et al. 1997 this volume). In many of these cases, it is not easy to discriminate between parental species and the hybrids.

Two opposite tendencies appeared in the literature on the genus *Acipenser*. (1) Recognizably different species have been considered to be the same species. This situation is illustrated below by two species pairs, *A. gueldenstaedtii* and *A. persicus* and *A. medirostris* and *A. mikadoi*. (2) Some authors elevated many forms to the rank of species. For instance, Duméril (1870) described six subgenera of *Acipenser* with more than 30 species of *Acipenser* in five of them (he considered *Huso* as the sixth subgenus of *Acipenser*). Most of the species described by Duméril (1870) have long since been recognized as conspecific with other well-known species.

We still do not know the number of species of *Acipenser*, and may never know it because of overfishing and habitat destruction in Europe and Asia, which have quickly eliminated sturgeons from certain river basins (see discussions in this volume by Bacalbaşa-Dobrovici 1997, Khodorevskaya et al. 1997, Krykhtin & Svirskii 1997, Wei et al. 1997). Therefore, we probably have already lost forever the opportunity to study some species of *Acipenser*. In the meantime, it is clear that genetic and molecular phylogenetic approaches are increasingly crucial for the recognition of sturgeon species and their relationships (for discussion, see Birstein et al. 1997 this volume).

In Eurasia, the genus *Acipenser* is centered upon three main basins: (1) the Black Sea and Sea of Azov, (2) Caspian Sea, and (3) the Aral Sea. Each of three main species of *Acipenser*, *A. gueldenstaedtii* Brandt, 1833, *A. stellatus* Pallas, 1771, and *A. nudiventris* Lovetsky, 1828 were described as having subspecies or forms in these basins (see Berg 1948, Shubina et al. 1989, Sokolov & Vasilev 1989a, Vlasenko et al. 1989a). If we follow the view on nomenclature of species discussed by Holčík & Jedlička (1994), then the concept of subspecies and trinomial nomenclature is inefficient. Therefore, we consider all intraspecies forms and subspecies of *A. gueldenstaedtii*, *A. stellatus*, and *A. nudiventris* invalid until detailed molecular and morphological studies of different forms within these species can be performed.[1] The same is true for *A. ruthenus* Linnaeus, 1758, for which a few intraspecies forms were described by different authors (see Berg 1948, Sokolov & Vasilev 1989b).

An example helps to illustrate the taxonomic frustration of sturgeon biologists. *Acipenser persicus* was described as a valid species by Borodin in 1897 (Borodin 1897, 1926), but it was later considered to be a subspecies (Berg 1934), and, still later, again regarded as a valid species (see Vlasenko et al. 1989b, Birstein & Bemis 1997 this volume, for discussion). Moreover, Artyukhin & Zarkua (1986) described two subspecies within *A. persicus*: the population inhabiting the Caspian Sea they named as *A. persicus persicus* Borodin, 1897, and the population inhabiting the Black Sea, as *A. persicus colchicus* Marti, 1940. Although some Russian authors follow this nomenclature (Pavlov et al. 1994), additional support from genetic and molecular data is desirable.

The validity of some Asian species and subspecies of *Acipenser* is questionable. For example, Ruban (1997 this volume) reviewed and presented new data on the Siberian sturgeon, *A. baerii* Brandt, 1869, which has an extremely wide range. Ruban's new work supports the traditionally recognized subspecies (*A.b. baerii, A.b. baicalensis* and *A.b. stenorrhynchus*, e.g., Sokolov & Vasiliev 1989c). No genetic study on the subspecies of *A. baerii* is yet available.

The three far eastern Asian species, *A. schrencki* Brand, 1869 of the Amur River, and *A. dabryanus* Duméril, 1868, and *A. sinensis* Gray, 1834 of the Yangtze River are certainly valid (see Krykhtin & Svirskii 1997, Wei et al. 1997, Zhuang et al. 1997, all this volume). Chinese sturgeon, *A. sinensis*, from the Pearl River differ morphologically from those of the Yangtze River, but whether this difference warrants separate species status is not clear (Wei et al. 1997).

The nomenclature and species status of the so-called 'green sturgeon' and 'Sakhalin sturgeon' of the Pacific Northwest of America and northeastern Pacific in Asia has been particularly confusing. Ayres (1854) described the American green sturgeon, *A. medirostris*. Nearly 40 years later, Hilgendorf (1892) described an Asian species caught in the northern waters of Japan as *A. mikadoi*, and Schmidt (1904) soon thereafter referred a sturgeon caught in the Aniwa Bay of Sakhalin Island to *A. mikadoi*. However, Berg (1911, 1948) considered this Sakhalin sturgeon to be conspecific with the American green sturgeon, *A. medirostris*. Schmidt (1950) eventually reconsidered his 1904 view, and named Sakhalin sturgeon as a subspecies of *A. medirostris*, *A. medirostris mikadoi* (Schmidt, 1950). Therefore, three names coexisted in the literature for the Sakhalin sturgeon: *A. mikadoi* (Okada & Matsubara

[1] In the literature on genetics, molecular phylogenetics and systematics, the taxonomic unit subspecies is often preserved (Avise 1994, Mallet 1995). Avise & Ball (1990) and Avise (1994, p 253) suggested that we recognize 'by the evidence of concordant phylogenetic partitions at multiple independent genetic attributes'. 'When phylogenetic concordance is exhibited across genetic characters solely because of extrinsic barriers to reproduction, subspecies status is suggested'. It is evident that according to these terminology, populations of the same species of sturgeon in disjunct sea basins (e.g., Caspian and Black seas), could be considered as subspecies.

1938, Matsubara 1955), *A. medirostris* (Berg 1948, Andriyashev & Panin 1953, Masuda et al. 1984, Houston 1988, Artyukhin & Andronov 1990, Pavlov et al. 1994), and *A. medirostris mikadoi* (Lindberg & Legeza 1965, Shilin 1995). Recently Birstein (Birstein et al. 1993, Birstein 1993) noted the difference in ploidy between the Sakhalin sturgeon and American green sturgeon, and suggested that they should be considered different species, *A. mikadoi* Hilgendorf, 1892, and *A. medirostris* Ayres, 1854, respectively. Molecular data on three mitochondrial genes presented in this volume (Birstein & DeSalle 1997) also show great differences between these two species. Other molecular data obtained show a close genetic relationship of *A. medirostris* to another American Pacific sturgeon species, *A. transmontanus* (Brown et al. 1996, Birstein et al. 1997). Therefore, *A. mikadoi* and *A. medirostris* should be considered as morphologically similar, but genetically different, species. The Sakhalin sturgeon inhabits the Sea of Japan up to the Korean Peninsula and waters to the north from Hokkaido Island (Berg 1948, Lindberg & Legeza 1965). It occurs in the mouths of small rivers of the Asian far east and Korean Peninsula, as well as the Amur River, and rivers of the Sakhalin Island. Now it spawns in the Tumnin (Datta) River in the Russian far east (Artuykhin & Andronov 1990), and historically it also spawned in the Ishikari and Teshio rivers of Hokkaido Island (Okada 1955). *Acipenser medirostris* ranges from the Gulf of Alaska to southern California (Houston 1988), with three known spawning rivers: the Sacramento and Klamath rivers in California and the Rogue River in Oregon (Moyle et al. 1994).

Two other species of sturgeons are usually mentioned in descriptions of the fish fauna of Japan, *A. kikuchii* Jordan & Snyder, 1901, and *A. multiscutatus* Tanaka, 1908 (Okada 1959–1960, Masuda et al. 1984, Rochard et al. 1991). Only one specimen of *A. kikuchii* is known (Jordan & Snyder 1901, 1906), and this species was re-identified as *A. sinensis* (Takeuchi 1979). Only a few specimens of *A. multiscutatus* were described (Tanaka 1908, Fowler 1941, Matsubara 1955). It seems that these specimens are morphologically similar to *A. schrenckii* (Lindberg & Legeza 1956) and are probably conspecific with *A. schrenckii*. It is most improbable that a sturgeon species could be restricted only to Japan and not inhabiting Asian continental waters (Artyukhin & Andronov 1990). There are no new reports on the catch of *A. multiscutatus* in Japanese literature (see a compilation of data in Honma 1988) since the review of Okada (1959-1960). Therefore, *A. multiscutatus* is most probably a synonym of *A. schrencki*.

It is easy to distinguish the second Pacific North American species, *A. transmontanus* Richardson, 1836, the freshwater North American *A. fulvescens* Rafinesque, 1817, and one of the two Atlantic North American sturgeons, *A. brevirostrum* Le Sueur, 1818 (Vladykov & Greeley 1963, Scott & Crossman 1973, Lee et al. 1980). Molecular data on the structure of the control region of mtDNA not only supported close relationships of two Pacific North American sturgeon species, *A. medirostris* and *A. transmontanus*, but also showed a significant genetic difference between these species, *A. fulvescens*, and the second Atlantic North American species, *A. oxyrinchus* (Brown et al. 1996).

American and the European Atlantic sturgeon were long considered to be one species, *A. sturio* Linnaeus, 1758. In this older terminology, the American Atlantic sturgeon was regarded as subspecies *A. sturio oxyrinchus*, with the European Atlantic sturgeon being known as *A. sturio sturio* (see Smith 1891, Vladykov & Greeley 1963). Magnin & Beaulieu (1963) suggested elevation of these subspecies to species ranks, with the European form retaining the name *A. sturio* Linnaeus, 1758, and American form named *A. oxyrinchus* Mitchill, 1815. Two subspecies, the Atlantic sturgeon, *A.o. oxyrinchus*, and the Gulf coast sturgeon, *A.o. desotoi*, were described within *A. oxyrinchus* (Vladykov 1955, Vladykov & Greeley 1963).[2] These two subspecies of *A. oxyrinchus* are morphologically similar, with the most significant known difference be-

[2] Since the description of the species, the name *A. oxyrinchus* has changed a few times. Mitchill described this species in 1815 under the name *A. oxyrinchus* (Mitchill, 1815). Later, the name was changed to *A. oxyrhynchus* and an incorrect date of publication (1814) began to be cited widely (e.g., Vladykov & Greely 1963). Also, *A. oxyrinchus desotoi* was first described under the name *A. oxyrhynchus de sotoi* (Vladykov 1955). In this volume we follow Smith & Clugston (1997) and use the names *A. o. oxyrinchus* and *A. oxyrinchus desotoi*.

ing the length of the spleen (in *A.o. oxyrinchus* the spleen is statistically smaller than it is in *A.o. de sotoi*, Wooley 1985). Molecular data are more informative for the discrimination between subspecies. Comparison of the control region of mtDNA sequences of both subspecies showed three fixed nucleotide changes in that region (Ong et al. 1996). Bowen & Avise (1990) suggested that there is genetic structuring among *A. oxyrinchus* from various drainages of the North American Atlantic coast. Recently, analyses of the control regions of mtDNA supported this hypothesis: Atlantic sturgeon populations in the Saint Lawrence and Saint John rivers (Canada), the Hudson River (U.S.A.), and rivers of Georgia (U.S A.) are genetically distinct (Waldman et al. 1996a,b).

Unpublished results of Birstein & DeSalle on the sequences of three more genes of mtDNA (cytochrome *b*, 12S rRNA, and 16S rRNA) also show a genetic difference between the two subspecies of *A. oxyrinchus* (one fixed nucleotide change in cytochrome *b* gene). The analysis of these genes demonstrated that the European *A. sturio* is the only sturgeon species closely related to *A. oxyrinchus*. Moreover, it appeared that there is a significant genetic differentiation within *A. sturio*. Birstein & DeSalle studied samples from two specimens of *A. sturio* caught in the Gironde estuary system (Dorgonne and Garonne rivers) and in the North Sea. The genetic difference between two individuals of *A. sturio* (6 nucleotide changes in the region of cytochrome *b* analyzed) was even more than the difference between subspecies of *A. oxyrinchus* (one change). These data seem to support the difference in some meristic characters between specimens from the Baltic Sea, from one side, and specimens from the Atlantic Ocean, Mediterranean and Black seas, from the other (Marti 1939, Magnin 1963, Ninua 1976, Holčík et al. 1989). Because *A. sturio* has almost disappeared in the wild (Holčík et al. 1989), more work should be done in museum collections on the comparison of specimens from different populations. This is especially important in terms of recovery projects for this species (Hochleithner 1995, Williot et al. 1997, this volume).

The last species in the genus *Acipenser* is the Adriatic sturgeon, *A. naccarii* Bonaparte, 1836. It is restricted to the Adriatic only and resembles *A. gueldenstaedtii* in meristic characters (Tortonese 1989).

Since Berg (1904), *Huso huso* Brandt, 1869 and *H. dauricus* Georgi, 1775 were considered as representatives of a distinct genus *Huso*, not *Acipenser* as they were usually considered in the 19th century (also see Findeis 1997, this volume). Results of recent molecular studies, however (see Birstein et al. 1997 this volume) showed that the two species of *Huso* do not form a separate monophyletic group, but are inserted among species of *Acipenser*. This result reactivates the old discussion on the validity of the genus *Huso*. In the absence of detailed work on this problem, it makes sense for now to regard *Huso* as a genus based on morphological and anatomical data (Findeis 1997 this volume). Also, a few subspecies were described within *H. huso* (reviewed in Pirogovskii et al. 1989). For instance, some authors still consider the Sea of Azov population of *H. huso* as *Huso huso maeoticus* Salnikov & Myatskii, 1934 (Pavlov et al. 1994). Until genetic differences can be shown in combination with morphology, we recommend the name *H. huso* for the Mediterranean, Black, Azov, and Caspian sea populations of beluga.

In conclusion, we recognize 17 valid extant species within *Acipenser*. For the moment, we accept that two species (*A. baerii* and *A. oxyrinchus*) contain subspecies. Further genetic and molecular studies will generate new data for correction of our contemporary knowledge about some of the species, including *A. sturio*.

A final note regarding the names of sturgeon species concerns the need to return to the originally published spellings for names of genera and species[3]. In addition to two recent clarifications on the correct spelling of species names for Siberian (*A. baerii*, see Ruban 1997, this volume) and American Atlantic sturgeon (*A. oxyrinchus*, see Gilbert 1992), we note the following correct spelling for two other

[3] Such decisions to use the originally published spellings of names, regardless of subsequent practices, are based on the International Code of Zoological Nomenclature (Ride et al. 1985). For a specific explanation of rules, see Chapters 31 and 33 of the International Code of Zoological Nomenclature. 1985, 3rd ed. International Trust for Zoological Nomenclature, London.

species of *Acipenser*. The scientific name of the Russian sturgeon should be spelled *Acipenser gueldenstaedtii* Brandt, 1833, and the scientific name of the Amur River sturgeon should be spelled *Acipenser schrenckii* Brandt, 1869.

References cited

Andriyashev, A.P. & K.I. Panin. 1953. Discovery of a Pacific sturgeon (*Acipenser medirostris* Ayres) in the Bering Sea. Zool. Zhurnal 32: 932–936 (in Russian).

Artyukhin, E.N. & A.E. Andronov. 1990. A morphological study of the green sturgeon, *Acipenser medirostris* (Chondrostei, Acipenseridae), from the Tumnin (Datta) River and some aspects of the ecology and zoogeography of Acipenseridae. Zool. Zhurnal 69: 81–91 (in Russian, English translation J. Ichthyol. 30 :11–21).

Artyukhin, E.N. & Z.G. Zarkua. 1986. On the question of taxonomic status of the sturgeon in the Rioni River (the Black Sea basin). Voprosy Ikhtiologii 26: 61–67 (in Russian).

Avise, J.C. 1994. Molecular markers, natural history and evolution. Chapman & Hall, New York. 511 pp.

Avise, J.C. & R.M. Ball, Jr. 1990. Principles of genealogical concordance in species concepts and biological taxonomy. Oxford Surv. Evol. Biol. 7: 45–67.

Ayres, W.O. 1854. Descriptions of the sturgeons. Proc. Calif. Acad. Sci. 1: 14–16.

Bacalbaşa-Dobrovici, N. 1997. Endangered migratory sturgeons of the lower Danube River and its delta. Env. Biol. Fish. (this volume).

Bemis, W.E., E.K. Findeis & L. Grande. 1997. An overview of Acipenseriformes. Env. Biol. Fish. (this volume).

Berg, L.S. 1904. Zur Systematic der Acipenseriden. Zool. Anz. 27: 665–667.

Berg, L.S. 1911. Fishes (Marsipobranchii and Pisces). Fauna of Russia and adjacent countries. Vol. 3, Vypusk 1. Izdatelstvo Akademii Nauk, St. Petersburg. 337 pp. (in Russian).

Berg, L.S. 1934. *Acipenser gueldenstaedti persicus*, a sturgeon from the south Caspian Sea. Ann. Mag. Nat. Hist., Ser. 10: 317–319.

Berg, L.S. 1948. The freshwater fishes of the USSR and adjacent countries, Vol. 1. Akademia Nauk USSR, Moscow & Leningrad. Part 1. (in Russian, English translation published by Israel Program for Scientific Translations, Jerusalem. 505 pp.)

Birstein, V. J. 1993. Is *Acipenser medirostris* one or two species? Sturgeon Quart. 1(2): 8.

Birstein, V.J. & W.E. Bemis. 1997. Leo Semenovich Berg and the biology of Acipenseriformes: a dedication. Env. Biol. Fish. (this volume).

Birstein, V.J., R. Hanner & R. DeSalle. 1997. Phylogeny of the Acipenseriformes: cytogenetic and molecular approaches. Env. Biol. Fish. (this volume).

Birstein, V.J., A.I. Poletaev & B.F. Goncharov. 1993. The DNA content in Eurasian sturgeon species determined by flow cytometry. Cytometry 14: 377–383.

Birstein, V.J. & R. DeSalle. 1997. Molecular phylogeny of Acipenserinae. Mol. Phylogenet. Evol. 7 (in press).

Bonaparte, C.L. 1836. Iconographia della fauna italica per le quatro classi degli animali vertebrate. 3. Pesci. Roma. Fasc. 15–18, 80–93.

Borodin, N.A. 1897. A report about a summer 1895 zoological expedition on board of the cruiser Uralets in the northern part of the Caspian Sea. Vestnik Rybopromyshlennosti 1: 1–31 (in Russian).

Borodin, N. 1936. *Acipenser persicus*, a sturgeon from the Caspian Sea. Ann. Mag. Nat. Hist., Ser. 9, 20: 26–28.

Bowen, B.W. & J.C. Avise. 1990. Genetic structure of Atlantic and Gulf of Mexico populations of sea bass, menhaden, and sturgeon: influence of zoogeographic factors and life history patterns. Mar. Biol. 107: 371–81.

Brandt, J.F. 1833. *In:* J.F. Brandt & J.T. Ratzeburg. Medizinische Zoologie. 2. Berlin. 364 pp.

Brandt, J.F. 1869. Einige Worte ber die europaischen-asiatischen Störarten (Sturionides). Bull. Acad. Imper. Sci. St.-Petersb. 14: 171–175.

Brown, J.R., K. Beckenbach, A.T. Beckenbach & M.J. Smith. 1996. Length variation, heteroplasmy and sequence divergence in the mitochondrial DNA of four species of sturgeon (*Acipenser*). Genetics 142: 525–535.

Duméril, A. 1868. Note sur trois poissons de la collection du museum, un esturgeon, un polyodonte et un malarmat. Nouv. Arch. Mus. Nat. Hist. Nat. Paris 4: 98–100.

Duméril, A. 1870. Histoire naturelle des poissons ou ichthyologie gnrale. T. 2. Librairie Encyclopdique de Roret, Paris. 624 pp.

Findeis, E.K. 1997. Osteology and phylogenetic interrelationships of sturgeons. Env. Biol. Fish. (this volume).

Fowler, H.W. 1941. The fishes of the groups Elasmobranchii, Holocephali, Isospondili and Ostariophysi obtained by the United States Bureau of Fisheries steamer Albatross in 1907 to 1910, chiefly in the Philippine Islands and adjacent seas. Bull. U. S. Nat. Mus. 100: 1–879.

Georgi, J.G. 1775. Bemerkungen einer Reise im russischen Reich. Vol. 1. St. Petersburg. 352 pp.

Gilbert, C.R. 1992. Atlantic sturgeon. pp. 31–39. *In:* R.A. Ashton (ed.) Rare and Endangered Biota of Florida, Vol. 2, Univ. Florida, Gainesville.

Grande, L. & W.E. Bemis. 1991. Osteology and phylogenetic relationships of fossil and recent paddlefishes (Polyodontidae) with comments on the interrelationships of Acipenseriformes. J. Vert. Paleo. 11, supplement 1: 1–121.

Grande, L. & W.E. Bemis. 1997. A comprehensive phylogenetic study of amiid fishes (Amiidae) based on comparative skeletal anatomy. J. Vert. Paleo. Supplement (in press).

Gray, J.E. 1834. Characters of two new species of sturgeons (*Acipenser*, Linn.). Proc. Zool. Soc. London 1834: 122–123.

Hilgendorf, F. 1892. Über eine neue Stör-Art aus Nord-Japan (*Acipenser mikadoi*). Sitzungsber. Ges. naturf. Freunde, Berlin 7: 98–100.

Hochleithner, M. 1995. Gesellschaft zur Rettung des Störs (*Acipenser sturio*) e.V.i.G. Österreichs Fisch. 48: 165–168.
Holčík, J. & L. Ledlička. 1994. Geographical variation of some taxonomically important characters in fishes: the case of the bitterling *Rhodeus sericeus*. Env. Fish. 41: 147–170.
Holčík, J., R. Kinzelbach, L.I. Sokolov & V.P. Vasilev. 1989. *Acipenser sturio* Linnaeus, 1758. pp. 367–394. *In:* J. Holčík (ed.) The Freshwater Fishes of Europe, Vol. 1, Pt. II, General Introduction to Fishes, Acipenseriformes, AULA-Verlag, Wiesbaden.
Honma, Y. 1988. Records and distributional notes on the sturgeons along the coast of Japanese Archipelago. Bull. Biogeogr. Soc. Japan 43: 51–55 (in Japanese).
Houston, J.J. 1988. Status of the green sturgeon, *Acipenser medirostris*, in Canada. Can. Field-Natur. 102: 286–290.
Jordan, D.S. & J.O. Snyder. 1901. Description of nine new species of fishes contained in museums of Japan. J. Coll. Sci. Imp. Univ. Tokyo 15: 301–311.
Jordan, D.S. & J.O. Snyder. 1906. A synopsis of the sturgeons (Acipenseridae) of Japan. Proc. U. S. Nat. Mus. 30: 397–398.
Khodorevskaya, R.P., G.F. Dovgopol, O.L. Zhuravleva & A.D. Vlasenko. 1997. Present status of commercial stocks of sturgeons in the Caspian Sea basin. Env. Biol. Fish. (this volume).
Krykhtin, M.L. & V.G. Svirskii. 1997. Endemic sturgeons of the Amur River: kaluga, *Huso dauricus* and Amur sturgeon, *Acipenser schrenckii*. Env. Biol. Fish. (this volume).
Lee, D.S., C.R. Gilbert, C.H. Hocutt, R.E. Jenkins, D.E. McAllister & J.R. Stauffer. Jr. 1980. Atlas of North American freshwater fishes. North Carolina Biol. Survey, Publ. # 1980–12. 867 pp.
Le Sueur, C.A. 1818. Description of several species of chondropterygious fishes of North America, with their varieties. Trans. Amer. Phil. Soc. 1: 383–394.
Lindberg, G.U. & M.J. Legeza. 1965. Fishes of the Sea of Japan and its adjacent waters, Sea of Okhotsk and Yellow Sea, Pt. 2. Izdatelstvo Akademii Nauk, Moscow. 391 pp. (in Russian).
Linnaeus, K. 1758. Systema naturae. pp. 230–238. *In:* Nantes Pisces. X ed., Vol. 1. Holmiae.
Lovetsky, A. 1828. On the fishes belonging to the sturgeon genus and inhabiting waters of the Russian Empire. Novyi Magazin Estestvennoi Istorii, Fiziki, Khimii i Svedenii Ekologicheskikh, Izdannyi I. Dvigubskim, Chast 6, Moscow. 3 pp. (in Russian).
Magnin, E. 1963. Recherches sur la systmatique et la biologie des Acipenserids *Acipenser sturio* L., *Acipenser oxyrhynchus* Mitch. et *Acipenser fulvescens* Raf. Ann. Stat. Centr. hydrobiol. appliq., Paris 9: 8–242.
Magnin, E. & G. Beaulieu. 1963. tude morphomtrique compare de *l'Acipenser oxyrhynchus* Mitchill du Saint-Laurent et de *l'Acipenser sturio* Linn de la Gironde. Natur. Can. 90: 5–38.
Mallet, J. 1995. A species definition for the Modern Synthesis. Trends Ecol. Evol. 10: 294–299.
Marti, V. Yu. 1939. Biology and harvest of *Acipenser sturio* in the Black Sea. Zool. Zhurnal 18: 435-442 (in Russian).
Marti, V.Yu. 1940. Systematics and biology of the Russian sturgeon from the Caucasian shore of the Black Sea. Zool. Zhurnal 19: 865–872 (in Russian).
Masuda, H., K. Amaoka, C. Araga, T. Uyeno & T. Yoshino. 1984. The fishes of the Japanese Archipelago. Tokai Univ. Press, Tokyo. 437 pp.
Matsubara, K. 1955. Fish morphology and hierarchy, Vol. I-III. Ishizaki Shoten, Tokyo. 1605 pp. (in Japanese).
Mitchill, S.L. 1815. The fishes of New York described and arranged. Trans. Lit. Philos. Soc. New York 1: 355–492.
Moyle, P.B., P.J. Foley & R.M. Yoshiyama. 1994. Status and biology of the green sturgeon, *Acipenser medirostris*. Sturgeon Quart. 2(1): 7.
Ninua, N.Sh. 1976. The Atlantic sturgeon from the Rioni River. Izdatelstvo Metsniereba, Tbilisi. 122 pp. (in Russian).
Okada, Y. 1959–1960. Studies on the freshwater fishes of Japan, Vol. I–II. J. Fac. Fish., Pref. Univ. Mie 4: 1–860.
Okada, Y. & K. Matsubara. 1938. Keys to the fishes and fish-like animals of Japan. Sanseido, Tokyo. 584 pp. (in Japanese).
Ong, T.-L., J. Stabile, I. Wirgin & J. R. Waldman. 1996. Genetic divergence between *Acipenser oxyrinchus oxyrinchus* and *A.o. desotoi* as assessed by mitochondrial DNA sequencing analysis. Copeia 1996 (in press).
Pallas, P.S. 1771. Reise durch verschiedene Provinzen des russischen Reiches, 1768 und 1769 sten Jahre. 1. St. Petersburg.
Pavlov, D.S, K.A. Savvaitova, L.I. Sokolov & S.S. Alekseev. 1994. Rare and endangered animals. Fishes. Vysshaya Shkola, Moscow. 334 pp. (in Russian).
Pirogovskii, M.I., L.I. Sokolov & V.P. Vasilev. 1989. *Huso huso* (Linnaeus, 1758). pp. 156-200. *In:* J. Holčík (ed.) The Freshwater Fishes of Europe, Vol. 1, Pt. II. General Introduction to Fishes, Acipenseriformes, AULA-Verlag, Wiesbaden.
Rafinesque, C.S. 1817. Additions to the observations on the sturgeons of North America. Amer. Month. Mag. Crit. Rev. 1: 128.
Richardson, J. 1836. The fish, Vol. 3. Fauna Boreali-Americana, or the zoology of the northern parts of British America: containing descriptions of the objects of natural history collected on the late northern land expeditions, under the command of Sir John Franklin, R.N. Richard Bentley, London. 327 pp.
Ride, W.D.L., C.W. Sabrosky, G. Bernardi & R.V. Melville. 1985. International code of zoological nomenclature. Univ. Calif. Press, Berkeley. 338 pp.
Rochard, E., P. Williot, G. Castelnaud & M. Lepage. 1991. lments de systmatique et de biologie des populations sauvages d'esturgeons. pp. 475–507. *In:* P. Williot (ed.) Acipenser, CEMAGREF Publ., Bordeaux.
Ruban, G.I. 1997. Species structure, contemporary distribution and status of the Siberian sturgeon, *Acipenser baerii*. Env. Biol. Fish. (this volume).
Salnikov, N.I. & S.M. Myatskii. 1934. On the systematics of the beluga from the Sea of Azov-Black Sea Basin. Trudy Nauchnoi Rybokhozyaistvennoi i Biologicheskoi Stantsii Gruzii 1: 31–50 (in Russian).
Schmidt, P.Yu. 1904. Fishes of the eastern seas of the Russian Empire. Izdanie Russkogo Geograficheskogo Obshchestva, St. Petersburg. 466 pp. (in Russian).

Schmidt, P.Yu. 1950. Fishes of the Sea of Okhotsk. Izdatelstvo Akademii Nauk, Moscow. 370 pp. (in Russian).

Scott, W.B. & E.J. Crossman. 1973. Freshwater fishes of Canada. Bull. Fish. Res. Board Can. 184. 966 pp.

Shilin, N.I. 1995. Programme for conservation of *Acipenser medirostris mikadoi* in the Russian Far East. pp. 262-267. *In:* A.D. Gershanovich & T.I.J. Smith (ed.) Proceedings of International Symposium on Sturgeons, 6–11 September 1993, Moscow-Kostroma-Moscow, VNIRO Publishing, Moscow.

Shubina, T.N., A.A. Popova & V.P. Vasilev. 1989. *Acipenser stellatus* Pallas, 1771. pp. 395–443. *In:* J. Holčík (ed.) The Freshwater Fishes of Europe, Vol. 1, Pt. II, General Introduction to Fishes, Acipenseriformes, AULA-Verlag, Wiesbaden.

Smith, H.M. 1891. Report on the fisheries of the South Atlantic States. Bull. U. S. Fish. Comm. 11: 269–356.

Smith, T.I.J. & J.P. Clugston. 1997. Status and management of Atlantic sturgeon, *Acipenser oxyrinchus*, in North America. Env. Biol. Fish. (this volume).

Sokolov, L.I. & V.P. Vasilev. 1989a. *Acipenser nudiventris* Lovetsky, 1828. pp. 206-226. *In:* J. Holčík (ed.) The Freshwater Fishes of Europe, Vol. 1, Pt. II, General Introduction to Fishes, Acipenseriformes, AULA-Verlag, Wiesbaden.

Sokolov, L.I. & V.P. Vasilev. 1989b. *Acipenser ruthenus Linnaeus*, 1758. pp. 227–262. *In:* J. Holčík (ed.) The Freshwater Fishes of Europe, Vol. 1, Pt. II, General Introduction to Fishes, Acipenseriformes, AULA-Verlag, Wiesbaden.

Sokolov, L.I. & V.P. Vasilev. 1989c. *Acipenser baeri* Brandt, 1869. pp. 263-284. *In:* J. Holčík (ed.) The Freshwater Fishes of Europe, Vol. 1, Pt. II, General Introduction to Fishes, Acipenseriformes, AULA-Verlag, Wiesbaden.

Takeuchi, T. 1979. Description of two acipenserid fishes, *Acipenser sinensis* Gray and *A. kikuchii* Jordan et Snyder, recorded from Sagamy Bay. Ann. Rep. Keikuy Aburatsubo Marine Parc. Aq. 10: 20–25 (in Japanese).

Tanaka, S. 1908. Notes on some Japanese fishes, with descriptions of fourteen new species. J. Coll. Sci. Imp. Univ. Tokyo 23: 1–54.

Tortonese, E. 1989. *Acipenser naccarii* Bonaparte, 1836. pp. 285-293. *In:* J. Holčík (ed.) The Freshwater Fishes of Europe, Vol. 1, Pt. II, General Introduction to Fishes, Acipenseriformes, AULA-Verlag, Wiesbaden.

Vladykov, V.D. 1955. A comparison of Atlantic Sea sturgeon with a new subspecies from the Gulf of Mexico (*Acipenser oxyrhynchus de sotoi*). J. Fish. Res. Board Can. 12: 754–761.

Vladykov, V.D. & J.R. Greeley. 1963. Order Acipenseroidei. pp. 24-60. *In:* H.B. Bigelow, C.M. Breder, D.M. Cohen, G.W. Mead, D. Merriman, Y.H. Olsen, W.C. Schroeder, L.P. Schultz & J. Tee-Van (ed.) Fishes of the Western North Atlantic, Mem. Sears Found. Mar. Res. 1.

Vlasenko, A.D., A.V. Pavlov, L.I. Sokolov & V.P. Vasilev. 1989a. *Acipenser gueldenstaedti* Brandt, 1833. pp. 294–344. *In:* J. Holčík (ed.) The Freshwater Fishes of Europe, Vol. 1, Pt. II, General Introduction to Fishes, Acipenseriformes, AULA-Verlag, Wiesbaden.

Vlasenko, A.D., A.V. Pavlov & V.P. Vasilev. 1989b. *Acipenser persicus* Borodin, 1897. pp. 345-366. *In:* J. Holčík (ed.) The Freshwater Fishes of Europe, Vol. 1, Pt. II, General Introduction to Fishes, Acipenseriformes, AULA-Verlag, Wiesbaden.

Waldman, J.R., J.T. Hart & I.I. Wirgin. 1996a. Stock composition of the New York Bight Atlantic sturgeon fishery based on analysis of mitochondrial DNA. Trans. Amer. Fish. Soc. 125: 364–371.

Waldman, J.R., K. Nolan, J. Hart & I.I. Wirgin. 1996b. Genetic differentiation of three key anadromous fish populations of the Hudson River. Estuaries 19: 759–768.

Wei, Q., F. Ke, J. Zhang, P. Zhuang, J. Luo, R. Zhou & W. Yang. 1997. Biology, fisheries, and conservation of sturgeons and paddlefish in China. Env. Biol. Fish. (this volume).

Williot, P., E. Rochard, G. Castelnaud, T. Rouault, R. Brun, M. Lepage & P. Elie. 1997. Biological characteristics of the European Atlantic sturgeon, *Acipenser sturio*, as the basis for a restoration program in France. Env. Biol. Fish. (this volume).

Wooley, C.M. 1985. Evaluation of morphometric characters used in taxonomic separation of Gulf of Mexico sturgeon, *Acipenser oxyrhynchus desotoi*. pp. 97–103. *In:* F.P. Binkowski & S.I. Doroshov (ed.) North American Sturgeons: Biology and Aquaculture Potential, Dr W. J. Junk Publishers, Dordrecht.

Zhuang, P., F. Ke, Q. Wei, X. He & Y. Cen. 1997. Biology and life history of Dabry's sturgeon, *Acipenser dabryanus*, in the Yangtze River. Env. Biol. Fish. (this volume).

Part 2: Biology and status reports on sturgeons and paddlefishes

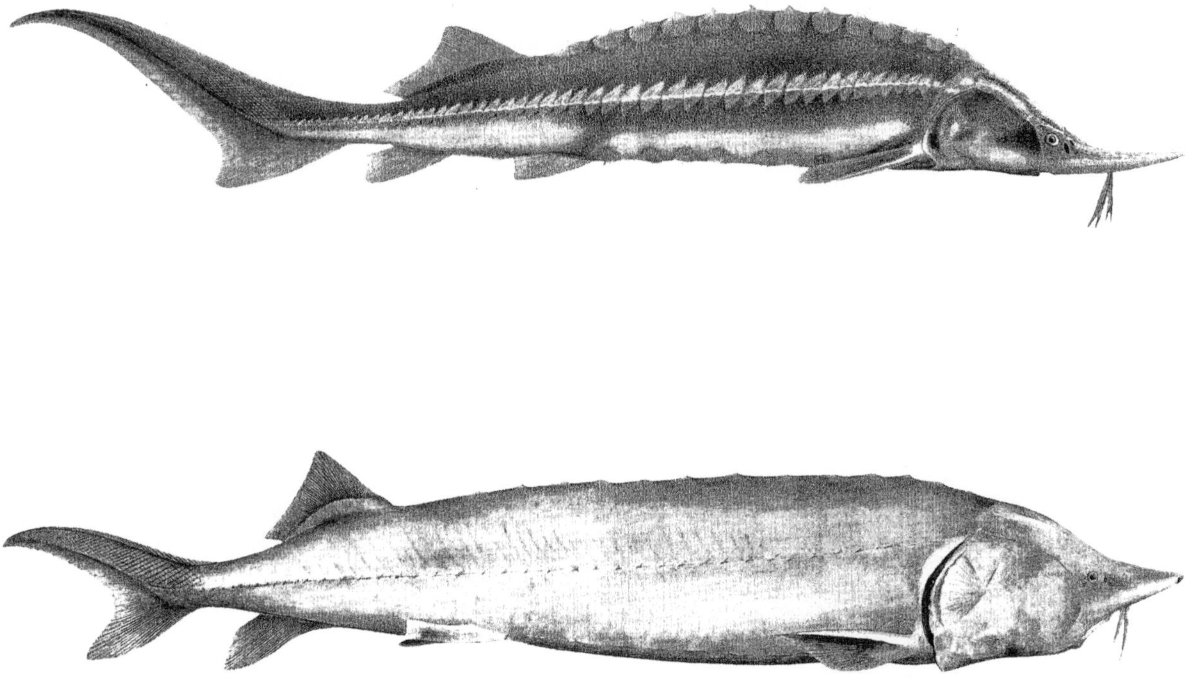

Huso huso juvenile ca. 37 cm long above a large adult 380 kg in weight (about 5 m long) from Antipa (1909, plate 24, fig. 120 and 123).

An early print of *Polyodon spathula* and *Scaphirhynchus platorynchus* from Wood (1863, p. 201).

Environmental Biology of Fishes **48**: 167–183, 1997.

Sturgeon rivers: an introduction to acipenseriform biogeography and life history

William E. Bemis[1] & Boyd Kynard[1,2]
[1] *Department of Biology and Graduate Program in Organismic and Evolutionary Biology, University of Massachusetts, Amherst, MA 01003, U.S.A.*
[2] *National Biological Service, Conte Anadromous Fish Research Center, 1 Migratory Way, Turners Falls, MA 01376, U.S.A.*

Received 22.5.1996 Accepted 3.6.1996

Key words: †Chondrosteidae, †Peipiaosteidae, Polyodontidae, Acipenseridae, Holarctic, anadromy, potamodromy

Synopsis

We present an overview of the global distribution of all 27 living species of Acipenseriformes in an attempt to understand their biogeographic history and the range of life history patterns displayed by different species. Our biogeographic analysis (based on the most recent phylogenetic analysis including fossil Acipenseriformes) suggests that Acipenseriformes originated in Europe, and that early diversification took place in Asia. Acipenseriformes do not have a common life history; variation within and between species is the rule rather than exception. The few relatively well-known case studies (e.g., Caspian Sea sturgeons, European Atlantic sturgeons in the Gironde system, and shortnose and North American Atlantic sturgeons in rivers of the east coast of America) greatly influence what we think we know about sturgeon biology. Our present level of phylogenetic understanding does not allow us to determine whether anadromy or potamodromy is the plesiomorphic life history pattern for Acipenseriformes. We propose that rivers in which spawning occurs must be the central unit for biogeographic analysis of living Acipenseriformes. After mapping these rivers, we recognized nine biogeographic provinces for acipenseriforms. Some repeated historical patterns emerge from this analysis, but, again, we are limited by our current understanding of phylogenetic relationships within the genus *Acipenser* in particular. Distribution and biogeographic data are central to deciding where to make new efforts to update existing status information for acipenseriform species. We single out a widely ranging and highly variable species, *Acipenser ruthenus*, as particularly intriguing, for it spans three of our nine biogeographic provinces, and apparently has different life history patterns in different river systems. Finally, we note new areas in need of basic research, particularly the need for more detailed descriptions and analyses of life histories of different populations of sturgeons.

Introduction

This paper attempts a new approach to the global biogeography of Acipenseriformes, although we admit from the outset that this is daunting topic because in its most comprehensive form, such an analysis concerns the history of the entire Holarctic region for the last 200 million years. Still, this topic is important because it lies at the interface between basic research on Acipenseriformes and practical steps needed to plan for sturgeon and paddlefish conservation (e.g. Rochard et al. 1990). Our review has several explicit purposes.

First, although Acipenseriformes has long been regarded as a biogeographically interesting group, the phylogeny necessary to study this question has yet to be assembled. Comprehensive phylogenetic understanding, particularly concerning relationships within the genus *Acipenser*, still eludes us. Complicating factors include high levels of ontogenetic and individual variation, hybridization, and extirpation of many populations within the historic ranges of certain species (Bemis et al. 1997b this volume). Some key intergeneric relationships are equally problematic. Results from karyological and molecular phylogenetic approaches (Birstein & DeSalle 1997) place the two species of *Huso* within *Acipenser* (as sister taxa to *A. ruthenus*) whereas osteological data place *Huso* as the sister taxon of all other sturgeons (Findeis 1997 this volume). Future updating of our interpretations is inevitable as our phylogenetic insight into Acipenseriformes improves.

Second, a detailed understanding of the geographic distribution of acipenseriforms is complicated by the wide ranges historically reported for adults of certain species, so that a more restrictive and useful definition of species ranges is required. We explicitly propose the concept that rivers in which spawning occurs should be the central unit of analysis for interpreting the biogeographic ranges of acipenseriform species (and other groups containing anadromous species). We provide a global summary of these rivers, but given the scope of the question (i.e., identify all rivers in the world in which sturgeons or paddlefish historically spawned) it is certain that more exhaustive analyses will yield additional rivers. It may also be necessary to find more restrictive ways to define our concept of spawning rivers, and perhaps we will succeed in provoking such a response.

Third, Acipenseriformes exhibit a broad array of spawning and feeding migratory patterns, with some species utilizing fresh water exclusively, others fresh water and estuarine environments, while others span the range from fresh water to fully marine environments. Different life history patterns have certainly influenced the biogeography of Acipenseriformes, although at present, we can do little more than catalogue them, because no one understands the genetic bases or adaptive significance of these patterns. An important but still neglected approach to studying the evolution of different life history patterns of acipenseriforms are ideas of Balon (1990 and references therein) that emphasize the importance of altricial or precocial patterns of development. This is likely to prove a productive approach, for size and yolk content differences are known for different species of sturgeons, though they have not yet been correlated with different patterns of life history.

Fourth, we want to introduce the remaining papers in the status part of this collection (Bemis et al. 1997a), which detail aspects of life history and biogeography for many of the extant species of Acipenseriformes. One need that emerges immediately is for more detailed river surveys and life history studies of virtually all species of sturgeons, particularly those from geographically remote regions in Asia and northern North America. These surveys need to be done with the most advanced technologies available, including in particular telemetry of individuals to determine life history patterns (Kynard 1997 this volume) and molecular based identification of populations within and between river systems (Wirgin et al. 1997 this volume).

Basic background

Biogeographic observations

We begin with three general biogeographic observations:

1. With the exception of the Pearl River in China, all spawning rivers used by Acipenseriformes lie entirely within the north temperate zone of Asia, Europe or North America, although individual adults have been taken at sea south of the Tropic of Cancer. All known fossil Acipenseriformes are also from north temperate localities (Grande & Bemis 1991, Jin 1995, Bemis et al. 1996 this volume, Grande & Bemis 1997). The absence of Acipenseriformes from tropical rivers is proba-

bly related to thermal requirements for maturation and early development, which generally need temperatures below 20° C (e.g., Artyukhin 1988, Dettlaff et al. 1993).

2. With the exception of *Acipenser ruthenus*, which lives in both Europe and Asia, no species within Acipenseriformes is known to spawn in rivers on two continents and few species spawn in more than two of the biogeographic provinces that we define below. This situation is unlike that, for example, for salmonids of the North Pacific Ocean, several of which spawn in both North American and Asian rivers (e.g., chum salmon, *Oncorhynchus keta* spawns in rivers along the east and north coast of Asia as well as the west coast of America, Salo 1991). Our interpretation concerning sturgeons and continents is subject to falsification, but all detailed work to date suggests that this pattern will hold true. For instance, *Acipenser medirostris* (west coast of North America) and *A. mikadoi* (Sea of Okhotsk and Sea of Japan) were at various times considered to be conspecific, but recent genetic and molecular data confirm that they are distinct species (Birstein 1993b, Birstein et al. 1997 this volume).
3. Much of the historic work concerning the distribution of different species of sturgeons (e.g., Berg 1948a, 1948b, 1959) predates contemporary concepts of continental drift. More recent accounts (e.g., Berra 1981, Hocutt & Wiley 1986, Bănărescu 1990, 1992, 1995) predate contemporary phylogenetic interpretations of acipenseriforms. If one restricts analysis to †Chondrosteidae, †Peipiaosteidae, Polyodontidae and the tribe Scaphirhynchini, all of which are small groups with intriguing but fairly simple biogeographic distributions (Grande & Bemis 1991, Jin 1995, Bemis et al. 1997b), then historical biogeography is easy to contemplate. The widely ranging genus *Acipenser*, however, imposes many difficult biogeographic questions, which is why it is a focus in our present analysis.

Life history observations

We make seven general observations about acipenseriform life history and spawning biology:

1. Acipenseriformes spawn repeatedly, but most females do not spawn annually. This pattern resembles that for anadromous fishes such as shad (*Alosa*; e.g., Leggett 1976) but is different from that typical for Pacific salmonids (*Oncorhynchus*; see Groot & Margolis 1991).
2. All acipenseriforms spawn in freshwaters of low salt content (0–0.1‰) even though adults of some species may migrate to feed in estuarine or brackish waters (approximately 14 to 27‰, Pearse & Gunter 1957) or seawater (35‰).
3. The timing of spawning for Acipenseriformes is highly variable, equaling or exceeding the variability found in any other group of fresh water or diadromous fishes. They spawn in all seasons and in highly variable conditions of water flow and temperature.
4. Characteristics of spawning migrations vary greatly among Acipenseriformes in total distance migrated, the distance upstream from salt water, etc. Several evolutionary scenarios and sets of terminology have been proposed to describe these variations in spawning migration pattern (reviewed below).
5. The few studies done to date indicate that the availability of suitable spawning habitat is critical to reproductive success. Spawning sites are characterized by areas with hard substrate of gravel to boulder size rocks containing many crevices. The water velocity near the bottom is typically moderate (Kynard 1997 this volume).
6. Annual spawning success and recruitment is highly unpredictable, and may be zero if river flows are too high during the brief reproductive window of females. High flows, whether caused by natural phenomena or controlled releases by dams, can create high bottom velocities that preclude or greatly reduce spawning success (Kynard 1997 this volume).
7. A particular spawning site is usually used from year to year. Such site fidelity might derive either from the particular characteristics of the site or from homing. Sturgeons are believed to have

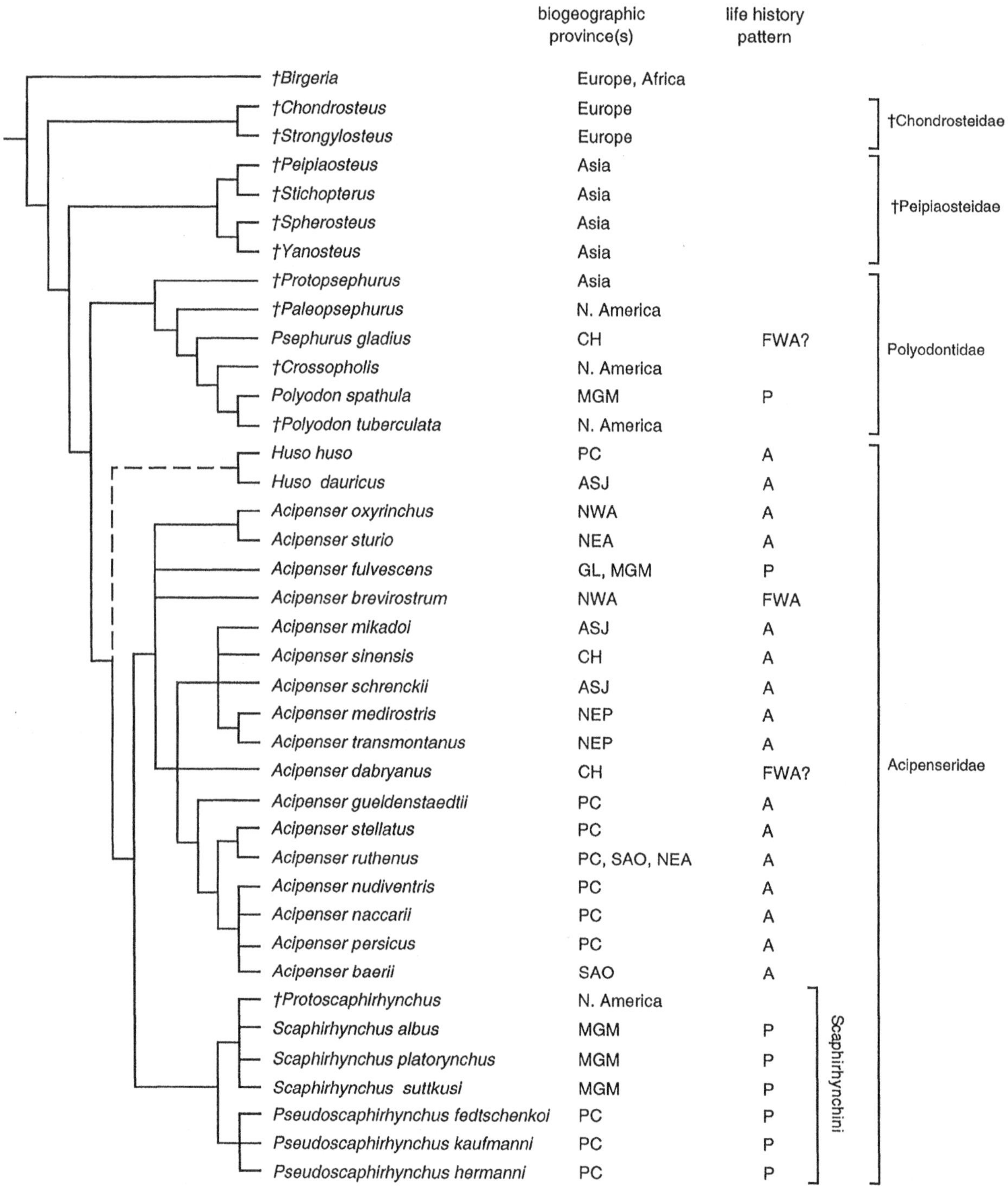

Figure 1. Tree suggesting possible evolutionary relationships among fossil and recent acipenseriforms. See text for explanation and discussion. Biogeographic areas are keyed to the map in Figure 3, and our scheme of provinces is explained in the biogeography section of the text. Fossils are preceded by dagger symbols; continents from which fossils were recovered are indicated. Life history pattern is keyed as follows: A – anadromous; P – potamodromous; FWA – freshwater amphidromous. These terms are defined in the life history section of the text. Data for relationships among acipenserines (*Acipenser* plus *Huso*) based on preliminary analyses by Birstein & DeSalle (1997).

strong homing capabilities, although direct evidence for this is only recently available, and the subject needs additional research (Waldman et al. 1996a,b, Wirgin et al. 1997 this volume). If homing proves to be as important as currently expected, then it might be the proximate explanation for the existence of different morphs or races within species, which is a particularly common pattern in the family Acipenseridae.

Species and evolutionary relationships

Figure 1 presents a tree of the well-preserved fossil and all living species of acipenseriforms. It includes *†Birgeria*, which was considered to be a closely related outgroup for Acipenseriformes by Bemis et al. (1997b this volume). Next to each extant taxon in Figure 1, we list the biogeographic province(s) in which it occurs and its supposed life history pattern (biogeographic provinces and life history patterns are described further below). For extinct taxa in Figure 1 (indicated with dagger symbols), we identify the continents from which the fossils were recovered. Life history cannot be assessed with certainty in fossils.

Our goal in presenting the tree in Figure 1 is not to present a single preferred hypothesis of relationships among acipenseriform taxa but rather to organize biogeographic and life history information. It should be regarded as a heuristic synthesis of formal phylogenetic analyses presented in this volume (Bemis et al. 1997b, Findeis 1997, Birstein et al. 1997) and elsewhere (Artyukhin 1995, Jin 1995, Grande & Bemis 1996). Some nodes in this tree are corroborated by all contemporary phylogenetic analyses. For example, we are now very confident about the placement of †Chondrosteidae as the sister taxon of all other Acipenseriformes (Grande & Bemis 1996). Both Polyodontidae and Acipenseridae are now considered to be monophyletic families (contrary to the view of Gardiner 1984; see Grande & Bemis 1991), and all available data support our concept of Acipenseroidei, a group containing Polyodontidae and Acipenseridae (see Grande & Bemis 1991, Bemis et al. 1997b for detailed comments on the strength of this node). Relationships within Polyodontidae also are well understood (Grande & Bemis 1991, Bemis et al. 1997b).

Other aspects of the tree in Figure 1 have been recently proposed on the basis of molecular sequence data (Birstein et al. 1997 this volume) and still others are decidedly controversial. A close comparison of formal phylogenetic hypotheses proposed by various current authors will reveal major differences in branching pattern within *Acipenser* as well as the placement of *Huso*. The largely unresolved pattern of relationships within *Acipenser* that we show in Figure 1 is derived from ongoing analyses of a growing molecular phylogenetic data set that is the basis for a separate formal phylogenetic analysis (Birstein & DeSalle 1997). A special problem is indicated on the tree by the dotted lines leading to *Huso huso* and *H. dauricus*. Based on a phylogenetic analysis of osteological and other morphological characters, Findeis (1997 this volume) proposed *Huso* as the sister taxon of all other species of Acipenseridae, and this is the placement we show in Figure 1. Birstein & DeSalle (1997), however, reported molecular characters that link *Huso* with *Acipenser ruthenus* (also see Berg 1948a,b).

Time and the biogeography of fossil acipenseriforms

Many relatively well-known Earth historical factors have impacted Acipenseriformes during their long (circa 200 Ma) history. Our intent is not to review these in detail but to outline the scope and time course of the changes. Divergence times are necessarily uncertain, given the relative paucity of well-preserved fossil taxa. To date, no one has used molecular phylogenetic data to estimate the times of divergence for major lineages within Acipenseriformes.

The outgroup for Acipenseriformes, *†Birgeria*, is known from the Triassic of Europe, North America and Madagascar (Nielsen 1949, Lehman 1952, Schwarz 1970). Two families of Acipenseriformes known only from fossils (†Chondrosteidae and Peipiaosteidae) are important for understanding the biogeography of the entire order (see discussion of

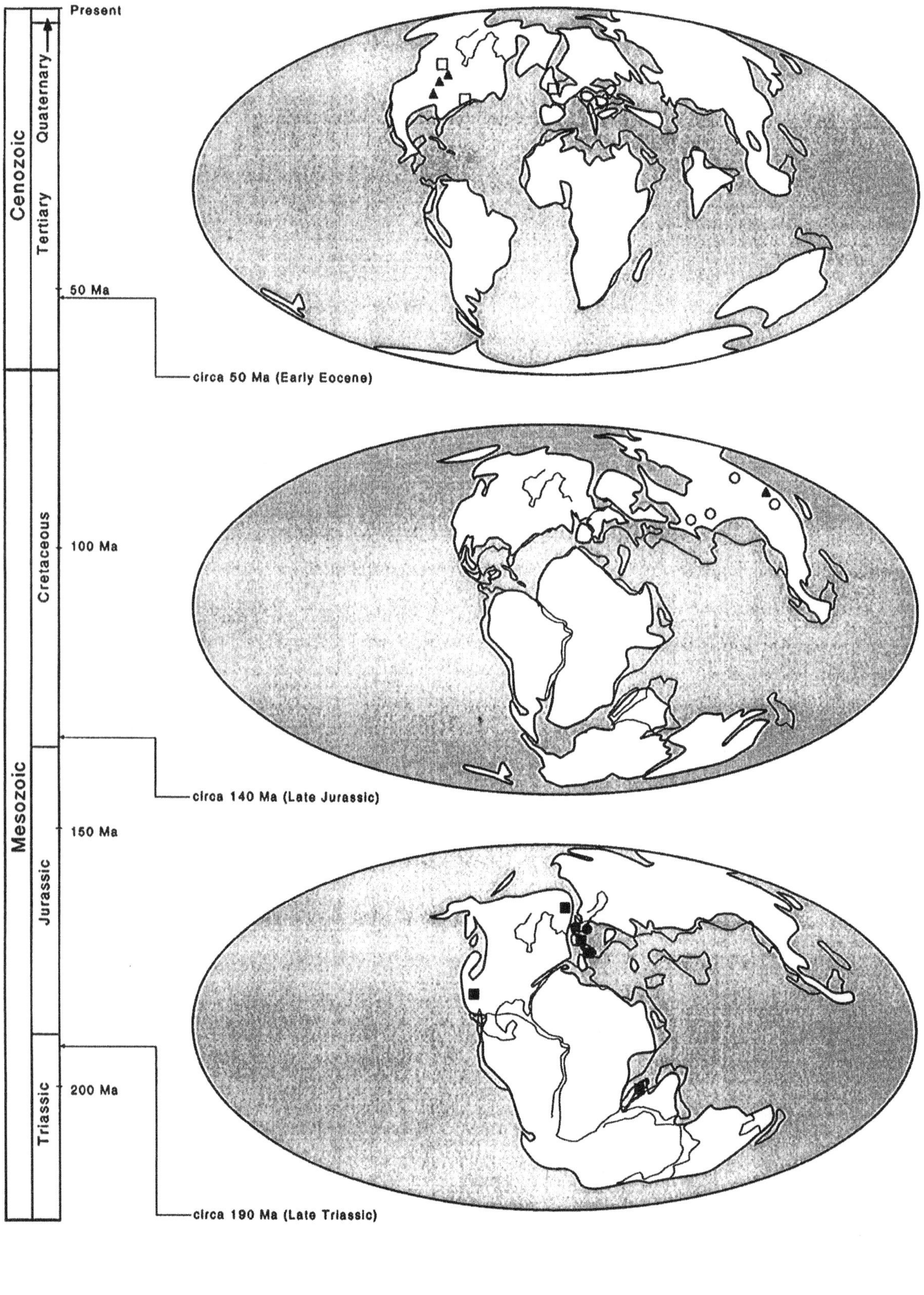

Present
Cenozoic
Quaternary
Tertiary
50 Ma
circa 50 Ma (Early Eocene)
Mesozoic
Cretaceous
100 Ma
circa 140 Ma (Late Jurassic)
Jurassic
150 Ma
Triassic
200 Ma
circa 190 Ma (Late Triassic)

←

Figure 2. Paleocoastline maps and the distribution of †Birgeriidae and Acipenseriformes. On the map representing the Late Triassic/Early Jurassic, the distribution of *†Birgeria* is indicated by solid squares, and the distribution of †Chondrosteidae is indicated by solid circles. On the map representing the Late Jurassic/Early Cretaceous, the distribution of †Peipiaosteidae is indicated by open circles, and the locality for the oldest fossil paddlefish, *†Protopsephurus*, is indicated by a solid triangle. On the map representing the Late Cretaceous/Early Tertiary, the distribution of three additional genera of paddlefishes (*†Paleopsephurus*, *†Crossopholis* and *Polyodon*) in western North America is indicated by solid triangles, and a few localities for fossil species assigned to *Acipenser* are marked with open squares.

the role of fossils in biogeographic studies in Grande 1985). Several fossil Polyodontidae and Acipenseridae are known. With the exception of the Green River paddlefish, †*Crossopholis magnicaudatus*, from the Early Eocene Green River Formation in southwestern Wyoming, the localities in which fossil paddlefishes and sturgeons occur lie within the historic ranges of the extant families. Some additional data about these fossil taxa are summarized in tabular form in Bemis et al. (1997b this volume) and Jin (1995). All well-preserved fossil genera of Acipenseriformes are included in the present study, but because species level distinctions within these genera are often problematic and have not been the subject of recent comprehensive reviews, generic level distinctions suffice for current purposes.

We organize our comments on the biogeography of fossil Acipenseriformes around three paleocoastline maps and a time scale (Figure 2; base maps were redrawn and simplified from Smith et al. 1994). The lowest map shows a reconstruction of the continents and their coastlines in the Late Triassic/Early Jurassic, with the localities of the outgroup taxon, †*Birgeria*, plotted in solid squares in Europe, North America and Madagascar. Also plotted on the lowest map (solid circles) are localities for †*Chondrosteus* and †*Strongylosteus*, from the Early

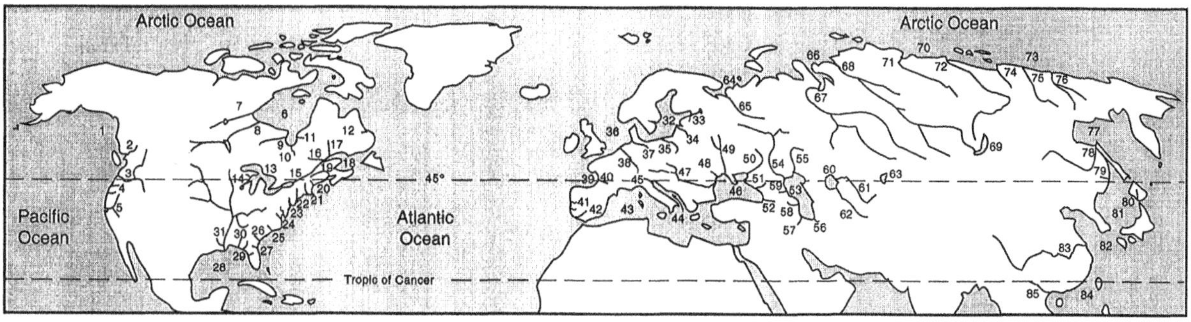

Figure 3. Major rivers, lakes, seas and oceans of the Holarctic relevant to the biogeographic ranges of recent Acipenseriformes. Base map redrawn from Bond (1996: Fig. 30–3); data used to assemble this figure are derived chiefly from secondary sources (Anonymous 1980, Vladykov & Greeley 1963, Scott & Crossman 1973, Hart 1973, Trautman 1981, Lee et al. 1980, Holčík 1989). Key: 1 – Gulf of Alaska; 2 – Fraser R.; 3 – Columbia R.; 4 – Rogue & Klamath R.; 5 – Sacramento R.; 6 – Hudson Bay; 7 – Churchill R.; 8 – Nelson R.; 9 – Albany R.; 10 – Moose R.; 11 – Rupert R.; 12 – Hamilton Inlet; 13 – Great Lakes (Superior, Huron, Michigan, Erie, & Ontario); 14 – L. Winnebago, Fox R., & Menominee R.; 15 – St. Lawrence R.; 16 – Ottawa R.; 17 – St. Maurice R; 18 – Gulf of St. Lawrence; 19 – St. John R.; 20 – Kennebec/Androscoggin R. & Merrimack R.; 21 – Connecticut R.; 22 – Hudson R.; 23 – Delaware R.; 24 – Chesapeake Bay system (includes Potomac & Susquehanna); 25 – Santee R.; 26 – Savannah R. & Altamaha R.; 27 – St. John's; 28 – Gulf of Mexico; 29 – Suwanee R. & Apalachicola R.; 30 – Alabama R.; 31 – Mississippi R. (includes Missouri, Ohio, & Tennessee rivers); 32 – Baltic Sea; 33 – Neva R., Nara R., & Luga R.; 34 – Wista R.; 35 – Oder R.; 36 – North Sea; 37 – Elbe R.; 38 – Rhine R.; 39 – Bay of Biscay; 40 – Gironde Estuary (Garrone & Dorgonne R.); 41 – Douro R. & Guadiana R.; 42 – Guadalquivir R.; 43 – Mediterranean Sea; 44 – Adriatic Sea; 45 – Po R.; 46 – Black Sea; 47 – Danube R.; 48 – Dnestr R.; 49 – Dniepr R.; 50 – Don R.; 51 – Kuban R.; 52 – Rioni R.; 53 – Caspian Sea; 54 – Volga R.; 55 – Ural R.; 56 – Gorgan R.; 57 – Qezel Owzan R.; 58 – Kura R.; 59 – Terek R.; 60 – Aral Sea; 61 – Syr Darya R.; 62 – Amu Darya R.; 63 – L. Balkash; 64 – White Sea; 65 – Severnaya-Dvina R.; 66 – Kara Sea; 67 – Ob R. (includes Irtysh R.); 68 – Yenesei R.; 69 – L. Baikal; 70 – Laptev Sea; 71 – Khatanga R.; 72 – Lena R.; 73 – East Siberian Sea; 74 – Yana R.; 75 – Indigirka R.; 76 – Kolyma R.; 77 – Sea of Okhotsk; 78 – Amur R.; 79 – Tumnin R.; 80 – Ishikari R.; 81 – Sea of Japan; 82 – East China Sea; 83 – Yangtze R.; 84 – South China Sea; 85 – Pearl R.

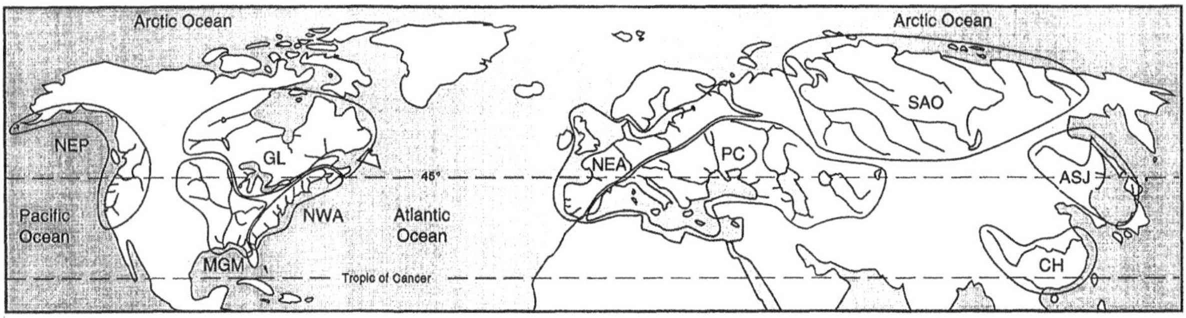

Figure 4. Nine biogeographic provinces for recent Acipenseriformes discussed in text. Also see Table 1. Key: NEP – North Eastern Pacific; GL – Great Lakes, Hudson Bay & St. Lawrence R.; NWA – North Western Atlantic; MGM – Mississippi R. & Gulf of Mexico; NEA – Northeastern Atlantic, including White, Baltic & North seas; PC – Ponto-Caspian Region, including Mediterranean, Aegean, Black, Caspian & Aral seas; SAO – Siberia & Arctic Ocean; ASJ – Amur R., Sea of Okhotsk & Sea of Japan; CH – China.

Jurassic of England and Germany, respectively. These are the diagnosable genera in the family †Chondrosteidae.

The middle map in Figure 2 shows a reconstruction of the continents and coastlines in Late Jurassic/Early Cretaceous times, with the localities plotted for the earliest known paddlefish, *†Protopsephurus* (solid triangle), and all four genera of the extinct family †Peipiaosteidae (*†Peipiaosteus* and *†Yanosteus* in China, *†Stichopterus* in Trans Baikal and Mongolia, and *†Spherosteus* in Kazakhstan indicated by open circles). The range of †Peipiaosteidae is restricted to Asia (Grande & Bemis 1996).

The top map in Figure 2 shows a reconstruction of the continents in Late Cretaceous/Early Tertiary times. Localities are plotted for three fossil paddlefishes: *†Paleopsephurus* from the Late Cretaceous Hell Creek Formation of Montana, *†Polyodon tuberculata* from the Early Paleocene Tullock Formation of Montana and *†Crossopholis* from the Early Eocene Green River Formation in Wyoming. A scaphirhynchine sturgeon, *†Protoscaphirhynchus* occurs in the Late Cretaceous Hell Creek Formation of Montana (and in fact was recovered from the same hadrosaur stomach as was the paddlefish *†Paleopsephurus*). Fossils assigned to the genus *Acipenser* are, for the most part, fragmentary, and have never been comprehensively reviewed or compared with the living species. For establishing the presence of *Acipenser* in North America, we plot in Figure 2 the locality of *†Acipenser albertensis* from the Late Cretaceous of Alberta and other fossil species assigned to *Acipenser* from the Early Eocene of England and Miocene of Virginia.

Biogeography of living acipenseriforms

Figure 3 maps selected major rivers of the world in which acipenseriforms spawn. The rivers, lakes and seas relevant to our analysis are coded by number to the list in the caption. Given the global scope of our survey, we followed a standard rule concerning the nomenclature of streams: we only name the relevant river that enters a particular ocean or sea basin. For example, the Mississippi River (# 31) is named in Figure 3 but not its major tributaries, which include the Missouri and Ohio rivers. Anadromous acipenserids are absent from the Mississippi River, and some acipenseriform species occur only in its upper tributaries (e.g., *Scaphirhynchus albus* lives in far upstream reaches of the Missouri River), but for purposes of our survey, the only river noted is the Mississippi. We also simplified many river systems and omitted many smaller rivers from our diagram.

We also found it convenient to define nine biogeographic provinces with which acipenseriforms are associated (Figure 4). Table 1 lists the provinces and species of acipenseriforms that currently live in each. Although some species occur in more than one province, we defined them for and primarily use them to discuss the biogeography of *Acipenser* and *Huso*. Most provinces are based on drainages feeding into distinct oceanic basins. Discrete geographic boundaries currently limit emigration of sturgeons from some of these provinces to adjacent provinces (e.g., around the lower one half of Florida there are no suitable spawning rivers). In other cases, the provinces are readily distinguishable based on geological history (e.g., the Mediterranean basin connected to the North Eastern Atlantic through the strait of Gibraltar in the Messinian; see Hsu 1972).

Life history

This section explains the life history patterns scored on the tree in Figure 1. Although some authors comment on evidence for anadromy in fossil acipenseriforms (e.g., Bai 1983), we consider that it is not possible to make meaningful comparisons based on such speculations. Thus, we restrict the analysis of life history patterns shown in Figure 1 to extant species.

Acipenseriforms migrate for two basic reasons: feeding and reproduction, and we illustrate some possible life history patterns in Figure 5. *Downstream migrations* of sturgeons are always associated with feeding. The interfaces between freshwater and saltwater or between rivers and large lakes can be nutrient rich, with abundant food. The shallow (< 100 m), near-shore continental shelf regions in which some species of sturgeons feed at sea are similarly productive environments. Sturgeons are not known to utilize deep environments while at sea, and do not in general make extensive offshore migrations. *Upstream migrations* are usually associat-

Table 1. Occurrence of species of acipenseriforms in nine biogeographic provinces (mapped in Figure 4).

NEP – North Eastern Pacific	*Acipenser medirostris*
	Acipenser transmontanus
GL – Great Lakes, Hudson Bay & St. Lawrence River	*Acipenser fulvescens*
	Acipenser o. oxyrinchus
NWA – North Western Atlantic	*Acipenser brevirostrum*
	Acipenser o. oxyrinchus
MGM – Mississippi R. & Gulf of Mexico	*Polyodon spathula*
	Acipenser oxyrinchus desotoi
	Scaphirhynchus albus
	Scaphirhynchus platorynchus
	Scaphirhynchus suttkusi
NEA – Northeastern Atlantic, including White, Baltic & North seas	*Acipenser ruthenus*
	Acipenser sturio
PC – Ponto-Caspian Region, including Mediterranean, Aegean, Black, Caspian & Aral seas	*Acipenser gueldenstaedtii*
	Acipenser nudiventris
	Acipenser naccarii
	Acipenser persicus
	Acipenser ruthenus
	Acipenser stellatus
	Acipenser sturio
	Huso huso
	Pseudoscaphirhynchus fedtschenkoi
	Pseudoscaphirhynchus hermanni
	Pseudoscaphirhynchus kaufmanni
SAO – Siberia & Arctic Ocean	*Acipenser baerii*
	Acipenser ruthenus
ASJ – Amur R., Sea of Okhotsk & Sea of Japan	*Acipenser mikadoi*
	Acipenser schrenckii
	Huso dauricus
CH – China	*Acipenser dabryanus*
	Acipenser sinensis
	Psephurus gladius

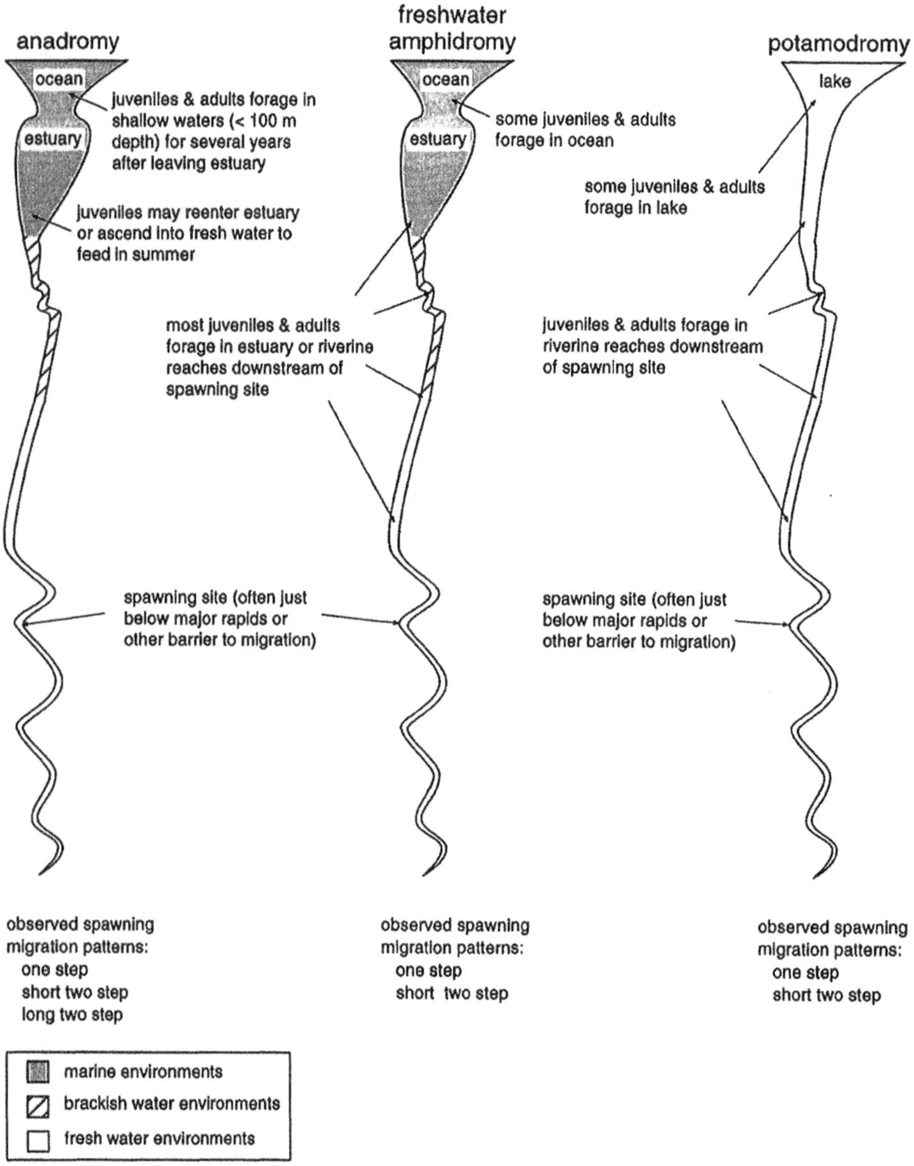

Figure 5. Concepts of anadromy, amphidromy and potamodromy in acipenseriforms. These terms are defined and further explained in the text. Patterns of spawning migrations used by fish employing these different life history patterns are indicated in text below each condition. Once a spawning migrant enters a river, it typically follows one of three spawning migration patterns. These are keyed to the drawing as follows: one step – short or long migration, spawning; short two step – migration, overwintering, short migration, spawning; long two step – migration, overwintering or oversummering (or both), long migration, spawning. See text for additional explanation.

ed with spawning activities, although in many cases, anadromous sturgeons may re-enter estuaries or even freshwater reaches of rivers during summer months to feed and amphidromous sturgeons may move from one riverine or estuarine foraging site to another. Fish with mature gonads that migrate upstream during the spawning season are commonly referred to as *spawning fish* or *spawners*, regardless of whether they successfully complete spawning. The age at first spawning migration is an important life history parameter for all species of sturgeons, although it is unknown in many cases. *Emigration* refers to an individual that leaves its river basin and migrates via a sea or lake. Emigrants may return to their natal river to spawn or colonize a new river basin (non-natal emigrants).

Thanks to McDowall (1987, 1988, 1992), other terms necessary to accurately describe migrations of Acipenseriformes have widely accepted definitions, which are:

Diadromous/diadromy – Fishes that migrate between salt water and fresh water (Myers 1949, McDowall 1988, 1992). Many, but not all, acipenseriforms are diadromous.

Anadromous/anadromy – Diadromous fishes that spend most of their lives at sea but return to fresh water to breed (Myers 1949, McDowall 1988, 1992). Most species in the genus *Acipenser* are anadromous, as are both species in the genus *Huso*. Surprisingly limited information is available about the physiological mechanisms that underlie anadromy in acipenseriforms (McEnroe & Cech 1985, 1987)

Amphidromous/amphidromy – Diadromous fishes whose migration from fresh water to the salt water, or vice-versa, is not for the purpose of breeding although it occurs regularly at some point(s) in the life cycle (McDowall 1988, 1992). McDowall (1992) defined two types of amphidromy: freshwater amphidromy, in which spawning is in fresh water and growth occurs during migrations into salt water, and marine amphidromy, in which spawning occurs in salt water and growth occurs during migrations into fresh water. All acipenseriforms spawn in fresh water, so that only freshwater amphidromy is relevant for the group. Only a few cases convincingly document freshwater amphidromy for any species of Acipenseriformes, because this requires detailed knowledge of the movements of individuals which can only be obtained from tagging and recapture or telemetric studies. The best documented of these species is the shortnose sturgeon, *Acipenser brevirostrum* (Bain 1997 this volume, Kynard 1997 this volume).

Potamodromous/potamodromy – Fishes that migrate within a river system to breed and forage (McDowall 1988, 1992). All shovelnose sturgeons (tribe Scaphirhynchini, *Scaphirhynchus* and *Pseudoscaphirhynchus*) are potamodromous. Paddlefishes (Polyodontidae) are potamodromous, for the few reports of *Polyodon* taken at sea seem to represent rare individuals, and all fossil polyodontids are from freshwater deposits (Grande & Bemis 1991, Bemis et al. 1997b this volume). The only possible exception is the Chinese paddlefish, *Psephurus gladius*, which was historically captured near the mouth of the Yangtze River and from the East China Sea, but which now is so rare that its true life history pattern will probably remain unknown. Based on the few data available, we think that it is probably fresh water amphidromous, because juvenile Chinese paddlefish were historically taken in the estuary of the Yangtze River (see Wei et al. 1997 this volume). Some species of *Acipenser*, such as *A. ruthenus*, are commonly considered to be potamodromous, such as the populations in upper reaches of the Danube River described by Hensel & Holčík (1997 this volume). Recent information suggests that *A. ruthenus* may prove to be amphidromous, because juveniles are commonly captured in salt water at the mouth of the Danube River. Other species, such as *A. schrenckii*, may be facultatively potamodromous, with some populations in upper reaches of the Amur River apparently never venturing near the estuary (Krykhtin & Svirskii 1997 this volume). Some authors refer to such populations as *residents*, meaning that the individual fish do not migrate to the sea. Poorly understood are other cases demonstrating the type of facultative potamodromy that occurs when dams obstruct passage of a formerly anadromous or amphidromous species, a condition referred to as *damlocked* (Kynard 1997 this volume). By itself, potamodromy can only provide negative evidence concerning a species' ability to cross large ocean basins.

In those marine coastal rivers that have sturgeons, usually at least two species (and sometimes as many as six) are present. If only two species are present, one is always anadromous, and the other is usually potamodromous (or amphidromous; see comments on the difficulties of detecting amphidromy above). Two clear examples of this are found in the Hudson River, which has *Acipenser oxyrinchus* (anadromous) and *A. brevirostrum* (amphidromous), and the Yangtze River, which has *A. sinensis* (anadromous) and *A. dabryanus* (amphidromous or potamodromous). If only one species is present, it is usually anadromous. The Northeast Atlantic region (Figure 4) has a high frequency of rivers, particularly in western France and the Iberian Peninsula, that have only one anadromous species, *A. sturio*. The absence of a second species in these rivers is not due to anthropogenic effects, but instead reflects the historical situation (Holčík 1989). It is unclear why the pattern that is so common elsewhere is not followed in the northeast Atlantic region.

In Figure 1, most taxa are scored with either an 'A' for anadromy or 'P' for potamodromy; only one species (*A. brevirostrum*) is marked 'FWA' to indicate freshwater amphidromy because without better telemetric and tagging studies, we cannot know how many seemingly freshwater species actually use patterns of freshwater amphidromy. Future work will almost certainly change some of our 'P' scores to 'FWA' scores. In some cases, a species may be potamodromous in one river basin and anadromous in another; in such cases, we scored the species 'A'.

Patterns of spawning migrations

Within acipenseriforms, variations in the pattern of spawning migration are found at the species, population, and individual levels. The genetic basis of spawning migration characteristics is well established for salmoniforms, for which extensive selection experiments have been done (also see papers in Groot & Margolis 1991). Virtually nothing is known about the heritability of spawning migration characteristics of acipenseriforms, but we expect a similar genetic basis to that known for salmonids. The existence of different spawning migration patterns in sturgeons has been discussed by many authors, including Berg (1934, 1959), Artyukhin (1988) and Kynard (1997 this volume). Berg (1934) introduced the terms *vernal* and *winter* races to describe groups of anadromous fishes migrating into rivers for spawning in the same year (vernal races) or next year (winter races). These terms stimulated a long discussion in the Russian literature concerning Eurasian sturgeons (reviewed by Barannikova 1957, Gerbilskiy 1957, Kazansky 1962, Artyukhin 1988). Some of the terms and discussions are contradictory and difficult to follow, particularly because it is not always possible to link migration times, spawning sites and specific migrants. Also, the terminology does not easily translate to conditions in North American rivers, many of which are shorter, smaller coastal streams than are the major rivers of the Black and Caspian Sea basins for which the terminology was originally developed. The simplified scheme summarized in the box in Figure 5 draws primarily from Kynard (1997) and Gerbilskiy (1957). It classifies spawning migrations as having one or two steps, with a variable length of time between the actual migration and the time of spawning. This scheme can be readily used to describe either anadromous, amphidromous or potamodromous acipenseriforms and individual variation within populations.

One step spawning migrations are those in which fish move directly upstream to the spawning site, spawn, and return downstream. Depending on the bioenergetic reserves of the fish, the migration may be short or long, and occur in winter or spring. This is usually thought to be the most common pattern for living acipenseriforms, although the few data available (mostly catch records) are conflicting. It corresponds to Gerbilskiy's (1957) migrant type I, in which the oocytes have reached their final size and spermatogensis is finished by the time migration starts. Fat deposits in connective tissue and muscles are depleted, and the stomach and digestive tract are empty and inactive indicating that feeding stopped some time before migration. Such migrants typically use spawning sites in the lower or middle reaches of rivers.

Short two step spawning migrations involve upstream migration, usually in the fall, followed by overwintering near the spawning site, followed by a very short migration to spawn the following spring. This pattern enables fish to use bioenergetic reserves gained during summer foraging for their initial long upstream migration. This corresponds roughly to Gerbilskiy's (1957) migrant type II, in which late stages of oogenesis are in progress, and the oocytes are still embedded in fatty tissue. Spermatogenesis is in the 'first wave' of divisions. There is abundant fat in connective tissue and dorsal muscles, and the hepatocytes are large because of lipid inclusions. Food remains in the stomach and digestive tract indicate that feeding took place just prior to the start of migration. These fish typically spawn in middle to upstream reaches of many rivers, such as the Volga, Ural, Danube, Hudson or Connecticut.

Long two step spawning migrations refer to fish that make an initial upstream migration, followed either by overwintering, oversummering, or both, then followed by a long upstream migration to the spawning site. Fish with this pattern may be in fresh water without feeding for 12 to 15 months, which effectively precludes this option for small to medium sized species because they lack sufficient bioenergetic capacities. Only very large species, such as *Huso huso* and *Acipenser sinensis* seem likely candidates for this pattern. This corresponds more or less to Gerbilskiy's (1957) migrant type III, which is characterized by late stages of oogenesis, and intermediate levels of fat in the ovary, connective tissues, and muscles at the start of migration. This type of migration is characteristic of some individuals of large species in the longest rivers, such as the Danube, Volga, Amur or Yangtze. The only place in North America where this pattern may have been present is the Columbia River, where very large white sturgeon, *A. transmontanus*, historically spawned in headwaters.

Explanations that have been offered concerning the adaptive significance of different spawning migratory patterns include river length, river gradient, temperature at the spawning site, and bioenergetics. For example, Artyukhin (1988) concluded that spawning in the spring is characteristic of most species of sturgeons in the Ponto-Caspian region that inhabit lowland rivers, whereas spawning during the summer is associated with rivers having a higher gradient. Kynard (1997 this volume) proposed a bioenergetic explanation of migratory patterns. As noted above, spawning migratory patterns can be variable within species or populations, and before conclusions are reached, it seems necessary to develop better understanding of sturgeon life histories. Factors such as spawning site fidelity and behavior between successive spawnings need to be broadly investigated across acipenseriforms, as well as the genetic bases of supposedly different stocks of the same species in particular river systems (Wirgin et al. 1997 this volume). Until then, it is safest to regard the three patterns of spawning migration as descriptive tools, rather than interpretive explanations for particular migration patterns, and we have not attempted to score the migratory patterns of species in Figure 1.

Discussion

Several events in Holarctic history stand out as influencing the contemporary distribution of sturgeons and paddlefishes. A complete review is far beyond our scope, so we note only a few highlights.

Place of origin of acipenseriforms and their early diversification

Some have asserted (e.g., Yakovlev 1977) that Acipenseriformes originated in northeastern Asia in the Triassic. This argument, however, cannot be based on either the earliest known occurrence of fossils or the greatest current diversity of taxa but must instead be consistent with the ranges of outgroup taxa (Nelson & Platnick 1981). Based on available phylogenetic and biogeographic evidence (Figure 1), the most plausible place and time of origin for Acipenseriformes is the Triassic of western Europe, for this is consistent with the range of †*Birgeria* (Europe, North America, and Madagascar) as well as with the range of †Chondrosteidae, which is interpreted by Grande & Bemis (1996) to be the sister taxon of all other Acipenseriformes (Figure 2). The localities of Late Jurassic/Early Cretaceous members of Peipiaosteidae in Central and Eastern Asia suggest (but do not provide definitive evidence for) early diversification of Acipenseriformes in central Asia. If true, this seems consistent with the greatest current species diversity of Acipenseridae in the Ponto-Caspian region and the relatively much later appearance of the group (Late Cretaceous) in North America (Figure 2). We should expect to find new acipenseriform fossils in Mesozoic deposits across Europe and Asia.

Were early acipenseriforms potamodromous or anadromous?

McDowall (1993) considered that sturgeons (and by extension, acipenseriforms in general) are unlikely to have had a recent marine ancestry, a conclusion that all fossil data and contemporary phylogenetic analyses support. This does not answer, however, whether anadromy originated within Acipenseriformes or was a plesiomorphic feature of the group. Unfortunately, we can only speculate about the answer, because we can never hope to understand much about the life history of †Birgeriidae, †Chondrosteidae or †Peipiaosteidae, and more distant outgroups are not helpful (Bemis et al. 1997b this volume). From the analysis in Figure 1, either potamodromy or anadromy could be the ancestral condition, because each condition appears twice on the tree (potamodromy in most or all species of Polyodontidae and Scaphirhynchini; anadromy in *Huso* and most species of *Acipenser*). If *Huso* is eventually nested within *Acipenser* (e.g., Birstein et al. 1997), then perhaps anadromy will emerge as a derived character of some clade that includes *Acipenser* and *Huso*. This would be very interesting, because anadromy seems likely to be linked to the great diversity within *Acipenser*.

Comments on biogeography of extant species

The Ponto-Caspian region currently has the greatest species diversity of Acipenseridae. Some sturgeons of the Ponto-Caspian region have striking morphological distinctions from all other species, such as the very elongate rostrum found in adult stellate sturgeon, *Acipenser stellatus*. There are some intriguing links of this Ponto-Caspian region both to the North Eastern Atlantic (*A. sturio*) and to the Amur River district (*Huso huso* and *H. dauricus*). One of the most intriguing links of the Ponto-Caspian region is to the Mississippi-Gulf of Mexico region in North America indicated by the scaphirhynchine sturgeons. All six extant species of *Scaphirhynchus* and *Pseudoscaphirhynchus* are considered to be potamodromous, and we know that Scaphirhynchini was present in North America in the Late Cretaceous (Figure 2). (A comparably intriguing and old link dating from at least the Late Cretaceous occurs between China and the Mississippi-Gulf of Mexico region as indicated by Polyodontidae; Grande & Bemis 1991.)

The Ponto-Caspian region has been very unstable over the last 150 million years, the period in which we suppose Acipenseridae has diversified. Some indication about the magnitude of the Earth historical changes in the Ponto-Caspian region is apparent in the diagrammatic maps in Figure 2, which represent only a small window on this part of the world (see Smith et al. 1994 for additional geographical and geological detail). The changes include major sea level variation, conversions of large bodies of water such as the Black Sea from freshwater lakes to marine environments, merging of island arcs with the southern continental borders of Europe and Asia, and major shifts in drainage patterns as mountain building occurred. The Black Sea has repeatedly been connected and disconnected with the Caspian and Aral seas. It is tempting to link the current diversity of acipenserids in this region to its extremely complex history.

Around the Pacific rim, we defined three biogeographic regions (NEP, ASJ, and CH, Figure 2) that together have six species of *Acipenser*. Based on the available phylogenetic interpretation (Figure 1), the five anadromous species (*A. transmontanus*, *A. medirostris*, *A. mikadoi*, *A. schrenckii* and *A. sinensis*) appear to form a monophyletic group. *Acipenser dabryanus*, a potamodromous and potentially amphidromous species believed to be restricted to the Yangtze River, lies outside the group of species from around the Pacific rim (see Wei et al. 1997 this volume and Zhuang et al. 1997 this volume for discussion of the ranges of sturgeons in China). One point concerning sturgeons of the Pacific rim is that we are unaware of any spawning in rivers north of the Fraser River, British Columbia. The explanation for this pattern in the Pacific is unknown, although other taxa of acipenserids, such as *A. fulves-*

cens in North America and *A. baerii* in Siberia, spawn in rivers at higher latitudes.

Some biogeographic patterns may be related to continental movements, such as the sister group relationship between *Acipenser sturio* in Europe and *A. oxyrinchus* in North America. These two species are anatomically similar and were long considered to be conspecific (Vladykov & Greeley 1963). They also share molecular sequence similarities (Birstein et al. 1997 this volume). Although these species live on opposite sides of the North Atlantic Ocean, and are presumably blocked from most interbreeding, we suspect that the separation between these taxa is actually much younger than the North Atlantic Ocean.

Species of acipenseriforms directly impacted by Pleistocene glaciation presumably include boreal taxa such as *A. baerii* (Ruban 1997, this volume). This species occurs in many of the northward flowing rivers of Siberia (SAO region, Figure 4). Although the extent of glaciation in Siberia during the last glacial maximum (18000 years bp) was not as extensive as one might suppose (Starkel 1991), *A. baerii* was certainly prevented from entering the Arctic Ocean. The location of its glacial refugia is uncertain. In North America, the lake sturgeon, *A. fulvescens*, may have occupied both Mississippian and Missourian refugia during glaciation, and recolonized its range in the northern United States and Canada by dispersal (Ferguson & Duckworth 1997, this volume).

Clearly, much is yet to be learned about biogeography and life history of Acipenseriformes. Synthesis should be the goal, particularly as new data on the status of poorly known populations become more available.

Acknowledgements

We are grateful to Eugene Balon for allowing us to prepare this manuscript for inclusion in the volume. Vadim Birstein and Rob DeSalle provided a preliminary analysis of their molecular phylogenetic data set to assist us in depicting acipenserine relationships shown in Figure 1 (thie complete data set, including a different placement of *Huso*, is analyzed in Birstein & DeSalle 1997). We also thank Vadim for extensive discussions of the main ideas presented here. In spite of a few minor differences of opinion, this paper would not exist were it not for his extremely helpful input. Paul Morris read and criticized a draft of the manuscript and helped with paleogeographic questions.

References cited

Anonymous. 1980. Rand McNally encyclopedia of world rivers. Bison Books, London. 352 pp.

Artyukhin, E.N. 1988. The nature and origin of summer spawning in sturgeons. Voprosy Ikhtiologii 5: 717–723 (in Russian, English translation J. Ichthyol. 5: 116–123).

Artyukhin, E.N. 1995. On biogeography and relationships within the genus *Acipenser*. Sturgeon Quarterly 3(2):6–8.

Bai, Y.-J. 1983. A new *Peipiaosteus* from Fengning of Hebei, China. Vert. PalAsiat. 21:341–346 (in Chinese).

Bain, M.B. 1997. Atlantic and shortnose sturgeons of the Hudson River: common and divergent life history attributes. Env. Biol. Fish. (this volume).

Balon, E.K. 1990. Epigenesis of an epigeneticist: the development of some alternative concepts on the early ontogeny and evolution of fishes. Guelph Ichthyol. Rev. 1:1–48.

Bănărescu, P. 1990. Zoogeography of fresh waters. Vol. 1 AULA-Verlag, Wiesbaden. pp. 1–511.

Bănărescu, P. 1992. Zoogeography of fresh waters. Vol. 2 AULA-Verlag, Wiesbaden. pp. 519–1091.

Bănărescu, P. 1995. Zoogeography of fresh waters. Vol. 3 AULA-Verlag, Wiesbaden. pp. 1099–1617.

Barannikova, I.A. 1957. Biological differentiation of the stocks of the Volga River-Caspian Sea sturgeons (in connection with the tasks of sturgeon farming in the Volga River delta). Uchenye Zapiski Leningradskogo Gosudarstvennogo Universiteta 228: 57–72 (in Russian).

Bemis, W.E., V.J. Birstein & J.R. Waldman. 1997a. Sturgeon biodiversity and conservation: an introduction. Env. Biol. Fish. (this volume).

Bemis, W.E., E.K. Findeis & L. Grande. 1997b. An overview of Acipenseriformes. Env. Biol. Fish. (this volume).

Berg, L.S. 1934. *Acipenser gaeldenstaedti persicus*, a sturgeon from the south Caspian Sea. Ann. Mag. Nat. Hist., Ser. 10: 317–319.

Berg, L.S. 1948a. On the position of the Acipenseriformes in the system of fishes. Trudy Zoologicheskogo Instituta 7: 7–57 (in Russian).

Berg, L.S. 1948b. The freshwater fishes of the USSR and adjacent countries, Vol. 1. Akademia Nauk USSR, Moscow & Leningrad. Part 1 (in Russian, English translation published by Israel Program for Scientific Translations, Jerusalem. 505 pp.)

Berg, L.S. 1959. Vernal and hiemal races among anadromous fishes. J. Fish Res. Board Can. 16: 515–537.

Berra, T.M. 1981. An atlas of distribution of the freshwater fish families of the world. University of Nebraska Press, Lincoln. 30 pp.

Birstein, V.J. 1993a. Sturgeons and paddlefishes: threatened fishes in need of conservation. Cons. Biol. 7: 773–787.

Birstein, V.J. 1993b. Is *Acipenser medirostris* one or two species? Sturgeon Quarterly 1(2): 8.

Birstein, V.J. & R. DeSalle. 1997. Molecular phylogeny of Acipenserinae. Mol. Phylogenet. Evol. 7 (in press).

Birstein, V.J., R. Hanner & R. DeSalle. 1997. Phylogeny of the Acipenseriformes: cytogenetic and molecular approaches. Env. Biol. Fish. (this volume).

Bond, C. E. 1996. Biology of fishes, 2nd ed. Saunders, Philadelphia. 750 pp.

Dettlaff, T.A., A.S. Ginsburg & O.I. Schmalhausen. 1993. Sturgeon fishes: developmental biology and aquaculture. Springer-Verlag, Berlin. 300 pp.

Findeis, E.K. 1997. Osteology and phylogenetic interrelationships of sturgeons (Acipenseridae). Env. Biol. Fish. (this volume).

Gardiner, B.G. 1984b. Sturgeons as living fossils. pp. 148–152. *In:* N. Eldredge & S.M. Stanley (ed.) Living Fossils, Springer Verlag, New York.

Gerbilskiy, N.L. 1957. Ways of development of the intraspecies biological differentiation, types of anadromous migrants, and a question of migration impulse in sturgeons. Uchenye Zapiski Leningradskogo Gosudarstvennogo Universiteta 228: 11–31 (in Russian).

Grande, L. 1985. The use of paleontology in systematics and biogeography, and a time control refinement for historical biogeography. Paleobiology 11:1–11.

Grande, L. & W.E. Bemis. 1991. Osteology and phylogenetic relationships of fossil and recent paddlefishes (Polyodontidae) with comments on the interrelationships of Acipenseriformes. J. Vert. Paleo. 11, supplement 1: 1–121.

Grande, L. & W.E. Bemis. 1996. Interrelationships of Acipenseriformes, with comments on "Chondrostei". pp. 85–15. *In:* M.L. J. Stiassny, L.R. Parenti & G.D. Johnson (ed.) Interrelationships of Fishes, Academic Press, San Diego.

Groot, C. & L. Margolis (ed.) 1991. Pacific salmon life histories. University of British Columbia Press, Vancouver. 564 pp.

Hart, J.L. 1973. Pacific fishes of Canada. Bull. Fish. Res. Board of Canada, Ottawa. 740 pp.

Hensel, K. & J. Holčík. 1997. Past and current status of sturgeons in the upper and middle Danube River. Env. Biol. Fish. (this volume).

Hilgendorf, F. 1892. Über eine neue Stör-Art aus Nord-Japan (*Acipenser mikadoi*). Sitzungsber. Ges. naturf. Freunde, Berlin 7: 98–100.

Hocutt, C.H. & E.O. Wiley. 1986. The zoogeography of North American freshwater fishes. John Wiley and Sons, New York. 866 pp.

Holčík, J. 1989. The freshwater fishes of Europe. Vol I. Part II. General introduction to fishes. Acipenseriformes. AULA-Verlag, Wiesbaden. 469 pp.

Hsu, K.J. 1972. When the Mediterranean dried up. Sci. Amer. 227: 27–36.

Hsu, K.J. 1978. When the Black Sea was drained. Sci. Amer. 238: 53–63.

Jin, F. 1995. Late Mesozoic Acipenseriformes (Osteichthyes: Actinopterygii) in Central Asia and their biogeographic implications. pp. 15–21. *In:*A. Sun & Y. Wang (ed.) Sixth Symposium on Mesozoic Terrestrial Ecosystems and Biota, Short Papers, China Ocean Press, Beijing.

Kazansky, B.N. 1962. Experimental analysis of the seasonal nature of the reproduction of the Volga River sturgeons in connection with the phenomenon of intraspecies biological differentiation. Uchenye Zapiski Leningradskogo Gosudarstvennogo Universiteta 311: 19–45 (in Russian).

Krykhtin, M.L. & V.G. Svirskii 1997. Endemic sturgeons of the Amur River: kaluga, *Huso dauricus* and Amur sturgeon, *Acipenser schrenckii*. Env. Biol. Fish. (this volume).

Kynard, B. 1997. Life history, latitudinal patterns, and status of the shortnose sturgeon, *Acipenser brevirostrum*. Env. Biol. Fish. (this volume).

Lee, D.S., C.R. Gilbert, C.H. Hocutt, R.E. Jenkins, D.E. McAllister & J.R. Stauffer, Jr. 1980. Atlas of North American freshwater fishes. North Carolina Biological Survey, Raleigh. 867 pp.

Leggett, W.C. 1976. The American shad *Alosa sapidissima*, with special reference to its migrations and population dynamics in the Connecticut River. Amer. Fish. Soc. Monogr. 1: 169–225.

Lehman, J.-P. 1952. Etude complmentaire des poissons de lEotrias de Madagascar. K. Svenska VetenskAkad. Handl. 2(4): 1–201.

McDowall, R.M. 1987. The occurrence and distribution of diadromy among fishes. Amer. Fish. Soc. Symp. 1: 1–13.

McDowall, R.M. 1988. Diadromy in fishes. Migrations between freshwater and marine environments. Croom Helm, London. 308 pp.

McDowall, R.M. 1992. Diadromy: origins and definition of terminology. Copeia 1992: 248–251.

McDowall, R.M. 1993. A recent marine ancestry for diadromous fishes? Sometimes yes, but mostly no! Env. Biol. Fish. 37: 329–335.

McEnroe, M. & J.J. Cech. 1985. Osmoregulation in juvenile and adult white sturgeon, *Acipenser transmontanus*. Env. Biol. Fish. 14: 23–40.

McEnroe, M. & J.J. Cech. 1987. Osmoregulation in white sturgeon: life history aspects. Amer. Fish. Soc. Symp. 1: 191–196.

Myers, G.S. 1949. Usage of anadromous, catadromous and allied terms for migratory fishes. Copeia 1949: 89–97.

Nelson, G. & N. Platnick. 1981. Systematics and biogeography. Cladistics and vicariance. Columbia University Press, New York. 567 pp.

Nielsen, E. 1949. Studies on Triassic fishes from East Greenland. 2. *Australosomus* and *Birgeria*. Meddr. Grönland. 146: 1–309.

Pearse, A.S. & G. Gunter. 1957. Salinity. Geol. Soc. Amer. Mem. 67:129–158.

Rochard, E., G. Castelnaud & M. Lepage. 1990. Sturgeons (Pis-

ces: Acipenseridae); threats and prospects. J. Fish Biol. (suppl. A) 37: 123–132.
Ruban, G.I. 1997. Species structure, contemporary distribution and status of the Siberian sturgeon, *Acipenser baerii*. Env. Biol. Fish. (this volume).
Salo, E.O. 1991. Life history of chum salmon (*Oncorhynchus keta*). pp. 231–310. *In:* C. Groot & L. Margolis (ed.) Pacific Salmon Life Histories, University of British Columbia Press, Vancouver.
Schwarz, W. 1970. Die Triasfauna der Tessiner Kalkalpen. XX. *Birgeria stensiöi* Aldinger. Schweiz Paläont. Abh. 1–93.
Scott, W.B. & E.J. Crossman. 1973. Freshwater fishes of Canada. Bull. 184, Fisheries Res. Board of Canada, Ottawa. 966 pp.
Smith, A.G., D.G. Smith & B.M. Funnell. 1994. Atlas of Mesozoic and Cenozoic coastlines. Cambridge University Press, Cambridge. 99 pp.
Starkel, L. 1991. Characteristics of the temperate zone and fluvial paleohydrology. pp. 3–12. *In:* L. Starkel, K.J. Gregory & J.B. Thornes (ed.) Temperate Paleohydrology, John Wiley and Sons, New York.
Trautman, M.B. 1981. The fishes of Ohio. Ohio State University Press, Columbus. 683 pp. [first printing in 1957]
Vladykov, V. & J.R. Greeley. 1963. Order Acipenseroidei. pp. 24–60. *In:* H.B. Bigelow & W.C. Schroeder (ed.) Fishes of the Western North Atlantic, Sears Foundation for Marine Research, Yale University, New Haven.
Waldman, J.R., J.T. Hart & I.I. Wirgin. 1996a. Stock composition of the New York Bight Atlantic sturgeon fishery based on analysis of mitochondrial DNA. Trans. Amer. Fish. Soc. (in press).
Waldman, J.R., K. Nolan, J. Hart & I.I. Wirgin. 1996b. Genetic differentiation of three key anadromous fish populations of the Hudson River. Estuaries 19: 759–768.
Wei, Q., F. Ke, J. Zhang, P. Zhuang, J. Luo, R. Zhou & W. Yang. 1997. Biology, fisheries, and conservation of sturgeons and paddlefish in China. Env. Biol. Fish. (this volume).
Williot, P. (ed.) 1991. Acipenser. CEMAGREF Publ., Bordeaux. 519 pp.
Wirgin, I.I., J.E. Stabile & J.R. Waldman. 1997. Molecular analysis in the conservation of sturgeons and paddlefish. Env. Biol. Fish. (this volume).
Yakovlev, V.N. 1977. Phylogenesis of acipenseriforms. pp. 116–143. *In:* V.V. Menner (ed.) Essays on Phylogeny and Systematics of Fossil Fishes and Agnathans. USSR, Akademy of Sciences, Moscow (in Russian).

Sturgeons of the western and eastern Atlantic: a – the shortnose sturgeon, *Acipenser brevirostrum* 143 cm TL from the outlet of Washademoak Lake, St. Johns River, New Brunswick, Canada, caught 2.1.1975 (ROM cat # 34310); b – American Atlantic sturgeon, *A. oxyrinchus* 150 cm TL from St. Lawrence River, Kamouraska, Quebec, now residing alive in the Montreal Biodome (hence the typical damage to the tip of the rostrum); c – European Atlantic sturgeon, *A. sturio* 107 cm TL from the Black Sea stock at the 'Grigore Antipa' Natural History Museum, Bucharest. Originals by Paul Vecsei, 1996.

Environmental Biology of Fishes **48**: 185–200, 1997.

Past and current status of sturgeons in the upper and middle Danube River

Karol Hensel[1] & Juraj Holčík[2]
[1] *Department of Zoology, Faculty of Natural Sciences, Comenius University, Mlynská dolina B-1, 842 15 Bratislava, Slovakia*
[2] *Department of Ichthyology, Institute of Zoology, Slovak Academy of Sciences, Dúbravská cesta 9, 842 06 Bratislava, Slovakia*

Received 2.8.1994 Accepted 25.3.1996

Key words: Huso huso, Acipenser nudiventris, A. stellatus, A. gueldenstaedti, A. ruthenus, anadromy, Romania, Serbia, Croatia, Hungary, Slovakia, Austria, Germany

Synopsis

Of the six species of sturgeons native to the Danube basin, five occurred in the upper and middle Danube. Among anadromous sturgeons were the large winter races of beluga, *Huso huso*, Russian sturgeon, *Acipenser gueldenstaedtii*, and stellate sturgeon, *A. stellatus*, which ascended the middle, and sometimes also the upper Danube, to spawn. Due to overfishing, followed by severe habitat alteration including damming and pollution, these anadromous sturgeons are critically endangered or extirpated from the upper and middle Danube. *Acipenser gueldenstaedtii* and *A. nudiventris* are represented only as resident non-migratory races with very small populations. The most abundant and widely distributed species is the sterlet, *A. ruthenus,* although it is presently limited to the middle Danube. Its population increased in some sections of the middle Danube during the past 15 years, presumably because of improving water quality, but this species remains at risk because of continuing habitat degradation.

Introduction

Six species of sturgeons historically occurred in the Danube River and some of its tributaries. The European Atlantic sturgeon, *Acipenser sturio,* was the rarest, and it only occasionally entered the Danube estuary. Beluga, *Huso huso* (Linnaeus, 1758), ship sturgeon, *A. nudiventris* Lovetski, 1828, stellate sturgeon, *A. stellatus* Pallas, 1771, Russian sturgeon, *A. gueldenstaedtii* Brandt, 1883, and sterlet, *A. ruthenus* (Linnaeus, 1758), however, were common to abundant (also see Bacalbaşa-Dobrovici 1997 this volume). Anadromous populations, especially winter races (= autumnal races of some authors; see Birstein & Bemis 1997 this volume, for discussion of this terminology) of beluga and Russian sturgeon, moved from the Black Sea into the Danube, ascending the middle and sometimes even the upper Danube and larger tributaries. Freshwater resident populations of some species of sturgeons also existed. Because sturgeons had such great economic importance, many historical records are available. However, overfishing and habitat alteration caused populations to collapse (Rochard et al. 1990, Birstein 1993). In particular, construction of the Đerdap I Dam (= Iron Gates Dam I) at the village of Sip (Iron Gate, river kilometer 942) in 1969 and later construction of the Đerdap II Dam (= Iron Gates Dam II) at Kusjak (river km 863) in 1984 blocked further upstream migration of anadromous sturgeons, and most species are now extirpated from the middle and upper Danube.

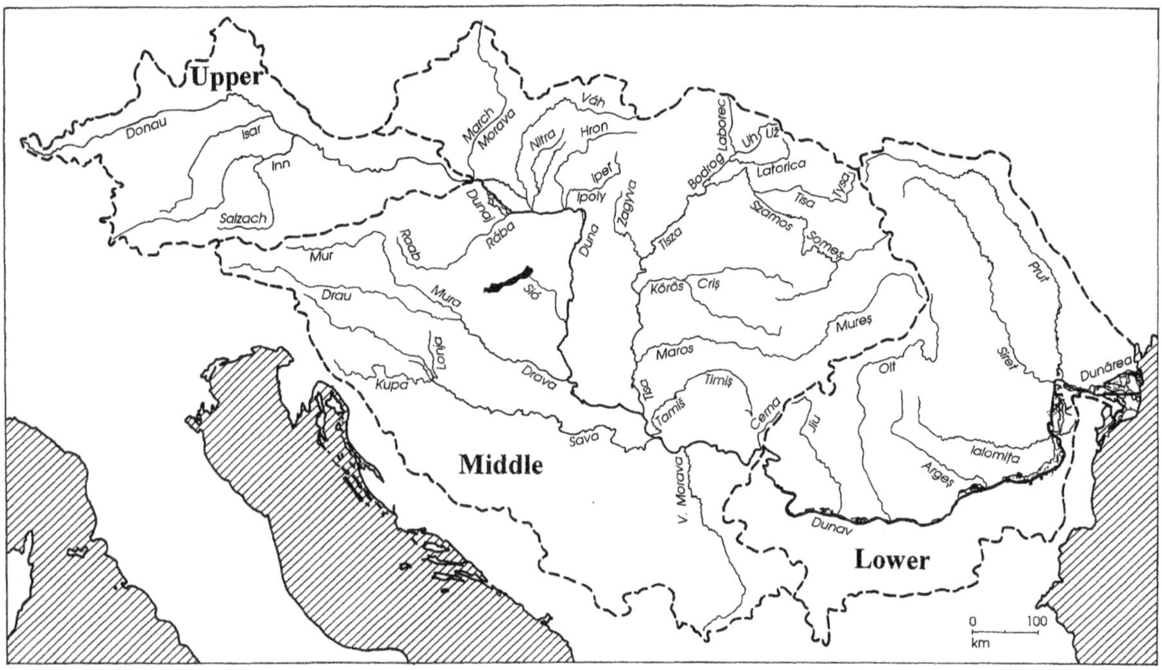

Figure 1. The Danube River basin showing rivers inhabitated by sturgeons (Acipenseridae). Original figure by K. Hensel.

Geographically the Danube is divided into three parts (Figure 1). The lower Danube (shared by Ukraine, Moldova, Romania, Bulgaria and Serbia) extends from the estuary up to the mouth of the Cerna River (river km 955) in the Iron Gates region. The middle Danube (shared by Romania, Serbia, Croatia, Hungary and Slovakia) runs from the Cerna up to the mouth of the Morava River (river km 1880), and the upper Danube flows through Austria and Germany (Balon et al. 1986).

This paper summarizes the history and status of five species of sturgeons in the middle and upper Danube. We report sizes as total length (TL, the distance between the anterior tip of the rostrum and the tip of the caudal fin) and weight (BW, total body weight). Data on catches of particular species are from proceedings of the JCIAFD.[1,2] We report locations of capture at either towns or particular river kilometers from the mouth.

Huso huso – beluga or great sturgeon

Vernal races (= spring races of some authors; see Birstein & Bemis 1997 this volume for terminology) and winter races of this anadromous species annually ascended the Danube River in large numbers (Figure 2). Although migrations of beluga continued year round, two peak periods were regularly observed, one for the winter and the second one for the spring strain. Upstream migration of the winter strain usually started in August and culminated in October or November. Migration of the spring strain lasted from January until April (Bănă-

[1] Joint Commission of the International Agreement on Fishing in the Danube River.

[2] Anonymous. 1983. Appendix. pp. 207–229. *In:* Documents from the 24. session of the Joint Commission of the International Agreement on the Fishing in the Danube River between the governments of the Soviet Union, People Republic of Bulgaria, People Republic of Hungary, Socialistic Republic of Romania, Czechoslovak Socialistic Republic and Socialistic Federative Republic of Yugoslavia. Moscow (in Russian).

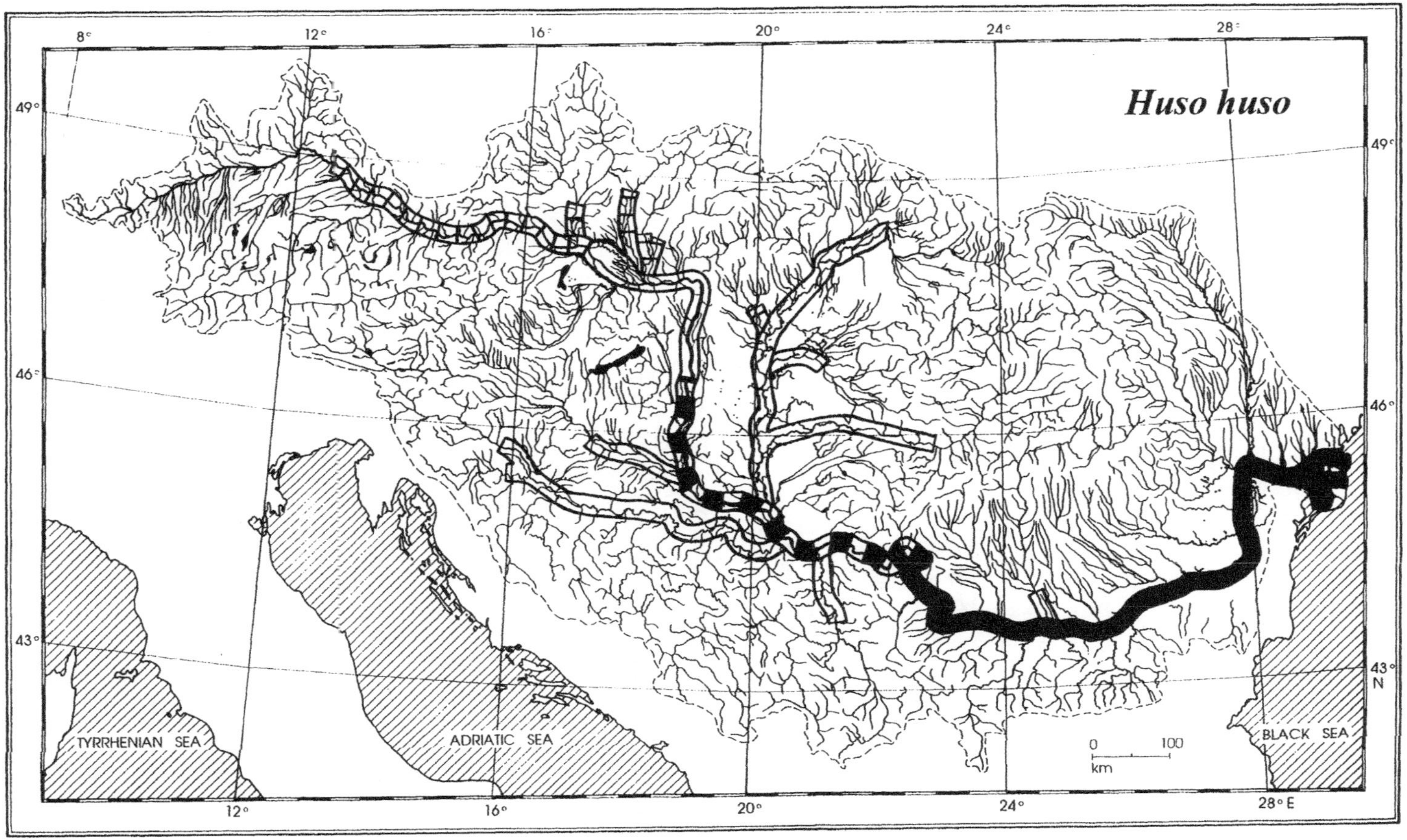

Figure 2. Distribution of the beluga, *Huso huso*, in the Danube drainage system. Regular (continuous black area) and occasional (black and white area) occurrence at present; regular (continuous white area) and occasional (striped white area) occurrence in the past. Original figure by K. Hensel.

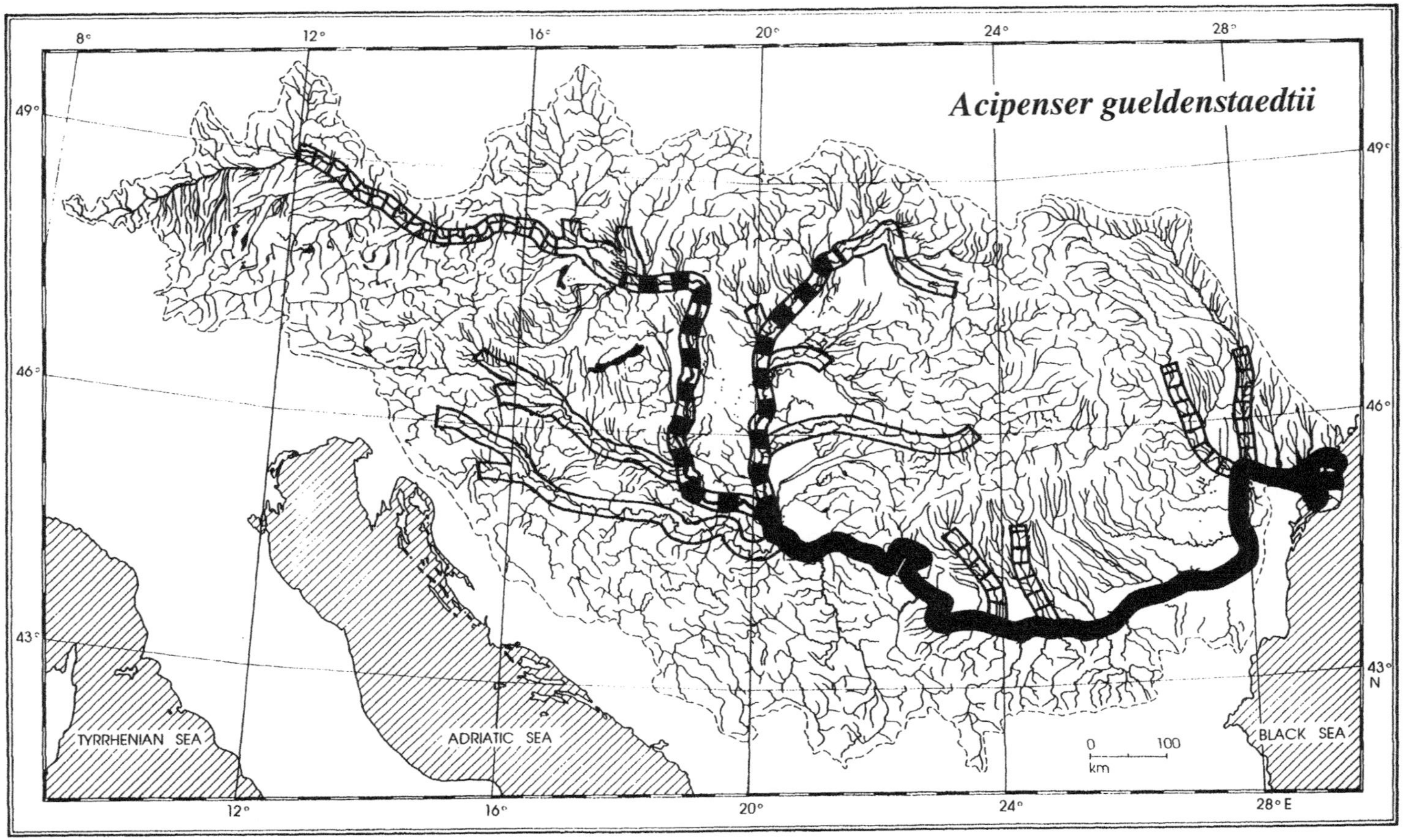

Figure 3. Distribution of the Russian sturgeon, *Acipenser gueldenstaedtii*, in the Danube drainage system. Regular (continuous black area) and occasional (black and white area) occurrence at present; regular (continuous white area) and occasional (striped white area) occurrence in the past. Information on distribution was compiled from Grossinger (1794), Fitzinger & Heckel (1835), Heckel & Kner (1858), Kornhuber (1863), Siebold (1863), Herman (1887), Ortvay (1902), Antipa (1909), Vutskits (1913), Munda (1926), Kähsbauer (1961), and Holčík (1995). Original figure by K. Hensel.

rescu 1964, Manea 1966, Kirilyuk & Rovnin[3]). The winter strain overwintered in the river and spawned in the following spring. Winter beluga ascended up to Bratislava (= Presburg, Preßburg or Pozsony, see Gesner 1575), rarely also entered the Austrian part of the Danube (Fitzinger & Heckel 1835) and occasionally even the Bavarian stretch up to Straubing (river km 2320; Siebold 1863). The main spawning grounds of beluga were located in the contemporary Slovak – Hungarian stretch of the river in the Žitný Ostrov reach below Bratislava (river km 1766–1866). The major fishery for beluga was concentrated in the Little Danube (which is the northern branch of the Danube River) near the mouth of the Váh River at the village of Kolárovo, see Figure 1) and in the Danube proper between Komárno and Sap (= Palkovičovo, see Balon 1967).

Beluga also entered other tributaries of the Danube, including the lower course of the Morava (= March) River (Jeitteles 1864), where a 2 m TL specimen was caught at Lanžhot (Zbořil & Absolon 1916). In the Váh River, beluga ascended up to Trnovec nad Váhom (Herman 1887) and exceptionally even up to Trenčín (Kornhuber 1861). Beluga also occurred in the Žitava River up to Nesvady (Holčík 1995), the Drava (= Drau) River (Taler 1953), the Tisa (= Tisza or Tysal) River (Heckel & Kner 1858) up to Trakany (Anonymous 1975) and its tributaries the Zagyva River, Körös (= Criş) River (Vutskits 1913) and Maros (= Mureş) River (Heckel & Kner 1858) where it occurred even at Hunedoara (Bănărescu 1964). In the Sava River, beluga were recorded at Zagreb (Glowacki 1896) and also in the Sava River's tributary, the Kupa River (Taler 1953). Beluga also entered the lower course of the Velika Morava River (Vutskits 1913) and the Olt River (Heckel & Kner 1858).

Beluga was among the most abundant of Danubian anadromous fishes, and it was the most valuable. Due to overfishing of brood fish during spawning migrations (Heckel 1851, Heckel & Kner 1858), catches of beluga started to decline after the 16th century (Balon 1967, 1968). Beluga were taken by means of special nets and particular hooks, called 'samolov'; however, the most effective method was the catching weir (Rohan-Csermák 1963). Because most of the fish migrating to spawn for the first time

[3] Kirilyuk, M.M. & A.A. Rovnin. 1983. The status of the brood stock, age structure and breeding conditions of sturgeons in 1981. pp. 28–36. *In:* Materials of the 24th Session of the joint Commission of the International Agreement on the Fishing in the Danube River between the Governments of the Soviet Union, Peoples Republic of Bulgaria, Peoples Republic of Hungary, Socialistic Republic of Romania, Czechoslovak Socialistic Republic and Socialistic Federative Republic of Yugoslavia, Moscow (in Russian).

Table 1. Specimens of *Acipenser gueldenstaedtii* recorded in Slovak and Hungarian segment of the Danube River since 1900[1,2].

1932:	lower stretch of the Morava River between its confluence with the Danube and Suchohrad (a 35 km long segment); 7 kg BW, estimated TL 950 mm.
1949:	mouth of the Little Danube at Komárno; 20.4 kg BW, estimated TL 1118 mm.
1960:	confluence of the Little Danube with the Nitra River, TL 375 mm, estimated BW 0,4 kg.
1962:	Danube at Zlatná na Ostrove (river km 1779); TL 850 mm, 10.5 kg BW.
1964:	Danube at Vel'ké Kosihy (river km 1787); two specimens 10.2 and 10.6 kg BW, estimated TL 1072 and 1085 mm.
1965:	same locality as above (Gunda 1966, erroneously writes 'Danube at Malé Kosihy' however, the latter village is on the right bank of the Ipel'River, where this species was never found); 10.4 kg BW, estimated TL 1079 mm.
1967:	Danube at Radvaň nad Dunajom (river km 1749); 355 mm, estimated BW 331 g; this specimen seems to be an anadromous form, as its calculated TL (295 mm) is substantially higher than that of the resident form (190 mm according to Lukin 1937).
1968:	Danube at Radvaň nad Dunajom, 424 mm TL; this was a hybrid between *A. ruthenus* and *A. gueldenstaedtii* (Hensel 1969).
1970:	Danube at Paks (river km 1827); 400 mm TL, estimated BW 643 g.
1980:	Tisza River at Tiszafüred; estimated TL 852 mm, 5 kg BW.

[1] Sturgeons caught in the Danube at Štúrovo in 1937 and 1957, BW 12 and 18 kg, respectively, where *A. guendelstaedtii* and not *H. huso* as said by Kux & Weisz (1962) because *H. huso* only matures at BW > 20 kg (Chugunov & Chugunova 1964). Moreover, beluga of such low weight have never been caught in the middle Danube (Khin 1957).

[2] Lengths or weights calculated from the GM regression: $BW = 0.000003994 \times TL^{3.10452}$; where BW = weight in grams, and TL = total length in mm (Holčík 1995).

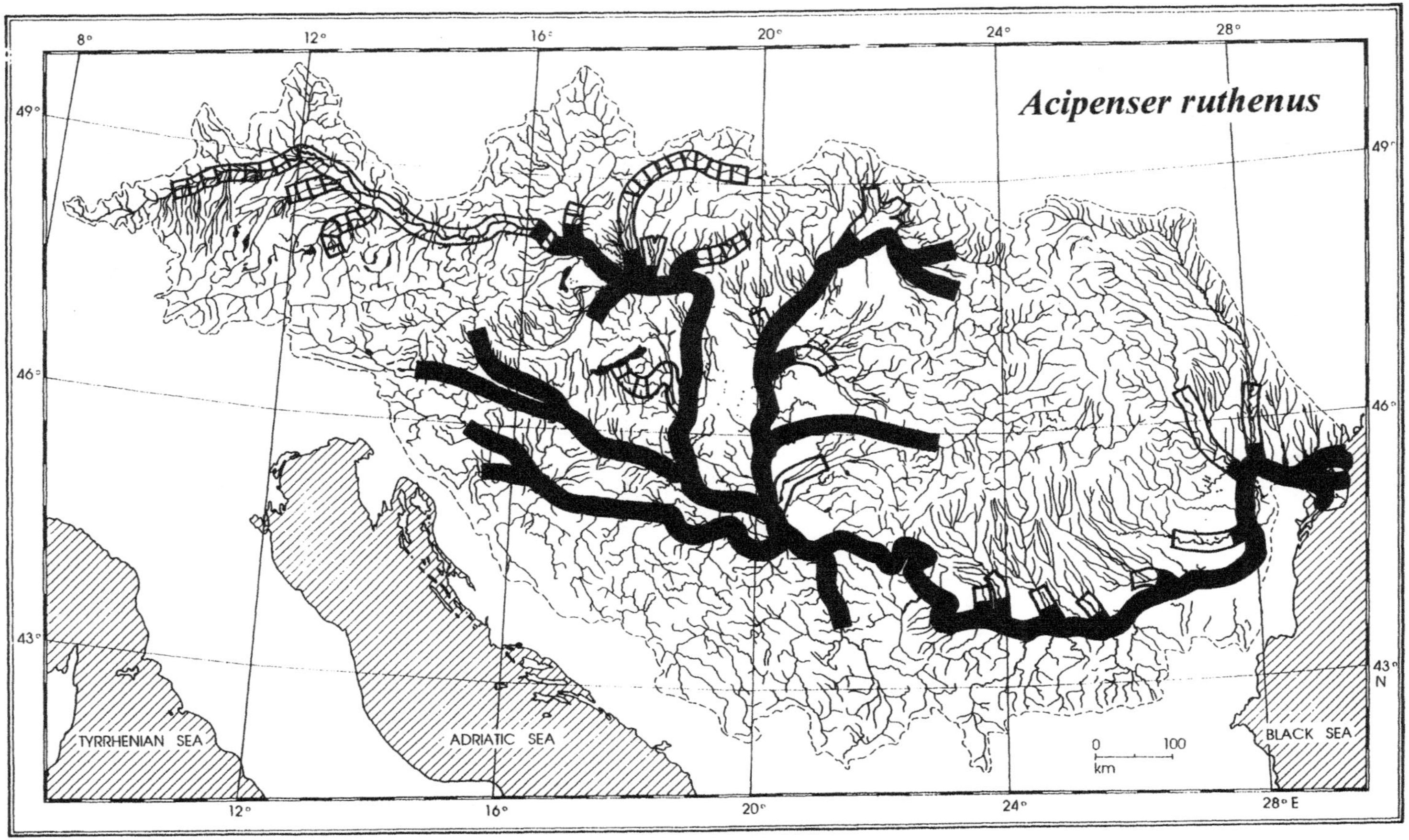

Figure 4. Distribution of the sterlet, *Acipenser ruthenus,* in the Danube drainage system. Regular (continuous black area) and occasional (black and white area) occurrence at present; regular (continuous white area) and occasional (striped white area) occurrence in the past. Information for this map was compiled from Grossinger (1754), Kornhuner (1861), Siebold (1863), Jeitteles (1864), Herman (1877), Moscáry (1877), Chyzer (1882), Malesevics (1892), Glowacki (1896), Antipa (1909), Vutskits (1913), Munda (1926), Mahen (1927), Vladykov (1931), Mihályi (1954), Kux (1956), Sedlár (1959, 1960, 1969), Žitňan (1963, 1965), Holčík (1968), Anonymous (1975), Sedlár et al. (1989), Sokolov & Vasil'ev (1989) and Holčík (1995). In the following list, towns or locations in parentheses give the farthest upriver record. Right hand tributaries include: the Isar River (Landshut), Inn River and its tributary Salzach River (Laufen), Sió River (Lake Balaton), Rába (= Raab) River, Drava River (Maribor), Mura River (Graz), Sava River (Sevnica) and its tributaries Kupa River (Karlovac) and Lonja River. Left hand tributaries include: the Morava River (Moravská Nová Ves), Váh River (Trenčín, exceptionally Liptovský Svätý Mikuláš) and its tributaries Nitra (Lándor) and Žitava rivers. Hron River (Kamenica and Hronom), Ipel' (= Ipoly) River, Tisa River (Sighetul Marmaţie) and its tributaries Bega River, Mureş River (Auid), Zagyva River and Bodrog River (Brehov) with tributaries Latorica River, Laborec River and Uh (= Už) River, Someş River (Dej), Tamiš (= Temes, Timiş) River.

Figure 5. Unusually large specimen of *Acipenser ruthenus* captured among over 100 sterlets in one seine haul on 9.6.1993 in the Danube River at Čenkov (river km 1730). Photograph by K. Hensel.

were caught, mortality surpassed recruitment. Beluga has a long life span and late sexual maturation (Pirogovskii et al. 1989), and the Danube population began to decrease rapidly (Rohan-Csermák 1963). Weir fishing disappeared from the middle Danube at the end of the sixteenth and from the Tisa River at the end of the seventeenth century, but Serbian fishermen employed it up to World War I at the Iron Gate, near the village of Sip. In the 17th and 18th centuries the last remnants of the beluga populations were so severely undermined that in the 19th century only a few beluga were caught in the foothills and in the lower Danube. The last beluga recorded in the Slovakian – Hungarian stretch of the Danube was a female, 3.1 m TL and 150 kg BW, taken at Štúrovo in 1925 (Khin 1957). According to Kornhuber (1901), Ortvay (1902) and Khin (1957), only 16 beluga were taken in this segment of the Danube between 1857 and 1957, of BW between 78 to 500 kg, and TL estimated to range from 2.2–7.4 m.[4] Beluga lose weight after the 1700 km migration up the Danube, as do other anadromous fishes (Nikol'skii 1974).

Construction of the Đerdap Dams (= Iron Gates Dams) greatly impacted the remaining beluga. Janković (1993) reported that catches of beluga and Russian sturgeons (separate data sets for the two species are not available) peaked during the five year period after construction of Iron Gates Dam I. In the period from 1972 to 1976, catches amounted to 115.7 metric tons, which is 23.1 tons higher than in the five years before construction of the dam. The higher catch was due to mass gathering of individuals below the dam, which allowed intensive fishing (see Wei et al. 1997 this volume for similar impact of construction of Gezhouba Dam on Yangtze River sturgeons). However, by a later five year period (1980 to 1984), the combined catch decreased to 78.2 tons. In the period from 1985 to 1989, the five years following construction of the Iron Gates Dam II, the combined catch dropped to 37.3 tons. Beluga only exceptionally overcome the dams via shiplocks: a male 3 m in TL weighing 181 kg was caught in Hungary at Paks (river km 1526–1528) on 16 May 1987, and this individual must have negotiated the locks at both dams (Pintér 1989).

According to the JCIAFD, the annual catch of beluga in the Danube between 1958 and 1981 varied from 19.7 to 240.4 tons, with a decrease in the last four years of the period. Most fish were taken by Romania (59.1%) and the former Soviet Union (30.7%) and the remainder were shared by Bulgaria and former Yugoslavia. Beluga is now extirpated from the upper Danube, critically endangered in the middle Danube and vulnerable in the lower Danube.

[4] The TL of these fish was estimated by Holčík (1994) using data from 9 specimens recorded by Khin (1957) to calculate a length-weight regression: $BW = -101975.47 + 81.69821\ TL$ (BW in grams TL in mm). This regression differs from that for beluga in the Sea of Azov calculated by Chugunov & Chugunova (1964; $BW = -4.41087 + 2.78706 \log TL$).

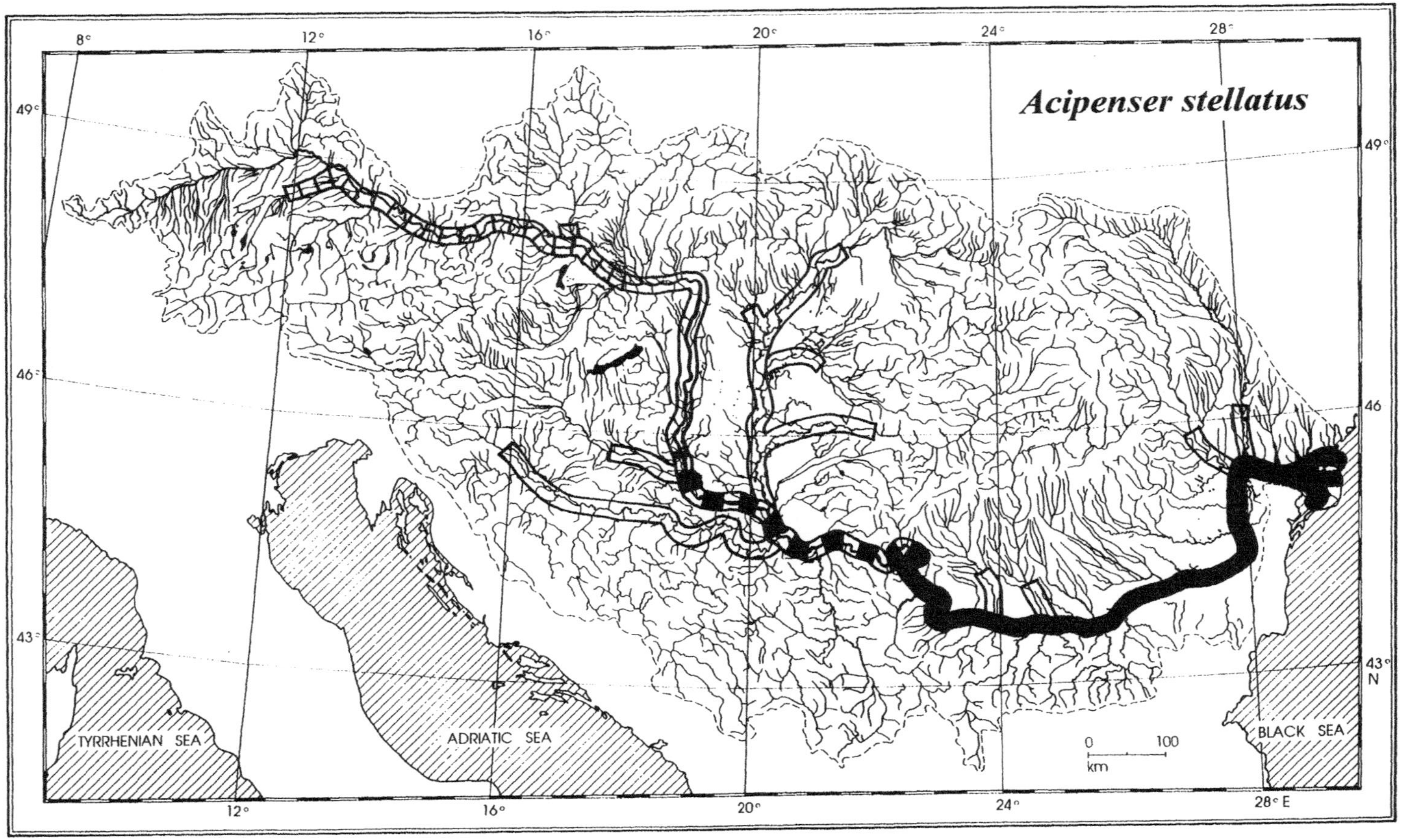

Figure 6. Distribution of the stellate sturgeon, *Acipenser stellatus*, in the Danube drainage system. Regular (continuous black area) and occasional (black and white area) occurrence at present; regular (continuous white area) and occasional (striped white area) occurrence in the past. Original figure by K. Hensel.

Acipenser gueldenstaedtii **– Russian sturgeon**

This is the largest Danubian species of the genus *Acipenser*, and was the most widely distributed anadromous species in the Danube River (Figure 3). According to Kornhuber (1863), Bănărescu (1964) and Manea (1966), the largest specimens reached 2–4 m in TL, with estimated BW of 70–600 kg. Anadromous Russian sturgeons weighing 60–90 kg regularly migrated upstream to Bratislava (river km 1869) and spawned in this section of the middle Danube in May and June. They rarely reached Vienna (river km 1925) and Regensburg (river km 2381).

In northern, or 'left bank' tributaries of the middle Danube, Russian sturgeons occurred in the Morava River (at Suchohrad), Váh River, Tisza River (up to Versényi) and its tributaries, the Szamos (= Someş) River, Zagyva River, Körös River, and the Mureş River (up to Mihalţ). It occasionally entered the tributaries of the lower Danube, including the Olt River, the Jiu River (up to Transylvania), the Prut River, and the Siret River. It occurred in southern, or 'right bank' tributaries of the Danube including the Drava River (and its tributary the Mura (= Mur) River, via which Russian sturgeon reached as far inland as Austria) and the Sava River up to Litija (as well as its tributary the Kupa River, up to Karlovac).

In the Volga River of Russia, *A. guendelstaedtii* occurred as both resident, non-migratory form and an anadromous migratory form (Lukin 1937). Heckel & Kner (1858) first noted that Russian sturgeon also occur in the Danube throughout the year, and the resident non-migratory form still occurs both in the lower (Manea 1966) and middle Danube (Sedlár 1960, Sedlár et al. 1989, Gunda 1966, Balon 1968a, Hensel 1969, Hárka 1980, and Holčík 1995). Table 1 lists all specimens of *A. gueldenstaedtii* recorded in the Slovak and Hungarian segment of the Danube River since 1900.

Holčík (1995) reported that until 1939, 10 to 15 Russian sturgeon weighing 2 to 3 kg were caught annually in the lower course of the Morava River. In the middle Danube, especially between river km 1749 and 1987, 3 to 4 specimens were caught annually until 1983. At present, this species is extremely rare in the middle Danube. In 1987 we learned that in the Danube near Bratislava one angler caught two large sturgeons, each about 1 m TL. Both specimens were released, but according to the description, these must have been *A. gueldenstaedtii*. Unverified records of Russian sturgeon exist also from the Hungarian stretch of the river (Pintér 1991).

According to the JCIAFD, annual catches from 1958 to 1981 varied from 7.1 to 42.3 metric tons (24.9 metric tons average). The greatest catch was recorded in Bulgaria (45% of the total) followed by former Yugoslavia (33.6%), former Soviet Union (13.1%) and Romania (8.3%). As already noted under the description of *H. huso*, the combined catch of beluga and Russian sturgeon dropped after construction of the dams at the Iron Gates (see Janković 1993). The Russian sturgeon is critically endangered in the Danube Basin.

Acipenser ruthenus **– sterlet**

The sterlet is the smallest species among Danube sturgeons. It is a potamodromous resident species. Tagging performed by Unger (1953) and Ristić (1970a) revealed maximum migration distances in the Danube of 322 km. In the Danube, sterlet regularly occurred up to Vienna, frequently to Linz, Passau and Regensburg, and even up to Ulm (Figure 4; Fitzinger & Heckel 1835, Heckel & Kner 1858, Siebold 1863). It was very abundant in the Danube near Bratislava (Kornhuber 1863, Ortvay 1902). According to Kinzelbach (1994), the large sterlet population in the upper Danube between Regensburg and Passau was autochthonous and not the result of migration as generally been thought. Sterlet also ascended or occurred in some of the Danube's tributaries (Figure 4).

Sterlet now has a very limited distribution in the middle and upper Danube. The species is extirpated from the German section of the Danube (Reichenbach-Klinke 1968, Balon et al. 1986), endangered in the Austrian section (Jungwirth 1975, Schiemer & Spindler 1989), and greatly diminished in the Slovakian section. Between 1962 and 1978, sterlet generally contracted in the Slovakian section to the 82 km stretch from Štúrovo to Čičov (Balon 1964, 1968b), and only occasionally was it found at Gab-

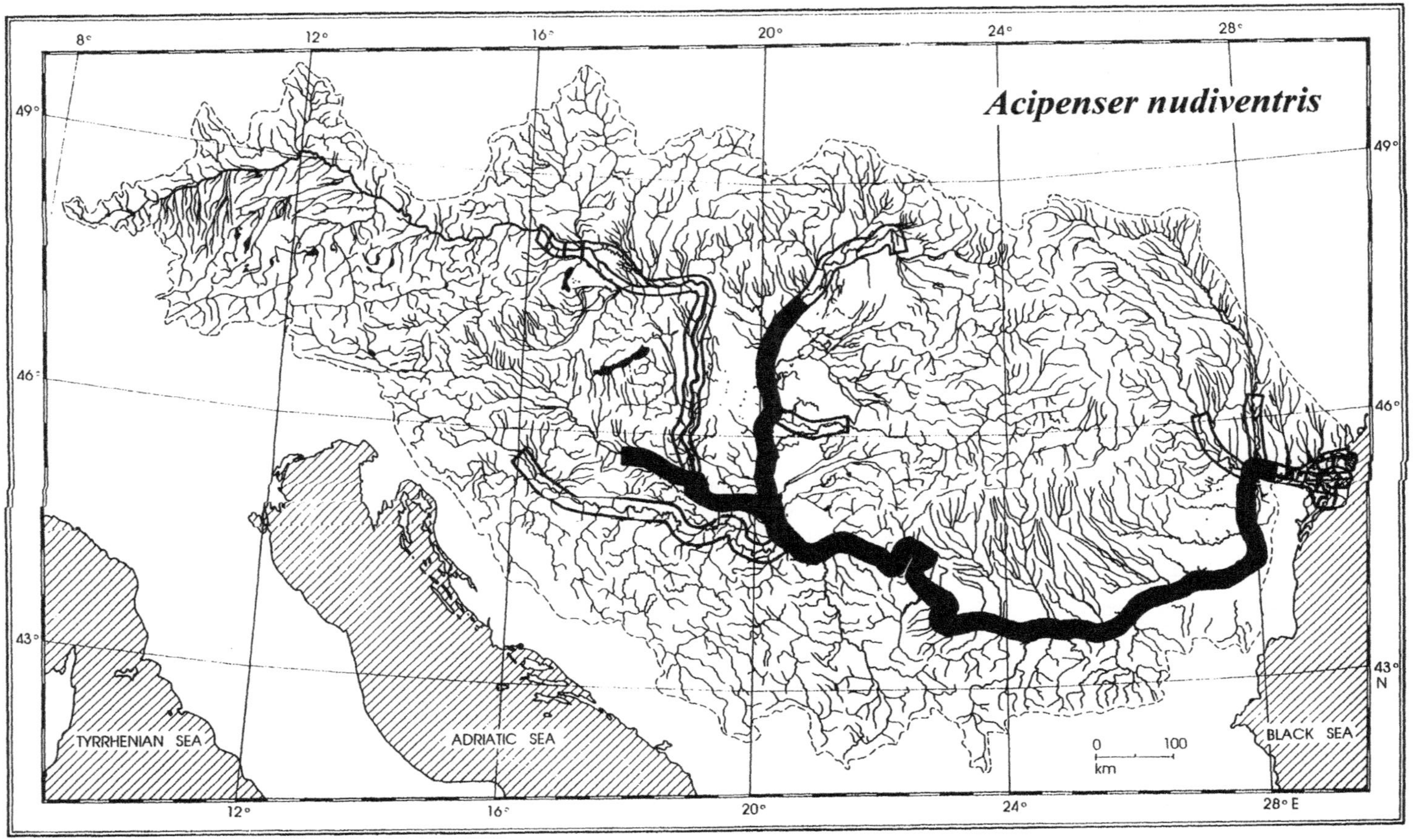

Figure 7. Distribution of the ship sturgeon, *Acipenser nudiventris*, in the Danube drainage system. Regular (continuous black area) and occasional (black and white area) occurrence at present; regular (continuous white area) and occasional (striped white area) occurrence in the past. Original figure by K. Hensel.

číkovo (Holčík et al. 1981). It disappeared from the Hron River (Sedlár et al. 1983), and now occurs only in the mouth of the Váh River. It also disappeared from the lower course of the Morava River sometime after 1966 (Holčík 1995). However, since 1975, water quality has improved and sterlet began to reappear above river km 1820 (Gabčíkovo, Slovakia). It is again found at Bratislava and in the lower course of the Morava, where 2 to 3 specimens have been caught annually by commercial fishermen since 1980 (Holčík 1995). Catches of sterlet are highest between river km 1749 and 1762 (Radvaň nad Dunajom – Iža), at Zlatná na Ostrove (river km 1778–709), at Vel'ké Kosihy (river km 1786–1789) and lowest at Čenkov (river km 1732–1733; but see Balon 1995a, b, figure 5) and Štúrovo (river km 1717). Population increases were presumably due to increasing water quality (Sedlár 1985, Sedlár et al. 1989) and stocking of juveniles from the Hungarian side of the Danube (Jaczó 1974, Tóth[5, 6, 7]). At the beginning of the 1980s, as many as 300 specimens of sterlet were caught by Slovak fishermen in one haul of a 300 m beach seine. Increases in sterlet catch in the Hungarian part of the Danube started in 1971 (Toth[8]), presumably caused by its emigration from the Tisza River where upstream migrations to spawning grounds were halted by dams. In the Serbian stretch of the Danube, the most abundant population of sterlet occurs near Belgrade and in the upstream section near Vojvodina, as well as in the lower (Serbian) parts of the Sava and Tisa rivers (Janković 1993). In the Slovak – Hungarian stretch of the Tisza River, the sterlet does not migrate (Holčík 1995). Its continuing presence was also recorded in the Rába River (Sokolov & Vasi'lev 1989), the Drava River (up to Carinthia, Honsig-Erlenburg & Schultz 1989), the Sava River (up to Sevnica), the Mura River (up to Mursko Središće) and the Kupa (= Kolpa) River up to Krasinec (see Povž & Sket 1990).

Annual catches of sterlet in the Danube between 1958 and 1981 varied between 36–117 metric tons (average 63.5 metric tons). The highest catches were former Yugoslavia (57.5%), followed by Bulgaria (28.0%), Romania (10.5%), Hungary (3.5%) and former Czechoslovakia (0.5%). The catch in the former USSR was so low that the JCIAFD did not record it. The Đerdap Dams are blamed for decreasing the catch of sterlet by 50% (Janković 1993). Ongoing construction activities of the Gabčíkovo hydropower station further threaten sterlet in the uppermost part of the middle Danube and the lower course of the Morava River. Tagging performed by the second author between 1992 and 1994 revealed that the barrages constructed at Čuňovo (river km 1840) and Gabčíkovo (river km 1820) are insurmountable obstacles for the upstream migration of any fishes, including sterlet.

Acipenser stellatus – stellate sturgeon

The stellate sturgeon was always rare in the middle

[5] Tóth, J. 1978. Information of the Hungarian part. pp. 278–287. *In:* Materials of the 19th Session of the Joint Commission of the International Agreement on the Fishing in the Danube River between the Governments of the Soviet Union, Peoples Republic of Bulgaria, Peoples Republic of Hungary, Socialistic Republic of Romania, Czechoslovak Socialistic Republic and Socialistic Federative Republic of Yugoslavia, Moscow (in Russian).

[6] Tóth, J. 1979. Information of the Hungarian part. pp. 125–150. *In:* Materials of the 19th Session of the Joint Commission of the International Agreement on the Fishing in the Danube River between the Governments of the Soviet Union, Peoples Republic of Bulgaria, Peoples Republic of Hungary, Socialistic Republic of Romania, Czechoslovak Socialistic Republic and Socialistic Federative Republic of Yugoslavia, Moscow (in Russian).

[7] Tóth, J. 1980. A Magyar fél tájekoztatója (Information of the Hungarian part). pp. 129–147. *In:* Materiály z 22. zasadania Zmiešanej komisie pre uplatňnovanie dohody o rybolove vo vodách Dunaja, Bratislava.

[8] Tóth, J. 1979b. Catch results changes of the sterlet (*Acipenser ruthenus* L.) in the Hungarian Danube. pp. 151–157. *In:* Materials of the 19th Session of the Joint Commission of the International Agreement on the Fishing in the Danube River between the Governments of the Soviet Union, Peoples Republic of Bulgaria, Peoples Republic of Hungary, Socialistic Republic of Romania, Czechoslovak Socialistic Republic and Socialistic Federative Republic of Yugoslavia, Moscow (in Russian).

Danube (Figure 6). It ascended upstream to Komárno (Grossinger 1794, Fitzinger & Heckel 1835, Kornhuber 1901), Bratislava (Kornhuber 1863, Orvay 1902), the Austrian part of the Danube (Fitzinger & Heckel 1835), and occasionally even reached the Bavarian stretch near Straubing (Gesner 1575) and the Isar River (Siebold 1863). During spawning migrations, stellate sturgeon entered tributaries of the lower Danube, such as the Prut, Siret, Olt and Jiul rivers (Antipa 1909); it was encountered also in some tributaries of the middle Danube, such as the Tisza River up to Tokaj (Heckel & Kner 1858) and in the lower courses of its tributaries the Maros and Körös rivers, in the mouth of the Zagyva River (Herman 1887) and in the lower course of the Drava and Sava rivers (Heckel & Kner 1858, Glowacki 1896, Vutskits 1913). Mahen (1927) mentioned stellate sturgeon from the mouth of the Morava River. However, this seems doubtful, as it has only rarely been recorded in the adjacent stretch of the Danube.

We consider *A. stellatus* to be extirpated not only from the upper Danube but also from the upper stretch of the middle Danube (the Slovakian and Hungarian section). The last known specimen from this section was caught at Komárno on 20 February 1926. The head of this specimen, measuring 325 mm, is at the Slovak National Museum in Bratislava (Holčík 1959), and is estimated to be from a specimen 1282 mm TL and 9.8 kg BW (Holčík 1995). The last stellate sturgeon in Hungary (100 cm TL) was caught in the Danube at Mohács in 1965 (Pintér 1991).

Construction of the Iron Gates dams blocked most migration of stellate sturgeon to the middle Danube, as few individuals succeed in passing through the shipping locks (Djisalov[9]). Janković (1993) analyzed catch of stellate sturgeon in the Serbian section of the Danube: from 1967–1970, the annual catch was around 1.4–2.0 tons but, in 1971, when the first dam was finished, the catch dropped to 184 kg. During the next 8 years this species was not recorded in the catch, except in 1975 when 284 kg was caught. In 1980, a catch of 80 kg was reported, but after construction of Iron Gates Dam II, the stellate sturgeon disappeared from the middle Danube catch. At present stellate sturgeon is only seldom taken, with an estimated annual catch < 100 kg. This species was never economically significant in the middle Danube, with a mean annual catch of only 7.8 tons in 1958–1981. Of this, 34.2% was shared by Bulgaria, 22.7% by the former USSR, 22.5% by former Yugoslavia and 20.6% by Romania (data from JCIAFD).

Acipenser nudiventris – ship sturgeon

The ship sturgeon forms both anadromous and resident populations, but in the Danube River, only the resident strain occurred (Bănărescu 1964, Manea 1966). This species was recorded in the lower Danube (occasionally in the Danube delta) and in the middle Danube, upstream to Bratislava (Figure 7, Kornhuber 1863). Only exceptionnally did it migrate to the Austrian segment of the river (Fitzginger & Heckel 1835). Ship sturgeon also occurred in some tributaries: the lower course of the Váh River (Heckel & Kner 1858, Herman 1887), the Tisza River at Mándok (Mihályi 1954), the Sava and Drava rivers (Heckel & Kner 1858, Vutskits 1913, Munda 1926), the Maros River (Hankó 1931), and also from two tributaries of the lower Danube, the Prut and Siret rivers (Bănărescu 1964).

Ship sturgeon was never abundant in the lower Danube (Manea 1970), although, as Pintér (1991) noted, it is difficult to verify this based on historical documents, particularly because fishermen did not always distinguish larger ship sturgeon from Russian sturgeon, and small ship sturgeon were confused with sterlet. The ship sturgeon is now very rare in the Danube, and only occasionally found in the catch of Romania and Serbia (Manea 1970, Bacalbaşa-Dobrovici 1991, Stamenković 1991, Janković 1993). Ship sturgeon completely disappeared from the Austrian and Slovak segment of the Da-

[9] Djisalov, N. 1983. Analysis of the migratory sturgeon fishery in the Yugoslavian part of the Danube in 1981. pp. 150–157. *In:* Materials of the 24th Session of the joint Commission of the International Agreement on the Fishing in the Danube River between the Governments of the Soviet Union, Peoples Republic of Bulgaria, Peoples Republic of Hungary, Socialistic Republic of Romania, Czechoslovak Socialistic Republic and Socialistic Federative Republic of Yugoslavia. Moscow (in Russian).

nube, and in the Hungarian section it is extremely rare. The largest specimen recorded (170 cm TL and 32 kg BW) was taken at Ercs in 1932 (Pintér 1989).

From tributaries of the middle Danube, ship sturgeon is known only from the Tisza and Drava rivers. Vásárhely (1957) reported many juveniles of ship sturgeon from the upper segment of the Tisza River at the Tiszalök Dam and suggested that adults are rare. Pintér (1991) recorded one specimen caught in August 1975 (about 70 cm in TL) in the Tisza River near Kiskör and other two specimens taken in the Tisza River at Tiszalök (however, their identification is questionable). From the Drava River one male 147 cm in TL, 20.5 kg BW was taken at Heresznye in August 1989 (Pintér 1991, 1994). The ship sturgeon in the Danube River basin is critically endangered.

Conclusions

Anadromous populations of beluga, Russian sturgeon and stellate sturgeon, represented by winter races, were heavily damaged by overfishing during previous centuries, and were then completely eliminated from the middle and upper Danube by construction of Đerdap I and Đerdap II dams (Iron Gates Dams I and II). Small stocks of the resident races of ship and Russian sturgeons occur in the middle Danube and some of its tributaries. The most abundant sturgeon in the Danube is the sterlet, but this species also experienced population declines. It disappeared from almost all of the upper Danube, where single specimens are now found only in its Austrian part. At present the sterlet is limited to the middle Danube and to lower courses of some tributaries.

Acknowledgements

We thank Vadim Birstein, John Waldman and Robert Boyle for arranging for one of us (Hensel) to travel to participate in the International Conference on Sturgeon Biodiversity and Conservation. Special thanks to E.K. Balon for his comments on the first draft of this paper and to D. Matichova for technical assistance. Two anonymous reviewers provided helpful comments, and W.E. Bemis extensively revised the text.

References cited

Anonymous [Speer, D.]. 1975. Uhorský Simplicissimus (Hungarian Simplicissimus). According to the Latin edition from 1683 translated by J. Vlachovič. Tatran, Bratislava 231 pp.

Antipa, G. 1909. Fauna ihtiologică a României (Fish fauna of Romania). Publ. fond. Adamachi, Bucureşti. 294 pp.

Bacalbaşa-Dobrovici, N. 1991. Status des differentes espèces d'esturgeon dans le Danube Roumain: problèmes lies a leur maintenance. pp. 185–192. *In:* P. Willot (ed.) *Acipenser,* Actes du premier colloque international sur l'esturgeon, Bordeaux 3–6 Octobre 1989, CEMAGREF, Bordeaux.

Bacalbaşa-Dobrovici, N. 1997. Endangered migratory sturgeons of the lower Danube River and its delta. Env. Biol. Fish. (this volume).

Balon, E.K. 1964. Predbežný zoznam kruhoústych a rýb Slovenska (Petromyzones et Teleostomi) (Preliminary list of lampreys and fishes of Slovakia (Petromyzones et Teleostomi). Biológia (Bratislava) 19: 343–358.

Balon, E.K. 1967. Vývoj ichtyofauny Dunaja, jej súčasný stav a pokus o prognózu d'alších zmien po výstavbe vodných diel (Development of the fish fauna of the Danube river, its present status, and an attempt for the prediction of further changes after the construction water works). Biologické práce 15: 5–121.

Balon, E.K. 1968a. Ďalší nález mlade *Acipenser güldenstädti colchicus* Marti, 1940 v československom úseku Dunaja (Further findings of the *Acipenser güldenstädti colchicus* Marti, 1940 in the Czechoslovak stretch of the Danube River). Acta Rer. Natur. Mus. Nat. Slovacae, Bratislava 14: 95–100.

Balon, E.K. 1968b. Ichtyofauna pozdl'žneho a priečneho profilu československého úseku Dunaja, druhové a početné zmeny rybích populácii a ich ochrana (Fishfauna of the longitudinal and cross profile of the Czechoslovak stretch of the Danube River, specific and density changes of the fish populations and their conservation). Československá ochrana prirody 1968: 203–229.

Balon, E.K. 1968c. Einfluß des Fischfangs aud die Fischgemeinschaften der Donau. Arch. Hydrobiol. (Suppl. Donauforschung 3) 34: 228–249.

Balon, E.K. 1995a. Origin and domestication of the wild carp, *Cyprinus carpio*: from Roman gourmets to the swimming flowers. Aquaculture 129: 3–48.

Balon, E.K. 1995b. The common carp, *Cyprinus carpio*: its wild origin, domestication in aquaculture, and selection as colored nishikigoi. Guelph Ichthyology Reviews 3: 1–55.

Balon, E.K., S.S. Crawford & A. Lelek. 1986. Fish communities of the upper Danube (Germany, Austria) prior to the new Rhein-Main-Donau connection. Env. Biol. Fish. 15: 243–271.

Bănărescu, P. 1964. Pisces – Osteichthyes. Fauna Republicii Populare Romîne, Vol. 13. Editura Acad. Republicii Populare Romîne, Bucureşti. 962 pp.

Birstein, V.J. 1993. Sturgeons and paddlefishes: threatened fishes in need of conservation. Cons. Biol. 7: 773–787.

Birstein, V.J. & W.E. Bemis. 1997. Leo Semenovich Berg and the biology of Acipenseriformes: a dedication. Env. Biol. Fish. (this volume).

Chugunov, N.L. & N.I. Chugunova. 1964. Comparative commercial and biological characteristics of the Azov Sea sturgeons. Trudy VNIRO 51: 87–182 (in Russian).

Chyzer, K. 1882. Die Fische des Zempliner Komitates. Jahrb. d. ungar. Karpaten Ver. 9: 1–16.

Fitzinger, L.J. & J. Heckel. 1835. Monographische Darstellung der Gattung *Acipenser*. Ann. Wiener Mus. Nat. 1: 269–326.

Gåßner, C. [= Gesner, K.] 1575. Fischbuch. Zürych, Ch. Froschower. x + 403 pp.

Głowacki J. 1896. Die Fischfauna der Save und der Isonzo. I. Jahresbericht d. k.k. Staats-Untergymnasiums in Cilli, 1895–1896: 1–37.

Grossinger, J.B. 1794. Universa historia physica Regni Hungariae secundum tria regna naturae digesta. Regni animalis pars III. Ichthyologia, sive historia piscium, et amphibiorum. S.P. Weber, Posonii et Comaromii, xiv + 400 pp.

Gunda, M. 1966. Lov jesetera v Dunaji (Sturgeon fishing in the Danube). Československé rybářství 1966: 182.

Hankó, B. 1931. Ursprung und Verbreitung der Fischfauna Ungarns. Archiv. fúr Hydrobiologie 23: 520–556.

Hárka, Á. 1980. Vágótok a Tiszában (Russian sturgeon from the Tisza River). Halászat 26: 82.

Heckel, J. 1951. Über die in den Seen Oberösterreichs vorkommenden Fische. Sitzungsberichte math. – naturh. Cl. kaiser. Akad. Wiss. in Wien 6: 145–149.

Heckel, J. & R. Kner. 1858. Die Süsswasserfische der Österreichischen Monarchie. W. Engelmann, Leipzig. 388 pp.

Hensel, K. 1969. The first find of natural sturgeon hybrid *Acipenser ruthenus* Linnaeus, 1758 × *Acipenser güldenstädti colchicus* Marti, 1940. Věstník Československé společnosti zoologické 33: 328–330.

Herman, O. 1887. A Magyar halászat könyve (Book of the Hungarian fisheries), Vol. I & II. Természettudományi Társulat, Budapest. 860 pp.

Holčík, J. 1968. Nález jesetera malého (*Acipenser ruthenus* Linnaeus, 1758) vo Váhu pri Liptovskom Mikuláši (Finding of the sterlet in the Váh river at Liptovský Mikuláš). Acta Rer. Natur. Mus. Nat. Slovacae, Bratislava 14: 109–112.

Holčík, J. 1959. About the occurrence of the stellate sturgeon (*Acipenser stellatus* Pall. 1771) in the River Danube near Komárno. Acta Rer. Natur. Mus. Nat. Slovenici 5: 44–48.

Holčík, J. 1995. Acipenseriformes. pp. 372–397. *In:* V. Baruš & O. Oliva (ed.) Mihulovci Petromyzontes a ryby Osteichthyes (1). (Lampreys Petromyzontes and bony fishes Osteichthyes), Fauna ČR a SR, Vol. 28/1, Academia, Praha.

Holčík, J., I. Bastl, M. Ertl & M. Vranovský. 1981. Hydrobiology and ichthyology of the Czechoslovak Danube in relation to predicted changes after the construction of the Gabčíkovo – Nagymaros river basin barage system. Práce Laboratória rybárstva a hydrobiológie 3: 19–158.

Honsig-Erlenburg, W. & N. Schultz. 1989. The fishes of Carinthia. Verl. Naturwiss. Verein f. Kärnten, Klagenfurt.

Jaczó, I. 1974. A kecsege mennyiségének változása folyóink-ban az 1947–1970. évi fogások és vizsgálatok alapján (Changes in the sterlet stocks in Hungarian rivers between 1947–1970 based on the investigations and catch statistics data). Halászat 20: 12.

Janković, D. 1993. Populations of Acipenseridae prior and after the construction of the HEPS Đerdap I and II. Acta Biologica Iugoslavica, Seria E. Ichthyologia 25: 29–34.

Jeitteles, L.H. 1864. Die Fische der March bei Olmütz. II. Abtheilung pp. 4–26. *In:* Bericht über das k. k. Gymnasium in Olmütz während des Schuljahres 1864, Olmütz.

Jungwirth M. 1975. Die Fischerei in Niederösterreich. Wiss. Schrift. Niederösterreich 6: 1–31.

Kähsbauer, P. 1961. Cyclostomata, Teleostomi (Pisces). Catalog-es Faunae Austriae, T. XXI aa. Springer Verlag, Wien 56 pp.

Khin, A. 1957. A Magyar vizák története (History of beluga sturgeon catches in Hungary). Mezögazadasági Múzeumi Füzetek 2: 124.

Kinzelbach, R. 1994. Ein weiterer alter Nachweis des Sterlets, *Acipenser ruthenus*, in der württembergischen Donau. pp. 281–291. *In:* R. Kinzelbach (ed.) Biologie der Donau, Limnologie aktuell 2, Gustav Fischer Verlag Stuttgart, Jena.

Kornhuber, G.A. 1861. Betrag zur Kenntnis der Fische in Waag Gebiet. Verhandlungen des Vereins für Naturkunde zu Presburg 5: 40–44.

Kornhuber, G.A. 1863. Bemerkungen über das Vorkommen der Fische um Presburg und an einigen anderen Orten Ungarns. Correspondezblatt der Vereins für Naturkunde zu Presburg 2: 205–213.

Kornhuber, A. 1901. Zoologische Bemerkungen, 2. Vom Scherg oder Sternhausen, *Acipenser stellatus* Pallas. Verhandlungen des Verereins für Natur- und Heilkunde zu Presburg, N.F. 12: 32–33.

Kux, Z. 1956. Příspěvek k ichthyofauně dolní Moravy a Dunaje (Contribution to the fish fauna of the lower reach of the Morava and the Danube rivers). Časopis Moravského musea, věd, přír. 41: 93–112.

Kux, Z. & T. Weisz. 1962. Ichtyofauna hlavního toku Dunaje a jeho některých přítoků v jihoslovenské nížině (Fish fauna of the main channel of the Danube and some of its tributaries in the south Slovakian lowland). Časopis Moravského musea, věd. přír. 47: 151–180.

Lukin, A.V. 1937. Biological analysis of the sturgeon catches in the middle course of the Volga River (Tartarian Republic). Izvestiya Akademii Nauk SSSR, Otdelenie matematicheskikh i estestvennykh nauk, seriya biologicheskaya 1: 211–220 (in Russian).

Mahen, J. 1927. Částečná revise ryb dunajské oblasti (Partial revision of the Danubian fish fauna). Sborník klubu přírodovědeckého v Brně 9: 56–69.

Malesevics, E. 1892. Loszoncz faunája (Fauna of the Losoncz (= Lučenec) region). Losoncz, Kárman – Társulat könyvnyomdája. 47 pp.

Manea, Gh. 1966. Contribuţii la studiul sturionilur din apele Romiîniei şi al reproducerii lor în legatura cu construcţiile hidroenergetice pe Dunarea inferioara. I. Une aspecte ale biologiei sturionilor (Contribution to the study on sturgeons in Rumanian waters and their reproduction in relation to the hydroenergetic construction in the Lower Danube. I. Aspects of the sturgeons biology). Bululetinul Institutului de Cercetări şi Proiectări Piscicole 25: 62–86.

Manea, G. 1970. Producţia şi reservele de sturioni de pe glob (Sturgeon world production and stocks). Buletinul de Cercetări Piscicole 29: 52–65.

Mihályi, F. 1954. Revision der Süsswasserfische von Ungarn und der angrenzenden Gebieten in der Sammlung der Ungarischen Naturwissenschaftlichen Museums. Annal. Hist.-Natur. Mus. Nat. Hungarici, Ser. nov. 5: 433–456.

Mocsáry, S. 1877. Adatok Zemplén és Ung megyék faunájahoz (Data on the fauna of the Zemplén and Ung counties). A magyar tud. Akad. 13: 131–185.

Munda, A. 1926. Ribe v Slovenskih vodah (Fishes in Slovenian waters). Izd. Slovensko Ribarsko Društvo v Ljubljani, Ljubljana. 63 pp.

Nikol'skii, G.V. 1974. Ecology of fishes, 3rd ed. Vysshaya shkola, Moscow. 365 pp. (in Russian).

Ortvay, T. 1902. Pozsonyvármegye és a területén fekvö Pozsony, Nagysyombat, Bazin, Modor s Szentgyörgy városok állatvilága (Animal world of Pozsony county and on its territory located cities, viz. Pozsony [= Bratislava], Nagyszombat [= Trnava], Basin [= Pezinok], Modor [= Modra], and Szentgyörgy [= Sväty Jur]). K. Stampfel, Pozsony, xvi + 448 pp.

Pintér, K. 1989. Magyarország halai, biológiájuk és hasznosításuk (Fishes of Hungary, their biology and management). Akademiai Kiadó, Budapest, 202 pp.

Pintér, K. 1991. Sturgeons in Hungary, past and present situation. pp. 173–178. *In:* P. Willot (ed.) *Acipenser*, Actes du premier colloque international sur l'esturgeon, Bordeaux 3–6 octobre 1989, *CEMAGREF*, Bordeaux.

Pintér, K. 1994. A tokalakúakkal (Acipenseriformes) foglalkozó magyar közlemények válogattot bibliográfiája (Selected bibliography of Hungarian papers on sturgeons (Acipenseriformes)). Halászat 87: 37–39.

Pirogovskii, M., L.I. Sokolov & V.P. Vasil'ev. 1989. *Huso huso* (Linnaeus, 1758). pp. 156–200. *In:* J. Holčík (ed.) The Freshwater Fishes of Europe, Vol. 1, pt. II, General Introduction to Fishes, Acipenseriformes, AULA-Verlag, Wiesbaden.

Povž, M. & B. Sket. 1990 Naše sladkovodne ribe (Our freshwater fishes). Mladinska knjiga, Ljubljana. 374 pp.

Reichenbach-Klinke, H. 1968. Fischfauna und Fischerei in der deutschen Donau. Archiv Hydrobiol., Suppl. 34 (Donauforschung): 12–23.

Ristić, M.D. 1970. Migracija riba u reci Dunav i njegovim pritokama, njen uticai na stanke i dinamiku populacija ekonomskih važnih riba kao i na ribolov (Fish migration in the Danube River and its tributaries, its influence on the status and population dynamics of the economically important fishes and on fisheries). Ribarstvo Jugoslavije 25: 1–15.

Rochard, E., G. Castelnaud & M. Lepage. 1990. Sturgeons (Pisces: Acipenseridae); treats and prospects. J. Fish Biology 37 (Suppl. A): 123–132.

Rohan-Csermák, G. de. 1963. Sturgeon hooks of Eurasia. Viking Fund Publications in Anthropology 35: i-xvi + 1–155.

Schiemer, F. & T. Spindler. 1989. Endangered fish species of the Danube River in Austria. Regulated Rivers 4: 397–407.

Schiemer, F. & H. Waidbacher. 1992. Strategies for conservation of a Danubian fish fauna. pp. 363–382. *In:* P.J. Boon, P. Calow & E. Petts (ed.) River Conservation and Management. John Wiley and Sons, New York.

Sedlár, J. 1959. Naše jeseterovité (Our sturgeons). Pol'ovníctvo a rybárstvo 11: 31.

Sedlár, J. 1960. Nález druhov *Acipenser ruthenus* Linneé 1758 a *Acipenser güldenstädti colchicus* Marti 1940 v Malom Dunaji (Finding of *Acipenser ruthenus* Linneé 1758 and *Acipenser güldenstädti colchicus* Marti 1940 in the Little Danube). Biológia (Bratislava) 15: 612–614.

Sedlár, J. 1969. Súčasný stav zarybnenia povodia rieky Nitry (Present status of the fish stock in the Nitra River). Biologické práce 15: 3–79.

Sedlár, J. 1985. Jeseter malý sa vracia (Sterlet is returning). Pol'ovnictvo a rybárstvo 37: 40.

Sedlár, J., A. Makara, I. Stráňai & J. Holčík. 1989. Atlas rýb (Fish atlas). Obzor, Bratislava. 372 pp.

Sedlár, J., I. Stráňai & A. Makara. 1983. Súčasný stav zarybnenia povodia Hrona. I. Druhové zloženie obsádky horskej a podhorskej zóny Hrona (Present status of fish stocks of the Hron River basin. I. Species composition of the stocks of the mountain and submountain zone of the Hron River). Pol'nohospodárstvo 29: 515–524.

Siebold, C.T.E. 1863. Süsswasserfische von Mitteleuropa. W. Engelmann, Leipzig. 430 pp.

Sokolov, L.I. & V.P. Vasil'ev. 1989. *Acipenser ruthenus* Linnaeus, 1758. pp. 227–262. *In:* J. Holčík (ed.) The Freshwater Fishes of Europe, Vol. 1, pt. II, General Introduction to Fishes, Acipenseriformes, AULA – Verlag, Wiesbaden.

Stamenković, S. 1991. Evolution des captures de trois espèces d'esturgeons *Huso huso*, *Acipenser gueldenstaedti* et *Acipenser stellatus* dans la partie Yougoslave du Danube. pp. 179–183. *In*: P. Willot (ed.) *Acipenser*, CEMAGREF, Bordeaux.

Taler, Z. 1953. Rasprostranjenje i popis slatkovodnih riba Yugoslavije (The distribution and description of the freshwater fishes of Yugoslavia). Glasnik Prirodnjačkog Muzeja Srpske Zemlje, ser. B 5/6: 425–455.

Unger, E. 1953. Die Fischmarkierung in den freien Gewässern Ungarns and die mit Gummiringen markierten Donau-Fische. Verh. Int. Ver. Limnol 7: 388–397.

Vásárhely, I. 1957. Felsö-Tiszai színtokok – és a tiszalöki vizlépsö (Ship sturgeon catches in the upper Tisza River and the Tiszalök water step). Halászat 50: 105.

Vladykov, V.D. 1931. Les Poissons de la Russie Souscarpathique (Tchécoslovaquie). Mém. Soc. Zool. France 29: 217–373.

Vutskits, G. 1913. Classis Pisces. Fauna Regni Hungariae. 1. Vertebrata Regia Soc. Sci. Natur. Hungarica, Budapest. 42 pp.
Wei, Q, F. Ke, J. Zhang, P. Zhuang, J. Luo, R. Zhou & W. Yang. 1997. Biology, fisheries, and conservation of sturgeons and paddlefish in China. Env. Biol. Fish. (this volume).
Zbořil, J. & K. Absolon. 1916. Zoologická pozorování z okolí hodonínského (Zoological observation from the Hodonín region). Časopis moravského musea zemského 15: 3–12.
Žitňan, R. 1963. Ryby rieky Hrona (Fishes of the Hron River). Pol'ovníctvo a rybárstvo 15: 16–17.
Žitňan, R. 1965. Ichtyofauna československého úseku Tisy (Fish fauna of the Czechoslovak reach of the Tisa River). Sbornik Východoslovenského múzea 6B: 61–67.

Environmental Biology of Fishes **48**: 201–207, 1997.

Endangered migratory sturgeons of the lower Danube River and its delta

Nicolae Bacalbaşa-Dobrovici
University 'Dunarea de jos' Galati, Str. Domeneasca 47, 6200 Galati, Romania

Received 2.8.1994 Accepted 9.3.1996

Key words: Acipenser, *Huso*, Black Sea, anthropogenic factors, Iron Gates Dam I, Iron Gates Dam II, eutrophication, hypoxia, *Mnemiopsis*

Synopsis

Historically, five acipenserid species migrated from the Black Sea into the Danube River: beluga *Huso huso*, Russian sturgeon *Acipenser gueldenstaedtii*, stellate sturgeon *A. stellatus*, ship sturgeon *A. nudiventris*, and perhaps European Atlantic sturgeon *A. sturio*. The freshwater sterlet *A. ruthenus* thrived in the Danube and its tributaries. Presently, only three anadromous species occur in the Romanian part of the Danube, *Huso huso*, *A. gueldenstaedtii*, and *A. stellatus*, while *A. ruthenus* lives in the Danube and its tributaries. Extreme depletion in the number of sturgeons was caused by many, primarily anthropogenic, factors which affected the Danube and the Black Sea shelf during recent last decades. Measures necessary for saving anadromous sturgeon species in the lower Danube are recommended.

Introduction

The Danube is the second longest river in Europe (2857 km). It is divided into three main regions: the upper Danube from the source to Vienna (890 km river length), the middle Danube from Vienna to Iron Gates Dam I (993 km river length), and the lower Danube from Iron Gates Dam I to the mouth (942 km river length). Thirty-five dams have been constructed on the upper Danube. The middle Danube was cut off from the lower Danube by construction of Iron Gates Dam I, built in 1970 (Figure 1). In 1984, the lower Danube was divided by the Iron Gates Dam II, located 80 km downstream from Iron Gates Dam I.

Until quite recently, five anadromous species of sturgeons migrated from the Black Sea into the Danube for spawning: the beluga, *Huso huso*, Russian sturgeon, *Acipenser gueldenstaedtii*, stellate sturgeon or sevruga, *A. stellatus*, ship sturgeon, *A. nudiventris*, and perhaps Atlantic sturgeon, *A. sturio* (Antipa 1916, 1933, Bănărescu 1964, Bacalbaşa-Dobrovici, 1989). In the 19th century, sturgeons swam upstream to Bavaria (Terofal 1980). The exclusively freshwater sterlet, *A. ruthenus*, also thrived in the Danube and its tributaries (Bănărescu 1964, Bacalbaşa-Dobrovici 1989). Presently, only *Huso huso*, *A. gueldenstaedtii*, and *A. stellatus* occur in the Romanian part of the Danube and their populations are impacted greatly by the dams and other installations (Bacalbaşa-Dobrovici 1991a, b). *Acipenser ruthenus* now lives primarily in the middle Danube and its tributaries. This paper reviews the depletion of populations of anadromous acipenserids in the lower Danube and discusses factors causing decreases in sturgeon populations in the Danube and Black Sea. The status of sturgeons in the upper and middle Danube is described by Hensel & Holčík (1997 this volume).

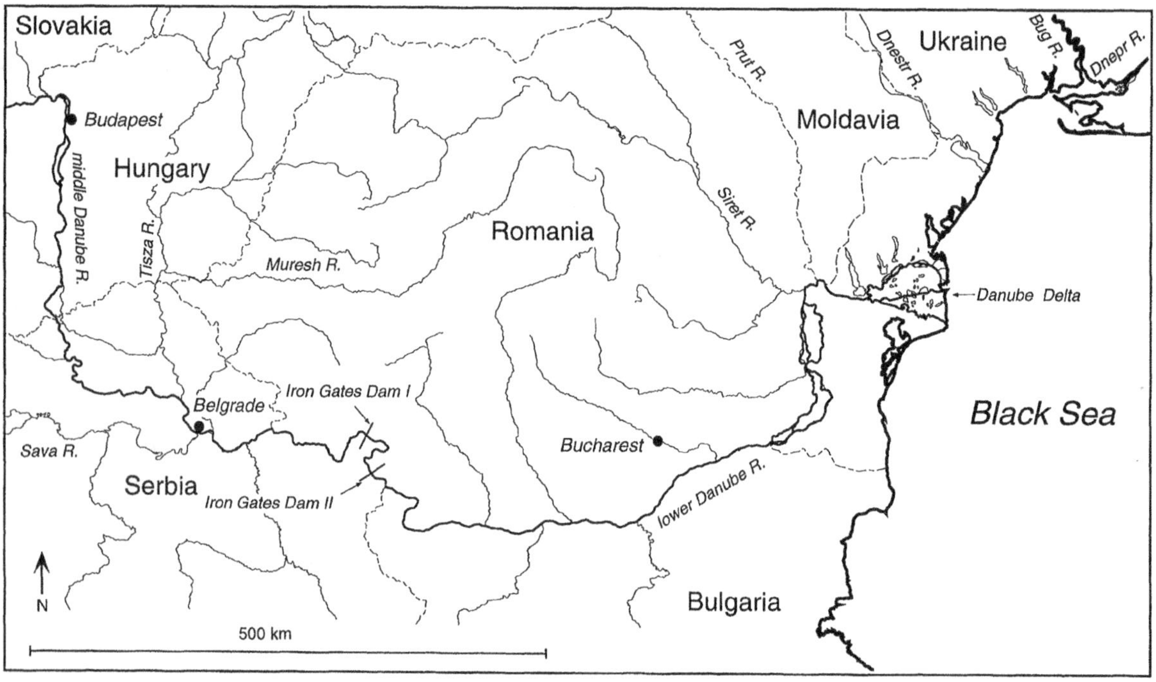

Figure 1. Map of the lower and portions of the middle Danube River and some major tributaries to show features related to sturgeon populations and migrations.

Sturgeon fisheries from ancient times to the 20th century

From ancient times, sturgeons had great economic importance in the Danube River region and were the basis of the population's wealth. Sturgeons of the Danube were mentioned by ancient Greek writers, Herodotus (484? – 425? B.C.) and Strabo (63? B.C. – 21? A.D.). Strabo wrote: 'In the Scythian north, sturgeon are caught and they are as big as dolphins' (translation from Strabo 1853). In Histria (a Greek port which existed 2200 years ago) the inhabitants were allowed to fish in the Danube mouth and to export salted fish to Greece and Rome without charge. Most of the exported fish were beluga and Russian sturgeon.

In the Middle Ages and until the end of the 18th century, sturgeons were an inexhaustible resource of the river (Giurescu 1964). Beluga were caught all along the Romanian part of the Danube (from the mouth to the Iron Gates), in the middle Danube and upstream up to Bavaria. Sturgeons thrived also in tributaries of the lower Danube including the Drava, Sava, Tizsa, Muresh, Siret, and Prut rivers (Figure 1).

In the 12–15th centuries, sturgeons were exported from the Danube area to Poland (Giurescu 1964). In 1409, Mircea the Great (prince of Vallachia) ordered all inhabitants of the villages located along the Danube to catch sturgeons three days a year for the court. The Italian monk Niccolo Barsari, who visited Moldavia in 1633–1639, mentioned that fishermen brought 1000–2000 sturgeons to Chilia every day. From here they were exported to Constantinopole, Poland, and Hungary. In 1690, the Austrian general Marsigli wrote that about 50–100 beluga were caught every day near Adakaleh Island (now submerged). In 1762, the French consul Peysonnel reported that about 25 000 beluga were caught annually in Chilia (Giurescu 1964).

Beginning in the 16th century, the town of Galati on the lower Danube was an important sturgeon fishing and market center. An Italian monk, Barsi, mentioned Galati as having a great abundance of different sturgeons and caviar. In 1646, another Italian, Bishop Bandini, wrote: 'Large beluga are caught here. You would not believe it if you have not seen it with your own eyes' (Giurescu 1964). In 1652, the traveler Robert Bargrave also noted that 'sometimes they catch such a big fish that they need 6–8 oxen to lift them together with a trap'.

Since the Middle Ages, sophisticated gear has been used for catching sturgeons and beluga in the lower Danube. Methods included fences made of wooden branches and provided with small gates for ships to pass through and big cage-like traps for sturgeons. The fence was attached to wooden poles placed into the river bed. Each installation lasted about seven months and was remade each spring because ice destroyed the fence and traps. Each site was operated by a team of 100–200 persons who lived nearby on special platforms.

Fourteen such installations near Chilia caught 1000–2000 sturgeons daily. In the 16th century, additional installations were located near the town of Ismail in the Danube delta (92 km from the mouth, Chilia Branch), in the Borcea Branch (from 248 to 370 km upstream), and in the Hungarian part of the river. In the Iron Gates zone sturgeons were caught by traps and iron baskets (Bacalbaşa-Dobrovici 1971).

Because of intensive fishing, declines in the populations of sturgeons were reported beginning in the early 19th century. In 1835, J. de Hagemeister wrote that beluga were much less abundant in the Danube (Giurescu 1964). In the 20th century, the catch of sturgeons in Romania (the lower reaches of the Danube River) dropped catastrophically and now the harvest is extremely small (Figure 2): only 11.5 metric tons in 1994 compared to about 200 metric tons per year in the 1960s (Bacalbaşa-Dobrovici 1991b). Not only the size, but also the structure of sturgeon populations in the Danube River changed dramatically. Individuals are much smaller and younger than in the past. *Acipenser sturio* disappeared from the sea catch, and there is a noticeable decrease in the numbers of *Huso huso*, *A. gueldenstaedtii* and especially *A. nudiventris*. Also, the population size of *A. stellatus* has decreased. Besides intensive fishing, other aspects of human activities have negatively impacted Danube River sturgeons, including deforestation, construction of hydrotechnical installations and dams, and pollution.

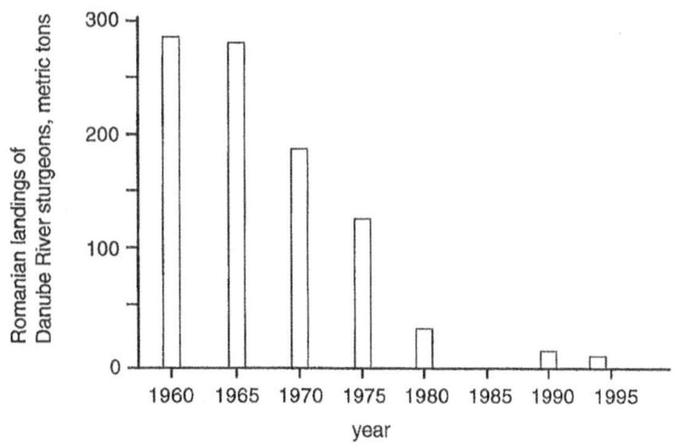

Figure 2. Decrease in the sturgeon catch in Romanian part of the Danube River. Data for 1960s through 1980s are from Bacalbaşa-Dobrovici (1991b).

Anthropogenic factors in the 20th century decline of Danube sturgeons

Deforestation

During the Middle Ages, forests located on the banks of the Danube River regulated the water level, and floods were rare. At the end of the 18th century, logging was officially encouraged, and persons who cut the trees were allowed to cultivate the cleared land. As a result, forested areas in Romania diminished from 55–60% in 1830 to 27% in 1930. Similar processes occurred in neighboring countries: in the Czech and Slovak Republics, only 34%, and in Bulgaria, only 29% of the historically forested areas now exist (Botzan 1984). Deforestation increased alluvial deposits, and water turbidity, which affected the sand, gravel and rock bottom of sturgeon spawning grounds.

Dikes and dams

Flood plains of the Danube changed drastically when dikes were built. Historically, the lower Danube flood plain included areas adjacent to the river (573 000 ha) and the delta (524 000 ha). At present, about 85% of the flood plains have been diked (Botzan 1984). The delta was diked to a lesser extent, and this stopped after the collapse of the communist regime (1989) for ecological reasons.

About 300 reservoirs in the Danube Basin were formed by damming. These lakes retain some alluvial deposits, especially large particles, and affect water levels in the Danube, which is 0.6 m higher than in historical times, and water velocity, which is slower in the riverine lakes region. In 1970, completion of the Iron Gates Dam I located 862 km upstream from the Danube mouth prevented sturgeons from reaching their historic spawning sites. Iron Gates Dam II, 80 km below the first project, shortened the possible migration to 862 km. A joint Bulgarian-Romanian dam at Turnu Magurele-Nicopol is planned, which would reduce the possible sturgeon migration to 265 km.

Pollution

Water pollution by heavy metals and pesticides in the lower Danube is very high (Oksiyuk et al. 1992) and it affects the entire biota (Pringle et al. 1993). No specific data are available on its impact on sturgeons, however.

Irrigation and gravel excavation

Water quality in the lower Danube is degraded by massive irrigation schemes. In Rumania, three million hectares of irrigated land decreases river flow and increases pollution by fertilizers and pesticides. Eutrophication now impacts the northwestern area of the Black Sea. Irrigation pumps kill fish larvae and juveniles. Sand and gravel taken from the Danube bed for construction work in the area near Calarasi (373 km) has destroyed sturgeon spawning grounds.

Water losses due to hydrotechnical constructions

Brackish water areas in the mouth of the Danube River and the northwestern part of the Black Sea depend primarily on flow in the Danube River, but natural flow is decreased by dams and irrigation. Other projects, such as shipping canals and junctions between different river branches in the delta, threaten to further alter the flow of the lower Danube. Fortunately, of all proposed projects (i.e., Levintov 1988), only the Rhine-Main-Danube Canal has been constructed. This canal uses water pumped into it from the Danube.

Diminishing role of the Danube Delta as a biofilter

The Danube Delta is an essential biofilter for the entire region. Also, it is the area of the contact of fresh riverine water with the brackish water of the Black Sea. Changes in water flow and circulation modified the whole ecosystem of the Danube Delta and affected its biofilter abilities. Poorly planned aquaculture and agriculture projects damaged the delta. Formerly, the Danube Delta was the largest area of reeds in Europe (almost 300 000 ha), a major component of its biofilter capacity. Unfortunately, 13.4% of the delta area (61 604 ha) was transformed into agricultural land during the communist regime (1948–1989). Attempts to cultivate reeds in a part of the delta created more problems because the heavy equipment which was used destroyed rhizomes of the natural reeds causing decreases in their area. Ill-conceived attempts at fish aquaculture on 53 000 ha of the delta failed, adding negative impacts on the biofilter capacity of the delta.

As a result of all these changes, eutrophication and turbidity increased in the delta, while biodiversity decreased, which in turn adversely affected the shelf area of the Black Sea. This shelf is crucial for sturgeons in the northwestern part of the Black Sea because this is where sturgeon live during the marine period of their life cycle.

Changes in the Black Sea

Geologic origin

The Black Sea has a long geological history, which has greatly impacted acipenserids. Originally a part of the Tethys Sea during the Upper Miocene (circa 20 MYBP), the Black Sea later, together with the Sea of Azov, Caspian and Aral seas, formed the Sarmatic (or Paratethys) Sea which covered the area from the Vienna basin to the Ural Mountains. In the Pliocene, the Sarmatic Sea was reduced to the smaller Pontic Sea, which included the three contemporary basins, the Aral, Caspian, and Black seas. During the Quaternary, these three basins separated. In the Pliocene (about 5 MYBP) the Strait of Gibraltar opened, and water from the Atlantic refilled the Mediterranean, which had been dry for about 1–2 million years. Eventually, the Mediterranean became reconnected with the Black Sea.

At present, water in the Black Sea is only half as salty as the Mediterranean Sea. The Black Sea water is divided into two strata. Due to the lack of vertical circulation, the lower stratum is abiotic. Within the surface stratum, the northwestern area is greatly impacted by three rivers: the Danube (which provides more than a half of the fresh water flowing into the Black Sea), the Dniester, and the Dnieper (together with the Bug River). This highly productive, low salinity zone is a good environment for the marine life of sturgeons.

The northwestern shelf of the Black Sea

Sturgeons migrating into the Danube River spend most of their life on the northwestern shelf of the Black Sea. The shelf is characterized by shallow water and a relatively flat bottom. Young sturgeon live and grow in this area, and adults return there after spawning in the Danube. Beluga feed mostly on fishes, while Russian and stellate sturgeons eat benthic invertebrates. During the last three decades, major changes in the biological equilibrium of the Black Sea occurred, affecting primarily the biota of the northwestern shelf.

Pollution

Pollution in the Black Sea is from tens to hundreds of times higher than that in the Atlantic or Pacific oceans and it is even higher than in the Mediterranean: 20 000 and 3775 kg km^{-3} of polluting agents in waters of the Black and Mediterranean seas respectively (Zaitsev 1992, 1993). Worse still, pollution is essentially perilittoral due to the perilittoral current in the Black Sea.

Eutrophication

Eutrophication of coastal waters had a serious impact on the Black Sea. Between the 1950s and the 1980s, the quantity of nutrient and organic substances brought by the Danube, Dnestr and Dnepr rivers into the Black Sea increased 400–500% (Table 1; see also Zhuravleva & Grubina 1993), causing intensive growth of phytoplankton from 670 mg m^{-3} (1950s) to 30 000 mg m^{-3} (Zaitsev 1991, 1993). The biomass of the jellyfish *Aurellia aurita* increased enormously from 1 million metric tons in the 1960s to 300–500 million metric tons in 1980. A simultaneous decline in the number of large planktonic crustaceans and planktophagous fishes occurred. Eutrophication also diminished water transparency by 50 to 80 percent and caused a drastic change in the benthic flora (Zaitsev 1992).

Effect of fish trawling

Bottom trawling devastated main areas of sturgeon habitat in the northwestern shelf and the Danube mouth. Over a 50 year period beginning in the

Table 1. Input of nutrient chemicals from the rivers entering into the northwestern part of the Black Sea (all values in parts million^{-1}, data from Zaitsev 1992).

	Danube River		Dnestr River		Dnepr River	
	1950	1986	1950	1986	1950	1986
Organic substances	2000	9800	100	246	250	664
Phosphates	13.00	50.00	0.14	1.00	0.80	4.00
Nitrates	97	238	2	13	55	89

1930s, the macrozoobenthic fauna near the Crimea decreased from 38 to 11 species, and their density diminished from 245 to 99 individuals m^{-2} (Zaitsev 1992).

Temporary hypoxic areas in the northwestern shelf

Eutrophication caused a new phenomenon, the appearance of temporary hypoxic areas, first noticed in August–September 1973. This hypoxic area affected 3500 km^2 between the Danube Delta and the Dnestr estuary (Zaitsev 1991, 1993). Since then, hypoxia has occurred periodically in the 10–40 m depth regions. Biological losses due to hypoxia between 1973 and 1990 were estimated at 60 million metric tons, including 5 million metric tons of fishes (Zaitsev 1993). Because of this destruction and periodic repopulation of the northwestern shelf area, molluscs and other benthic food organisms are represented by mostly young individuals. Besides threatening the survival of young sturgeons, hypoxia caused changes in populations of prey species inhabiting their feeding grounds.

Explosive growth of the ctenophore Mnemiopsis leidyi

The American comb jellyfish *Mnemiopsis leidyi* (or *M. maccradyi* according to Zaika & Sergeeva 1990) seems to have been introduced into the Black Sea in 1982 with the ballast water from ships; before this, it inhabited the North American Atlantic waters (Vinogradov et al. 1989, Travis 1993). In such an isolated marine basin with a depauperate fauna as the Black Sea, *M. leidyi* does not have competitors and its biomass grew very fast so that in August–September 1989 there were 800 million metric tons of this predator, which feeds on zooplankton, pelagic fish eggs, embryos and larvae (Zaitsev 1992). The abundance of *M. leidyi* produced a complete collapse of the anchovy fishery in the Sea of Azov in 1989. Changes in the faunal structure and distribution of invertebrates caused by *M. leidyi* (Kovalev et al. 1994) indirectly impact sturgeons.

Sturgeon survival in the lower Danube and Black Sea: the social context

Since 1878 (the year of the Berlin Peace agreement), the lower Danube and the Danube Delta have been under the control of the Romanian state. For a long time, overfishing of sturgeons in this area was extensive. During the communist regime (1948–1989), the centralized economy did not consider ecological criteria in the sturgeon fishery. The situation, however, has not yet improved in the post-communist period. Moreover, fishing permits have increased, resulting in a lack of information about the extent of the catch, which is now extremely extensive in Romania. The Danube Delta Biosphere Reserve (Gâstescu 1993), in which fish harvest is controlled, is a lucky exception to the generally uncontrolled situation. It is very difficult to organize protective measures for sturgeons in the lower Danube, and all species, especially *Huso huso* and *Acipenser gueldenstaedtii*, should be considered as threatened or endangered in Romania.

Considering all of the negative conditions for sturgeons migrating into the Danube River, the following conservation measures are recommended: (1) an end to fishing in the lower Danube; (2) research on the survival of young sturgeons in the contemporary conditions of the lower Danube and the Black Sea; (3) restocking of the endangered sturgeon species; (4) conservation (cryopreservation) of genetic materials of sturgeons from the Danube populations.[1]

Acknowledgements

I thank Vadim Birstein and John Waldman for inviting me to participate in the International Conference on Sturgeon Biodiversity and Conservation. Two anonymous reviewers commented on my original draft. The English text was substantially revised

[1] Since this paper was written, a new five-year project entitled 'Recovery Program of the Danubian Anadromous Migratory Sturgeons' began in 1995 funded by the Romanian Ministry of Research and Technology and a grant of the Global Environment Trust Fund administrated by the World Bank [The Editors].

and improved by Vadim Birstein and William E. Bemis, who also drew the map and figure.

References cited

Antipa, G. 1916. Pescaria si pescuitul în România. Acad. Româna, Fond V. Adamachi 8(46). Bucuresti. 794 pp. (in Romanian).

Antipa, G. 1933. Les esturgeons de la Mer Noire, leur biologie et les measures nécessaires pour leur protection. Comiss. Internat. Explor. Scient. de la Mer Méditerranée, Rapports et procès-verbaux des réunions 8 (Nouv. Sér.): 61–75.

Bacalbaşa-Dobrovici, N. 1971. Eigentümliche Fischereimethoden mit Stationären Fanggeräten vom rumänischen Ufer des neuen Donaustausees 'Eisernes Tor'. Protokolle z. Fischereitechnik 12: 401–407.

Bacalbaşa-Dobrovici, N. 1989. The Danube River and its fisheries. pp. 455–468. *In*: D.P. Dodge (ed.) Proceedings of the International Large River Symposium, Can. Spec. Publ. Fish. Aquat. Sci. 106: 455–468.

Bacalbaşa-Dobrovici, N. 1991a. Fish population evolution (1985–1988) in the last Danube built Reservoir Iron Gates II. Verh. Internat. Verein. Limnol. 24: 2524–2527.

Bacalbaşa-Dobrovici, N. 1991b. Statut des differantes especes d'esturgeons dans le Danube Roumain: problemes lies a leur maintanence. pp. 185–192. *In*: P. Williot (ed.) Acipenser, Cemagref Publ., Bordeaux.

Bănărescu, P. 1964. Pisces-Osteichthyes. Fauna Republicii Populare Romîne, 13, Editura Academiei Republicii Populare Romîne, Bucuresti. 962 pp.

Botzan, M. 1984. Apele în viata poporului român. Edit. CERES, Bucuresti. 410 pp.

Gâstescu, P. 1993. The Danube Delta: geographical characteristics and ecological recovery. GeoJournal 29: 57–67.

Giurescu, C.C. 1964. Istoria pescuitului si a pisciculturii in România. Editura Academiei Rep. Pop. Rom., Bucuresti. 390 pp.

Hensel, K. & J. Holčík. 1997. Past and current status of sturgeons in the upper and middle Danube River. Env. Biol. Fish. (this volume).

Kovalev, A.V., V.E. Zaika, N.A. Ostrovskaya, N.G. Sergeeva, V.V. Melnikov, I.Yu. Tamoikin, N.I. Ivanova & L.S. Svetlichnyi. 1994. *Mnemiopsis mccradyi* Mayer, 1900 is a new inhabitant of the Black Sea. Gidrobiologicheskiy Zhurnal 30: 104–107 (in Russian).

Levintov, A.E. 1988. What will be the future of the Danube River Delta? Priroda 1: 48–53 (in Russian).

Oksiyuk, O. P., L. A. Zhuravlev, A. V. Lyashenko, I. Kh. Bashmakova, Yu. I. Karpezo & A. I. Ivanov. 1992. Water pollution of the Danube River in the Ukraine: general indices. Gidrobiologicheskiy Zhurnal 28: 3–11. (in Russian; English translation: Hydrobiol. J. 29: 1–10 (1993)).

Pringle, C., G. Vellidis, F. Heliotis, D. Bandacu & S. Cristofor. 1993. Environmental problems of the Danube Delta. Amer. Sci. 81: 350–360.

Strabo, 1853. Geographica. C. Müller & F. Dübner (ed.), Paris.

Terofal, F. 1980. Ausgestorbene und seltene Fische sowie Neueinbürgerung in bayrische Gewässer. pp. 104–131. *In*: Im Dienste der bayrischen Fischerei, Landesfischereiverband Bayer, München.

Travis, J. 1993. Invader threatens Black, Azov seas. Science 262: 1856–1857.

Vinogradov, M.E., E.A. Shushkina, E.I. Musaeva & P.Yu. Sorokin. 1989. A new invader into the Black Sea, the ctenophore *Mnemiopsis leidyi* (A. Agassiz) (Ctenophora: Lobata). Okeanologiya 29: 293–299 (in Russian).

Zaika, V.E. & N.G. Sergeeva. 1990. Morphology and development of the ctenophore-invador *Mnemiopsis* (Ctenophora, Lobata) in the Black Sea environment. Zoologicheskii Zhurnal 69: 5–11 (in Russian).

Zaitsev, Yu.P. 1991. Cultural euthrophication of the Black Sea and other south European seas. La mer (Tokyo) 29: 1–7.

Zaitzev, Yu.P. 1992. Ecological status of the Ukranian zone of the Black Sea shelf: a survey. Gidrobiologicheskiy Zhurnal 28: 3–18 (in Russian; English translation: Hydrobiol. J. 29: 4–22 (1993)).

Zaitsev, Yu.P. 1993. Impact of eutrophication on the Black Sea fauna. pp. 63–86. *In*: Studies and Reviews 64, Part 2. General Fisheries Council for the Mediterranean.

Zhuravleva, L.A. & N.A. Grubina. 1993. Phosphorus regime of the lower Danube and addition of phosphorus to the Black Sea. Gidrobiologicheskiy Zhurnal 29: 81–88 (in Russian; English translation: Hydrobiol. J. 31: 92–101 (1995)).

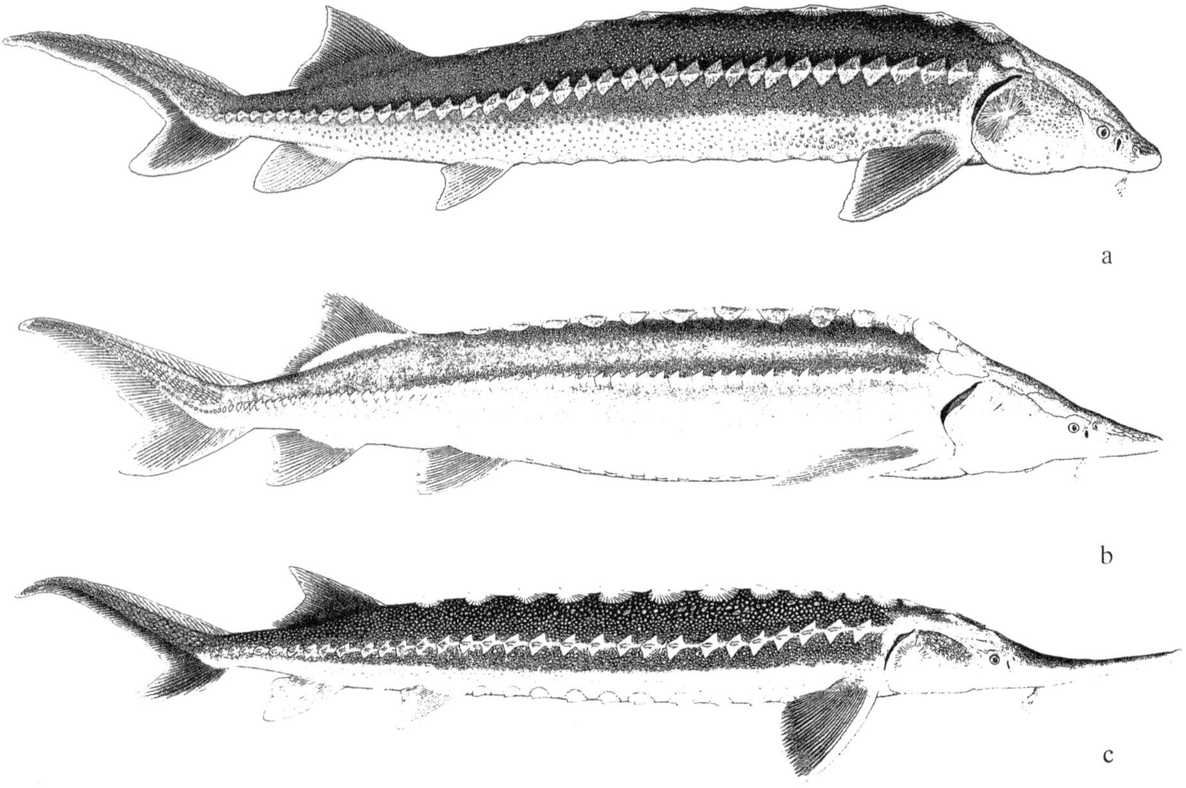

Sturgeons from the Danube river (Black Sea) collection at the 'Grigore Antipa' Natural History Museum, Bucharest: a – *Acipenser gueldenstaedtii* 146 cm TL from Antipa's personal working depository assembled at the turn of the century. Note the rare lack of medium sized denticles in the dorso-lateral region. b – *A. nudiventris* 104 cm TL, and c – *A. stellatus* 109 cm TL from 1940 in the same collection (all wet preserved). Originals by Paul Vecsei, 1996.

Environmental Biology of Fishes **48**: 209–219, 1997.

Present status of commercial stocks of sturgeons in the Caspian Sea basin

Raissa P. Khodorevskaya, Galina F. Dovgopol, Olga L. Zhuravleva & Anatolii D. Vlasenko
Caspian Fisheries Research Institute, 1 Savushkina st., Astrakhan 414056, Russia

Received 2.8.1994 Accepted 19.3.1996

Key words: beluga sturgeon, *Huso huso*, Russian sturgeon, *Acipenser gueldenstaedtii*, stellate sturgeon, *Acipenser stellatus*, population size, artificial propagation, pollution, poaching

Synopsis

Catches for the last 25 years are analyzed for beluga *Huso huso*, stellate sturgeon *A. stellatus* and Russian sturgeon *Acipenser gueldenstaedtii*, which are the three commercially important species of sturgeons found in the Caspian Sea Basin. Population sizes for generations born between 1961 and 1970 are estimated, and found to depend on natural reproduction and the number of young fish stocked annually from sturgeon hatcheries located in the Volga River Delta. A ban on sea fishing from 1962 to 1991 positively impacted the number and total biomass of commercial stocks. Sturgeon growth rates depend on water levels in the Caspian Sea. In order to preserve Caspian Sea sturgeon populations, it will be necessary to coordinate efforts of all countries surrounding the Caspian Sea to achieve rational harvests, preserve juveniles, and produce at least 100 million juveniles annually from hatcheries.

Introduction

Recently, 80 to 90 percent of the world's sturgeon catch was taken from the Caspian Sea Basin, mainly from the Volga River (Barannikova et al. 1995). The Volga River and Caspian Sea are home to three commercial species, the beluga (*Huso huso*, see Pirogovskii et al. 1989), Russian sturgeon (*Acipenser gueldenstaedtii*, see Vlasenko et al. 1989a), and stellate sturgeon (*A. stellatus*, see Shubina et al. 1989), in the Russian (northern) part of this basin. A fourth commercial species, the Persian sturgeon (*A. persicus*, see Vlasenko et al. 1989b), inhabits mostly the southern (Iranian) part of the Caspian Sea and the rivers entering into it. *Acipenser persicus* is not discussed in this paper because of lack of data.

The natural reproduction of commercial sturgeon species decreased in the Volga River after the Volgograd Dam was built between 1958 and 1960 (Figure 1). The dam prevents sturgeons from reaching their main spawning grounds. At present, the Ural River is the only large river entering the northern part of the Caspian Sea in which natural reproduction still occurs. Sturgeons no longer use the Kura and Terek rivers, where spawning previously occurred (Berg 1948). In this paper, we report on population changes in sturgeon stocks in the Caspian Sea from the early 1960s until 1994 based on data published in Dyuzhikov (1960), Shilov (1966), Khoroshko (1967, 1970), Vlasenko (1979, 1990), Slivka et al. (1982), Khodorevskaya (1986, 1992), Veshchev & Novikova (1988), Veshchev (1991a, b), Veshchev et al. (1992), Raspopov (1992, 1993), Dovgopol et al. (1993), Novikova (1993), Raspopov et al. (1994), Khodorevskaya et al. (1995), and Levin (1995). Also, the present status of natural reproduction in the Volga and the Ural rivers is described.

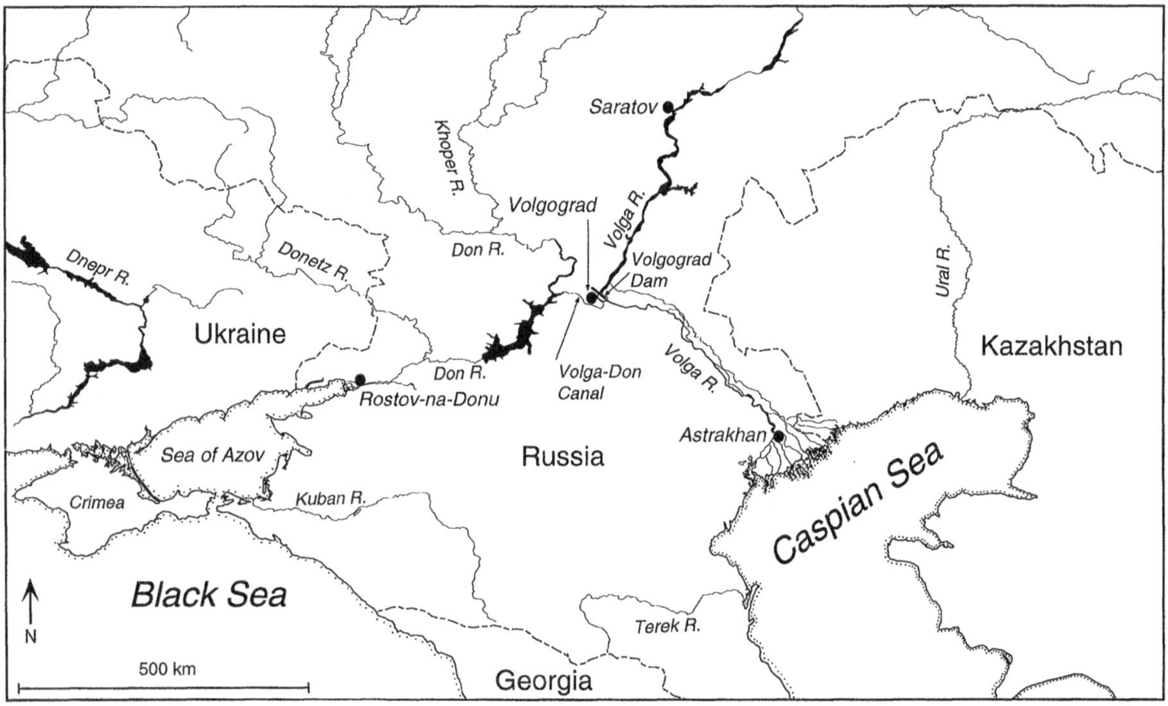

Figure 1. Map showing lower portions of the Volga and Ural rivers, northern part of the Caspian Sea, and adjacent regions referred to in text. Other river systems historically used by the three commercially important species of sturgeons (*Huso huso*, *Acipenser stellatus*, and *A. gueldenstaedtii*) are labeled.

Harvest trends during the last 40 years

Prior to 1951, commercial sturgeon fishing concentrated in the Caspian Sea (Korobochkina 1964). Following recommendations to concentrate harvest to only the lower reaches of the Volga River (Derzhavin 1947), a ban on sea harvest was instituted. The only sturgeon taken in the Caspian Sea were as a by-catch to other fishes. The introduction of plastic nets in the late 1950s for harvest in the Caspian Sea greatly increased in the number of young sturgeons caught as by-catch. In 1957, of the total 2.6 million sturgeons harvested in the northern Caspian Sea, 1.8 million were young sturgeons, and in 1959–1961, the by-catch of young sturgeons reached 2–3 million (Korobochkina 1964). From 1962 until 1991, sturgeons were not harvested legally in the northern part of the Caspian Sea.

For the last 35 years, natural and artificial reproduction contributed to the total commercial sturgeon stock in the Caspian Sea (Barannikova 1995). Sturgeon population sizes also depend on the volume of harvest, the construction and operation of dams, water consumption for irrigation and industry, and impacts of pollution. All three commercially important species of sturgeons now have fewer spawning fish migrating into the Volga and Ural rivers than in the past. Russian and beluga sturgeons no longer use the Kura, Terek and Sulak rivers, although small runs of stellate sturgeons still enter into the Terek and Sulak rivers to spawn.

The Volga River and its delta are the most important areas in terms of commercial harvest. About 75% of the total sturgeon catch in the Caspian Sea basin comes from this area, with Russian sturgeon providing 60–70% of this total. Until recently, stellate sturgeon made up about 30% of the catch in the Volga River. Beluga constituted 5.0–6.0% of the catch. Between 1976 and 1981, Russian sturgeon were taken about four times more frequently than

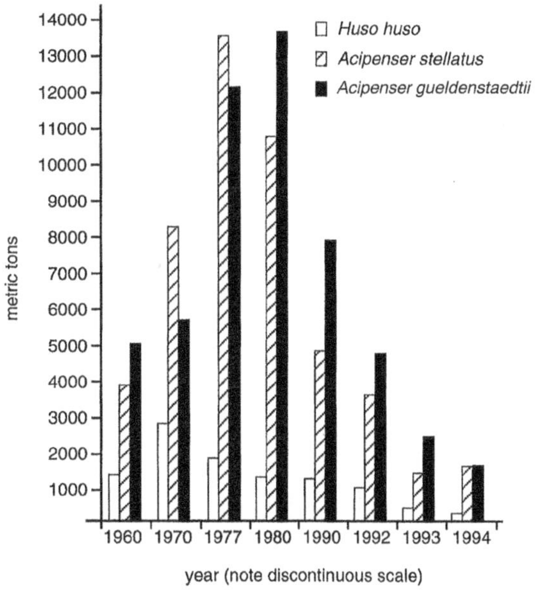

Figure 2. Sturgeon catch in the northern part of the Caspian Sea in thousands of metric tons. Catches for stellate sturgeon (*Acipenser stellatus*) and Russian sturgeon (*A. gueldenstaedtii*) peaked in the 1970s, when catches of beluga (*Huso huso*) were already in decline.

beluga and stellate sturgeons, whereas in the early 1960s, this was only one and one half times. Harvest figures (in thousands of metric tons) are shown for selected years for each of the three species (Figure 2).

Population changes and status

Beluga

Historically, beluga were harvested more intensively than were Russian and stellate sturgeons (Korobochkina 1964). For the last 35 years, the number of spawning adult beluga has decreased. In the early 1970s, about 25 000 individuals with a total weight 2600 metric tons migrated into the Volga River. In recent years, the number of spawning fish harvested did not exceed 11 700 individuals weighing a total 750 metric tons. Commercial catch of beluga decreased from 2000 metric tons in the early 1970s to less than 500 metric tons at present. In the early 1970s, 21% of spawning beluga sturgeon migrating into the Volga River reached the spawning grounds. In 1976, 36–40% of spawning beluga reached the spawning grounds, but the number of fish reaching the grounds remained at 4000–6000 individuals.

Declines in the number of belugas occurred after construction of a series of hydroelectric dams on the Volga River in the late 1950s (Frantsuzov 1960). Beluga no longer can reach their historic spawning grounds because their movements are blocked by the Volgograd Dam (Figure 1). Beluga now spawn downriver from the city of Volgograd, in reaches in which they did not spawn previously. In the past, the migration distance for juveniles was much longer (originating from spawning grounds far above Volgograd), and juveniles grew to a larger size in the

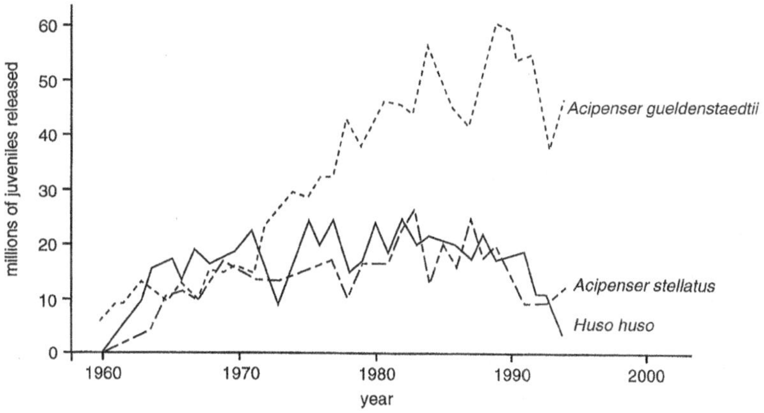

Figure 3. Releases of sturgeon juveniles produced by hatcheries located in the northern part of the Caspian Sea.

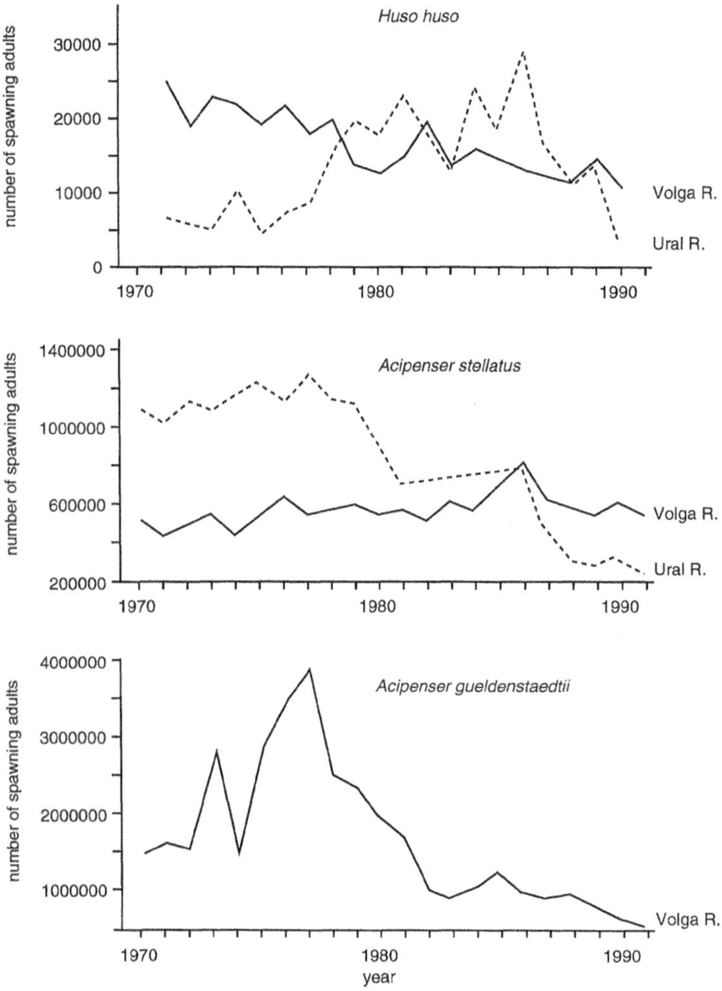

Figure 4. Estimated numbers of sturgeons in the Northern Caspian Sea from the Volga and Ural river populations.

river than they do now. These changes resulted in decreases in natural reproduction in the Volga River and in changes in the population structure of this species.

In response to this decline in natural reproduction, the Soviet government began a program in the early 1960s to enhance sturgeons through artificial propagation (Barannikova 1995). Throughout the 1960s, more than 3.9 million beluga juveniles were released from hatcheries annually. In the 1970s, the annual release reached more than 12.9 million young, and by the early 1980s, the average number of the young belugas released into the Volga River was 19.4 million (Figure 3). At present, practically all beluga (96.3%) in the Volga River consist of hatchery propagated fish (Khodorevskaya 1986, 1992). However, artificial propagation does not completely compensate for the loss of natural reproduction of beluga in the Volga River. The population of beluga continues to decline even though the number of beluga harvested does not exceed 0.1% of the number of individuals released.

Spawning sites for beluga in the Ural River remain intact and the Ural River stock of beluga is replenished by natural reproduction. Until the late 1970s, the number and biomass of beluga migrating into the Ural River was considerably smaller than that migrating into the Volga River (Figure 4).

However, since 1979, the number of beluga entering the Ural River has exceeded the number entering the Volga River. The ban on sturgeon harvest from the Caspian Sea enacted in the 1960s contributed greatly to the increase in the number of spawning beluga that migrate into the Ural River.

The biomass of beluga from the 1952 through 1976 generations, which took part in spawning during the years 1978 through 1987, grew from 2800 to 4000 metric tons. In 1990, about 14000 individuals migrated into the Volga River from the sea. Approximately 35% were females and males which were spawning for the first time (11–17 years old), 60% of the fish were spawning for the second time (18–30 years old), and the remaining 5% were 31–52 year old individuals (Table 1). The present commercial catch targets 14–22 year old individuals from 1966 to 1974 spawnings. Catches in the near future will be based on fish spawned between 1970 and 1978.

The size of the beluga population in the Caspian Sea is small, and the number of fish migrating into the Volga River is low. The commercial catch of beluga is expected to remain low because conditions in the Caspian Sea are unfavorable (see below).

Stellate sturgeon

Stellate sturgeon still have natural spawning grounds below the Volgograd Dam, and 60% of their historic spawning sites remain intact. The number of stellate sturgeon harvested varies from 194 000 in 1967 to 884 000 in 1986. From 85 000 to 388 000 individuals (20–44% of the total number of spawning stellate sturgeons migrating into the Volga River) reached the spawning grounds below the Volgograd Dam. Many stellate sturgeon reached spawning grounds in 1978, 1979, 1983, and 1985–1988 (616 000–884 000 individuals, Figure 4). Harvests varied from 3870 metric tons in 1960 to 4550 metric tons in 1986.

The spawning stock of stellate sturgeon migrating into the Volga River in the late 1980s consisted of fish either spawned after regulation of the Volga River flow or from fish released from hatcheries. At present, the generations hatched between 1972 and 1978 dominate the catch. The largest stock reported consisted of 15.7% first-time-spawners, 72.9% second-time-spawners, and 11.4% older individuals (Table 1). An increase in the number of spawning stellate sturgeon migrating into the Volga River in the late 1980s could be due to stabilization of natural reproduction in this species.

From the late 1960s until 1985, fewer stellate sturgeon have entered the Volga River (approximately 500 000 on average) than the Ural River (800 000–1 300 000, Figure 4). The number of stellate sturgeon decreased in the Ural River because of overfishing and insufficient annual recruitment, and between 1986 and 1992, stocks have declined sharply in both rivers. In 1992, the Ural River population was only one-third of the number observed in 1986.

Table 1. Changes in the percent composition of spawning populations of sturgeons in the Volga River during the last 35 years.

Species and characteristics	1965	1980	1985	1991
Beluga sturgeon				
1st-time-spawners	30	34	41	35
2nd-time-spawners	67	60	50	60
Older individuals	3	6	7	5
Stellate sturgeon				
1st-time-spawners	20	4	6	12
2nd-time-spawners	70	80	72	51
Older individuals	10	16	22	37
Russian sturgeon				
1st-time-spawners	9	14	10	14
2nd-time-spawners	63	74	76	74
Older individuals	28	12	14	12

The number of spawning fish in the Ural River is critical for the survival of this population.

Russian sturgeon

The tonnage of Russian sturgeon harvested from the Volga-Caspian stock increased from 1961 to 1977. The increase in harvest followed the 1962 ban on sturgeon fishing in the Caspian Sea. The number of Russian sturgeon harvested increased from 480 000 in 1950 to 3 746 800 in 1974, and tonnage increased more than 8 times. For 20 years, from 1966 until 1985, the number of spawning fish harvested remained more than one million individuals. During that period, the catch was based on fish hatched between 1935 and 1961, before Volga River flow became regulated by dams. Beginning in 1978, the number of fish harvested decreased to 766 600 (Figure 4c) and their tonnage declined to 16 300 metric tons. The decline followed a sharp decrease in natural reproduction, because as much as 80% of the spawning grounds for Russian sturgeon became unavailable to the fish after the Volgograd Dam was built in 1958–1960. The number of individuals harvested from the 1959–1960 generations was 691 500–730 000, and that from the 1965–1968 generations was 461 000–600 000.

The age structure of the fish migrating in the Volga River has also changed (Table 1). In the early 1960s, the run consisted of 8–12% first-time-spawners, 62–75% of second-time-spawners, and 16–28% of older fish. During peak years (1966–1985), second-time-spawners (68–78%) and older individuals (10–18%) prevailed. This phenomena continued until the early 1990s (Table 1). At present, older individuals dominate the run (approximately 50%), while first time spawning fish constitute around 11%.

Natural reproduction plays an important role in the formation of the present Russian sturgeon stock. After the Volga River flow became regulated in 1959, natural reproduction decreased from 7500 metric tons in 1960 to 3000 metric tons in 1981–1985. In the early 1990s, natural reproduction decreased to 830 metric tons because of the low number of sturgeons reaching the spawning grounds.

Factors affecting sturgeon stocks

Effects of sea level on sturgeon growth

Fluctuations in water levels in the Caspian Sea and consequent changes in salinity impacted sturgeon stocks. Fluctuations impact accessibility to feeding sites, the abundance of food organisms at these sites, and the concentration and distribution of sturgeons in the sea. We used the rate of weight gain to estimate the effect of sea level fluctuations.

Changes were seen in beluga that returned to rivers to spawn. Beginning in 1970, the sea level decreased until, in 1977, it reached a minimum at 29 m, a previous lowest level which also occurred in 1936–1937. Starting in 1972, the relative rate of weight gain in beluga began to decrease. The relative weight gain decreased more in males than in females. Once sea level began to rise in 1978, beluga spawners continued to show a decrease in relative weight. In the early 1970s, the average weight of beluga females was 110 kg, and in 1990–1991, it was only 57 kg. The sex structure of the spawning stock of beluga also has changed, with females now (1991) constituting 21 to 24% of the stock as opposed to twice that percent in the 1960s.

The rise in sea water level since 1978 was correlated with an increase in the growth rate of stellate sturgeon. A sharp drop in growth rate in 1989 may be due to high levels of pollution in the Volga River and the Caspian Sea. The discharge of toxic chemicals caused deaths of stellate sturgeon in the river and negatively impacted sturgeon feeding in the Caspian Sea (Khodorevskaya et al. 1995).

Russian sturgeon are less tolerant of high water salinity than are stellate sturgeon. Optimal conditions for growth of Russian sturgeon are: salt concentration of not more than 10, an abundance of brackish water prey organisms, and extended opportunity for juveniles to live in the river. Between 1967 and 1978, the relative rate of growth of Russian sturgeon returning as spawners decreased sharply with minimum weight gains from the 1970–1980s. Lukyanenko et al. (1986) considered that the relative increase in salinity negatively impacted the growth rate of Russian sturgeon in the 1970s.

Influence of dams on natural reproduction

Construction of a series of hydroelectric dams in the middle reaches of the Volga River in the 1950s blocked sturgeons from reaching their primary spawning sites. Only 372 from a total of 3390 ha of suitable spawning area is left intact (Khoroshko 1970). The success of natural reproduction depends on the volume of water during spring flood, runoff during the summer, water temperatures during the spawning, amount of suitable spawning grounds available, the condition of the substrate of the spawning grounds, the number of fish reaching the spawning grounds, and the quality or condition of spawners reaching the spawning grounds. As a result of new fishery rules introduced in 1986 in the Volga River Delta, the period of legal harvest in the river was shortened to increase the number of spawning fish reaching the spawning grounds. This, and favorable ecological conditions (i.e., an increase in water levels in the river and the Caspian Sea), allowed natural reproduction to increase in the late 1980s. However, levels of poaching in the Caspian Sea and in the Volga River have increased immensely in the last few years, which has decreased the efficiency of natural reproduction of all three commercial sturgeon species.

Artificial propagation

The Volga River sturgeon hatcheries are situated next to the river and began to release juveniles into the Volga River first in 1957; trends since then are shown in Figure 3. Juveniles are released into the river at age of 30–50 days (Lukyanenko et al. 1984). The number of beluga juveniles increased from 0.5 million in 1959 to 16.0 million in 1970. However, we found that this did not increase the number of beluga harvested in later years. The release of stellate sturgeon juveniles, along with a ban on sturgeon catch from the sea in 1962, resulted in increases in the number of stellate sturgeon caught. In the late 1980s, 30% of the stellate sturgeon harvest was from hatchery produced fish. Unfortunately, increases in the number of juveniles released from hatcheries did not seem to affect the number of spawning beluga or stellate sturgeon, that later entered the river.

The number of Russian sturgeon juveniles released from sturgeon hatcheries increased from 0.7 million in 1955 to 20–40 million in 1980–1983. However, these numbers also failed to stabilize the size of the Russian sturgeon population in the Volga River and Caspian Sea. Artificial reproduction of Russian sturgeon has not compensated for stock losses caused by overfishing, pollution and other anthropogenic factors. As previously mentioned, decreases in water levels in the Caspian Sea in the 1980s may also have negatively impacted both growth and survival of juveniles in the Volga River delta during their first winter, while fish are adapting to more saline conditions. To offset a continuing decrease in the size of stock, 40 to 60 million juvenile Russian sturgeon were released annually from 1986 to 1990 (Figure 3). In the late 1980s, hatchery propagated fish represented an estimated 25–30% of the catch.

Water pollution and illegal harvest as threats

In 1984, specimens of Russian sturgeon with degenerated muscles began to appear in the Volga River and the Caspian Sea. In 1987, muscle degeneration was noted on a massive scale in all three commercial sturgeon species covered in this paper (Altufev et al. 1989, Lukyanenko 1990). The phenomenon was called 'muscle atrophy' and was intensively studied from 1987 through 1992 (Evgeneva et al. 1990, Altufev et al. 1992, Kuzmina et al. 1992). Results indicated that fibrils of the striated muscle tissues degenerated and were replaced by fat and connective tissues. It was suggested that muscle atrophy was caused by cumulative toxicosis resulting from increasing pollution levels in the Caspian Sea basin (Altufev et al. 1989, Lukyanenko 1989). Volga River water pollution also was extremely high by the late 1970s. In 1979, the concentration of the pesticide hexachlorane ranged between 0.0003 mg l^{-1} in the Volga River delta and 0.0025 mg l^{-1} near the city of Volgograd (Lukyanenko 1990). Water pollution has increased since then. Recent experiments support the hypothesis of pollution-caused etiology of muscle atrophy. The most common oil products, diesel fuel, and hexachlorocyclohexane, are known to cause anomalies in muscles of juvenile Russian sturgeon and beluga that are similar to those seen in sturgeons with muscle atrophy (Altufiev 1994).

Accumulation of heavy metals and pesticides in gonads, livers, and muscles was discovered recently (Andreev et al. 1989, Gapeeva et al. 1990, Golovin et al. 1990, Kirillov et al. 1990, Moroz 1990, Paveleva et al. 1990). Since the late 1980s, high levels of tumors, abnormalities in gonad development and gametogenesis, and disturbances in the morphogenesis of organs have been found in all three species of sturgeons (Romanov et al. 1989, Romanov & Sheveleva 1992, Romanov & Altufev 1990). In 1990, 100% of eggs taken from females of all three species of sturgeons caught in the lower reaches of the Volga River showed abnormalities and 100% of the embryos were nonviable (Shagaeva et al. 1993). Therefore, the impact of pollution on present populations of sturgeons in the Volga River and Caspian Sea is very high and its effect will intensify in the near future.

Another serious threat to survival of sturgeons in the Caspian Sea basin is uncontrolled overfishing and an enormously increased level of poaching. After dissolution of the Soviet Union in 1991, sea harvest, which had been prohibited for 30 years (since 1962) under Soviet law, began again. The absence of common fishery agreements among the states bordering the Caspian Sea (Russia, Azerbaidjan, Turkmenistan, Kazakhstan, and Iran)[1] aggravated the already grave situation of the commercial species of sturgeons. Political and economic instability in Russia, along with inflationary pressures in the region, caused a sharp rise in poaching in the Volga River during the early 1990s, concentrated at feeding and spawning areas. Unfortunately, the unprecedented scale of poaching in the Volga River occurred simultaneously with decreases in efforts to replenish sturgeon stocks by stocking of artificially reared juveniles. Together, these factors threaten the survival of commercial sturgeon species in the northern part of the Caspian Sea basin.

[1] On 14 November 1996, Russia, Azerbaijan, Turkmenistan, and Iran signed an agreement to ban all sturgeon fishery in the Caspian Sea in 1997. Under the agreement, sturgeon fishery will only be allowed in the deepest waters of the Volga and Ural rivers (Editors' note, February 1997).

Dynamics of the loss of Caspian Sea sturgeon fisheries

(1) Before Volga River flows were regulated in 1958, sturgeon stocks in the Volga-Caspian system were supported by natural reproduction only. Between 1950–1958, returning spawning populations consisted of about 20 000 individuals of beluga, 400 000 stellate sturgeon, and 700 000 Russian sturgeon.

(2) From 1959 through 1972, recruitment was mostly natural. A ban on the catch of sturgeon from the Caspian Sea in 1962 had a positive effect on returning stocks. Also, beginning in 1957, sturgeon hatcheries began to release juveniles to maximize the size of sturgeon populations. The number of surgeons returning to the Volga River to spawn was from 5700 to 11 000 beluga, from 600 000 to 907 000 stellate sturgeon, and from 334 000 to 450 000 Russian sturgeon. All three commercial species were cut off from their historic spawning areas by construction of dams in 1958–1960, and the length of the migration path was reduced. This initiated a gradual decrease in natural reproduction so that by the 1970s, the beluga stock was consisted primarily of hatchery propagated fish.

(3) The period of 1973 through 1977 saw a sharp reduction in natural reproduction, a worsening environmental situation caused by a drop in the sea level which increased water salinity, decreased the area of feeding grounds, and reduced the deltaic area where juveniles overwinter. This period was particularly critical for the survival of juveniles of all three species, especially for the Russian sturgeon, and many eggs laid on spawning grounds did not develop.

(4) From 1978 through 1989, environmental conditions changed. Water levels rose in the Caspian Sea and salinity decreased. The number of juveniles released in the delta of the Volga River increased to 19 million beluga juveniles, 18 million stellate sturgeon, and 45.7 million Russian sturgeon. But, beginning in 1985, high levels of water pollution began to affect sturgeons, and their natural reproduction decreased sharply. The recruitment of individuals from these generations have not been estimated yet since these fish are still too young to be caught.

(5) Since 1990, declines in sturgeon populations have occurred due to poaching and overfishing in the Volga River and Caspian Sea. The number of juveniles released from hatcheries decreased sharply because of a worsening economic situation in Russia. In order to preserve sturgeon stocks in the northern Caspian Sea, an international agreement is needed on fishery regulations and control of fishing in the sea. Strict measures against illegal fishing are urgently needed. It will be necessary to restore the propagation efforts of hatcheries and improve the technology for artificial reproduction of sturgeons.[1]

Acknowledgements

We are grateful to Vadim Birstein, John Waldman and Robert Boyle for inviting one of us (Anatolii Vlasenko) to present our paper at the International Conference on Sturgeon Biodiversity and Conservation. We thank Vadim Birstein, Kent Keenlyne, Todd Georgi and William Bemis for editing our manuscript to improve the English. William E. Bemis drew the map and figures.

References cited

Altufiev, Yu. V. 1994. Morphofunctional state of muscle tissue and liver of Russian sturgeon and beluga juveniles in experiments on chronic intoxication. Voprosy Ikhtiologii 34: 135–138 (in Russian, English translation J. Ichthyol. 34: 134–139).

Altufiev, Yu. V., A. A. Romanov & N. N. Sheveleva. 1989. On the histopathology of muscle tissue of some acipenserids of the Caspian Sea basin. pp. 4–5. *In*: Sturgeon Industry in the Waters of the USSR, Pt. 1, Astrakhan (in Russian).

Altufiev, Yu. V., A. A. Romanov & N. N. Sheveleva. 1992. Histology of the striated muscle tissue and liver in Caspian Sea sturgeons. Voprosy Ikhtiologii 32: 157–171 (in Russian, English translation J. Ichthyol. 32: 100–115).

Andreev, V. V., V. N. Kryuchkov & V. A. Grigoriev. 1989. Accumulation of heavy metals in water ecosystems and their influence on sturgeons. pp. 6–7. *In*: Sturgeon Industry in the Water Bodies of the USSR, Pt. 1, Astrakhan (in Russian).

Barannikova, I. A. 1995. Measures to maintain sturgeon fisheries under conditions of ecosystem changes. pp. 131–136. *In*: A. D. Gershanovich & T. I. J. Smith (ed.) Proceedings of the International Symposium on Sturgeons, 6–11 September 1993, VNIRO Publishing, Moscow.

Barannikova, I. A., I. A. Burtsev, A. D. Vlasenko, A. D. Gershanovich, E. V. Makarov & M. S. Chebanov. 1995. Sturgeon fisheries in Russia. pp. 124–136. *In*: A. D. Gershanovich & T. I. J. Smith (ed.) Proceedings of the International Symposium on Sturgeons, 6–11 September 1993, VNIRO Publishing, Moscow.

Berg, L. S. 1948. The freshwater fishes of the USSR and adjacent countries, Vol. 1, Part 1. Akademia Nauk USSR, Moscow & Leningrad (in Russian, English translation published by Israel Program for Scientific Translations, Jerusalem. 505 pp.)

Derzhavin, A. N. 1922. Stellate sturgeon. A biological essay. Izvestiya Bakinskoi Ikhtiologicheskoi Laboratorii 1: 1–204 (in Russian).

Derzhavin, A. N. 1947. Sturgeon stock reproduction. Izdatelstvo Akademii Nauk Azerbaijanskoi SSR, Baku. 247 pp. (in Russian).

Dovgopol, G. F., P. V. Veshchev & T. V. Ozeryanskaya. 1993. Estimation of population size of generations of the stellate sturgeon, *Acipenser stellatus*, and the major factors affecting its population structure. Voprosy Ikhtiologii 33: 93–99 (in Russian).

Dyuzhikov, A. T. 1960. Stock composition and reproduction of Russian sturgeon in the Volga River below the Volgograd hydroelectric station. Trudy Saratovskogo Otdeleniya GosNIORKh 6: 76–116 (in Russian).

Evgeneva, T. P., O. K. Basurmanova & A. B. Shekhter. 1989. Degenerative changes in white muscle of the Russian sturgeon (*Acipenser gueldenstaedti*). Doklady Akademii Nauk USSR 307: 462–466 (in Russian, English translation Doklady Biol. Sci. 307: 485–489).

Frantsuzov, N. N. 1960. The number of spawning Russian sturgeons in the Volga River populations and regulation of the Russian sturgeon harvest in the present conditions. Nauchno-Tekhnicheskii Bulletin GosNIORKh 12: 11–29 (in Russian).

Gapeeva, M. V., O. L. Tselmovich & O. L. Shirokova. 1990. Heavy metals in acipenserid fishes of the lower Volga River. pp. 55–61. *In*: V. I. Lukyanenko (ed.) Physiological-Biochem-

[1] In 1995, the prospects for the three commercial sturgeon species of the Volga River worsened. Only 35 female beluga were captured for artificial breeding at the hatcheries located in the delta, and only 86 female beluga were legally taken in the entire northern area of the Caspian Sea (in the 1960s, about 2000 female beluga were caught annually in the same region). Practically all Russian and stellate sturgeons migrating to spawning grounds below the Volgograd Dam were harvested in 1995 by poachers, who operated between Astrakhan and Volgograd (see Figure 1). Fewer than five of the sturgeon hatcheries in the Volga Delta produced juveniles for stocking during 1995, a great decrease from the more than 12 hatcheries operated historically. Juveniles from the hatcheries were released directly into the Volga River, where predation losses are high, instead of being released directly onto sturgeon feeding grounds in the Caspian Sea, the preferred method used throughout the 1980s (editors note, February 1996).

ical Status of the Volga River-Caspian Sea Acipenserids in Normal Individuals and in the Individuals with Muscle Degeneration (Cumulative Polytoxicosis), Institute of Biology of Inland Waters Press, Rybinsk (in Russian).

Golovin, A. N., S. G. Kirichenko, O. A. Galutva, E. N. Konishcheva & L. D. Kurlapova. 1990. Content of some chemical elements and chlororganic pesticides in muscles and organs of the Russian sturgeon. pp. 52–54. *In*: V. I. Lukyanenko (ed.). Physiological-Biochemical Status of the Volga River-Caspian Sea Acipenserids in Normal Individuals and in the Individuals with Muscle Degeneration (Cumulative Polytoxicosis), Institute of Biology of Inland Waters Press, Rybinsk (in Russian).

Khodorevskaya, R. P. 1986. Commercial sturgeon stocks and prediction of their possible harvest in the Volga-Caspian area. pp. 189–199. *In*: Dynamics of the Number of Commercial Fishes, Nauka Press, Moscow (in Russian).

Khodorevskaya, R. P. 1992. Stocks formation of spawning sturgeon populations in the Volga River. pp. 445–448. *In*: Biological Resources of the Caspian Sea. Astrakhan (in Russian).

Khodorevskaya, R. P., G. F. Dovgopol & O. L. Zhuravleva. 1995. Formation of commercial sturgeon (Acipenseridae) stocks. pp. 137–150. *In*: A. D. Gershanovich & T. I. J. Smith (ed.) Proceedings of the International Symposium on Sturgeons, 6–11 September 1993, VNIRO Publishing, Moscow.

Khoroshko, P. N. 1967. Spawning of the Russian and stellate sturgeons in the lower Volga River. Trudy TsNIORKh 1: 95 (in Russian).

Khoroshko, P. N. 1970. On the spawning ecology of the Russian sturgeon in the changed conditions of the Volga River. Trudy TsNIORKh 2: 105–111 (in Russian).

Kirillov, V. N., N. A. Kanieva & R. A. Davletyarova. 1990. Comparative biochemical characteristics and levels of chlororganic pesticide content in the organs of Russian sturgeon having pathological changes and not having them. pp. 69–74. *In*: V. I. Lukyanenko (ed.). Physiological-Biochemical Status of the Volga River-Caspian Sea Acipenserids in Normal Individuals and in the Individuals with Muscle Degeneration (Cumulative Polytoxicosis), Institute of Biology of Inland Waters Press, Rybinsk (in Russian).

Korobochkina, Z. S. 1964. Main historical stages of sturgeon fishery in the Caspian Sea basin. Trudy VNIRO 52, Pt. 1: 59–86 (in Russian).

Kuzmina, O. Yu., V. I. Lukyanenko, Ye. I. Shakhmatova, Ye. A. Lavrova & Yu. V. Natochin. 1992. Specific features of water and salt homeostasis in sturgeon during muscle degeneration. Voprosy Ikhtiologii 32: 138–143 (in Russian, English translation J. Ichthyol. 33: 93–100).

Levin, A. V. 1995. Russian sturgeon, *Acipenser gueldenstaedti* Brandt, stocking in the Volgo-Caspian basin. pp. 178–188. *In*: A. D. Gershanovich & T. I. J. Smith (ed.) Proceedings of the International Symposium on Sturgeons, 6–11 September 1993, VNIRO Publishing, Moscow.

Lukyanenko, V. I. 1989. The phenomenon of stratification of muscle tissue and loosening of egg membrane in sturgeons. pp. 207–209. *In*: Sturgeon Industry in the Water Bodies of the USSR, Pt. 1, Astrakhan (in Russian).

Lukyanenko, V. I. 1990. Impact of the multiple factor anthropogenic pressure on the habitat conditions, reproduction, number, and harvest of acipenserids. pp. 25–44. *In*: V. I. Lukyanenko (ed.) Physiological-Biochemical Status of the Volga River-Caspian Sea Acipenserids in Normal Individuals and in the Individuals with Muscle Degeneration (Cumulative Polytoxicosis), Institute of Biology of Inland Waters Press, Rybinsk (in Russian).

Lukyanenko, V. I., R. Yu. Kasimov & A. A. Kokoza. 1984. Age and weight standards of hatchery raised juveniles of Caspian Sea acipenserids. Institute of Biology of Inland Waters Press, Volgograd. 229 pp. (in Russian).

Lukyanenko, V. I., P. P. Geraskin & A. I. Surial. 1986. Hemoglobin fraction compositions in anadromous and fresh water sturgeons. Ecologiya 6: 71–77 (in Russian).

Moroz, I. E. 1990. Accumulation of chlororganic pesticides in the sturgeon bodies and the extant of manifestation of toxic effects at different stages of their ontogenesis. pp. 62–68. *In*: V. I. Lukyanenko (ed.) Physiological-Biochemical Status of the Volga River-Caspian Sea Acipenserids in Normal Individuals and in the Individuals with Muscle Degeneration (Cumulative Polytoxicosis), Institute of Biology of Inland Waters Press, Rybinsk (in Russian).

Novikova, A. S. 1993. Current status of natural reproduction of beluga, *Huso huso*, in the lower Volga River. Voprosy Ikhtiologii 33: 556–560 (in Russian, English translation J. Ichthyol. 34: 68–75).

Paveleva, L. G., I. E. Zimakov, A. V. Komarova & E. M. Golik. 1990. Some aspects of the effect of anthropogenic pollution on the acipenserids of the Volgo-Caspian basin. pp. 45–51. *In*: V. I. Lukyanenko (ed.) Physiological-Biochemical Status of the Volga River-Caspian Sea Acipenserids in Normal Individuals and in the Individuals with Muscle Degeneration (Cumulative Polytoxicosis), Institute of Biology of Inland Waters Press, Rybinsk (in Russian).

Pirogovskii, M. I., L. I. Sokolov & V. P. Vasilev. 1989. *Huso huso* (Linnaeus, 1758). pp. 156–200. *In*: J. Holčík (ed.) The Freshwater Fishes of Europe, Vol. 1, Pt. II, General Introduction to Fishes, Acipenseriformes, AULA-Verlag, Wiesbaden.

Raspopov, V. M. 1992. Age structure and population dynamics of the beluga, *Huso huso*, migrating into the Volga River. Voprosy Ikhtiologii 32: 74–80 (in Russian, English translation J. Ichthyol. 33: 105–112).

Raspopov, V. M. 1993. Growth rate of Caspian Sea beluga sturgeon. Voprosy Ikhtiologii 33: 417–426 (in Russian, English translation J. Ichthyol. 33: 72–84).

Raspopov, V. M., A. S. Novikova, O. L. Zhuravleva, I. N. Lepilina & A. E. Egorova. 1994. Effectiveness of natural reproduction of the Russian sturgeon, *Acipenser gueldenstaedti*, during regulation of the Volga River flow. Voprosy Ikhtiologii 34: 348–352 (in Russian, English translation J. Ichthyol. 34: 9–17).

Romanov, A. A. & Yu. V. Altufiev. 1990. Tumors in gonads and liver of the Caspian Sea sturgeons (Acipenseridae). Voprosy Ikhtiologii 30: 1040–1044 (in Russian, English translation J. Ichthyol. 30: 44–49).

Romanov, A. A. & N. N. Sheveleva. 1992. Disruption of gonadogenesis in Caspian sturgeons. Voprosy Ikhtiologii 32:176–180 (in Russian, English translation J. Ichthyol. 33: 127–133).

Romanov A. A., N. N. Sheveleva & Yu. V. Altufev. 1989. Disturbances in the liver morphogenesis in the Caspian Sea sturgeons. pp. 282–284. *In*: Sturgeon Industry in the Water Bodies of the USSR, Pt. 1, Astrakhan (in Russian).

Shagaeva, V. G., M. P. Nikolskaya, N. V. Akimova, K. P. Markov & N. G. Nikolskaya. 1993. A study of the early ontogeny of Volga River sturgeons (Acipenseridae) subjected to human activity. Voprosy Ikhtiologii 33: 230–240 (in Russian, English translation J. Ichthyol. 33: 23–41).

Shilov, V. I. 1966. Sturgeon reproduction in the lower reservoir of the Volgograd hydroelectric power station. Voprosy Ikhtiologii 6: 663–672 (in Russian).

Shubina, T. N., A. A. Popova & V. P. Vasilev. 1989. *Acipenser stellatus* Pallas, 1771. pp. 395–443. *In*: J. Holčík (ed.) The Freshwater Fishes of Europe, Vol. 1, Pt. II, General Introduction to Fishes, Acipenseriformes, AULA-Verlag, Wiesbaden.

Slivka, A. P., G. F. Dovgopol & S. S. Zakharov. 1982. The dynamics of the stellate sturgeon number in the Northern Caspian Sea. pp. 65–71. *In*: Biological Productivity of the Caspian Sea and Sea of Azov, VNIRO Press, Moscow (in Russian).

Veshchev, P. V. 1991a. Effectiveness of natural reproduction in the stellate sturgeon, *Acipenser stellatus*, under conditions of regulated flow of the Volga River. Voprosy Ikhtiologii 31: 222–227 (in Russian, English translation J. Ichthyol. 31: 76–83).

Veshchev, P. V. 1991b. Characteristics of spawning stocks and reproduction of the Volga River stellate sturgeon, *Acipenser stellatus*, under new ecological conditions. Voprosy Ikhtiologii 31: 442–450 (in Russian, English translation J. Ichthyol. 31: 121–132).

Veshchev, P. V. & A. S. Novikova. 1987. Reproduction of sevruga, *Acipenser stellatus*, in the lower Volga River. Voprosy Ikhtiologii 27: 801–808 (in Russian, English translation J. Ichthyol. 28: 39–47).

Veshchev, P. V., A. D. Vlasenko & G. F. Dovgopol. 1992. Analysis of the commercial return coefficients of the stellate sturgeon, *Acipenser stellatus*. Voprosy Ikhtiologii 32: 78–836 (in Russian, English translation J. Ichthyol. 33: 56–62).

Vlasenko, A. D. 1979. Estimation of the size of Russian sturgeon stock replenishment from natural reproduction in the Volga River. pp. 38–40. *In*: Sturgeon Aquaculture in the USSR Inland Reservoirs, Astrakhan (in Russian).

Vlasenko, A. D. 1990. Formation of the number of acipenserids in the Caspian Sea. Rybnoe Khozyaistvo 7:53–56 (in Russian).

Vlasenko, A. D., A. V. Pavlov, L. I. Sokolov & V. P. Vasilev. 1989a. *Acipenser gueldenstaedti* Brandt, 1833. pp. 294–344. *In*: J. Holčík (ed.) The Freshwater Fishes of Europe, Vol. 1, Pt. II, General Introduction to Fishes, Acipenseriformes, AULA-Verlag, Wiesbaden.

Vlasenko, A. D., A. V. Pavlov & V. P. Vasilev. 1989b. *Acipenser persicus* Borodin, 1897. pp. 345–366. *In*: J. Holčík (ed.) The Freshwater Fishes of Europe, Vol. 1, Pt. II, General Introduction to Fishes, Acipenseriformes, AULA-Verlag, Wiesbaden.

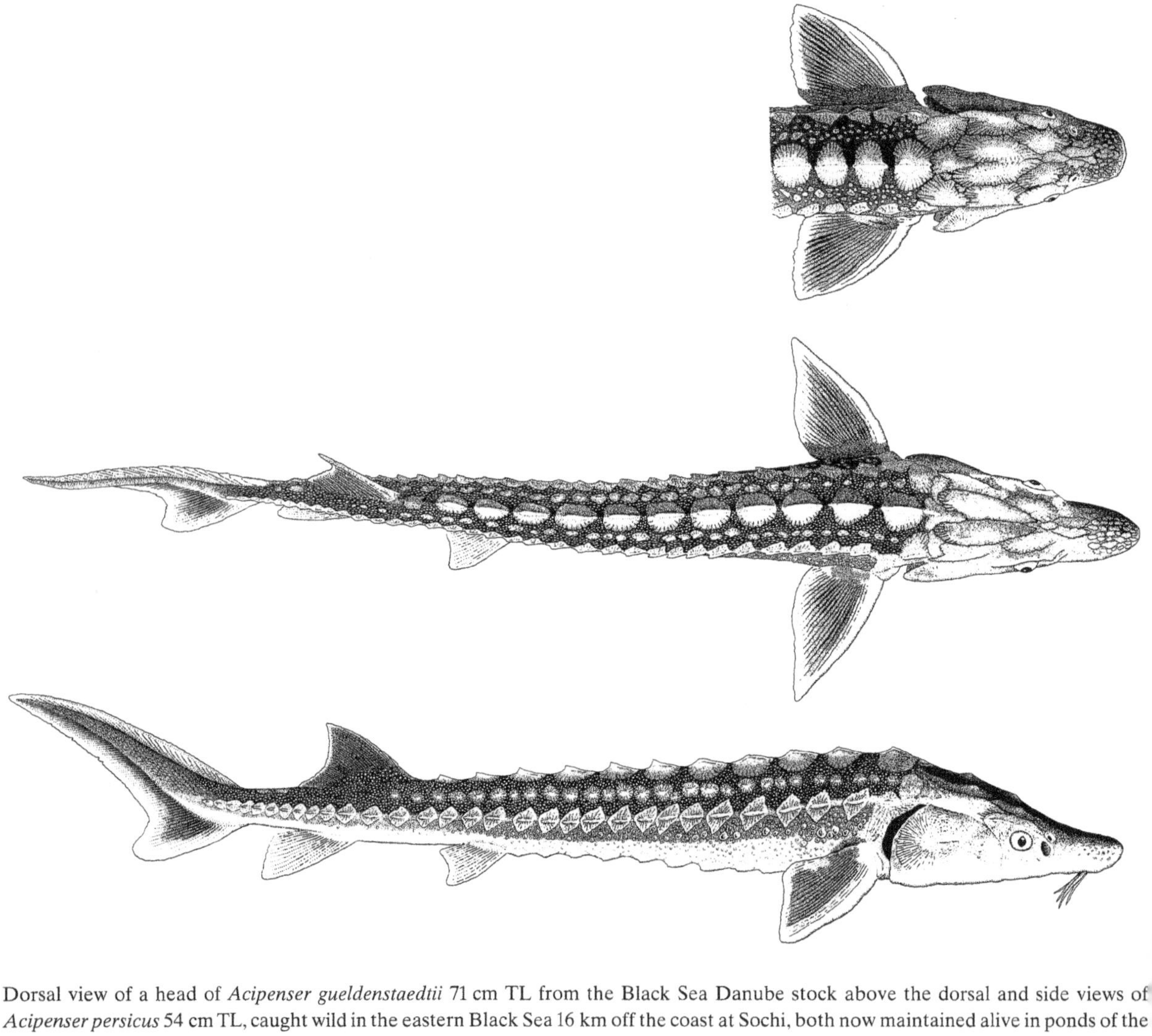

Dorsal view of a head of *Acipenser gueldenstaedtii* 71 cm TL from the Black Sea Danube stock above the dorsal and side views of *Acipenser persicus* 54 cm TL, caught wild in the eastern Black Sea 16 km off the coast at Sochi, both now maintained alive in ponds of the Propa-Gen International, Komadi, Hungary. Originals by Paul Vecsei, 1996.

Environmental Biology of Fishes **48**: 221–230, 1997.

Species structure, contemporary distribution and status of the Siberian sturgeon, *Acipenser baerii*

Georgii I. Ruban
A. N. Severtsov Institute of Ecology and Evolution, Russian Academy of Sciences, Moscow, 117071, Russia

Received 2.8. 1994 Accepted 12.3.1996

Key words: Ob River, Lena River, Yenisey River, Kolyma River, Indigirka River, Lake Baikal, taxonomy, subspecies, range, pollution, histological anomalies

Synopsis

A detailed analysis of the historical and contemporary range of the Siberian sturgeon, *Acipenser baerii*, shows that the contemporary status of its populations and forms can be described as threatened or endangered. Recently, the abundance of the nominal subspecies, *A. b. baerii*, which inhabits mainly the Ob River basin, decreased sharply. Due to construction of hydroelectric dams, up to 40% of the spawning grounds became inaccessible for migrating sturgeon of this subspecies. The Lake Baikal subspecies, *A. baerii baicalensis*, is extremely rare and was included in the Russian Federation Red Data Book in 1983. The abundance of the east Siberian subspecies, *A. baerii stenorrhynchus*, inhabiting the basins of the east Siberia rivers, has also significantly decreased during the last few years. Its range in the Yenisey and Lena River basins is gradually being reduced. Gametogenesis is anomalous in a high number of females from all populations of this subspecies (in the Kolyma and Indigirka river stocks 80–100% of females were anomalous in 1987–1989). These anomalies seem to be caused by high levels of water pollution.

Introduction

The range of Siberian sturgeon, *Acipenser baerii*, is very large. However, because the rivers it inhabits are not easily accessible to researchers (it lives in practically all large Siberian rivers, many of which are difficult to reach) and because of its low abundance, studies of the Siberian sturgeon are few. Recent intensive anthropogenic impact on the north Siberian water bodies (fisheries, construction of dams, and pollution) makes it necessary to evaluate the contemporary status of different forms and populations of this species. Moreover, until recently, the populations of *A. baerii* from the northeastern part of its range were little known and were not included in the recent review on this species (Sokolov & Vasiliev 1989). This paper offers a revision of the information on the range of *A. baerii*, including its taxonomy, and evaluation of the contemporary status of different subspecies and populations. The northeastern populations are the primary focus.

Historic and contemporary range of *Acipenser baerii*

The range of the Siberian sturgeon extends in the meridional direction from 73–74°n. l. (Lena River, Ob Bay) to 48–49°n. l. (Chernyi Irtysh and Selenga rivers) (Dryagin 1948a, Votinov et al. 1975), and in the longitudinal direction from the Ob River basin to the Kolyma River, up to 97° (Dryagin 1948a). The data on the range of *A. baerii* are included in Table 1 and Figure 1, and a detailed description of the range is given below.

In the Ob-Irtysh basin, the northern boundary of the Siberian sturgeon range is located at the Ob Bay near the Drovyanoy Cape (Dryagin 1948b, 1949). The sturgeon occurs in the Ob River within its entire length, 3680 km from the confluence of the Biya and Katun rivers (which form the Ob River), to its delta. Sturgeon migrate upstream to the Katun River for 50–70 km and were found in the mouth of the Biya River (Dryagin 1949, Petkevich et al. 1950). They were also caught in Lake Teletskoe (Berg 1948). In addition, the Siberian sturgeon inhabited the tributaries of the Ob River: the Chulym, Charysh, Nadym, and Irtysh rivers; sometimes they were found in the mouth of two other large tributaries, the Polui and Synya rivers (Dryagin 1948b, 1949). Sturgeon ranged throughout the entire length of the Irtysh River, up to Zaisan Lake and Chernyi Irtysh River (Sedelnikov 1910, Bogan 1939, Dryagin 1948b, 1949, Petkevich et al. 1950, Votinov 1963, Votinov et al. 1975). Siberian sturgeon were also caught in the Chinese part of the Irtysh River (Tchernyi Irtysch River) up to the Kren River. *Acipenser baerii* inhabited also the Irtysh River tributary, the Tobol River, and its tributaries, the Tura and Tavda rivers (Dryagin 1948b, 1949).

Table 1. Ranges of the subspecies of the Siberian sturgeon, *Acipenser baerii*.

Main rivers or lakes	Tributaries
a. ***Acipenser baerii baerii***	
(1) Ob River, from the Ob Bay (the northern boundary) up to the confluence of the Biya and Katun rivers (3680 km); Teletskoe Lake	Chulym, Charysh, Nadym, and Irtysh rivers
(2) Taz River (300 km upstream) and Taz Bay	Messo-Yakha, Anti-Payuta, Ader-Payuta rivers
(3) Pur River (100 km upstream)	
(4) Gyda and Yuribey rivers and Gyda Bay	
b. ***A. baerii stenorrhynchus***	
(1) Yenisey River, from the Yenisey Gulf (the northern boundary) up to the city of Krasnoyarsk (2450 km, upstream the contemporary southern boundary) or the Oznachennaya village (3100 km upstream, historical range): (a) migrating stock (throughout the whole river length); (b) non-migrating stock (historically, from the city of Igarka up to the Oznachennaya village; 2300 km upstream)	(a) Tuba and Abakan rivers (historically); (b) Angara, Podkamennaya and Nizhnyaya Tunguska rivers (non-migrating populations); (c) tributaries of the Nizhnyaya Tunguska River: the Kochechumo, Vivi, and Tutokchan rivers
(2) Pyasina River	Pyasina River basin lakes, Lama and Melkoe
(3) Khatanga River (up to the confluence of the Kheta and Kotui rivers)	Kheta River (350–460 km up from the confluence with the Kotuy River)
(4) Anabar River	(a) Vitim River (860 km upstream); (b) Olekma River (30 km upstream); (c) Aldan River and its tributary, the Amga River; (d) Vilyuy River and its tributaries, the Chona, Chirkuo, and Akhtaranda rivers
(5) Olenek River (1020 km upstreasm)	
(6) Indigirka River (850 km upstream)	
(7) Alazeya River	Bor-Yuryakh River
(8) Kolyma River (1500 km upstream)	Korkodon and Ozhogina rivers
c. ***A. baerii baicalensis***	
Baikal Lake and its trubutaries	(a) Selenga River (1000 km upstream) and its tributaries, Chikoy and Orhon rivers; (b) Tula and Delger-Muren rivers; (c) Barguzin River (300 km upstream); (d) Verkhnyaya Angara (100–150 km upstream) and Kichera rivers

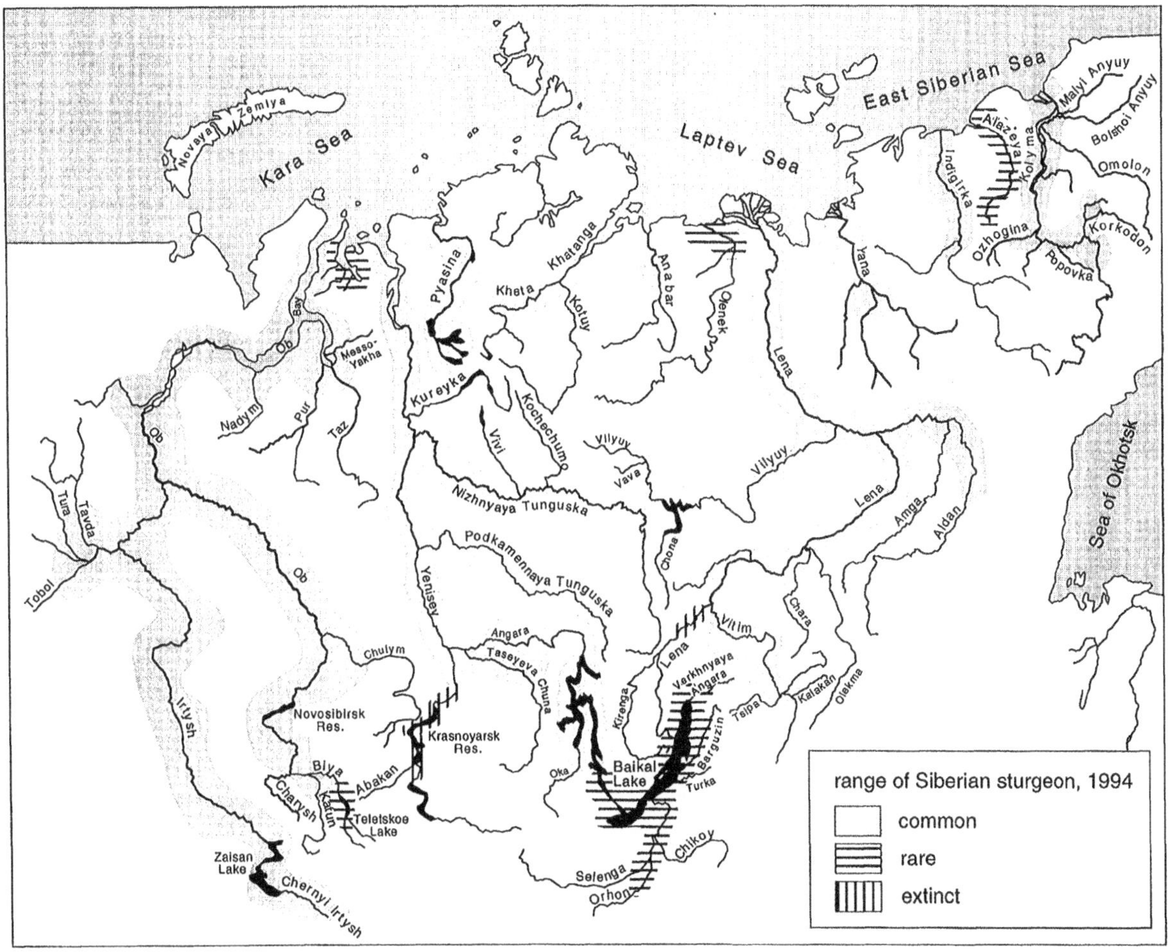

Figure 1. The range of the Siberian sturgeon, *Acipenser baerii*. Areas where sturgeon are still common, rare, or extinct are indicated by shading.

Sturgeon became cut off from approximately 40% of their spawning grounds in the Ob River after the Novosibirsk hydroelectric dam was constructed in 1957, and in the Irtysh River after building the Ust-Kamenogorsk hydroelectric dam in 1952 (Votinov et al. 1975). The situation became even worse after the Shulbinsk hydroelectric dam was constructed in 1985 on the Irtysh River downstream from the Ust-Kamenogorsk dam. The migrating form of the sturgeon cannot swim to the spawning grounds located upstream.

In the Taz River, the Siberian sturgeon was found upstream up to 300 km from the mouth. It also occurs throughout Taz Bay and in the mouths of Taz Bay tributaries: the Messo-Yakha, Anti-Payuta, and Ader-Payuta rivers. It inhabits the lower reaches of the Pur River up to 100 km from the mouth (Chupretov & Slepokurov 1979). Historically, Siberian sturgeon also occurred in the Gyda Bay and Gyda and Yuribey rivers (Burmakin 1941, Dryagin 1949).

The Shirokaya bay of the Yenisey Gulf is the northern boundary of the Siberian sturgeons range in the Yenisey River basin, while Oznachennoe village was considered as the southern boundary of the range before construction of the hydroelectric dam. The distance between these two geographic points is more than 3100 km (Podlesnyi 1955, 1958,

1963). It was assumed that the non-migrating form lived in the Yenisey River over 2300 km from the Oznachennoe village to the city of Igarka (Podlesnyi 1955). Sturgeon did not, apparently, migrate upstream from the Yartsevo village (1759 km from the river mouth) (Podlesnyi 1955) or the town of Yeniseysk (Podlesnyi 1958). They were caught rarely upstream from the city of Krasnoyarsk (2454 km from the Yenisey mouth) (Podlesnyi 1963). In the past, sturgeon migrated to the near-mouth parts of tributaries of the middle Yenisey River, the Tuba and Abakan rivers, but now, according to the information received from local people, it is completely absent upstream from the Krasnoyarsk hydroelectric dam (built in 1967). Therefore, the sturgeon range in the Yenisey River has been reduced, and its southern boundary moved to the north by 500–600 km.

In the large Yenisey tributaries, such as the Angara, Podkamennaya and Nizhnyaya Tunguska rivers, there were small populations of non-migrating sturgeon (Podlesnyi 1955, 1958); the upper boundary of its range in the two latter rivers has not been determined. The sturgeon occurred also in the lower tributaries of the Nizhnyaya Tunguska River, the Kochechum, Vivi, Kuchumdek, and Tutokchan rivers (Dryagin 1949). It was also found in the Kureika River in the 19th century (Tretyakov 1869).

In the Angara River the sturgeon inhabited an area from the mouth up to the estuary of the Belaya River (Dryagin 1949). Also, juveniles were caught several times near the Angara River source and near the city of Irkutsk (Yegorov 1941, 1961). The sturgeon occurred in the Angara River tributaries, the Taseyeva (and its tributary Chuna) and Oka rivers (Yegorov 1963).

In Lake Baikal, Siberian sturgeon were most abundant in the area near the delta of the Selenga River, as well as in the Barguzinskii and Chivyrkuiskii bays. Sturgeon moved from these main habitat regions along the coast in the shallow-water zone of the lake to the mouths of the large tributaries of the lake, from which they migrated into these tributaries. They were rare in the northern part of Lake Baikal at the mouth of the Verkhnyaya Angara and Kichera rivers. They migrated into the Selenga River up to 1000 km, including its tributaries Chikoy and Orhon rivers (Yegorov 1961), as well as into the Tula and Delger-Muren rivers (Sokolov & Shatunovsky 1983). Also, they migrated into the Barguzin River more than 300 km upstream, and into the Verkhnyaya Angara River, up to 100–150 km (Yegorov 1961). Siberian sturgeon were also found in smaller tributaries such as the Turka River.

In the Pyasina River Siberian sturgeon were rather rare (Dryagin 1949). They were not caught in the Pyasina Bay (Ostroumov 1937), but were found in the Pyasina basin lakes, the Lama and Melkoe (Belykh 1940, Logashov 1940).

In the Khatanga River basin sturgeon were found both in the estuary (Tretyakov 1869, Berg 1926) and upstream from the Khatanga River to the confluence of the Kheta and Kotuy rivers, which form the Khatanga River. In the Kheta River, they were caught over 460 km from the mouth to Volochanka village, although their main habitat in this river is from the mouth to 350 km upstream (Lukyanchikov 1967). Sometimes sturgeon were noticed in the flood plain lakes of the Kotuy River, where they migrated with the spring water. Sturgeon were found in the middle and low reaches of the Anabar River (Kirillov 1972).

In the Olenek River the sturgeon was a rare fish (Kirillov 1972). Usually it migrated upstream up to the Pur River mouth and, exceptionally, to the mouth of the Chemudakh brook located 1020 km from the river mouth.

Siberian sturgeon inhabited the Lena River basin up from the river mouth and the Neyelova Bay (Dryagin 1948a, 1949). During high-water years they migrated into the Tiksi Bay and coastal regions of the Bulunkan and Sogo bays (Kirillov 1950). Within the Lena River, sturgeon moved up to the Korshunovo village (Borisov 1928, Dryagin 1933, 1949, Karantonis et al. 1956, Dormidontov 1963, Kirillov 1972), and the total length of its range was approximately 3300 km in this river. The town of Kirensk is considered as the southern boundary of the sturgeon range in the Lena River. Earlier, in the 1840s, the southern boundary was at Makaryevskoe village located upstream from the town of Kirensk (Maak 1886). At the end of the 19th century, sturgeon also inhabited the right tributary of the Lena River, the Kirenga River (Borisov 1928). Therefore, the Siberian sturgeon range in the Lena River was reduced during the last 150 years by 300 km at the expense of the upper reaches of the river (Figure 1).

The Siberian sturgeon also occurred in some Lena River tributaries: the Vitim, Olekma, Aldan, and Vilyuy rivers (Dryagin 1949, Karantonis et al. 1956, Kirillov 1972). In the Vitim River they were found both in the lower (Kirillov 1972) and upper reaches from the Tsipa River mouth (860 km upstream from the Vitim mouth) (Kozhov 1950), and sometimes even farther upstream, up to the mouth of the Kalakan River (Kalashnikov 1978). Sturgeon were also found in the Olekma River (Dryagin 1949, Kirillov 1972). According to my own observation, sturgeon now migrate upstream the Olekma River approximately 30 km.

In the Aldan River, sturgeon inhabited the lower and middle reaches and they were especially abundant in the left tributary, the Amga River (Kirillov 1964). The upper boundaries of the range in the Aldan and Amga rivers are not known. According to information from local people, sturgeon occur in the Aldan River up to Ust-Mil village (Sokolov et al. 1986). In the Vilyuy River sturgeon were found from the mouth up to the Vava River, and they migrated into its tributaries, the Chona, Malaya and Bolshaya Botuobiya, Tyung, and Markha rivers (Kirillov 1972). Before the flow regulation in the late 1960s, it was most abundant in the Chona, Chirkuo, and Akhtaranda river mouths. After the impoundment of the Vilyuy water reservoir (the Vilyuyskaya dam was built in 1965), the sturgeon moved upstream (Kirillov & Solomonov 1979).

In the Yana River, sturgeon occurred from the delta up to the Verkhoyansk settlement (Kirillov 1972). In the Indigirka River they were caught upstream up to the Krest-Maior settlement (850 km upstream) and some individuals were found up to the Zashiversk settlement (Kirillov 1953, 1972). Only single specimens were caught in the Alazeya River; they migrated to the mouth of its right tributary, the Bor-Yuryakh River (Dryagin 1933).

In the Kolyma River, sturgeon occurred from the delta region up to the Seimchan settlement (Dryagin 1933), i.e., within 1500 km, but mainly up to the Verkhnekolymsk settlement (Dryagin 1948a). In 1988–1989 we found that sturgeon were abundant in the Kolyma River upstream from this town, near the mouth of the Popovka River (1085 km from the Kolyma River mouth) (Ruban & Akimova 1993). Sturgeon were caught only in two tributaries of the Kolyma River, the Korkodon and Ozhogina River (Dryagin 1948a).

Taxonomy and species structure of *A. baerii*

The first description of the Siberian sturgeon, *Acipenser baerii* Brandt 1869, was based on specimens caught in the Ob and Lena rivers (Brandt 1869). According to Chapters 31 and 33 of the International Code of Zoological Nomenclature (1988)[1], the initial spelling of this species name should be preserved and the widely used name *A. baeri* (for instance, Sokolov & Vasilev 1989) is incorrect (Ruban & Panaiotidi 1994). Nikolskii (1896) described the sturgeon from the Yenisey River as *A. stenorrhynchus*, and a form from Lake Baikal as a variety of this species, *A. stenorrhynchus* var. *baicalensis*. Later Menshikov (1947) reduced the species, *A. stenorrhynchus*, to the rank of subspecies, *A. baerii stenorrhynchus*. According to Chapter 45 of the International Code of Zoological Nomenclature, the Baikal variety of the sturgeon described by Nikolskii (1896) should be considered as a subspecies, *A. baerii baicalensis* Nikolsky, 1896 (Ruban & Panaiotidi 1994).

The taxonomic status of another subspecies, the Yakut sturgeon, *A. baerii chatys* Dryagin, 1948a, remained unclear until recently (see, for instance, Sokolov & Vasilev 1989). This form inhabits the rivers of Yakutiya[2] from the Khatanga River in the west to the Kolyma River in the east. Many ichthyologists did not consider this form as a subspecies (Nikolskii 1939, Berg 1948, Andriyashev 1954). Comparative analysis of morphological characters in large number of individuals from the Yenisey (*A. baerii ste-*

[1] International Code of Zoological Nomenclature. 1988, 3rd ed. Nauka Press, Leningrad. 202 pp. (in Russian).

[2] Yakutiya (now sometimes referred to as Saha) is a huge north central Siberian automomous republic within Russia which is roughly bounded by the Anabar River on the west, the Kolyma River on the east and the upper Lena and Aldan rivers at the south.

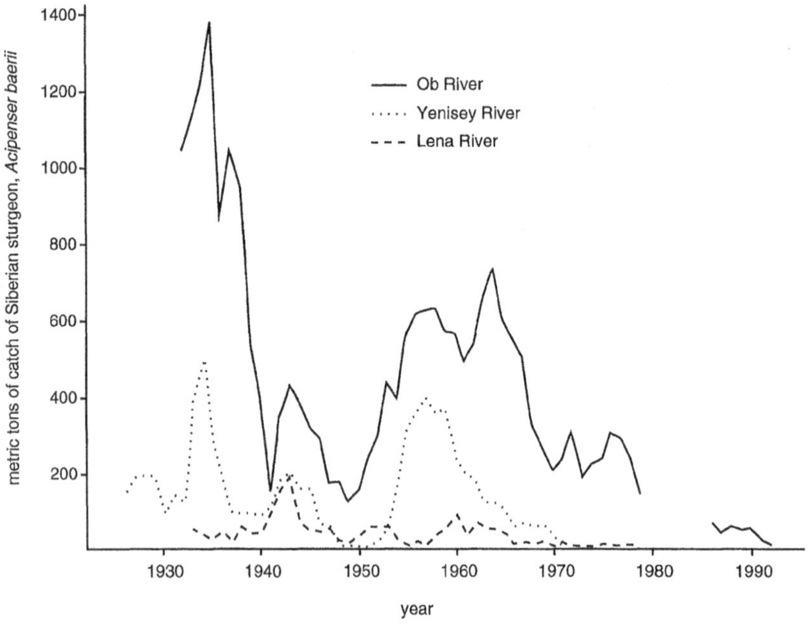

Figure 2. Siberian sturgeon catches in the 1980–1990s in the Ob, Yenisey, and Lena rivers. Data from Kirillov (1972), Yegorov (1988), Gundrizer et al. (1983), Votinov et al. (1975), Dryagin (1949) and own.

norrhynchus) and Lena (*A. baerii chatys*) rivers showed that differences between the fish from these populations do not reach a subspecific level (Ruban & Panaiotidi 1994). Comparison of our results with previous data obtained by the other authors (Menshikov 1947, Dryagin 1948b, Podlesnyi 1955, Sokolov & Vasilev 1989, Ruban 1989, 1992) points to clinal variation in a number of meristic characters of Siberian sturgeon (Ruban & Panaiotidi 1994). Usually it is considered that when the geographic variation within a species is clinal, it is not appropriate to name such forms as a subspecies (Mayr 1969, Holcik & Jedlicka 1994). Therefore, it is not right to consider *A. baerii chatys* as a subspecies of the Siberian sturgeon and this form should be included in the subspecies *A. baerii stenorrhynchus* described by Menshikov (1947) for the fish from populations of the Yenisey, Lena, and Kolyma rivers (Ruban & Panaiotidi 1994). Additionally, because of the absence of sturgeon specimens from the Ob River basin in museum collections, the problem of *A. b. stenorrhynchus* also cannot be solved definitively. At present one can consider the Siberian sturgeon to consist of three subspecies: the nominal *A. baerii baerii* Brandt, 1869 from the Ob River basin, *A. baerii baicalensis* Nikolsky, 1896 from the Lake Baikal basin, and *A. baerii stenorrhynchus* Nikolsky, 1896 from other Siberian waters.

Contemporary status of *A. baerii*

As described above, the range of the Siberian sturgeon is constantly being reduced; the number of individuals in each population is also decreasing. Depletion is caused by three main factors: the elimination of spawning grounds after the construction of dams, overfishing, and water pollution.

Fishing

Although the catch of Siberian sturgeon has always been relatively small and never exceeded 1769 metric tons per year, its impact on the status of different populations was devastating. The highest catches were recorded in the Ob-Irtysh and Yenisey rivers basins: 1401 and 504 metric tons per year, respectively (Figure 2). In other Siberian rivers the sturgeon catch was much smaller and there was no specialized sturgeon fishery. The highest catch in the Lena River was 189.9 metric tons in 1943. The populations of two sturgeon subspecies, the nominal *A. baerii baerii* of the Ob River basin and the East-Siberian *A. baerii stenorrhynchus* of the Yenisey River, were especially affected by fishing. The third subspecies, *A. baerii baicalensis*, is at present extremely rare and has been included in the Red Data Book of the Russian Federation (Kolosov 1983).

Histological data on the effect of pollution

Disturbances in the reproductive system are often correlated with worsening conditions of the environment, especially high levels of chemical pollution. Our long-term histological studies (methods in Roskin & Levinson 1957) on the development and functioning of the reproductive system in Siberian sturgeon inhabiting East Siberia showed abnormal gametogenesis in individuals from the Lena, Indigirka, Kolyma, and Yenisey rivers.

Between 1964 and 1977, only single cases of degeneration of oocytes during the period of cytoplasmic growth were reported in females of the Lena River population (Akimova & Ruban 1993). But in 1986 the number of females with such defects was close to 59% (23 of 39 females studied). New types of anomalies appeared in these females: some of them had many unspawned eggs, and amitotic division of sex cells and degeneration of the nuclear membrane in oocytes during vitellogenesis were characteristic for many females (Akimova & Ruban 1995).

Similarly, but in a shorter period of time and to a greater extent, the state of the reproductive system in females of the Indigirka River population also changed (Ruban & Akimova 1991, Akimova & Ruban 1993). The percent of females of different ages with partially degenerated oocytes during the period of cytoplasmic growth increased within four years (1984–1987) from 77 to 100%. In some females degeneration occurred repeatedly. The growth period of the oocytes was characterized by asynchronous development (which is uncommon for sturgeons), decreases in the strength of their membranes, and by degeneration of 15% of the germ cells. Simultaneously, the remaining oocytes undergoing cytoplasmic growth continued to degenerate.

In the Kolyma River population up to 81–83% of all females had various defects in germ cells in 1988 and 1989 (Akimova & Ruban 1993, Ruban & Akimova 1993). Just as in the Indigirka population, degeneration of some oocytes was observed during the period of cytoplasmic growth, in some females repeatedly, and amitotic divisions of oocytes were also found. Degeneration of oocytes at the time of growth was higher (up to 20% of oocytes) in females of the Kolyma River population than in females of other populations. Defects in oocyte membranes resulted in 50–100% of mature eggs having a deformed, angular shape. The gelatinous envelopes contained inclusions of uncertain content, which were absent in egg envelopes of sturgeon from the Lena and Indigirka populations. In some females of the Kolyma River population, destruction of the nuclear membrane was observed in still immature oocytes. This destruction caused the disintegration of the nuclei.

Some oocytes had evidently degenerated during the period of cytoplasmic growth in females from the Yenisey River population. Also an amitotic division of oocytes, which caused the degeneration of the dividing cells, was observed. In some females the oocyte envelopes became locally thinner during the period of growth in almost 50% of oocytes. We also observed extensive deformation of oocytes, appearance of cavities with foreign body inclusions under the oocyte envelopes and among yolk granules, and mass resorption of mature eggs in some females. Unspawned eggs were present in gonads of some females after spawning. In general, the discovered abnormalities of gametogenesis in the Yenisey population were not so numerous as in the Kolyma or Indigirka River populations. This enabled us to declare this phenomenon as an initial state of pathological development and functioning of the reproductive system (Akimova et al. 1995). Disturbances in gametogenesis of females from the Yenisey River are probably caused by a high level of water pollution, especially, by pesticides (Tchuprov 1986).

Acknowledgements

Vadim Birstein, John Waldman and Robert Boyle invited me to the International Conference on Sturgeon Biodiversity and Conservation. I am thankful to Vadim Birstein and anonymous reviewers for their editorial notes on the draft of the manuscript and to William E. Bemis for drawing the map and figure.

References cited

Akimova, N. V. & G. I. Ruban. 1993. The condition of the reproductive system of the Siberian sturgeon, *Acipenser baeri*, as a bioindicator. Voprosy Ikhtiologii 32: 102–109 (in Russian, English translation: J. Ichthyol. 33: 15–24).

Akimova, N. V. & G. I. Ruban. 1995. Disturbances of Siberian sturgeons generative system resulted from anthropogenic influence. pp. 74–79. *In*: A. D. Gershanovich & T. I. J. Smith (ed.) Proceedings of the International Symposium on Sturgeons, 6–11 September 1993, Moscow-Kostroma-Moscow, VNIRO Publishing, Moscow.

Akimova, N. V., G. I. Ruban & Yu. V. Mikhalyev. 1995. Analysis of the state of reproductive system of the Siberian sturgeon in Central Siberia. pp. 93–98. *In*: I. A. Shilov (ed.) Northern Ecosystems: Structure, Adaptations, Stability, Izdatelstvo Moskovskogo Gosudarstvennogo Universiteta, Moscow (in Russian).

Andriyashev, A. P. 1954. Fishes of the northern seas of the USSR. Izdatelstvo Akademii Nauk USSR, Moscow-Leningrad. 566 pp. (in Russian).

Belykh, F. I. 1940. Lake Lama and its fishery use. Trudy Nauchno-Issledovatelskogo Instituta Polyarnogo Zemledeliya, Zhivotnovodstva i Promyslovogo Khozyaistva, Seriya Promyslovoe Khozyaistvo 11: 73–100 (in Russian).

Berg, L. S. 1926. Fishes of the Khatanga Basin. Materialy Komissii AN SSSR po Izucheniyu Yakutskoi ASSR 2: 1–22 (in Russian).

Berg, L. S. 1948. The freshwater fishes of the USSR and adjacent countries, Vol. 1, Part 1. Akademia Nauk SSSR, Moscow & Leningrad (in Russian, English translation published by Israel Program for Scientific Translations, Jerusalem. 505 pp.)

Bogan, F. E. 1939. Contribution to the biology of the Siberian sturgeon (*Acipenser baeri* Brandt) of the Irtysh River basin. Uchenye Zapiski Permskogo Gosudarstvennogo Universiteta 3: 145–163 (in Russian).

Borisov, P. G. 1928. Fishes of the Lena River. Trudy Komissii AN SSSR po Izucheniyu Yakutskoi Respubliki 9: 1–181 (in Russian).

Brandt, J. G. 1869/1870. Einige Worte ber die europisch-asiatischen Strarten (Sturionides). Bull. Acad. Imper. Sci. St.-Petersbourg 14: 171–175.

Burmakin, E. V. 1941. Some little commercial and noncommercial fishes of the Gyda Bay system. Trudy Nauchno-Issledovatelskogo Instituta Polyarnogo Zemledeliya, Zhivotnovodstva I Promyslovogo Khozyaistva, Seriya Promyslovoe Khozyaistvo 15: 149–158 (in Russian).

Chupretov, V. M. & V. A. Slepokurov. 1979. On winter distribution of the Siberian sturgeon in the Ob and Taz bays. pp. 270–271. *In*: V. I. Lukyanenko (ed.) Sturgeon Fishery in Inland Water Bodies of the USSR, Abstracts of Papers at the 2nd All-Union Conference, Astrakhan (in Russian).

Dormidontov, A. S. 1963. Fishery utilization of the Lena River sturgeon. pp. 182–187. *In:* Sturgeon Fishery in the Water Bodies of the USSR, Izdatelstvo Akademii Nauk SSSR, Moscow (in Russian).

Dryagin, P. A. 1933. Fish resources of Yakutiya. Trudy Soveta po Izucheniyu Proisvoditelnykh Sil 5: 3–94 (in Russian).

Dryagin, P. A. 1947. Sturgeon catches in the water bodies of Siberia. Rybnoe Khozyaistvo 1: 34–38 (in Russian).

Dryagin, P. A. 1948a. On some morphological and biological distinctions between sturgeon occurring in the rivers of Yakutiya and the Siberian sturgeon, *Acipenser baerii* Brandt. Zoologicheskii Zhurnal 27: 525–534 (in Russian).

Dryagin, P. A. 1948b. Commercial fishes of the Ob-Irtysh River Basin. Izvestiya Vsesouznogo Instituta Ozernogo i Rechnogo Rybnogo Khozyaistva 25(2): 3–104 (in Russian).

Dryagin, P. A.1949. Biology of the Siberian sturgeon, its reserves and rational utilization. Izvestiya Vsesouznogo Instituta Ozernogo i Rechnogo Rybnogo Khozyaistva 29: 3–51 (in Russian).

Gundrizer, A. N., A. G. Egorov, V. G. Afanaseva, S. A. Enshina, Yu. V. Mikhalev, R. I. Setsko & A. A. Khakimullin. 1983. Perspectives of reproduction of Siberian sturgeons. pp. 241–253. *In:* I.A. Barannikova & L. S. Berditchevsky (ed.) Biological Foundations of Sturgeon Management, Nauka, Moscow (in Russian).

Holčík, J. & L. Jedlička. 1994. Geographical variation of some taxonomically important characters in fishes: the case of the bitterling *Rhodeus sericeus*. Env. Biol. Fish. 41: 147–170.

Kalashnikov, Yu. E. 1978. Fishes of the Vitim River Basin. Nauka Press, Novosibirsk. 190 pp. (in Russian).

Karantonis, F. E., F. N. Kirillov & F. B. Mukhomediyarov. 1956. Fishes of the middle Lena reaches. Trudy Instituta Biologii Yakutskogo Filiala AN USSR 2: 3–144 (in Russian).

Kirillov, F. N. 1950. Fishes of the Tiksi Bay. Uchenye Zapiski Tomskogo Universiteta 15: 155–162 (in Russian).

Kirillov, F. N. 1953. Fishes of the Indigirka River and their fishery. Candidate of Biol. Sci. Dissertation Thesis, Yakutsk. 13 pp. (in Russian).

Kirillov, F. N. 1964. Species composition of fishes of the Aldan River. pp. 73–82. *In*: Vertebrate Animals of Yakutiya, Yakutsk (in Russian).

Kirillov, F. N. 1972. Fishes of Yakutiya. Nauka Press, Moscow. 360 pp. (in Russian).

Kirillov, F. N. & N. G. Solomonov (ed.) 1979. Biology of the Vilyui Water Reservoir. Nauka Press, Novosibirsk. 270 pp. (in Russian).

Kolosov, A. M. (ed.) 1983. RSFSR [Russian Federation] Red Data Book. Animals. Rosselkhozizdat, Moscow. 455 pp. (in Russian).

Kozhov, M. M. 1950. Fresh waters of East Siberia. OGIZ Press, Irkutsk. 367 pp. (in Russian).

Logashov, M. V. 1940. Lake Melkoe and its utilization for fishery. Trudy Nauchno-Issledovatelskogo Instituta Polyarnogo Zemledeliya, Zhivotnovodstva i Promyslovogo Khozyaistva, Seriya Promyslovoe Khozyaistvo 11: 7–71 (in Russian).

Lukyanchikov, F. V. 1967. Fishes of the Khatanga River system. Trudy Krasnoyarskogo Otdeleniya SibNIIRKh 9: 11–93 (in Russian).

Maak, R. U. 1886. Vilyui Region of the Yakutsk District, Part 2. St.-Petersburg. 366 pp. (in Russian).

Mayr, E. 1969. Principles of systematic zoology. McGraw-Hill, New York. 428 pp.

Menshikov, M. I. 1947. On geographical variation of the Siberian sturgeon, *Acipenser baerii* Brandt. Doklady AN USSR 55: 371–374 (in Russian).

Nikolskii, A. M. 1896. Siberian sturgeon (*Acipenser stenorrhynchus* sp. nov.). Ezhegodnik Zoologicheskogo Museya Akademii Nauk 1: 400–405 (in Russian).

Nikolskii, G. V. 1939. Materials on taxonomy of the Siberian sturgeon, *Acipenser baerii* Brandt. Sbornik Trudov Gosudarstvennogo Zoologicheskogo Muzeya pri MGU 5: 136–148 (in Russian).

Ostroumov, N. A. 1937. Fishes and fishery of the Pyasina River. Trudy Polyarnoi Komissii AN USSR 30: 1–115 (in Russian).

Petkevich, A. N., V. N. Bashmakov & A. Ya. Bashmakova. 1950. Sturgeons of the middle and upper reaches of the Ob River. Trudy Barabinskogo Otdeleniya Vsesoyuznogo Nauchno-Issledovatelskogo Otdeleniya Instituta Ozernogo i Rechnogo Rybnogo Khozyaistva 4: 3–54 (in Russian).

Podlesnyi, A. V. 1955. Sturgeon (*Acipenser baerii stenorrhynchus* A. Nikolsky) of the Yenisey River. Voprosy Ikhtiologii 4: 21–40 (in Russian).

Podlesnyi, A. V. 1958. Fishes of the Yenisey River, conditions of their life and utilization of them. Trudy Vsesoyuznogo Nauchno-Issledovatelskogo Instituta Ozernogo i Rechnogo Rybnogo Khozyaistva 44: 97–178 (in Russian).

Podlesnyi, A. V. 1963. State of reserves of sturgeons in the Yenisey River and ways to increase them. pp. 200–205. *In*: E. N. Pavlovskii (ed.) Sturgeon Fishery in Water Bodies of the USSR, Izdatelstvo Akademii Nauk USSR, Moscow (in Russian).

Roskin, G. I. & L. V. Levinson. 1957. Microscopic techniques. Sovetskaya Nauka, Moscow. 468 pp. (in Russian).

Ruban, G. I. 1989. Clinal variation of morphological characters in the Siberian sturgeon, *Acipenser baeri*, of the Lena River basin. Voprosy Ikhtiologii 29: 48–55 (in Russian, English translation: J. Ichthyol. 29: 48–55).

Ruban, G. I. 1992. Plasticity of development in natural and experimental populations of Siberian sturgeon, *Acipenser baeri* Brandt. Acta Zool. Fennica 191: 43–46.

Ruban, G. I. & N. V. Akimova. 1991. Notes on the ecology of the Siberian sturgeon, *Acipenser baeri*, in the Indigirka River. Voprosy Ikhtiologii 31: 596–605 (in Russian, English translation: J. Ichthyol. 31: 118–129).

Ruban, G. I. & N. V. Akimova. 1993. Ecological charactetristics of Siberian sturgeon, *Acipenser baeri*, from the Kolyma River. Voprosy Ikhtiologii 33: 84–92 (in Russian, English translation: J. Ichthyol. 33: 66–80).

Ruban, G. I. & A. I. Panaiotidi. 1994. Comparative morphological analysis of subspecies of the Siberian sturgeon, *Acipenser baerii stenorrhynchus* A. Nikolsky and *Acipenser baerii chatys* Drjagin (Acipenseriformes, Acipenseridae), in the Yenisey and Lena rivers. Voprosy Ikhtiologii 34: 469–478 (in Russian, English translation: J. Ichthyol. 34: 58–71).

Sedelnikov, A. K. 1910. Lake Zaisan. Zapiski Zapadno-Sibirskogo Otdela Imperatorskogo Russkogo Geograficheskogo Obshchestva 35: 1–253 (in Russian).

Sokolov, L. I., B. V. Koshelev, O. V. Khalatyan, G. I. Ruban, N. V. Akimova & E. L. Sokolova. 1986. Ecomorphological characteristics of the Siberian sturgeon – *Acipenser baeri*, from the Aldan River. Voprosy Ikhtiologii 26: 55–64 (in Russian, English translation: J. Ichthyol. 26: 55–64).

Sokolov, L. I. & V. P. Vasilev. 1989. *Acipenser baeri* Brandt, 1869. pp. 263–284. *In*: J. Holčík (ed.) The Freshwater Fishes of Europe, Vol. 1, Pt. II, General Introduction to Fishes, Acipenseriformes, AULA-Verlag, Wiesbaden.

Sokolov, V. E. & M. I. Shatunovsky (ed.) 1983. Fishes of the Mongolian Peoples Republic. Nauka Press, Moscow. 277 pp. (in Russian).

Tchuprov, S. M. 1986. Ecological and morphological characteristics of fishes from eastern Siberian reservoirs (Krasnoyarskoe and Sayanskoe reservoirs). Candidate Biol. Sci. Dissertation Thesis, Moscow. 27 pp. (in Russian).

Tretyakov, P. 1869. Turukhanskii District. Zapiski Imperatorskogo Russkogo Obshchestva po Obshchei Geografii 2: 215–530 (in Russian).

Votinov, N. P. 1963. Biological foundations of artificial reproduction of the Ob River sturgeon. Trudy Ob-Tazovskogo Otdeleniya Gosudarstvennogo Nauchno-Issledovatelskogo Instituta Ozernogo i Rechnogo Rybnogo Khozyaistva, Novaya Seria 3: 5–102 (in Russian).

Votinov, N. P., V. N. Zlokazov, V. P. Kasyanov & R. I. Setsko. 1975. Status of sturgeon reserves in the rivers of Siberia and measures aimed to increase these reserves. Sredneuralskoe Knizhnoe Izdatelstvo, Sverdlovsk. 94 pp. (in Russian).

Yegorov, A. G. 1941. Baikal sturgeon. Rybnoe Khozyaistvo 5: 22–23 (in Russian).

Yegorov, A. G. 1961. Baikal sturgeon, *Acipenser baerii stenorrhynchus* natio *baicalensis* A. Nikolsky (taxonomy, biology, fishery, and reproduction of reserves), Ulan-Ude (USSR). 121 pp. (in Russian).

Yegorov, A. G. 1963. State and perspectives of development of the sturgeon fishery in the Baikal Lake and Angara River sys-

tem. pp. 188–195. *In*: Sturgeon Fishery in Water Bodies of the USSR, Izdatelstvo Akademii Nauk USSR, Moscow (in Russian).
Yegorov, A. G. 1988. Perspectives of reproduction of sturgeons in water bodies of East Siberia and Far East. pp. 7–17. *In*: P. Ya. Tugarina (ed.) Studies of Fishes of East Siberia, Irkutsk State University, Irkutsk (in Russian).

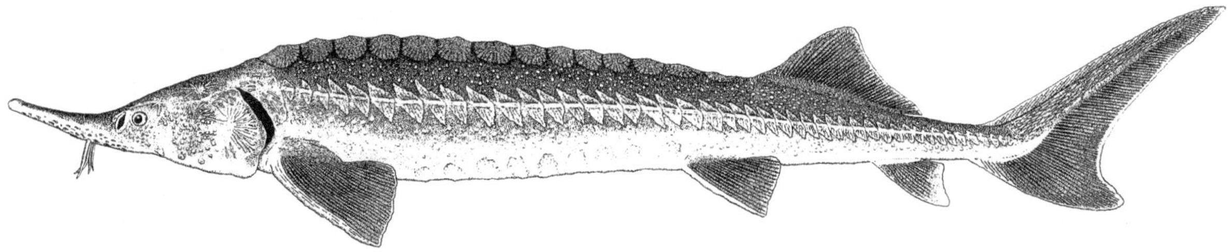

Acipenser baerii 47 cm TL offspring from Lena River parents bred at the Fish Culture Research Institute, Szarvas, Hungary. The long rostrum is characteristic of specimens from the Lena and Yenisei rivers. Original by Paul Vecsei, 1996.

Environmental Biology of Fishes **48**: 231–239, 1997.

Endemic sturgeons of the Amur River: kaluga, *Huso dauricus*, and Amur sturgeon, *Acipenser schrenckii*

Mikhail L. Krykhtin & Victor G. Svirskii
Pacific Research Institute of Fisheries and Oceanography, 4 Shevchenko Alley, Vladivostok 690600, Russia

Received 2.8.1994 Accepted 8.3.1996

Key words: anadromy, population, hybridization, endangered species, poaching

Synopsis

General biological characteristics and the contemporary status of the kaluga, *Huso dauricus*, and Amur sturgeon, *Acipenser schrenckii*, are described. Both inhabit the Amur River basin. Kaluga is the largest freshwater fish in this river system reaching more than 5.6 m in length and more than 1000 kg in weight. We recognize four populations of kaluga: the first is from the estuary of the Amur River and coastal brackish waters of the Sea of Okhotsk and Sea of Japan, the second is from the lower Amur River, the third is from the middle-Amur, and the fourth occurs in lower reaches of the Zeya and Bureya rivers. Freshwater and brackish water morphs exist in the estuary population, with the freshwater morph predominating in number. The number of individuals in the lower Amur River population at age 2 or greater was recently estimated to be 40 000, and in the middle Amur, 30 000. The population will continue to decline because of rampant overfishing. The Amur sturgeon is represented in the Amur River basin by two morphs: brown and gray. Brown morphs occur in the middle and lower parts of the Amur River; they grow more slowly than the gray ones. Today, the lower Amur River population of Amur sturgeon is made up of 95 000 fish at age 2 or greater and is approximately half as large as the population in the middle Amur River. Populations of kaluga and Amur sturgeon in the Zeya and Bureya rivers are extremely small and on the verge of extinction.

Introduction

The Amur River in the Russian far east is home to four species of the family Acipenseridae, kaluga *Huso dauricus*, Amur sturgeon *Acipenser schrenckii*, Sakhalin sturgeon *A. mikadoi*, and sterlet *A. ruthenus*. Only kaluga and Amur sturgeon are endemic to this river (Berg 1948, Nikolskii 1956). Sakhalin sturgeon was recorded in the Sea of Okhotsk from the Amur River estuary to northern Japan and the Korean Peninsular (Berg 1948, Artyukhin & Andronov 1990, Shilin 1995), and five to ten Sakhalin sturgeon are caught annually in the Amur River estuary. Sterlet was introduced into the Amur River from the Ob River in 1956–1959, and since then only a few sexually mature individuals have been caught in the Amur.

Historically, kaluga and Amur sturgeon were commercial species. In 1891, 595 metric tons of kaluga and 607 metric tons of Amur sturgeon were caught in the Amur River, constituting 42.5% of the total catch of all fishes in the Amur River that year (Kryukov 1894). As in many other places, the survival of sturgeon populations in the Amur River became problematic after the turn of the 20th century. By 1909, the catch of kaluga decreased to less than one third and that of the Amur sturgeon to about one fifth of 1900 levels. From 1915 until 1917, Rus-

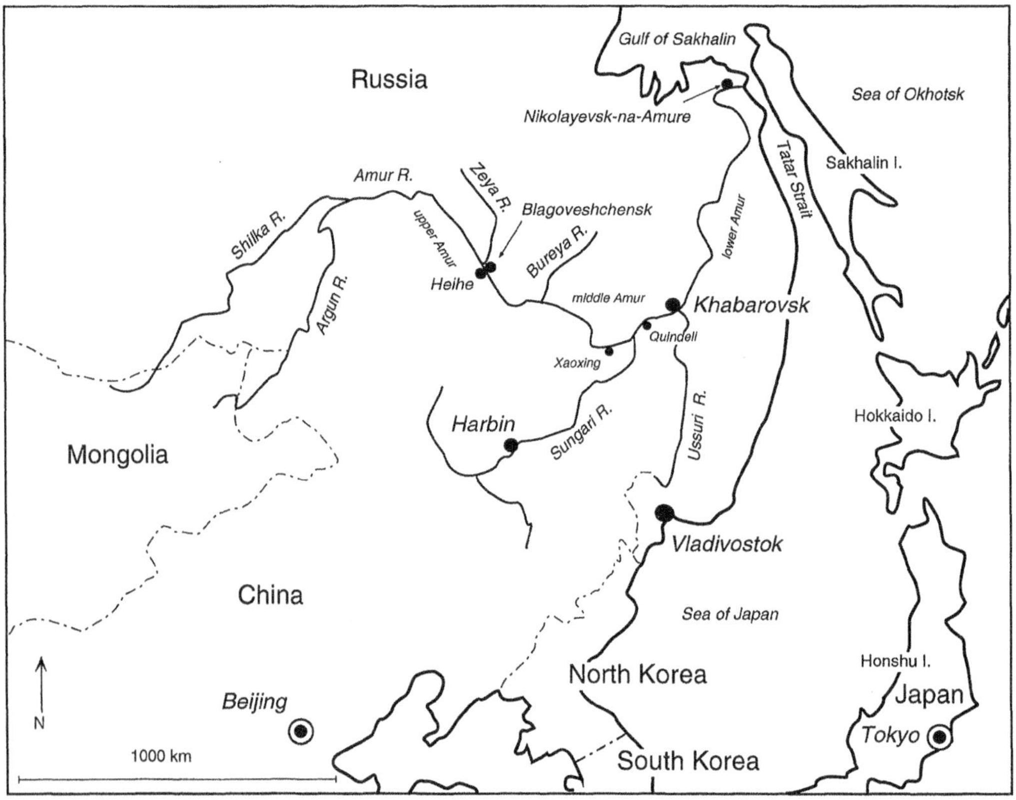

Figure 1. Map of the Amur River System showing the upper, middle and lower reaches of the Amur River. The Ussuri River is known as the Wusulijang, and the Sungari River is known as the Songhuajing in Chinese. The upper Amur is a portion of the river above Blagoveshchensk, the middle Amur extends from Blagoveshchensk to Khabarovsk, and the lower Amur extends to the mouth. Only tributaries currently or historically important for acipenserids are shown.

sian authorities prohibited fishing of sturgeons during the spawning period. In 1923, authorities issued a ban on the catch of sturgeons across the USSR; this ban was withdrawn in 1930. Three years after the second world war in 1948, 61 metric tons of kaluga and 4.2 metric tons of Amur sturgeon were caught, i.e., the catch of kaluga was one tenth of the 1891 level, whereas that of the Amur sturgeon was less than one-hundredth of the 1891 level (Svirskii 1971). In 1958, USSR authorities banned the catch of kaluga and Amur sturgeon. A ban is formally still in effect. Both species are on the IUCN Red List (1994), with *H. dauricus* considered rare and *A. schrenckii* vulnerable.

Although aspects of morphology of kaluga and Amur sturgeon have been described before (Berg 1948, Nikolsky 1956), the biology and population structure of these species have not been studied extensively (but see Wei et al. 1997 this volume). This paper describes general characteristics and the contemporary status of both species.

Materials and methods

Fishes were collected during government surveys of the Amur River estuary, where 30–40 tons of kaluga were caught annually, as well as in the upper and middle reaches of the Amur River. Data on development were obtained at the hatcheries located in the lower (1962–1968) and middle (1992–1993) reaches of the Amur River. Live fish were observed at the Vladivostok Oceanarium. Standard methods

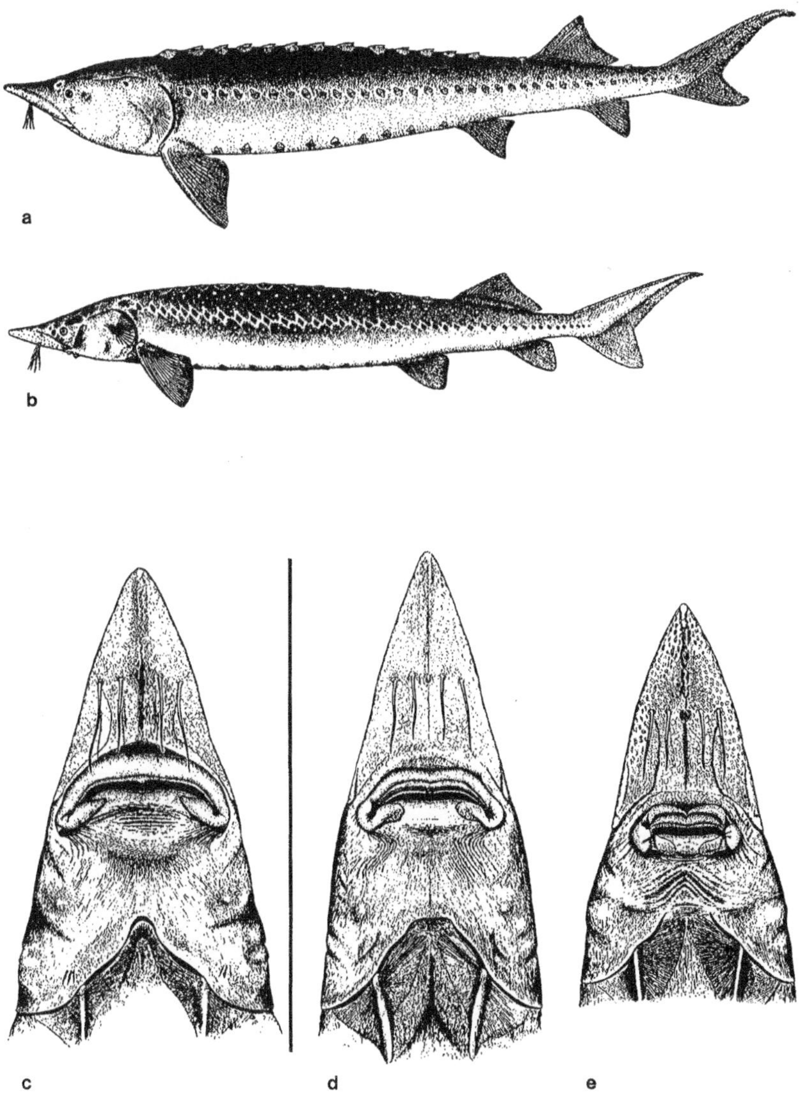

Figure 2. Acipenserid fishes of the Amur River: a - kaluga, *Huso dauricus*, b - Amur sturgeon, *Acipenser schrenckii*, c - ventral surface of the head of kaluga, *H. dauricus*, d - ventral surface of the head of a kaluga/Amur sturgeon hybrid, and e - ventral surface of the head of Amur sturgeon, *A. schrenckii*. Note difference in head and mouth shape of *H. dauricus* and *A. schrenckii*.

of measuring morphological characters and determining age were used (Romeis 1954, Pravdin 1966).

General characteristics of the Amur River

The Amur River is formed by the confluence of the Argun and Shilka rivers (Figure 1). It enters into the Amur estuary of the Tatar Strait. The Amur estuary is 48 km long and 16 km wide at the mouth of the river. The Amur River is 4092 km long if its longest tributary, the Shilka River, is included. The total size of the basin is 1 856 000 km^2. For much of its length, it forms the border between Russia and China.

According to the structure of its valley, bed and flow characteristics, the Amur River can be divided into three parts (Figure 1). The upper reach of the

Amur extends down to the city of Blagoveshchensk (upper Amur, 883 km); the middle reach continues down to the mouth of the Ussuri River, opposite the city of Khabarovsk (middle Amur, 975 km); and the lower reach continues down to the estuary (lower Amur, 966 km). Its hydrology is characterized by spring floods. The difference between the highest and lowest (winter) water levels is different in various parts of the river: about 10 m in the upper Amur, 11 m in the middle Amur, 7–8 m in the lower Amur, and up to 3 m near the estuary. The current velocity ranges from 0.5 to 2.0 m sec^{-1}. During the Quaternary, there were times when the Amur River bed was either elongated 1.5–2.0 times, or when sea water penetrated deeply into the continent, reaching the area of Khabarovsk. Therefore, a seawater body with a different gradient of salinity existed at the location of the contemporary lower Amur and estuary. These environmental changes evidently affected sturgeon populations during their history in this region (Svirskii 1968).

Kaluga sturgeon

Biology and population structure

Kaluga is the largest freshwater fish in the Amur River basin, reaching more than 5.6 m in length, more than 1000 kg in weight, and an age of more than 80 years (Figure 2a, c). It inhabits the Amur River Basin from the estuary to its upper reaches, including several large tributaries and lakes (Nikolskii 1956). Young kaluga have been caught at the sea during summer in coastal waters of the Sea of Okhotsk (Kostarev & Tyurnin 1970), near the northeastern part of Sakhalin Island (Gritsenko & Kostyunin 1979), in the northern part of the Tatar Strait (Krykhtin 1984a,b), and in the Sea of Japan near the islands of Hokkaido (Amaoka & Nakaya 1975) and Honshu (Honma & Itano 1994). During the last decade, the number of young fish increased considerably in coastal waters of the northern part of the Tatar Strait and in the southwestern part of the Gulf of Sakhalin.

We recognize four populations of kaluga in the Amur River basin. The first lives in the estuary and coastal brackish waters of the Sea of Okhotsk and Sea of Japan, the second lives in the lower Amur, the third in the middle Amur, and the fourth in the lower reaches of the Zeya and Bureya rivers. We know more about the estuary population than we do about the other three, and most of our observations in this paper concern this group. Two ecological morphs exist in the estuary population, which we term the freshwater and brackish water forms (Lukyanenko et al. 1979, Krykhtin 1985). The freshwater morph predominates, making up 75–80% of the estuary population. They feed only in fresh water. The brackish water form spends winter in the river or estuary and, in late June–early July, migrates downstream to the brackish water of the estuary and northern part of the Tatar Strait, as well as to the southwestern part of the Sakhalin gulf, to salinities of 12–16‰.

In autumn, when the salinity of the estuary increases, brackish water kaluga return to the river, where they overwinter together with freshwater kaluga. If storms rapidly fill the estuary with cold sea water from the Sea of Okhotsk in late November–early December, then some of the brackish water individuals, mostly juveniles, cannot reach the fresh water zones and die in the sea water at salinities of about 29–30‰ and water temperatures below 0°C (Krykhtin 1984a).

Kaluga consume mostly invertebrates in the first year of life, later switching to juveniles of pelagic fishes such as chum salmon, *Oncorhynchus keta*. At age three to four years, kaluga start to feed on adult fishes. In estuaries and coastal sea regions kaluga catch saffron cod, *Eleginus gracilis*, and ocean perch, *Sebastes alutus*. Cannibalism is frequent. Kaluga do not feed during winter nor do broodstock eat during spawning migrations (Soldatov 1915, Yukhimenko 1963, Svirskii 1971, Krykhtin 1979, Krykhtin & Gorbach 1986).

Maturation, spawning migration, and breeding

Males from the estuary population spawn for the first time at age 14–21 years, and females, at age 17–23. Water temperature affects the time of maturation of females: during warm years, females of the same generation mature and spawn a year earlier than they would during cold years. Males spawn once every three to four years, and females, every four to five years (Svirskii 1971, Krykhtin 1986). Total degeneration of previtellogenic oocytes (40–470 µm in diameter) was observed in a few adult females at age 18–24 years (Svirskii 1979). Maturation in these females can be delayed for two years and intervals between spawnings may increase to six to eight years.

Fecundity in 411 females ranging from 16 to 30 years from the estuary population ranged from 186 000 to 4 225 000 eggs (mean = 977 465 ± 23 692). Relative fecundity was 3300 to 15 100 eggs kg^{-1} of body weight. In some females, the coelenterate parasite *Polypodium hydriforme* caused a decrease in individual fecundity by 19% (Svirskii 1984). Gonads of most of the future spawners were at stage IV of maturation (according to the stages of Nedoshivin 1928). In autumn and early winter, most of these future spawning fish migrate from the estuary into the Amur River, where they spend the winter in preparation for spawning during the next spring (Krykhtin 1986). A smaller group (about 5%) of all of the spring spawners migrates into the Amur River during the spring, spawning soon after migration. Thus, there are two seasonal forms of kaluga in the estuary, which we call winter and spring. The winter form predominates.

Most of the mature fishes from the estuary population spawn 50–150 km upstream from the town of Nikolayevsk-na-Amure, while a small portion spawns on sites located less than 500 km upstream from the mouth of the river. Some individuals migrate as far as Khabarovsk, nearly 1000 km from the mouth, and spawn in the middle Amur.

Spawning takes place during a small increase in the water level, from the end of May until the beginning of July, when the water temperature ranges from 12 to 21° C. The peak of spawning usually occurs in the middle of June. Pebble deposits in the main river bed and some large side channels serve as spawning grounds (Svirskii 1976a). After spawning, the fish return to the estuary to feed.

Embryonic development until hatching lasts 82–112 h, and development of free embryos until the transition to active feeding takes 7.5–14 days at 14–22° C (Svirskii 1971). Free embryos and larvae of the estuary drift downstream from the spawning grounds to the lower reaches of the Amur River and estuary. Some juveniles from the estuary population remain in the lower reaches until age two to five years, where they feed together with juveniles from the population of the lower Amur.

Kaluga from the lower Amur population spawn at the same spawning grounds and at the same time as individuals from the estuary population. These fish migrate from the lower Amur to the spawning grounds from May to the first half of June. The process of maturation and intervals between successive spawnings appear to be similar in both populations. Individuals from the lower Amur population grow more slowly and are probably adapted to the higher water temperature of the lower Amur.

The middle Amur River population of kaluga extends approximately 900 km from the river mouth and includes the upper part of the lower reach of the Amur River and lower part of the middle reach of the Amur River. Many individuals of the middle Amur population reach sexual maturity much earlier than do those of the estuary population. From 1988 to 1992, 11–16 year old females constituted up to 25% of the females migrating to spawning grounds from the lower region of the middle reach of the Amur. Such early maturation was never observed in the estuary population. Individual fecundity varied from 238 000 to 4 868 000 eggs in 126 females studied at 11–16 years of age. Their relative fecundity varied from 5000 to 11 000 eggs kg^{-1} body weight.

Major spawning grounds of the middle Amur River population of kaluga are located in the lower region of the middle reach of the Amur. Smaller spawning grounds are located in the Sungari and Ussuri rivers. The spawning migration of the middle Amur population takes place from May to the first half of June. Their free embryos drift downstream from the spawning grounds to the regions with well-developed flood plains.

The Zeya and Bureya populations are now represented by rare individuals in the upper region of the middle reach of the Amur, in the upper reach of the Amur, and in the lower regions of the Zeya, Shilka, and Argun rivers. They migrate to spawning grounds located in the upper Amur River and in a region 250 km long downstream from Blagoveshchensk, in the second half of May and early June. Spawning takes place in June. The biology of this population has not been studied.

Contemporary status

The status of kaluga populations in the Amur basin has changed since 1900. At the end of the 19th century, when the highest catches were recorded (more than 595 metric tons per year), the middle Amur population was the most abundant. This is evident from the analysis of absolute and relative catches throughout the Amur River. According to data for 1891, fish from the middle Amur population constituted 87% of the annual kaluga catch in the Amur River Basin. Fish from the estuary and lower Amur populations, however, constituted 2% of the catch, with remaining 11% coming from the Zeya and Bureya population (Kryukov 1894).

At present, the estuary population is most abundant. Due to the strictly limited catch since 1976, the total number of fish in the estuary population increased by approximately one third, and the number of large fish of more than 100 kg in weight increased by 2.5 times in comparison with the early 1970s (Krykhtin 1979), so that there were about 70 000 fish greater than one year old at the end of 1980s. Approximately 5000 of these fish weighed more than 100 kg and were potentially sexually mature.

However, by 1993, as a result of an illegal fishery in the lower Amur during the spawning migration, the number of sexually mature fish in the estuary population was reduced by approximately 30–35%. The current population of kaluga consists predominantly of young fish, with only 2–3% of the population weighing more than 100 kg and classified as adults. Thus, according calculations using the area method and based on irregular control catches in the lower and middle Amur, the number of individuals older than 2 years in the lower Amur population is approximately 40 000 and in the middle Amur population, 30 000. The decrease in the kaluga population which appeared at the end of the 1960s continues. Further decrease in population size should be expected, especially in the middle part of the Amur River. The size of the Zeya and Bureya population, if evaluated on the basis of the very low catch in the Amur River within the Amur district (0.09–1.03 metric tons), is so small that the population is on the verge of disappearing.

The efficiency of natural breeding of kaluga is very low, as can be seen from the slow rate of restoration of the estuary population: until the beginning of the 1990s, its size increased only 35%, or less than 2% per year.

Amur sturgeon

Biology and population structure

The Amur sturgeon (Figure 2b,d) is represented in the Amur River basin by two morphs: brown and gray. Young and adults of the brown morph inhabit the middle and lower reaches of the Amur River. Brown morphs are rare and grow more slowly than do the gray ones, with females maturing 96–117 cm long and weights of 3.5–5.6 kg, whereas female gray morphs are 125–142 cm long and weigh 8.3–16.4 kg by maturity at age 12 years. There are a few small local concentrations of brown morph in the Amur River.

The maximum length of the gray morph is about 3 m, and the weight is 190 kg at an age of more than 60 years. The distribution, mode of life, and population structure of the gray morph resemble those of kaluga, but they do not enter the sea. They feed on benthos and cannibalism rarely occurs. Freshwater mollusks and larvae of the Arctic lamprey, *Lampetra japonica*, are usually present in the stomachs of the gray morph individuals (Yukhimenko 1963, Svirskii 1971).

Maturation, spawning migration, and breeding

Most gray morph individuals mature at an age of 10–14 years, being 105–125 cm in length and weighing 6.0–18.5 kg. Females reproduce at least every four years. If previtellogenic and vitellogenic oocytes degenerate, then the age of the first maturation, as in kaluga, can be delayed by two to four years (Svirskii 1979). Total fecundity of Amur sturgeons is approximately one-fourth that of kaluga: in 317 females 8 to 45 years old, the mean fecundity per female was 287 780 ± 24 489 eggs (from 41 000 to 1 057 000 eggs in 388 females). The relative fecundity, however, is much higher than that of kaluga, from 4600 to 17 300 eggs kg^{-1} of body weight.

Both morphs begin to migrate to the spawning grounds in autumn. During winter, the gonads of 80% of the future spawners have not yet reached stage IV. A few sexually immature fish migrate from the estuary into the Amur River, where they live until the completion of sexual maturation. They return to the estuary only after spawning, one or two years later.

Amur sturgeon migrate to the spawning grounds in small groups of 3–5 fish. Spawning takes place in the spring within 25–30 days at all spawning grounds of the lower Amur. Gray morphs of the Amur River spawn at the same spawning grounds as kaluga, and during spawning, a small number of kaluga-Amur sturgeon hybrids are produced (2–5% of all larvae; Figure 2d). The hybrids are predominantly males (up to 79%). Some hybrids may reach 1.9 m in length and 70 kg in weight.

The embryo period lasts from 83 to 295 hours at 24°C and 12°C, respectively (about the same time as kaluga). The transition of free embryos to active feeding takes 8.5 days at 18–20°C, i.e., 1.0–1.5 days earlier than in kaluga (Svirskii 1976a). Amur sturgeon survive early development two or three times more successfully than do kaluga.

Contemporary status

The current size of the estuary population of Amur sturgeon is relatively low: about 3000 fish are older than 2 years in the estuary. In 1891, when the catch of the Amur sturgeon in the Amur River Basin reached 607 metric tons, the fish caught in the lower Amur and estuary constituted only about 3% of the whole annual catch (Kryukov 1894). Most fish (89%) were caught in the middle Amur, from Ekaterino-Nikolskaya village to Tambovskaya village. The rest of the catch (8%) was taken upstream from Ekaterino-Nikolskaya village. Differences in the catch were due mostly to regional differences in the abundance of sturgeons.

Now, the lower Amur population of the Amur sturgeon comprises about 95 000 fish greater than age two years and is approximately half as large as the population in the middle Amur. The Zeya and Bureya population of the Amur sturgeon is extremely small and on the verge of extinction.

If the decrease in Amur sturgeon populations continues, then the survival of this species is in doubt, especially the population inhabiting the middle Amur.

General comments on the status of kaluga and Amur River sturgeon

Since the fall of the Soviet Union, poaching in the Amur River has increased enormously. Fishermen catch fish now not only for food but also for sale. In the lower Amur, intensive fishing of kaluga and Amur sturgeon migrating to spawning grounds started in 1991. The catch increased both on the Russian and Chinese banks of the Amur River, despite the ban on fishing issued in 1958.

The middle Amur populations of both kaluga and Amur sturgeon are especially overfished during their spawning migrations and on the spawning grounds. In the upper region of the lower Amur River, illegal fishing is carried out not only by single fishermen but also by organized groups. Chinese fishermen alone caught approximately 410 metric tons of kaluga and Amur sturgeon in 1989, and 170 metric tons in 1993, taking migrating kaluga and Amur sturgeon at the prespawning state. Caviar produced was exported to the United States and other countries. There is no agreement between Russia and China regarding sturgeon fishing in the Amur River. Russians started a sturgeon fishery in the lower region of the middle Amur in 1991, and by 1993, had taken more than 200 metric tons of kaluga and Amur sturgeon migrating into the middle reach of the Amur River for spawning, although official records indicated 64.4 metric tons in 1991, 62.6 metric tons in 1992, and 47.8 metric tons in 1993. In the near future this type of uncontrolled fishery will decimate kaluga and Amur sturgeon populations in the estuary and the lower Amur. Contamination of the Amur River and its tributaries with heavy metals, oil products, phenols, organic substances, and other pollutants is a potential but unknown threat.

Because they spawn on the river bed, the low water level in the Amur River during some years does not apparently affect the breeding of sturgeons. Most young and adult fish also feed in the same area of the river. Juveniles migrating for food to the small tributaries of the Amur River will not overwinter in these areas and, therefore, are not affected by the low water levels during autumn-winter periods. The mass death of the kaluga and Amur sturgeon juveniles that sometimes was observed in the Amur River estuary (due to cold, saline water) affected only the estuary populations. Predation on kaluga and Amur sturgeon eggs and juveniles in the Amur River did not increase within the last two decades but was, on the contrary, markedly reduced.

Kaluga and Amur sturgeon populations in the Amur River seem to be in better shape than do beluga and sturgeon populations in the European part of Russia (Khodorevskaya et al. 1997 this volume). Taking into consideration the existing and expected status of the kaluga and Amur sturgeon populations, their very low rate of breeding and low efficiency of natural reproduction in the Amur River basin, we consider it necessary to build hatcheries for artificial breeding and restocking of these species[1].

Acknowledgements

We thank Vadim Birstein and John Waldman for the invitation to participate in the International Conference on Sturgeon Biodiversity and Conservation. Two anonymous reviewers commented on our original draft. The English text was substantially revised and improved by Vadim Birstein. William E. Bemis drew the map.

[1] In 1995, sturgeon poaching from the Russian and Chinese banks of the Amur River and its tributary, the Ussuri River, intensified. Increasingly, poaching in the Russian part of the lower Amur is carried out by well-organized and well-equipped groups (editors' note, February 1996).

References cited

Amaoka, K. & K. Nakaya. 1975. First record of the kaluga sturgeon, *Huso dauricus*, from Japan. Japan. J. Ichthyol. 22: 164–166.

Artyukhin, E.N. & A.E. Andronov. 1990. A morphological study of the green sturgeon, *Acipenser medirostris* (Chondrostei, Acipenseridae), from the Tumnin (Datta) River and some aspects of the ecology and zoogeography of Acipenseridae. Zoologicheskii Zhurnal 69: 81–91 (in Russian, English translation J. Ichthyol. 30: 11–21).

Gritsenko, O.F. & G.M. Kostyunin. 1979. The Amur River white fish, *Coregonus ussuriensis* Berg, and kaluga, *Huso dauricus* (Georgi), in the Sakhalin island waters. Voprosy Ikhtiologii 19: 1125–1128 (in Russian).

Honma, Y. & H. Itano. 1994. A record of a great Siberian sturgeon, *Huso dauricus*, off Niigata, Sea of Japan (Osteichthyes: Acipenseridae). Japan. J. Ichthyol. 41: 317–321.

Khodorevskaya, R.P., G.F. Dovgopol, O.L. Zhuravleva & A.D. Vlasenko. 1997. Present status of commercial stocks of sturgeons in the Caspian Sea basin. Env. Biol. Fish. (this volume).

Kostarev, V.L. & B.V. Tyurnin. 1970. Kaluga in the waters of the north-western part of the Sea of Okhotsk. Izvestiya Tikhookeanskogo Nauchno-Issledovatelskogo Instituta Rybnogo Khozyaistva i Okeanografii 74: 346–247 (in Russian).

Krykhtin, M.L. 1972. Changes in the composition and abundance of populations of the kaluga, *Huso dauricus* (Georgi), and sturgeon, *Acipenser schrencki* Brandt, during the period of fishery closure in the Amur River Basin. Voprosy Ikhtiologii 12: 2–12 (in Russian).

Krykhtin, M.L. 1975. Growth of kaluga, *Huso dauricus* (Georgi), in the Amur River estuary. Izvestiya Tikhookeanskogo Nauchno-Issledovatelskogo Insituta Rybnogo Khoziaistva i Okenograpfii 98: 129–136 (in Russian).

Krykhtin, M.L. 1979. The modern state and perspectives of development of the sturgeon fishery in the Amur River Basin. pp. 68–74. *In*: Biological Foundations of Development of the Sturgeon Fishery in the Water Bodies of the USSR, Nauka Press, Moscow (in Russian).

Krykhtin, M.L. 1984a. On the causes of kaluga death in the Amur River estuary. pp. 163–164. *In*: Sturgeon Fishery in the Water Bodies of the USSR, Astrakhan (in Russian).

Krykhtin, M.L. 1984b. Comparative estimate of the rate of reproduction, natural loss, and size of populations of the Amur River fresh water fishes. Voprosy Ikhtiologii 24: 204–211 (in Russian).

Krykhtin, M.L. 1985. Ecological-physiological variability in the kaluga population of the Amur River estuary. pp. 105–196. *In*: Ecological Physiology and Biochemistry of Fishes, Vilnus (in Russian).

Krykhtin, M.L. 1986. The rate of sexual maturation and reproduction cycle of the kaluga, *Huso dauricus* (Georgi), in the Amur River estuary. Voprosy Ikhtiologii 26: 945–954 (in Russian).

Krykhtin, M.L. & E.I. Gorbach. 1986. Distribution and feeding of the kaluga in the Amur River estuary. pp. 161–162. *In*: Formation of Sturgeon Resources Under the Conditions of Complex Utilization of Water Resources, Astrakhan (in Russian).

Kryukov, N.A. 1894. Some data on the fishery in the Amur River Basin. Zapiski Priamurskogo Otdela Imperatorskogo Russkogo Geografitcheskogo Obshchestva, Vol. 1, Vypusk 1, St. Petersburg. 87 pp. (in Russian).

Lukyanenko, V.I., P.P. Geraskin, N.V. Bal & G.F. Metallov. 1979. Physiological-biochemical characteristics of the kaluga from the Amur River as an object of acclimatization in the southern regions of the USSR. Gidrobiologicheskii Zhurnal 15: 71–77 (in Russian).

Nedoshivin, A.Ya. 1928. Materials on the Don River fishery. Trudy Azovo-Chernomorskoi Nauchno-Promyslovoi Ekspeditsii 4: 1–175 (in Russian).

Nikolskii, G.V. 1956. Fishes of the Amur River Basin. USSR Academy of Sciences Press, Moscow. 552 pp. (in Russian).

Pravdin, I.F. 1966. Manual for studies of fishes. Pishchevaya Promyshlennost, Moscow. 376 pp. (in Russian).

Romeis, B. 1954. Microscopic technique. Izdatelstvo Inostrannoi Literatury, Moscow. 718 pp. (in Russian).

Shilin, N.I. 1995. Programme for conservation of *Acipenser medirostris mikadoi* in the Russian far east. pp. 262–267. *In*: A.D. Gershanovich & T.I.J. Smith (ed.) Proceedings of the International Symposium on Sturgeons, 6–11 September 1993, VNIRO Publishing, Moscow.

Soldatov, V.K. 1915. A study on the Amur River acipenserids. Materialy k Poznaniyu Russkogo Rybolovstva, Vol. 3, Vypusk 12, Petrograd. 415 pp. (in Russian).

Svirskii, V.G. 1967. The Amur River sturgeon and kaluga (taxonomy, biology, perspectives of recovery). Candidate Biol. Sci. Thesis, Far Eastern State University Press, Vladivostok. 31 pp. (in Russian).

Svirskii, V.G. 1968. Possible ways of the modern sturgeon status formation in connection with the ocean level fluctuations during the Quaternary. pp. 203–208. *In*: Problems of Oceanology, Far Eastern State University Press, Vladivostok (in Russian).

Svirskii, V.G. 1971. The Amur River sturgeon and kaluga. Uchenye Zapiski Dalnevostochnogo Gosudarstvennogo Universiteta 15: 19–33 (in Russian).

Svirskii, V.G. 1976a. Embryonic and postembryonic development of the Amur River acipenserids. pp. 8–20. *In*: Biology of Fishes in the Far East. Far Eastern State University Press, Vladivostok (in Russian).

Svirskii, V.G. 1976b. Hydrological characteristics of the Low Amur River and their role in reproduction of the Amur River sturgeon and kaluga. pp. 70–74. *In*: Biology of the Far Eastern Fishes, Far Eastern State University Press, Vladivostok (in Russian).

Svirskii, V.G. 1979. Degeneration of germ cells in the ovaries of the Amur River sturgeon and kaluga as the reflection of adaptive resistance of fishes to the extreme environmental factors. p. 170. *In*: Ecology, Physiology and Biochemistry of Fishes, Vol. 2, Astrakhan (in Russian).

Svirskii, V.G. 1984. *Polypodium hydriforme* (Coelenterata) in the Amur River acipenserids. Parasitology 18: 362–366 (in Russian).

Wei, Q., F. Ke, J. Zhang, P. Zhuang, J. Luo, R. Zhou & W. Yang. 1997. Biology, fisheries, and conservation of sturgeons and paddlefish in China. Env. Biol. Fish. (this volume).

Yukhimenko, S.S. 1963. Feeding of the Amur sturgeon, *Acipenser schrencki* Brandt, and kaluga, *Huso dauricus* (Georgi), in the lower reaches of the Amur River. Voprosy Ikhtiologii 3: 311–318 (in Russian).

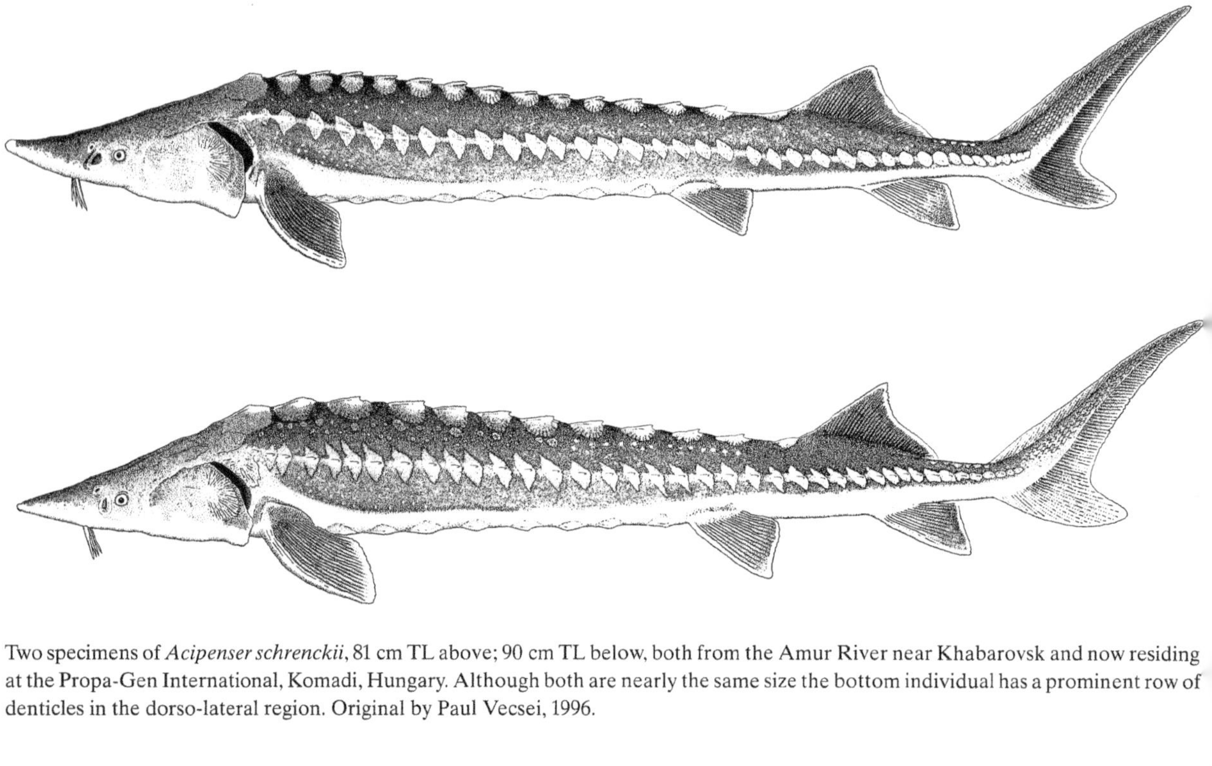

Two specimens of *Acipenser schrenckii*, 81 cm TL above; 90 cm TL below, both from the Amur River near Khabarovsk and now residing at the Propa-Gen International, Komadi, Hungary. Although both are nearly the same size the bottom individual has a prominent row of denticles in the dorso-lateral region. Original by Paul Vecsei, 1996.

Environmental Biology of Fishes **48**: 241–255, 1997.

Biology, fisheries, and conservation of sturgeons and paddlefish in China

Qiwei Wei[1,2], Fu'en Ke[2], Jueming Zhang[3], Ping Zhuang[2], Junde Luo[2], Rueqiong Zhou[2] & Wenhua Yang[2]
[1] *State Key Laboratory of Freshwater Fish Germplasm Resources and Biotechnology, Shashi 434000, China*
[2] *Yangtze River Fisheries Research Institute, Chinese Academy of Fishery Sciences, Shashi 434000, China*
[3] *Amur River Fisheries Research Institute, Chinese Academy of Fishery Sciences, Harbin 150070, China*

Received 2.8.1994 Accepted 10.3.1996

Key words: Acipenseriformes, *Acipenser*, *Huso*, *Psephurus*, Yangtze River, Amur River, Gezhouba Dam, Three Gorges Project, spawning areas

Synopsis

This paper reviews five of the eight species of acipenseriforms that occur in China, chiefly those of the Amur and Yangtze rivers. Kaluga *Huso dauricus* and Amur sturgeon *Acipenser schrenckii* are endemic to the Amur River. Both species still support fisheries, but stocks are declining due to overfishing. Acipenseriformes of the Yangtze River are primarily threatened by hydroelectric dams that block free passage to spawning and feeding areas. The Chinese paddlefish *Psephurus gladius* now is rare in the Yangtze River system, and its spawning activities were severely limited by completion of the Gezhouba Dam in 1981. Since 1988, only 3–10 adult paddlefishes per year have been found below the dam. Limited spawning still exists above the dam, but when the new Three Gorges Dam is complete, it will further threaten the paddlefish. Artificial propagation appears to be the only hope for preventing extinction of *P. gladius*, but it has yet to be successfully bred in captivity. Dabry's sturgeon *A. dabryanus* is a small, exclusively freshwater sturgeon found only in the Yangtze River system. It is concentrated today in reaches of the main stream above Gezhouba Dam. The fishery has been closed since 1983, but populations continue to decline. *Acipenser dabryanus* has been cultured since the 1970s, and holds promise for commercial aquaculture; availability of aquacultural methods offers hope for enhancing natural populations. The Chinese sturgeon *A. sinensis* occurs in the Yangtze and Pearl rivers and seas of east Asia. There is still disagreement about the taxonomy of the Pearl and Yangtze River populations. The Yangtze River population is anadromous. Adults begin spawning at about age 14 years (males) and 21 years (females), and adults spend over 15 months in the river for reproduction. Spawning sites of *A. sinensis* were found every year since 1982 below the Gezhouba Dam, but it seems that insufficient suitable ground is available for spawning. Since 1983, commercial fishing has been prohibited but more measures need to be taken such as establishing protected areas and characterizing critical spawning, summering and wintering habitats.

Introduction

Eight species of Acipenseriformes are native to China. Kaluga *Huso dauricus* and Amur sturgeon *Acipenser schrenckii* are shared with Russia in the Amur River system, and Siberian sturgeon *A. baerii* and sterlet *A. ruthenus* with Kazakhstan and Russia in the Irtysh River. *Acipenser nudiventris* occurs in the Ili River, which is a tributary of Lake Balkhash in Kazakhstan. The remaining three species, Chinese sturgeon *A. sinensis*, Dabry's sturgeon *A. dabryanus* and Chinese paddlefish *Psephurus gladius*, are endemic to China, particularly the Yangtze River and China Sea. In 1988, North American paddle-

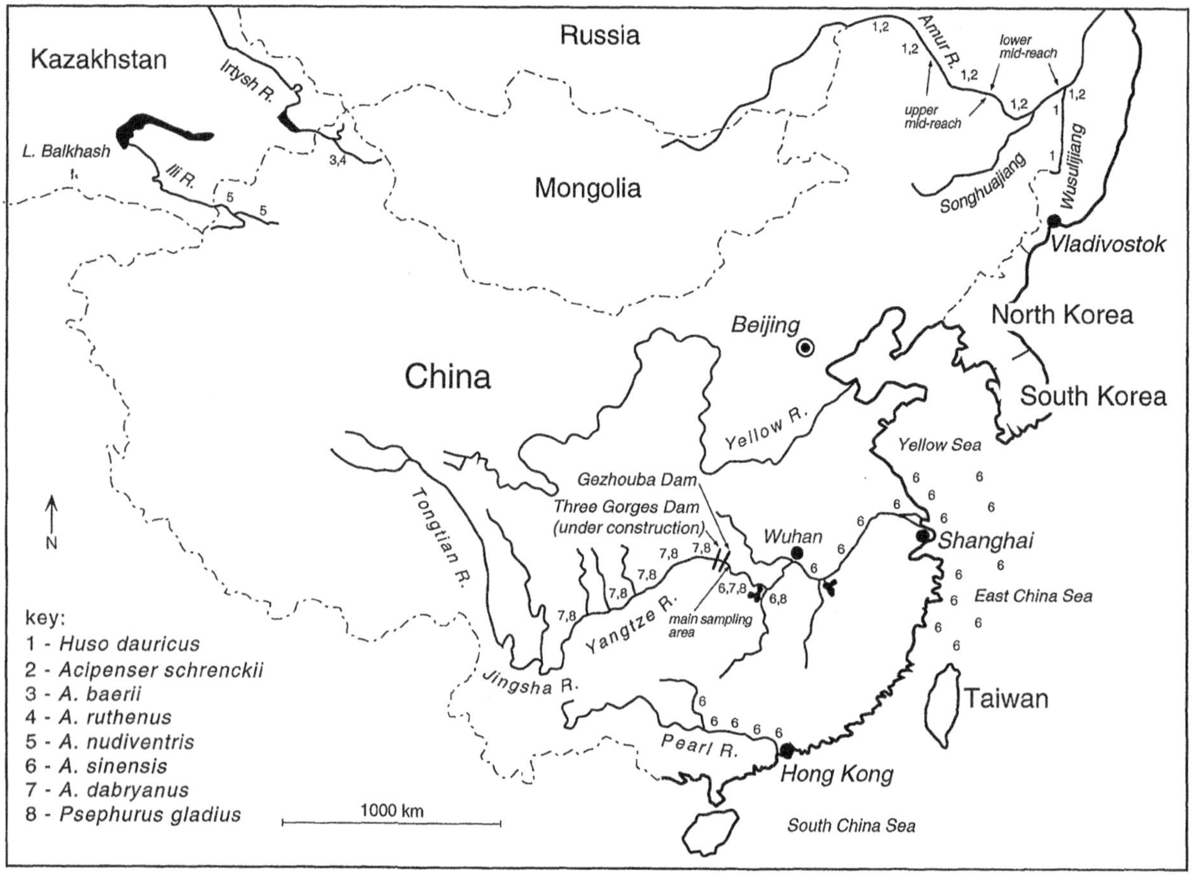

Figure 1. Map of China and adjacent countries showing the Amur, Yellow, Yangtze and Pearl rivers and current ranges of sturgeons and paddlefish. Much of the Amur River forms the border between China and Russia. The Amur is traditionally divided into upper, middle and lower reaches. Most of our data for this system concerns conditions in the middle reach, which we in turn divide into upper middle reach and lower middle reach as shown. The Wusulijiang is also known as the Ussuri River, and the Songhuajiang is also known as the Sungari. Also see Krykhtin & Svirskii (1997) for additional details on Amur River acipenserids. The Yangtze River is also divided into upper, middle and lower reaches. Most of our data concern sturgeons and paddlefish of the middle reach, near Gezhouba Dam. Also see Zhuang et al. (1997) for additional details on Yangtze River Acipenseriformes.

fish, *Polyodon spathula*, were introduced in Hubei Province, China, where they are being successfully reared in ponds.

Documentation of sturgeon in China dates to at least 1104 BC, when sturgeon were described as 'wang-wei', which means 'king of fishes'. More than ten ancient Chinese books mention sturgeons, including descriptions of morphology, habits, distribution, fishing and utilization (Anonymous 1988). These early descriptions did not distinguish between sturgeons and paddlefish. Sturgeons were depicted as fishes related to god, such as the sturgeon dragon in Guangdong China, and as precious fishes, they were sent to emperors as a tribute. Many parts of sturgeons were considered to have medicinal value.

We still have relatively poor knowledge of the biology of Chinese sturgeons and paddlefish. Until the 1970s, most research and publications concerned taxonomy. Between 1972 and 1979, we studied sturgeon biology and fisheries on the Chinese side of the Amur River (Zhang 1985), and in 1988, established a propagation station for *Huso dauricus* and *Acipenser schrenckii* at Qingdeli, Heilongjiang

Province. Also starting in the 1970s were two long term investigations of the two species of sturgeons and the paddlefish in the Yangtze River. The first project was done between 1972–1975, directed by the Yangtze River Fisheries Research Institute (Anonymous 1988). The second project began in 1980, prompted by the construction of Gezhouba Dam (Fu 1985, Ke et al. 1984, 1985). These research programs provide the basis for fisheries management and conservation of the three species of Acipenseriformes native to the Yangtze River. So far, *A. baerii*, *A. ruthenus* and *A. nudiventris* from Xingjiang Province in northwestern China have hardly been studied.

In the last thirty years, all stocks of sturgeons declined due to overfishing, construction of hydroelectric dams, and pollution. The Pearl River population of *A. sinensis* and populations of *A. dabryanus* and *Psephurus gladius* in the Yangtze River have been severely impacted. This paper reviews the sturgeons and paddlefish in the Amur and Yangtze rivers, and discusses conservation strategies.

Materials and methods

Investigations on sturgeons of the Amur River were conducted from 1978 to 1979. Studies on the Yangtze River were made each fall (October and November) from 1981 through 1993, and in the spring of 1984. Sturgeons are caught by fishermen in accordance with our requirements. In the Amur River, the fishermen use three-layer gill nets (see Zhang 1985 for description). In the Yangtze River, sturgeons are taken using row hooks (Deng et al. 1991). Chinese paddlefish are caught incidentally in gill nets set for copperfish, *Coreius heterodom*. Our main sampling location for *Acipenser sinensis* is in reaches just below the Gezhouba Dam (Figure 1). Specimens were measured to the nearest cm (total length, TL, body length, BL, fork length, FL) and weighed to the nearest 0.25 kg (body weight, BW).

Since 1983, we have taken Chinese sturgeons, *A. sinensis*, from the Yangtze River for artificial propagation and population analysis. We removed the pectoral fin rays, clavicles and cleithra for age determination, and examined the condition of the gonads. Each aging method was read at least twice, and the results were usually in agreement (Anonymous 1988, the age based on annular counts of the first pectoral fin ray was used in cases of disagreement). Methods used to determine the degree of gonadal development were taken from Anonymous (1988) and Deng et al. (1991). Several specimens in each sample collected since 1982 exhibited abnormally developed gonads, and we calculated gonadosomatic indices (GSI = gonadal weight/BW) to study this. We identified the time and general location of spawning for Chinese sturgeon by exam-

Table 1. Composition of spawning stock of *Acipenser schrenckii* in the middle reach of the Amur River as related to river location.

TL range (cm)	Upper middle reach [1] (N=198)			Lower middle reach [2] (N=125)		
	x ± SD (cm)	N	%	x±SD (cm)	N	%
100–120	116.0 ± 5.7	73	36.8	113.7 ± 5.3	16	12.8
120–140	129.8 ± 6.6	73	36.8	130.8 ± 5.8	21	16.8
140–160	148.8 ± 6.0	22	11.1	150.5 ± 6.1	32	25.6
160–180	166.2 ± 5.0	20	10.1	168.9 ± 5.1	30	24
180–200	189.2 ± 5.5	7	3.5	189.3 ± 5.6	20	16
200–220	203.5 ± 2.1	2	1.0	208.5 ± 6.1	4	3.2
260–280	267.0	1	0.4	242.5 ± 2.1	2	1.6

[1] From Heihe to Luobei; see Figure 1.
[2] From Zhaoxing to Qingdeli; see Figure 1.
SD = standard deviation.
N = number of samples (individuals).

ining stomachs of copperfish, *Coreius heterodom*, for the presence of sturgeon eggs (Deng & Xu 1991). Hydrological data were provided by local monitoring stations. Catches were determined from our census or by local fisheries managers. All data were analyzed using SAS and MSA (Beijing University 1993) software.

Results and discussion

Amur River

The Amur River (also known as the Heilong Jiang) arises from the confluence of the Shilka and Argun rivers. From there, it flows nearly 3000 km to its entry into the Sea of Okhotsk. For about one third of its length, the Amur River forms the border between China and Russia. Geographers distinguish three portions of the Amur River usually referred to as upper, middle, and lower Amur. Our studies focus on the sturgeons of the middle Amur, a section of river about 1005 km long, that we subdivide into upper and lower middle reaches (Figure 1), with the upper middle reach extending between the towns of Heihe and Luobei and the lower middle reach extending from Zhaoxing to Qingdeli. Additional information about the Amur and its sturgeons, particularly populations from the lower reach and estuary of the Amur River, is given by Krykhtin & Svirskii (1997 this volume). Two species of sturgeons are native to the middle reach of the Amur River.

Amur sturgeon, *Acipenser schrenckii*
The Amur sturgeon is mostly restricted to the main channel of the Amur River from the upper middle reach to the mouth (Figure 1). It also occurs in the Wusulijiang River (also known as the Ussuri River), a tributary to the Amur River, but it is now severely depleted in the Songhuajiang River (also known as the Sungari River). The species consists of a few distinct populations. In general, there is a gradient of increasing mean total length (TL) and body weight (BW) of spawning stocks from the upper to the lower middle reaches (Table 1). Males first spawn at age 7 to 8 years, at sizes of about 103 cm TL and 4 kg BW. First spawning for females occurs at age 9 to 10 years, at about 105 cm TL and 6 kg BW. Spawning individuals in the lower middle reach are older than those in the upper middle reach. In the lower middle reach, 72.6% of fish were between age 15 and 28 years, whereas in the upper middle reach, 75.3% of fish were between 13 and 24 years of age (Table 2). The oldest individual was 45 years. The sex ratio also varied between reaches, with a skewed sex ratio in the lower middle reach (2.0 : 1, N = 251). Between late May and early June, we examined gonads of 24 fish aged 15 to 38 years. The GSI of these specimens ranged from 12.7% to 34.7%, averaging 23.5 ± 5.6%. Fecundity ranged from 114 000 to 1 292 000 with mean of 385 000. The number of eggs per gram of body weight ranged from 31.0 to 64.4, with a mean of 44.6. Fecundity was positively correlated with total length and age. Ripe eggs ranged from 3.0 to 3.5 mm in diameter.

Kaluga, *Huso dauricus*
The kaluga is a large freshwater sturgeon found only in the Amur River. They tend to be solitary and non-migratory. Fisheries exist throughout the mainstream of the middle reach of the Amur River.

Kaluga ranged from 160 to 400 cm TL in the lower middle reach. Fish between 200 and 320 cm TL ac-

Table 2. Age composition of spawning stock of *Acipenser schrenckii* in the middle reach of the Amur River.

Age (years)	7–8	9–10	11–12	13–14	15–16	17–18	19–20	21–22	23–24	25–26	27–28	29–30	31–32	35–36	37–38	45
H-L: Nt	12	4	2	9	12	16	14	10	9	3	0	1	0	1	0	0
Z-Q: Nt	2	3	4	7	10	23	10	12	12	25	11	6	8	3	3	1
Z-Q: Nm	2	3	4	7	6	16	7	8	1	5	3	1	0	0	0	0
Z-Q: Nf	0	0	0	0	4	7	3	4	11	15	8	5	8	3	3	1

H-L = upper middle reach from Heihe to Luobei; Z-Q = lower middle reach from Zhaoxing to Qingdeli; Nt = total number of samples; Nm = number of males; Nf = number of females; and values = individuals.

counted for 87.4% near Zhaoxing, whereas fish between 180 and 300 cm TL accounted for 87.4% near Qingdeli (Table 3). Body weight ranged from 40–501 kg. In a sample of 79 specimens, 74.7% were between 40 and 165 kg BW. Females tended to be larger than males. In a sample of 53 females, 52.8% ranged from 90 kg to 190 kg, and 41.5% ranged from 190 kg to 501 kg. The spawning population ranged from 12 to 54 years of age (Table 4). Age is positively correlated with length and weight. The oldest individual (54 years) was also the largest at 390 cm TL and 501 kg BW. Sex ratios differed between the upper and lower middle reaches. The ratio of males to females was 1.4 : 1 near Zhaoxing (N = 113) and 3.7 : 1 near Qingdeli (N = 48). The average GSI for females was 13.9% in May (maximum of 17.3%); by June, the GSI was 14.1% (maximum of 18.4%). Fecundity ranged from 383 400 to 3 280 000 (N = 22; 224–390 cm TL). Mean fecundity was 600 000, 1 500 000, and 3 000 000 for individuals with TL 224–274 cm, 281–327 cm and 350–390 cm, respectively. The number of eggs per gram of body weight ranged from 27.8 to 53.0, with a mean of 41.4. Ripe eggs ranged from 2.5 to 3.5 mm.

Reproduction. – Both Amur sturgeon and kaluga usually spawn between May and July when water temperatures range from 15° to 20°C. Spawning habitat is characterized by calm water and sand and gravel substrates. Kaluga spawn at a depth of 2 to 3 m. Appropriate spawning habitat is available from the upper to the lower middle reaches of the Amur River. Known spawning sites are located at the mouth of the Pingyanghe and Xueshuiwen rivers in Xunke County, in the Yadanhe and Zhaoxing rivers within Luobei County, and in the Qingdeli and Xiabaca rivers within Tongjiang County. We have not found any spawning sites in the Songhuajiang and Wusulijiang rivers.

Natural hybridization occurs between kaluga and Amur sturgeon in the Amur and Songhuajiang rivers (Ren 1981). Historically, the hybrids were considered a distinct species (Gong 1940). They have intermediate characteristics and resemble artificially produced hybrids (Krykhtin & Svirskii 1986, fig. 2d).

Fisheries. – The pull nets and row hooks used to capture sturgeons in the Amur River until the 1950s were replaced by three-layer gill nets. Fishing impact on sturgeons was low before the 1970s because few people lived along the Amur River. However, with increasing population and the high profit of sturgeon fishing, catches increased. There are now 2 to 5 sturgeon fishing boats per kilometer along the mid reaches of the Chinese side of the Amur River, and annual exports of caviar increased from 3 metric tons in the 1970s to 12 tons in 1990. Separate catch data for the two species are not available. Incomplete statistics from 9 sites show that the annual catch from 1952 to 1956 and 1959 was commonly 70 to 80 tons year^{-1} in the entire middle reaches of the Chinese side, with a minimum of 30 tons year^{-1}. In 1978, 90 tons were harvested, 141 tons in 1981, 175 tons in 1985, and 200 tons by 1987. In recent years, production decreased at some locations, especially in the upper middle reaches, despite comparable fishing effort. Fishing has moved to the lower part of the mid reaches, and harvests in these areas are now declining. Annual catches at Qingdeli, for instance, were 36.3, 31.7, 27, 26.7 and 21.6 tons in the years from 1985 through 1989.

Most of the fish taken are spawning-sized adults. This has grave consequences for the future of the fishery. We sampled 256 Amur sturgeon from fishermen at Luobei, which included 123 young fish ranging from 20 to 100 cm TL, and 133 adults ranging from 100 to 180 cm TL. In another large sample of Amur sturgeon (407 individuals from Zhaoxing and Qingdeli), young fish accounted for only 5.7% of the total sample. For kaluga from Zhaoxing and Qingdeli, young accounted for 29.3% of the total sample.

Conservation efforts. – The government of Heilongjiang Province issued specific regulations for the protection and management of sturgeons in the 1950s, and renewed them in 1982. Current regulations are not fully implemented because of insufficient management. Furthermore, some regulations are based on insufficient understanding of the stocks. As the stocks declined, a propagation station for Amur sturgeon was set up at Qingdeli in 1988. About 900 000 larvae (0.2 to 0.4 g BW) and

168 000 juveniles (1.0 to 1.5 or 20 to 30 g BW) were stocked into the Amur river from 1988 to 1991, and the number of young sturgeon in the Amur River seemed to be increasing through 1991 (Chen & Zhou 1992).

Yangtze River

The Yangtze (also known as the Chang Jiang) arises in southwestern China and flows 5500 km across the country to reach the East China Sea near Shanghai (Figure 1). Based on its outflow, the Yangtze is the fourth largest river in the world, surpassed only by the Amazon, Congo, and Indus rivers. Near the headwaters, the mainstream of the Yangtze is known as the Tongtian River; its eastward course toward Sichuan is known as the Jingsha River; as it passes through Sichuan it becomes known as the Yangtze. It passes through some of the most densely populated areas of the planet, yet it retains three native species of Acipenseriformes. This situation is now deteriorating rapidly, and all three species in the Yangtze River have been listed as state protected animals in Category I.

Dabry's sturgeon, *Acipenser dabryanus*
Dabry's sturgeon is a freshwater species that occurs only in the middle and upper reaches of the Yangtze River (Figure 1). It reaches at least 130 cm TL and 16.0 kg BW. Major research was done from 1972 to 1975. This species is easily propagated, and has excellent commercial aquaculture potential. The native population, however, sharply declined in the last two decades. Some fish still occur in the upper middle reach but few occurred in reaches below the Gezhouba Dam in recent years. See Zhuang et al. (1996, this volume) for a review of this species.

Chinese paddlefish, *Psephurus gladius*
The Chinese paddlefish is characterized by its sword-like rostrum and several osteological features that distinguish it from other fossil and living paddlefishes (Grande & Bemis 1991, 1996, Bemis et al. 1997 this volume). Rostral length (RL) is as much as 1/4 to 1/3 times the total length (TL = –17.134 + 3.8957 RL, N = 16, R = 0.9841, S = 9.9181, F = 249.366). Detailed biological investigations are impossible because *Psephurus* is now very rare. Migration patterns and spawning sites of the Chinese paddlefish are unknown, although it historically occurred in the East China Sea, so it is presumably anadromous (Anonymous 1988). It is also recorded from the Yellow River and Yellow Sea (Zhu et al. 1963, Li 1965).

Within the Yangtze River, Chinese paddlefish were commonly found in the mainstream, and sometimes in its tributaries including Tuojiang, Mingjiang, Jialinjiang, Qiantangjiang and Yongjiang rivers, and Dongting and Poyang lakes (Anon-

Table 3. Composition of spawning stock of *Huso dauricus* in the middle reach of the Amur River by sex.

TL range (cm)	Male			Female		
	x±SD	N	%	x±SD	N	%
160–180		0	0	174.0±1.0	3	5.3
180–200	192.3±0.6	3	2.7	190.1±6.4	8	14.0
200–220	212.3±4.1	11	10.1	214.4±3.8	7	12.3
220–240	231.7±5.8	17	15.4	230.7±4.8	10	17.5
240–260	250.5±5.9	18	16.4	253.0±3.8	12	21.1
260–280	271.6±5.7	27	24.5	269.5±5.8	6	10.5
280–300	288.7±6.7	11	10.1	292.9±7.6	7	12.2
300–320	308.6±5.6	12	10.9	307.5±6.4	2	3.5
320–340	329.6±5.8	6	5.4		0	0
340–360	358.0	1	0.9	342.0	1	1.8
360–380	368.3±2.9	3	2.7	375.0	1	1.8
380–400	390.0	1	0.9		0	0

SD = standard deviation; N = number of samples.

ymous 1988, Lin & Zeng 1987, also see Zhuang et al. 1996 for a map of the Yangtze River). Prior to 1980, mature individuals only occurred in the upper reaches of the Yangtze River, juveniles (4 cm TL) occurred in the Jiangan-Fongdu section of the upper Yangtze River, and fingerlings (75 to 250 mm TL, 5 to 95 g BW) occurred in the Xupu-Chongming section of the lower Yangtze River (Anonymous 1988). Juveniles ranging 80 to 530 mm TL were found on the eastern tidal beach of Chongming Island (i.e., the Yangtze estuary) from 1983 to 1985.

No report on the population structure of paddlefish is available. According to Lin & Zeng (1987), the maximum weight is 500 kg. The total length of 17 specimens, which were incidental-catches or carcasses found just below the Gezhouba Dam from 1981 to 1986, ranged from 148.8 cm to 262 cm TL (Figure 2). Individuals 182 to 244 cm TL accounted for 88.2% of this sample. The mean weight of 11 ripe individuals taken in 1973 from upper reaches of the Yangtze River was 37.5 kg (Anonymous 1988, average body weight 26.9 kg, range from 13.9 to 53.5 kg). The relationship between TL and BW was: $BW = 1.577 \times 10^{-7}\ TL^{3.5250}$ (N = 19, R = 0.9863, FF = 607.702). Ages of spawning individuals ranged from 8 to 12 years. Males with gonads at stage IV accounted for 75% of the sample, and the remaining 25% had testes at stage III (see Anonymous 1988). A single female found on 20 November, 1984, was 10 years old, 244 cm TL and 41.5 kg BW; its ovaries were at stage IV of development. It carried 338 000 eggs with average diameters of 2.45 mm. Yu et al. (1986) described a female paddlefish found just below the Gezhouba Dam, with gonads at stage IV and egg diameters ranging from 2.8 to 3.8 mm.

Reproduction and present status. – The spawning grounds for Chinese paddlefish are unknown. They are probably scattered and limited to reaches above Chongqing (Anonymous 1988). The Sichuan Fisheries Research Institute (1980) reported a spawning ground near Yibing as evidenced by the capture of a female with running eggs on 6 April 1980. The mean diameter of the eggs, which were fully distended with water, was 2.8 mm. The spawning site was described as 500 m in length, consisting of sand in the upper part and gravel or cobble in the lower part. Spawning presumably occurred on 4 April 1980, at a water temperature between 18.3 and 20.0°C, surface water velocity of 0.72–0.94 m s^{-1} and at a maximum water depth of about 10 m. They suggested that the spawning period may be from late March to early April. We think that natural spawning still exists in the upper reaches of Yangtze River because juveniles have been found there since the construction of Gezhouba Dam (Liu 1992). Spawning in reaches below Gezhouba Dam must have occurred from 1983 to 1985, as evidenced by the occurrence of juveniles in the Yangtze River estuary in the same period (Zhu & Yu 1987, Yu et al. 1986). However, no juveniles have been found below the Gezhouba Dam since 1986, and we think that perhaps these spawning grounds disappeared after 1986. The total number of individual Chinese paddlefish recovered below Gezhouba Dam has been declining since the middle of the last decade (Figure 3), which suggests that no recruitment to this population is occurring.

Artificial propagation of Chinese paddlefish has not been successful. This is in marked contrast to the ready availability of eggs and embryos of North American paddlefish, *Polyodon spathula* (e.g., Bemis & Grande 1992). We have difficulties catching broodstock because of the rarity of *Psephurus*. Moreover, captive ripe males and females have never been available at the same time. We also find adult *Psephurus* difficult to keep in captivity.

Conservation efforts. – Before 1983, it was permissible to take *Psephurus* throughout the river. The species has been fully protected since 1983, and protected stations for Chinese sturgeon and paddlefish were set up along reaches in Hubei and Sichuan provinces. Fishermen now usually release their incidental catches. Several attempts to artificially spawn paddlefish were unsuccessful, and we have

Table 4. Age composition of spawning stock of *Huso dauricus* in the middle reach of the Amur River.

Age range (years)	12–20	22–40	42–44	47	54
N	15	108	6	1	1
Percentage	11.5	82.4	4.6	0.75	0.75

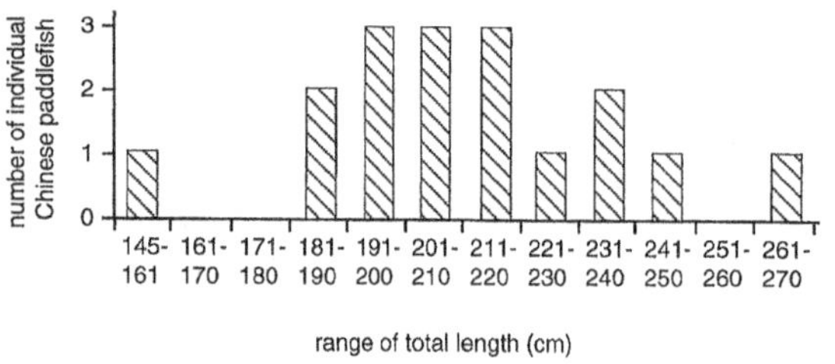

Figure 2. Size frequency distribution of seventeen adult Chinese paddlefish, *Psephurus gladius*, found below Gezhouba Dam between 1982 and 1986.

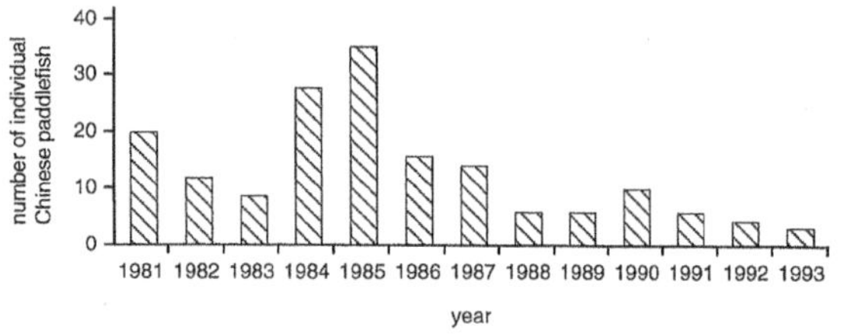

Figure 3. Total number of occurrences of adult Chinese paddlefish, *Psephurus gladius*, below Gezhouba Dam from 1981 to 1993.

been unable to hold brood fish for more than a month.

Chinese sturgeon, *Acipenser sinensis*

As presently understood, the Chinese sturgeon is restricted to the main channel of the Yangtze River, the Pearl River, and the East and South China Seas (Figure 1). There is still uncertainty about the taxonomy of the two populations in the Pearl and Yangtze rivers. The original specimen of *A. sinensis* (32 cm TL), collected by Reeves, may have been from the Pearl River (Anonymous 1988). Wu et al. (1963) considered the larger anadromous sturgeon from the Yangtze River to be *A. sinensis* [note that Zhu (1963), regarded all sturgeon in the Yangtze River and in the East China Sea to be *A. dabryanus*, an opinion not shared by later workers]. It has been suggested that the large anadromous sturgeons of the Pearl and Yangtze rivers should be considered as two isolated species (Anonymous 1988). The large anadromous sturgeon from the Pearl River seem to differ in morphology, spawning time and migration pattern from those in the Yangtze River (Zhou et al. 1994), but much more evidence needs to be collected to examine this taxonomic problem, particularly because very little information is available about the Pearl River population (Zhen 1989). The Pearl River population appears to be even more endangered than the one in the Yangtze River. A new hydroelectric project, the Changzhou Dam, will block spawning migrations (Chu et al. 1994).

Migration of the Yangtze River population. – Adults with gonads approaching maturity (in early stage III), arrive from the sea in June or July at the mouth of the Yangtze River to ascend its main channel (Anonymous 1988). The adults do not feed while in the river. They arrive at the Jingjiang reach, not far below the Gezhouba Dam, in September or October, where they overwinter. Ripe individuals were formerly found as far inland as the Jingsha River (Figure 1) during the following October and November, where they spawned. Prior to construction of the Gezhouba Dam, the migration distance was as long as 2500 to 3300 km.

Juvenile *A. sinensis* 7 to 38 cm TL occur in the Yangtze River estuary from the middle of April through early October. These are presumably one year old individuals (Wei et al. 1994, Anonymous 1988). Juveniles weighing a few kilograms can be found in coastal waters near the river mouth. Individuals from 25 to 250 kg in weight were registered in some fishing grounds of East China Sea and Yellow Sea (Anonymous 1988).

Structure of the Yangtze River population. – Some investigations on the structure of the spawning population have been performed (Ke & Wei 1992, Deng et al. 1991, Anonymous 1988, Deng et al. 1985). Females are larger than males (Figure 4) and the sex ratio is consistently about 1:1. We recorded the total lengths (TL) of 475 individuals taken between 1981 and 1993, and found it to range from 189 cm to 389 cm (mean = 273 cm). The 266 males in the sample ranged from 189 to 305 cm TL (mean = 242 cm), and the 209 females ranged from 253 to 389 cm TL (mean = 313 cm). Fish weighed 42.5 to 420 kg (mean of 213.7 kg). Males weighed 42.5 to 167.5 kg (mean of 85.4 kg), and females weighed 104.5 to 420 kg (mean of 217.3 kg). Total length is significantly cor-

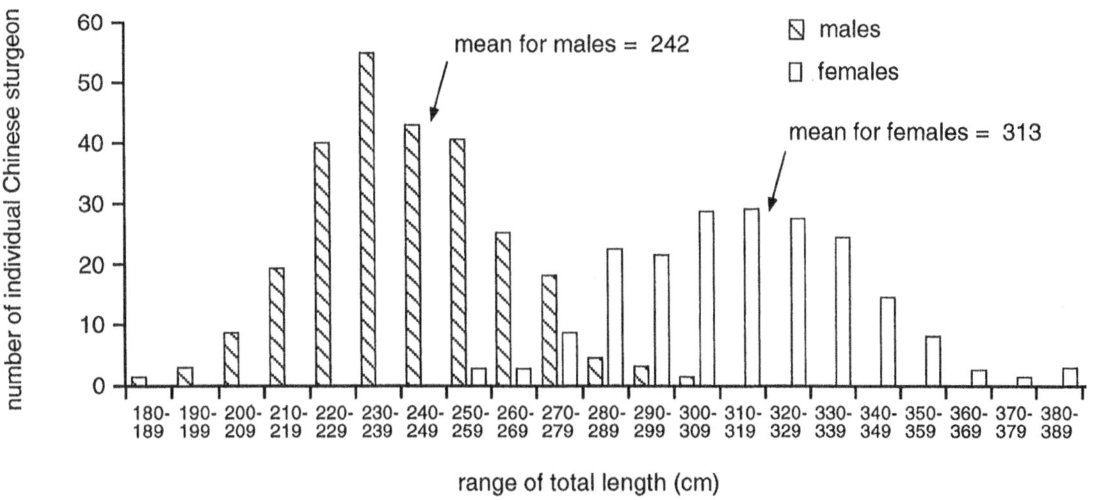

Figure 4. Size frequency distribution of spawning population of Chinese sturgeon, *A. sinensis*, from 1981 to 1993. Males and females are coded separately to show sexual dimorphism of this population. N = 475.

related with weight: $BW = 1.1969 \times 10^{-6}TL^{3.2981}$ (N = 269, r = 0.9452, $S_{BW.TL} = 25.58$, F = 2239.39).

From 1981 through 1993, 384 fish were used for age determination. This sample included 214 males, 169 females and 1 inter-sexual. Fish ranged from age 8 to 34 years (N = 383). The mean age of the spawning population was 17.0 years (14.0 years for males and 20.7 years for females, Figure 5). Males ranged between 8 and 27 years, with the majority between 10 and 18 years (90.2%); females ranged between 13 and 34 years with the majority between 16 and 27 years (86.4%). Deng et al. (1991) reported that fish making an initial spawning accounted for 84% of all males and 76% of all females. Our results differ. We determined whether reproductive marks occurred in all three structures used for aging (the first pectoral fin ray, the clavicle and cleithrum) in a subsample of 341 of the 384 specimens. Reproductive marks were indistinct in the remaining 43 specimens, in which the average ages were 14.3 years for males (N = 18) and 21.4 years for females (N = 25). The first column of Table 5 shows that individuals making an initial spawning run accounted for 66.3% of males and 44.4% of females. We found that the first spawning mark exactly recorded the age of first reproductive migrations, allowing us to

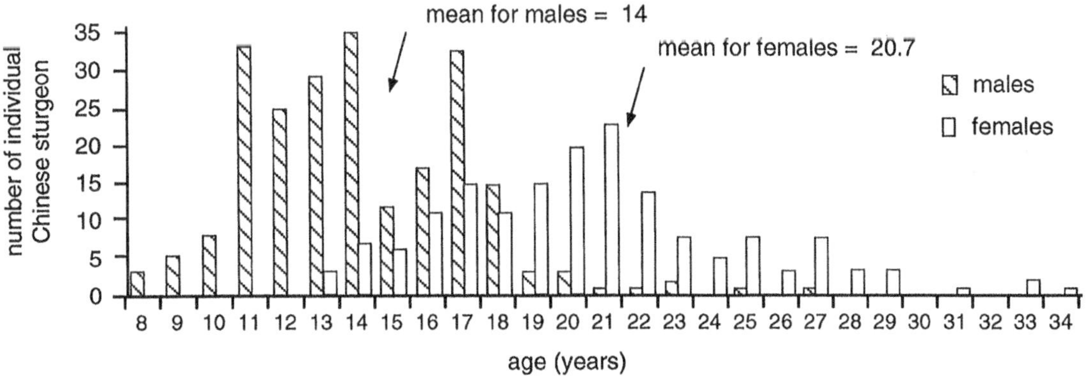

Figure 5. Age frequency distribution of spawning population of Chinese sturgeon, *A. sinensis*, from 1981 to 1993. N = 383.

determine the age at first reproduction for the entire subsample of 341. The ages of individuals which belong to the class of initial spawners are not significantly different from the age of first spawning for repeated spawners ($t = 0.485 < t_{0.005} = 2.358$). This demonstrates the validity of our determination of initial reproductive marks. It also suggests that the average age at initial reproduction of *A. sinensis* is 14.3 years. Because of the blurring of structures following the initial spawning mark, and of insufficient sample, we cannot verify the time intervals between repeated spawning migrations.

Chinese sturgeon exhibit different individual growth rates. Commonly, females are larger than males of the same age. For instance, 14 year old males had a mean TL of 248.0 cm (N = 35, SD = 13.00, range 221–274 cm) whereas females had a mean TL of 289.6 cm (N = 7, SD = 11.76, range 272–306). By age 18 years, males had a mean TL of 256.9 cm (N=15, SD=14.85, range 234–285 cm) and females had a mean TL of 300.4 cm (N=11, SD=26.00, range 263–340 cm). The growth rate of *A. sinensis* is one of the highest known among the species of Acipenseriformes (Holčík 1989, Smith 1985).

Some changes in age composition in recent years. – We were interested in examining age composition of the population as it may have changed since completion of the Gezhouba Dam in 1981. With very limited samples available, we classified them into four groups: group I (1981–1983), group II (1984–1986), group III (1987–1989) and group IV (1990–1993) (in 1990 only 3 specimens were available). ANOVA shows that the differences in age are not significant between each group for females and also for males before 1990. Males in group IV, however, are significantly older than those in groups I, II and III ($p < 0.01$). The mean age of group IV males is 17.1 years while those of group I, II and III are 13.0, 12.8 and 12.6 years, respectively. Furthermore, males younger than 12 years are very underrepresented (percentages younger than 13 years were 47.1%, 47.5%, 43.5% and 4.7% in groups I to IV, respectively). No fish born in 1982 and 1983 were found, and only two fish born in 1981 and 1984 were found. On the other hand, the number of repeated spawners was higher for group IV males (26.2, 28.6, 34.4 and 46.5% in group I, II, III and IV, respectively) but did not increase for females (52.0, 69.2, 65.0 and 42.6% in group I, II, III and IV, respectively). Taken together, these results suggest that recruitment of males to the population is declining at a dangerous rate.

Gonadosomatic index and its alteration after construction of Gezhouba Dam. – As mentioned above, two groups of spawning adults can be found in the river at the same time, one with gonads at stage III (or rarely II) and another with gonads at stage IV or V (or rarely VI). Before completion of the Gezhouba Dam, females stage III gonads exhibited these characteristics: gonadal fat 60–70% (September and October) or 20–60% (November and December); GSI 2.8–7.1% (mean 4.8%, September and October) or 2.4–9.0% (mean 7.3%, November and December); and egg diameter 2.0–2.5 mm (September and October) or 2.5–3.5 mm (November and December; summarized from Y. Z. Leng in Anony-

Table 5. Ages of *Acipenser sinensis* at initial and repeated reproduction (1981–1993).

	Actual ages of fish at initial reproduction				Ages of fish at subsequent reproduction				Combined estimate of age at initial reproduction [1]		
	N	%	x	Range (years)	N	%	x	Range (years)	N	x ± SD	Range (years)
Total	195	57.2	14.8	8–26	146	42.8	19.6	8–34	341	14.3 ± 3.4	8–26
Male	130	66.3	12.2	8–25	66	33.7	15.9	8–26	196	12.5 ± 2.6	8–25
Female	64	44.4	18.0	14–26	80	55.6	22.5	16–34	144	16.8 ± 2.9	14–26

[1] As explained in text, this estimate compiles data from the first column in the table together with data from analysis of spawning marks detected in fish that made a subsequent spawning migration. N = number of samples, x = mean age, SD = standard deviation.

mous 1988). A sample of 54 females with stage IV gonads had: no gonadal fat; GSI 11.77–25.95% (mean 19.11%); and egg diameter 4.0 × 4.2–4.5 × 5.0 mm. However, since the damming, this situation has changed. In Table 6 we report data on the developmental status of the ovary for 10 females sampled in October 1984. Only in individual 4 the ovary was in stage IV of development. For the other 9 specimens, GSI and egg diameter still had not attained conditions typical of stage IV even though the gonadal fat was exhausted. The shape and color of eggs were different, and sections showed that the yolk was being re-absorbed (Ke et al. 1985), so that the ovaries were regressing or degenerating. A similar phenomenon has been observed in other sturgeons (Kozlovsky 1968). In 1984, males also exhibited gonadal degeneration. Since 1984, we have investigated further the gonadal degeneration of spawning males and females (Figure 6). The absolute number of individuals exhibiting gonadal degeneration varies from year to year but the percentage of fish showing gonadal degeneration has decreased from the 1984 high.

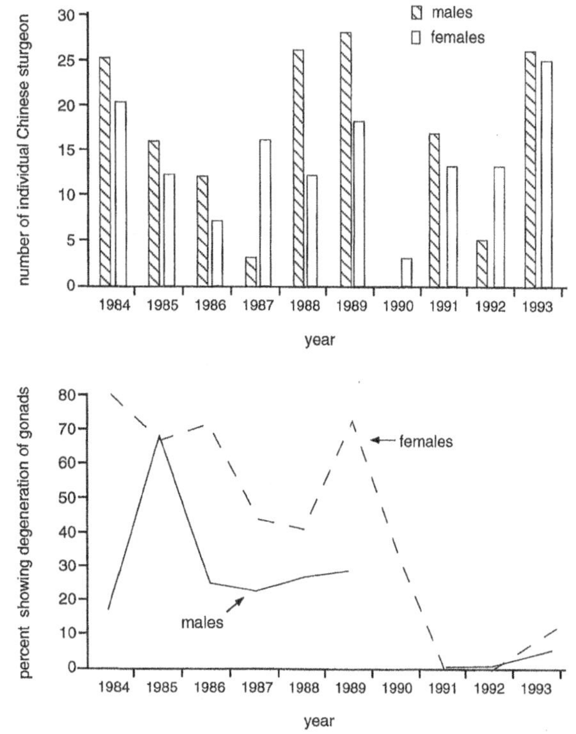

Figure 6. Annual variation in gonadal degeneration of the spawning population of Chinese sturgeon, *A. sinensis*.

Spawning. – Before the damming of the Yangtze River by Gezhouba Dam in January 1981, spawning areas for *A. sinensis* were distributed in the section from the upper Yangtze above Huling to the lower Jingsha River below Xingshi, covering at least 800 km of river length. At least 16 historical spawning sites are known (Anonymous 1988). However, only one major spawning site has been found below the dam. It is in a narrow area about 5 km long, just below the dam. A minor spawning site was also found at Huyiatan, 15 km below the dam, on 23 October 1986 and 14 November 1987. Spawning sites are usually located at a bend in the river where the river bottom consists of gravel or rocks. Typical spawning sites have a complex substrate, with water depth ranging from 3 to 40 m, although we believe that shallow areas are the preferred spawning sites. In 1993, an area 4 to 10 m deep was used for spawning (Kynard et al. 1995). Spawning occurs in a short period from middle of October to middle of November when water temperature is between 15.2–20.2°C (Anonymous 1988). In general, spawning occurs twice during the period.

We documented 19 spawning that took place be-

Table 6. Reproductive condition of ten female *Acipenser sinensis* from the Yichang section sampled in October 1984.

	1	2	3	4	5	6	7	8	9	10
Date	10/19	10/21	10/25	10/29	10/29	10/30	10/31	10/31	10/31	10/31
Body weight, kg	348.5	235.0	220.0	250.5	135.5	222.5	171.0	177.0	232.5	301.5
GSI, % [1]	4.9	5.0	8.5	23.3	5.4	5.3	5.7	8.8	8.3	9.0
Egg diameter, mm	2.2	2.1	2.2	4.0	2.7	2.4	2.6	2.6	2.7	2.1
Gonadal fat, %	0	0	0	0	0	0	0	0	0	0

[1] GSI, % = gonad weight/body weight × 100.

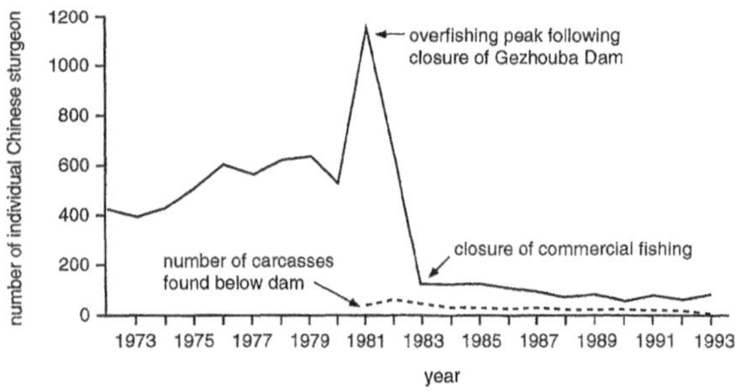

Figure 7. Annual landings of the migratory population of Chinese sturgeon, *A. sinensis*, in the Yangtze River.

low the Gezhouba Dam since it closed in January 1981. Except in 1981, we found spawning activities each year in the major spawning site. Because the spawning population was overfished below the dam in 1981 and 1982 (Figure 7), few fish were available to spawn in 1982, and as a result, spawning activities were scattered and abnormal. Table 7 shows that spawning dates after the damming are similar to those before the damming. In half of the years sampled, spawning only occurred once at the major spawning site, but in 1987 we found eggs in stomachs of copperfish for 17 days, which we interpret as evidence of two spawnings during this period. We are not sure what factors trigger spawning, although some researchers consider that water level (Anonymous 1988) or water temperature (Deng et al. 1991) are critical factors. Buckley & Kynard (1985) reported that shortnose sturgeon *A. brevirostrum* require a specific water velocity to spawn, and Chinese sturgeon may be similar. We believe that the available spawning sites are insufficient because ripe females can still be taken even after the spawning period is over.

Harvest and stock size of the migratory population in the Yangtze River. – Before 1981, sturgeon fisheries mainly occurred in the middle and upper reaches of Yangtze. Fishing was unlimited, but seasonal, and mostly in the fall. Gear include gill nets (upper reach) and row hooks (middle reach). Catch data are not available prior to 1972. Between 1972 and 1980, the annual mean catch of migratory adults was 517 individuals or 77 550 kg (150 kg per individual) for the entire river. When the river was dammed in January 1981, almost all of the fish trapped in a small area below the dam were caught in 1981 and 1982. As a consequence, the catch reached a peak of 1163,

Table 7. Hydrological characteristics of spawning areas used by *A. sinensis* at Yichang below Gezhouba Dam from 1983–1993.

		Water temperature (°C)		Water level (m)		Silt content (kg m^{-3})		Water velocity (m s^{-1})[2]	
Date	Days[1]	Start	Fluctuation	Start	Fluctuation	Start	Fluctuation	Start	Fluctuation
7.11.83	5	17.9	17.9–17.2	43.01	42.90–43.55	0.35	0.35–0.43	0.89	0.88–1.07
16.10.84	7	18.9	18.9–18.6	45.00	45.45–44.36	0.66	0.33–0.66	1.42	1.30–1.49
13.11.84	5	17.1	17.1–16.8	41.71	41.69–41.90	0.10	0.10–0.08	0.81	0.80–0.84
13.10.85	7	19.8	19.8–18.8	44.75	44.75–44.22	0.69	0.69–0.60	1.39	1.39–1.28
7.11.85	7	17.5	17.5–15.7	43.73	44.31–43.41	0.29	0.56–0.21	1.15	1.39–1.07
21.10.86	5	18.2	18.3–18.0	45.38	45.38–44.35	0.87	0.87–0.51	1.39	1.39–1.28
31.10.87	17	17.2	17.4–15.8	43.00	43.29–41.49	0.44	0.44–0.09	1.16	1.16–0.90
13.10.88	4	18.8	18.8–18.3	46.33	46.33–45.41	0.59	0.59–0.45	1.19	1.19–1.72
3.11.88	3	18.6	18.6–18.0	43.46	43.46–42.84	0.45	0.45–0.35	1.26	1.26–1.18
27.10.89	8	17.7	18.0–17.7	46.59	46.59–44.52	1.00	1.00–0.53	1.66	1.66–1.38
15.10.90	7	18.9	19.0–18.8	47.29	47.29–45.49	1.32	1.91–0.67	1.98	1.98–1.49
31.10.90	5	17.8	17.8–17.2	44.23	44.23–43.11	0.30	1.28–1.09	1.28	1.28–1.09
23.10.91	7	19.4	19.4–18.0	44.24	44.24–43.44	0.62	1.82–1.32	1.82	1.82–1.32
17.10.92	8	19.0	19.2–18.6	43.71	43.71–42.78	0.70	0.98–0.38	1.27	1.27–1.13
17.10.93	6	20.0	20.0–18.8	45.20	46.31–44.70	0.83	1.24–0.79	2.07	2.20–1.81
30.10.93	7	18.0	18.0–17.2	44.24	44.24–43.54	0.59	0.67–0.49	1.73	1.93–1.29

[1] Number of days during which sturgeon eggs were found in the stomach of *Coreius guichenoti*.

[2] Mean water velocity at the Moji monitoring section; other hydrological data provided by Yichang Hydrological Monitoring Station.

including 161 from the reaches above the dam (Fig. 7). Commercial fishing has been closed since 1983, and fishing for scientific or propagation purposes produces annual catches of about 100 individuals. In addition to fishing mortality, many individuals are killed by the dam, and are discovered as carcasses below it. The number of carcasses has declined as the number of adult sized fish trapped above the dam continues to decrease.

We tagged 57 adults collected from sites at Shashi (51) and at Yichang (6) in 1983 and 1984 in order to estimate the stock size. By the following year, we had recaptured 6 individuals, including 4 from the areas near the release sites and 2 from the sea near the mouth of the Yangtze. Using Peterson's method, we estimated that the mean stock size of the spawning stock of Chinese sturgeon was 2176 (946 and 4169 as 95% confidential intervals) in 1983 and 1984. Given alternate year spawning migrations, the annual stock size migrating into the river was about 1000 individuals.

No exact data on juveniles are available for years prior 1981. However, the stock of juveniles has decreased at the Yangtze estuary. In the 1960s, this stock supported a major commercial fishery, but it has declined to about 5000 fish per year (Wei et al. 1994).

Conservation efforts. – All commercial fishing has been banned since 1983 and a series of protective measures have been taken, including setting up protection stations along the river and a station at Yichang for artificial propagation and research. In 1988, *A. sinensis* was listed as a state protected animal in class I. In 1983, the first successful artificial spawning was made by the Yangtze River Fisheries Institute (Fu et al. 1985). Since then, an average of 250 000 larvae and some juveniles have been stocked into the river annually. The total number stocked was 2.8 million larvae and 17 000 juveniles (size 2–10 g) from 1983 through 1993. Both the general public and fishermen are now protective of sturgeon, and willingly release incidental catches. For instance, in just the Yichang section, we are aware of 148 incidental catches that were released between 1986 and 1993.

Recommendations

Amur River

Stocks of *Huso dauricus* and *Acipenser schrenckii* in the Amur River are declining dramatically due to overfishing. If we ignore this situation, then we will loose the sturgeon fishery and perhaps even both species. Loss of sturgeon fisheries has occurred in many other places (Holčík et al. 1989, Smith 1985, Galbreath 1985, Folz & Meyers 1985, Pinter 1991). We recommend that the following measures be immediately taken: (1) modify existing regulations for sturgeon fishery management; (2) reduce fishing effort, including restrictions on the number of boats, the type of gear used, and the maximum allowable catch; (3) strictly forbid fishing juveniles and to strengthen artificial propagation; (4) strengthen scientific investigations on the stock to establish a rational maximum allowable catch; (5) work out a joint pledge for sturgeon conservation in the shared waters of China and Russia.

Yangtze River

The construction of Gezhouba Dam greatly affected stocks of *P. gladius* and *A. sinensis*. If no measures are taken, paddlefish will disappear in the middle and lower reaches of Yangtze in the very near future, placing the species on the verge of extinction. Wild populations of *A. dabryanus* are declining despite protection. The typical anadromous species, *A. sinensis*, is increasingly threatened. As the largest dam ever constructed, The Three Gorges Project, nears completion (1997), the habitat for both species of sturgeons and the paddlefish will be further harmed. This huge dam will be 175 m high and is only 47 km upstream from the Gezhouba Dam. Furthermore, it will use 1.3 million tons of gravel and rocks from the reach below Gezhouba Dam as construction material (Ke & Wei 1993), directly damaging the largest known existing spawning ground for Chinese sturgeon and many other commercial fishes. Despite these serious impacts, some measures can and should be taken, such as: (1) establishing protected areas for juvenile Chinese sturgeon at the Yangtze estuary to permit effective recruitment of Chinese sturgeon including the Yichang Protected Area for spawning and the Chongming Protected Area (Wei et al. 1994); (2) characterizing habitats for Chinese sturgeon including spawning sites, summer and winter grounds using telemetry and related techniques (Buckley & Kynard 1985) to enable agencies to protect these sites or, possibly, to re-create additional spawning areas below the dam; (3) conducting studies of paddlefish, particularly on artificial culture, which may be the only way to maintain this species.

Acknowledgements

The Yichang Fisheries Management Station coordinated the field surveys, and the Bureau of East China Sea Fisheries Management assisted our work in the Yangtze Estuary. We thank Boyd Kynard for critical review of the initial manuscript, and James D. Williams, Theodore I. J. Smith, Frank M. Parauka and Kim Graham for providing us with valuable literature. We extend special thanks to Vadim Birstein, Robert H. Boyle and John Waldman for organizing the international sturgeon conference. William E. Bemis substantially revised the text of our original manuscript and drew the map and figures.

References cited

Anonymous. 1988. The biology of the sturgeon in Yangtze and their artificial propagation. The Yangtze Aquatic Resources Survey Group, Sichuan Scientific & Technical Publishing House. 284 pp. (in Chinese).

Bemis, W.E. & L. Grande. 1992. Early development of the actinopterygian head. I. External development and staging of the paddlefish *Polyodon spathula*. J. Morphol. 213: 47–83.

Bemis, W. E., E.K. Findeis & L. Grande. 1997. An overview of Acipenseriformes. Env. Biol. Fish. (this volume).

Buckley, J. & B. Kynard. 1985. Habitat use and behavior of prespawning and spawning shortnose sturgeon, *Acipenser brevirostrum*, in the Connecticut River. pp. 111–117. *In*: F.P. Binkowski & S.I. Doroshov (ed.) North American Sturgeons: Biology and Aquaculture Potential, Dr W. Junk Publishers, Dordrecht.

Chen, S.D. & X.G. Zhou. 1992. A study of the artificial propagation of *Acipenser schrencki*. Heilongjiang Fisheries 4: 8–12 (in Chinese).

Deng, X., Z.L. Deng & M.Y. Cai. 1991. Spawning population characteristics of *Acipenser sinensis* in Yangtze River just below Gezhouba Dam. pp. 235–242. *In*: P. Williot (ed.) Acipenser, CEMAGREF Publ., Bordeaux.

Deng, Z.L., Y.G. Xu & Y. Zhao. 1991. Analysis on *Acipenser sinensis* spawning ground and spawning scales below Gezhouba Hydroelectric Dam by the means of examining the digestive contents of benthic fishes. pp. 243–250. *In*: P. Williot (ed.) Acipenser, CEMAGREF Publ., Bordeaux.

Folz, D.J. & L.S. Meyers. 1985. Management of the lake sturgeon, *Acipenser fulvescens*, population in the Lake Winnebago system, Wisconsin. pp. 135–146. *In*: F.P. Binkowski & S.I. Doroshov (ed.) North American Sturgeons: Biology and Aquaculture Potential, Dr W. Junk Publishers, Dordrecht.

Fu, Z.J. 1985. The artificial propagation of *Acipenser sinensis* Gray below Gezhouba Dam. Freshwater Fisheries 1: 1–5 (in Chinese).

Galbreath, J.L. 1985. Status, life history, and management of Columbia River white sturgeon, *Acipenser transmontanus*. pp. 119–125. *In*: F.P. Binkowski & S.I. Doroshov (eds.) North American Sturgeons: Biology and Aquaculture Potential, Dr W. Junk Publishers, Dordrecht.

Grande, L. & W. E. Bemis. 1991. Osteology and phylogenetic relationships of fossil and Recent paddlefishes (Polyodontidae) with comments on the interrelationships of Acipenseriformes. J. Vert. Paleo. 11, supplement 1: 1–121.

Grande, L. & W.E. Bemis. 1966. Interrelationships of Acipenseriformes with comments on "Chondrostei". pp. 85–115. *In*: M.L.J. Stiassny, L.R. Parenti & G.D. Johnson (eds.) Interrelationships of Fishes. Academic Press, New York.

Holčík J. 1989. The freshwater fishes of Europe, Vol. 1, Part II, General introduction to fishes, Acipenseriformes. AULA-Verlag, Wiesbaden. 470 pp.

June, F.C. 1977. Reproductive patterns in seventeen species of warmwater fishes in a Mississippi reservoir. Env. Biol. Fish. 2: 285–296

Ke, F.E. 1984. Effects of Gezhouba Dam on Chinese sturgeon: investigation of the dynamics of the sturgeon. Freshwater Fisheries 3: 16–19 (in Chinese).

Ke, F.E., D.G. Hu, G.L. Zhang & Q.W. Wei. 1985. An observation of gonadal degeneration of the spawning population of Chinese sturgeon below Gezhouba Dam. Freshwater Fisheries 4: 38–41 (in Chinese).

Ke, F.E., Q.W. Wei, D.G. Hu, G.L. Zhang & P. Zhuang. 1992. Investigations of ecological effect of Gezhouba Hydroelectric Project on Chinese sturgeon stock. pp. 3–4. *In:* Papers Presented at the International Symposium on Sturgeon and Paddlefishes, Chongqing.

Ke, F.E. & Q.W. Wei. 1992. Investigations on the structure of spawning population of Chinese sturgeon (*Acipenser sinensis* Gray) and the estimate of its stock. Freshwater Fisheries 4: 7–11 (in Chinese).

Ke, F.E. & Q.W. Wei. 1993. Effects of Yangtze Gorges Project on fishery resources and remedy measures. Freshwater Fisheries 6: 15–18 (in Chinese).

Kozlovsky, A.D. 1968. Resorption of sexual products in fishes as a stimulus to biological modification. Prob. Ichthyol. 8: 803–807.

Krykhtin, M. & V. Svirskii. 1996. Endemic sturgeons of the Amur River: kaluga, *Huso dauricus* and Amur sturgeon, *Acipenser schrencki*. Env. Biol. Fish. (this volume).

Kynard B., Q.W. Wei & F.E. Ke. 1995. Use of ultrasonic telemetry to locate the spawning area of Chinese sturgeon. Chinese Sci. Bull. 40: 668–671.

Li, S.Z. 1965. Discussion on the ichthyofauna in Yellow River. Chinese J. Zool. 7: 217–222.

Lin, F.S. & X.Z. Zeng. 1987. Precious and rare aquatic animals in China. pp. 28–31. Zhejiang Sci. & Tech. Press, Hongzhou (in Chinese).

Pinter, K. 1991. Sturgeon in Hungary, past and present situation. pp. 173–178. *In*: F.P. Binkowski & S.I. Doroshov (eds.) North American Sturgeons: Biology and Aquaculture Potential, Dr W. Junk Publishers, Dordrecht.

Ren, M.L. 1981. The fishes of the Amur River. pp. 5–10. Heilongjiang People's Publishing House (in Chinese).

Sichuan Fisheries Research Institute, 1980. A preliminary report on a spawning ground for *Psephurus gladius*. Sichuan Fisheries 3: 39–40 (in Chinese).

Smith, T.I.J. 1985. The fishery, biology, and management of Atlantic sturgeon, *Acipenser oxyrhynchus*, in North America. pp. 61–72. *In*: F.P. Binkowski & S.I. Doroshov (eds.) North American Sturgeons: Biology and Aquaculture Potential, Dr W. Junk Publishers, Dordrecht.

Wei, Q.W., F.E. Ke, P. Zhuang, J.D. Luo & W.H. Yang. 1994. On protection of juvenile Chinese sturgeon (*Acipenser sinensis* Gray) in the Yangtze Estuary. pp. 435–444. *In*: The Proceedings of the First Conference on National Parks and Protected Areas in East Asia, Environmental Science Press, Beijing.

Wu, X.W. 1963. The economic animals of China – the freshwater fishes. pp. 12–16. Science Press, Beijing (in Chinese).

Yu, Z.T., Z.L. Deng, Y. Zhao & X. Huang. 1986. Observations on the gonad development of *Psephurus gladius* (Martens) in Changjiang River below Gezhouba Dam. Acta Hydrobiol. Sinica 10: 293–294 (in Chinese).

Zhou, J., D.K. Si & Z.L. Deng. 1994. An investigation of *Acipenser sinensis* in Xijiang (Pearl River). Guangxi Freshwater Fisheries 2: 5–9 (in Chinese).

Zhen, C.Y. 1989. Ichthyography of Pearl River. pp. 18–20. Science Press, Beijing (in Chinese).

Zhang, J.M. 1985. Fishery resources of Heilongjiang Province. pp. 352–389. The Chaoxian Nationality Publishing House of Heilongjiang (in Chinese).

Zhu, C.D & N. Yu. 1987. A preliminary study on the morphology, growth and feeding habits of young *Psephurus gladius* in the estuary region of the Changjiang (Yangtze) River. Acta Hydrobiol. Sinica 11: 289–298 (in Chinese).

Zhuang, P., F. Ke, Q. Wei, X. He & Y. Cen. 1996. Biology and life history of Dabry's sturgeon, *Acipenser dabryanus*, in the Yangtze River. Env. Biol. Fish. (this volume).

Zhu, Y.D. 1963. Ichthyography of the East China Sea. pp. 90–94. Science Press, Beijing (in Chinese).

Zoology Institute of Chinese Academy of Science. 1979. The Fishes of Xingjiang Province. pp. 6–8, The Xingjiang People's Press (in Chinese).

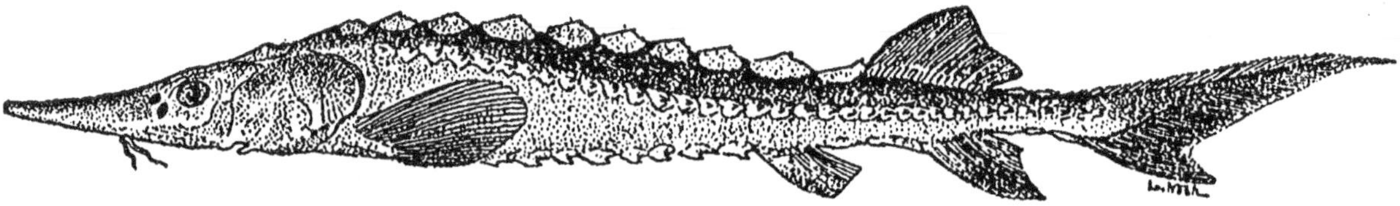

Reproduction of *Acipenser dabryanus* Duméril, 1868, 31.8 cm SL, probably from the Dongting Lake, Yangtze River, in J.T. Nichols (1943) 'The Fresh-water Fishes of China'.

Environmental Biology of Fishes **48**: 257–264, 1997.

Biology and life history of Dabry's sturgeon, *Acipenser dabryanus*, in the Yangtze River

Ping Zhuang, Fu'en Ke, Qiwei Wei, Xuefu He[1] & Yuji Cen
Yangtze River Fisheries Research Institute, Chinese Academy of Fishery Sciences, Shashi, Hubei, 434000, China
[1] *Department of Biology, Southwest Normal University, Chongqing, Sichuan, 630715, China*

Received 2.8.1994 Accepted 11.3.1996

Key words: Gezhouba dam, population size, conservation

Synopsis

Dabry's sturgeon, *Acipenser dabryanus*, is a relatively small (130 cm, 16 kg) and now rare sturgeon restricted to the Yangtze River Basin. It behaves as a resident freshwater fish, does not undertake long distance migrations (except for spawning), and lives in a variety of habitats. It historically spawned in the upper Yangtze River, but the spawning sites are unknown. *Acipenser dabryanus* reaches maturity earlier than do other Chinese sturgeons, which gives the species aquaculture potential, and artificial spawning has been carried out. However, the native population in the Yangtze has sharply declined in the last two decades due to overfishing, pollution and habitat alteration and destruction, especially since the construction of the Gezhouba Dam, which was built in 1981 across the Yangtze River at Yichang, Hubei Province. Since 1981, Dabry's sturgeon rarely occurs below the Gezhouba Dam because downstream movements are blocked. Clearly, conservation of Dabry's sturgeon must be emphasized. Conservation methods may include protecting habitats, controlling capture and stock replenishment.

Introduction

Dabry's sturgeon, *Acipenser dabryanus*, is one of two species of acipenserids found in the Yangtze (= Changjiang) River, China. The second species is the Chinese sturgeon, *A. sinensis*. More than a thousand years before the Christian era, external anatomy, general habits and characteristics, fishing methods and supposed medicinal value of sturgeons were described in ancient Chinese literature. However, the two species of sturgeons in the Yangtze River were not scientifically distinguished until the middle of the 19th century. *Acipenser dabryanus* was described by Duméril (1868) based on a type specimen collected from the Yangtze River by M. Dabry and now kept in the Museum National D'Histoire Naturelle, Paris, France (Zhang et al. 1993). Günther (1873), Wu (1930), Nichols (1943), Wu et al. (1963), and Fu et al. (1988) subsequently studied the morphology of this species using additional specimens from the Yangtze River. Until the 1960s, there was some taxonomic confusion between Dabry's sturgeon and Chinese sturgeon (Zhu et al. 1963, Wu et al. 1963). Now, we accept that Dabry's sturgeon is a freshwater species restricted to the Yangtze River, whereas the Chinese sturgeon is an anadromous species that inhabits both the Yangtze and Pearl rivers (see Wei et al. 1997 this volume for data on the Chinese sturgeon, *A. sinensis*).

Dabry's sturgeon is also called the Yangtze sturgeon, Changjiang sturgeon or river sturgeon, and it was an important commercial fish in the middle and

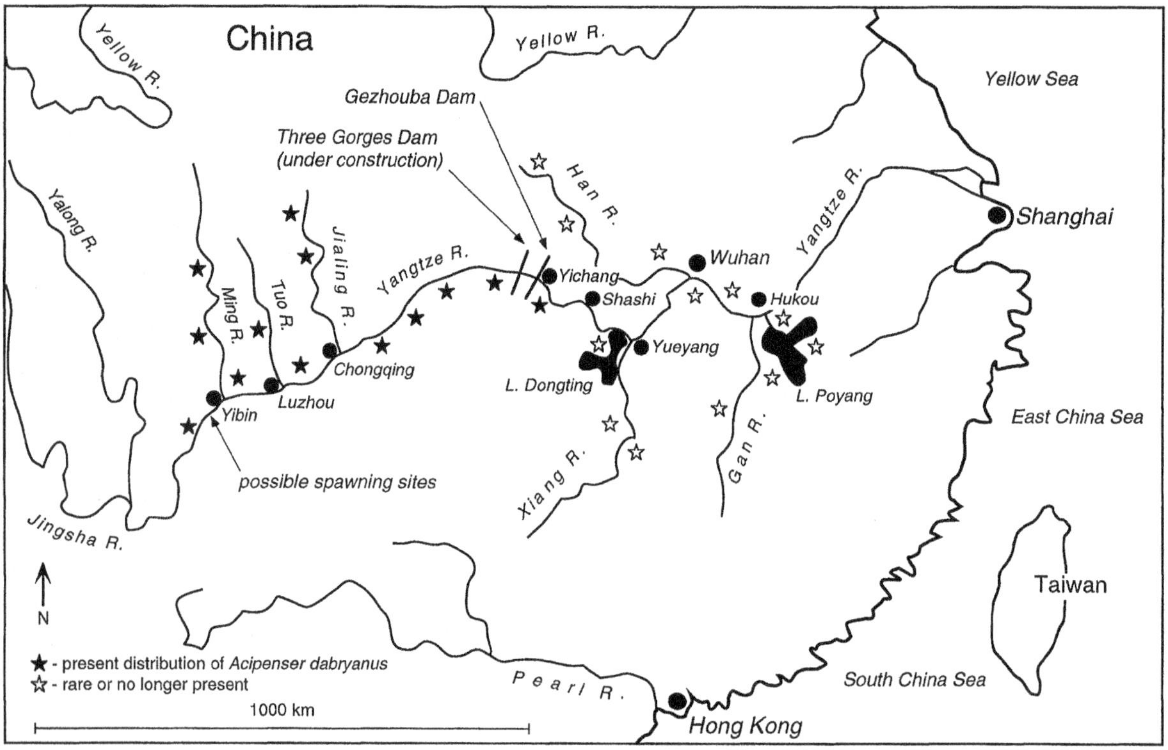

Figure 1. Past and present ranges of *Acipenser dabryanus*. Data from Nichols (1943), Wu et al. (1963), Yang (1986), Zhang (1988), Ke et al. (1989), He (1990), Zen et al. (1990) and the research of the authors. Names of places on the map are consistent with those used in the New York Times Atlas of the World, Third Revised Concise Edition.

upper reaches of the Yangtze River (Yang 1986, Zen et al. 1990, He 1990). As a result of overfishing, pollution and habitat alteration and destruction, Dabry's sturgeon populations have been declining for the last two decades (Wu 1990). The species now rarely occurs in reaches below the Gezhouba Dam, which was built in 1981 across the Yangtze River at Yichang, Hubei Province (Figure 1). Commercial capture of Dabry's sturgeon in the Yangtze River has been banned since the early 1980s when it was listed as an endangered species under the state's special protection in Category 1. Unfortunately, this listing seems to have occurred too late to significantly impact conservation of wild populations of Dabry's sturgeon.

This paper reviews the general biology and life history of Dabry's sturgeon and provides a source

Table 1. Meristic features of adult Dabry's sturgeon (modified from Fu 1988).

Character	Range	Mean	SD	CV	N
Gill rakers	32–55	45.6	0.95	3.99	18
Dorsal fin rays	44–57	51.3	0.89	3.56	16
Anal fin rays	25–36	30.4	0.68	2.79	17
Dorsal scutes	8–13	10.3	0.25	1.11	19
Lateral scutes	26–38	32.6	0.31	1.46	22
Ventral scutes	9–13	10.6	0.17	0.79	22

of pertinent information that can be quickly located in order to make conservation recommendations.

Description

Dabry's sturgeon is a freshwater fish, attaining a length of more than 130 cm, and weight of more than 16 kg. It is very similar in shape to similar sized Chinese sturgeon, but Chinese sturgeon achieve much larger sizes than do Dabry's sturgeon. The body of young Dabry's sturgeon is very rough because of many bony plates on the skin. This may provide a good distinction between young Dabry's sturgeon and Chinese sturgeon (Fu 1988). In addition, at similar total lengths, the rostrum of Dabry's sturgeon is slightly shorter than that of Chinese sturgeon. Adult Dabry's sturgeon have more than 30 gill rakers whereas Chinese sturgeon have fewer than 28.

Above the lateral row of scutes, the body of Dabry's sturgeon is dark gray, brown gray or yellow gray in color; the rest of the body is milky white in color. The lateral row of scutes is a distinct color demarcation line until the fish reaches adult size. Table 1 shows some meristic characteristics of adults.

Distribution and habitat

Dabry's sturgeon is restricted to the Yangtze River system. Historically, under the natural, unaltered conditions that existed until the middle of the 20th century, Dabry's sturgeon widely inhabited the upper and middle reaches of the Yangtze River and its large tributaries, including the Ming, Tuo, Jialing, Xiang and Han rivers, as well as the large lakes linking up with the Yangtze River (Figure 1). At present, Dabry's sturgeon is mainly distributed in the upper main stream of the Yangtze River as it passes through Sichuan Province. It also enters major tributaries, including the Ming, Tuo, and Jialing rivers. Occasionally, it occurs in the lower and middle sections of the Yangtze River. It is now rare in Dongting Lake (Hunan Province) and Poyang Lake (Jiangxi Province), although it was formerly reported from these places.

Dabry's sturgeon is a potamodromous freshwater fish that inhabits sandy shoals, with silt ground and gentle water flow. When the water level rises in the mainstream in the spring, the fish move into the tributaries to feed. Young individuals often stay in sandy shallows, and occur frequently in stretches between Luzhou and Jiangjing, Sichuan Province, where the current velocity is not rapid (Zhang et al. 1988). Dabry's sturgeon is often found in areas where there are drainage pipelines of waste water that promote growth of prey foods. Although Da-

Table 2. Percent occurrence of different foods in different life intervals of Dabry's sturgeon (modified from Zen et al. 1990).

Food category	Standard length ranges (cm)			
	Below 10	10–15	15–40	Above 40
Oligochaeta	100.0	50.0	10.5	29.8
Plecoptera	0.0	8.3	18.4	0.0
Ephemeroptera	0.0	16.7	28.9	14.8
Odonata	0.0	0.0	34.2	29.8
Chironomidae	0.0	33.3	31.6	22.2
Other insects	0.0	8.3	15.6	31.0
Plankton	50.0	0.0	0.0	3. 87
Bacillariophyta	50.0	8.3	5.3	0.0
Chlorophyta	50.0	8.3	13.2	3.7
Other plants	0.0	16.7	10.5	37.0
Shrimps	0.0	50.0	13.2	29.8
Gobiidae	0.0	66.7	36.8	18.5

bry's sturgeon is adaptable to a variety of habitats, individuals prefer sublittoral areas 10–20 m from the river bank, with current velocity of about 1 m s^{-1}, water depth of 8–10 m, sandy silt ground and an abundance of detritus and benthic organisms during most of the year.

Feeding habits

Feeding habits of Dabry's sturgeon vary with age, season and habitat. Their food items are wide-ranging. Although Dabry's sturgeon prefer living animal prey, seeds, leaves and stems of plant remains are frequently found in analyses of stomach contents. Dietary variation during different life stages is shown in Table 2. Young fish initially feed almost exclusively on zooplankton and oligochaetes. Older individuals eat oligochaetes and small fishes, such as gobies, as well as many other food items, including chironomids, odonates, and aquatic plants (Zen et al. 1990). Tu (1980) reported that a 12 cm Dabry's sturgeon ate gobies up to 2 cm in length. From May to July, the diet is broadest. Dabry's sturgeon feeds throughout the winter. Typically, feeding is more active at night than during the day.

Feeding habits of Dabry's sturgeon are influenced by environmental factors, such as water level fluctuations. Because benthic food organisms are concentrated in sublittoral areas, these are the preferred feeding grounds.

Tu (1980) studied the feeding intensity (F) of Dabry's sturgeon as a function of size. Feeding intensity is calculated by the following equation: $F = DW/BW \times 10000$, where F is the feeding intensity, DW is the diet weight in stomach, and BW is the total body weight. Feeding intensity varies indirectly with standard body length according to the following relationship: $F=3337L^{-1.0213}$.

Age and growth

We usually age specimens of Dabry's sturgeon by grinding sections of either a dorsal scute or an operculum and counting the annuli. Table 3 shows the results of a study of individuals from upper reaches of the Yangtze River. In this sample, 92% of the individuals were younger than 3 years (Fu 1988). This is consistent with other studies that suggest that Dabry's sturgeon achieves maturity faster than other sturgeons. Three year old fish can reach 3 kg in weight. Between 3–4 years and 7–8 years, growth rates in wild populations usually decline at sexual maturation. Males reach sexual maturity at 4–6 years of age and females at 6–8 years of age. In pond culture, Dabry's sturgeon can grow very rapidly (up to 3. 5 kg each year, L. Ling personal communication). Females generally grow slightly faster in length than males, and at maturity, they usually weigh more than do males of the same length or age because of their large ovaries.

The relationship between age (a) and length (L) for the ages contained in the data set is $L = 36 + 9.9a$. This relationship does not follow the von Bertalanffy growth model. The relationships between body

Table 3. Growth of wild Dabry's sturgeon in upper reaches of Yangtze River (modified from Zen et al. 1990 and Fu 1988).

Age	Weight (kg)		Length (cm)		N
	Mean	Annual increase	Mean	Annual increase	
0	0.08		21.3		66
1	0.65	0.55	44.1	22.8	24
2	1.30	0.68	55.0	10.9	8
3	3.05	1.75	68.2	13.2	6
4	4.25	1.20	77.5	9.3	4
5	6.60	2.35	86.5	9.3	4
6	8.60	2.00	95.5	9.0	3
7	11.85	3.25	101.0	5.0	3
8	14.70	2.85	106.0	5.0	2

weight (BW) and length (L) before and after sexual maturity are as follows:

$$lgBW = 6.005 + 2.72lgL$$
(males and females before sexual maturity);
$$lgBW = 6.175 + 2.72lgL$$
(males after sexual maturity);
$$lgBW = 6.219 + 2.72lgL$$
(females after sexual maturity).

Movement

Dabry's sturgeon generally behaves as a resident fish and does not undertake long migrations. This behavior resembles that of Amur sturgeon, *A. schrenckii*, in the Amur River (Nikolski 1960). Historically, some individuals of Dabry's sturgeon reached the lower-middle section of the Yangtze River below Yichang (Figure 1), but this was before the construction of the Gezhouba Dam, and such migrations are now impossible. Most of the available information about movements is based upon recapture of tagged fish. Zhang (1988) reported that a tagged fish swam 97 km downstream for 6 days before recapture, but this is the only report of movement over such a long distance. Most tagged individuals only moved several kilometers either up or downstream before recapture in 2 days to 8 months after releasing.

Dabry's sturgeon swim upstream for spawning during spring floods. Spawning fishes do not aggregate to swim upstream together as a group but instead move individually. After spawning, the spent fish move slowly back downstream to the sandy shoals where they feed intensively. Sometime they enter large lakes to feed. There is a record that Dabry's sturgeon was historically found far downstream in Anhui Province, more than 2000 km from the probable spawning areas (He 1990).

Reproductive biology

Males start to mature at 4 years, and all are mature by 7 years of age. At maturity, males weigh more than 4. 5 kg. Females usually do not mature until at least 6 years of age; all are mature by 8 years. Mature females generally weigh at least 9 kg. In natural populations, only 6. 7% of individuals reached sexual maturity (4. 8% of males and 1. 9% of females, Zen et al. 1988). It is hard to identify the sex of the fish from external characteristics. When the body weight of female Dabry's sturgeon reaches 2. 5–5. 0 kg, the ovary can reach developmental stage II; by 5 kg in weight, the ovary is at stage III (stages according to classification of Conte et al. 1988). The mature ovary accounts for 2/3 of the volume of abdominal cavity and the ovary of gravid females may comprise 10.0–18.8% of the body weight. Gravid females can contain from 57 000 to 102 000 eggs. Mature eggs are gray to black and range from 2.7–3.4 mm in diameter. Eggs are sticky, firmly adhere to stones on the bottom of the river after breeding, and are not readily eaten by other fishes. We have never found that copperfish and yellow catfish eat Dabry's sturgeon eggs although those species usually eat many eggs of Chinese sturgeon during their spawning period. Male Dabry's sturgeon can spawn annually, but most females cannot. The time required to develop mature eggs is unknown, but apparently it is longer than one year.

Natural and artificial spawning

Based on the capture of ripe individuals, we conclude that spawning occurs in the spring, although some individuals may spawn in the fall (Yang 1986, Zen et al. 1990). Since the late 1950s, researchers have searched for spawning sites, but they remain unknown despite considerable effort. Mature fish are unknown in the reaches below Yibin, Sichuan Province, whereas young are often caught in the reach below Yibin. Thus we infer that the spawning areas are in upper mainstream of the Yangtze, above Yibin (Figure 1). The spawning environment probably has a rubble, cobble and gravel bottom, clear water, water velocity of 1.2–1.5 m s^{-1}, water depth of 5–15 m.

Xie (1979) reported on early development (see Bemis & Grande 1992 for SEM stages of paddlefishes, which are very similar). At 17–18° C, the first cleavage occurs 3–4 hours after fertilization; the mid-gastrula stage occurs 28–30 hours after fertilization, neurulation is complete 43–45 hours after fertilization, and hatching occurs 115–117 hours after fertilization. Newly hatched embryos, about 4 mm long, swim erratically and vertically due to their large yolksac. Three days after hatching, the larvae can balance themselves. Four days after hatching, the pectoral fin appears, with dorsal fin appearance on day 5 and anal fin appearance on day 6. In 10–11 days, most of the yolk is absorbed and the mouth parts are developed so the young are ready to feed. Yolk is depleted 12 days after hatching. Young fish immediately swim downstream. They can be found in reaches at Luxian, Hejiang and Jiangjing, about 100–200 km below the supposed spawning areas.

Artificial stripping of Dabry's sturgeon was reported by Xie et al. (1976) using mature fish caught in the Jingsha River. Subsequently, Tian et al. (1986) reported artificial stripping of Dabry's sturgeon domesticated in ponds from juvenile to adult. Early workers used injections of pituitary glands of Chinese or Dabry's sturgeon to induce ovulation. Induction included two injections, with the initial (or primary) injection separated from the second (or resolving) injection by 8 hours. Ovulation occurred within 14–24 hours after administration of the second injection. More recently, luteinizing hormone releasing hormone analog (LHRH-A) has been used to induce ovulation of Dabry's sturgeon.

Artificially propagated interspecific hybrids of Dabry's sturgeon (male) and Chinese sturgeon (female) exhibit significant hybrid vigor, prove to be fertile, and seem to be a promising candidate for commercial production. Hybrid progeny add weight much faster than do Dabry's sturgeon, and about as fast as Chinese sturgeon, before 3 years of age. At 6 years of age, female hybrids weigh more than 40 kg, much larger than Dabry's sturgeon. Furthermore, female hybrids reach maturity at 6 years of age, 2 years earlier than most female Dabry's sturgeon, and male hybrids reach maturity at 3 years of age, which is also earlier than male Dabry's sturgeon. Some of the hybrids are hermaphrodites (D.J. Xie personal communication).

Aquaculture, fisheries, and conservation

Domestication of Dabry's sturgeon started in the early 1970s (Xie 1979). Because it matures rapidly, it was expected that the species had potential both for intensive aquaculture in ponds and for extensive aquaculture in lakes and reservoirs by artificial stocking (Jia et al. 1981, Yang 1986, He 1990). However, limited experience with production techniques and poorly developed markets prevented early success in this industry. Only in the last 10–15 years with the improvement of propagation techniques has commercial culture of Dabry's sturgeon been possible (Xie 1992).

Dabry's sturgeon was an important commercial species in the upper Yangtze River, near Yibin, Sichuan Province (Figure 1). In the 1960's, landings of Dabry's sturgeon reached 10% of the total harvest of Hejiang Fishing Squad (He 1990). In the 1970s, up to 5000 kg of Dabry's sturgeon could be caught in a spring season from the Yibin reach. Because adult Dabry's sturgeon are uncommon and do not like to aggregate, young individuals, weighing several hundred grams, made up most of the harvest. Since the 1970s, however, total landings of Dabry's sturgeon sharply declined, and it now seldom occurs below Gezhouba Dam.

Possible causes of decline in Dabry's sturgeon populations are: (1) *Unsuitable fishing methods*. Fishermen repeatedly have decreased the mesh size of nets, so that they take many small fish. This method is harmful to the small fish, especially during the flood season, when many young Dabry's sturgeon concentrate to feed. As a result, the sturgeon taken from the Luzhou reach usually weigh less than 50 g. (2) *Overfishing*. For example, in the Neijiang reach of the Tuo River there were only 500 fishing boats in 1950s, but this number increased to about 2000 by 1985. In the Leshan Reach of the Ming River, drift gill nets are crowded together from day to night. (3) *Unsuitable fishing season*. The traditional busy fishing season in the main stream of Jingsha River is between March and August, with more than 30% of the catch processed in April and May. However, this is also spawning season of Dabry's sturgeon in the main stream, so it threatens the spawning stock. (4) *Pollution from factories and pesticides*. With economic development, more and more factories are built in the Yangtze Valley. Much untreated waste water discharges into the river each year. For instance, Wenjiang district in upper Ming River pours out 10 million tons of industrial waste water to the Ming River each year. Also, many agricultural chemicals are used in Yangtze Valley. (5) *Habitat alterations and destruction*. The upper Yangtze Valley is seriously deforested, so that vast amounts of silt are washed to river, muddying its waters and impacting fishes and their food organisms. Perhaps most important are the negative impacts of hydroelectric projects. Since the construction of Gezhouba Dam in 1981, adult Dabry's sturgeon trapped below the dam cannot migrate upstream to the original spawning areas, and no suitable spawning grounds have been found below the dam. Young fish cannot pass the dam to reach the feeding grounds in the middle reaches of the Yangtze River and the lakes linking up with it. The Three Gorges Dam, currently under construction, will inundate many habitats of Dabry's sturgeon. Other dams have been built, are being built, or will be built in upper Jingsha River and its tributaries. These structures may alter the populations of Dabry's sturgeon in the main stream system by restricting their natural movements or by influencing spawning behavior (Wu et al. 1991).

In conclusion, we propose some tactics for the conservation of Dabry's sturgeon. (1) Commercial fishing of Dabry's sturgeon must be strictly forbidden. Although Dabry's sturgeon has been listed as an endangered species and is under special protection, net fishermen still catch them accidentally. Caught fish are usually hurt, cannot survive easily, even if they are immediately released to the river again (L. Ling personal communication). Small mesh nets and detrimental fishing methods must be prohibited. (2) Spawning areas and feeding grounds need to be properly protected. When we plan hydroelectric projects, methods for conserving Dabry's sturgeon and other fishes must be built into the plan. Waste water must be properly treated before being poured into the river. (3) Stock enhancement of Dabry's sturgeon should be urgently carried out. Successful efforts to replenish stocks of other Acipenseriformes have been made in the United States and Russia (Graham 1986, Smith 1990, 1991, Khodorevskaya et al. 1997 this volume). Perhaps Dabry's sturgeon could also be maintained by artificial stocking, particularly because culture methods for the fish has already been developed. Enhancement stations could be established along the Jingsha River at Yibin (Sichuan Province) and below Gezhouba Dam at Yichang (Hubei Province) to release juveniles for stock enhancement. (4) Research on Dabry's sturgeon should be increased. Up to now, very limited information is available on the biology and life history of Dabry's sturgeon, and many more topics need study. A particularly critical step in the conservation of Dabry's sturgeon is to determine where the populations spawn so that those areas can be precisely protected.

Acknowledgements

We thank Vadim Birstein, Robert H. Boyle and John Waldman for organizing the International Sturgeon Conference. We thank Boyd Kynard for constructive suggestions in preparing our draft. We are also grateful to Ling Ling and Mu Tianrong for help collecting new data on population dynamics. James D. Williams, Frank M. Parauka, Kim Graham and Theodore I. J. Smith provided valuable foreign reference materials. William E. Bemis reviewed and substantially revised the original draft of our manuscript and drew the map.

References cited

Bemis, W.E. & L. Grande. 1992. Early development of the actinopterygian head. I. External development and staging of the paddlefish *Polyodon spathula*. J. Morphol. 213: 47–83.

Buckley, J. & B. Kynard. 1985. Habitat use and behavior of prespawning and spawning shortnose sturgeon, *Acipenser brevirostrum*, in the Connecticut River. pp. 111–117. *In*: F.P. Binkowski & S.I. Doroshov (ed.) North American Sturgeons: Biology and Aquaculture Potential, Dr W. Junk Publishers, Dordrecht.

Conte, F.S., S.I. Doroshov & P.B. Lutes. 1988. Hatchery manual for white sturgeon, University of California Press, Oakland. 120 pp.

Deng, Q.X. & J.S. Hu. 1988. The skeletal system of Dabry's sturgeon, *Acipenser dabryanus*. pp. 179–187. *In*: The Biology of Sturgeons in Changjiang and Their Artificial Propagation, Sichuan Scientific & Technical Publishing House, Chengdu (in Chinese).

Duméril, A.H.A. 1868. Note sur trois poissons: *Acipenser dabryanus*,*Polyodon gladius* et *Peristhethidion prionocephalum*, accompagné de quelques considérations générales sur les groupes auxquels ces espèces appartiennent. Nouv. Arch. Mus. Nat. Hist. Nat. Paris 4: 93–116.

Fu, C.R. 1988. Age and growth of Dabry's sturgeon, *Acipenser dabryanus*. pp. 220–226. *In*: The Biology of Sturgeons in Changjiang and Their Artificial Propagation, Sichuan Scientific & Technical Publishing House, Chengdu (in Chinese).

Graham, L.K. 1986. Establishing and maintaining of paddlefish population by stocking. pp. 96–104. *In*: J.G. Dillard, L.K. Graham & T.R. Russell (ed.) The Paddlefish: Status, Management and Propagation, North Central Division, Special Publication Number 7, American Fisheries Society, Bethesda.

Günther, A. 1873. Report on a collection of fishes from China. Ann. Mag. Nat. Hist. 12 (series 4): 239–250.

He, X.F. 1990. Biology of rare and precious fishes of Sichuan Province. pp. 125–157. *In*: B.N. Shi (ed.) Ichthyofauna and Fisheries Resources of Sichuan Province, Southwest Normal University Press, Chongqing (in Chinese).

Jia, J.D. & Z.L. Wang. 1981. Ecology of Dabry's sturgeon, *Acipenser dabryanus*. J. Freshwater Fisheries 6: 12–13 (in Chinese).

Ke, F.E., P. Zhuang & Q.W. Wei. 1989. Status and management of fisheries in the Yangtze Valley. pp. 313–320. *In*: T.C. Guan & X.Q. Wang (ed.) Development and Management of Rivers in China, China Science & Technology Press, Beijing (in Chinese).

Khodorevskaya, R.P., G.F. Dovgopol, O.L. Zhuravleva & A.D. Vlasenko. 1996. Present status of commercial stocks of sturgeons in the Caspian Sea Basin. Env. Biol. Fish. (this volume).

Nichols, J.T. 1943. The fresh-water fishes of China. American Museum of Natural History, New York, Natural History of Central Asia 9: 1–322.

Nikolski, G.V. 1960. Fishes of the Amur Valley. China Science Press, Beijing. 472 pp. (in Chinese).

Smith, T.I.J. 1986. Culture of North American sturgeons for fishery enhancement. pp. 19–27. *In*: A. K. Sparks (ed.) Marine Farming and Enhancement, Proceedings of the Fifteenth U.S.-Japan Meeting on Aquaculture, October 22–23, 1986, Kyoto.

Smith, T.I.J. 1991. Development of shortnose sturgeon, *Acipenser brevirostrum*, stock enhancement program in North America. pp. 329–337. *In*: P. Williot (ed.) Acipenser, CEMAGREF Publ., Bordeaux.

Tian, Y.D. & D.Y. Chen. 1986. The first success of artificial propagation of Dabry's sturgeon, *Acipenser dabryanus*, cultured in pond. J. Today's Fisheries 12: 13 (in Chinese).

Tu, M.R. 1980. Some information on the nutrition of Dabry's sturgeon, *Acipenser dabryanus*. J. Fisheries Sci. 1: 11–14 (in Chinese).

Wei, Q., F. Ke, J. Zhang, P. Zhuang, J. Luo, R. Zhou & W. Yang. 1996. Biology, fisheries, and conservation of sturgeons and paddlefish in China. Env. Biol. Fish. (this volume).

Wu, H.W. 1930. On some fishes collected from the Yangtze Valley. Sinensia 1: 65–86.

Wu, J. & M.S. Wu. 1990. Status and utilization of fisheries in Jingsha River. pp. 53–61. *In*: W. L. Huang (ed.) Utilization and Conservation of Potamodromous Fisheries in Sichuan Province, Sichuan Scientific & Technical Publishing House, Chengdu (in Chinese).

Wu, X.W. 1963. Economic animals of China (freshwater fishes). pp. 12–16. China Science Press, Beijing (in Chinese).

Xie, D.J. 1979. Artificial propagation and rearing of Dabry's sturgeon, *Acipenser dabryanus*. J. Sichuan Fisheries 1: 2–8 (in Chinese).

Xie, D.J., Y.P. Tian & D.Y. Chen. 1981. Artificially induced maturity of male Dabry's sturgeon, *Acipenser dabryanus*, cultured in ponds. J. Freshwater Fisheries 5: 14–17 (in Chinese).

Xie, D.J., Y.P. Tian & D.Y. Chen. 1992. Study on domestication and propagation of the Changjiang sturgeon in ponds. p. 6. *In*: Proceedings of the International Symposium on Sturgeons and Paddlefishes, June 5–8, 1992, Chongqing.

Yang, G.R. 1986. Ichthyography of Hubei Province. Hubei Science & Technology Press, Wuhan. 242 pp. (in Chinese).

Zen, Q.Z. 1990. Fisheries resources in Yangtze valley, China Ocean Press, Beijing. 281 pp. (in Chinese).

Zhang, M.K. 1988. Movements and distribution of Dabry's sturgeon, *Acipenser dabryanus*. pp. 230–231. *In*: The Biology of Sturgeons in Changjiang and Their Artificial Propagation, Sichuan Scientific & Technical Publishing House, Chengdu (in Chinese).

Zhang, Y.L. & C.R. Fu. 1993. Redescription of a sturgeon, *Acipenser dabryanus* Duméril 1868 (Pisces, Acipenseriformes). Acta Zool. Taiwanica 4: 59–61.

Zhu, Y.D. 1963. Ichthyography of Eastern China. pp. 90–94. China Science Press, Beijing (in Chinese).

Zhuang, P. 1993. Impacts of the three gorges project on the aquatic animal resources in Yangtze Valley and the tactics for their conservation. pp. 26–29. *In*: J.D. Jia (ed.) Proceedings of the Annual Academic Meeting of Chinese Fisheries Society, 15–19 December 1993, Beijing (in Chinese).

Environmental Biology of Fishes **48**: 265–278, 1997.

Observations on the reproductive cycle of cultured white sturgeon, *Acipenser transmontanus*

Serge I. Doroshov, Gary P. Moberg & Joel P. Van Eenennaam
Department of Animal Science, University of California, Davis, CA 95616, U.S.A.

Received 10.1.1995 Accepted 8.4.1996

Key words: gametogenesis, neuroendocrine control, pubertal age and body size, breeding intervals

Synopsis

Males and females of cultured white sturgeon, *Acipenser transmontanus*, mature at an average age of 4 and 8 years, respectively. However, the onset of ovarian vitellogenesis and puberty are highly asynchronous in the female stock. Gonadal cycles are annual in males and biennial in females, and gametogenesis is influenced by season. Neuroendocrine regulation of reproduction appears to involve a dual gonadotropin system controlling gonadal development and spawning. Labile puberty and sex-specific duration of the gonadal cycle are distinct characteristics of cultured and wild sturgeon. Photoperiod and temperature play a significant role in environmental regulation of the reproductive cycle, but further studies are necessary to elucidate the roles of endogenous and environmental factors in sturgeon reproduction which is critically important for both aquaculture and conservation of endangered wild stocks.

Introduction

Sturgeons and paddlefishes, order Acipenseriformes, have attracted century-long attention of developmental biologists (recent reviews by Dettlaff et al. 1993, Bolker 1993, Bemis & Grande 1992), but studies of their reproductive physiology were limited by the lack of cultured animals. The complexity of the reproductive cycle of sturgeon has been revealed through fishery and hatchery studies (Holčík 1989, Barannikova 1991, Dettlaff et al. 1993). Hormonal induction of ovulation and spermiation in wild sturgeon, developed by Russian scientists, made hatchery propagation feasible and contributed to early advances in fish endocrinology (Pickford & Atz 1957). However, difficulties in obtaining experimental animals and growing them to maturity were major obstacles in reproductive research on Acipenseriformes.

Current development of sturgeon and paddlefish culture provides opportunities for studies of reproduction in Acipenseriformes. Gametogenesis and gonadal cycles were elucidated in cultured Siberian sturgeon *A. baerii* (Akimova et al. 1979, Le Menn & Pelissero 1991, Williot et al. 1991), and the hybrid *Huso huso* × *A. ruthenus* (Burtsev 1983, Fujii et al. 1991). Structure and function of reproductive hormones have also been investigated (Burzawa-Gerard et al. 1975a, 1975b, Kuznetsov et al. 1983, Goncharov et al. 1991, Sherwood et al. 1991, Lescheid et al. 1995, Moberg et al. 1995). Nevertheless, information on gonadal cycles and reproductive physiology of Acipenseriformes remains fragmentary.

This paper summarizes preliminary findings on reproductive development in cultured white sturgeon, *A. transmontanus*. As this species is long lived and slow maturing, our observations are incomplete and investigations continue. The material was obtained from sturgeon broodstock raised at five commercial farms in Northern California and at the University of California, Davis research facilities.

Broodfish and methods

White sturgeon live along the Pacific coast of North America and in watersheds of the Fraser, Columbia, and Sacramento-San Joaquin rivers (PSMFC[1]). It is a semi-anadromous species, with landlocked populations in freshwater reservoirs and tributaries of the Columbia River basin. In the wild, males reach first sexual maturity at age 10–12 years and females at 15–32 years (PSMFC[1]). River migrations and spawning exhibit seasonal patterns. In the Sacramento River peak spawning was observed in April, at water temperature 14–15°C (Kohlhorst 1976), while in the lower Columbia River major spawning activity occurred in May at water temperature 12–14°C (McCabe & Tracy 1994). The commercial fishery and hydroconstruction at the turn of the century severely depleted and fragmented wild stocks (Skinner[2], Semakula & Larkin 1968, Galbreath 1985). The restricted sport fishery is currently the major harvester of stocks in the Columbia River and San Francisco Bay (PSMFC[1]). Artificial propagation of white sturgeon was established during the 1980's (Conte et al. 1988), and the offspring were used by aquaculturists for raising domestic broodstocks and production of sturgeon for the food market (Logan et al. 1995).

The broodfish studied were 3–14 year old first-hatchery generations of wild sturgeon originated from the Sacramento and Columbia rivers. They were raised in freshwater tanks and fed a commercial salmonid diet at a rate of 0.5–2.0% body weight per day. Most rearing facilities were large (6–9 m diameter and 1.5 m depth) outdoor tanks exposed to natural photoperiod. In the research facility supplied by irrigation water, rearing temperature fluctuated seasonally (10–12°C in the winter and 16–18°C in the summer). Commercial facilities were supplied by underground water with a constant temperature of 20°C. Ripe broodfish raised on commercial farms were transferred to cooler water (10–14°C) 3–6 months before spawning, to prevent ovarian atresia and testicular regression. Fish were held in artificially chilled tanks or transported to other facilities with a supply of cool reservoir water. To examine reproductive cycles, gonadal development of marked fish was monitored by tissue biopsy starting at age 3–4 years. Commercial broodstock were sampled annually in the fall, and the ripe females were re-sampled before spawning (March-May). In the experiments conducted at the university, fish were sampled several times per year. Animals were anesthetized in a 100 ppm MS-222 bath, 7 ml of blood was drawn with a vacutainer from the caudal vein, centrifuged, and plasma was stored frozen for the analyses of reproductive hormones and metabolites. Gonads were biopsied (1 cm^3 fragment) through a small abdominal incision, and tissue was fixed in 10% buffered formalin. As fish reached full sexual maturity, ovulation and spermiation were induced by the administration of mammalian GnRH analog, GnRHa ([D-Ala^6, Pro^9 N-Et]-GnRH, Sigma) and extracts of dried common carp pituitary glands (Stoller Fisheries). Although optimal hormonal induction protocols are still being investigated, the most common hormonal treatment for ovulation was a 'priming' dose of 10 $\mu g\ kg^{-1}$ of GnRHa followed by intramuscular injection of 4.5 $mg\ kg^{-1}$ carp pituitary material 12 hours later. To induce spermiation, sturgeon were given a single injection of 1.5 $mg\ kg^{-1}$ carp pituitary, and semen was collected by catheter 30–32 hours later. Additional procedures for hatchery breeding of sturgeon were described by Conte et al. (1988) and Doroshov et al. (1994).

Stages of gametogenesis were examined on paraffin sections stained by periodic acid Schiff (PAS) or by hematoxylin and eosin (H&E) stains (Doroshov et al. 1991). The pre-ovulatory ovarian stage was detected by germinal vesicle migration (measuring distance of the nucleus from the animal pole of boiled and bisected eggs) and by the in vitro oo-

[1] Pacific States Marine Fisheries Commission (PSMFC). 1992. White sturgeon management framework plan. Portland, Oregon. 201 pp.

[2] Skinner, J.E. 1962. An historical review of the fish and wildlife resources of the San Francisco Bay area. California Department of Fish & Game, Water Project Branch Report No. 1. 226 pp.

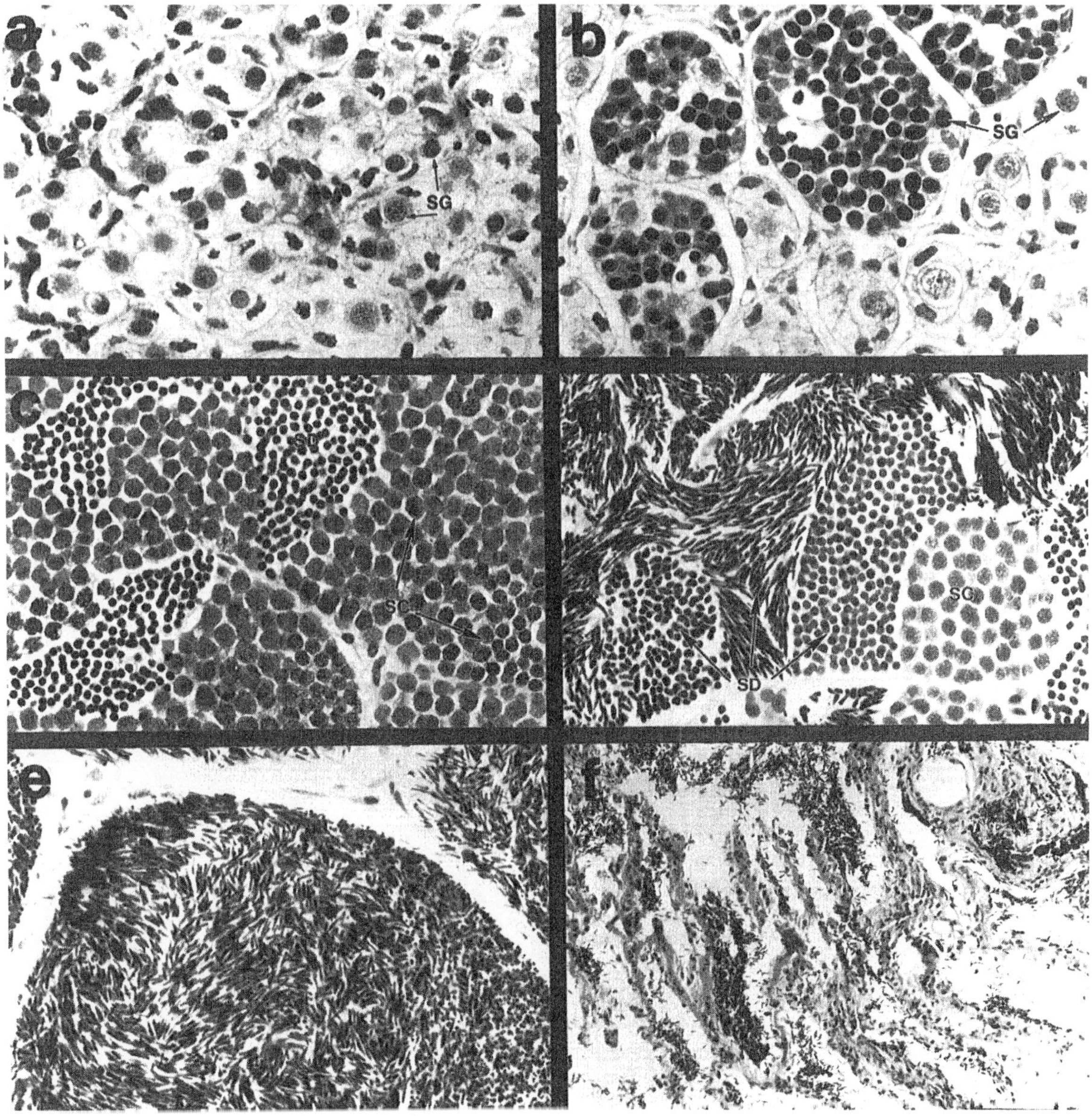

Figure 1. Different states of gametogenesis in cultured white sturgeon male. Paraffin sections stained by H&E (scale at bottom right 0.1 mm): a – primary spermatogonia enclosed in cysts, b – proliferation of spermatogonia, c – meiosis (primary, secondary spermatocytes, and spermatids), d – late meiosis and spermiogenesis, e – cysts with differentiated spermatozoa, f – postspawning testis (SG = spermatogonia, SC = spermatocytes, SD = spermatids, SZ = spermatozoa).

cyte maturation response after incubation with 5 μg ml^{-1} progesterone for 16 h at 16° C (Doroshov et al. 1994). Relative concentrations of plasma vitellogenin were estimated by measuring total plasma calcium using atomic absorption spectrophotometry. In vitellogenic females, plasma calcium concentrations exhibited a significant linear relationship (r^2 = 0.96, n = 72) with plasma vitellogenin, measured by an enzyme immunoassay employing polyclonal antibody (Linares-Casenave et al. 1994). Basal level of total plasma calcium in males from the same rearing facilities was 90 ± 0.5 μg ml^{-1} (mean ± s.e.m., n = 208, Kroll 1990). Gonadotropins (stGTH I and stGTH II) were measured by radioimmunoassays (RIA)

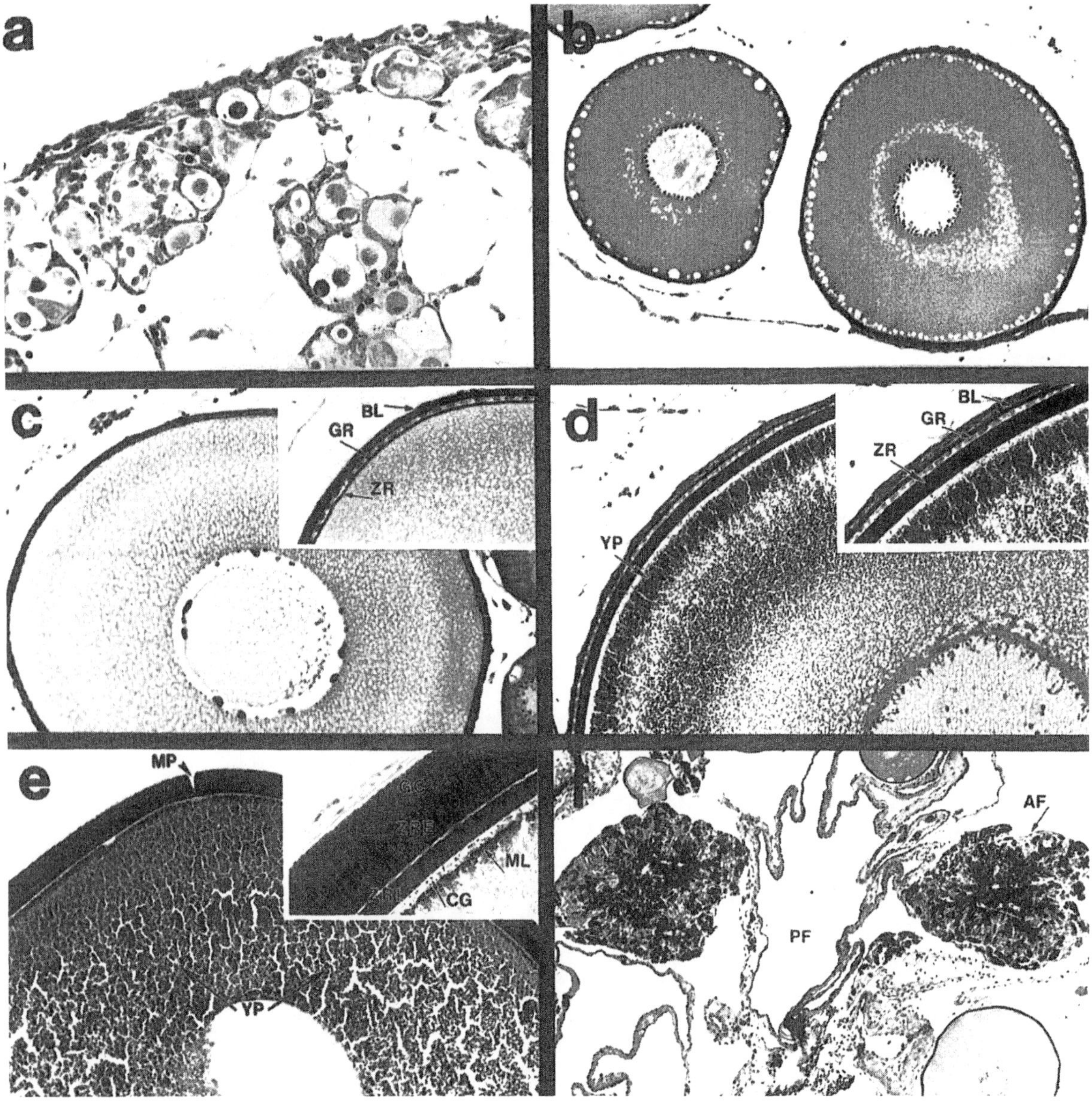

Figure 2. Different states of gametogenesis in cultured white sturgeon female. Paraffin sections stained by PAS: a – early phase of oogenesis, oogonia and oocytes initiating meiosis, b – differentiated ovarian follicles with the oocytes in the late primary growth phase, c – previtellogenic follicle with differentiated granulosa layer and vitelline envelope, d – mid-vitellogenesis, deposition of yolk bodies, e – ripe follicle with polarized oocyte near completion of vitellogenic growth, f – ovarian tissue after ovulation, containing post-ovulatory follicles, atretic follicles, and new generation of oocytes (BL = basal lamina, GR = granulosa cells, ZR = zona radiata, YP = yolk platelets, MP = micropyle, GC = gelatinous coat, ZRE = zona radiata externa, ZRI = zona radiata interna, CG = cortical granules, ML = melanin granules, PF = post-ovulatory follicle, AF = atretic follicle).

described by Moberg et al. (1991b, 1995). Briefly, the RIA's were based on specific polyclonal antibodies to purified stGTH's fractions. The cross-reactivity of stGTH I antibody with stGTH II was 2%, and that of stGTH II antibody with stGTH I was 9.3%. Minimum detectable concentrations for the stGTH I and stGTH II were 0.84 ng ml^{-1} and 1.25 ng ml^{-1}, respectively. Inter- and intra-assay variability was

10.2 and 7.3% for stGTH I, and 9.7 and 6.1% for stGTH II. Testosterone (Gay & Kerlan 1978), estradiol-17β (England et al. 1974) and 17α, 20β-dihydroxy-4-pregnen-3-one (Scott et al. 1982) were analyzed by previously described RIA procedures. Minimum detectable concentrations for these steroids were 0.52 ng ml^{-1}, 0.13 ng ml^{-1}, and 0.10 ng ml^{-1}, respectively.

In the following description of gonadal development we applied the term 'puberty' to the gonadal condition or age of animals that have completed the first gonadal cycles. The term 'vitellogenesis' was used to define a period of oocyte growth associated with deposition of yolk platelets in the cytoplasm. The three layers of the oocyte envelope were termed zona radiata interna, zona radiata externa and gelatinous (jelly) coat, following the classification of Dettlaff et al. (1993).

Gonadal development and gametogenesis

While gonadal sex differentiation in cultured white sturgeon has not been investigated in detail, our observations on gonadal development in domestic (F_2) offspring suggest that anatomical and cytological differentiation of sex glands were completed by the age 18 months, at fork length 58–72 cm and body weight 1.1–2.3 kg. The females had distinct 'ovarian grooves' on the lateral side of the ovary, with lamellae containing the nests of primary oogonia and the oocytes in meiotic prophase. The germinal portion of the testis was seen as a narrow layer of solid tissue on the dorsal part of the gonads, containing spermatogonia enclosed by a fibrous wall. The gonads of both sexes were largely composed of adipose fatty tissue surrounded by the peritoneal epithelium. Further observations at age 3–4 years, conducted on hundreds of marked fish on several commercial farms, indicated that domestic offspring had a 1:1 sex ratio.

Full differentiation of testicular follicles and proliferation of spermatogonia were observed at age 3 years (Figure 1a, b), and were followed by meiosis and spermiogenesis (Figure 1c, d). Cell proliferation was accompanied by the enlargement of the germinal portion of testes and by resorption of adipose tissue. Meiosis starts in October-November and continues for 3–4 months. Ripe males that have completed the first cycle of spermatogenesis are typically 4 years old and 10–15 kg body weight. They have enlarged white testes containing cysts with differentiated spermatozoa (Figure 1e). Individual cycles are asynchronous in a stock, but the majority of males mature from February to June. In summer, the unshed spermatozoa and remaining meiotic cells are reabsorbed (Figure 1f) and only primary spermatogonia remain in the cyst walls. Elevated water temperature (above 15° C) accelerates spermatogenesis but induces rapid testicular regression in the post-meiotic phase. Spontaneous spermiation was occasionally observed, but for scheduled spawning spermiation was induced by treatment with carp pituitary extracts. Males are capable of at least three consecutive spermiations at biweekly intervals, producing up to 200 ml of milt at each spermiation. As in other Acipenseriformes, white sturgeon spermatozoa possess an acrosome and undergo acrosome reaction during insemination (Cherr & Clark 1985). White sturgeon sperm contains the enzyme amidase, which is associated with sperm penetration through the egg envelope in non-teleostean vertebrates (Ciereszko et al. 1994). However, the functions of this enzyme and the details of the acrosomal reaction in insemination of eggs in sturgeon are unknown.

Gonadal development of females is long and asynchronous, relative to individual age and body size (see 'Age and body size at puberty' below). Oocyte meiosis and differentiation of the ovarian follicle start at age 2–3 years. The narrow ovarian groove (formed on the lateral side of the gonad) contains clusters of oogonia and small oocytes surrounded by squamous follicular cells (Figure 2a). By age 4–5 years gonial mitosis appears complete, and the ovigerous lamellae contain only oocytes in the primary growth phase, reaching 100–300 μm diameter and surrounded by a few granulosa cells, PAS-positive basement membrane and the thecal layer. The oocyte cytoplasm is strongly basophilic and contains unstained spherical vesicles in the cortex. Numerous small nucleoli appear in the periphery of the nucleus (Figure 2b). Oocytes are visible to the unaided eye as small transparent spheres and

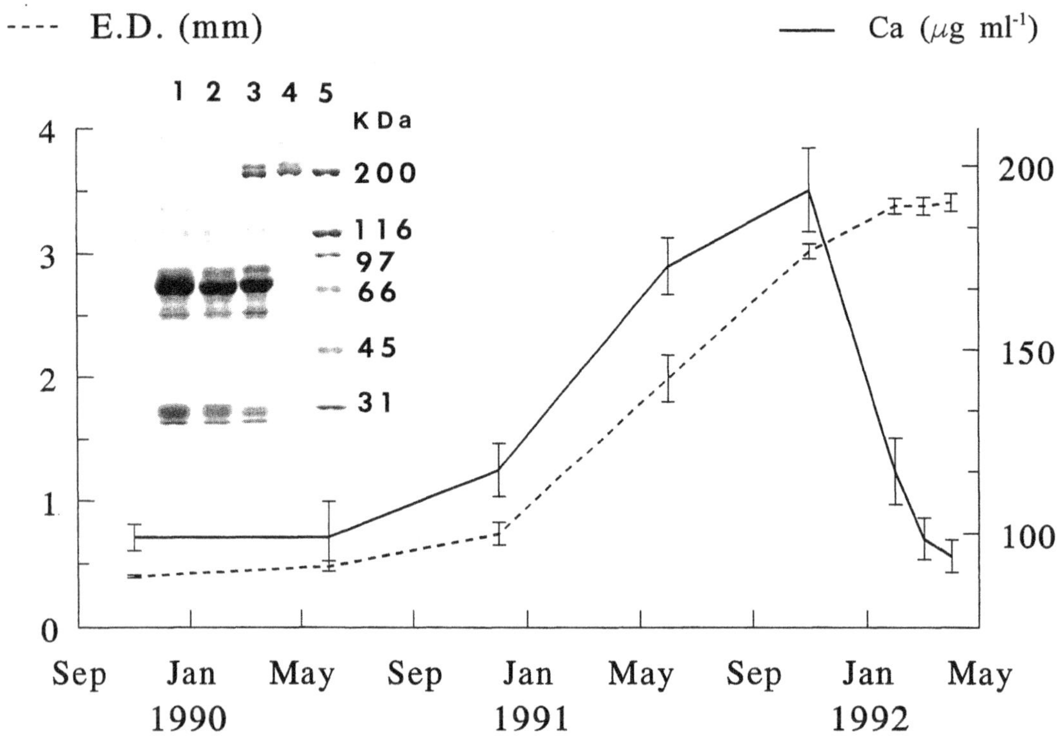

Figure 3. Total plasma calcium (Ca) concentrations and diameter (ED) of vitellogenic ovarian follicles in eight females repeatedly sampled during the first ovarian cycle. Bars are x ± se. Plasma calcium has a linear relationship with plasma vitellogenin (Ca = 98.94 + 0.015 Vg μg ml^{-1}), and elevation of total plasma calcium above 100 μg ml^{-1} indicates the presence of vitellogenin (Linares-Casenave et al. 1994). Inset: SDS-PAGE of plasma from mature male (lane 1), immature female (lane 2), late vitellogenic female (lane 3), purified white sturgeon vitellogenin (lane 4), and molecular markers (lane 5). Vitellogenic female plasma and purified vitellogenin exhibit two polypeptides with an approximate molecular mass of 190 and 210 KDa. Courtesy of J. Linares-Casenave, University of California, Davis.

are present in the ovary throughout consecutive reproductive cycles. As oogenesis advances, some of the oocytes will increase in diameter to 400–600 μm and appear as enlarged translucent spheres. Their cytoplasm is less basophilic and exhibits vesicular structure. However, crystalline yolk bodies are not detectable with light microscopy (Figure 2c). The distinct features of this stage are the appearance of PAS-positive envelope and differentiation of the granulosa cell layer (Figure 2c, inset). Deposition of crystalline yolk usually follows this stage, but can be delayed in some fish for one or two years. Low concentrations of vitellogenin (< 1 μg ml^{-1}) are detected in plasma of white sturgeon at this stage (Linares-Casenave et al. 1994); morphological evidence of endocytosis at similar stage was reported in Siberian sturgeon (LeMenn & Pelissero 1991), suggesting the initiation of yolk precursor uptake by the oocyte.

Vitellogenic growth of the oocyte starts at age 4–8 years and a body weight of 15–30 kg, and continues for 16–18 months. Oocytes increase in diameter from 0.6 to 3.5 mm and change in color from light yellow to brown, grey, and finally to black, due to synthesis and deposition of melanin granules in the cortical cytoplasm. The crystalline yolk bodies first appear in the peripheral cytoplasm and, while increasing in size, exhibit centripetal progression (Figure 2d) similar to yolk deposition in the amphibian oocyte (Dumont 1972). Gradually, the yolk platelets and oil globules fill the entire cytoplasm except for the perinuclear area. When vitellogenic growth subsides (diameter > 3 mm), the oocyte undergoes major cytoarchitectural changes and becomes ovoid in shape with a slightly pointed animal pole. The germinal vesicle migrates to the animal pole and the large yolk bodies and oil globules aggregate in the vegetal hemisphere (Figure 2e, vege-

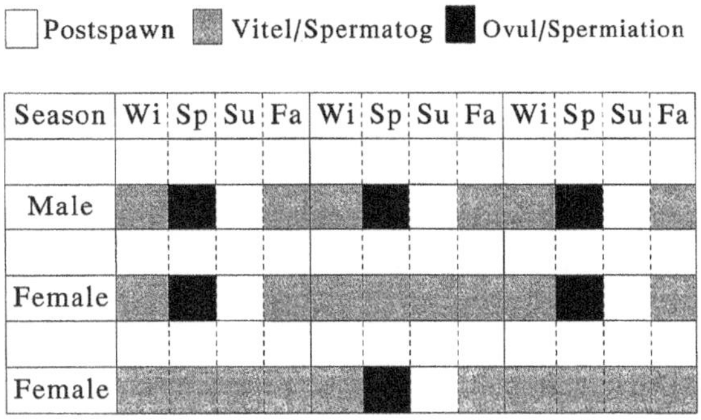

Figure 4. Approximate length and seasonality of the testicular and ovarian cycles in domestic white sturgeon, based on observations of iteroparous broodfish raised in outdoor tanks. The biennial ovarian cycle is shown for odd and even years.

tal portion not shown). The egg envelope reaches maximum thickness and consists of three PAS-positive layers: zona radiata interna, zona radiata externa and gelatinous coat. The cortical alveoli form a distinct layer beneath the oolemma.

During the vitellogenic phase of ovarian development, plasma concentrations of vitellogenin significantly increase. Hepatic synthesis of vitellogenin is stimulated by estrogen (Moberg et al. 1991a), which causes expression of the vitellogenin gene in the hepatocytes (Bidwell et al. 1991, Bidwell & Carlson 1995). Oocyte growth and changes in plasma vitellogenin level (measured as total plasma calcium) in females raised at the university facility are shown in Figure 3. Mean plasma vitellogenin concentration reaches 7 mg ml^{-1} (200 μg ml^{-1} calcium) during the late phase of vitellogenic growth and decreases significantly before spawning. Vitellogenesis in these females was first observed in November of 1990, and ovulation was successfully induced in April 1992. Two putative vitellogenin proteins were detected in the plasma of vitellogenic females by SDS-PAGE, with estimated molecular masses of 190 and 210 kDa (Figure 3, inset).

Final ovarian maturation in white sturgeon (oocyte maturation and ovulation) appears to be similar to that described by Dettlaff et al. (1993) in *A. gueldenstaedtii*. Captive white sturgeon females can be induced to ovulate by administration of GnRHa or carp pituitary extracts when they reach a responsive stage (Lutes et al. 1987, Doroshov et al. 1994). At spawning, the number of ova removed by caesarean section ranged from 100 000–200 000, and the average fecundity measured in fish sacrificed for caviar production was 209 000 eggs at a mean body weight of 29 kg (n = 67). At ovulation, the ovary contains post-ovulatory follicles, previtellogenic oocytes, and atretic follicles with dark pigmentation (Figure 2f). The late phase of ovarian development is highly sensitive to environmental temperature, and a prolonged exposure to water temperature of 19° C induces ovarian atresia before completion of germinal vesicle migration and acquisition of the meiotic response by the oocytes (Webb et al. 1994). These observations and studies with Russian sturgeon, *A. gueldenstaedtii*, and Siberian sturgeon, *A. baerii*, suggest that exposure to low temperature during the late phase of vitellogenesis is required for some sturgeon species to complete a normal ovarian cycle (Williot et al. 1991, Dettlaff et al. 1993).

Gonadal cycles

Repeatedly spawned males of cultured sturgeon exhibit annual seasonal cycles of spermatogenesis, with meiosis occurring during the fall and winter, and spermiation during the spring (Figure 4). Sperm that was successfully used for insemination was obtained each spring from the same males during 4 consecutive years. Additional evidence for an annual cycle in males was provided by observations on seasonal changes in the pituitary content of gonadotropins correlated with histological stages of gametogenesis (Moberg et al. 1995).

The duration and seasonality of the ovarian cycles are currently being investigated. We collected data from 147 virgin females that were sampled on farms each fall for several years. The majority (67%) exhibited patterns of a biennial cycle, similar to that in Figure 3, but some (29%) had a shorter duration of vitellogenesis (the stages as depicted in Figure 2c the first fall and in Figure 2e the following fall). Furthermore in small numbers of fish (4%) vitellogenesis continued for a period of 3–4 years. In approximately 23% of all females the onset of vitellogenic growth was arrested, and the advanced oocytes with differentiated envelopes and granulosa

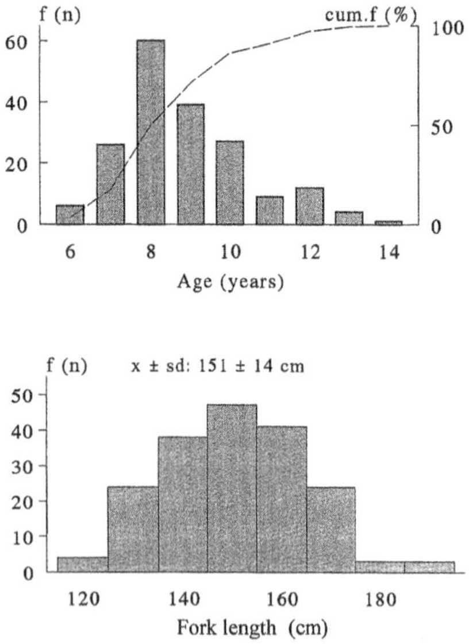

Figure 5. The age and body size at puberty in cultured white sturgeon females. Data are pooled observations on the broodstock of five commercial farms (N = 184).

layer (Figure 2c) persisted over a period of 1–3 years before the initiation of yolk deposition. The reabsorption of previtellogenic follicles and development of new oocytes could potentially occur within the annual sampling interval, but no atretic follicles were observed on the histological slides from any of these fish.

Observations on the second ovarian cycle in repeatedly spawned females (n = 5) revealed a biennial cycle, with three of these fish spawning on odd years, 1991 and 1993, and two on even years, 1992 and 1994 (Figure 4). All fish started the second vitellogenesis during the first fall after spring spawning, and their oocytes reached the late phase of vitellogenic growth in the fall of the following year. They were spawned a second time during a period from March to June, and the calendar date of the second ovulation in individual females was very similar to that of the first spawning.

These preliminary observations reveal annual testicular and biennial ovarian cycles in cultured white sturgeon. Gonadal development in both sexes exhibits seasonal patterns, with testicular meiosis and ovarian vitellogenesis starting in the fall, and final gonadal maturation completed during the spring. The observations on seasonal changes in the pituitary and plasma concentrations of two sturgeon gonadotropins (Moberg et al. 1995) also revealed that gonadal cycles in white sturgeon are controlled by seasonal factors.

Age and body size at puberty

Most (80–90%) cultured white sturgeon males reach puberty at age 3–4 years and body weight of 7–14 kg. In commercial farms, sturgeon males are selected for spawning from the production stock at age 4 years and body weight 10–15 kg. In the wild, white sturgeon males reach puberty at age 10–12 years, and body weight of approximately 12 kg (PSMFC[1]). Thus, the body size at first sexual maturity appears to be similar in wild and cultured males, but the latter reach puberty at a younger age.

Cultured females reach puberty at an age from 6 to 14 years, with fifty percent maturing by age 8, at a mean body weight of 32 kg and a mean fork length of 151 cm (Figure 5). These data are based on a large number of gravid females pooled from different age cohorts raised on farms, and the maturation rates may be affected by farm husbandry and the age groups life history. However, the proportions of mature females within each age group showed very similar trends (Doroshov et al. 1994, Logan et al. 1995). The body weight of females that completed their first sexual maturation was highly variable, 32 ± 10 kg (× ± sd, range 12–61 kg). Similar variability was found in wild-caught females used for farm spawning during the past four years (36 ± 13 kg, n = 43), although the age and iteroparity of these fish could not be examined. The smallest mature wild female weighed 12 kg, and the largest ‘virgin’ female caught in San Francisco Bay (stage similar to Figure 2a) weighed 53 kg (S. Doroshov unpublished observations). A recent report indicates that the pubertal age of wild white sturgeon females ranges from 15 to 32 years (PSMFC[1]). This suggests that a highly variable pubertal age and body size in cultured sturgeon may reflect the patterns existing in natural population. However, cultured females reach puberty at a considerably younger age compared to wild fish. Similar observations were reported for cultured Siberian sturgeon, *A. baerii* (Akimova et al. 1979, Williot et al. 1991).

Neuroendocrine control

Knowledge of the neuroendocrine regulation of reproduction in Acipenseriformes remains inadequate, but our recent observations on cultured white sturgeon appear to indicate general similarity of regulatory mechanisms with those of teleosts. Based on these preliminary observations, we propose a model for the neuroendocrine regulation of reproduction in white sturgeon (Figure 6). We believe the gonadotropin-releasing hormone (GnRH), and possibly the neurotransmitter dopamine (DA) interact on the pituitary gonadotropes to regulate the synthesis and release of two putative gonadotropins. The gonadotropins control gonadal development and gamete release, and stimulate the synthesis of gonadal steroids, androgen (A) and estrogen (E). The gonadal steroids, especially testosterone, appear to feedback on the pituitary, and possibly the hypothalamus, to influence the synthesis and secretion of gonadotropins. The function of the hypothalamic-pituitary-gonadal axis is further modulated by a number of exogenous and endogenous factors, including season, body size and age of fish, as previously discussed. The proposed model has been developed from the following data.

The brains of white, *A. transmontanus*, and Russian sturgeon, *A. gueldenstaedtii*, contain the mammalian form of GnRH (mGnRH) and a small amount of chicken GnRH II (Sherwood et al. 1991, Lescheid et al. 1995). High concentrations of mGnRH in brains of mature Russian sturgeon caught during their spawning migrations, as well as its localization in the forebrain of Siberian sturgeon (Lepretre et al. 1993), suggest that this form of GnRH is responsible for pituitary gonadotropin release (Lescheid et al. 1995). High sensitivity of sturgeon pituitary gonadotropes to exogenous stimulation by mGnRH analog (D-Ala^6, Pro^9 N-Et]GnRH, has been demonstrated by experiments on spawning induction in European sturgeon species (Goncharov et al. 1991) and studies on the effect of GnRHa injections on plasma concentrations of gonadotropins in stellate

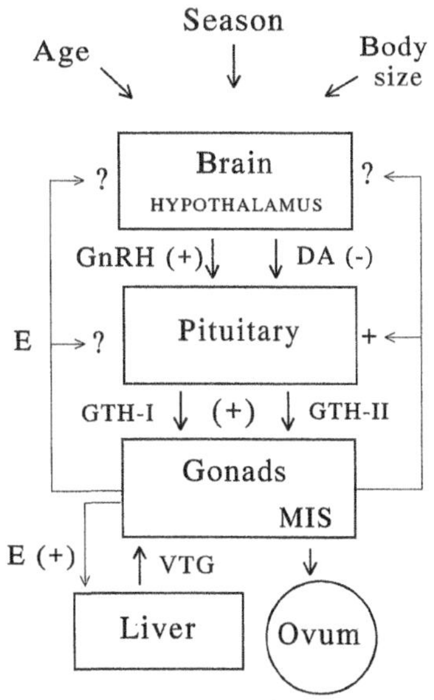

Figure 6. Hypothetical neuroendocrine reproductive axis of cultured white sturgeon based on current studies (see text for description).

sturgeon, *A. stellatus* (Barannikova & Bukovskaya 1990), and white sturgeon (Moberg et al. 1995).

The presence of a dual gonadotropin control of gonadal development in sturgeon is being investigated. Gonadotropic hormone was first purified from pituitary glands of stellate sturgeon, and it was initially believed that Acipenseriformes possessed only a single form of GTH that regulated all aspects of reproductive development (Burzawa-Gerard et al. 1975a, 1975b). Recently, two potential gonadotropins have been isolated from Russian sturgeon pituitaries, designated sturgeon gonadotropin I, or stGTH I, and sturgeon gonadotropin II, or stGTH II (Moberg et al. 1995). Based on physiological evidence, these two gonadotropins appear to be functional analogs of the salmonid gonadotropins GTH I and GTH II (Kawauchi et al. 1989, Swanson et al. 1989), possessing many of the same biological functions. Like GTH I in salmonids, stGTH I appears to induce and maintain follicular development and vitellogenesis, while stGTH II is instrumental in inducing ovarian maturation and ovulation (Moberg et al. 1995).

Plasma levels of both stGTH I and stGTH II are low (< 1–2 ng ml^{-1}) in reproductively immature sturgeon. With the onset of meiosis in males and vitellogenesis in females, the pituitary concentrations of both gonadotropins increase, with the stGTH I being the predominant gonadotropin in the pituitary. Plasma concentration of stGTH I also increases during vitellogenesis, similar to the increases of GTH I reported during the same state of development in salmonids (Suzuki et al. 1988). These increases in the pituitary and plasma concentrations of stGTH I suggest that this hormone regulates the onset of vitellogenesis as has been observed in trout where GTH I stimulates the uptake of vitellogenin by the developing oocytes (Tyler et al. 1991). While it is not known whether stGTH I has a similar role, we have observed that female sturgeon in the previtellogenic state will not sequester circulating vitellogenin, even in the presence of elevated concentrations of plasma estrogen and vitellogenin (Moberg et al. 1991a), when the concentrations of stGTH I are low in both the pituitary and plasma. Furthermore, the administration of a gonadotropin releasing hormone analog, GnRHa, will not stimulate the secretion of the gonadotropin in these females even though GnRHa administration will stimulate the secretion of stGTH I in mature male sturgeon (Moberg & Doroshov 1992) and induce ovulation and spermiation in ripe fish (Doroshov & Lutes 1984, Fujii et al. 1991, Goncharov et al. 1991). These findings are consistent with the hypothesized role of stGTH I in initiating ovarian development and the onset of vitellogenesis.

Prior to the onset of final ovarian maturation and ovulation, pituitary concentration of stGTH II rises sharply to a level greater than 100 $\mu g\ mg^{-1}$ in the preovulatory white sturgeon female (Moberg et al. 1995). It is during this period of reproduction that stGTH II becomes the predominant gonadotropin in the pituitary. Comparable changes in gonadotropins have also been observed during the reproductive cycle of salmonids (Swanson et al. 1989). As was observed for GTH II in salmonids (Suzuki et al. 1988), the plasma concentration of stGTH II in white sturgeon is not elevated until the time of final gonadal maturation and spawning, when it increases to more than 20 $ng\ ml^{-1}$ (Moberg et al. 1995). We have also observed a functional difference between the stGTHs in an in vitro oocyte maturation bioassay using ripe ovarian follicles of white sturgeon. It was found that stGTH II was more potent than stGTH I in inducing germinal vesicle breakdown, GVBD (Moberg et al. 1991b). This effect, in conjunction with the increased pituitary and plasma concentrations of stGTH II prior to final ovarian maturation, suggests that stGTH II gonadotropin regulates the final reproductive events leading to spawning.

In males, similar changes in the secretion of stGTHs occur during their annual reproductive cycle. In the winter during the meiotic state of spermatogenesis, the mature white sturgeon male has greater amounts of stGTH I than stGTH II in the pituitary. If GnRHa is administered during spermatogenesis, stGTH I is also released in greater amounts than stGTH II, consistent with stGTH I being responsible for regulating the meiotic phase of spermatogenesis. In the spring, during spermiation, the pituitary concentration of stGTH II exceeds the levels of stGTH I and the GnRHa-induced release of this gonadotropin is maximal, implying that stGTH II may be responsible for controlling spermiation. In the summer, during reproductive quiescence, relative low amounts of both stGTHs are released following the administration of GnRHa, indicating that the regulation of the neuroendocrine axis is modulated by such environmental factors as season (Moberg et al. 1995).

In some species of teleosts, the neurotransmitter dopamine acts as an endogenous inhibitor of GnRH-induced pituitary secretion of GTH (Peter et al. 1991). A comparable effect has been found in sturgeon (Pavlick 1995). Administration of dopamine to mature white sturgeon males effectively inhibited the GnRHa-induced elevation of both stGTHs in blood plasma as well as spermiation, while the administration of the dopamine antagonist pimozide potentiated stGTH I and stGTH II release when used in combination with GnRHa. These results indicate a dopaminergic inhibition of GnRH action in white sturgeon, similar to what has been observed in several teleost species.

Little is known about the feedback effects of the gonadal steroids on the regulation of the hypothalamic-pituitary-gonadal axis of sturgeon. From the onset of the meiotic phase in gametogenesis, both sexes in sturgeon exhibit high concentrations of plasma androgens (Moberg et al. 1995, Cuisset et al. 1995). Recently, we demonstrated that the implants of exogenous testosterone in sexually undifferentiated white sturgeon will significantly elevate the pituitary concentrations of both stGTHs (Pavlick et al. 1995). Likewise, pre-vitellogenic females implanted with testosterone have higher levels of pituitary stGTHs than controls, suggesting that the pituitaries of adult sturgeon are also sensitive to the positive feedback of testosterone. This positive feedback effect of testosterone in sturgeon is comparable to the reported effect of testosterone in modern teleost species (Crim & Evans 1983). It still remains to be determined whether or not testosterone can also have a negative feedback effect on the secretion of the gonadotropins and what the potential feedback effects of estrogen are.

In vitro oocyte maturation is readily induced in white sturgeon by C-21 steroids, with 17α,20β-Dihydroxy-4-pregnen-3-one (17α,20β-DP) being most potent (Lutes 1985). Our observations on ovulating domestic females showed consistent and significant increases in plasma concentrations of 17α,20β-DP correlated with an increase of stGTH II. Wild-caught pre-ovulatory white sturgeon females also had significantly higher plasma concentrations of 17α,20β-DP, compared to late vitellogenic females that did not reach the responsive state of gonadal development (Lutes et al. 1987). Similar data were obtained with migratory females of wild Atlantic sturgeon, *A. oxyrinchus* (Van Eenennaam et al. 1996). While the identity of the putative maturation-inducing substance (MIS) in white sturgeon has not been established, the potential similarity in the native MIS between sturgeon and salmonids (Young et al. 1986) may lend further support to a generally similar endocrine control of reproduction in the Acipenseriformes and modern teleosts.

Conclusions

Our data support several previous studies with Siberian sturgeon and the sturgeon hybrid (*H. huso* × *A. ruthenus*), which suggested that puberty and sexual maturation of Acipenseriformes are accelerated by intensive culture utilizing artificial feed and warm water. In the wild, most species of sturgeons reach first sexual maturity at age 10 to 20 years, while in culture the age interval 3–10 years is most frequently reported. Thus, Acipenseriformes appear to have a labile pubertal age that may be influenced by environmental factors and growth rates. One interesting aspect of oogenesis observed in cultured white sturgeon was the ability of females to maintain ovarian follicles in an arrested but advanced previtellogenic state for one or more years before the onset of yolk deposition. This suggests that ovarian recrudescence may be affected by some endogenous and exogenous factors, resulting in highly asynchronous female puberty. High variability in ovarian development rates of captive fish may be caused by culture stress and competition related to high rearing densities, commonly utilized on the commercial farms.

Sex-related differences in pubertal age and length of the gonadal cycle were previously reported in wild and cultured Acipenseriformes (see Holčík 1989 for review). Males mature at a younger age and usually have annual gonadal cycles whereas the females mature late and have ovarian cycles lasting longer than one year. Estimates of breeding frequency in wild sturgeon females range from one to ten years (Holčík 1989), but these assumptions are usually inferred from the age-body size population structure and the presence of spawning marks on finray sections (Roussow 1957). Observations on cultured sturgeon suggest a predominantly biennial ovarian cycle (Williot et al. 1991, Doroshov et al. 1994). Variations in the length of individual cycles maybe driven by exogenous (environmental) and endogenous (genetic) factors, but the current knowledge of reproductive physiology in Acipenseriformes is inadequate to answer these questions. It is possible that the length of the reproductive cycle in wild and cultured sturgeon is labile, as is their pubertal age. However, it is more likely that Acipenseriformes have developed endogenous reproductive rhythms of specific duration, as indicated by their relatively advanced neuroendocrine control of reproduction and the clearly apparent seasonality of the gonadal cycle. As in teleosts (Bromage et al. 1993), annual photoperiod is likely to control gametogenesis in sturgeon, although no experimental evidence is available, to our knowledge. The effect of seasonal factors on reproduction of Acipenseriformes may play a role in differentiation of wild stocks of Russian sturgeon and stellate sturgeon into different spawning ecotypes, generally similar to spawning races in salmonids ('vernal' and 'hiemal' ecotypes, distinguished by the season and stage of maturity when they initiate their spawning migration in the river, Barannikova 1991, Dettlaff et al. 1993).

Thermal environment plays an important role in the reproduction of Acipenseriformes, as it does in teleosts (de Vlaming 1972, Sundararaj & Vasal 1976) and other ectothermic vertebrates (Licht et al. 1969). In agreement with observations on *A. stellatus* and *A. gueldenstaedti* (Dettlaff et al. 1993), we observed high sensitivity of gonadal development to environmental temperature in white sturgeon. The vitellogenic phase of ovarian development appears to be accelerated by warm temperature (18–22° C), whereas the late stages of gonadal development (postmeiotic in males and germinal vesicle migration in females) are negatively affected by elevated temperature. Further research on reproductive development and endogenous regulation of the reproductive cycle in cultured Acipenseriformes may elucidate the roles of environmental and genetic factors, an understanding of which will be critically important for both aquaculture and conservation of endangered sturgeon species.

Acknowledgements

This work was supported by the Western Regional Aquaculture Consortium, USDA, Development of Sturgeon Broodstocks project No. 459497 and the NOAA, National Sea Grant College Program, U.S. Department of Commerce, under grant number NA85AA-D-SG140, project number R/A 85, through the California Sea Grant College Program, and in part by the California State Resource Agency. We gratefully acknowledge the members of the California Sturgeon Broodstock Development Committee and the graduate students who have assisted in the many years of field research.

References cited

Akimova, N.V., V.S. Malyutin, I.I. Smolyanov & L.I. Sokolov. 1979. Growth and gametogenesis of the Siberian sturgeon (*Acipenser baeri* Brandt) under experimental and natural conditions. pp. 179–183. *In*: Proceedings 7th Japan-Soviet Joint Symposium on Aquaculture, Sept. 1978, Tokyo.

Barannikova, I.A. 1991. Peculiarities of intrapopulation differentiation of sturgeon (*Acipenser gueldenstaedti* Brandt) under present day conditions. pp. 137–142. *In*: P. Williot (ed.) Acipenser, CEMAGREF Publ., Bordeaux.

Barannikova, I.A. & O.S. Bukovskaya. 1990. Influence of luliberin on pituitary gonadotropocytes state and dynamics of serum gonadotropin concentrations in stellate sturgeon (*Acipenser stellatus* Pallas). Vopr. Ikhtiol. 30: 126–136 (in Russian).

Bemis, W.E. & L. Grande. 1992. Early development of the actinopterygian head. I. External development and staging of the paddlefish *Polyodon spathula*. J. Morphol. 213: 47–83.

Bidwell, C.A. & D.M. Carlson. 1995. Characterization of vitellogenin from white sturgeon, *Acipenser transmontanus*. J. Mol. Evol. 41: 104–112.

Bidwell, C.A., K.J. Kroll, E.I. Severud, S.I. Doroshov & D.M. Carlson. 1991. Identification and preliminary characterization of white sturgeon vitellogenin mRNA. Gen. Comp. Endocrinol. 83: 415–424.

Bolker, J.A. 1993. The mechanisms of gastrulation in the white sturgeon. J. Exp. Zool. 266: 132–145.

Bromage, N., C. Randall, D. Davies, M. Thrush, J. Duston, M. Carillo & S. Zanuy. 1993. Photoperiodism and the control of reproduction and development in farmed fish. pp. 81–102. *In*: B. Lallou & P. Vitello (ed.) Coastal and Estuarine Studies, Vol. 43, American Geophysical Union, Washington D.C.

Burtsev, I.A. 1983. Hybridization and selection of sturgeon during full cycle breeding and domestication. pp. 102–113. *In*: V.S. Kirpichnikov (ed.) Biological Foundations of Fish Culture: Problems of Genetics and Selection, Nauka Press, Leningrad (in Russian).

Burzawa-Gerard, E., B.F. Goncharov & Y.A. Fontaine. 1975a. L'hormone gonadotrope hypophysaire d'un poisson chondrosteen, l'esturgeon (*Acipenser stellatus* Pall.). 1. Purification. Gen. Comp. Endocrinol. 27: 289–295.

Burzawa-Gerard, E., B.F. Goncharov & Y.A. Fontainb. L'hormone gonadotrope hypophysaire d'un poisson chondrosteen, l'esturgeon (*Acipenser stellatus* Pall.). 2. Proprietes biochimiques. Gen. Comp. Endocrinol. 27: 296–304.

Cherr, G.N. & W.H. Clark Jr. 1985. Gamete interaction in the white sturgeon *Acipenser transmontanus*: a morphological and physiological review. Env. Biol. Fish. 14: 11–22.

Ciereszko, A., K. Dabrowski, F. Lin & S. Doroshov. 1994. Identification of trypsin-like activity in sturgeon spermatozoa. J. Exp. Zool. 268: 486–491.

Conte, F.S., S.I. Doroshov, P.B. Lutes & E.M. Strange. 1988. Hatchery manual for the white sturgeon. Publication #3322, University of California Press, Oakland. 104 pp.

Crim, L.W. & D.M. Evans. 1983. Influence of testosterone and/or luteinizing hormone releasing hormone analogue on precocious sexual development in the juvenile rainbow trout. Biol. Reprod. 29: 137–142.

Cuisset, B., A. Fostier, P. Williot, C. Bennetau-Pelissero & F. Le Menn. 1995. Occurrence and *in vitro* biosynthesis of 11-ketotestosterone in Siberian sturgeon, *Acipenser baeri* Brandt, maturing females. Fish Physiol. Biochem. 14: 313–332.

Dettlaff, T.A., A.S. Ginsburg & O.I. Schmalhausen. 1993. Sturgeon fishes: developmental biology and aquaculture. Springer-Verlag, New York. 300 pp.

de Vlaming, V.L. 1972. Environmental control of teleost reproductive cycles: a brief review. J. Fish Biol. 4: 131–140.

Doroshov, S.I. & P.B. Lutes. 1984. Preliminary data on the induction of ovulation in white sturgeon (*Acipenser transmontanus* Richardson). Aquaculture 38: 221–227.

Doroshov, J.N., J.P. Van Eenennaam, F.A. Chapman & S.I. Doroshov. 1991. Histological study of the ovarian development in wild white sturgeon, *Acipenser transmontanus*. pp. 129–135. *In*: P. Williot (ed.) Acipenser, CEMAGREF Publ., Bordeaux.

Doroshov, S.I., J.P. Van Eenennaam & G.P. Moberg. 1994. Reproductive management of cultured white sturgeon (*Acipenser transmontanus*). pp. 156–161. *In*: D.D. Mackinlay (ed.) High Performance Fish, Proceedings of an International Fish Physiology Symposium, July 1994, Fish Physiology Association, Vancouver.

Dumont, J.N. 1972. Oogenesis in *Xenopus laevis* (Daudin). 1. Stages of oocyte development in laboratory maintained animals. J. Morphol. 136: 153–180.

England, B.G., G.D. Niswender & A.R. Midgley Jr. 1974. Radioimmunoassay of estradiol-17β without chromatography. J. Clinical Endocrinol. Metab. 38: 42–50.

Fujii, K., K. Hirose, A. Hara, M. Shiraishi & T. Maruyama. 1991. Use of vitellogenin level as a maturational indicator for artificial spawning of cultured hybrid sturgeon *Huso huso* × *Acipenser ruthenus*. pp. 381–388. *In*: P. Williot (ed.) Acipenser, CEMAGREF Publ., Bordeaux.

Galbreath, J.L. 1985. Status, life history, and management of Columbia river white sturgeon, *Acipenser transmontanus*. pp. 119–126. *In*: F.P. Binkowski & S.I. Doroshov (ed.) North American Sturgeons: Biology and Aquaculture Potential, Dr W. Junk Publishers, Dordrecht.

Gay, V.L. & J.T. Kerlan. 1978. Serum LH and FSH following passive immunization against circulating testosterone in the intact male rat and in orchidectomized rat bearing subcutaneous silastic implants of testosterone. Arch. Androl. 1: 257–266.

Goncharov, B.F., L.V. Igumnova, I.S. Polupan & E.A. Savelieva. 1991. Induced oocyte maturation, ovulation and spermiation in sturgeons (Acipenseridae) using synthetic analogue of gonadotropin-releasing hormone. pp. 351–364. *In*: P. Williot (ed.) Acipenser, CEMAGREF Publ., Bordeaux.

Holčík, J. 1989. The freshwater fishes of Europe, Vol. 1, Part II. Acipenseriformes. AULA-Verlag, Wiesbaden.

Kawauchi, H., K. Suzuki, H. Itoh, P. Swanson, N. Naito, Y. Nagahama & S. Itoh. 1989. The duality of teleost gonadotropins. Fish Physiol. Biochem. 7: 29–38.

Kohlhorst, D.W. 1976. Sturgeon spawning in the Sacramento River in 1973, as determined by the distribution of larvae. California Fish and Game 62: 32–40.

Kroll, K.J. 1990. Preliminary characterization of vitellogenin and yolk proteins of white sturgeon (*Acipenser transmontanus*). M.S. Thesis. University of California, Davis. 84 pp.

Kuznetsov, A.A., B.F. Goncharov & E. Burzawa-Gerard. 1983. Pituitary gonadotropic hormone from a chondrostean fish, starred sturgeon (*Acipenser stellatus* Pall.). 3. Polymorphism. Gen. Comp. Endocrinol. 49: 364–374.

Le Menn, F. & C. Pelissero. 1991. Histological and ultrastructural studies of oogenesis of siberian sturgeon. pp. 113–128. *In*: P. Williot (ed.) Acipenser, CEMAGREF Publ., Bordeaux.

Lepretre, E., I. Anglade, P. Williot, F. Vandesande, G. Tramu & O. Kah. 1993. Comparative distribution of mammalian GnRH (gonadotropin-releasing hormone) and chicken GnRH-II in the brain of the immature Siberian sturgeon (*Acipenser baeri*). J. Comp. Neurol. 337: 568–583.

Lescheid, D.W., J.F.F. Powell, W.H. Fischer, M. Park, A. Craig, O. Bukovskaya, I.A. Barannikova & N.M. Sherwood. 1995. Mammalian gonadotropin-releasing hormone (GnRH) identified by primary structure in Russian sturgeon, *Acipenser gueldenstaedti*. Regulatory Peptides 55: 299–309.

Licht, P., H.E. Hoyer & P.G.W.J. Van Oordt. 1969. Influence of photoperiod and temperature on testicular recrudescence and body growth in the lizards, *Lacerta sicula* and *Lacerta muralis*. J. Zool. 157: 469–501.

Linares-Casenave, J., K.J. Kroll, J.P. Van Eenennaam & S.I. Doroshov. 1994. Development and application of an enzyme-linked immunosorbent assay (ELISA) for the detection of plasma vitellogenin in white sturgeon (*Acipenser transmontanus*). pp. 165–169. *In*: D.D. Mackinlay (ed.) High Performance Fish, Proceedings of an International Fish Physiology Symposium, July 1994, Fish Physiology Association, Vancouver.

Logan, S.H., W.E. Johnston & S.I. Doroshov. 1995. Economics of joint production of sturgeon (*Acipenser transmontanus* Richardson) and roe for caviar. Aquaculture 130: 299–316.

Lutes, P.B. 1985. Oocyte maturation in white sturgeon, *Acipenser transmontanus*: some mechanisms and applications. Env. Biol. Fish. 14: 87–92.

Lutes, P.B., S.I. Doroshov, F. Chapman, J. Harrah, R. Fitzgerald & M. Fitzpatrick. 1987. Morpho-physiological predictors of ovulatory success in white sturgeon, *Acipenser transmontanus* Richardson. Aquaculture 66: 43–52.

McCabe, G.T. Jr. & C.A. Tracy. 1994. Spawning and early life history of white sturgeon, *Acipenser transmontanus*, in the lower Columbia River. U.S. Fish. Bull. 92: 760–772.

Moberg, G.P., J.G. Watson, S. Doroshov, H. Papkoff & R.J. Pavlick Jr. 1995. Physiological evidence for two sturgeon gonadotropins in *Acipenser transmontanus*. Aquaculture 135: 27–39.

Moberg, G.P. & S.I. Doroshov. 1992. Reproduction in cultured white sturgeon, *Acipenser transmontanus*. NOAA Technical Report, NMFS 106: 99–104.

Moberg, G.P., S.I. Doroshov, J.P. Van Eenennaam & J.G. Watson. 1991a. Effects of various hormone implants on vitellogenin synthesis and ovarian development in cultured white sturgeon. pp. 389–399. *In*: P. Williot (ed.) Acipenser, CEMAGREF Publ., Bordeaux.

Moberg, G.P., J.G. Watson, H. Papkoff, K.J. Kroll & S.I. Doroshov. 1991b. Development of radioimmunoassays for two sturgeon gonadotropins. pp. 11–12. *In*: A.P. Scott, J.P. Sumpter, D.E. Kinne & M.S. Rolfe (ed.) Proceedings of 4th International Symposium on the Reproductive Physiology of Fish, Fish Symposium 91, Sheffield.

Pavlick, R.J. Jr. 1995. Endocrine regulation of two sturgeon pituitary gonadotropins in *Acipenser transmontanus*. Ph.D. Dissertation, University of California, Davis. 146 pp.

Pavlick, R.J. Jr., J.G. Watson & G.P. Moberg. 1995. Influence of exogenous testosterone treatment on two sturgeon gonadotropins in sexually undifferentiated *Acipenser transmontanus*. pp. 86–94. *In*: A.D. Gershanovich & T.I.J. Smith (ed.) Proceedings of the International Symposium on Sturgeons, 6–11 September 1993, VNIRO Publ., Moscow.

Peter, R.E., B.D. Trudeaux & B.D. Stoley. 1991. Brain regulation of reproduction in teleosts. Bull. Instit. Zool. Acad. Sinica, Monograph 16: 89–118.

Pickford, G.E. & J.W. Atz. 1957. The physiology of the pituitary gland of fishes. New York Zoological Society, New York, 613 pp.

Roussow, G. 1957. Some considerations concerning sturgeon spawning periodicity. J. Can. Fish. Res. Board 14: 553–572.

Scott, A.P., E.L. Sheldrick & A.P.F. Flint. 1982. Measurement of 17, 20β-Dihydroxy-4pregnen-3-one in plasma of trout (*Salmo gairdneri* Richardson): seasonal changes and response to salmon pituitary extract. Gen. Comp. Endocrinol. 46: 444–451.

Semakula, S.N. & P.A. Larkin. 1968. Age, growth, food, and yield of the white sturgeon (*Acipenser transmontanus*) of the Fraser River, British Columbia. J. Fish. Res. Board. Can. 25: 2589–2602.

Sherwood, N.M., S.I. Doroshov & V. Lance. 1991. Gonadotropin-releasing hormone (GnRH) in bony fish that are phylogenetically ancient: reedfish (*Calamoichthys calabaricus*), sturgeon (*Acipenser transmontanus*), and alligator gar (*Lepisosteus spatula*). Gen. Comp. Endocrinol. 84: 44–57.

Sundararaj, B. & S. Vasal. 1976. Photoperiod and temperature control in the regulation of reproduction in the female catfish. J. Fish. Res. Board. Can. 33: 959–973.

Suzuki, K., A. Kanamori, Y. Nagahama & H. Kawauchi. 1988. Development of salmon GTH I and GTH II radioimmunoassays. Gen. Comp. Endocrinol. 71: 459–467.

Swanson, P., M. Bernard, M. Nozki, K. Suzuki, H. Kawauchi & W.W. Dickhoff. 1989. Gonadotropins I and II in juvenile coho salmon. Fish Physiol. Biochem. 7: 169–176.

Tyler, C.R., J.P. Sumpter, H. Kawuachi & P. Swanson. 1991. Involvement of gonadotropin in the uptake of vitellogenic oocytes of the rainbow trout, *Oncorhynchus mykiss*. Gen. Comp. Endocrinol. 84: 291–299.

Van Eenennaam, J.P., S.I. Doroshov, G.P. Moberg, J.G. Watson, D.S. Moore & J. Linares. 1996. Reproductive conditions of the Atlantic sturgeon (*Acipenser oxyrinchus*) in the Hudson River. Estuaries (in press).

Webb, M.A., S.I. Doroshov & J.P. Van Eenennaam. 1994. The effects of temperature on final ovarian maturation of the white sturgeon. pp. 186–190. *In*: D.D. Mackinlay (ed.) High Performance Fish, Proceedings of an International Fish Physiology Symposium, July 1994, Fish Physiology Association, Vancouver.

Williot, P., R. Brun, T. Rouault & O. Rooryck. 1991. Management of female spawners of the siberian sturgeon, *Acipenser baeri* Brandt: first results. pp. 365–380. *In*: P. Williot (ed.) Acipenser, CEMAGREF Publ., Bordeaux.

Young, G., S. Adachi & Y. Nagahama. 1986. Role of ovarian thecal and granulosa layers in gonadotropin-induced synthesis of a salmonid maturation-inducing substance (17α,20β-dihydroxy-4-pregnen-3-one). Dev. Biol. 118: 1–8.

Environmental Biology of Fishes **48**: 279–289, 1997.

Contemporary status of the North American paddlefish, *Polyodon spathula*

Kim Graham
Missouri Department of Conservation, 1110 South College Ave., Columbia, MO 65201, U.S.A.

Received 28.7.1994 Accepted 13.3.1996

Key words: Polyodontidae, commercial fishing, sport fishing, distribution, endangered status

Synopsis

North American paddlefish, *Polyodon spathula*, were once abundant in most large rivers and tributaries of the Mississippi River basin, but numbers have declined dramatically in most areas during the past 100 years. Habitat destruction and river modification are the most obvious changes affecting their distribution and abundance. Although peripheral range has dwindled, paddlefish still occur over most of their historic range and are still found in 22 states. Populations are currently increasing in 3 states, stable in 14, declining in 2, unknown in 3, and extirpated in 4. Sport harvests presently occur in 14 states, however two states with traditionally important sport fisheries report decreased recruitment into the population and are planning more restrictive regulations. Commercial fisheries are reported in only six states. During the past 10 years, five states have removed paddlefish from their commercial list primarily because of declines in adult stocks due to overfishing or illegal fishing. Ten states are currently stocking paddlefish to supplement existing populations or to recover paddlefish populations in the periphery of its native range.

Introduction

The North American paddlefish, *Polyodon spathula*, is one of two living species of paddlefishes, the other being the Chinese paddlefish, *Psephurus gladius* (for additional basic information on Polyodontidae, see Russell 1986, Grande & Bemis 1991, Bemis et al. 1997 this volume, and Wei et al. 1997 this volume). Paddlefish once were abundant in most large rivers and major tributaries of the Mississippi River basin (Carlson & Bonislawsky 1981), however since the turn of the century, these populations have declined dramatically in most areas (Gengerke 1986). Habitat destruction and river modification are the most obvious changes affecting the abundance and distribution of paddlefish. Construction and operation of dams on mainstem streams has had severe impacts (Sparrowe 1986, Unkenholz 1986). Dams eliminated traditional spawning sites, interrupted natural spawning migrations, altered water flow regimes, dewatered streams, and eliminated backwater areas that were important as nursery and feeding areas. To a lesser degree, industrial pollution, poaching adults for caviar, and overfishing by commercial and sport fishermen have adversely affected paddlefish populations (Pflieger 1975, Carlson & Bonislawsky 1981, Pasch & Alexander 1986).

In 1989, the U.S. Fish and Wildlife Service was petitioned to include paddlefish on the list of Threatened and Endangered Species under provisions of the Endangered Species Act of 1973. The U.S. Fish and Wildlife Service, after collecting supplemental information from all 22 states, agreed that the listing of paddlefish as 'threatened' was not warranted. Because of the uncertainty of the species' status in several portions of its range, the U.S. Fish and Wildlife Service recommended a reclassifi-

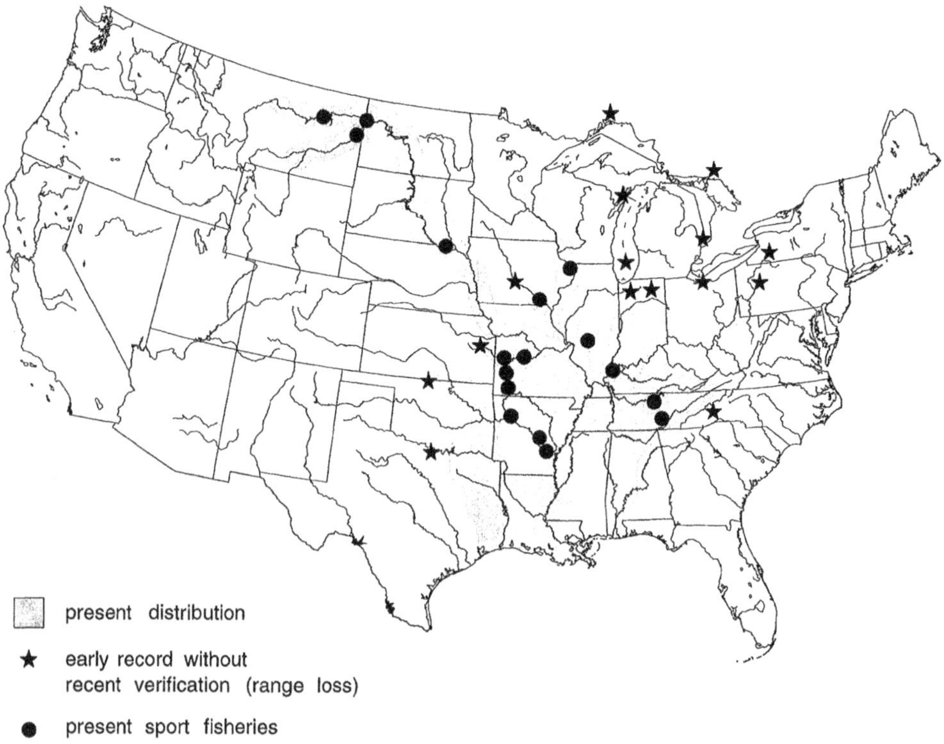

Figure 1. Past and present distribution of North American paddlefish, *Polyodon spathula*, and locations of sport fisheries (adapted from Carlson & Bonislawsky 1981). Map drawn by W. E. Bemis.

cation from category 3C to a category 2 under authority of the Endangered Species Act of 1973, as amended. Category 3C is intended for taxa that have proven to be more abundant or widespread than previously believed and/or those that are not subject to any identifiable threat. Category 2 indicates taxa for which information now in the possession of the U.S. Fish and Wildlife Service indicates that proposing to list as endangered or threatened is possibly appropriate, but for which conclusive data on biological vulnerability and threat are not currently available to support proposed rules. The U.S. Fish and Wildlife Service believes this classification will encourage further investigation and biological research of the species' status throughout its range. Additionally, paddlefish were added to the list of Appendix II of the Convention on International Trade in Endangered Species of Wild Fauna and Flora (CITES) in 1992 primarily because of concern about illegal poaching for the international caviar trade.

This paper reviews major changes in paddlefish status in the United States since Gengerke's (1986) report, illustrates historical changes since the turn of the 20th century, discusses major reasons for declines, defines current range of paddlefish in the United States, describes current status of paddlefish in the United States, and predicts their future as a fisheries resource.

Distribution

Historically, paddlefish were abundant throughout the Mississippi River basin and adjacent Gulf drainages, with a few records from the Great Lakes (Gengerke 1986). Neill et al.[1] reported that several

[1] Neill, W.H., B.R. Murphy, C.R. Vignali, P.W. Dorsett & V.M. Pitman. 1994. Salinity responses of paddlefish. Texas Parks and Wildlife Department. Dingell-Johnson Project F-31-R-20, Project 81. Final Report. 38 pp.

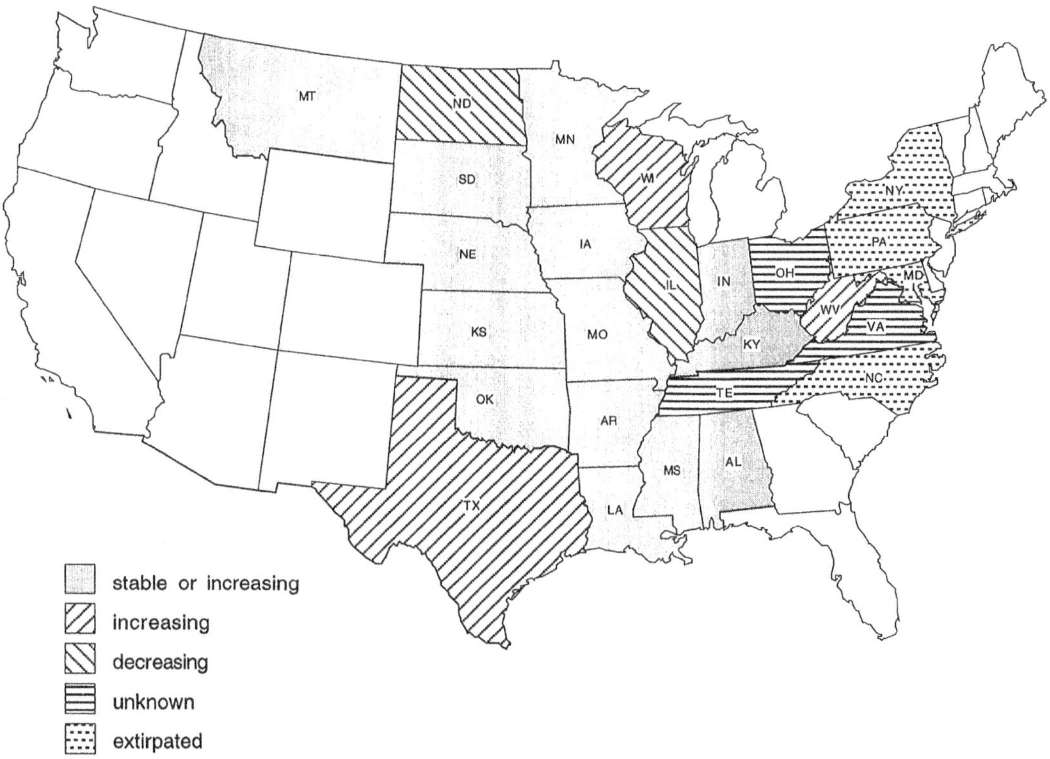

Figure 2. Status of North American paddlefish stocks in the United States. AL, Alabama; AR, Arkansas; IA, Iowa; IL, Illinois; IN, Indiana; KS, Kansas; KY, Kentucky; LA, Louisiana; MD, Maryland; MN, Minnesota; MO, Missouri; MS, Mississippi; MT, Montana; NC, North Carolina; ND, North Dakota; NE, Nebraska; NY, New York; OH, Ohio; OK, Oklahoma; PA, Pennsylvania; SD, South Dakota; TE, Tennesee; TX, Texas; VA, Virginia; WI, Wisconsin; WV, West Virginia. Map drawn by W. E. Bemis.

paddlefish stocked into large river systems of east Texas were reported captured in Galveston Bay. This ability to survive in brackish water probably explains the occurrence of several specimens captured by shrimp trawlers, and of two paddlefish originally tagged and released in Toledo Bend Reservoir in Louisiana that were recaptured in the Neches River in Texas.

The former range of paddlefish in the United States encompassed 26 states (Figure 1). They have been extirpated in four states on the periphery of their range (Maryland, New York, North Carolina, and Pennsylvania). Even in states long considered strongholds for paddlefish (Iowa, Nebraska, Oklahoma, and Alabama), portions of their historical range has diminished. During the last 100 years, significant declines in major paddlefish populations have occurred in the Mississippi, Missouri, Ohio, and Red rivers.

Status by state analysis

In 1983, Gengerke (1986) contacted all states known or suspected to have paddlefish to gather information pertaining distribution, abundance and status of present day paddlefish populations. I contacted those same states to determine changes during the past 10 years. I will not repeat the 1983 information for individual states and will only discuss major changes since that time.

Paddlefish populations are considered by resource agency personnel to be increasing in three states, stable in fourteen, declining in two, unknown in three, and extirpated in four (Figure 2). Since the last survey (1983), the number of states reporting paddlefish as stable or stable/increasing remains at fourteen. Fifteen of the twenty-two states recorded changes in the status of their paddlefish stocks. Of these, seven were positive changes, and eight indicated negative changes in status since 1983 (Table 1). Significant positive changes were reported in Iowa, which changed status from declining to stable/increasing; Kansas, which changed from declining to stable/increasing; South Dakota, which changed from declining to stable; Wisconsin, which changed from stable to increasing; and Texas and West Virginia, which changed from declining to increasing. All but two of these states (Iowa and Wisconsin) implemented stocking programs to supplement existing stocks or recover historic populations.

North Dakota reported the most significant negative change in status since 1983 by changing from stable/increasing to declining (Table 1). Additionally, Montana and Nebraska changed their status from stable to stable/declining. Most of the other declining shifts in status since 1983 are the result of states reporting that their stock inventories are unknown at this time.

Several significant changes in the management and regulation of paddlefish during the last 10 years were prompted by recognition of continued degradation of paddlefish habitat, threats of industrial, commercial, or agricultural contaminants in paddlefish, and an increase in the demand for paddlefish caviar. Since 1983, 86% of the states where paddlefish still occur have changed their regulations of sport and/or commercial paddlefish fisheries (Tables 1 and 2). Alabama, Virginia, and West Virginia no longer allow a sport harvest, and Alabama, Io-

Table 1. Classification and population status of paddlefish in all states containing paddlefish for 1994, as compared to 1983.[1]

State	Classification		Status	
	1983[2]	1994	1983[2]	1994
Alabama	Commercial	Special Concern	Decline	Stable/Decline
Arkansas	Commercial	Sport/Commercial	Increase	Stable
Illinois	Commercial	Sport/Commercial	Decline	Decline
Indiana	Sport	Sport	Stable	Stable
Iowa	Commercial	Sport	Stable	Stable
Kansas	Sport	Sport	Decline	Stable/Increase
Kentucky	Commercial	Commercial	Stable	Stable
Louisiana	Commercial	Special concern	Stable	Stable
Maryland	Threatened	Extirpated	Extirpated	Extirpated
Minnesota	Protected	Threatened	Stable	Stable
Mississippi	Commercial	Commercial	Stable/Increase	Stable
Missouri	Game	Game	Stable	Stable
Montana	Sport	Sport/Special concern	Stable	Stable/Decline
Nebraska	Sport	Sport	Stable	Stable/Decline
New York	Extirpated	Extirpated	Extirpated	Extirpated[3]
North Carolina	Not Classified	Endangered	Extirpated	Extirpated[4]
North Dakota	Commercial	Sport/Special concern	Stable/Increase	Decline
Ohio	Endangered	Threatened	Decline	Unknown
Oklahoma	Commercial	Non-game	Unknown	Stable
Pennsylvania	Extirpated	Extirpated	Extirpated	Extirpated
South Dakota	Sport	Sport	Decline	Stable
Tennessee	Commercial	Sport/Commercial	Stable	Unknown
Texas	Endangered	Endangered	Decline	Increase
Virginia	Non-Game	Threatened	Stable	Unknown
West Virginia	Sport	Special concern	Decline	Increase
Wisconsin	Watch list	Watch list	Stable	Increase

[1] Paddlefish are considered to be extirpated from Maryland, New York, North Carolina and Pennsylvania.

[2] From Gengerke (1986).

[3] Sole report of a paddlefish in New York was one fish in 1800s.

[4] There are two unconfirmed reports of paddlefish being taken during the last 19 years.

wa, Louisiana, Oklahoma, and Virginia now prohibit commercial harvests (Table 2). Alabama, Louisiana, and West Virginia now consider paddlefish as a species of special concern and Virginia classifies them as threatened. Some states currently use quotas and length limits, while others utilize creel limits or protected zones to regulate their sport harvests. Some of the more important changes in classification and status of paddlefish stocks since 1983 are as follows:

Alabama (no sport or commercial fisheries, classified as special concern, status stable/declining)
A no harvest regulation (sport and commercial) was implemented in 1989. Paddlefish in the Tennessee River are considered to be 'very reduced in numbers'. Overharvest and habitat alterations are considered main threats.

Arkansas (sport and commercial fisheries, classified as sport/commercial, status stable)
All populations in major river systems of the state are regarded as self-sustaining. Arkansas currently utilizes seasonal restrictions on the upper White River and border waters to protect mature females from commercial harvest. Commercial fishermen are also restricted by a 76 cm eye-to-fork-of-tail length limit from fall until late spring as additional

Table 2. Type of fishery allowed in all states containing paddlefish for 1994, compared to 1983, and current stocking programs.[1]

State	Sport fishery		Commercial fishery		Stocking program
	1983[2]	1994	1983[2]	1994	1994
Alabama	Yes	No	Yes	No	No
Arkansas	Yes	Yes	Yes	Yes	Yes
Illinois	Yes	Yes	Yes	Yes	No
Indiana	Yes	Yes[3]	No	No	No
Iowa	Yes	Yes	Yes	No	No
Kansas	Yes	Yes	No	No	Yes
Kentucky	Yes	Yes	Yes	Yes	No
Louisiana	No	No	Yes	No	Yes
Maryland	No	No	No	No	No
Minnesota	No	No	No	No	No
Mississippi	Yes	Yes	Yes	Yes	No
Missouri	Yes	Yes	Yes	Yes[4]	Yes
Montana	Yes	Yes	No	No	No
Nebraska	Yes	Yes[5]	No	No	No
New York	No	No	No	No	No
North Carolina	No	No	No	No	No
North Dakota	Yes	Yes	No	No	No
Ohio	No	No	No	No	No
Oklahoma	Yes	Yes	Yes	No	Yes
Pennsylvania	No	No	No	No	Yes
South Dakota	Yes	Yes[6]	No	No	Yes
Tennessee	Yes	Yes	Yes	Yes	Yes
Texas	No	No	No	No	Yes
Virginia	Yes	No	Yes	No	No
West Virginia	Yes	No	No	No	Yes
Wisconsin	No	No	No	No	No

[1] Paddlefish are considered to be extirpated from Maryland, New York, North Carolina and Pennsylvania.
[2] From Gengerke (1986).
[3] Paddlefish may be taken only by hook and line. Snagging is not a legal method of take.
[4] Commercial fishing only allowed on Mississippi and lower St. Francis rivers.
[5] Share tailwater fishery (1600-fish quota) below Gavins Point Dam with South Dakota.
[6] Share tailwater fishery (1600-fish quota) below Gavins Point Dam with Nebraska.

protection for females. The statewide commercial harvest during 1993 was estimated to be 136 000–181 000 kg. Most sport fisheries are located below dams on major tributaries to the Mississippi River. Arkansas is currently stocking juveniles into the White River (Beaver Reservoir) to establish another paddlefish stock and hopefully a fishery. Habitat alteration, contaminants, and sand and gravel dredging are considered major threats.

Illinois (sport and commercial fisheries, classified as sport/commercial, status declining)
Nearly 90 percent of the reported commercial harvest occurs in the Mississippi River. The recent decline in commercial harvest (21 319 kg in 1993 compared to 24 500 kg in 1983) may be related to poor reporting by commercial fishermen. Although there is a shortage of information, overfishing and habitat loss are considered threats.

Indiana (sport fishing only, classified as sport, status stable)
An insignificant legal sport harvest is only by hook and line, no snagging. Major concerns include illegal harvest, habitat alteration and contaminants.

Iowa (sport fishing only, classified as sport, status stable)
Iowa closed its commercial fishery on the Missouri River in 1986 and the Mississippi River in 1987. Stocks in the pooled portion of the Mississippi River are thought to be declining slightly, whereas below Lock and Dam 26, the stock is increasing. Iowa's sport harvest occurs below locks and dams on the pooled portion of the Mississippi River. Illegal harvest, habitat alteration, and contaminants are of concern.

Kansas (sport fishing only, classified as sport, status stable/increasing)
Two sport fisheries below low dams on the Neosho and Marais des Cygnes rivers flowing into Oklahoma and Missouri, respectively, depend upon high spring flows. In 1993, the Neosho River fishery yielded 87 paddlefish, while the Marais des Cygnes River fishery produced between 500–550 fish. Kansas is currently attempting to restore the historic distribution by stocking juveniles into the Arkansas, Blue and upper Neosho rivers. Habitat alteration, contaminants, and illegal harvest are major management concerns.

Kentucky (sport and commercial fisheries, classified as commercial, status stable)
A commercial fishery is allowed on Kentucky and Barkley lakes and the Ohio River only. In 1993, the commercial harvest from Kentucky and Barkley lakes was 3906 kg. Kentucky's major sport fishery is below Kentucky Dam, where fewer than 1000 paddlefish were harvested in 1993. Habitat destruction and contaminants in the entire Ohio River are considered major threats.

Louisiana (no sport or commercial fisheries, classified as special concern, status stable)
In 1992, a three year moratorium on paddlefish harvest, including commercial and sport fishing, became permanent. There is general concern regarding the impacts of overfishing of adult paddlefish stocks throughout the state. Illegal caviar trade may be a problem at times. Habitat alterations and contaminants are considered threats. In cooperation with Texas, juvenile paddlefish are currently being stocked into Toledo Bend Reservoir to establish a population.

Maryland (no sport or commercial fisheries, classified as extirpated, status extirpated)
Paddlefish are currently considered extirpated from the state.

Minnesota (no sport or commercial fisheries, classified as threatened, status stable)
Although paddlefish remain rare, the largest stock is found in the Mississippi River. Minnesota currently has a long-range management plan that includes a potential for stocking juvenile paddlefish into the Minnesota River below Granite Falls. Habitat alteration appears to be the most serious problem.

Mississippi (sport and commercial fisheries, classified as commercial, status stable)
Paddlefish stocks in Pascagoula River are declining,

presumably as a result of high commercial harvest for eggs in the mid-1980s. The remaining stocks in the state appear to be stable. The commercial season is closed state-wide in early spring and closed on the Pascagoula River system and border waters with Louisiana during late fall to early spring to protect mature females. Sport fisheries occur predominantly below dams on major tributaries to the Mississippi River. There is little information regarding the annual harvests of paddlefish by sport and commercial methods. Major management concerns are habitat destruction and possible overfishing.

Missouri (sport and commercial fisheries, classified as game, status stable)
Sport fisheries occur in the upper portions of Lake of the Ozarks, Harry S. Truman Lake, and Table Rock Lake. The 1992 sport harvest above Harry S. Truman Lake was 4041 paddlefish. The most recent creel census at Lake of the Ozarks (1988) indicated that approximately 2000 paddlefish were harvested, and 350 paddlefish were estimated harvested during the last year of a creel census (1990) at Table Rock Lake. Lake of the Ozarks, Table Rock Lake, and Harry S. Truman Lake currently receive annual, supplemental stocking of early juveniles. No juveniles were stocked into Table Rock Lake from 1991 to 1994 because of an apparent lack of interest in sport snagging due to chlordane advisories. In 1994, the chlordane advisories were lifted and sport snagging interest increased to near previous levels. Major concerns are habitat alteration and poaching for eggs. Commercial fishing was closed on the Missouri River in 1990. The 1992 commercial harvest from the Mississippi River was estimated at 2188 kg.

Montana (sport fishery only, classified as sport, but designated as a species of special concern, status stable/declining)
Sport fisheries are located on the Missouri River above Fort Peck Reservoir and in the Yellowstone River at Intake, Montana, and downstream to the confluence with the Missouri River. Recent information suggests the Yellowstone River stock (from Lake Sakakawea in North Dakota) is suffering from decreased natural recruitment. Declining reservoir productivity and poor survival of young are believed to be problems. In 1989, the Montana legislature passed an act which established a procedure for legally collecting and marketing paddlefish eggs taken during the Yellowstone River sport snagging season. An estimated 1360–4536 kg of paddlefish eggs have been collected each spring for caviar. A Montana/North Dakota paddlefish management plan suggests a reduced harvest quota for the Garrison Dam paddlefish beginning in 1996.

Nebraska (sport fishery only, classified as sport, status stable/declining)
Nebraska currently shares a paddlefish fishery with South Dakota in the Gavins Point Dam tailwater. A 1600-fish quota with a 88–114 cm eye-to-fork-of-tail slot length limit was established in 1989 and is routinely reached in less than one week. The most important concern is the need to restore part of the natural hydrography of the Missouri River, habitat alteration, waste management, and harvest management.

New York (no sport or commercial fisheries, classified as extirpated, status extirpated)
The sole report of a paddlefish is from Chautauqua Lake during the late 1800s and was likely the result of unusual movement resulting from flooding in the Ohio River Valley.

North Carolina (no sport or commercial fisheries, classified as endangered, status extirpated)
There are only two unsubstantiated records for the state; both are from Madison County.

North Dakota (sport fishing only, classified as sport, but designated as a species of special concern, status declining)
North Dakota's major sport fishery is in the Missouri River above Lake Sakakawea and the Yellowstone River. Recent information suggests the paddlefish population in Lake Sakakawea is suffering from low natural recruitment. Declining reservoir productivity and predation of young by walleye and sauger may be a problem. In 1993, state administrative authorization allowed a procedure for legally collecting and marketing paddlefish eggs from this paddlefish population during the sport snagging

season. During 1993–1994, 2268–2726 kg of paddlefish eggs were collected for caviar. A Montana/North Dakota paddlefish management plan suggests a reduced harvest quota, beginning in 1996.

Ohio (no sport or commercial fisheries, classified threatened, status unknown)
Ohio prohibits snagging but allows pole and line harvest. Fewer than 10 fish were harvested statewide in 1993. Ohio is currently conducting radio telemetry studies in the lower Scioto and Ohio rivers. Management concerns include dams and contaminants.

Oklahoma (sport fishery only, classified as nongame, status stable)
Little information exists regarding paddlefish abundance outside the Grand Lake system, however a growing population is believed to exist in Keystone Reservoir. Oklahoma's major sport fishery occurs on the Grand River above Grand Lake of the Cherokees. Smaller fisheries are located below several Arkansas River dams. Habitat alteration and overfishing (legal and poaching for eggs) are of concern. Early juveniles are being stocked into Kaw Reservoir. Commercial harvest was prohibited in 1992.

Pennsylvania (no sport or commercial fisheries, classified as extirpated, status extirpated)
Pennsylvania initiated a recovery program in 1991 to increase stocks in the upper Ohio and Allegheny rivers. Habitat alteration and water quality are believed to be most responsible for past declines; however, water quality in these streams has improved.

South Dakota (sport fishery only, classified as sport, status stable)
South Dakota shares responsibility with Nebraska for a 1600-fish quota and a 88–114 cm (eye-to-fork-of-tail) slot length limit below Gavins Point Dam. They began a stocking program in mid-1970s in Lake Francis Case to guarantee a broodstock source. Since 1990, South Dakota has been stocking juveniles into Lake Francis Case where natural reproduction is suspected to no longer occur, and into the Fort Randall Dam tailwaters where natural reproduction is limited.

Tennessee (sport and commercial fisheries, classified as sport/commercial, status unknown)
A modest sport harvest of about 2500 paddlefish annually is spread among several tailwaters, however this harvest is considered insignificant compared to the commercial harvest. All waters in the

Table 3. Sport harvest of paddlefish in 1993 and commercial harvest of paddlefish in 1992.

State	Sport harvest numbers (1993)	Commercial harvest in kg (1992)
Arkansas	Not available	136 000–181 000
Illinois	Not available	21 145
Indiana	Insignificant	Prohibited
Iowa	200	Prohibited
Kansas	< 600	Prohibited
Kentucky	< 1000	3906
Mississippi	Not available	Not available
Missouri	4041[1]	2188
Montana	3000	Prohibited
Nebraska	1600[2]	Prohibited
North Dakota	2000	Prohibited
Ohio	< 10	Prohibited
Oklahoma	5000	Prohibited
South Dakota	1600[2]	Prohibited
Tennessee	2500	60 328

[1] Creel Census information is for 1992 at Harry S. Truman Lake. No creel census available from Lake of the Ozarks or Table Rock Lake. Most recent Lake of the Ozarks harvest estimated at near 2000 paddlefish and Table Rock Lake at about 350.
[2] South Dakota and Nebraska share Gavins Point tailwater with a 1600 fish quota and an 88–114 cm eye-to-fork-of-tail slot length limit.

state are closed to commercial fishing except those designated as open. Commercial harvests are also controlled by various gear restrictions. Most of the commercially harvested paddlefish are taken from Tennessee and Cumberland river reservoirs. Commercial harvest is high (60 328 kg in 1992), but considerably less than 197 768 kg reported in 1975. Sport and commercial seasons are closed during a two-month period in the spring to protect adults during the spawning season. Concerns include overfishing, habitat destruction and contaminants. Several reservoirs receive maintenance stockings annually.

Texas (no sport or commercial fisheries, classified as endangered, status increasing)
Texas is currently utilizing an aggressive stocking program to recover paddlefish populations in six east-Texas streams. Early indications are that populations are increasing in most of these streams. Major concerns continue to be habitat destruction and water quality.

Virginia (no sport or commercial fisheries, classified as threatened, status unknown)
Paddlefish occur in Powell and Clinch rivers but in low numbers. Major reasons for decline include contaminants and siltation (soil and coal fines).

West Virginia (no sport or commercial fisheries, classified as species of special concern, status increasing).
West Virginia began stocking juveniles into the Ohio and Kanawha rivers in 1992. Prior to 1992, one or two paddlefish were reported annually. Navigational dams and habitat alterations are of major concern.

Wisconsin (no sport or commercial fisheries, classified as watch-listed, status increasing)
Wisconsin is presently conducting research on paddlefish populations in the Wisconsin and Mississippi rivers. A reintroduction plan (including stocking juveniles) is being considered above Prairie du Sac Dam on the lower Wisconsin River. Habitat degradation, water quality, and illegal harvest are major concerns.

Sport harvests

Sport harvest occurs throughout most of the existing range for paddlefish; however, many of these fisheries are small and dependent upon unpredictable river flows. Large sport fisheries, supported by self-sustaining or augmented stocks, exist only in the upper and central portions of the United States (Figure 1 and Table 3). Even these traditional sport fisheries are being challenged by increasing fishing pressure, continued habitat degradation and occasional mismanagement. Montana and North Dakota, once thought to have stable populations, are beginning to see effects of overfishing of adult stock presumably because of low natural recruitment. Slot length limits in South Dakota and Nebraska (on a shared tailwaters fishery) protect mature paddlefish, yet allow some harvest of large fish. Missouri is considering annual quotas and Oklahoma and Arkansas have reduced their daily bag limit. Since Gengerke's (1986) report, Alabama, Louisiana, Virginia, and West Virginia no longer allow sport harvest of paddlefish.

The most important sport fisheries are located on the Tennessee River at Kentucky Dam in Kentucky; the Yellowstone River at Intake, Montana; the Osage River at Warsaw, Missouri and the upper Osage River above Truman Lake, Missouri; Gavins Point Dam tailwaters on the Missouri River in Nebraska and South Dakota; the Missouri River at Williston, North Dakota (above Lake Sakakawea); and Highway Bridge 171 on the Missouri River in Montana (above Ft. Peck Reservoir).

Commercial harvests

It is apparent that commercial harvest of paddlefish is now less than reported by Gengerke (1986). It is not possible to obtain accurate information for commercial harvests from several states. Many states do not require commercial fishermen to report their catch and in some states where reporting is mandatory, results are suspect. Additionally, commercial fishermen routinely fish several drainages in several states and quantitative assessment is difficult.

Since Gengerke's (1986) report, five states (Alabama, Iowa, Louisiana, Oklahoma, and Virginia) have prohibited commercial harvest of paddlefish (Table 3). Most of these states were concerned about overfishing, illegal fishing or declines in adult stocks.

Arkansas and Tennessee report the largest commercial fisheries (Table 3). Arkansas estimates 136 000–181 000 kg of paddlefish are taken commercially each year from its several river systems. These figures are reduced slightly since 1983 because of a continued decline in some populations. Tennessee reports 60 328 kg harvested commercially in 1992, compared to 197 768 kg in 1975. All states bordering the pooled portion of the Mississippi River report declines in their commercial fisheries since 1983.

All states bordering the Missouri River now prohibit commercial fishing for paddlefish because harvests were small and size of fish was decreasing. Commercial exploitation, particularly in southern reservoirs, continues to be a major factor affecting the viability of paddlefish populations throughout their range. The incentive for illegal harvest has increased tremendously in the past 10 years because paddlefish eggs for caviar routinely sell for $100 to $200 per kg. The vulnerability of paddlefish to commercial (legal and illegal) harvest, behavioral characteristics, and low recruitment rate (slow maturation) is well documented by Pasch & Alexander (1986; also see Boreman 1997 this volume). Because of these problems, several states have removed paddlefish from their commercial list or implemented regulations to protect paddlefish from overexploitation.

Stocking programs

Ten states currently stock paddlefish juveniles to supplement existing stocks where natural recruitment is lacking or insufficient to maintain populations, or are stocking to recover paddlefish stocks in the periphery of its native range (Table 3). Wisconsin and Minnesota have not yet stocked fish but have drafted plans to guide recovery efforts for paddlefish in the lower Wisconsin and Minnesota rivers, respectively. It is encouraging that Kansas, Oklahoma, Pennsylvania, Texas, and West Virginia are attempting to recover remnant stocks in selected watersheds.

Conclusions

(1) The primary difficulty in assessing the current status or trends of paddlefish populations in most states is a lack of information about population sizes, age structure, growth, or harvest rates. While several paddlefish populations appear to be stable or increasing, *Polyodon spathula* is in decline in much of its current range because of continued habitat modification and degradation, increased contamination, and overfishing.

(2) Declines in paddlefish populations were identified by five of twenty-two states where paddlefish currently exist. Most of these states consider the loss or alteration of habitat as significant. Because of these serious resource impacts, states should continue seeking mitigation for habitat losses in select rivers to restore paddlefish habitat. Efforts should include restoration or maintenance of habitat diversity, including purchase or long-term easement on selected private lands so that old river channels and oxbows can be reconnected to main channels. Potential and known spawning areas should be protected from degradation.

(3) Resource agencies must remain alert for illegal harvests of eggs for caviar, for paddlefish populations can be damaged quickly if illegal activities are not recognized and stopped.

(4) Although the peripheral range of paddlefish has diminished slightly, they still occur in the majority of their original range, and in come cases are reinvading areas previously occupied. Aggressive stocking programs are currently in planning stages or actually underway to restore paddlefish to historic ranges. Water quality has improved in many previously polluted rivers and stocking should be successful. While states wait for stocked paddlefish populations to develop, they must quickly inventory available and potential spawning and nursery habitat and determine if paddlefish can prosper naturally. It has taken less than 100 years to destroy the future for many paddlefish populations. If state

and federal resource agencies work together to correct mistakes made in the last century, North American paddlefish can have a guaranteed future.

Acknowledgements

Two anonymous reviewers provided comments on the manuscript, and I also thank K. DeiSanti and T. Russell who provided a critical review of the manuscript, and the many state fisheries biologists, who compiled status, harvest, and present distribution information for the questionnaire. William E. Bemis drew the maps.

References cited

Bemis, W., E. Findeis & L. Grande. 1997. An overview of Acipenseriformes. Env. Biol. Fish. (this volume).

Boreman, J. 1997. Sensitivity of North American sturgeons and paddlefish to fishing mortality. Env. Biol. Fish. (this volume).

Carlson, D.M. & P.S. Bonislawsky. 1981. The paddlefish, *Polyodon spathula*, fisheries of the midwestern United States. Fisheries 6(2): 17–22, 26–27.

Gengerke, T.A. 1986. Distribution and abundance of paddlefish in the United States. pp. 22–35. *In*: J.G. Dillard, L.K. Graham & T.R. Russell (ed.) The Paddlefish: Status, Management and Propagation, N. Cent. Div., Amer. Fish. Soc. Spec. Pub. No. 7.

Grande, L. & W.E. Bemis. 1991. Osteology and phylogenetic relationships of fossil and recent paddlefishes (Polyodontidae) with comments on the interrelationships of Acipenseriformes. J. Vert. Paleo. 11, supplement 1: 1–121.

Pasch R.W. & C.M. Alexander. 1986. Effects of commercial fishing on paddlefish populations. pp. 46–53. *In*: J.G. Dillard, L.K. Graham & T.R. Russell (ed.) The Paddlefish: Status, Management and Propagation, N. Cent. Div., Amer. Fish. Soc. Spec. Pub. No. 7.

Pflieger, W.L. 1975. The fishes of Missouri. Missouri Department of Conservation, Jefferson City. 343 pp.

Russell, T.R. 1986. Biology and life history of paddlefish – a review. pp. 2–20. *In*: J.G. Dillard, L.K. Graham & T.R. Russell (ed.) The Paddlefish: Status, Management and Propagation, N. Cent. Div., Amer. Fish. Soc. Spec. Pub. No. 7.

Sparrowe, R.D. 1986. Threats to paddlefish habitat. pp. 36–45. *In*: J.G. Dillard, L.K. Graham & T.R. Russell (ed.) The Paddlefish: Status, Management and Propagation, N. Cent. Div., Amer. Fish. Soc. Spec. Pub. No. 7.

Unkenholz, D.G. 1986. Effects of dams and other habitat alterations on paddlefish sport fisheries. pp. 54–61. *In*: J.G. Dillard, L.K. Graham & T.R. Russell (ed.) The Paddlefish: Status, Management and Propagation, N. Cent. Div., Amer. Fish. Soc. Spec. Pub. No. 7.

Wei, Q., F. Ke, J. Zhang, P. Zhuang, J. Luo, R. Zhou & W. Yang. 1997. Biology, fisheries, and conservation of sturgeons and paddlefish in China. Env. Biol. Fish. (this volume).

Right side portrait of *Scaphirhynchus platorynchus* 98 cm TL from the Missouri River downstream of Great Falls, Fort Benton, Montana. Original by Paul Vecsei, 1996.

Environmental Biology of Fishes **48**: 291–298, 1997.

Life history and status of the shovelnose sturgeon, *Scaphirhynchus platorynchus*

Kent D. Keenlyne
U.S. Fish and Wildlife Service, 420 South Garfield Ave., Pierre, SD 57501–5408, U.S.A.

Received 20.8.1994 Accepted 14.3.1996

Key words: range, commercial harvest, market value, Mississippi, Missouri River

Synopsis

The shovelnose sturgeon, *Scaphirhynchus platorynchus*, is a freshwater sturgeon of the Mississippi and Missouri rivers and their tributaries. It is one of the smaller North American sturgeons, seldom weighing more than 2.5 kg over most of its range except in the upper Missouri River, where individuals of over 7 kg have been found. Spawning occurs in spring at temperatures between 17 and 21 °C over rock or gravel substrate downstream from dams, near rock structures, or in tributaries. most males reach sexual maturity at 5 years, most females at 7 years. Adults do not spawn every year. Shovelnose sturgeon are found in large, turbid rivers and frequently concentrate in areas downstream from dams or at the mouths of tributaries. Population densities range up to 2500 fish per km. They are commonly found in areas of current over sandy bottoms or near rocky points or bars, where they feed primarily on aquatic invertebrates. The shovelnose sturgeon is classified as a sport species in 12 of 24 states where it occurs. Commercial harvest is allowed in seven states, where fresh shovelnose sturgeon sell for 55 to 88 cents per kg, smoked shovelnose for about $5.75 per kg, and roe from 33 to 110 dollars per kg. About 25 tons of shovelnose sturgeon are harvested commercially each year. Shovelnose sturgeon are considered extirpated in three states, fully protected in four states, and rare, threatened, or of special concern in eight states. Populations are considered stable throughout most of the upper Mississippi, lower Missouri, Red, and Atchafalaya rivers. Three states, Wyoming, West Virginia, and New Mexico, have developed plans to reintroduce the species into rivers where it has been extirpated.

Introduction

The shovelnose sturgeon, *Scaphirhynchus platorynchus*, is indigenous to the Mississippi River drainage. The genus has occurred in this region for nearly 100 million years (Bailey & Cross 1954). Although one of the most abundant sturgeons in North America, its distribution has diminished in the last 100 years, and population numbers have been reduced throughout most of its range due to habitat alteration, overharvest, and water contamination. Although not as abundant as it once was, the shovelnose sturgeon is still one of the few sturgeons that can be harvested commercially in the United States (Helms 1974, Carlson et al. 1985).

Considerable information has been published on the biology and life history of the shovelnose sturgeon, but no status review has been compiled for this species in 10 years, and no compilation of commercial harvest data exists on this species over its entire range. The purposes of this paper are to summarize the general biology and life history of the shovelnose sturgeon and to present information on its current status. Status information was obtained

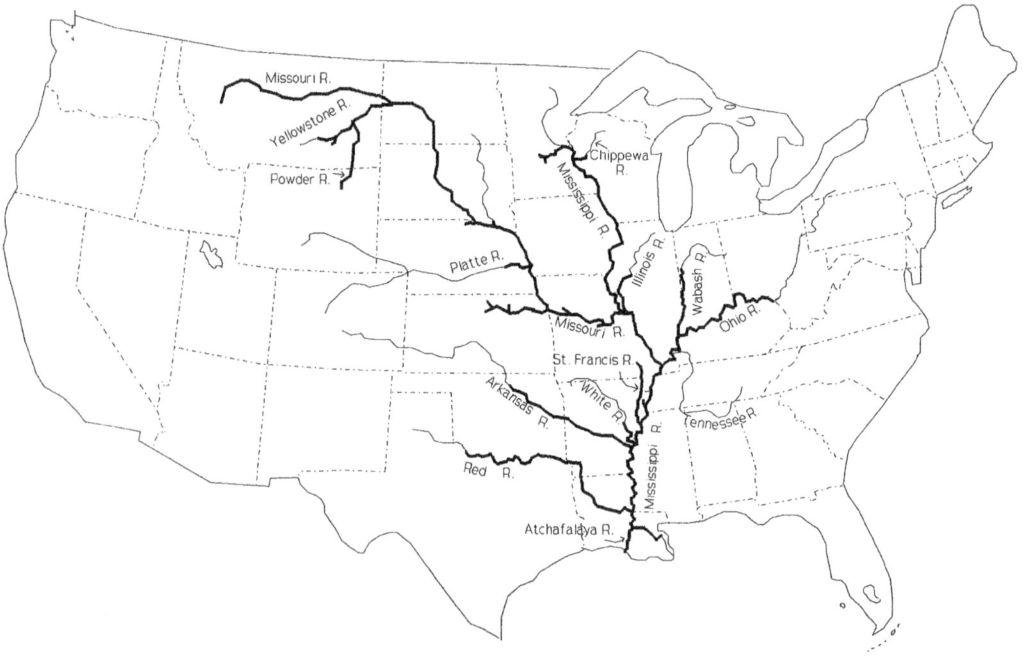

Figure 1. The current distribution of the shovelnose sturgeon.

with a questionnaire sent to all 24 states within the range of this species; biology and life history information were obtained from a review of literature.

Distribution, habitat, and abundance

Species of the genus *Scaphirhynchus* represent a distinctive group of freshwater sturgeons confined to the larger rivers of the interior United States (Bailey & Cross 1954). The pallid sturgeon, *S. albus*, is sympatric with the shovelnose sturgeon throughout the Missouri and Atchafalya rivers and the lower Mississippi River downstream of its confluence with the Missouri. The Alabama shovelnose sturgeon, *S. suttkusi*, is a rare species recently proposed from the Mobile River basin of Alabama and Mississippi (Williams & Clemmer 1991). The shovelnose sturgeon is also known as sand sturgeon, hackleback, or switchtail and is the most common of the three species. The current distribution of the shovelnose sturgeon, as obtained from the questionnaires used in this study, is presented in Figure 1.

Shovelnose sturgeon are primarily a bottom dwelling species (Barnickol & Starrett 1951, Moos 1978, Curtis 1990). They live where there is a current of 20–40 cm sec^{-1} (Hurley et al. 1987, Curtis 1990). Shovelnose sturgeon are usually found in pools downstream from sandbars in unchannelized rivers (Schmulbach et al. 1975, Moos 1978, Durkee et al. 1979); along the main channel border, downstream from dams, or in association with wing dams in rivers with navigational training structures (Hurley 1983, Pennington et al. 1983, Carlson et al. 1985, Hurley et al. 1987, Curtis 1990); and in the riverine habitat upstream of reservoirs in impounded rivers (Held 1969, Curtis 1990). They are usually found in association with sand substrate, often near rock or gravel (Christenson 1975, Hurley 1983, Carlson et al. 1985), where a current exists (Coker 1930, Schmulbach et al. 1975, Carlson et al. 1985, Curtis 1990).

Abundance of shovelnose sturgeon seems to be related to the size of the rivers in which they live and to human activities. Coker (1930) stated that the fate of Mississippi River sturgeon was a tragedy of shortsightedness in the conduct of the fishery. Shovelnose sturgeon were once so common that they were considered a nuisance to commercial fishermen and were destroyed when caught (Coker 1930). Barnickol & Starrett (1951) indicated that the decline of the shovelnose sturgeon in the Mississippi River also coincided with the development of the river as a navigation canal. The abundance of shovelnose sturgeon has been estimated for rivers of several sizes with a variety of habitats and varying degrees of modification: Schmulbach (1974) estimated 2500 fish km^{-1} for the unchannelized Missouri River, Helms (1972) estimated 1030 fish km^{-1} for the navigation-altered Mississippi River, Christenson (1975) estimated 100 fish km^{-1} for the small Red Cedar River in Wisconsin, and Elser et al. (1977) estimated 403 to 537 fish km^{-1} for the Tongue River in Montana.

Food, growth, and reproduction

Shovelnose sturgeon are opportunistic feeders that prey on aquatic invertebrates, primarily immature insects (Hoopes 1960, Held 1969, Ranthum 1969, Elser et al. 1977, Modde & Schmulbach 1977, Durkee et al. 1979, Carlson et al. 1985). Several studies indicate that the abundance of food affects growth rates of shovelnose sturgeon. Altered stream flows affect both the ability of shovelnose sturgeon to find food (Modde & Schmulbach 1977) and the abundance of prey organisms (Elser et al. 1977).

Carlander (1969) summarized shovelnose sturgeon growth data from fish collected from the Missouri River as the reservoirs were filling, and found them to be 213 mm total length (TL) at age 1, 274 mm at age 2, 399 mm at age 5, and 503 mm at age 10. In a study on the Mississippi River, Helms (1973) found shovelnose sturgeon to be 211 mm fork length (FL) at age 1, 328 mm at age 2, 495 mm at age 4, 592 mm at age 6, and 701 mm at age 10. Even though lengths of the Missouri River fish were in total length and the Mississippi River fish in fork length, the Missouri River fish were much shorter at given ages. Zweiacker (1967) found that Missouri River shovelnose sturgeon nearly ceased growing and reproducing after the Missouri River dams were constructed. Shovelnose sturgeon from 8 to 27 years of age were all about the same length, even though their ages varied by as much as 19 years (Zweiacker 1967).

The shovelnose sturgeon is the smallest of the North American sturgeons. Carlander (1969) indicates a maximum weight for the shovelnose sturgeon of 4.5 kg with most specimens being less than 2.5 kg. Shovelnose sturgeon in the upper Missouri River, however, are notably larger than specimens throughout most of its range. Peterman & Haddix (1975) examined 427 shovelnose in Montana and found an average weight of 2.4 kg; 11% of the individuals weighed more than 3.6 kg, 5% more than 4.5 kg, and specimens up to 7 kg were found. Keenlyne et. al. (1994) found significant morphometric differences between upper Missouri River fish compared to downriver populations which suggests that a different strain of shovelnose sturgeon exists in the upper Missouri River.

Spawning habitat of shovelnose sturgeon has not been described. Spawning is believed to occur over hard substrate in primary tributary streams to the main rivers (Cross 1967, Peterman & Haddix 1975, Christenson 1975, Elser et al. 1977) or along the borders of the main river channels (Coker 1930, Moos 1978). Studies conducted before major modifications were made to river channels and tributary flows concluded that shovelnose sturgeon swam up tributaries to spawn (Forbes & Richardson 1920, Coker 1930). Moos (1978) stated that shovelnose sturgeon only used tributaries infrequently and chose to remain in the larger rivers. Cross (1967) suggested that shovelnose sturgeon may seek tributary streams for spawning when flows are high. Moos (1978) observed that shovelnose sturgeon moved up to 540 km and stated that dams have probably prevented movement to traditional spawning areas, contributing to a lack of recruitment since damming of the upper Missouri River.

Although actual spawning has not been described, the capture of fish in spawning condition indicates that shovelnose sturgeon spawn from late April to June (Forbes & Richardson 1920, Coker 1930, Eddy & Surber 1947, Barnickol & Starrett 1951, Christenson 1975, Elser et al. 1977, Moos 1978) at water temperatures between 16.9 and 20.5 °C (Christenson 1975, Elser et al. 1977). Most males become sexually mature at age 5, while most females do not mature until age 7 (Helms 1973). In areas of poor food supply, males and females become sexually mature at a smaller size (Moos 1978). Females do not spawn every year with the frequency of spawning influenced by food supply and ability to store adequate fat to produce mature gametes (Christenson 1975, Moos 1978). Gonads of mature males represent 2–6% of body weight and 10–22% of females (Zweiacker 1967, Helms 1973, Christenson 1975, Moos 1978).

Several authors have identified what are believed to be reproduction problems with this species. June (1977) reported finding shovelnose sturgeon with mature eggs into July in Missouri River reservoirs with up to 52% of the females showing massive follicular atresia and concluded that this demonstrated unfavorable spawning conditions for the sturgeon as a result of river impoundment. He also reported high levels (2.1% incidence rate) of hermaphroditism in shovelnose sturgeon; a phemonenon also noted on other Missouri River studies by Carlson et al. (1985), with a 3% incidence rate, and Moos (1978), with a 1.6% incidence rate. Carlson et al. (1985) also discovered hybridization with *S. albus*. Several authors have noted a lack of natural reproduction in areas of the Missouri River and have attributed it to man-made alterations to shovelnose sturgeon habitat (Bailey & Cross 1954, June 1977, Moos 1978).

Current harvest

The shovelnose sturgeon is presently classified as a sport fish in 12 states and a commercial species in 7 (Table 1). It is considered extirpated in 3 states, is fully protected in 4, and is considered rare, threatened, or of special concern in 8 states. Some states have dual classifications and some classify it differently in various waters.

Table 1. Classification, status, and type of fishery allowed in states within the range of the shovelnose sturgeon.

State	Classification	Status		Type of fishery	
		Since 1940	Since 1990	Sport	Commercial
Alabama	Extirpated	Decline	Extinct	No	No
Arkansas	Commercial	Decline	Unknown	Yes	Yes
Illinois	Sport/commercial	Unknown	Unknown	Yes	Yes
Indiana	None	Unknown	Unknown	Yes	Yes
Iowa	Sport/commercial	Unknown	Stable	Yes	Yes
Kansas	Sport	Decline	Unknown	Yes	No
Kentucky	Sport	Unknown	Unknown	Yes	Yes
Louisiana	Special concern	Unknown	Unknown	No	No
Minnesota	Sport/concern	Unknown	Stable	Yes	No
Mississippi	Rare	Unknown	Unknown	No	No
Missouri	Sport/commercial	Decline	Unknown	Yes	Yes
Montana	Sport	Unknown	Stable	Yes	No
Nebraska	Sport	Decline	Stable	Yes	No
New Mexico	Extirpated	Extinct	Extinct	No	No
North Dakota	Protected	Decline	Stable	No	No
Ohio	Endangered	Decline	Unknown	No	No
Oklahoma	Special concern	Decline	Unknown	No	No
Pennsylvania	Extirpated	Unknown	Unknown	No	No
South Dakota	Protected	Unknown	Unknown	No	No
Tennessee	Extirpated	Unknown	Unknown	No	No
Texas	Endangered	Decline	Unknown	No	No
West Virginia	Extirpated	Decline	Extinct	No	No
Wisconsin	Sport/commercial	Decline	Unknown	Yes	Yes
Wyoming	Sport/concern	Decline	Stable	Yes	No

Sport and commercial harvests of shovelnose sturgeon are difficult to compare or analyze due to absence of similar data across its range. According to responses to a questionnaire used to obtain information for this project, sport harvest of shovelnose sturgeon generally is considered low with most of the harvest coinciding with the spring spawning season. Few anglers fish specifically for shovelnose sturgeon; much of the catch is incidental. Estimates of sport fishing exploitation rates are limited to the work of Christenson (1975) on the Red Cedar-Chippewa rivers in Wisconsin (2% annual exploitation rate) and Elser et al. (1977) on the Yellowstone and Tongue rivers in Montana (a 1% rate). No data are available on recent exploitation rates as a result of commercial harvest.

Today, there is a significant commercial harvest from the Mississippi River upstream from the mouth of the Ohio River, from the lower Missouri River, and from the Wabash River. In states where commercial harvest data were available (Table 2), information was provided by commercial anglers as part of their licensing requirement. Helms (1972) considered such data to be conservative. Presently, about 25 tons of shovelnose sturgeon are harvested annually; 60% comes from the Mississippi River upstream of St. Louis, Missouri. Limited information on the values of shovelnose sturgeon products exists.

Most harvest occurs in May and June and is believed to be associated with movement that coincides with spawning activity (Hurley 1983). Coker (1930) stated that shovelnose sturgeon tend to swim near the surface during the spawning run and Becker (1983) indicated that floating nets can be used by fishermen to catch sturgeon at that time. The fish are also vulnerable to propeller strikes by pleasure or commercial craft while swimming near the surface for several authors have noted finding fish with anterior ends of rostrums missing or severed caudal peduncles (Helms 1973, Christenson 1975, Moos 1978). Helms (1972) reported that 42% of the shovelnose sturgeon are harvested during May and June.

Most of the commercial harvest is taken with trammel nets, although some operators use traps. During spring, traps are often baited with sexually mature females, which are believed to attract more shovelnose sturgeon. Helms (1972) described the two major means of setting nets as dead or stationary sets in currents along the channel borders or drifting, where bottom conditions allow. Drifting of trammel nets can be very effective during periods when shovelnose sturgeon are relatively inactive and concentrated downstream from dams or near wing dams, or are located in deeper pools downstream from sand bars. When flows are reduced in winter and river waters drop in regulated rivers, shovelnose sturgeon concentrate and can become very vulnerable to netting. At this time, a technique that involves weighting one end of the trammel net and setting it near the edge of a scour hole and then sweeping the other end through the hole like a seine is effective.

Some fish are sold locally or directly to the public by commercial anglers as hog dressed or eviscerated fish, while others are eviscerated fish with heads removed. In distant markets or where the market

Table 2. Commercial harvest and prices (in dollars per kg) of shovelnose sturgeon by state in 1992. N.A. indicates data were not available.

State	Market price			Harvest (kg)
	Fresh	Smoked	Roe	
Arkansas	N.A.	N.A.	88.20–110.25	14 500
Illinois	0.80	5.75	77.20	22 335
Indiana	N.A.	N.A.	55.10	N.A.
Iowa	1.00	5.51	55.10	34 603
Kentucky	N.A.	N.A.	88.20	N.A.
Missouri	0.66	N.A.	66.15–88.20	26 460
Wisconsin	0.66	N.A.	33.10	11 025

exceeds the supply, fish are often sold as smoked sturgeon.

Shovelnose sturgeon are also harvested in late fall and early winter for both meat and the highly valued roe (Coker 1930). Moos (1978) reported that shovelnose sturgeon eggs are uniformly dark by the first of January and change little through the winter. Some of the best caviar is produced from fish taken at this time because the eggs are uniform in size and color, firm, relatively easy to process, and of high quality and taste before fat is incorporated into the eggs in spring during the final egg maturation process (Moos 1978). The general decline in prices of roe from north to south (Table 2) reflects proximity to market and quality of eggs. Southern states enjoy the advantage of having a longer ice-free period on the rivers, allowing commercial operators the opportunity to harvest sturgeon later in fall and earlier in spring, when the roe is best for producing quality caviar.

Current status and outlook

The shovelnose sturgeon is the widest-ranging freshwater sturgeon in North America. There is little question that its range and many populations have been reduced as a result of human actions, either through overharvest early in the 20th century or through modification of riverine habitats by dams and river-training structures (Coker 1930, Barnickol & Starrett 1951, Carlander 1954, Modde & Schmulbach 1977). A comparison of the historic range (Lee et al. 1980) to the present range (Figure 1) indicates that the species is now absent from the Rio Grande River and from upstream reaches of several large western rivers where movement has been blocked by dams and stream flow has been altered.

In the questionnaire developed for this study, 19 states responded that habitat alteration is a concern in regard to the welfare of the shovelnose sturgeon, six mentioned pollution as a concern, one mentioned overharvest, one mentioned hybridization, and three expressed no issues of concern. Flow alteration and habitat fragmentation, as a result of damming of many of the rivers within the range of this species, is a continuing problem to the long-term health of the species, especially as damming and fragmentation may be affecting replacement, reproduction, and gene flow.

In this survey, 12 states indicated that shovelnose sturgeon populations have declined in the last 50 years, one state reported it to have become extinct, and 11 states did not have information necessary to make an assessment of population trends for that period. Six states considered shovelnose sturgeon populations to be stable since 1990, three states indicated that the species is now considered extirpated within their state, and 15 states did not have sufficient data to make trend analyses on this species. Three states, Wyoming, West Virginia, and New Mexico, are developing plans to restock shovelnose sturgeon into waters that they once inhabited.

Hybridization of the shovelnose sturgeon with the pallid sturgeon is an emerging concern among sturgeon fishery managers. The possible introgression of genes from the more common shovelnose is viewed as a threat to the rare pallid sturgeon (Carlson et al. 1985, Keenlyne et al. 1994). Molecular technologies have been unable to differentiate among shovelnose species and their hybrids (Phelps & Allendorf 1983). Species like the shovelnose sturgeon that likely evolved with polyploidy (Blacklidge & Bidwell 1993) are difficult to study through normal genetic testing procedures. Imposition of recent introgressive hybridization may contribute to the present state of confusion about integrity of species and will continue to remain a problem for scientists and administrators who attempt to manage for shovelnose species in the future.

The welfare of the shovelnose sturgeon may have future implications to our large rivers in the central United States. Becker (1983) lists the shovelnose sturgeon as a host for glochidia of the commercially valuable yellow sand-shell *Lampsilis teres*, pimpleback *Quadrula pustulosa*, and hickory-nut *Obovaria olivaria* pearly mussels. Shovelnose sturgeon also are the only known host for the parasitic larvae of the hickory-nut mussel (Coker 1930).

Acknowledgements

I thank S. Whitmore, who developed the figure for this report, J. Zuboy and J. Schmulbach, who provided critical reviews of the manuscript, and the many state fishery biologists, who compiled status, harvest, and present range information for the questionnaire.

References cited

Bailey, R.M. & F.B. Cross. 1954. River sturgeons of the American genus *Scaphirhynchus*: characters, distribution, and synonymy. Pap. Mich. Acad. Sci., Arts, Let. 39: 169–208.

Barnickol, P.G. & W. Starrett. 1951. Commercial and sport fishes of the Mississippi River between Caruthersville, Missouri and Dubuque, Iowa. Bull. Ill. Nat. Hist. Surv. 25: 267–350.

Becker, G.C. 1983. Fishes of Wisconsin. University of Wisconsin Press, Madison. 1052 pp.

Blacklidge, K.H. & C.A. Bidwell. 1993. Three ploidy levels indicated by genome quantification in Acipenseriformes of North America. J. Hered. 84: 427–430.

Carlander, H.B. 1954. History of fish and fishing in the upper Mississippi River. Upper Miss. River Cons. Comm., Rock Island. 96 pp.

Carlander, K.D. 1969. Handbook of freshwater fishery biology. Iowa State University Press, Ames. 752 pp.

Carlson, D.M., W.L. Pflieger, L. Trial & P.S. Haverland. 1985. Distribution, biology, and hybridization of *Scaphirhynchus albus* and *S. platorynchus* in the Missouri and Mississippi rivers. Env. Biol. Fish. 14: 51–59.

Christenson, L.M. 1975. The shovelnose sturgeon *Scaphirhynchus platorynchus* (Rafinesque) in the Red Cedar – Chippewa rivers system Wisconsin. Wis. Dep. Nat. Resour., Res. Rep. No. 82. 23 pp.

Coker, R.E. 1930. Studies of common fishes of the Mississippi River at Keokuk. U. S. Bur. Fish. Bull. 45: 141–225.

Cross, F.B. 1967. Handbook of fishes of Kansas. Mus. Nat. Hist. Univ. Kans., Misc. Publ. 45: 1–357.

Curtis, G.L. 1990. Habitat use by shovelnose sturgeon in Pool 13, upper Mississippi River, Iowa. M.S. Thesis, Iowa State University, Ames. 79 pp.

Durkee, P., B. Paulson & R. Bellig. 1979. Shovelnose sturgeon (*Scaphirhynchus platorynchus*) in the Minnesota River. Minn. Acad. Sci. 45: 18–20.

Eddy, S. & T. Surber. 1947. Northern fishes. Univ. Minn. Press, Minneapolis. 267 pp.

Elser, A.A., R.C. McFarland & D. Schwehr. 1977. The effect of altered streamflow on fish of the Yellowstone and Tongue rivers, Montana. Mont. Dept. Fish and Game Tech. Rept. 8: 1–180.

Forbes, S.A. & R.E. Richardson. 1920. The fishes of Illinois. Ill. Nat. Hist. Surv. 3: 1–357.

Held, J.W. 1969. Some early summer foods of the shovelnose sturgeon in the Missouri River. Trans. Amer. Fish. Soc. 98: 514–517.

Helms, D.R. 1972. Progress report on the first year study of shovelnose sturgeon in the Mississippi River. Iowa Conservation Commission, Des Moines. 22 pp.

Helms, D.R. 1973. Progress report on the second year of study of shovelnose sturgeon in the Mississippi River. Iowa Conservation Commission, Des Moines. 33 pp.

Helms, D.R. 1974. Age and growth of the shovelnose sturgeon, *Scaphirhynchus platorynchus* (Rafinesque), in the Mississippi River. Proc. Iowa Acad. Sci. 81: 73–75.

Hoopes, D.T. 1960. Utilization of mayflies and caddisflies by some Mississippi River fish. Trans. Amer. Fish. Soc. 89: 32–34.

Hurley, S.T. 1983. Habitat associations and movements of shovelnose sturgeon in Pool 13 of the upper Mississippi River. M.S. Thesis, Iowa State University, Ames. 82 pp.

Hurley, S.T., W.A. Hubert & J.G. Nickum. 1987. Habitats and movements of shovelnose sturgeons in the upper Mississippi River. Trans. Amer. Fish. Soc. 116: 655–662.

June, F.C. 1977. Reproductive patterns in seventeen species of warmwater fishes in a Missouri River reservoir. Env. Biol. Fish. 2: 285–296.

Keenlyne, K.D., C.J. Henry, A. Tews & P. Clancey. 1994. Morphometric comparisons of Upper Missouri River sturgeons. Trans. Amer. Fish. Soc. 123: 779–785.

Lee, D.S., C.R. Gilbert, C.H. Hocutt, R.E. Jenkins, D.E. McAllister & J.R. Stauffer Jr. 1980. Atlas of North American freshwater fishes. North Carolina Biol. Surv., Pub. 1980–12, Raleigh. 867 pp.

Modde, T. & J.C. Schmulbach. 1977. Food and feeding behavior of the shovelnose sturgeon, *Scaphirhynchus platorynchus*, in the unchannelized Missouri River, South Dakota. Trans. Amer. Fish. Soc. 106: 602–608.

Moos, R.E. 1978. Movement and reproduction of shovelnose sturgeon, *Scaphirhynchus platorynchus*, in the Missouri River, South Dakota. Ph.D. Dissertation. University of South Dakota, Vermillion. 216 pp.

Pennington, C.H., J.A. Baker & M.E. Potter. 1983. Fish populations along natural and revetted banks on the lower Mississippi River. N. Amer. J. Fish. Manag. 3: 204–211.

Peterman, L.G. & M.H. Haddix. 1975. Lower Yellowstone River fishery study. Montana Dept. Fish and Game, Prog. Report 1: 1–56.

Phelps, S.R. & F. Allendorf. 1983. Genetic identity of pallid and shovelnose sturgeon (*Scaphirhynchus albus* and *S. platorynchus*). Copeia 1983: 696–700.

Ranthum, R.G. 1969. Distribution and food habits of several species of fish in Pool 19, Mississippi River. M.S. Thesis, Iowa State University, Ames. 207 pp.

Schmulbach, J.C. 1974. An ecological study of the Missouri River prior to channelization. University of South Dakota, Vermillion. 34 pp.

Schmulbach, J.C., G. Gould & C.L. Groen. 1975. Relative abundance and distribution of fishes in the Missouri River, Gavins Point Dam to Rulo, Nebraska. S. D. Acad. Sci. 54: 194–222.

Williams, J.D. & G.H. Clemmer. 1991. *Scaphirhynchus suttkusi*, a new sturgeon (Pisces: Acipenseridae) from the Mobile Basin of Alabama and Mississippi. Bull. Alabama Mus. Nat. Hist. 10: 17–31.

Zweiacker, P. 1967. Aspects of the life history of the shovelnose sturgeon, *Scaphirhynchus platorynchus* (Rafinesque), in the Missouri River. M.A. Thesis, University of South Dakota, Vermillion. 46 pp.

Environmental Biology of Fishes **48**: 299–309, 1997.

The status and distribution of lake sturgeon, *Acipenser fulvescens*, in the Canadian provinces of Manitoba, Ontario and Quebec: a genetic perspective

Moira M. Ferguson[1] & George A. Duckworth[2]
[1] *Department of Zoology and Institute of Ichthyology, University of Guelph, Guelph, Ontario N1G 2W1, Canada*
[2] *Ontario Ministry of Natural Resources, Northwest Region, 140–4th Ave., Cochrane, Ontario P0L 1C0, Canada*

Received 18.11.1994 Accepted 8.4.1996

Key words: postglacial recolonization, anthropogenic influences, fish, Acipenseridae, conservation

Synopsis

Genetic analysis of mitochondrial DNA sequence variation indicates that most of a sample of 396 lake sturgeon, *Acipenser fulvescens*, from the northern part of their range belonged to either one of two haplotypes. The vast majority of fish from the Great Lakes/St. Lawrence and Mississippi drainages were of a single haplotype while those from the Hudson/James Bay were composed of both haplotypes. This haplotypic distribution suggests that fish from one refugium (possibly Missourian) recolonized the Hudson-James Bay drainage while those from a second (possibly Mississippian) recolonized the Laurentian Great Lakes and St. Lawrence River. Lake sturgeon still inhabit much of their native postglacial distribution in Manitoba, Ontario and Quebec. However, the stresses of commercial overexploitation and habitat alteration, usually through hydroelectric dam construction and operation, have either singly or in tandem brought about the reduction, if not extirpation, of some populations within the range. The largest zone of extirpation and population reduction has occurred in the Lake Winnipeg drainage area, which covers more than one-third of Manitoba. Other areas where populations have been reduced to remnant levels, if not extirpated, include the lower Laurentian Great Lakes of Lake Ontario and Lake Erie. In northern Ontario, lake sturgeon populations whose riverine habitats have been fragmented by two or more dams are substantially reduced from their former levels. In Quebec, more attention has been paid to limiting the exploitive stresses on lake sturgeon populations. Combination of the genetic and status data suggests that both northern and southern populations of lake sturgeon (possibly from two glacial refugia) have been impacted severely from anthropogenic influences.

Introduction

The lake sturgeon, *Acipenser fulvescens*, is one of the most widely distributed members of the North American fish fauna. Its native range includes three major watersheds: the Laurentian Great Lakes, Hudson-James Bay, and the Mississippi River (Houston 1987). The Canadian distribution of lake sturgeon includes the five provinces of Alberta, Saskatchewan, Manitoba, Ontario, and Quebec. It exists as far west as Edmonton on the North Saskatchewan River, as far east as St. Roch de Aulnaires on the St. Lawrence River, as far north as the Seal River, a tributary on the west coast of Hudson Bay, and as far south as the main Mississippi River and in most of its larger tributaries southward to southern Arkansas.

The current distribution of lake sturgeon has been impacted by historical processes such as postglacial recolonization of the northern part of the range and anthropogenic influences including overfishing and habitat alteration. We discuss these factors and then evaluate the current status of this species in the Canadian provinces of Manitoba, Ontario, and Quebec from a genetic perspective. We also summarize and contrast the overall status in the United States.

Influences on distribution

Postglacial recolonization

Wisconsinan glaciation was a major feature of the recent zoogeographic past of North America (Hocutt & Wiley 1986). At its maximum about 18 000 years ago, the Wisconsinan ice sheet covered much of the extant Canadian range of lake sturgeon. Those populations that occurred in Canada prior to the Wisconsinan glacial period either perished or moved into refuges south of the ice sheet. Enormous amounts of meltwater formed rivers and large lakes, providing dispersal routes for reinvasion. Paleogeographic evidence suggests that lake sturgeon reinvaded the northern part of its range via the Warren, Brule, and Chicago dispersal routes from a Mississippian Refugium (Mandrak & Crossman 1992).

Distribution of mitochondrial DNA (mtDNA) genetic types (haplotypes) has been used to infer postglacial recolonization from refugia and test conclusions based on paleogeographic evidence (Billington & Hebert 1991). MtDNA is a closed circular molecule which is maternally inherited and not subject to recombination (Moritz et al. 1987). Nucleotide variation in mtDNA is usually detected indirectly by cutting the molecule with restriction endonucleases or by determining nucleotide sequences directly. MtDNA retains a history of past isolation for a longer period relative to nuclear DNA because of its transmission (Billington & Hebert 1991). Gene frequencies will diverge in both the nuclear and mitochondrial genomes once a single population becomes subdivided into two isolates because of the evolutionary processes of genetic drift and mutation. This divergence may be enhanced if the isolates become fixed for different mtDNA types because of stochastic lineage extinction (Avise et al. 1987). If the descendants of two populations originating from different refugia come into secondary contact after glacial recession, then gene frequencies at variable loci in the nuclear genome will homogenize until there is no evidence of past isolation. However, this process occurs more slowly in the mitochondrial genome because of its smaller genetic population size (1/4) relative to the nuclear genome, which means that four times as

Table 1. Sampling locations and number of lake sturgeon analyzed for mtDNA haplotype variation. Fish were collected between 1989–1993 unless otherwise indicated.

Sampling location	n
Hudson Bay-James Bay	
Moose River basin	129
Waswanipi River	12[a]
	12
Lower Nelson R.	10[b]
Rainy River	16
Total	179
Great Lakes-St. Lawrence	
Lake St. Clair	12
Lake Temiskaming	18
Lake Winnebago	18
Lake Ontario (1873)[c]	1
Ottawa River (Rockland, Quyon, Gatineau)	17[a]
	10
Lake of Two Mountains	11[a]
St. Lawrence River	
Lac Saint-Louis	11[a,d]
Montreal (1866)[c]	1
Lac Saint-Pierre	20[a]
	22
Gentilly	11[a]
Quebec City	22
Sturgeon River (Lake Superior)	21
Total	195
Mississippi	
Mississippi River (1870)[c]	1
Flambeau River (Mississippi)	21
Total	22

[a] Data from Guenette et al. (1993), otherwise from Ferguson et al. (1993) and Ferguson (unpublished).

[b] Sampled a single individual with a different haplotype at a sequence position other than 54.

[c] Preserved specimen housed in the British Museum of Natural History, London.

[d] Sampled a single individual with a different haplotype as determined by digestion with *Bcl* 1.

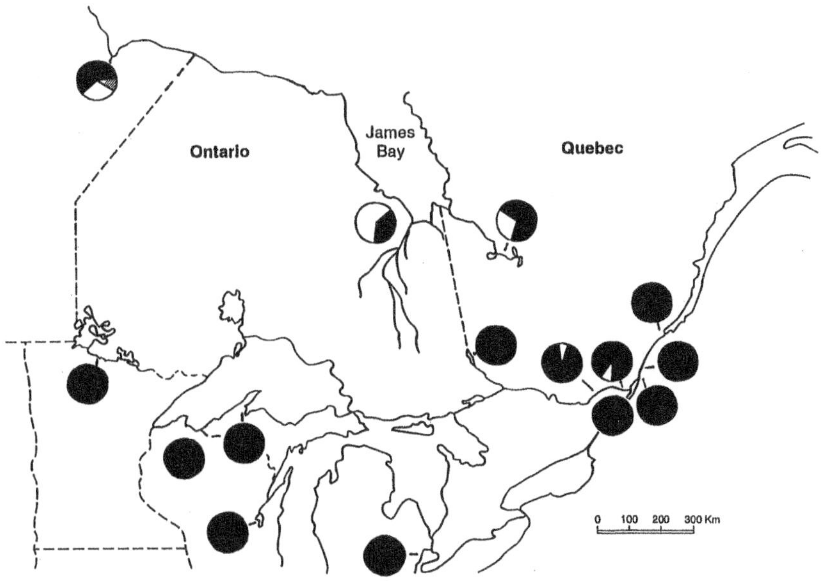

Figure 1. Mitochondrial DNA haplotype frequencies in lake sturgeon from the northern part of their range. See Table 1 for sample sizes. Haplotype 1 = black, haplotype 2 = clear.

much migration is required to erode mtDNA haplotype divergence than nuclear gene divergence. Thus, the basic premise is that fish originating from different refugia will differ genetically in mtDNA haplotype frequencies (Billington & Hebert 1991). Ward et al. (1989) found that mtDNA haplotypes of walleye, *Stizostedion vitreum*, cluster into three main groups, each reflecting recolonization of Canadian waters from each of three glacial refugia. A similar approach has been used for lake whitefish, *Coregonus clupeaformis* (Bernatchez & Dodson 1990), lake charr *Salvelinus namaycush*, arctic charr *S. alpinus* (Wilson 1995), and brook charr *S. fontinalis* (Danzmann & Ihssen 1995).

We analyzed mtDNA variation of lake sturgeon from the northern part of their range to assess factors which may influence the distribution of mtDNA haplotype lineages (Ferguson et al. 1993, Guenette et al. 1993). Most of the 396 lake sturgeon analyzed were characterized by only two mtDNA haplotypes based on a restriction fragment length polymorphism (RFLP) analysis or direct sequencing of 275 nucleotides in the mtDNA control region (Table 1, Figure 1). Fish could be categorized into two major haplotypes using either of two restriction enzymes which detected polymorphisms (*Ava* II, *Hinc* II) or nucleotide variation at position 54 in the 275 base pairs sequenced (details in Ferguson et al. 1993). Both analyses provided equivalent haplotypic designations. Three populations from the Hudson-James Bay watershed in northern Manitoba, Ontario, and Quebec shared the same two common haplotypes. However, the vast majority of more southerly fish from the Great Lakes/St. Lawrence and Mississippi watersheds were of a single haplotype. This included three specimens housed at the British Museum of Natural History and collected in the late 1800's. The only apparent disparity in this pattern was the collection of a very small number of haplotype 2 fish near the confluence of the Ottawa and St. Lawrence Rivers by Guenette et al. (1993).

The distribution of mtDNA haplotypes (Figure 1) can be explained by two alternative scenarios corresponding to a single refugium hypothesis versus a two refugium hypothesis. The resolution of the controversy depends on the ability of each scenario to explain the commonality of haplotype 2 fish in northern samples and rarity in southern samples. In the first scenario, the disjunct distribution of haplotype 2 is because lake sturgeon used two distinct routes of colonization from two refugia (possibly Missourian and Mississippian) to reinvade the contemporary range of distribution. Fish from a Missourian refugium would have recolonized the headwaters of the James Bay and Hudson Bay areas of Ontario and Quebec through the connection between Lake Agassiz and Lake Ojibway-Barlow (about 9500 years ago) (Crossman & McAllister 1986) resulting in a northerly distribution of descendants. As mentioned previously, Mississippian fish would have used the Warren, Brule, and Chicago dispersal routes to colonize the southern part of the range (Mandrak & Crossman 1992). The large number of haplotype 2 fish sampled from northern sites (Nelson River, Manitoba; Moose River basin, Ontario; Waswanipi River, East Megiscane region, Quebec) coupled with the very high predominance of haplotype 1 fish in the south is consistent with the two refugium hypothesis. The observation that all 21 fish sampled from the Flambeau River, a tributary of the Mississippi River were haplotype 1 lends further credence to the idea that populations derived from a Mississippian refugium were haplotype 1. According to the two refuge hypothesis, the three haplotype 2 fish collected by Guenette et al. (1993) in the Ottawa River and Lake of Two Mountains could be the result of secondary contact among the two refugial groups; Lake Ojibway-Barlow was connected with the Ottawa River up until about 8000 years ago (Crossman & McAllister 1986). According to the single refugium model, lake sturgeon recolonized Canada from a single Mississippian refugium (Guenette et al. 1993) a scenario compatible with the paleogeographic evidence (Crossman & McAllister 1986, Mandrak & Crossman 1992). Mississippian refugium fish would have contained both haplotypes, one of which was largely eliminated in the southern part of the range via stochastic lineage extinctions (Avise et al. 1987). The few haplotype 2 fish observed in the Lake of Two Mountains and the Ottawa River would be the remnants of a historically more common lineage.

We favor the two refuge hypothesis given the distribution of haplotypes over the broader geographic scale illustrated in Figure 1. Stochastic lineage extinction is expected to be random with respect to the extinction of haplotypes in specific populations. It is difficult to envisage that all the southerly populations sampled by us and Guenette et al. (1993) from throughout the Great Lakes and Mississippi regions would have had the same haplotype eliminated by chance, and it is difficult to imagine differential selection of mtDNA haplotypes.

Exploitation

Since the existence of aboriginal culture in North America, lake sturgeon have been a key food source, especially during spring ceremonial festivities at lake sturgeon spawning sites. During the early 1800's, lake sturgeon was also sought after as a trade item since the isinglass (a form of gelatin obtained from the inner lining of the swimbladder) could be used as a clarifying agent in wine, beer and jelly making. Records of its harvest were first kept by the Hudson Bay Company which provided aboriginal people access to markets (Holzshamm & McCarthy 1988). There is no evidence that the harvest levels of lake sturgeon prior to the 1860's had any influence on population levels.

Early settlers to North America did not value lake sturgeon as a food source. This changed about 1855 when a market for caviar developed at Sandusky, Ohio, on Lake Erie, followed by the sale of smoked flesh in 1860. Spurred by these market demands, local markets and fisheries spread to lakes Huron, Ontario, Superior, Nipissing, and Nipigon and Lake of the Woods. In every case, after an initial high yield, the fisheries displayed a rapid and permanent decline to very low levels (Harkness & Diamond 1961). For example, the Lake Erie catch fell from over 2500 tonnes per year to less than 500 tonnes between 1885–1895.

The development of commercial fisheries for lake sturgeon in Manitoba were delayed relative to those in Ontario. The harvest from Lake Winnipeg and its tributaries, the Red and Assiniboine rivers, peaked at 445 tonnes in 1900 and crashed to 13 tonnes by 1910 when the fishery was closed (Houston 1987). The fishery reopened after 6 years with fluctuating landings until 1928 when it was again closed. Declining catches in Lake Winnipeg spurred interest in more northern locations such as the Nelson and Churchill rivers where commercial fishing began in 1907 and 1924, respectively. Successive closures and openings characterized commercial fishing in these locations. In 1961, commercial fishing for sturgeon was closed in Manitoba but it resumed again in 1970, albeit at low levels.

After fisheries in the larger lakes had declined, smaller northern inland waters became the subject of commercial interest. Impacts on these populations are masked by the aggregate reporting of statistics. For example, in Ontario, commercially harvested fish from all northern inland waters (rivers and lakes) were recorded in one category making it impossible to trace results from one particular waterbody. This is further exacerbated by incomplete reporting by commercial fisherman and the relatively low priority that managers place on ensuring compliance in this fishery. By the late 1980's, the combined northern inland harvest in Ontario was twice that of the Great Lakes (Duckworth et al.[1]).

Habitat alteration

Habitat protection is considered to be the key factor in the conservation and rehabilitation of the remaining lake sturgeon stocks in Ontario (Duckworth et al.[1]). Maintenance and enhancement of existing lake sturgeon habitat is considered the third highest priority in Manitoba after the maintenance of genetic integrity and the protection of existing stocks (Anonymous[2]). Lake sturgeon require swift current and large rough substrate for spawning and embryo incubation. This dependence on riverine environments makes them vulnerable to development on rivers that alters habitat.

Hydroelectric generation facilities affect both periodic and seasonal water level fluctuations, causing decreased production and loss of fish (Payne 1987). Low water conditions after spawning can affect success of embryo survival as embryos experience variable water temperatures, low oxygen concentrations, and desiccation. Young fish can become trapped in shallow pools and subjected to heavy mortality through predation, temperature, and oxygen stress. Adults have become stranded in shallow pools and mortality occurs when pools become anoxic or freeze (Duckworth et al.[1]).

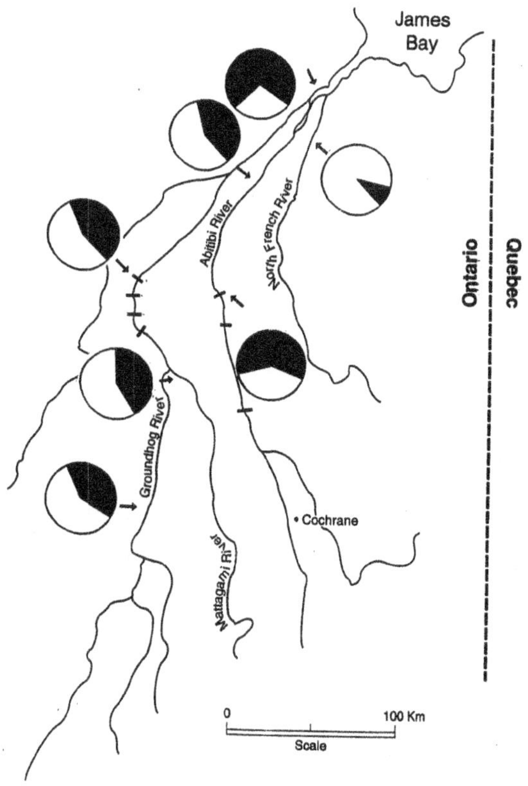

Figure 2. Mitochondrial DNA haplotype frequencies in lake sturgeon from the Moose River basin, Ontario. Hydroelectric generating stations are represented by slashes (data from Ferguson et al. 1993). Haplotype 1 = black, haplotype 2 = clear.

Dams also restrict movements of lake sturgeon, preventing fish from reaching critical habitat such as spawning sites, and by stranding fish between barriers. High water conditions associated with dams can also flood and eliminate rapids and chutes previously used by spawning fish (Duckworth et al.[1]). A population genetics study of lake sturgeon within the Moose River Basin in northeastern Ontario suggested that there is significant gene flow among most sites in the watershed (Ferguson et al. 1993, Figure 2). Dam construction or other artificial

[1] Duckworth, G., T. Mosindy, E. Armstrong, G. Goodchild, G. Preston, M. Hart & C. Jessop. 1992. A draft management strategy for lake sturgeon in Ontario. Edition 6, 31 July 1992. Ministry of Natural Resources Unpublished Manuscript.

[2] Anonymous. 1992. A sturgeon management strategy for Manitoba: a discussion paper (draft). October 1992. Manitoba Natural Resources Department, Fisheries Branch Unpublished Manuscript.

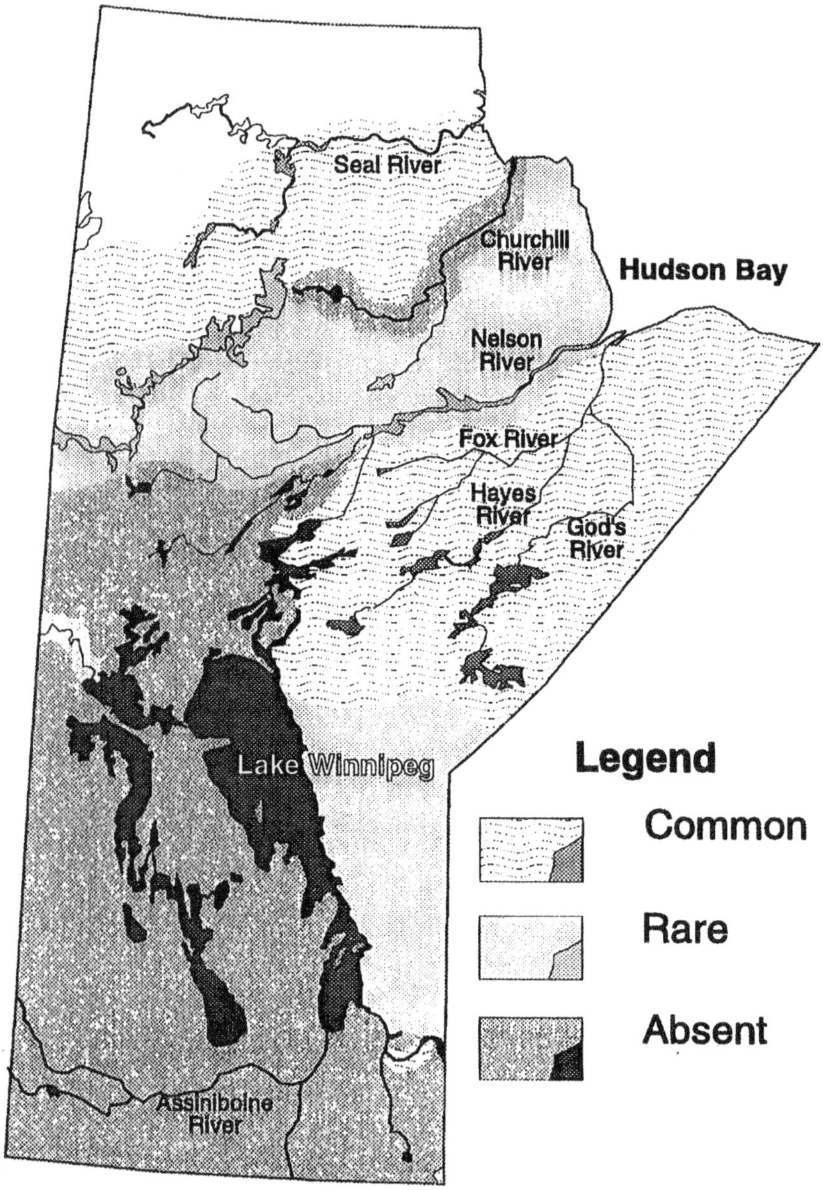

Figure 3. Status of lake sturgeon in the province of Manitoba, Canada.

barriers to migration could impact the genetic integrity of the species by fragmenting it into isolated stocks and causing the loss of genetic variability through evolutionary processes such as random genetic drift and inbreeding. Populations with lowered genetic variability may be less able to withstand future stresses. Genetic impacts of such anthropogenic influences are illustrated by the reduced mtDNA haplotypic diversity in white sturgeon, *Acipenser transmontanus*, from isolated sections of the Columbia River (Brown et al. 1992).

The other impact on Canadian lake sturgeon comes from the effects of the forest products industry including construction of forest access roads, log driving, and pulp mill effluent. Roads often cross rivers at spawning sites due to the presence of rock

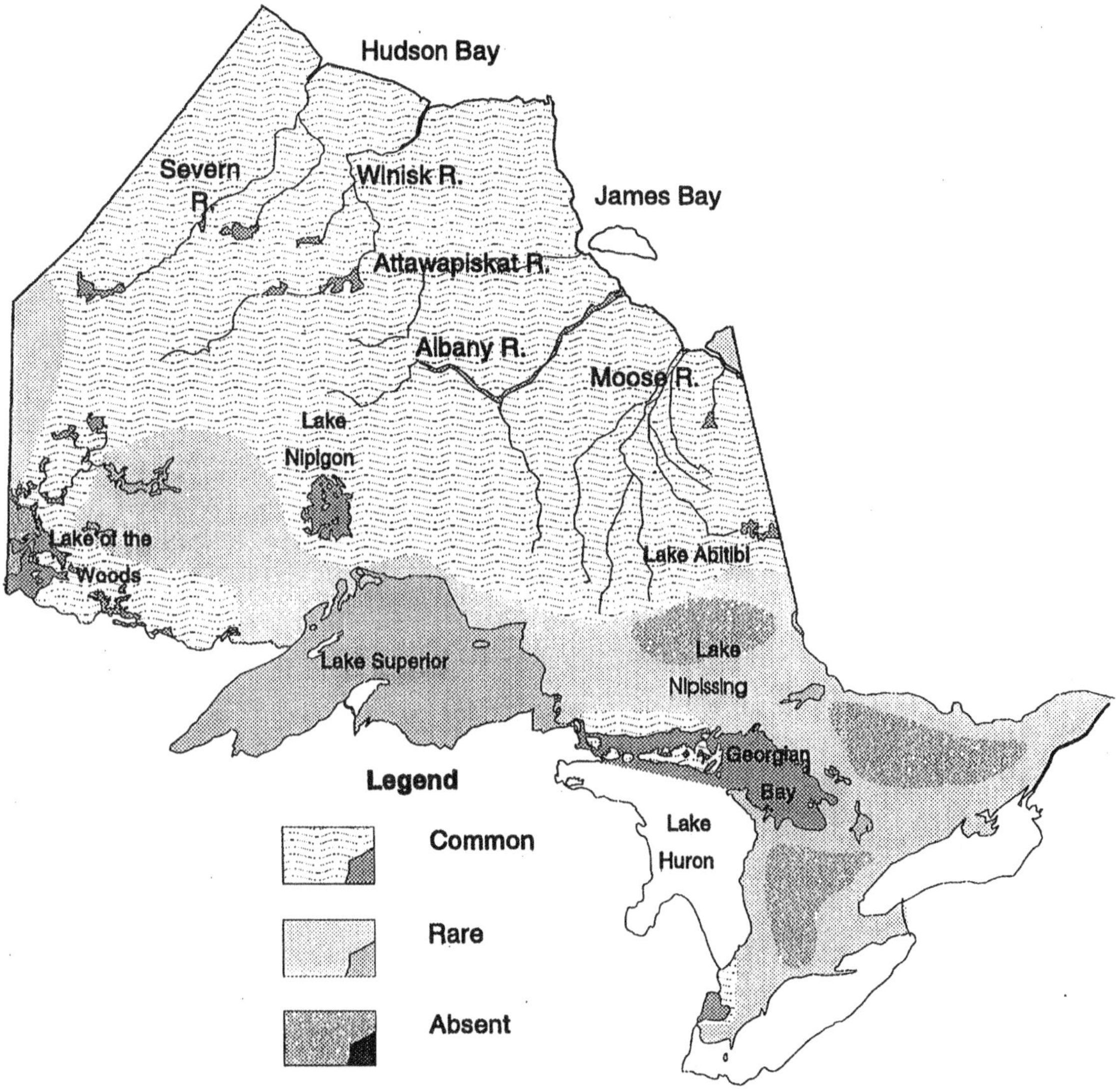

Figure 4. Status of lake sturgeon in the province of Ontario, Canada.

at the narrower cross sections. These can obstruct fish passage and reduce the amount of spawning habitat available. Erosion and siltation during and after road construction impairs water quality. Impaired reproduction in such areas has also been suggested (Duckworth et al. 1992). Pulp and paper mill effluent deposits organic debris and decreases water quality by reducing dissolved oxygen and disrupting benthic communities.

Current status of lake sturgeon

We categorized regions of the postglacial distribution of lake sturgeon into areas where the species was either common, rare, or absent (extirpated) (Figure 3–5). Data sets describing the distribution and abundance of lake sturgeon in the provinces of Ontario, Manitoba, and Quebec differ considerably, and interpretation was required to categorize geographical regions into the above categories. Data for Ontario are from the Ontario Ministry of Nat-

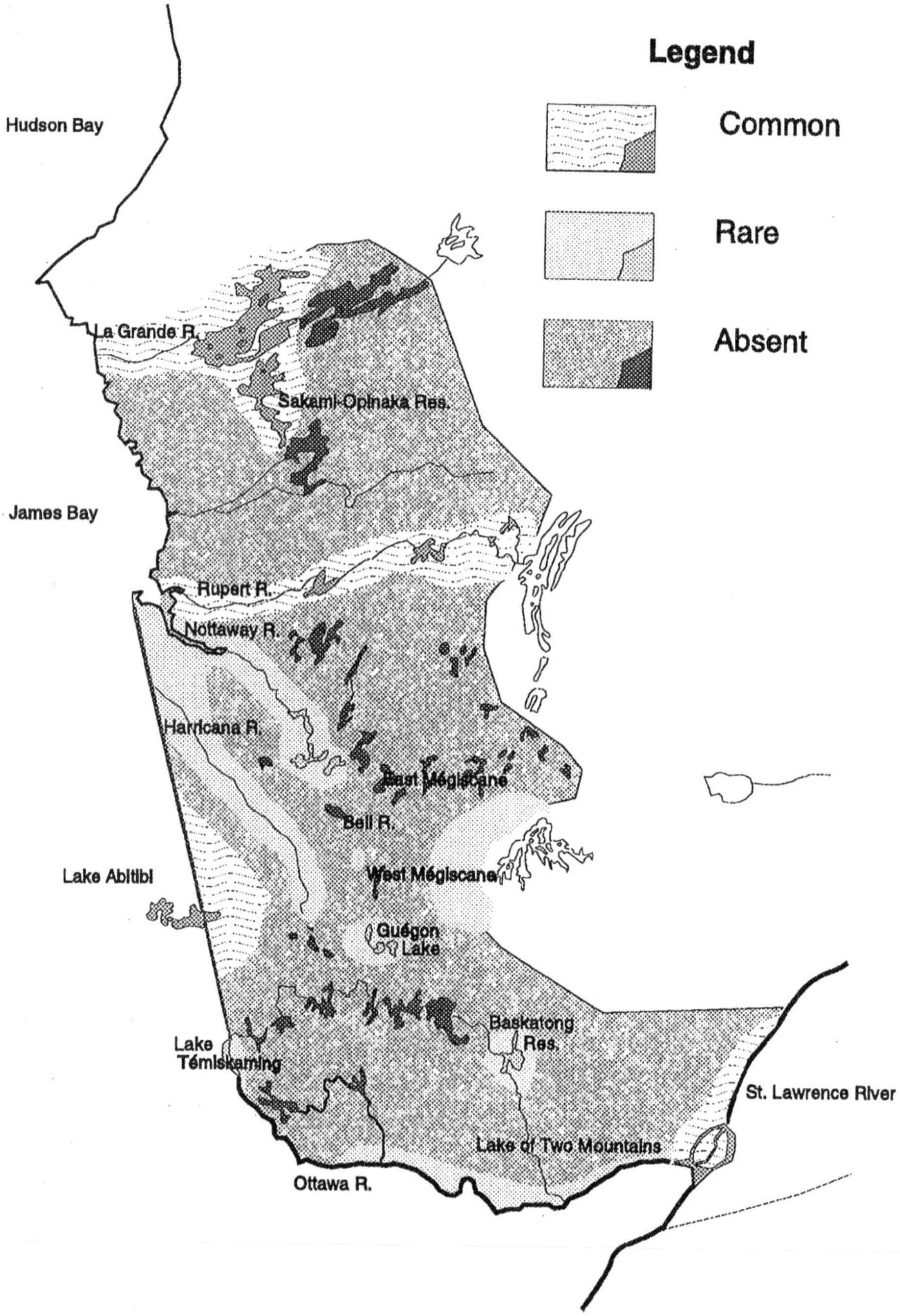

Figure 5. Status of lake sturgeon in the province of Quebec, Canada.

ural Resources files on individual locations where fish have been found during various lake and river surveys. These data were represented as individual dots on the map and could be translated into our absent, rare, and common designations by drawing lines on the map to enclose areas of similar dot density (Figure 3). Areas that contained no known populations (i.e., no dots on the map) were designated as 'absent'. Very few lake sturgeon were found in 'rare' areas, whereas areas designated as 'common' contained many populations (i.e., dot density was high). Data for Manitoba (Rob Cann personal communication) was of similar scope to the Ontario data but differed in category designation (Figure 4). Our class of 'absent' was equivalent to their remnant/extirpated, our 'rare' was equivalent to their poor/fair and 'common' was equivalent to good/unimpacted. Fortin et al.[3] compiled the data base for Quebec (Figure 5). Lake sturgeon distribution in Quebec is limited to the southeast corner of the province and Fortin et al. reviewed the status of 14 populations within this distribution. These populations were designated as having either 'common' or 'rare' status with the intervening areas having 'absent' categorization. Quebec populations qualifying for rare status were those where commercial fisheries have been closed due to overexploitation and, in one case, habitat alteration. The remaining populations continue to support commercial and subsistence native fisheries and were considered to have common status.

In Manitoba, lake sturgeon were plentiful and widely distributed as recently as one hundred years ago (Anonymous[2]). There are now very few waters where lake sturgeon may still be commercially fished; the species has been identified as vulnerable under the Manitoba Endangered Species Act. Manitoba is considering the elimination of both sport and commercial harvests until stocks can be rehabilitated or stabilized. Populations in the Nelson River system vary from remnant to poor to good, depending on the section of the river (Figure 3). Stock status also varies within the Winnipeg River system. Remnant populations occur near the Ontario boundary. Populations in the next section west are fair to good even though they occur between two dams and aboriginal domestic harvest is considered to be beyond sustainable yield levels. The final section of the Winnipeg River near Lake Winnipeg also lies between two dams and is considered to have remnant populations with habitat impacts and exploitation having taken their toll. The Churchill River populations are poor. The most inland reaches of the river have remnant nonsustainable populations severely impacted by changes to the hydraulic regime. Little is known about the status of downstream populations in the Churchill system; the last section close to Hudson Bay supports only a remnant population due to habitat changes associated with a major water diversion. Finally, the Hayes River stocks are less impacted, which might be attributable to minimal habitat alteration. Despite being fished commercially, the populations appear to be sustaining themselves.

In Ontario, robust, healthy populations are generally limited to more remote northern rivers and lakes (Figure 4). The Mattagami-Groundhog, Frederick House, Moose, Albany and Attawapiskat rivers are prime examples (Duckworth et al.[1]). Populations from Lake of the Woods (Mosindy & Rusak[4]) and possibly, the north channel of Lake Huron and southern Lake Huron appear to be increasing from recent depressed levels and appear capable of sustaining very modest fisheries (Duckworth et al.[1]). Existing evidence suggests that populations in the Ottawa River, Lake Nipissing, Abitibi River, St. Lawrence River, Lake Erie, Lake Ontario, and Lake Superior are severely depressed and cannot sustain further harvest if they are to be rehabilitated. Although not formally adopted as policy, the Ontario Ministry of Natural Resources is taking a conservative approach in allowing the commercial and sport harvest of lake sturgeon stocks. Legal fishing has often been eliminated where stocks are low. The quotas on lake sturgeon were removed from commercial licenses in 1984 for Lake Erie, in 1990 for Lake Superior and the Ottawa River, and 1991 for Lake Nipissing and the Attawapiskat River. The current implementation of the Canadian federal habitat policy is protecting valuable lake sturgeon habitat.

[3] Fortin, R., S. Guenette & P. Dumont. 1992. Biologie, exploitation, modelisation et gestion des populations d'esturgeon jaune (*Acipenser fulvescens*) dans 14 reseaux de lacs et de rivieres du Quebec. Rapport de Recherche Commanditee, Presente au Gouvernement du Quebec, Ministere du Loisir, de la Chasse et de la Peche. Direction de la gestion des especes et des habitats.

[4] Mosindy, T. & J. Rusak. 1991. An assessment of lake sturgeon populations in Lake of the Woods and the Rainy River, 1987–1990. Lake of the Woods Fisheries Assessment Unit Report 1991:01.

Lake sturgeon are more restricted in their distribution in Quebec than they are in either Manitoba and Ontario, such that their range covers the southwest corner of the province from La Grande River on James Bay to the St. Lawrence River in the south (Figure 5). The eastern limit occurs on the St. Lawrence at about the Saguenay River because of increasing salinity. Even though no populations in Quebec are known to have been extirpated, some have declined (Rejean Fortin personal communication). After reviewing the available information on the 14 populations referred to previously (Fortin et al.[3]), biologists recommended that four fisheries be closed. These include the Ottawa River in the south, the Hurricana and Gueguen rivers in the southcentral area of the distribution and the Baskatong Reservoir lying north of the Ottawa River. The Ottawa and Hurricana River populations are considered overexploited. The Baskatong population is showing signs of stress through variable year-class strength and may be restricted to suboptimal spawning habitat due to deposition of wood and debris. The Gueguen population does not appear to be reproducing successfully as current harvesting yields very few small fish. Lake sturgeon in the Megiscane area, roughly centrally located in the distribution, are considered vulnerable despite having a high harvest level. Closure of the Megiscane populations has not been recommended but restrictions on harvest have been suggested.

The St. Lawrence River populations of Quebec are showing signs of overexploitation from the intensive commercial fishing that has occurred for several generations (Figure 5). In recent years, the commercial harvest from this system has been over 100 tonnes, probably the largest commercial lake sturgeon fishery in North America (Dumont et al. 1987). Lake sturgeon are abundant in the St. Lawrence River near Montreal but populations are considered poor in the upstream Lac St. Francois area and downstream from Lac St. Pierre (Trois Rivieres area) towards the Saguenay River. Analysis of data collected in the 1980's led to restrictions on the fishery in the river upstream from Trois-Rivieres. Elimination of certain permits and reducing the length of the season reduced the harvest while protection of mature breeding fish was sought through restricting the use of seine nets and set lines and limiting gill net mesh from 190 to 203 mm. Commercial fisheries in the Lake Temiskaming and northern areas are operated on an annual quota basis. Quotas on Lake Temiskaming are set at 0.08 kg ha^{-1}. This quota is considered to be a sustainable harvest level and is used as a model for the management of lake sturgeon fisheries in northern Quebec. Recommendations have been made to cut 0.2 kg ha^{-1} quotas for other northern locations in half. Furthermore, the quotas for such large systems should be determined by available sturgeon habitat and not by the total waterbody area. Finally, pulse fishing, where the total allowable harvest for several years is removed in one year and then the fishery is closed in successive years, has been suggested in Quebec.

In contrast to the Canadian situation where lake sturgeon are considered common in many parts of the country, their distribution within the United States is fragmented. Lake sturgeon have been classed as a rare species over much of its range in the United States by the U.S. Fish and Wildlife Service. The species is considered threatened in Nebraska, Illinois, and Kentucky, rare in Minnesota, Missouri, Arkansas, and Alabama, endangered in South Dakota, Iowa, Indiana, Ohio, Vermont, Pennsylvania, West Virginia, and Tennessee and depleted in Georgia by local state authorities (G. Priegel & T. Thuemler personal communication). Wisconsin is the only state where lake sturgeon can be considered common.

Combination of the mtDNA haplotype and status data suggests that both northern and southern populations of lake sturgeon have been impacted by anthropogenic influences. For example, a population with multiple haplotypes and possibly representing the descendants of a Missourian refugial group (e.g., lower Nelson River) has been designated rare. Similarly, rare designation has been assigned to populations with single haplotypes (Lake St. Clair, Ottawa River), possibly from a Mississippian refugium. Although our genetic analysis provides important information on the glacial history of lake sturgeon relevant to the conservation of this species it does not have the resolution (genetic variation) to identify management units. For instance, we cannot determine whether management units correspond to local populations or fish within particular drainages. This will require the utilization of more hypervariable marker systems such as simple sequence repeats (microsatellites).

Acknowledgements

We thank Rob Cann and Rejean Fortin for providing us with status data, Jim Brown for the generous gift of Nelson River lake sturgeon mtDNA, and many individuals for providing lake sturgeon tissue. Louis Bernatchez, Michael Gatt, Blake Konkle, Susan Lee, Michelle Malott, and Scot McKinley participated in the genetic analysis and Jim Finnigan produced the status maps. Neil Billington and an anonymous referee read the manuscript and provided many helpful comments. The genetic research was supported by funds from Ontario Hydro, Ontario Ministry of Natural Resources (Environmental Youth Corps), and the Natural Sciences and Engineering Research Council of Canada. We thank Vadim Birstein, John Waldman, and the Hudson River Foundation for giving us the opportunity to present this paper.

References cited

Avise, J.C., J. Arnold, R.M. Ball, E. Bermingham, T. Lamb, J.E. Neigel, C.A. Reeb & N.C. Saunders. 1987. Intraspecific phylogeography: the mitochondrial DNA bridge between population genetics and systematics. Ann. Rev. Ecol. Syst. 18: 489–522.

Bernatchez, L. & J.J. Dodson. 1990. Allopatric origin of sympatric populations of lake whitefish (*Coregonus clupeaformis*) as revealed by mitochondrial DNA restriction analysis. Evolution 44: 1263–1271.

Billington, N. & P.D.N. Hebert. 1991. Mitochondrial DNA diversity in fishes and its implications for introductions. Can. J. Fish. Aquat. Sci. 48 (Suppl. 1): 80–94.

Brown, J.R., A.T. Beckenbach & M.J. Smith. 1992. Influences of Pleistocene glaciations and human intervention upon mitochondrial DNA diversity in white sturgeon (*Acipenser transmontanus*) populations. Can. J. Fish. Aquat. Sci. 49: 358–367.

Crossman, E.J. & D.E. McAllister. 1986. Zoogeography of freshwater fishes of the Hudson Bay drainage, Ungava Bay and the Arctic archipelago. pp. 53–104. *In*: C.H. Hocutt & E.O. Wiley (ed.) The Zoogeography of North American Freshwater Fishes, Wiley-Interscience, New York.

Danzmann, R.G. & P.E. Ihssen. 1995. A phylogeographic survey of brook charr (*Salvelinus fontinalis*) in Algonquin Park, Ontario based upon mitochondrial DNA variation. Molecular Ecology 4: 681–697.

Dumont, P., R. Fortin, G. Desjardins & M. Bernard. 1987. Biology and exploitation of the lake sturgeon (*Acipenser fulvescens*) in the Quebec waters of the Saint-Laurent River. pp. 57–76. *In*: C.H. Olver (ed.) Proceedings of a Workshop on Lake Sturgeon (*Acipenser fulvescens*). Ont. Fish. Tech. Rep. Ser. No. 23.

Ferguson, M.M., L. Bernatchez, M. Gatt, B.R. Konkle, S. Lee, M.L. Malott & R.S. McKinley. 1993. Distribution of mitochondrial DNA variation in lake sturgeon (*Acipenser fulvescens*) from the Moose River basin, Ontario, Canada. J. Fish Biol. 43 (Suppl. A): 91–101.

Guenette, S., R. Fortin & R. Rassart. 1993. Mitochondrial DNA in lake sturgeon (*Acipenser fulvescens*) from the St. Lawrence River and James Bay drainage basins in Quebec, Canada. Can. J. Fish. Aquat. Sci. 50: 659–664.

Harkness, W.J. & J.R. Dymond. 1961. The lake sturgeon. Ontario Department of Lands and Forests, Fisheries and Wildlife Branch, Toronto. 121 pp.

Hocutt, H. & E.O. Wiley. 1986. The zoogeography of North American fishes. Wiley-Interscience, New York. 866 pp.

Holzhamm, T.E. & M. McCarthy. 1988. Potential fishery for lake sturgeon (*Acipenser fulvescens*) as indicated by the returns of the Hudson's Bay Company Lac la Pluie District. Can. J. Fish. Aquat. Sci. 45: 921–923.

Houston, J.J. 1987. Status of the lake sturgeon (*Acipenser fulvescens*) in Canada. Can. Field-Naturalist 101: 171–185.

Mandrak, N.E. & E.J. Crossman. 1992. Postglacial dispersal of freshwater fishes into Ontario. Can. J. Zool. 70: 2247–2259.

Moritz, C., T.E. Dowling & W.M. Brown. 1987. Evolution of animal mitochondrial DNA: relevance for population biology and systematics. Ann. Rev. Ecol. Syst. 18: 269–292.

Payne, D.A. 1987. Biology and population dynamics of lake sturgeon (*Acipenser fulvescens*) from the Frederick House, Abitibi and Mattagami rivers, Ontario. pp. 10–19. *In*: C.H. Olver (ed.) Proceedings of a Workshop on the Lake Sturgeon (*Acipenser fulvescens*). Ont. Fish. Tech. Rep. Ser. No. 23.

Ward, R.D., N. Billington & P.D.N. Hebert. 1989. Allozyme and mitochondrial DNA variation in populations of walleye, *Stizostedion vitreum*. Can. J. Fish. Aquat. Sci. 47: 2074–2084.

Wilson, C.C. 1995. Phylogeography and postglacial dispersal of two North American *Salvelinus* species. Ph.D. Thesis, University of Guelph, Guelph. 171 pp.

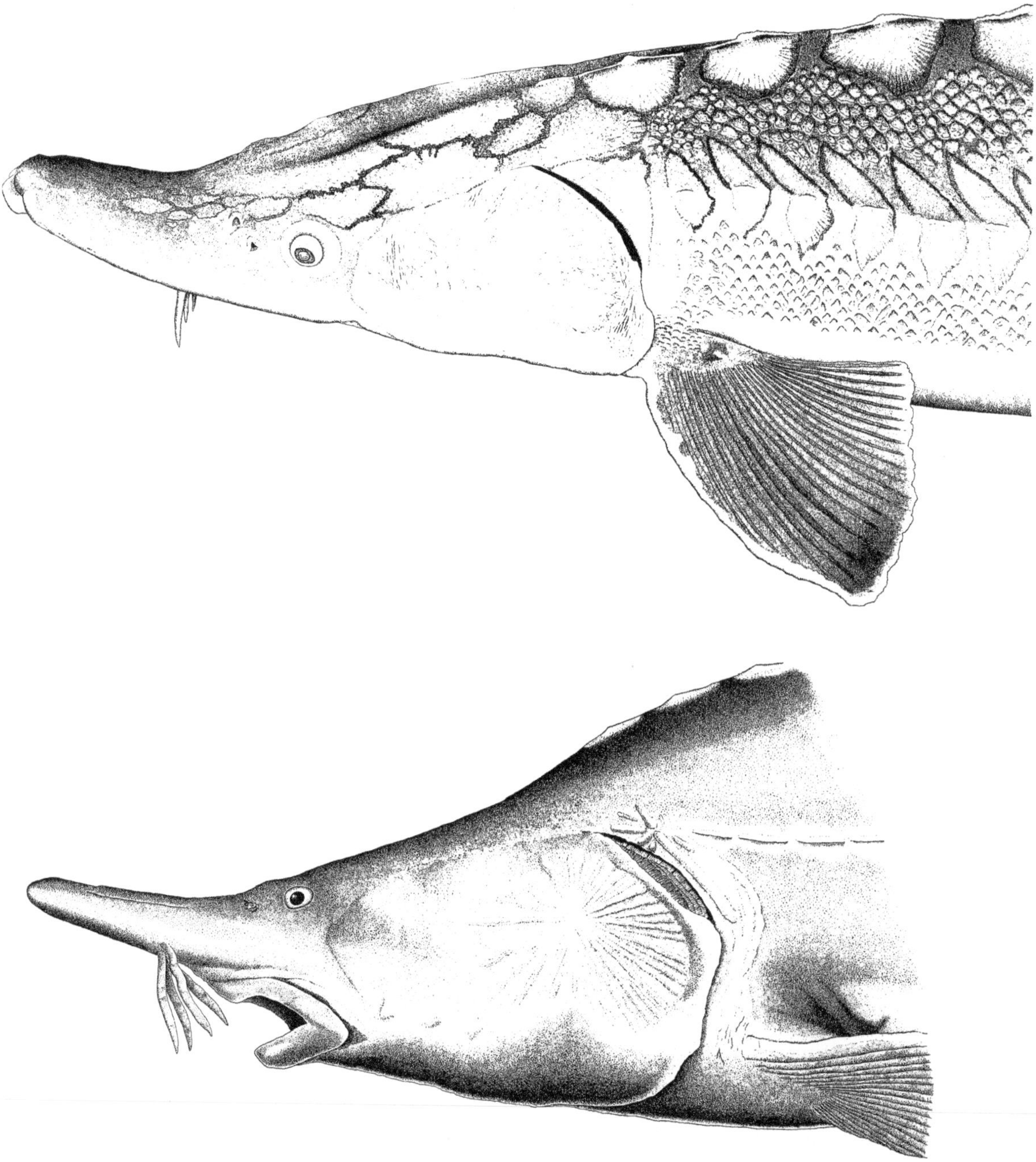

Left side portrait tilted dorsally of *Acipenser oxyrinchus* 150 cm TL from St. Lawrence River at Kamouraska, Quebec, which now resides in the Montreal Biodome (hence the aquarium damage to the rostrum), the same specimen as on p. 184. The irregular rows of denticles above and below lateral scutes are all depressions (below). The 35 cm long head, above, of a large *Huso huso* from the Caspian Sea (1967) donated for the Montreal Expo and then kept in the Toronto Metropolitan Zoo is now preserved in the royal Ontario Museum collection. Originals by Paul Vecsei, 1996.

Environmental Biology of Fishes **48**: 311–317, 1997.

Lake sturgeon management in the Menominee River, a Wisconsin-Michigan boundary water

Thomas F. Thuemler
Wisconsin Department of Natural Resources, 1636 Industrial Parkway, Marinette, WI 54143, U.S.A.

Received 2.8.1994 Accepted 17.3.1996

Key words: Acipenser fulvescens, regulations, exploitation, population estimates, standing stocks

Synopsis

The Menominee River, a boundary water between northeastern Wisconsin and the upper peninsula of Michigan, contains a sport fishery for lake sturgeon, *Acipenser fulvescens*, which is jointly managed by both states. Previous studies indicated that overfishing of this sturgeon population was occurring, and this investigation examined the impact of new angling regulations. The sturgeon population is fragmented into sections by hydroelectric dams. Stocks from the three main sections of the river were compared before and after implementation of the new angling regulations. Records of the legal harvest of lake sturgeon from each river section were obtained through a registration system, which has been in effect since 1983, and estimates of exploitation were derived from these data. Overfishing of lake sturgeon stocks in two of the three sections of the Menominee River is still occurring. Management recommendations are made which would allow for a continued fishery by providing further protection to the stocks.

Introduction

Lake sturgeon, *Acipenser fulvescens*, historically occurred in larger rivers and lakes in three major watersheds, the Great Lakes, the Mississippi River and the Hudson Bay, throughout the eastern portion of North America (Scott & Crossman 1973, Lee et al. 1980, Houston 1987). They are now considered to be threatened over most of their former range in the United States (Miller 1972). In Canada, this species is less abundant than formerly (see Ferguson & Duckworth 1997 this volume). Many populations are sufficient to provide modest managed fisheries (Houston 1987). In Wisconsin, several self-sustaining lake sturgeon populations support limited sport fisheries (Larson[1], Thuemler 1985, Folz & Meyers 1985). These fisheries are closely monitored to guard against overexploitation. There are no commercial fisheries for lake sturgeon in Wisconsin waters.

The Menominee River forms the boundary between northeastern Wisconsin and the upper peninsula of Michigan, and flows into Green Bay. Being a border river, its fish populations are managed jointly by the states of Wisconsin and Michigan. Studies conducted in 1969 and 1970 and again in 1978 and 1979 on the lake sturgeon fishery in the Menominee River (Priegel[2], Thuemler 1985) suggested that exploitation was too high. Regulations were changed to protect the stock and prevent closure of the stur-

[1] Larson, T.R. 1988. The lake sturgeon fishery of Lake Wisconsin, 1978–1985. Fish Management Report #136, Wisc. Dept. Nat. Res., Madison. 34 pp.

[2] Priegel, G.R. 1973. Lake sturgeon management on the Menominee River. Tech. Bull. 67, Wisc. Dept. Nat. Res., Madison. 24 pp.

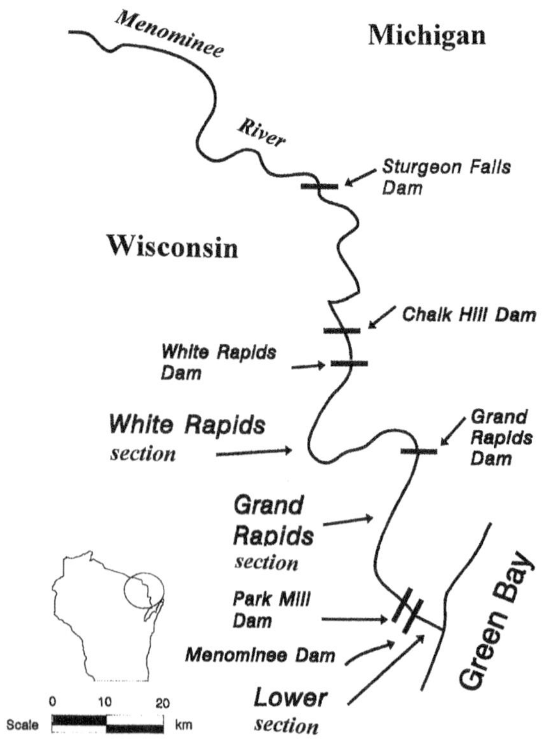

Figure 1. Map of the Menominee River lake sturgeon study area showing hydroelectric dams that fragment the river into sections. Sturgeons in the three major sections – White Rapids section, Grand Rapids section, and Lower section – are managed as separate stocks.

geon fishery. The current study examines the effectiveness of the new regulations.

Geography of the Menominee River

The Menominee River, formed by the confluence of the Brule and Michigammie rivers, flows for 154 km in a southeasterly direction before it joins the waters of Green Bay near the cities of Marinette, Wisconsin and Menominee, Michigan (Figure 1). The Menominee River Basin encompasses just over 1544 square kilometers in the states of Wisconsin and Michigan and has a mean annual discharge of 113 280 liters $second^{-1}$ (Holmstrom et al.[3]). It is a hardwater river having slightly alkaline, lightly stained brown water (Carlson et al.[4]). The Menominee River Basin lies within the southern portion of the Canadian Precambrian shield. It has an average gradient of 1.3 m km^{-1} and it flows over a mixed substrate of sand, rubble and bedrock.

Ten hydroelectric dams on the river fragment its fish populations, as there are no fish passage facilities at these sites. Lake sturgeon were historically found throughout the river up to the present site of the Sturgeon Falls Dam. This study concentrates on sturgeon populations in three distinct sections of river between the White Rapids Dam and the river mouth (Figure 1). The lowermost section of the river encompasses 3.9 km and is bounded by the Menominee Dam at its upper end and the waters of Green Bay at the lower end (Lower section). Fish in the lower section of river can move freely into Green Bay, but the Menominee Dam blocks any upstream movement. About 2 km above the Menominee Dam is the Park Mill Dam. The entire section of river between these two dams is impounded water called the Lower Scott Flowage. The next section of river is 34 km long and is bounded by the Grand Rapids Dam at its upper end and the Park Mill Dam at the lower end (Grand Rapids section). The uppermost section of river that contains lake sturgeon is 42 km long and is bounded by the Grand Rapids and White Rapids dams (White Rapids section).

Methods

Sturgeon were sampled from the Menominee River using boomshocker electrofishing units (Thuemler 1985). Surveys were conducted in 1970, 1978, and 1990 on the White Rapids section, in 1979 and 1990 in the Grand Rapids section and in 1991 in the Lower section of river. All surveys were conducted in

[3] Holmstrom, B.K., C.A. Han & R.M. Erickson. 1983. Water resources data, Wisconsin water year 1982. U.S. Geological Survey Water Report WI-82-1. 426 pp.

[4] Carlson, H.C., L.M. Andrews & C.W. Threinen. 1975. Surface water resources of Marinette County. Wisc. Dept. Nat. Res., Madison. 110 pp.

Table 1. Estimates of the size of the lake sturgeon population from different sections of the Menominee River. Shown in parentheses are the 95% confidence intervals.

River section	Year of population estimate				
	1970	1978	1979	1990	1991
Lower section					893 (457–1329)
Grand Rapids section			2834 (1729–3939)	3201 (2526–3876)	
White Rapids section	2865 (2299–3430)	2749 (1997–3501)		3156 (2512–3800)	

July or August. All sturgeon captured in 1990 and 1991 were measured in both total and fork length (TL and FL, respectively) to the nearest centimeter, but prior to 1990 only total length measurements were taken. A representative sample of sturgeon were weighed. Capture-mark and recapture population estimates were made using the adjusted Peterson estimator (Ricker 1975). Separate population estimates were made for each of the three sections of river. Fish from both the initial and recapture runs were collected with the electrofishing gear. Recapture took place seven to ten days after the initial run. All sturgeon ≥ 25 cm TL were included and were tagged with either a Monel metal tag in the base of the dorsal fin or with a Floy dart tag in the caudal fin or with both. Because the number of fish recaptured in the various size groups was not large enough to make unbiased estimates, the number of fish actually sampled in each 2.5 cm size group was used to estimate the fraction of the total population in each size group. This method assumes that capture efficiencies are equal among the various size groups. Although this method does not allow one to calculate confidence intervals around the estimates, it is the best estimate possible given the data available.

Legal harvest of lake sturgeon by sport anglers on the Menominee River only occurs during a two month fall season. Since 1983, licensed anglers have needed a special permit to fish for lake sturgeon in all Wisconsin waters. In 1974, a minimum size limit of 127 cm was imposed; prior to that time a 107 cm size limit had been in effect. In 1992, the bag limit for an angler was reduced from 2 fish season^{-1} to 1 fish season^{-1}. All legal fish taken in the sport fishery must be registered and tagged at a Department of Natural Resources approved station. All fish harvested are measured, weighed and checked for tags at that time. Harvest information from the various sections of river was obtained through this registration system.

Angler exploitation for the various sections of river was obtained by calculating the average annual harvest from registration data and dividing it by the estimated number of sturgeon over 127 cm

Table 2. Estimated number of lake sturgeon in the White Rapids section of the Menominee River, by size group.

TL (cm)	1970 survey	1978 survey	1990 survey
< 107	2680	2543	2423
> 107	185	206	733
> 127	115	105	320
> 140	57	45	121
> 152	20	9	53
> 165	2	0	9

Table 3. Estimated number of lake sturgeon in the Grand Rapids section of the Menominee River, by size group.

TL (cm)	1979 survey	1990 survey
< 107	2584	2867
> 107	250	334
> 127	125	91
> 140	37	24
> 152	22	3
> 165	0	0

Table 4. Estimated number of lake sturgeon in the Lower section of the Menominee River, by size group.

TL (cm)	1991 survey
< 107	456
> 107	437
> 127	108
> 140	29
> 152	0
> 165	0

from the most recent population estimates (1990 and 1991).

Results and discussion

Various studies report sturgeon lengths in either fork length or total length, but not in both measurements. This makes it difficult to compare age and growth information between studies. In this study the relationship between these two measurements was calculated. We measured both FL and TL for 1839 lake sturgeon, between 25 and 175 cm TL. The relationship between these variables for lake sturgeon in the Menominee River is:

$$TL = 3.52688 + 1.05682\ FL;$$
$$N = 1839;\ r = 0.998.$$

All population estimates satisfied the requirements to insure unbiased estimates (MC – 4N) (Robson & Regier 1964). Lake sturgeon population estimates for sections of the Menominee River with 95% confidence limits are given in Table 1. The estimated number of sturgeon over selected sizes in the three sections of river are shown in Tables 2, 3 and 4. Standing stock estimates for lake sturgeon from the different sections of river were calculated from the 1990 and 1991 estimates. The Lower section had the highest standing stock of 67.8 kg hectare^{-1}, followed by the White Rapids section at 23.8 kg hectare^{-1}, and then the Grand Rapids section at 15.6 kg hectare^{-1}.

Nearly 5000 lake sturgeon have been tagged in the different sections of the Menominee River since 1969. In various surveys and through the registration process, 255 tagged sturgeon have been recaptured. Most of these fish were recaptured in the same section of river in which they were tagged. Of 194 recaptured sturgeon originally tagged in the White Rapids section of the river, only four fish or 2% were found to have moved, and they moved downstream into the Grand Rapids section. Sixty one sturgeon originally tagged in the Grand Rapids section of the river were recaptured and only 1 or 1.6%, was recovered in the Lower section.

Table 5 shows the annual harvest of lake sturgeon from the Menominee River since the inception of the registration system in 1983, Between 1500 and 2000 anglers participate in the sturgeon fishery on the Menominee River each year. The number of anglers has increased steadily since the inception of the registration system. Roughly five percent of anglers successfully harvested a legal sturgeon. In the White Rapids section of the river the harvest has ranged from 8 to 33 fish annually, with an average of just under 19 fish year^{-1}. The average weight of an individual fish taken in this section was 16.1 kg, and the average weight of all sturgeon harvested was 0.50 kg hectare^{-1} year $^{-1}$. In the Grand Rapids section of the river the average annual harvest has been 10.9 fish, with a range of 1 to 28 fish taken annually. These fish had an average weight of 15.8 kg and the average weight of all sturgeon harvested from the section was 0.35 kg hectare^{-1} year^{-1}. In the

Table 5. Annual harvest of lake sturgeon from the three sections of the Menominee River.

Year	White Rapids section	Grand Rapids section	Lower section	Total
1983	9	9	1	19
1984	8	4	1	13
1985	12	1	2	15
1986	13	10	12	35
1987	20	15	23	58
1988	21	14	22	57
1989	32	28	20	80
1990	18	13	22	53
1991	19	9	33	61
1992	23	3	46	72
1993	33	14	40	87
Total	208	120	222	550

Lower section of river the harvest has ranged between 1 and 46 fish year^{-1}. Average annual harvest in this section has been 20.2 fish at an average weight of 16.6 kg and an average weight of all sturgeon harvested of 3.37 kg hectare^{-1} year^{-1}. All harvest estimates were calculated based on the annual average harvest between 1983 and 1993.

Average annual exploitation rates on lake sturgeon from the three sections of river were: Lower section 18.7%, Grand Rapids section, 12%, and White Rapids section, 8.5%.

The lake sturgeon population in the Menominee River is fragmented by numerous hydroelectric dams. Tagging studies showed that there is little movement of juvenile or adult fish from one section of river to another. Dams create a barrier to all upstream movement and very few marked fish showed any inclination to move downstream. We have not tagged lake sturgeon less than 25 cm TL, so if downstream movement and dispersal occur, it may take place at sizes > 25 cm TL. Lack of easy movement by adults between the three sections of river necessitates that they be managed as distinct stocks.

White Rapids section

The sturgeon stock in the White Rapids section responded positively to regulation changes over the past nineteen years. Although population size in this section has not increased markedly since the 1970 estimate, its size structure has improved. The number of sturgeon over the minimum size of 127 cm has more than doubled since 1970, while the number of undersized fish has shown a slight increase. There is no statistical difference ($p < 0.05$) in the mean length of sturgeon harvested when comparing data from the first six years of the registration system (1983–1988) with data from the last five years (1989–1993).

The exploitation rate of 8.5%, is the lowest of the three sections of river studied and it is lower than previously reported. Priegel[2] calculated exploitation rates of 13% in 1969 and 17% in 1970 for sturgeon in this section, and he suggested that exploitation rates no higher than 5% would be more desirable for this fishery. There are few published data on exploitation rates on lake sturgeon fisheries. In the Saint Laurent River in Quebec, exploitation rates ranged from 10 to 20% on different sections (Dumont et al.[5]). Exploitation rates of 15 to 20% in the Lac Saint-Pierre section of the Saint Laurent impacted the average length, weight and age of this stock (Fortin et al. 1993). The exploitation rate for the Lake Winnebago sturgeon population, between 1976 and 1983, has averaged 2.5%, and between 1953 and 1959 it averaged 4.3% (Folz & Meyers 1985). The adult sturgeon stock in Lake Winnebago has more than doubled over the past thirty years under that type of exploitation.

Table 5 indicates an increasing annual harvest of sturgeon in this section of river. This probably results from an increased number of anglers and the greater number of legal size sturgeon available in the White Rapids section. It appears that the current harvest regulations are maintaining the sturgeon stock in this section of the Menominee River, however if exploitation rates increase much over the present level, then further restrictions will have to be applied. Continual monitoring of this stock is recommended.

Grand Rapids section

Standing stock of lake sturgeon in this section of river is lower than in either of the other two sections. Although the 1990 estimate of the population is similar to the 1979 estimate, the number of legal size sturgeon (> 127 cm) decreased by 27%. The average exploitation rate was 12% for the 1983 through 1993 period. There is no statistical difference ($p < 0.05$) in the mean length of sturgeon harvested when comparing the first six years of registration data (1983–1988) with the last five years (1989–1993). Table 5 shows the annual harvest of sturgeon from this section has been fairly stable over the period of the reg-

[5] Dumont, P., R. Fortin, G. Desjardins & M. Bernard. 1987. Biology and exploitation of lake sturgeon (*Acipenser fulvescens*) in the Quebec waters of the Saint-Laurent River. pp. 57–76. *In*: C.H. Olver (ed.) Proceedings of a Workshop on the Lake Sturgeon (*Acipenser fulvescens*). Ontario Fisheries Tech. Rep. Ser. No. 23.

istration system. The drop in the number of legal size sturgeon in this section and the high exploitation rate are cause for concern, especially in a stock that has so few spawning size fish. Additional harvest restrictions are needed.

Lower section

The Lower section of the river currently supports the highest standing stock of lake sturgeon of any section of the Menominee River. This stock expanded as water quality improved over the last two decades. Sturgeon move freely between the river and the waters of Green Bay. The population estimate in 1990 was made during July, and it is not known how the standing stock changes seasonally. The number of sturgeon found in this section may be quite different during the fall fishing season. The exploitation rate for this section was 18.7%. This is higher than on either of the other sections of river. The size structure of the stock in July showed a lack of fish over 152 cm in length. The harvest continued to increase. There is no statistical difference ($p < 0.05$) in the mean length of sturgeon harvested when comparing the first six years of registration data (1983–1988) with the last five years (1989–1993).

Further investigations are needed to determine the standing stock of sturgeon in this section of river. However the apparent high exploitation rate and the lack of any fish > 152 cm TL is a cause for concern. Current regulations should be modified to further restrict the harvest.

Table 6. Portion of the season in which harvest of lake sturgeon has taken place on the Menominee River, 1983–1993.

Period of season	Percentage of the harvest
1 –14 September	43
15–30 September	29
1–14 October	14
15 October–1 November	14

Management recommendations

Lake sturgeon stocks of the Menominee River, with the exception of those fish in the Lower section, spend their entire life in riverine habitat (Priegel[2], Thuemler 1988). In many parts of their range lake sturgeon spawn in river systems, but spend most of their life in lacustrine habitat (Lyons & Kempinger[6], Baker[7], Houston 1987). Lake sturgeon are currently found and were historically abundant only in lakes or in large rivers with extensive shallow water areas (Harkness & Dymond 1961, Priegel & Wirth 1977). In the Menominee River, lake sturgeon could historically move freely into and out of the waters of Green Bay. With the construction of dams on the river in the 1800s this population was fragmented into separate stocks. The long term goal of the Wisconsin and Michigan Departments of Natural Resources is to have free passage of lake sturgeon throughout their former range in the Menominee River (Thuemler & Schnicke[8]). This goal will take many years to achieve and interim measures are needed to protect the current sturgeon stocks in the river.

Two alternatives could be used to decrease exploitation, especially in the lower two sections of river. Imposition of a higher size limit is not being considered because it would tend to further bias the harvest towards larger female fish. Shortening the current two month fishing season would be possible, however the season would have to be shortened to about two weeks to effect a 50% reduction in exploitation (Table 6). The second alternative would be complete closure of the season every other year. This should halve the exploitation rates and yet still allow some harvest, and might be acceptable if a catch and release only season operated in the year when harvest was prohibited. This would permit an-

[6] Lyons, J. & J.J. Kempinger. 1992. Movements of adult lake sturgeon in the Lake Winnebago system. Wisc. Dept. Nat. Res., Res. Report 156, Madison.

[7] Baker, J.P. 1980. The distribution, ecology, and management of the lake sturgeon (*Acipenser fulvescens* Rafinesque) in Michigan. Mich. Dep. Nat. Res., Fish. Res. Report No. 1883. Ann Arbor. 95 pp.

[8] Thuemler, T.F. & G. Schnicke. 1993. Menominee River fisheries plan. File Report. Wisc. Dept. Nat. Res., Madison. 51 pp.

glers continued opportunity to fish for lake sturgeon and yet provide the additional needed protection for the lake sturgeon stocks.

Acknowledgements

I wish to thank the Wisconsin Public Service Corporation, who donated the funds necessary to conduct the investigations on lake sturgeon in 1990 and 1991. I would also like to thank Gary Schnicke, Michigan Department of Natural Resources, Greg Kornely and Brian Belonger, Wisconsin Department of Natural Resources, for their aid in collecting survey data and their valuable advice on data analysis and review of this manuscript. Various personnel at the privately owned registration stations also played an important role by collecting accurate harvest data. The helpful suggestions of two anonymous reviewers was appreciated. William E. Bemis was corresponding editor for the manuscript.

References cited

Ferguson, M.M. & G.A. Duckworth. 1997. The status and distribution of lake sturgeon, *Acipenser fulvescens*, in the Canadian province of Manitoba, Ontario and Quebec: a genetic perspective. Env. Biol. Fish. (this volume).

Folz, D.J. & L.S. Meyers. 1985. Management of the lake sturgeon, *Acipenser fulvescens*, population in the Lake Winnebago system, Wisconsin. pp. 135–146. *In*: F.P. Binkowski & S.I. Doroshov (ed.) North American Sturgeons: Biology and Aquaculture Potential, Dr W.J. Junk Publishers, Dordrecht.

Fortin, R., J.R. Mongeau, G. Desjardins & P. Dumont. 1993. Movements and biological statistics of lake sturgeon (*Acipenser fulvescens*) populations from the St. Lawrence and Ottawa River system, Quebec. Can. J. Zool. 71: 638–650.

Harkness, W.J.K. & J.R. Dymond. 1961. The lake sturgeon. Ontario Department of Lands and Forests, Fish and Wildlife Branch, Maple. 97 pp.

Houston, J.J. 1987. Status of the lake sturgeon, *Acipenser fulvescens*, in Canada. Can. Field Nat. 101: 171–185.

Lee, D.S., C.R. Gilbert, C.H. Hocutt, R.E. Jenkins, D.E. McAllister & J.R. Stauffer, Jr. 1980. Atlas of North American freshwater fishes. North Carolina Biol. Surv., Publication #1980–12. 867 pp.

Miller, R.R. 1972. Threatened freshwater fishes of the United States. Trans. Amer. Fish. Soc. 101: 239–252.

Priegel, G.R. & T.L. Wirth. 1977. The lake sturgeon: its life history, ecology and management. Wisc. Dept. Nat. Res. Publication 4–3600(77), Madison. 20 pp.

Ricker, W.E. 1975. Computation and interpretation of biological statistics of fish populations. Bull. Fish. Res. Board Can. 191. 382 pp.

Robson, D.S. & H.A. Regier. 1964. Sample size in Peterson mark-recapture estimates. Trans. Amer. Fish. Soc. 93: 215–226.

Scott, W.B. & E.J. Crossman. 1973. Freshwater fishes of Canada. Bull. Fish. Res. Board Can. 184. 966 pp.

Thuemler, T.F. 1985. The lake sturgeon, *Acipenser fulvescens*, in the Menominee River, Wisconsin-Michigan. Env. Biol. Fish. 14: 73–78.

Thuemler, T.F. 1988. Movements of young lake sturgeons stocked in the Menominee River, Wisconsin. Amer. Fish. Soc. Symp. 5: 104–109.

Landsat image of the lower reach of the Connecticut River in western Massachusetts and Connecticut. The location of the Holyoke Dam at South Hadley Falls is indicated with the white arrow. Holyoke Dam creates an impoundment (barely discernible in this Landsat image) that is known as the Holyoke Pool. Despite the presence of a fish lifting facility at Holyoke Dam, the Connecticut River population of shortnose sturgeon, *Acipenser brevirostrum*, has been effectively divided since construction of the first dam at this site in the 1840s. The spawning site for the downstream stock of sturgeon is located immediately below Holyoke Dam. Sturgeons restricted to the Holyoke Pool spawn several kilometers above the dam at what is probably the historic spawning site for the entire Connecticut River population (see Kynard 1997 this volume). Image courtesy of Curtice Griffin and the New England Gap Analysis Program (scale bar = 30 km). Comments by W.E. Bemis; also see Bemis & Kynard (1997 this volume).

Environmental Biology of Fishes **48**: 319–334, 1997.

Life history, latitudinal patterns, and status of the shortnose sturgeon, *Acipenser brevirostrum*

Boyd Kynard
National Biological Service, Conte Anadromous Fish Research Center, PO Box 796, Turners Falls, MA 01376, U.S.A. and Graduate Program in Organismic and Evolutionary Biology, University of Massachusetts, Amherst, MA 01003, U.S.A.

Received 31.10.1994 Accepted 26.4.1996

Key words: adaptation, spawning migration, dam impacts, behavior, anadromy, Acipenseridae

Synopsis

Historically, shortnose sturgeon inhabited most major rivers on the Atlantic coast of North America south of the Saint John River, Canada. Today, only 16 populations may remain. Major anthropogenic impacts on shortnose sturgeon are blockage of spawning runs by dams, harvest of adults (bycatch and poaching), dredging of fresh/saltwater riverine reaches, regulation of river flows, and pollution. The pattern of anadromy (adult use of salt water) varies with latitude. The pattern may reflect bioenergetic adaptations to latitudinal differences between fresh and salt water habitats for thermal and foraging suitability. The greater adult abundance in northern and north-central populations likely reflects a historical difference with southern populations that is currently accentuated by increased anthropogenic impacts on southern populations. Adult abundance is less than the minimum estimated viable population abundance of 1000 adults for 5 of 11 surveyed populations, and all natural southern populations. Across the latitudinal range, spawning adults typically travel to about river km 200 or farther upstream. Dams built downstream of spawning reaches block spawning runs, and can divide amphidromous populations into up- and downstream segments. Conservation efforts should correct environmental and harvest impacts, not stock cultured fish into wild populations.

Introduction

Shortnose sturgeon, *Acipenser brevirostrum*, is a small sturgeon that attains a maximum length of about 120 cm, weight of 24 kg, and lives a maximum of 50-60 years (Dadswell et al. 1984, Figure 1). Adults resemble similar-sized juvenile Atlantic sturgeon, *A. oxyrinchus oxyrinchus* that historically co-occurred in the lower mainstems of major Atlantic coast rivers from the Saint John River, New Brunswick, Canada, to the St. Johns River, Florida, United States. Fisheries data and surveys during the 22 years since listing of shortnose sturgeon under the Endangered Species Act of 1973 indicated populations are likely present in 16 rivers (Table 1). Additional surveys may yet find a few undiscovered populations.

Since Dadswell et al. (1984) compiled biological information on shortnose sturgeon, additional studies on location and characteristics of foraging and spawning habitats, adult abundance, and migration patterns have been conducted throughout the species' range. This is the best information on latitudinal variation of these life history traits for any sturgeon species. A comparison among intraspecific populations for life history characteristics

Figure 1. Lateral view of adult male shortnose sturgeon *Acipenser brevirostrum* (94 cm TL) from the Connecticut River, Massachusetts, U.S.A.

often reveals patterns that are not obvious, and can lead to identification of adaptations and evolutionary processes (Frank & Leggett 1994). This report summarizes much of the information on shortnose sturgeon life history, examines latitudinal variation in anadromy and spawning migrations, investigates impacts of dams on spawning migration and abundance, and reviews threats to the species. Explanations that are proposed for behavioral patterns will hopefully stimulate discussion and testing of hypotheses.

Range and colonization

The present range of shortnose sturgeon is disjunct, with northern populations separated from southern populations by a distance of about 400 km near the geographic center. No known populations occur from the Delaware River, New Jersey, to the Cape Fear River, North Carolina (Table 1). Historically, populations were likely present in all large rivers in this area, which includes Chesapeake Bay (Dadswell et al. 1984). Because the separation distance is great between northern and southern populations, there may be no interchange of adults.

Unlike adult Atlantic sturgeon that range widely along the coast (Murawski & Pacheco[1]), most shortnose sturgeon adults remain in their natal river or estuary. The lack of marine movements by most adults suggests that the recolonization rate of shortnose sturgeon to rivers where they have been extirpated would be slow. The lack of evidence for successful recolonization of any river where shortnose sturgeon was extirpated by Atlantic sturgeon fisheries and dams 100 years ago in the mid-Atlantic (Ryder 1890, Smith 1985), supports the hypothesis of a slow recolonization rate.

Rare individuals that are occasionally captured at sea near the coast (review by Dadswell et al. 1984) could represent emigrants that colonize new rivers and maintain gene flow among populations. The phenomenon of marine migrants is not understood, but available information on marine captures in northern and southern parts of the range presented by Dadswell et al. (1984), and corrected for misidentified North Carolina captures by Ross et al. (1988), indicates a greater incidence of marine emigrants in the north compared to the south. Because northern populations are also larger than southern populations, there may be a relationship between population size and number of marine emigrants. If this is so, only a large population like the Hudson River population may be providing emigrants.

Density dependent regulation of population abundance that depends on emigration of fish in relation to food abundance is found in stream dwelling salmonids, and may be present in other fishes that have limited movements and a restricted range (Frank & Leggett 1994). Populations of shortnose sturgeon have limited movements and a restricted home range within their river and estuary. Further, young sturgeon have a size dependent dominance

[1] Murawski, S.A. & A.L. Pacheco. 1977. Biological and fisheries data on Atlantic sturgeon, *Acipenser oxyrhynchus* (Mitchill). Nat. Mar. Fish. Serv., Sandy Hook Lab., Sandy Hook. Tech. Report No. 10. 78 pp.

hierarchy that determines use of foraging habitat (C. Cauthron & B. Kynard unpublished data). Thus, a behavioral mechanism may be present in adults that could regulate density and emigration in each river relative to resource abundance.

Abundance of adults

Commercial harvest data are not a useful indicator of the historical abundance of shortnose sturgeon because catches of Atlantic and shortnose sturgeons were combined (Smith 1985). Consequently, recent trends in abundance are only available for two rivers – the Hudson and Connecticut. The Hudson River population has increased since the 1970s concurrent with the decline of Atlantic sturgeon from about 30 000 to 38 000 (Bain 1997 this volume), and the Connecticut River population seems unchanged since the 1970s (Taubert 1980a, Buckley & Kynard[2], Savoy & Shake[3], M. Kieffer & B. Kynard

[2] Buckley, J. & B. Kynard. 1983. Studies on shortnose sturgeon. Massachusetts Cooperative Fisheries Research Unit, University of Massachusetts, Amherst. Report to National Marine Fisheries Service, Gloucester and Northeast Utilities Service Co., Hartford. 40 pp.

[3] Savoy, T. & D. Shake. 1992. Sturgeon status in Connecticut waters. Proj. AFC-20-1, Conn. Dept. Env. Prot., Report to U.S. Fish & Wildl. Serv., Newton Corner. 51 pp.

Table 1. Twenty Atlantic coast rivers showing the north to south distribution of shortnose sturgeon. Each river is characterized for studies done since 1984 and best estimate of adult abundance.

River	Province/State	Studies[a]	Abundance	Source
Saint John R.	New Brunswick	none	18 000	Dadswell (1979)
Penobscot R.	Maine	P	0	Squiers (pers. comm.)
Kennebec R.	Maine	none	7000	Squiers et al.[6]
Androscoggin R.	Maine	M, A, S	3.000	Squiers[7], Squiers et al.[14], spawner extrapolation
Merrimack R.	Massachusetts	P, A, M, H, S	< 100[c]	Kieffer & Kynard (1996)
Taunton R.	Massachusetts	P	0	Buerkett & Kynard (1993)[b]
Connecticut R.	Massachusetts & Connecticut	A, M, H, S, D, C, CU, B, PA	1200	Savoy & Shake[3], Kieffer & Kynard unpublished data
Hudson R.	New York	A, M, H, D, F	38 000	Bain (1997)
Delaware R.	New Jersey	A, M, H, S	13 000	O'Herron et al. (1993)
Cape Fear R.	North Carolina	P, M, S, B, PA	< 100[c]	Moser & Ross (1994)
Pee Dee R.	South Carolina	none	1000[c]	Marchette & Smiley[8]
Santee-Cooper R.	South Carolina	P, M, PA	no data	Cooke unpublished data
Edisto R.	South Carolina	none	no data	Dadswell et al. (1984)
Ashepoo R.	South Carolina	none	no data	Dadswell et al. (1984)
Savannah R.	South Carolina	A, M, H, C, CU, ST, B	1676[d]	Smith et al. (1994), Collins et al. (1991)
Ogeechee R.	Georgia	P, A, M, H, S, B	216	Rogers & Weber (1994)[e]
Altamaha R.	Georgia	A, M, H, S, B	650	Rogers & Weber[9], unpublished data
Satilla R.	Georgia	P	0	Rogers unpublished data
St. Marys R.	Georgia	P	0	Rogers & Weber[4]
St. Johns R.	Florida	none	no data	Dadswell et al. (1984)

[a] Studies key: P – presence, A – abundance/size frequencies, M – movements, H – summering/wintering habitat, S – spawning migrations/habitat, D – sex determination, C – contamination, CU – culture, ST – stocking, B – bycatch, PA – passage, P – pollution.

[b] Buerkett, C. & B. Kynard. 1993. Survey for sturgeons in the Taunton River, Massachusetts. Report to Massachusetts Div. Mar. Fish., Boston. 23 pp.

[c] Estimate based on best judgement of experts.

[d] Estimate includes adults produced from wild and cultured fish introduced 1984–1992 (Smith et al. 1995).

[e] Rogers, G.S. & W. Weber. 1994. Occurrence of shortnose sturgeon (*Acipenser brevirostrum*) in the Ogeechee – Canoochee River system, Georgia during the summer of 1993. Report of Georgia Dept. Nat. Res. 13 pp.

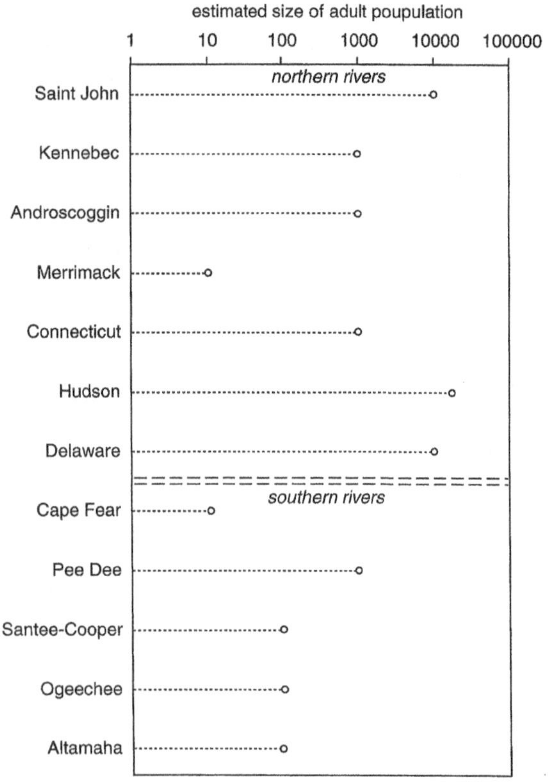

Figure 2. Comparison of seven northern and north-central populations with five southern populations for adult abundance. Estimates for populations in the Cape Fear and Santee-Cooper rivers are expert opinions.

unpublished data). Additional study may find the Hudson River population even larger than 38 000.

Abundance of adults has been estimated for a total of seven populations in northern or north-central rivers and for five southern populations. Abundance varies from tens of thousands (Saint John, Hudson, and Delaware rivers) to tens in the Merrimack and Cape Fear rivers (Table 1). The great abundance of adults in the Delaware River suggests that large populations were likely present in other large rivers, like the Potomac and Susquehanna rivers, located at the center of the range. For the past 100 years, a dissolved oxygen block at Philadelphia has prevented most spring and summer up- or downstream migration of fishes (Chittenden 1974). This pollution block may also have prevented shortnose sturgeon from using the estuary and emigrating to nearby rivers in Chesapeake Bay. Adult shortnose sturgeon in the Delaware River are presently restricted to only 75 km (Brundage & Meadows 1982, Hastings et al. 1987, O'Herron et al. 1993). Recent improvement in water quality may result in renewed access of shortnose sturgeon to the lower Delaware River and tidal interface (an additional 100 km), and could result in a dramatic increase in abundance. It may also provide emigrants to recolonize those Chesapeake Bay rivers that have suitable habitat.

Adult abundance is significantly higher in northern populations than in southern populations (Mann Whitney Test, one tail, $p < 0.02$, Figure 2). This difference may reflect a historical pattern of smaller southern populations due to intrinsic differences in river characteristics. But Rogers & Weber[4] believe that a relatively pristine large southern river like the Altamaha River, Georgia, should support more than 650 adults. All natural southern populations are below or near the minimum viable population level of 1000 adults suggested for vertebrates (Thompson 1991).

Anthropogenic factors are likely responsible for the present low abundance of shortnose sturgeon in southern rivers. All surveyed southern populations are exposed to three or more of the following impacts: harvest (bycatch and poaching), dams, river flow regulation, pollution (particularly, paper mill effluent), and dredging of fresh/saltwater interface reaches. Most southern rivers have never been surveyed well for shortnose sturgeon, or not in many years. Recent surveys of two Georgia rivers found degraded water quality and other impacts; shortnose sturgeon were absent in both (Table 1). The status of shortnose sturgeon in the St. Johns River, Florida, is unknown, but a remnant population may be present. The southernmost limit of shortnose sturgeon distribution that has been recently documented is the Altamaha River, Georgia.

The Savannah River was stocked with more than

[4] Rogers, S.G. & W. Weber. 1995. Status and restoration of Atlantic and shortnose sturgeons in Georgia. Report of Georgia Dept. Nat. Resources. 28 pp.

[5] Even the most well-designed and carefully managed fish culturing methods inherently (and often unintentionally) select certain types of individuals that may or may not be adapted for conditions in the wild.

97 000 cultured[5] shortnose sturgeon less than 9 weeks old from 1984 to 1992 (Smith et al. 1995). Only 19% were tagged, so identification of wild and cultured individuals is impossible. These fish mature at age 4–5, so stocking has already contributed to the estimated total number of 1676 adults (T. Smith personal communication). Stocking cultured fish into the wild population may directly threaten the long-term existence of the Savannah River population by disrupting the population's genetic adaptations to local environmental conditions (Waples 1991, Fleming 1994). Stocking cultured fish is not a permanent solution to environmental or harvest problems that cause low population abundance; these problems still remain in the Savannah River. Some of the stocked young have recruited to the adult cohort, so abundance of adults has temporarily increased. Because the estimate of adults includes wild and stocked fish, the Savannah River population was not included in analyses of abundance trends of the species.

Latitudinal pattern of anadromy

Capture-tagging-recapture and telemetry studies of adult shortnose sturgeon indicate a latitudinal pattern in the amount of time spent in salt water. Adults at the extreme northern part of the range (northern rivers – Saint John, Androscoggin, and Kennebec rivers, Maine; Table 1) use saline water for all or most of the year (Dadswell 1979, Squiers et al.[6], Squiers[7]). Adults leave the estuary and forage in Saint John River fresh water for only a few months in spring and summer (e.g. June-August) when river temperature is warmest. Some adults overwinter in 30% seawater. Adults in less northerly rivers to the center of the range (north-central rivers – Merrimack River, Massachusetts, to the Delaware River) use saline water the least. Adults forage and spawn in fresh water for years, then enter the fresh/saltwater reach, usually only briefly (Buckley & Kynard 1985a, Dovel et al. 1992, O'Herron et al. 1993, Kieffer & Kynard 1993). This pattern of fresh and saltwater use is termed freshwater amphidromy (Kieffer & Kynard 1993). Because feeding occurs when temperature exceeds about 7° C (Dadswell 1979), north-central rivers provide suitable thermal foraging conditions in fresh water for about 7 months (April - October). This is likely the longest continuous period of a suitable thermal regime for foraging in fresh water. Adults in some southern rivers (Cape Fear, Pee Dee, Savannah, and Altamaha rivers) forage in or just upstream of the fresh/saltwater interface moving to high salinity water either briefly, or for a long period, during fall and winter (Marchette & Smiley[8], Hall et al. 1991, Moser & Ross 1994, Rogers & Weber[9]). Some adults in the Altamaha River remain in freshwater reaches after spawning, but extent of this freshwater use is not known. River temperatures that exceed 28-30° C during the summer create unsuitable thermal conditions in most freshwater reaches. During summer, most adults and juveniles 1-year old and older remain at the fresh/saltwater interface in a few deep, cool water refuges (Hall et al. 1991, Flournoy et al.[10], Rogers & Weber[4,9]). These southern populations seem to use fresh water environment less than any populations known.

An exception to the previous pattern of salt water use by southern adults is found in the dammed Santee-Cooper River system, South Carolina, where adults remain upstream of dams all year in fresh water (e.g. one adult captured in 1984 at river km 190, D. Cooke personal communication; 19 adults cap-

[6] Squiers, T.S., L.S Flagg, M. Smith, K. Sherman & D. Ricker. 1981. American shad enhancement and status of sturgeon stocks in selected Maine waters. Maine Dept. Mar. Resour., Report to Nat. Mar. Fish. Serv., Gloucester, pp. 20-64.

[7] Squiers, T.S. 1982. Evaluation of the 1980-82 spawning run of shortnose sturgeon (*Acipenser brevirostrum*) in the Androscoggin River, Maine. Maine Dept. Mar. Resour., Final Report to Central Maine Power Co., Augusta. 15 pp.

[8] Marchette, D.E & R. Smiley. 1982. Biology and life history of incidentally captured shortnose sturgeon, *Acipenser brevirostrum*, in South Carolina. Report of South Carolina Wildl. Mar. Res., Brunswick. 57 pp.

[9] Rogers, S.G. & W. Weber. 1995. Movements of shortnose sturgeon in the Altamaha River system, Georgia. Georgia Dept. Nat. Resour., Contrib. 57. 78 pp.

[10] Flournoy, P.H., S.G. Rogers & P.S. Crawford. 1992. Restoration of shortnose sturgeon in the Altamaha River, Georgia. Georgia Dept. Nat. Resour., Final Report to U.S. Fish & Wildl. Serv., Atlanta. Proj. AFS-2. 54 pp.

tured in 1991 by poachers in the Congaree River, headwaters of the Santee River, M. Collins personal communication). Also, the presence of migrating ripe adults in the tidal reaches of both rivers (D. Cooke personal communication) and adults in headwaters above dams is strong evidence that adults in these populations are amphidromous, like those in north-central rivers. It seems likely that before the rivers were dammed, adults moved from fresh/saltwater foraging areas to discrete headwater reaches unknown to biologists. This could be done by spawned fish, which would be far upriver after spawning. This situation may also occur undetected in other southern rivers. The extent of fresh water use in all southern rivers needs further examination.

The latitudinal pattern of saltwater use may reflect bioenergetic adaptations for obtaining the optimal environment for foraging and growth. Important components of shortnose sturgeon bioenergetics are likely the spatio-temporal pattern of forage abundance in river and estuarine areas and the length of time a suitable thermal regime is available for foraging in fresh water. An acceptable thermal regime in fresh water is shortest in northern rivers, longest in north-central rivers where fish spend the most time in fresh water, and may be intermediate in southern rivers. Southern rivers have a suitable thermal regime in fresh water during the fall, winter, and spring, but available data indicate most fish do not enter fresh water to forage; instead, they use the fresh/saltwater interface. If so, perhaps forage abundance in southern rivers is higher in the fresh/saltwater interface than in fresh water.

Life history

Growth

Dadswell et al. (1984) reviewed growth throughout the latitudinal range. Growth of juvenile fish in all populations is rapid, with growth of 14-30 cm during the first year. Southern fish grow the fastest, but do not reach the larger size of northern fish, which continue to grow throughout life. This phenomenon may be related to different bioenergetic styles of southern and northern shortnose sturgeon, but sufficient data are not available for conclusions. Shortnose sturgeon located upstream of Holyoke Dam at river km 140 of the Connecticut River have the slowest growth rate of any surveyed (Taubert 1980a). This suggests growth advantages are associated with foraging in the lower river or fresh/saltwater interface.

Length at maturity (45-55 cm FL) is similar throughout the latitudinal range of shortnose sturgeon, but southern fish mature at younger ages than do northern fish (Dadswell et al. 1984). Males mature at 2-3 years in Georgia, 3-5 years in South Carolina, and 10-11 years in the Saint John River; females mature at 4-5 years in Georgia, 7-10 years in the Hudson River, and 12-18 years in the Saint John River. Northern adults live 30-67 years; southern adults live 10-25 years. These data are based on studies that used pectoral rays to age fish, a procedure that is not verified with fish of known age.

Spawning

Spawning periodicity of males and females is poorly understood. Males spawn more frequently than females in all northern populations, but perhaps not in all southern populations. Dadswell (1979) estimated that the Saint John River males spawned at 2-year intervals and females at 3 to 5-year intervals. Annual spawning by some males is documented in the Merrimack, Connecticut, and Hudson rivers (Dovel et al. 1992, Kieffer & Kynard 1996, M. Kieffer & B. Kynard unpublished data). Collins & Smith (1996) indicate that some Savannah River fish of both sexes spawn during successive years. Abundance of spawning females can vary annually by an order of magnitude in both northern and southern rivers (Buckley & Kynard[2], Smith et al.[11]). Moreover, fecundity is highly variable among populations. Fecundity of females from the Saint John River ranged from 27 000 to 208 000 eggs (average,

[11] Smith, T.I.J., E. Kennedy & M.R. Collins. 1992. Identification of critical habitat requirements of shortnose sturgeon in South Carolina. South Carolina Wildl. & Mar. Resour. Dept., Report to U.S. Fish & Wildl. Serv., Proj. AFS-17. 105 pp.

11 568 eggs per kg body weight; Dadswell et al. 1984). Females in the Saint John River have the highest mean potential lifetime fecundity, 197 000 eggs, compared to 111 000 eggs for Pee Dee River females (Boreman et al.[12]). Males dominate the sex ratio of spawners (not the total population) in all rivers. Sex ratios of male:female spawners in 2 northern and 1 southern river are for Hudson River – 2.5:1 (Pekovitch[13]); Connecticut River – 3.5:1 (Taubert 1980b) and 3 to 7:1 (Buckley & Kynard 1985b, M. Kieffer & B. Kynard unpublished data); and Savannah River – 3.5:1 (Collins & Smith 1996).

Shortnose sturgeon vary in prespawning migration pattern and the type present may reflect energetic adaptations to migration distance, river discharge and temperature, and physiological condition of fish (Kieffer & Kynard 1993). The three patterns of migrations are (1) a short 1-step migration done in spring only a few weeks before spawning, (2) a long 1-step migration done many weeks in late winter and spring before spawning, and (3) a short 2-step migration composed of a long fall migration, which places fish near the spawning site for overwintering, then a short migration like the short 1-step in spring to spawn. Amphidromous adults in the Merrimack, Connecticut, and Delaware rivers that overwinter only a short distance (< 25 km) downstream of the spawning area use a short 1-step migration (Buckley & Kynard 1985a, Kieffer & Kynard 1993, O'Herron et al. 1993, M. Kieffer & B. Kynard unpublished data). Of the Connecticut River prespawning adults that in the fall prior to spawning need to move a total of 80 km or more to spawn, 70% use a 2-step migration and 30% use a long 1-step spring migration (Buckley & Kynard 1985a). No long 1-step spring migration was found in Connecticut River prespawners that, as of their fall location, had to move 140 km or farther to spawn.

Although total migration distance is unclear, prespawning Saint John River adults have a 2-step migration and fish must likely move about 45-65 km in spring (Dadswell 1979). The 2-step migration pattern used by most Connecticut River adults that need to move 80 km may be an energetic adaptation to take advantage of warmer river temperature, reduced river discharge, and superior physiological condition of fish in the fall compared to spring. Only adults in southern rivers (Savannah, Altamaha, and Pee Dee rivers) have a long 1-step migration that exceeds 80 km. Most southern fish migrate to about river km 200 or farther in late winter (Marchette & Smiley[8], Hall et al. 1991, Rogers & Weber[4,9]). Although details are lacking, both fall and spring migrations were found in the Altamaha River (Rogers & Weber 1995[4]). This indicates a greater diversity of migration patterns than found previously in southern rivers. The presence of long 1-step winter migrations that exceed 200 km in southern rivers may be related to warmer river temperatures or the ability of southern adults to continue feeding in winter, or both. These conditions provide energy resources for a long migration that are not available for northern adults. Northern adults cease foraging in November, about 5 months before most fish initiate a short 1-step spawning migration. Although some adult shortnose sturgeon in a population migrate upstream in fall and some migrate in spring, all spawn together. The pre-spawning migrations of winter and spring 'races' of European sturgeon species (Berg 1959) likely also reflect different migration styles of adults that eventually spawn together.

Shortnose sturgeon spawn in late winter (southern rivers) to mid-spring (northern rivers) when river temperature increases to about 9° C. Spawning usually ceases at 12-15° C (Dadswell et al. 1984, Buckley & Kynard 1985b, Hall et al. 1991, O'Herron et al. 1993, Squiers et al.[14], Kieffer & Kynard 1996). When high Connecticut River discharge delayed spawning in 1994, a few females spawned successfully at 18° C (M. Kieffer & B. Kynard unpublished

[12] Boreman, J., W.J. Overholtz, & M.P. Sissenwine. 1984. A preliminary analysis of the effects of fishing on shortnose sturgeon. Nat. Mar. Fish. Serv., Woods Hole, Mass., Ref. Doc. No. 84-17. 19 pp.

[13] Pekovitch, A. W. 1979. Distribution and some life history aspects of the shortnose sturgeon (*Acipenser brevirostrum*) in the upper Hudson River estuary. Report of Hazleton Environ. Sci. Corp., Ill. 23 pp.

[14] Squiers, T.S., M. Robillard & N. Gray. 1993. Assessment of potential shortnose sturgeon spawning sites in the upper tidal reach of the Androscoggin River. Report of Maine Dept. Mar. Resour., Augusta, Maine. 43 pp.

data). Savannah River fish held for culture also spawned at 17-18° C (Smith et al. 1985).

Spawning occurs at moderate river discharge levels and high discharge can deleteriously affect spawning. In the Connecticut and Merrimack rivers, spawning occurs after peak spring flows, when only rain events or regulated flows create high discharge (Taubert 1980b, Buckley & Kynard 1985b, Kieffer & Kynard 1996). For adults that spawn directly below a hydropower dam in tailrace flows, the facility's operation controls the suitability of water velocities for spawning and rearing of eggs and embryos. High river discharge in May of 1983, 1991, 1992, and 1996 during the normal spawning period of Connecticut River adults likely inhibited females from spawning by creating unacceptably fast water velocities at or near the bottom (Buckley & Kynard 1985b, M. Kieffer & B. Kynard unpublished data). All information supports Buckley & Kynard (1985b) that acceptable river conditions (flows) must be available before endogenous factors trigger ovulation and spawning.

Although channels are used for spawning in many rivers, Connecticut River females used a wide range of water depths, so depth may be less important that water velocity. Channel with gravel substrate was likely used for spawning in the Saint John River (Dadswell 1979), channel with gravel, rubble, and ledge bottom in the Androscoggin River (Squiers et al.[14]), channel with rubble in the Merrimack River (Kieffer & Kynard 1996), shallow riffle channels in the Delaware River (O'Herron et al. 1993), channel curves with rocks, gravel/sand/logs in the Savannah River (Hall et al. 1991, Collins & Smith 1995), and channel with gravel, cobble, and large rocks adjacent to bluff formations in the Altamaha River (Rogers & Weber[9]). Connecticut River females tracked for 3 years during spawning used water depths of 1.2-10.4 m and bottom velocities of 0.4-1.8 m sec^{-1} (mean, 0.7 m sec^{-1}; Kieffer & Kynard 1996, P. Vinogradov, M. Kieffer & B. Kynard unpublished data).

Telemetry of spawning fish throughout the species' range indicates that spawning occurs during a few days to 2-3 weeks (Androscoggin River - Squiers et al.[14], Merrimack River - Kieffer & Kynard 1996, Connecticut River - Buckley & Kynard 1985b, M. Kieffer & B. Kynard unpublished data, Delaware River - O'Herron et al. 1993; and Savannah River - Hall et al. 1991). The end of spawning is easily determined because fish leave the spawning area and move downstream, some at the rapid rate of 32 km d^{-1} (Buckley & Kynard 1985b, Hall et al. 1991, Kieffer & Kynard 1996). When shortnose sturgeon move up- or downstream, they follow the channel (M. Kieffer & B. Kynard unpublished data).

Eggs, embryos, and larvae

Early life history is complex in this lithophilous species (Balon 1975, 1985). Females deposit brown to black demersal eggs (approximately 3.5 mm diameter) that quickly adhere to bottom material and increase to approximately 4 mm diameter. Embryos hatch in 111 hours at 18-20° C or about 200 hours at 12° C (Buckley & Kynard 1981, Smith et al. 1995). Free embryos (eleutheroembryos) 1-8 days old are 7-11 mm long, black in color, photonegative, and seek cover during final development (Richmond & Kynard 1995). Embryos have large yolk-sacs, poorly developed sensory systems and can only swim using swim-up and drift behavior. The photonegative behavior of embryos suggests that they would hide under any available cover in spawning areas. Eleutheroembryos develop into feeding larvae (about 15 mm TL) in 8-12 days at 15-17° C, and, as Bemis & Grande (1992) pointed out, this is a period of rapid change in the acquisition of sensory, feeding, and locomotor systems. Larvae have well-developed eyes, open electrosensory (ampullary) organs, a mouth with teeth, and fins that enable them to swim normally. In laboratory tests, larvae were photopositive, nocturnally active, and preferred white substrate and the deepest water available (Richmond & Kynard 1995). The selection of deep water by larvae was also indicated by their capture in river channels (Taubert & Dadswell 1980, Bath et al. 1981).

Recent laboratory studies of Connecticut River larvae found most ceased downstream migration after 2 days, although some emigration continued for 14 days (C. Cauthron & B. Kynard unpublished data). This is sufficient time to move many kilometers downstream, but not sufficient time to move to the estuary from any known unobstructed spawning location.

Tolerance of early life stages to increasing salinity and low dissolved oxygen increases with age. Twenty-two day old larvae from the Savannah River tolerated a maximum of 9 ppt salinity and required more than 3 mg l^{-1} oxygen, while fish about 300 days old tolerated 25 ppt salinity for 18 hours and most survived short periods of 3 mg l^{-1} oxygen (Jenkins et al. 1993).

Young-of-the-year

Behavior and movements of young-of-the-year (YOY) are not fully understood in any river, but data from rivers throughout the range suggest that fish remain upriver in fresh water for about 1 year. In the Saint John River, YOY and older juveniles remain in fresh water, growing slowly for several years (Dadswell 1979). Some Hudson River YOY occur within 18 km of Troy Dam, the upstream limit of spawning. They apparently remain in fresh water for about 1 year before moving downstream to the fresh/saltwater interface (Carlson & Simpson 1987, Dovel et al. 1992).

Sparse information is available on feeding and habitat use of YOY. Carlson & Simpson (1987) found YOY foraged in channel habitat on amphipods and dipteran larvae on mud, not sand substrate. Saint John River YOY used deep water and intermediate depth areas (Pottle & Dadswell).[15]

One-year old juveniles and adults

With the exception of the Saint John River, where all juveniles remain in fresh water for several years (Dadswell et al. 1984), 1-year old and older juveniles in other populations join adults and show similar spatio-temporal patterns of habitat use. During all year in the Connecticut, Hudson, Savannah, and Altamaha rivers, and in the Saint John River during summer, juveniles use the same fresh water or fresh/saltwater interface as adults (Dadswell 1979, Hall et al. 1991, Savoy & Shake[3], Dovel et al. 1992, Flournoy et al.[10], Rogers & Weber[9]). Juveniles 2-3 year old and adults occur together in freshwater concentration areas in the Connecticut River (Taubert 1980a, M. Kieffer & B. Kynard unpublished data). Juveniles 1-year and older and adults in the Connecticut River had similar summer home ranges (respective means, 6.3 and 4.9 km) and winter home ranges (respective means, 2.7 and 2.6 km; B. Kynard, D. Seibel, M. Kieffer & M. Horgan unpublished data). The similarity of home ranges further indicates that juveniles use space similar to adults.

Concentration areas used by juveniles and adults in fresh water are often located upstream of natural constrictions or in headwaters of dammed reaches, locations where river velocity slows and creates large sandy shoals (Buckley & Kynard 1985a, Kieffer & Kynard 1993). Perhaps, these geomorphological features create hydraulic conditions that favor substrate for freshwater mussels, a major food item of adults (Dadswell et al. 1984). Adults also feed on introduced bivalves, including the introduced zebra mussel, *Dressena polymorpha* (M. Bain personal communication).

Habitat use of juveniles and adults has been studied in several rivers. Juveniles in the Saint John, Hudson, and Savannah rivers use sand and mud substrate in deep channels (Pottle & Dadswell[15], Hall et al. 1991, Dovel et al. 1992). Some Saint John River adults foraged during summer in backwaters of fresh to low salinity lakes with aquatic vegetation or on mud substrate along river banks (Dadswell 1979). During summer and winter in freshwater reaches of the Connecticut River, juveniles and adults selected similar geomorphological reaches of curves and runs with island, but not straight runs (Seibel 1993). Juveniles and adults forage in channel and shoal areas (Dadswell 1979, O'Herron et al. 1993), but Connecticut River juveniles used shoals more at night than adults (Seibel 1993). Although use of channel and shoal areas was highly variable among individuals during summer and fall, all Connecticut River juveniles and adults overwintered in

[15] Pottle, R. & M. J. Dadswell. 1979. Studies on larval and juvenile shortnose sturgeon (*Acipenser brevirostrum*). Report to Northeast Utilities Service Co., Hartford. 79 pp.

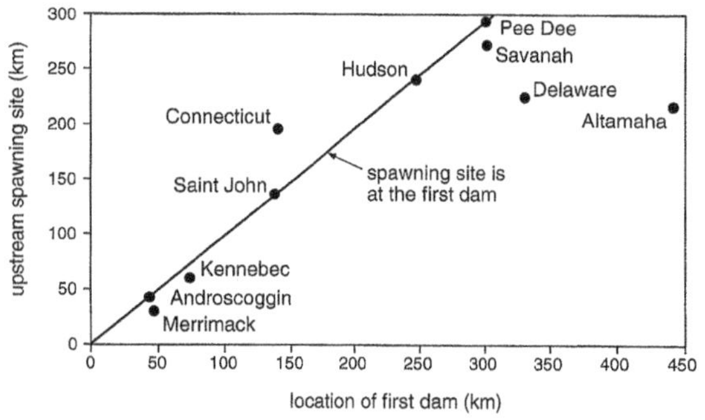

Figure 3. Relationship in 10 populations between the maximum upstream spawning location of shortnose sturgeon and location of the first (lowermost) dam. A 1:1 relationship of spawning location to dam location is indicated by the 45° line.

deep water channel habitat within or directly downstream of the summer range. No telemetered juvenile or adult tracked in the Connecticut River ever used the Holyoke Dam reservoir area except when moving up- or downstream to areas of use.

In estuarine environments, most juveniles and adults forage together in the fresh/saltwater interface where salinity is variable, substrate is usually mud and sand, and vegetation is often present. Adults in the Saint John estuary foraged over sand/mud or mud substrate with emergent macrophyte vegetation in 5-10 m depths in summer and overwintered in the lower estuary in deep water with mud substrate. Kennebec and Androscoggin river adults foraged on tidal mud flats with patchy macrophyte vegetation and 18-25 ppt salinity (McCleave et al. 1977), or in shallow or deep tidal channels with salinity of 0-21 ppt that often had vegetation (Squiers et al.[6]). Adults overwintered in deep water in the lower estuary (Squires et al.[7]). Pee Dee River adults foraged in the fresh/saltwater interface (0.5-1 ppt salinity) during spring and summer, then overwintered downstream in the lower estuary in 15 ppt salinity (Dadswell et al. 1984). A similar pattern was found in the Savannah River (Hall et al. 1991), but not in the Altamaha River, where fish occur less in high salinity water (Flournoy et al.[10], Rogers & Weber[9]).

All telemetry studies of foraging behavior of juveniles and adults cited previously and diet studies cited in Dadswell et al. (1984) found much individual variability. This suggests that shortnose sturgeon have a broad niche breadth; thus, the species should be capable of occupying a wide geographic range, many habitats, and producing large populations (Brown 1984). The great individual variability for foraging has likely contributed to the extraordinary persistence of shortnose sturgeon during the many habitat changes they have experienced during their long existence.

Spawning migration and impact of dams

Spawning site locations in three unobstructed rivers and two rivers with the first dam located more than 240 km upstream provide information on the natural pattern of shortnose sturgeon migratory distance. The maximum upstream spawning site location of fish in unobstructed rivers is river km 220-225 in the Delaware River, river km 275-278 in the Savannah River, and river km 210-220 in the Altamaha River. This indicates a common pattern of migrating upstream to about river km 200 or farther (Hall et al. 1991, O'Herron et al. 1993, Rogers & Weber[9], Collins & Smith 1996). Some Savannah River fish stopped at river km 179-190, but even these fish migrated almost to river km 200. Also, some Altamaha River migrants stopped at river km 50-125, the lowermost potential spawning location found in any unobstructed river. If these fish spawned, then there is more variation for spawning location in the Altamaha River than elsewhere in the species' range. Because this pattern is not supported by observations from other rivers, and spawning was not verified, the Altamaha River migrants need further study. In addition, adults in the Hudson River spawn near Troy Dam (river km 246, Dovel et al. 1992). Pee Dee River adults likely spawn in the reach downstream of Blewett Falls Dam at river km 298 (Ross et al. 1988), although some spawned at river km 192 (Marchette & Smiley[8]) similar to the pattern observed in the Savannah River. Thus, all adults in four rivers (two north-central and two southern) migrate to about river km 200 or farther to spawn. No data were collected from the far northern rivers because all have dams in the lower river that block upstream migration. Saint John River adults may spawn in the reach below Mactaquac Dam (river km 137-120, Dadswell 1979).

Adults likely have a behavioral drive to reach a historical spawning area that is located at about river km 200 km or farther. When a dam blocks the spawning migration, females apparently move as far upstream as they can, then may or may not spawn in the reach below the dam. Comparison of maximum upstream spawning locations in rivers throughout the range shows an almost 1:1 relationship of known or suspected spawning locations and location of the first dam (Figure 3). Exceptions are rivers with dams located 300 km or farther upstream, i.e., the Savannah River (dam at river km 300), the Delaware River (dam at river 331), and the Altamaha River (dam at river km 441). Only runs in these rivers likely escaped blockage, either partially or totally, by dams. Although Connecticut River fish appear to be an exception (Figure 3), they also spawn near river km 200 downstream of a dam (the second dam located at river km 198, not the first dam at river km 140). Merrimack River fish do not attempt to move farther upstream than river km 32, and do not spawn directly downstream of Essex Dam at river km 46 (Kieffer & Kynard 1996). This different migratory behavior is not likely related to the small number of fish present because rare Cape Fear River shortnose sturgeon continue to migrate as far upstream as possible (e.g. to Lock & Dam No. 1 at river km 96, Moser & Ross 1994). All evidence from other north-central rivers indicates the present spawning pattern of the Merrimack River fish is not likely the historical pattern. Closer examination of Kennebec River adults, where ripe males were captured at river km 58 (Squiers et al.[7]), may show that prespawning adults gather at river km 58, but spawn as far upstream as they can go (e.g., below the dam at river km 69).

A dam built downstream of a spawning reach will block the migration of anadromous spawners, but it may divide amphidromous populations into an upriver segment, with access to the spawning site, and a lower river segment whose upstream spawning and foraging migrations are blocked by the dam. This is likely the situation in the Connecticut River with Holyoke Dam at river km 140. It also may be the situation in the Santee (Wilson Dam at river km 140) and Cooper rivers (Pinopilis Dam at river km 80) which currently share the upriver reach of the Santee River. In the Santee-Cooper river system, the occurrence of ripe lower river adults and headwater adults indicates that the dams divided the original population, which was likely amphidromous, not anadromous. Although only two divided populations are known, other rivers may contain or have contained in the past, an undetected upriver population segment. Persistence of an upriver population segment is likely determined by many factors, particularly fishing mortality and availability of spawning conditions. In particular, the Holyoke Dam was built on the site of a large rapids near South Hadley Falls, Massachusetts. This was not a high waterfall, and historical records document spearfishing for sturgeon in the rapids, evidence that they were able to pass this potential obstruction. After Holyoke Dam was built in 1849, however, the upriver segment of the Connecticut River population persisted unknown to biologists for almost 100 years.

In a divided population, upriver fish can move downstream past the dam and join lower river fish, but the reverse is difficult, if not impossible. Downstream movement of juveniles and adults has been documented in the Connecticut River, where it appears to be a natural movement pattern timed to occur with increased river discharge (Seibel 1993, M. Kieffer & B. Kynard unpublished data). Although some fish from the upriver segment move downstream pass the dam, the pattern is variable (as with other movement patterns of the species), and will take years of study to decipher. Lower river migrants can enter fishlifts at Holyoke Dam, but passage is infrequent with 81 fish lifted from 1975 to 1995 (Holyoke Fishlift data). Three adults/year (range, 0-13) are lifted during April-October with most fish lifted singly on one day in the spillway lift (B. Kynard unpublished data). Although this level of passage insures gene flow, the small number of fish lifted annually contributes little to total reproduction.

All evidence indicates spawning by the Connecticut River upriver segment is the main source of recruitment for the entire population. A spawning run of lower river fish annually migrates upstream

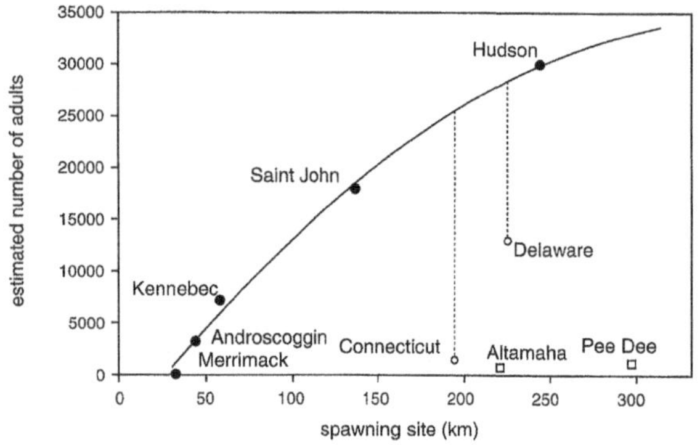

Figure 4. Relationship in nine populations between the abundance of adult shortnose sturgeon and maximum upriver spawning location. The regression line represents the equation: population size = 14435 ln(river km) - 50987, $r^2 = 0.99$, $p < 0.001$.

to Holyoke Dam; these fish are ripe and artificially cultured eggs develop normally (Buckley & Kynard 1981, 1985b, Richmond & Kynard 1995). However, intense netting for eggs and embryos showed that no females spawned in 1993-1994, and probably only 1 of 50 estimated females spawned in 1995 (P. Vinogradov unpublished data). These results just below Holyoke Dam were concurrent in all years with successful spawning and production of many eggs and embryos by tagged females 54 km upstream near Turners Falls Dam (river km 198). Thus, the successful breeding portion of the Connecticut River population is mainly the estimated 300 or so upriver adults, of which about 25% spawn annually (M. Kieffer & B. Kynard unpublished data). The estimated 850 lower river adults (Buckley & Kynard[2], Savoy & Shake[3]) likely contribute little to reproduction. Because only about one-third of the total Connecticut River adults are able to spawn successfully, low population abundance should be expected. Similar results of variable spawning success among divided population segments of white sturgeon, *A. transmontanus*, was found in the Columbia River (Parsley & Beckman 1994).

There is only one naturally landlocked sturgeon population (e.g., resident population above a natural barrier) in the coastal rivers of North America – the federally endangered Kootenai River white sturgeon, which occurs in the headwaters of the Columbia River (Anders[16]). The phenomena of migratory sturgeons establishing freshwater residency upstream of dams is not understood, but it occurs under similar circumstances with other sturgeons – white sturgeon, *A. transmontanus*, in the mid-Columbia River (North et al. 1993) and lake sturgeon, *A. fulvescens*, in rivers of the Great Lakes (Thuemler 1985). The Connecticut and Santee/Cooper river populations have been referred to as partially landlocked (Dadswell et al. 1984), but no evidence of a natural historical division exists. To clarify terminology, perhaps naturally landlocked populations should be designated landlocked, and segments of migratory populations upstream of dams designated damlocked.

Biological significance of spawning location

Upstream spawning location may be an important component of reproductive success for shortnose sturgeon. A comparison of nine populations across the species' range for the relationship between maximum upstream spawning location and adult abundance used the following equation:

$$\text{population size} = 14435 \ln(\text{river km}) - 50987.$$

The regression line fits the data well ($r^2 = 0.99$) and was significant (< 0.001, Figure 4). Spawning location has a positive relationship to population size and this relationship is best represented when scaled with natural logs. Spawning location was positively related to abundance in northern and north-central populations, with only the abundance of adults in the Connecticut and Delaware rivers located far below the regression line. Reasons for the low adult abundance in both these rivers were discussed previously. Recent results from the Hudson River received after the analysis indicates the population has increased to 38 000 (Bain 1997 this volume). This would cause the regression line to be located above the present position on Figure 4, but

[16] Anders, P. 1993. Kootenai River white sturgeon studies. Report to U.S. Army Corps of Engineers, Bonneville Power Admin., Portland. 16 pp.

would not change the basic relationships. Abundance of the Saint John River population, whose migration is likely blocked, would be located well below the new regression line. In general, the two southern rivers (Altamaha and Pee Dee) do not fit the pattern of the northern and north-central rivers, and neither would the Savannah River, even with enhanced abundance due to stocking. Perhaps, abundance of southern populations is not related to spawning location, but their spawning locations are far upriver, like northern populations. The low abundance of southern adults may be affected more by anthropogenic impacts than by spawning location. More research is needed to understand factors that affect abundance of southern populations.

Why do females have a strong behavioral drive to move to river km 200-300 km to spawn? Perhaps, the behavior is related to finding suitable substrate and bottom velocity for spawning. However, in the Connecticut River these habitat conditions are available at downstream reaches, particularly Enfield Rapids, Connecticut, at river km 110. But prespawning fish continue upstream until blocked by Holyoke Dam. Migration patterns previously discussed suggest that three behavioral patterns of shortnose sturgeon insure that early life stages rear far upstream: (1) spawning females migrate far upstream, (2) larvae migrate a short distance, and (3) YOY are non-migratory for about 1 year. The behavior of females, larvae, and YOY may be adaptations that insure young fish do not contact salt water until salinity tolerance develops. Larvae may also gain a survival advantage from predators by delaying downstream migration until they are larger.

Superior survival of fish reared upstream of saltwater is supported by two kinds of information. Least squares linear regression comparisons of 10 rivers for relationships between adult abundance and two independent variables, (1) distance from spawning location to head of tide, and (2) distance from spawning location to salt water (maximum intrusion of salinity) found significant relationships with both factors (respectively, $r^2 = 0.88$, $p < 0.0005$, $r^2 = 0.75$, $p < 0.025$). The relationship with head of tide was particularly strong. Both analyses support the hypothesis of increased survival associated with spawning distance upstream, but both independent variables are closely correlated and provide no conclusions regarding salinity. However, results of stocking 2-9 week old fish in the Savannah River also indicated rearing distance upstream may affect survival. Fish released and presumably reared between river km 103-273, upstream of head-of-tide (river km 83), had an adult return rate of 1.3%, while fish stocked at river km 65-58, downstream of head-of-tide, had a 0.4% return rate (Smith et al. 1995). In the Savannah River, salinity penetrates in summer to river km 36, but contact of stocked YOY with salinity was not monitored. Smith et al. (1995) speculated that different acclimation and imprinting times produced the different returns. There is insufficient information for firm conclusions, but evidence in the present report indicates that rearing location upstream is important for survival of early life stages and population abundance.

Survival and recruitment

There is only limited information on survival of early life stages. Richmond & Kynard (1995) found that substrate with abundant crevices is likely critical for survival of eggs and embryos. Mortality of eggs due to the fungus *Saprolegnia*, although high in cultured eggs, was only 8% (mean of five daily samples of mid-developed eggs) at a Connecticut River spawning site (M. Horgan & B. Kynard unpublished data). Fish predation on eggs at the same spawning site also was insignificant. No data are available on mortality of embryos or larvae, but losses due to predation and unsuccessful initiation of feeding are probably high in these motile life periods. Year-class strength is likely established early in life, probably within 1-2 months.

Dredging in fresh/saltwater reaches of rivers containing shortnose sturgeon may destroy or alter juvenile and adult habitat. Fish may also be impinged. No adults were impinged during evaluations done in the Delaware River (Hastings[17]), but nothing is

[17] Hastings, R.W. 1983. A study of the shortnose sturgeon (*Acipenser brevirostrum* population in the upper tidal Delaware River: assessment of impacts of maintenance dredging. Final Report to U.S. Army Corps of Engineers, Philadelphia. 29 pp.

known about small juveniles that may be less able to avoid impingement. Smith et al.[11] believe the low abundance of Savannah River juveniles was caused by dredging in the fresh/saltwater interface. A similar situation may exist in other rivers.

Shortnose sturgeon are protected from directed fisheries, but they are captured across their range as bycatch, mainly in gill net fisheries for American shad, *Alosa sapidissima*. Dadswell (1979) estimated incidental fishing mortality of adults in the Saint John River gill net fishery was less than 10% of total mortality. In the Connecticut River, Savoy & Shake[3] estimated 2-25 adults were taken annually by the American shad fishery, and some fish are also caught by sport fishers angling for catfish, *Ictalurus* spp. Marchette & Smiley[8] also reported catch of shortnose sturgeon by sport fishers in South Carolina. Recent evaluation of sturgeon bycatch in American shad gill net fisheries and shrimp trawl fisheries in South Carolina and Georgia found that gill net fisheries captured about 2% of the adults. Of those captured, 11% died and 15% were injured (Collins & Smith 1996). Even if spawning migrants are released after capture by commercial fishers, handling greatly disrupts their migration (Moser & Ross 1994).

Poaching of adults in southern rivers using gill nets and traps seems widespread. Although no estimate of poaching exists, poachers captured 8 of 10 radio-tagged migrants below Pinopolis Dam on the Cooper River (D. Cooke personal communication) and 11 adults were trapped in the upper Santee River (M. Collins personal communication). Adults are extremely vulnerable to poaching because they group together in concentration areas and a commonly available inexpensive gear (gill nets) captures them.

Gill net fisheries bycatch and poaching are likely having a significant impact on southern, but not on northern populations. The last attempt to estimate the effects of harvest on shortnose sturgeon was done more than 10 years ago using data from populations in the Saint John, Hudson, and Pee Dee rivers (Boreman et al.[12]). Given the new information on losses of fish to bycatch and poaching in southern rivers, the effect of fishing mortality on southern populations should be reevaluated.

Conservation

The primary need of shortnose sturgeon conservation is completion of a recovery plan for the species. This plan will guide the direction of future conservation efforts. Generally, conservation should insure that existing populations survive and increase to carrying capacity. All aspects of the species' life history indicate that shortnose sturgeon should be abundant in most rivers. Thus, the expected abundance of adults in northern and north-central populations should be thousands to tens of thousands of adults (Figure 4). Expected abundance in southern rivers is uncertain, but large rivers should likely have thousands of adults. In small southern rivers, periodic removal of even small numbers of adults by poachers will probably make all other conservation measures unsuccessful. Impacts of fishing bycatch should be identified and reduced. In rivers where spawning migrations are blocked by dams, long-term solutions should be found for passing migrants (if spawning habitat remains upriver). Because spawning habitat requirements are known, the usefulness of creating artificial spawning areas should be investigated for spawning runs blocked by dams. Migratory movement patterns and river reaches used for foraging and spawning of all populations should be identified and protected. Spawning location and habitat should continue to be identified, particularly in southern rivers. Acceptable spawning conditions for fish blocked by dams should be maintained. The effect of contaminants on shortnose sturgeon reproduction and survival are unknown and should be investigated in these long-lived fish. Finally, long-term monitoring of population dynamics, abundance, and recruitment is needed in all populations to establish trends.

Using cultured fish for supplementing population abundance or restoring extirpated populations should be done carefully. Restoration should be attempted only after suitable habitat conditions for all life intervals exist and a suitable donor stock is available. The Delaware River stock should probably be used to restore Chesapeake Bay rivers. This restoration may be important to avoid long term genetic isolation between northern and southern segments of the range. Supplementary stocking of cultured fish into existing populations should only be used when wild populations are near extirpation and anthropogenic impacts cannot be corrected before the population is extirpated. Enhancement stocking may temporarily increase adult abundance, but it may alter the genetic basis of local adaptations of the wild population, possibly resulting in long-term reduction of individual fitness and decline of the population (Waples 1991, Fleming 1994).

Acknowledgements

Thanks to Vadim J. Birstein, John Waldman and Robert Boyle for the invitation to attend the International Conference on Sturgeon Biodiversity and Conservation. I thank members of the Shortnose Sturgeon Recovery Team (Mary Moser, Gordon Rogers and Tom Squiers) and Mark Collins and Doug Cooke of South Carolina for providing additional information, reviewing the manuscript, or both. John Waldman, Matt Droud, Vadim J. Birstein and William E. Bemis contributed editorial reviews and suggestions. Martin Horgan performed statistical analyses.

References cited

Bain, M.B. 1997. Atlantic and shortnose sturgeons of the Hudson River: common and divergent life history attributes. Env. Biol. Fish. (this volume).

Balon, E.K. 1975. Reproductive guilds of fishes: a proposal and definition. J. Fish. Res. Board Can. 32: 821–864.

Balon, E.K. 1985. Early life histories of fishes: new developmental, ecological and evolutionary perspectives. Developments in Env. Biol. Fish. 5, Dr W. Junk Publishers, Dordrecht. 280 pp.

Bath, D.W., J.M. O'Connor, J.B. Albert & L.G. Arvidson. 1981. Development and identification of larval Atlantic sturgeon (*Acipenser oxyrhynchus*) and shortnose sturgeon (*A. brevirostrum*) from the Hudson River estuary, New York. Copeia 1981: 711-717.

Bemis, W.E. & L. Grande. 1992. Early development of the actinopterygian head. I. External development and staging of the paddlefish *Polyodon spathula*. J. Morphol. 213: 47-83.

Berg, L.S. 1959. Vernal and hiemal races among anadromous fishes. J. Fish. Res. Board Can. 16: 515-537.

Brown, J.H. 1984. On the relationship between abundance and distribution of species. Amer. Nat. 124: 155-279.

Brundage, H.M. & R.E. Meadows. 1982. Occurrence of the endangered shortnose sturgeon, *Acipenser brevirostrum*, in the Delaware estuary. Estuaries 5: 203-208.

Buckley, J. & B. Kynard. 1981. Spawning and rearing of shortnose sturgeon from the Connecticut River. Prog. Fish Cult. 43: 74-76.

Buckley, J. & B. Kynard. 1985a. Yearly movements of shortnose sturgeon in the Connecticut River. Trans. Amer. Fish. Soc. 114: 813-820.

Buckley, J. & B. Kynard. 1985b. Habitat use and behavior of prespawning and spawning shortnose sturgeon, *Acipenser brevirostrum*, in the Connecticut River. pp. 111-117. *In:* F.P. Binkowski & S.I. Doroshov (ed.) North American Sturgeons, Dr W. Junk Publisher, Dordrecht.

Carlson, D.M. & K.W. Simpson. 1987. Gut contents of juvenile shortnose sturgeons in the upper Hudson estuary. Copeia 1987: 796-802.

Chittenden, M.E. 1974. Trends in the abundance of American shad *Alosa sapidissima* in the Delaware River basin. Chesapeake Science 15: 96-103.

Collins, M.R. & T.I.J. Smith. 1996. Characteristics of the adult segment of the Savannah River population of shortnose sturgeon, *Acipenser brevirostrum*. Proc. SE Assoc. Fish & Wildl. Agencies (in press).

Dadswell, M.J. 1979. Biology and population characteristics of the shortnose sturgeon, *Acipenser brevirostrum* LeSueur 1818 (Osteichthyes: Acipenseridae), in the Saint John River estuary, New Brunswick, Canada. Can. J. Zool. 57: 2186-2210.

Dadswell, M.J., B.D. Taubert, T.S. Squiers, D. Marchette & J. Buckley. 1984. Synopsis of biological data on shortnose sturgeon, *Acipenser brevirostrum* LeSueur 1818. FAO Fish. Synop. 140: 1-45.

Dovel, W.L., A.W. Pekovitch & T.J. Berggren. 1992. Biology of the shortnose sturgeon (*Acipenser brevirostrum* Leseur 1818) in the Hudson River estuary, New York. pp. 187-216. *In:* C.L. Smith (ed.) Estuarine Research of the 1980s: Seventh Symposium on Hudson River Research, State Univ. New York Press, Albany.

Fleming, I.A. 1994. Captive breeding and the conservation of wild salmon populations. Cons. Biol. 8: 886-888.

Frank, K.T. & W.C. Leggett. 1994. Fisheries ecology in the context of ecological and evolutionary theory. Ann. Rev. Ecol. Syst. 25: 401-422.

Hall, W.J., T.I.J. Smith & S.D. Lamprecht. 1991. Movements and habitats of shortnose sturgeon *Acipenser brevirostrum* in the Savannah River. Copeia 1991: 695-702.

Hastings, R.W., J.C. O'Herron, K. Schick & M.A. Lazzari. 1987. Occurrence and distribution of shortnose sturgeon, *Acipenser brevirostrum*, in the upper tidal Delaware River. Estuaries 10: 337-341.

Jenkins, W.E., T. Smith, L. Heyward & D.M. Knott. 1993. Tolerance of shortnose sturgeon, *Acipenser brevirostrum*, juveniles to different salinity and dissolved oxygen concentrations. Proc. Ann. Conf. SE Assoc. Fish. & Wildl. Agen. 47: 476-484.

Kieffer, M. & B. Kynard. 1993. Annual movements of shortnose and Atlantic sturgeons in the Merrimack River, Massachusetts. Trans. Amer. Fish. Soc. 122: 1088-1103.

Kieffer, M. & B. Kynard. 1996. Spawning of shortnose sturgeon in the Merrimack River, Massachusetts. Trans. Amer. Fish. Soc. 125: 179-186.

McCleave, J.D., S.M. Fried & A.K. Towt. 1977. Daily movements of shortnose sturgeon, *Acipenser brevirostrum*, in a Maine estuary. Copeia 1977: 149-157.

Moser, M.L. & S.W. Ross. 1994. Habitat use and behavior of shortnose and Atlantic sturgeons in the lower Cape Fear River, North Carolina. Trans. Amer. Fish. Soc. 124: 225-234.

North, J.A., R.C. Beamesderfer & T. Rien. 1993. Distribution and movements of white sturgeon in three lower Columbia River reservoirs. NW Science 67: 105-111.

O'Herron, J.C., K.W. Able & R.W. Hastings. 1993. Movements of shortnose sturgeon (*Acipenser brevirostrum*) in the Delaware River. Estuaries 16: 235-240.

Parsley, M.J. & L.G. Beckman. 1994. White sturgeon spawning and rearing habitat in the lower Columbia River. North Amer. J. Fish. Manag. 14: 812-827.

Richmond, A. & B. Kynard. 1995. Ontogenic behavior of shortnose sturgeon. Copeia 1995: 172-182.

Ross, S.W., F.C. Rohde & D.G. Lindquist. 1988. Endangered, threatened, and rare fauna of North Carolina. Part I of II. A reevaluation of the marine and estuarine fishes. Occ. Pap NC Biol. Surv. 1988-7.

Ryder, J.A. 1890. The sturgeon and sturgeon industry of the eastern coast of the United States with a recount of experiments bearing upon sturgeon culture. Bull. US Fish. Comm. 8: 231-328.

Seibel, D. 1993. Habitat selection, movements, and response to illumination of shortnose sturgeons in the Connecticut River. MS Thesis, University of Massachusetts, Amherst. 47 pp.

Smith, T.I.J. 1985. The fishery, biology, and management of Atlantic sturgeon, *Acipenser oxyrhynchus*, in North America. Env. Biol. Fish. 14: 61-72.

Smith, T.I.J., E.K. Dingley, R.D. Lindsey, S.B. Van Sant, R.A. Smiley & A.D. Stokes. 1985. Spawning and culture of shortnose sturgeon, *Acipenser brevirostrum*. J. World Maricul. Soc. 16: 104-113.

Smith, T.I.J, L.D. Heyward, W.E. Jenkins & M.R. Collins. 1995. Culture and stock enhancement of shortnose sturgeon, *Acipenser brevirostrum*, in the southern United States. pp. 204-214. *In:* A.D. Gershanovich & T.I.J. Smith (ed.) Proceedings of the International Symposium on Sturgeons, 6-11 September 1993, VNIRO Publishing, Moscow.

Taubert, B.D. 1980a. Biology of the shortnose sturgeon (*Acipenser brevirostrum*) in the Holyoke Pool, Connecticut River, Massachusetts. Ph.D Dissertation, University of Massachusetts, Amherst. 136 pp.

Taubert, B.D. 1980b. Reproduction of shortnose sturgeon, *Acipenser brevirostrum*, in the Holyoke Pool, Connecticut River, Massachusetts. Copeia 1980: 114-117.

Taubert, B.D. & M.J. Dadswell. 1980. Description of some larval shortnose sturgeon (*Acipenser brevirostrum*) from the Holyoke Pool, Connecticut River, Massachusetts, U.S.A., and the Saint John River, New Brunswick, Canada. Can. J. Zool. 58: 1125-1128.

Thompson, G.G. 1991. Determining minimum viable populations under the Endangered Species Act. Nat. Mar. Fish. Serv., Tech. Memorandum NMFS F/NWC-198. 78 pp.

Thuemler, T.F. 1985. The lake sturgeon, *Acipenser fulvescens*, in the Menominee River, Wisconsin-Michigan. Env. Biol. Fish. 14: 73-78.

Waples, R.S. 1991. Genetic interactions between hatchery and wild salmonids: lessons from the Pacific Northwest. Can. J. Fish. & Aquat. Sci. 48: 124-133.

Environmental Biology of Fishes **48**: 335–346, 1997.

Status and management of Atlantic sturgeon, *Acipenser oxyrinchus*, in North America

Theodore I. J. Smith[1] & James P. Clugston[2]
[1] *South Carolina Department of Natural Resources, P.O. Box 12559, Charleston, SC 29422, U.S.A.*
[2] *National Biological Service, 7920 NW 71st St., Gainesville, FL 32653, U.S.A.*

Received 2.1.1995 Accepted 30.3.1996

Key words: Acipenseridae, gulf sturgeon, fisheries, life history, culture, regulations, value

Synopsis

The Atlantic sturgeon, *Acipenser oxyrinchus*, consists of two subspecies distributed along the Atlantic coast of North America from Labrador to the east coast of Florida (Atlantic sturgeon subspecies – *A. o. oxyrinchus*) and along the Gulf of Mexico from Florida Bay, Florida to the mouth of the Mississippi River (Gulf sturgeon subspecies – *A. o. desotoi*). The species has been exploited throughout its range with landings peaking around the turn of the 20th century followed by drastic declines shortly thereafter. During recent years, landings in Canadian waters have increased substantially (approximately 129 metric tons in 1993) while in the United States landings are more controlled or prohibited (approximately 22–24 metric tons in 1993). Recently, the Atlantic States Marine Fisheries Commission developed a Fishery management plan for Atlantic sturgeon, and the United States Fish & Wildlife Service and Gulf States Marine Fisheries Commission drafted a Gulf Sturgeon Recovery/Management Plan. Fishery managers in Canada are in the process of establishing more stringent fishery regulations for sturgeon. Thus, the impact on populations due to harvesting should be substantially reduced. Current research focus includes: life history and population status studies, stock delineation, and development of culture and stock enhancement techniques. Implementation of the findings of such studies may be helpful in the restoration of depleted stocks.

Introduction

As one of North America's prized fish species, the Atlantic sturgeon, *Acipenser oxyrinchus*, served not only as food but also as an item of commerce for early European settlers. However, due to a combination of unregulated harvesting, damming of spawning rivers, and water pollution, the abundance of this species has dramatically declined, and harvest is now prohibited throughout much of its range. In the few areas where it is harvested, landings are relatively low, totaling about 200 metric tons (mt). This large anadromous species matures at an advanced age and utilizes rivers, bays, estuaries, coastal and continental shelf waters during its life cycle. Thus, it is highly susceptible to fishing and to human-induced habitat perturbations. During recent years, there has been increased interest to protect this species and restore populations. This concern is exemplified through development of fishery recovery and management plans.

Acipenser oxyrinchus has a broad distribution in major coastal river systems and estuarine and marine waters of eastern North America. Recently, it was reported that the species name '*oxyrhynchus*' has been misspelled for over 100 years and that the original correct spelling is '*oxyrinchus*' (Gilbert 1992). Thus, the original spelling is used in this paper. The species is represented by two subspecies (Vladykov 1955, Wooley 1985, Birstein 1993). Along the Canadian coast, the northern subspecies, *A. o. oxyrinchus*, (referred to as Atlantic sturgeon throughout the remainder of the paper), occurs in Hamilton Inlet on the Atlantic coast of Labrador (Bachus 1951), and is common in the Gulf of St. Lawrence, the St. Lawrence River, the Saint John River, New Brunswick, and in the Bay of Fundy (Murawski & Pacheco[1]). In United States waters, this subspecies occurs along the entire Atlantic coast to the St. Johns River in eastern Florida (Vladykov & Greeley 1963). In eight of fifteen coastal states, harvesting Atlantic sturgeon is prohibited.

The southern subspecies, *A. o. desotoi*, (referred to as 'Gulf sturgeon' throughout this paper) has a more restricted range and occurs in most river systems of the northern Gulf of Mexico from the mouth of the Mississippi River to the Suwannee River and in coastal waters as far south as Florida Bay, Florida (Wooley & Crateau 1985). The Gulf sturgeon differs from the Atlantic sturgeon in relative head length and pectoral fin length, shape of dorsal scutes, and length of spleen (Vladykov 1955, Vladykov & Greeley 1963). Wooley (1985) reexamined these differences and determined that the ratio of spleen length to fork length was the only statistically reliable characteristic to distinguish between subspecies. The Gulf sturgeon was listed as 'threatened' under the Endangered Species Act in 1991 and is no longer harvested.

Genetic tools are now being used to examine the population structure of Atlantic and Gulf sturgeons. Bowen & Avise (1990) used restrictive fragment length polymorphism (RFLP) analysis of the entire mitochondrial genome and reported that sturgeon exhibited low genotypic diversity, small sequence differences between mtDNA genotypes, and limited sharing of genotypes between the two coasts. More recently, Ong et al. (1996) utilized direct sequence analysis of the mtDNA control region and provided genetic evidence supportive of the subspecies designations for Atlantic and Gulf sturgeon.

Life history and ecology

There is substantial information on the life history and ecology of Atlantic sturgeon and detailed reviews have been provided by Murawski & Pacheco[1], Hoff[2], Rulifson & Huish (1982), Van Den Avyle[3], Smith & Dingley (1984), Smith (1985), Gilbert[4], and Taub[5]. The Atlantic sturgeon is the second largest acipenserid fish in North America with a maximum total length of 4.3 m (Scott & Crossman 1973, by comparison, white sturgeon, *A. transmontanus*, achieve maximum 6.1 m TL). Atlantic sturgeon undertake upriver spawning migrations beginning in February/March in the southern rivers, April/May in the mid-Atlantic region, and May–July in Canadian waters (Smith 1985). Female Atlantic sturgeon mature at about 7–19 years in SC (Smith et al.[6]) and 27–28 years in the St. Lawrence River (Scott & Crossman 1973). Female Gulf sturgeon mature at about 8–12 years in Florida (Huff[7]); females do not spawn every year so recruitment is very low. In

[1] Murawski, S. A. & A. L. Pacheco. 1977. Biological and fisheries data on Atlantic sturgeon, *Acipenser oxyrhynchus* (Mitchill). Nat. Mar. Fish. Ser., Tech. Ser. Rep. 10: 1–69.

[2] Hoff, J. G. 1980. Review of the present status of the stocks of Atlantic sturgeon, *Acipenser oxyrhynchus* (Mitchill). Southeast. Mass. Univ., Report to Nat. Mar. Fish. Ser., North Dartmouth. 136 pp.

[3] Van den Avyle, M. J. 1983. Species profiles: life histories and environmental requirements (South Atlantic) – Atlantic sturgeon. U.S. Fish Wildl. Ser., Div. Biol. Ser. FWS/OBS–82/11. U.S. Army Corps Eng., TREL–82–4. 38 pp.

[4] Gilbert, C. R. 1989. Species profiles: life histories and environmental requirements of coastal fishes and invertebrates (Mid-Atlantic Bight) – Atlantic and shortnose sturgeons. U.S. Fish Wildl. Ser. Biol. Rep. 82(11.122), U.S. Army Corps of Engineers TR EL–82–4. 28 pp.

[5] Taub, S. H. 1990. Fishery management plan for Atlantic sturgeon (*Acipenser oxyrhynchus*). Fisheries Management Rep. No. 17 of Atlantic States Marine Fisheries Commission. 73 pp.

[6] Smith, T. I. J., D. E. Marchette & R. A. Smiley. 1982. Life history, ecology, culture and management of Atlantic sturgeon, *Acipenser oxyrhynchus oxyrhynchus* Mitchill, in South Carolina. S.C. Wildl. Mar. Resour. Res. Dep., Final Rep. to U.S. Fish Wildl. Ser. Proj. AFS–9. 75 pp.

[7] Huff, J. A. 1975. Life history of Gulf of Mexico sturgeon, *Acipenser oxyrhynchus desotoi*, in Suwannee River, Florida. Florida Dep. Nat. Resour., Mar. Resour. Publ. 16, St. Petersburg. 32 pp.

some areas, a small fall spawning migration consisting of ripe Atlantic sturgeon adults has also been reported (Smith et al. 1984). Although actual spawning locations are not well known they are believed to include flowing water over hard substrates (rocks, rubble, shale, and sand).

Information on the early life history for both subspecies is scarce and based primarily on culture studies. For Atlantic sturgeon, the highly adhesive eggs require incubation times of 94 h (20° C) to about 140 h (18° C) (Smith et al. 1980). The yolksac is absorbed in about 10 days, and the small fish begin a demersal existence. Hatching time for artificially reared Gulf sturgeon eggs ranges from 54 h at 22.7–23.3° C to 85 h at 18.4° C (Parauka et al. 1991). Little is known of the behavior and habitat requirements of these small fish and it is assumed that they slowly move downriver from the spawning sites. Once Atlantic sturgeon subadults attain a size of ≥ 30 cm, they are regularly captured in tidally influenced lower river and estuarine areas (Dovel & Berggren 1983, Lazzari et al. 1986, Collins et al. 1996). Some movement of juveniles between river systems occurs (Dovel & Berggren 1983). Most juvenile Atlantic sturgeon remain in slightly brackish water near the river mouth/estuarine zone for a number of months or years (Kieffer & Kynard 1993) and then move into coastal and continental shelf waters where they grow and mature.

Information on the characteristics of Atlantic sturgeon populations is being developed. Based on tagging studies, Atlantic sturgeon are known to undertake extensive coastal migrations (Holland & Yelverton[8], Murawski & Pacheco[1], Hoff[2], Rulifson & Huish 1982, Smith et al.[6], Dovel & Berggren 1983). However, tagging studies provide little information on the source of recruits to various stocks and such information is basic to effective management. Recently, using mitochondrial DNA analysis, Waldman et al. (1996a) found that the Hudson River contributes most of the Atlantic sturgeon captured in the New Jersey intercept fishery in the spring and also a substantial portion of the fish harvested in the autumn fishery. In addition, it is believed that the Hudson River provides a substantial number of the juveniles that aggregate in the Delaware Bay. In 1990, studies suggested that there may be genetic structuring among Atlantic sturgeon in various Atlantic coast drainages (Bowen & Avise 1990). This hypothesis was supported when Atlantic sturgeon from the Saint Lawrence and Saint John rivers, Canada, the Hudson River, and rivers in Georgia were identified as genetically distinct populations (Waldman et al. 1996a, b).

Much information exists on the biology of Gulf sturgeon in the Apalachicola and Suwannee rivers of Florida (Huff (1975, footnote 7), Wooley & Crateau 1985, Odenkirk 1989, Foster 1993, Mason & Clugston 1993, Clugston et al. 1995). Gulf sturgeon usually begin to migrate into coastal rivers from the Gulf of Mexico during mid to late February as the rivers warm to 16–19° C. Migration continues through May and peaks during late March or early April in the Suwannee River, when the river temperature reaches about 20° C (Foster 1993, Clugston et al. 1995). Fish of all sizes (1.0–75.0 kg) move upriver and settle in various reaches of the river from river km 24 to km 215. Downstream movement usually begins in mid-November as the water cools to 20° C. All sturgeon in the Suwannee River, except young of the year, return to the Gulf of Mexico by early December. Small sturgeon (0.2–2.5 kg) remain in and near the river mouth during winter and spring. Larger fish move to unidentified areas in the Gulf of Mexico.

Gulf sturgeon over one year of age do not eat during the summer and fall despite an abundant supply of benthic organisms in the Suwannee River (Mason & Clugston 1993). In spring, stomachs of subadults and adults that migrate from the Gulf of Mexico into the Suwannee River and small sturgeon that remain near the river mouth are full of benthic invertebrates. Based on capture, mark and recapture studies, juvenile and adult Gulf sturgeon decrease in weight during the summer while in freshwater (Wooley & Crateau 1985, Clugston et al. 1995). Net growth of Gulf sturgeon results from a series of weight gains while they are in the Gulf of Mexico during winter and spring, and weight losses

[8] Holland, B. F., Jr. & G. F. Yelverton. 1973. Distribution and biological studies of anadromous fishes offshore North Carolina. N.C. Dep. Nat. Econ. Res. Spec. Sci. Rep. 24, Morehead City. 132 pp.

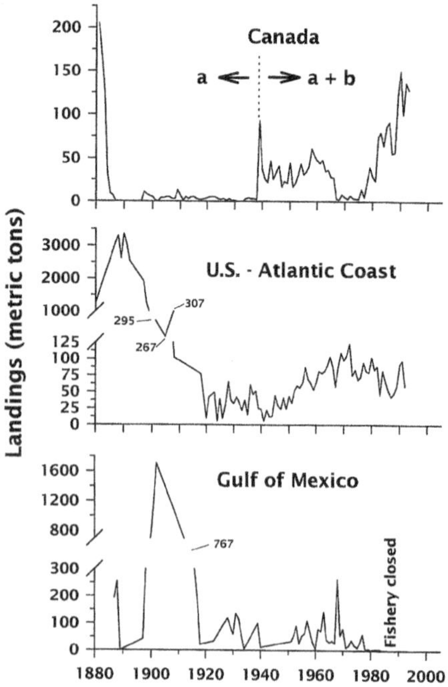

Figure 1. Reported landings of Atlantic and Gulf sturgeon. Data for Canada prior to 1940 show only landings from the Saint John River, New Brunswick (a, M. Dadswell personal communication); after 1940, the figure shows combined landings for the Saint John and St. Lawrence (Quebec) rivers (a + b, G. Trencia personal communication). U.S.-Atlantic Coast landings are based on NMFS data and may include shortnose sturgeon prior to 1972. Gulf of Mexico landings are based on J. M. Barkuloo[10].

during summer and fall when they are in coastal rivers. High water temperature was suggested as the main cause of feeding inactivity in the Suwannee River (Mason & Clugston 1993). Although this river remained cooler than near-shore Gulf of Mexico waters during summer, river temperatures were still believed to exceed optimum temperature for feeding and growth. Foster (1993), located juvenile and adult Gulf sturgeon in 29° C water in the Suwannee River using radio transmitters.

Gulf sturgeon spawn in the upper reaches of the Suwannee River (Clugston et al. 1995). The smallest sturgeon (76 g and 85 g) ever reported from this river were captured at river km 215 and 237 during the winter of 1991. Verification of spawning in this area was accomplished with the collection of sturgeon eggs on artificial substrates near river km 215 in April 1993 (Marchant & Shutters 1996). The river was in flood, 8.1 m and 5.5 m deep, respectively, where eggs were collected. Temperature at both locations was 18.3° C. Additional eggs were collected during late March and April 1994 at river km 201 to 221 when water temperatures ranged from 18.8 to 20.1° C.

Studies are underway to genetically characterize the Gulf sturgeon. Results to date show definitive inter-regional differentiation of mtDNA genotypes indicating that fish from rivers in the eastern, central and western Gulf are genetically distinct (Isaac Wirgin personal communication). Further, it is suggested that the homing fidelity of Gulf sturgeon may be quite high and that mtDNA differentiation among fish from geographically proximal rivers may be possible.

Reasons for decline

Exploitation

Early utilization of Atlantic sturgeon can be traced to 2190 B.C. in New England (Ritchie 1969), however, major fisheries for Atlantic sturgeon began during the last quarter of the 19th century. These fisheries focused on the spring migrations when the sturgeon moved into coastal rivers to spawn. Although most historical sturgeon landings data are probably inaccurate and do not include fishing effort, they do reflect major trends in harvest. Some Atlantic sturgeon landings may have included the smaller (maximum size 1.4 m TL) shortnose sturgeon, *A. brevirostrum* up until 1973 when this species was listed as endangered. However, most landings are probably based on the much larger Atlantic sturgeon. U.S. landings, recorded initially in 1880 by the US Fish Commission, peaked about 1890 when approximately 3350 mt were landed (Figure 1). Because the fish were so vigorously pursued, all major fisheries collapsed within about 10 years with landings in 1901 less than 10% (295 mt) of the former peak. Today, fisheries are still depressed and more closely regulated. During 1990–1992, mean total U.S. landings were 82.4 mt or 2% of that reported in 1890.

The major U.S. historical fisheries for the Atlantic sturgeon occurred during the period 1870–1920. They were initially centered on the Delaware River and the Chesapeake Bay system with most landings reported by New Jersey (NJ) and Delaware (DE) and, to a lesser degree, by Virginia (VA) (Murawski & Pacheco[1]). During the early period of peak exploitation, substantial landings were also reported by North Carolina (NC), South Carolina (SC), and Georgia (GA). By the late 1970s and early 1980s, SC, GA, and NC accounted for approximately 80% of the U.S. landings (Table 1). In recent years, landings have again shifted and are now centered in the Hudson River, and coastal New York (NY) and NJ. During 1990–1992, NY and NJ reported approximately 93% of U.S. landings.

Some sturgeon harvested in directed fisheries were taken using pound nets, weirs, stake row nets, trammel nets, trawls, harpoons, and snares. However, large floating and anchored gill nets with stretch mesh sizes of 33–41 cm and depths of 4–8 m were the most commonly used fishing gear (Cobb[9], Galligan 1960, Huff[7], Smith et al. 1984). In a recent review of U.S. Atlantic sturgeon fisheries, Taub[5] determined that incidental captures of Atlantic sturgeon contributed more to total landings than did directed sturgeon fisheries. For example, in 1987 nearly 77% of the total reported landings were incidental to other commercial fisheries. Such fisheries used a variety of gear (e.g., trawls, haul seines, pound nets, gill nets) to catch various groundfishes, shad, shrimp, and lobster bait. Sturgeon landed in these fisheries were usually subadults. Recent studies in SC and GA confirm that sturgeon bycatch still occurs and reported the incidental capture of sturgeons in the shad gill net and penaeid shrimp trawl fisheries (Collins et al. 1996). This trend of harvesting both subadults and adults continues in the remaining major U.S. fisheries. Harvests in NY are primarily mature adult Atlantic sturgeon from the Hudson River and juveniles and young adults from the coastal fisheries. The NJ fishery lands mostly immature fish in its directed and incidental fisheries.

The Atlantic sturgeon also has a long history of exploitation in Canadian waters. During the period of peak exploitation, the only long term harvest record is from the Saint John River fishery. Harvest pattern for this fishery was similar to that in the U.S.,

[9] Cobb, J. N. 1900. The sturgeon fishery of Delaware River and Bay. Rep. U.S. Comm. Fish and Fisheries for 1899, Part 25: 369–380.

Table 1. Reported landings (metric tons, mt) of Atlantic sturgeon in the United States by state in 1982, 1987, and 1992 (based on NMFS data).

State	1982		1987		1992	
	Landings	Rank*	Landings	Rank*	Landings	Rank*
Maine	1.2		0.4		0.0	
New Hampshire	0.0		0.4		0.0	
Massachusetts	1.0		3.1	4	< 0.1	
Rhode Island	0.5		1.9		1.2	
Connecticut	0.0		0.0		0.0	
New York	9.9	4	17.1	1	17.8	2
New Jersey	3.3	5	9.1	2	38.1	1
Delaware	0.6		< 0.1		0.0	
Maryland	0.5		0.5		0.6	
Virginia	1.7		0.0		0.0	
North Carolina	10.6	3	6.2	3	0.0	
South Carolina	45.4	1	0.0		0.0	
Georgia	12.8	2	2.9	5	1.0	
Florida (E. coast)	< 0.1		0.0		0.0	
Total	87.6		41.6		58.8	

* Rank assigned for states with reported landings 2 ≥ mt.

with landings substantially reduced from those reported before the turn of the century (Figure 1). From about 1951–1982, U.S. landings generally exceeded those reported from Canada. However, there has been a consistent trend of increased catches from Canada with 1990–1992 landings (mean 130 mt) near peak levels.

There are two major Canadian fisheries: one centered in the St. Lawrence River between St. Joachim and Cacouna, Quebec (Guy Trencia personal communication); and, another in the Saint John River, New Brunswick (Michael Dadswell personal communication). Both are directed fisheries. The increase in Canadian landings during the past 12 years is primarily from the St. Lawrence River fishery, which supports about 35 fishermen. Fish taken from the St. Lawrence River are rarely mature. In recent years, only 1–2 fisherman fish for Atlantic sturgeon in the Saint John River fishery with annual landings on the order of 5–10 mt.

The Gulf sturgeon fishery also experienced a major decline since about the turn of the century (Figure 1) based on commercial landing statistics compiled by Barkuloo[10]. Directed fisheries occurred only along the Florida (FL) and Alabama (AL) coasts. There are periods of increased harvest but, as with the Atlantic sturgeon, no effort information is available to calculate catch per unit effort.

Over-harvesting is believed to be the single major cause of the precipitous declines in abundance of Atlantic sturgeon (Ryder 1890, Vladykov & Greeley 1963, Hoff[2], Taub[5]). Vulnerability to overfishing is clear as mature fish were relatively easily captured during spring spawning migrations and juveniles were harvested from estuarine nursery habitats.

Damming of spawning rivers

Construction of dams in some northeastern and southeastern rivers excluded Atlantic sturgeon from historical spawning sites. In the northeast, such dams include those at the head of the Androscoggin River (1807) and Kennebec River (1837) in Maine, the dam at Lawrence (1847) on the Merrimack River in New Hampshire, and the Enfield Rapids Dam on the Connecticut River (Hoover[11], Galligan 1960, Murawski & Pacheco[1]). In South Carolina, Atlantic sturgeon have been excluded from historic spawning sites since about 1870 when mill dams and water supply dams were constructed on the Peedee, Wateree, Congaree and Savannah rivers (Leland 1968). Similarly, dams have limited Gulf sturgeon access to migration routes and historic spawning areas throughout the Gulf of Mexico (Murawski & Pacheco[1], Wooley & Crateau 1985, USFWS & Gulf States Marine Fisheries Commission[12]). A dam constructed in 1962 across North Bay of the St. Andrew Bay system in FL prevents passage of all anadromous species as does the Ross Barnett dam on the Pearl River. The Jim Woodruff Lock and Dam constructed in 1957 on the Apalachicola River also appears to provide complete restriction as no tagged sturgeon have been taken upriver of this structure and no evidence exists which indicates that the Gulf sturgeon can pass through the lock system (USFWS & Gulf States Marine Fisheries Commission[12]).

Pollution and industrialization

Much of the Atlantic sturgeon's historic riverine habitat has been degraded by water pollution and extensive dredging (Taub[5]). Along the middle Atlantic coast and the Gulf of Mexico, water degradation from industrial and domestic discharges impacted spawning and nursery habitats (Hoover[11], Galligan 1960, Leland 1968, Murawski & Pacheco[1], Barkuloo[10]). The 1970 National Estuary Study indicated that dredging and filling activities were particularly destructive to fish habitat and reported that 73% of the U.S. estuaries have been moderate-

[10] Barkuloo, J. M. 1988. Report on the conservation status of the Gulf of Mexico sturgeon, *Acipenser oxyrhynchus desotoi*. U.S. Fish and Wildlife Service, Panama City. 33 pp.

[11] Hoover, E. E. 1938. Biological survey of the Merrimack watershed. Fish Game Comm., Concord. 238 pp.

[12] U.S. Fish and Wildlife Service & Gulf States Marine Fisheries Commission. Gulf Sturgeon Recovery/Management Plan. Atlanta, Georgia. 170 pp.

ly to severely degraded. Dredging and filling disturbs benthic fauna, eliminates deep holes and alters rock substrates, all important to sturgeon. Anecdotal evidence indicates that old river bottom not subjected to maintenance dredging is preferred by Atlantic sturgeon (Taub[5]). Contaminants have not been intensively examined in the Atlantic sturgeon but an early study indicated that concentrations of polychlorinated biphenyls (PCBs) in St. Lawrence and Hudson River Atlantic sturgeon generally exceeded Food and Drug Administration guidelines (2 mg l^{-1}) for human consumption (Murawski & Pacheco[1]). In 1986, New York Department of Environmental Conservation analyzed Atlantic sturgeon from the Hudson River and found low levels (0.15 to 1.70 mg l^{-1}) in all tissues except the brain which contained an average concentration of 7.92 mg l^{-1} PCBs (Sloan 1987). Gulf sturgeon collected from a number of rivers from 1985 to 1991 were analyzed for pesticides and heavy metals (Bateman & Brim[13]). Concentrations of arsenic, mercury, DDT metabolites, toxaphene, polycyclic aromatic hydrocarbons, and aliphatic hydrocarbons were sufficiently high to warrant concern. Such products are known to cause reproductive failure, reduced survival, and various physiological alterations in fish.

[13] Bateman, D. H. & M. S. Brim. 1994. Environmental contaminants in Gulf sturgeon of northwest Florida 1985–1991. USFWS, Pub. No. PCFO-EC 94–09, Panama City. 23 pp.

Table 2. Ex-vessel price ($ per kg whole weight) for Atlantic sturgeon landed in New York, New Jersey and Georgia during 1988–1992*.

Year	Price ($ per kg)		
	New York	New Jersey	Georgia
1988	2.46	2.70	8.13
1989	3.03	3.02	7.86
1990	3.98	3.99	10.00
1991	4.02	2.36	12.50
1992	4.44	3.53	13.37
Mean	3.59	3.12	10.37

* Based on NMFS data and G. Rogers personal communication.

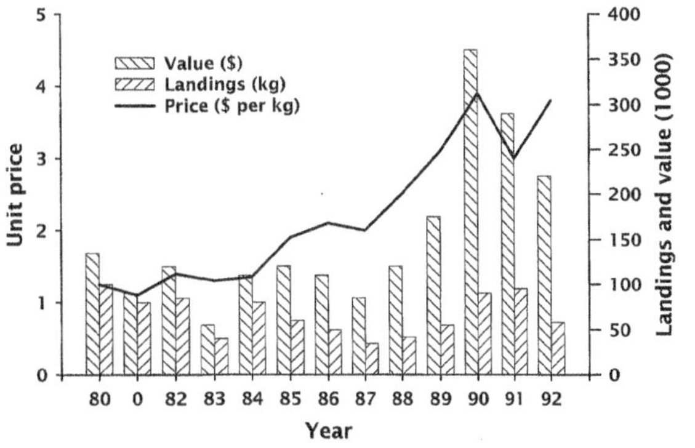

Figure 2. Ex-vessel value and landings for the Atlantic sturgeon fishery from 1980–1992.

Value and employment

The former sturgeon fisheries were economically important and provided substantial direct (fishermen) and indirect employment such as net manufacturing, boat building, food processing, and shipping (Murawski & Pacheco[1]). Today, however, sturgeon fisheries have an insignificant impact on coastal economies and provide no full time employment. Sturgeon fishermen now number approximately 100 (35 in Canada; 65 in U.S.) and all use the fishery as a source of supplemental income.

The reported ex-vessel value of U.S. landings of Atlantic sturgeon has fluctuated from $59 000 to $365 000 (mean $155 000) during the period 1980 to 1992. During this time, landings have ranged from about 40 to 100 mt while ex-vessel price and value have increased (Figure 2). Average price during 1980–1982 was $1.27 per kg whole weight and $3.58 per kg during 1990–1992. When the ex-vessel prices and values of the fishery are adjusted for inflation during this period (deflated using Consumer Price Index for fishery products), there has been a 53% increase in real ex-vessel price and a 41% increase in real value of the Atlantic sturgeon fishery since 1980–1982. States reporting landings based partially on incidental captures and landings of juveniles have a lower mean ex-vessel price than those with directed fisheries where adults are harvested. For example, ex-vessel prices during 1988–1992 for landings from NY and NJ (where much of the harvest is subadults) averaged $3.59 per kg and $3.12

per kg, respectively (Table 2). In GA, where there is a directed gill net fishery focused on capture of roe females, the ex-vessel average price was $10.37 per kg during the same period.

A detailed analysis of the South Carolina fishery was performed for 1977 to 1982 and the total value of the caviar exceeded the total value of the flesh (Smith et al. 1984). In SC, the ex-vessel price of processed caviar was $66 per kg versus $2.42 per kg for the dressed carcass in 1982. The current prices for caviar in GA are about $121 per kg ex-vessel, $278 per kg wholesale, and $350 per kg retail (Bertha Boone, Walters Caviar Co. personal communication). In New York, prices as high as $220 per kg ex-vessel have been reported (Holloway 1994). During 1993, ex-vessel dressed carcass (headless, finless and gutted) price was about $5–7 per kg in GA but after further processing and smoking in NY, the product retailed for $33–40 per kg (Bertha Boone, Walter's Caviar Co., personal communication).

Management

Until recently, management of Atlantic sturgeon was the responsibility of the individual states and other regional jurisdictional entities. This resulted in a wide diversity of regulations involving licensing, harvest size, fishing gear, seasons, and reporting requirements (Smith 1985, Taub[5]). In U.S. waters, lack of uniform regulations is no longer a problem but in Canadian waters, sturgeon fishery regulations are different for the St. Lawrence and Saint John Rivers (Table 3).

In 1990, the Atlantic States Marine Fisheries

Table 3. Management regulations for taking Atlantic sturgeon, *Acipenser o. oxyrinchus*, in 1994. Harvesting of Gulf sturgeon, *Acipenser oxyrinchus desotoi*, is prohibited.

Area	Type of fishery	Laws and regulations
Canada		
St. Lawrence R.	directed	amount of gill net limited by fishing zone; minimum gill net stretch mesh 17.8 cm; net owner identification
St. John R.	directed	minimum fish length 122 cm TL; minimum gill net stretch mesh 33 cm; open season except June
United States		
Maine	closed	landings prohibited
New Hampshire	closed	landings prohibited
Massachusetts	closed	landings prohibited
Rhode Island	incidental	minimum fish size 213 cm TL, license required in freshwater
Connecticut	incidental	minimum fish size 213 cm TL; trawl license required; maximum of 3 fish per day per vessel
New York	directed and incidental	minimum fish size 152 cm TL; special sturgeon license required; open seasons: 15 May–15 Jun in Hudson R. and marine district; 1 Oct–30 Nov in marine district; landed fish must be tagged, reported, and biological data collected
New Jersey	directed and incidental	minimum fish size 152 cm TL; special sturgeon license required; individual quotas; logbook records; allow access to fish for collection of biological data
Pennsylvania	closed	landings prohibited
Maryland	incidental	minimum fish size 213 cm TL; commercial finfish license required; maximum gill net stretch mesh size 8.9 cm
Delaware	directed	minimum fish size 213 cm TL; commercial gill net license required
Virginia	closed	landings prohibited
North Carolina	closed	landings prohibited
South Carolina	closed	landings prohibited
Georgia	directed	minimum fish size 213 cm TL; minimum gill net stretch mesh 30 cm; gill nets only; open season 15 Feb–15 Apr, commercial saltwater fishing license required
Florida	closed	landings prohibited

Commission (ASMFC) completed development of a Fishery Management Plan (FMP) for Atlantic sturgeon (Taub[5]). The goal of the FMP is 'to provide the framework to allow restoration of the Atlantic sturgeon resource to fishable abundance throughout its range'. Fishable abundance was defined as an annual harvest of about 317 mt or about 10% of the historic landings in 1890. The FMP has several management objectives which include: protection from further stock depletion; expansion of knowledge concerning the stock(s); enhancement and restoration; and coordination of research and management activities throughout the species'range. To achieve these objectives, 13 specific management recommendations were identified. The recommendation of greatest immediate importance was that each state limit harvest by establishing a fishing moratorium; or, by establishing a minimum harvest size of 213 cm TL (191 FL) coupled with a monitoring program; or, by developing a 'conservation equivalency' plan which is acceptable to ASMFC. Other recommendations include identification and protection of critical habitats, documentation of movements, evaluation of status and genetic characteristics of populations, development of propagation techniques, and evaluation of hatchery fish for stock restoration purposes.

Development of the FMP focused attention on the Atlantic sturgeon and has resulted in efforts to protect and conserve the remaining populations in U.S. waters. In 1994, NJ submitted an acceptable plan to ASMFC, thereby bringing all Atlantic states into compliance with the primary recommendation to control harvest (Table 3).

In Canada, there are plans to improve management and prevent over-harvest of Atlantic sturgeon in the St. Lawrence River. In addition to regulation of total amount of netting and minimum mesh size (17.8 cm stretch), all nets must be identified with the owner's name beginning in 1994. In subsequent years, there are plans to regulate fishing season and fish harvest size to prevent overfishing (Guy Trencia personal communication).

Prior to its classification as a 'threatened species' in 1991, Gulf sturgeon harvest regulations varied among the coastal states along the Gulf of Mexico. Alabama (AL) permitted commercial harvest in coastal waters but harvest in freshwater has been prohibited since 1972. Louisiana (LA) permitted commercial harvest until 1990 while the Mississippi fishery has been closed since 1974. Florida classified both subspecies as 'species of special concern' and prohibited harvest on both coasts beginning in 1984.

Under the Endangered Species Act, it is illegal for anyone to take, kill, possess or sell the Gulf sturgeon. Classification as 'threatened' prompted the establishment of a Gulf Sturgeon Recovery/Management Task Team which recently completed a Gulf Sturgeon Recovery/Management Plan (USFWS & Gulf States Marine Fisheries Commission[12]). A major short-term objective of the Gulf sturgeon plan is to prevent further reduction of existing wild populations of this subspecies. The long-term objective is to establish population levels that would allow delisting the Gulf sturgeon by discrete management units.

Culture and stock enhancement research

Early fishery workers were aware of the catastrophic decline in sturgeon landings. After detailed study of the fisheries, they concluded that 'the only means of maintaining and increasing the industry was through artificial propagation which should be successfully accomplished at a comparatively insignificant outlay' (Ryder 1890). At the time, sturgeon were considered second in economic importance to the lobster (Stone 1900). The first artificial spawning was achieved by the New York State Fish Commission using fish from the Hudson River in 1875 (Harkness & Dymond 1961). Based on this limited success, the U.S. Fish Commission initiated artificial propagation studies on the Delaware River in 1888. This work and that of other culturists was thwarted by difficulties associated with collecting ripe males and females simultaneously and fungal infection of the incubating eggs. No substantial successes were achieved, and by 1912 most work on sturgeon culture had been abandoned in the U.S.

and Canada (Dean 1894, Leach[14], Harkness & Dymond 1961, Hoff[2], Smith & Dingley 1984).

In recent years, there has been renewed interest in culturing North American sturgeons (Smith 1990). A cooperative USFWS effort was established in SC focused on propagation of Atlantic sturgeon (Smith et al.[6]). In 1979 and 1981, Atlantic sturgeon were successfully stripped using injection of acetone-dried pituitary glands (Smith et al. 1980, 1981). In 1993 and 1994, success was obtained in stripping Hudson River Atlantic sturgeon using LHRHa (Richard A. St. Pierre personal communication). Studies are underway to develop culture techniques at USFWS and National Biological Service (NBS) facilities in PA. Such studies include diet research on all life stages, pathology, and sperm cryopreservation. After 20 months, cultured Atlantic sturgeon grew to 66 cm TL and 1.2 kg. In fall 1994, 5000 cultured juveniles (approximately 10 cm TL) were tagged with coded wire tags and stocked in the Hudson River. To date, several of these fish have been reported captured.

The first artificial spawning of Gulf sturgeon was achieved with fish from the Suwannee River on 22 March 1989 (Parauka et al. 1991) and efforts to develop culture techniques are continuing. To examine the potential for stock enhancement, about 1200 of the 20–30 cm TL sturgeon were tagged using passive integrated transponders (PIT tags). These fish were released by NBS and University of Florida personnel at 10 sites in the Suwannee River during December 1992. By February 1995, 6.2% of the stocked fish have been captured by NBS as part of this continuing effort.

Due to the interest in stocking for restoration and management purposes, the ASMFC recently requested the Atlantic Sturgeon Aquaculture and Stocking Committee to recommend stocking guidelines. A series of recommendations were developed which addressed a variety of issues including: origin and numbers of broodstock; genetic and ecological considerations; stocking numbers, sizes and locations; and, planning and evaluation (St. Pierre[15]).

Prospects

In the U.S., the Gulf sturgeon is fully protected from harvest while fishing for Atlantic sturgeon has been severely curtailed or halted in all states. In addition, there is national focus on all eight North American sturgeons and the paddlefish as evidenced by the recent development of a document titled 'Framework for the Management and Conservation of Paddlefish and Sturgeon Species in the United States'. This document, prepared by the National Paddlefish and Sturgeon Steering Committee[16], identifies problems and provides guidance for needed research.

The recent adoption of ASMFC recommendations to control minimum harvest size (213 cm TL) or to establish a 'conservation equivalency' measure is a major first step for preservation of Atlantic sturgeon in U.S. waters. Not only will landings be greatly reduced but juveniles will also be protected. Based on current quotas and fishery regulations, total 1994 harvest in the U.S. is expected to be about 20 mt with a minimum fish harvest size of 152 cm TL. In Canada, there is substantial concern about the recent increase in landings in the St. Lawrence River fishery. Consequently, fishery managers are analyzing landings data and planning implementation of strict harvest regulations during the next several years to protect the sturgeon from overfishing.

In addition to regulatory issues, emphasis has been placed on answering the basic, life history questions needed for effective management of the species. Culture and marking techniques are being developed for possible use in stock enhancement efforts for both subspecies and guidelines have

[14] Leach, G. C. 1920. Artificial propagation of sturgeon, review of sturgeon culture in the United States. Rep. U.S. Fish Comm. 1919: 3–5.

[15] St. Pierre, R. A. 1995. Breeding and stocking protocol for cultured Atlantic sturgeon. Rep. Atlantic Sturgeon Aquaculture and Stocking Committee, Atlantic States Marine Fisheries Commission. 17 pp.

[16] National Paddlefish & Sturgeon Steering Committee. 1993. Framework for the management and conservation of paddlefish and sturgeon species in the United States. USFWS, Division of Fish Hatcheries, Washington, DC. 41 pp.

been prepared by the Atlantic Sturgeon Aquaculture and Stocking Committee (Smith[17], St. Pierre[15]). Preliminary stock enhancement work with Atlantic and Gulf sturgeons, and shortnose sturgeon (Smith et al. 1995) is encouraging, but substantial additional work will be required to fully test the concept. Atlantic sturgeon movement and recruitment data are being collected through USFWS and ASMFC cooperation. Tagging and reporting of recaptured fish is encouraged by the establishment of a central data repository in MD. Tags are provided free and anyone encountering sturgeon during various research and monitoring activities is requested to tag and release sturgeon as well as collect basic biological data. In states with 'conservation equivalency' status, detailed biological information is collected from all fish landed. Further, incidental capture and fate of sturgeons caught in select fisheries in SC and GA was documented to provide information on possible impacts to the stocks (Collins et al. 1996).

The USFWS and NBS continue to collect sturgeon distribution, abundance, and life history information in the coastal rivers of the Gulf of Mexico. Research is focused on collection of young Gulf sturgeon and identification of spawning habitat. Little is known of estuarine and marine habitat requirements and work will be directed to these areas. The Caribbean Conservation Corporation, a private, non-profit research organization, is conducting a long-term study to estimate the growth and number of sturgeon entering the Suwannee River each spring. Recently, a number of Gulf states initiated status and life history studies in their respective river systems. With the controlled harvest of sturgeon now in place, it is essential to continue investigations to determine critical habitat requirements, to establish culture and stocking parameters, and to document the status and origin of stocks. Once such studies are completed, it may be possible to restore some stocks to a level that will support sustainable fishery harvests.

[17] Smith, T. I. J. 1992. Recommendations concerning the culture and stocking of Atlantic sturgeon. Rep. Atlantic Sturgeon Aquaculture and Stocking Committee, Atlantic States Marine Fisheries Commission, Spec. Rep. No. 22. 20 pp.

Acknowledgements

This paper is a joint contribution of the South Carolina Marine Resources Center (contribution number 343) and the Southeastern Biological Science Center, National Biological Services. We greatly appreciate the information and comments provided by Guy Trencia, Direction Regionale de Quebec, and Michael Dadswell, Acadia University. We thank the many reviewers, especially Gordon Rogers, Wally Jenkins, Mark Collins, Lorna Patrick, Frank Parauka, and Ken Sulak.

References cited

Bachus, R. H. 1951. New and rare records of fishes from Labrador. Copeia 1951: 288–294.

Birstein, V. J. 1993. Sturgeons and paddlefishes: threatened fishes in need of conservation. Conserv. Biol. 7: 773–787.

Bowen, B. W. & J. C. Avise. 1990. Genetic structure of Atlantic and Gulf of Mexico populations of sea bass, menhaden, and sturgeon: influence of zoogeographic factors and life history patterns. Marine Biology 107: 371–381.

Clugston, J. P., A. M. Foster & S. H. Carr. 1995. Gulf sturgeon, *Acipenser oxyrinchus desotoi*, in the Suwannee River, Florida, USA. pp. 215–224. *In:* A. D. Gershanovich & T. I. J. Smith (ed.) Proceedings of the International Symposium on Sturgeons, VNIRO Publishing, Moscow.

Collins, M. R., S. G. Rogers & T. I. J. Smith. 1996. Bycatch of sturgeons along the southern Atlantic coast of the U.S. N. Amer. J. Fish. Manag. 16: 24–29.

Dean, B. 1894. Recent experiments in sturgeon hatching on the Delaware River. U.S. Fish Comm. Bull. (1893) 13: 335–339.

Dovel, W. L. & T. J. Berggren. 1983. Atlantic sturgeon of the Hudson estuary, New York. J. New York Fish & Game 30(2): 140–172.

Foster, A. M. 1993. Movement of Gulf sturgeon, *Acipenser oxyrinchus desotoi*, in the Suwannee River, Florida. Masters Thesis, University of Florida, Gainesville. 130 pp.

Galligan, J. P. 1960. History of the Connecticut River sturgeon fishery. Conn. Wildl. Cons. Bull. 6: 1, 5–6.

Gilbert, C. R. 1992. Atlantic sturgeon. pp. 31–39. *In:* R. A. Ashton (ed.) Rare and Endangered Biota of Florida, Volume 2, Fishes, University of Florida, Gainesville.

Harkness, W. J. K. & J. R. Dymond. 1961. The lake sturgeon, the history of its fishery and problems of conservation. Ontario Dept. Lands & Forests, Toronto. 121 pp.

Holloway, M. 1994. Caviar on the Hudson. Audubon 96(3): 72–75.

Kieffer, M. C. & B. Kynard. 1993. Annual movements of shortnose and Atlantic sturgeons in the Merrimack River, Massachusetts. Trans. Amer. Fish. Soc. 122: 1088–1103.

Lazzari, M. A., J. C. OHerron, II & R. W. Hastings. 1986. Occurrence of juvenile Atlantic sturgeon, *Acipenser oxyrhynchus*, in the upper tidal Delaware River. Estuaries 9: 356–361.

Leland, J. G., III. 1968. A survey of the sturgeon fishery of South Carolina. Contr. Bears Bluff Lab. 47: 1–27.

Marchant, S. R. & M. K. Shutters. 1996. Artificial substrates collect the first Gulf sturgeon eggs. N. Amer. J. Fish. Manag. 16: 445–447.

Mason, W. T., Jr. & J. P. Clugston. 1993. Foods of the Gulf sturgeon in the Suwannee River, Florida. Trans. Amer. Fish. Soc. 122: 378–385.

Odenkirk, J. S. 1989. Movements of Gulf of Mexico sturgeon in the Apalachicola River, Florida. Southeast. Assoc. Fish and Wildl. Ag. 43: 230–238.

Ong, T.-L., J. Stabile, I. Wirgin & J. R. Waldman. 1996. Genetic divergence between *Acipenser oxyrinchus oxyrinchus* and *A. o. desotoi* as assessed by mitochondrial DNA sequencing analysis. Copeia 1996: 464–469.

Parauka, F. M., W. J. Troxel, F. A. Chapman & L. G. McBay. 1991. Hormone induced ovulation and artificial spawning of the Gulf of Mexico sturgeon (*Acipenser oxyrhynchus desotoi*). Prog. Fish-Cult. 53: 113–117.

Ritchie, W. A. 1969. The archaeology of Marthas Vineyard. The National History Press, Garden City. 253 pp.

Rulifson, R. A. & M. T. Huish. 1982. Anadromous fish in the southeastern United States and recommendations for development of a management plan. U.S. Fish Wild. Ser., Fish. Resour. Reg. 4. 525 pp.

Ryder, J. A. 1890. The sturgeon and sturgeon industries of the eastern coast of the United States, with an account of experiments bearing upon sturgeon culture. U.S. Fish Comm., Bull. (1888)8: 231–238.

Scott, W. B. & E. J. Crossman. 1973. Freshwater fishes of Canada. Fish. Res. Board Can. Bull. 184. 966 pp.

Sloan, R. 1987. Toxic substances in fish and wildlife: analyses since May 1, 1982. Vol. 6, New York State Dept. Envir. Conserv., Div. Fish and Wildl., Albany. 182 pp.

Smith, T. I. J. 1985. The fishery, biology and management of Atlantic sturgeon, *Acipenser oxyrhynchus*, in North America. Env. Biol. Fish. 14: 61–72.

Smith, T. I. J. 1990. Culture of North American sturgeons for fishery enhancement. pp. 19–27. *In:* A. K. Sparks (ed.) Marine Farming and Enhancement, Proceedings of the 15th U.S.-Japan Meeting on Aquaculture, Kyoto, Japan 1986, NOAA Tech. Rep. NMFS 85.

Smith, T. I. J. & E. K. Dingley. 1984. Review of biology and culture of Atlantic (*Acipenser oxyrhynchus*) and shortnose sturgeon (*A. brevirostrum*). J. World Maricul. Soc. 15: 210–218.

Smith, T. I. J., E. K. Dingley & D. E. Marchette. 1980. Induced spawning and culture of Atlantic sturgeon. Prog. Fish Cult. 42: 147–151.

Smith, T. I. J., E. K. Dingley & D. E. Marchette. 1981. Culture trials with Atlantic sturgeon, *Acipenser oxyrhynchus* in the U.S.A. J. World Maricul. Soc. 12: 78–87.

Smith, T. I. J., D. E. Marchette & G. F. Ulrich. 1984. The Atlantic sturgeon fishery in South Carolina. N. Amer. J. Fish. Manag. 4: 164–176.

Smith, T. I. J., L. D. Heyward, W. E. Jenkins & M. R. Collins. 1995. Culture and stock enhancement of shortnose sturgeon, *Acipenser brevirostrum*, in the southern United States. pp. 204–214. *In:* A. D. Gershanovich & T. I. J. Smith (ed.) Proceedings of the International Symposium on Sturgeons, VNIRO Publishing, Moscow.

Stone, L. 1900. The spawning habits of the lake sturgeon (*Acipenser rubincundus*). Trans. Amer. Fish. Soc. 29: 118–128.

Vladykov, V. 1955. A comparison of Atlantic sea sturgeon with a new subspecies from the Gulf of Mexico (*Acipenser oxyrhynchus desotoi*). J. Fish Res. Board Can. 12: 754–761.

Vladykov, V. D. & J. R. Greeley. 1963. Order Acipenseroidei. Fishes of the western North Atlantic. Sears Found. Mar. Res., Yale University, New Haven. 630 pp.

Waldman, J. R., J. T. Hart & I. I. Wirgin. 1996a. Stock composition of the New York Bight Atlantic sturgeon fishery based on analysis of mitochondrial DNA. Trans. Amer. Fish. Soc. 125: 364–371.

Waldman, J. R., K. Nolan, J. Hart & I. I. Wirgin. 1996b. Genetic differentiation of three key anadromous fish populations of the Hudson River. Estuaries (in press).

Wooley, C. M. 1985. Evaluation of morphometric characters used in taxonomic separation of Gulf of Mexico sturgeon, *Acipenser oxryhynchus desotoi*. pp. 97–103. *In:* F. P. Binkowski & S. I. Doroshov (ed.) North American Sturgeons: Biology and Aquaculture Potential, Developments in Environmental Biology of Fishes 6, Dr W. Junk Publishers, The Hague.

Wooley, C. M. & E. J. Crateau. 1985. Movement, microhabitat, exploitation, and management of Gulf of Mexico sturgeon, Apalachicola River, Florida. N. Amer. J. Fish. Manag. 5: 590–605.

Environmental Biology of Fishes **48**: 347–358, 1997.

Atlantic and shortnose sturgeons of the Hudson River: common and divergent life history attributes

Mark B. Bain
New York Cooperative Fish and Wildlife Research Unit, Department of Natural Resources, Fernow Hall, Cornell University, Ithaca, NY 14853, U.S.A.

Received 10.1.1995 Accepted 1.5.1996

Key words: Acipenseridae, *Acipenser oxyrinchus, Acipenser brevirostrum*, distribution, habitat use, movements, size, maturity, age, growth

Synopsis

The Hudson River estuary supports substantial number of Atlantic sturgeon, *Acipenser oxyrinchus*, and shortnose sturgeon, *Acipenser brevirostrum*. Both species have complex life cycles that have been studied sporadically in the past 50 years. The life cycle of the shortnose sturgeon may be divided into four life intervals: non-spawning adults, spawning adults, eggs and larvae, and juveniles. The life cycle of the Atlantic sturgeon is reviewed in six intervals: non-spawning adults, female spawners, male spawners, eggs and larvae, early juveniles, and late juveniles. Both species are long-lived, mature at advanced age, have rapid and similar growth during the first few years of life, feed on generally similar taxa, use deep channel habitats for all life intervals, and have complex migratory patterns with distinct, seasonal, concentration areas. Atlantic and shortnose sturgeons differ, however, in ages and sizes at maturity, maximum size, timing and location of spawning, migratory behaviors, and management. Use of marine habitats and long-distance coastal migrations are restricted to Atlantic sturgeon, but some evidence indicates that large Atlantic sturgeon juveniles reside in riverine habitats along the Atlantic coast during warm months. Movements and habitat use by both sturgeons in the Hudson River estuary contrasts with the spatial segregation of the species reported in other river systems. Juvenile shortnose sturgeon and early juvenile Atlantic sturgeon have almost the same distributions in the Hudson River estuary during all seasons. During this period of co-occurrence, both species are very similar in size, grow at about the same rate, feed on similar foods, and share deep, channel habitats. Adult shortnose sturgeon distribution overlaps with the distribution of juvenile Atlantic sturgeon, and the latter commence river emigration at a size comparable to co-occurring adult shortnose sturgeon. Life history information on the Hudson River sturgeons substantiates the need to carefully conserve these species because of vulnerability to exploitation and habitat disruption.

Introduction

The Hudson River supports substantial populations of Atlantic sturgeon, *Acipenser oxyrinchus*, and shortnose sturgeon, *Acipenser brevirostrum*. The Atlantic sturgeon is one of North America's largest fishes, and an important commercial species in the Hudson River and Atlantic coast waters (species reviewed in Smith & Clugston 1997, this volume). In contrast, the shortnose sturgeon is the smallest species of *Acipenser* in North America, and a charter member (included in the original US

Endangered Species Act, 1973) of the U. S. endangered species list (species reviewed in Kynard 1997 this volume).

Observations of sturgeon in the Hudson River date back to the earliest historical accounts of human settlement in the region. Both species were very abundant, often captured for food, and easily observed by people during some part of the year. The first scientific accounts of the Hudson River sturgeons emerged from the New York State Biological Survey conducted in the mid-1930s (Curran & Ries[1], Greeley 1937[2], Townes[3]. These studies documented some basic life history attributes such as distribution in the river, sizes and ages of mature fish, and diet. Almost no additional information was collected on the Hudson River sturgeons for 40 years, but then in the 1970s major concerns emerged about the effect of electric generating stations on fish resources of the Hudson River (Barnthouse et al. 1984). William Dovel led extensive studies (Dovel & Berggren 1983, Dovel et al. 1992) which provide most of our current knowledge on the Hudson River sturgeons. Electric utilities that operate power plants along the Hudson River initiated comprehensive environmental monitoring programs that continue today. Some biologists (Carlson & Simpson 1987, Geoghegan et al. 1992, Hoff et al. 1988, Young et al. 1988) working with monitoring program samples and data provided relatively recent reports of sturgeon distributions and life history attributes.

Understanding the complex life cycles of the sturgeons has challenged biologists for more than 50 years. Until recently, the shortnose sturgeon was believed to be an anadromous fish, and therefore the responsibility for recovering this endangered species was assigned to the National Marine Fisheries Service[4]. Dadswell (1979) provided the first thorough study of the life history of shortnose sturgeon, and a comparably detailed analysis of the biology of any Atlantic sturgeon population has not been reported. Despite numerous and varied reports on the biology of both Hudson River sturgeons, life history reviews within the last 10 years have concluded that important life cycle attributes remain poorly known or unknown (Gilbert 1989, T.I.J. Smith 1985). In only one case (Saint John River and estuary, New Brunswick, Canada) has abundant populations of both species been studied (Dadswell 1979), and they were found to segregate on the basis of habitat, presumably to minimize competition.

In this paper, I review the knowledge of Atlantic and shortnose sturgeon in the Hudson River estuary by summarizing information from biologists going back to 1937. This summary is organized around distinct life intervals of each sturgeon in an effort to present a complete picture for both species. Finally, the potential interactions between the two species will be considered because the Hudson River has sizable populations of both species, and some evidence (Dadswell 1979, Dadswell et al.[5], Dovel et al. 1992) suggests that competition between them may influence habitat use.

Sturgeon are limited to the lower 246 km of the Hudson River (Figure 1) where habitats range from a typical freshwater river channel to a brackish water fjord (for physicochemical and morphological reviews see Coch & Bokuniewicz 1986 and others in the same volume, Limburg et al. 1989, Smith 1992). This estuary system is nearly straight and oriented in a north-south direction from the New York City harbor (southern tip of Manhattan Island; km 0 [km for river location in kilometers above mouth]) to the Troy Dam (Federal Green Island Dam; km 246) near Albany, New York. The normal tidal ampli-

[1] Curran, H.W. & D.T. Ries. 1937. Fisheries investigations in the lower Hudson River. pp. 124–145. *In:* A Biological Survey of the Lower Hudson Watershed, Supplement to the 26th Annual Report of the New York State Conservation Department, Albany.

[2] Greeley, J.R. 1937. Fishes of the area with annotated list. pp. 45–103. *In:* A Biological Survey of the Lower Hudson Watershed, Supplement to the 24th Annual Report of the New York State Conservation Department, Albany

[3] Townes, H.K., Jr. 1937. pp. 217–230. *In:* A Biological Survey of the Lower Hudson Watershed. Supplement to the 26th Annual Report of the New York State Conservation Department, Albany.

[4] U.S. Federal Register Vol. 39, No. 230, Pages 41367–41377; 27 November 1974.

[5] Dadswell, M.J., B.D. Taubert, T.S. Squires, D. Marcette & J. Buckley. 1984. Synopsis of biological data on the shortnose sturgeon, *Acipenser brevirostrum* LeSueur, 1818. NOAA Technical Report NMFS 14, National Marine Fisheries Service, Washington, D.C.

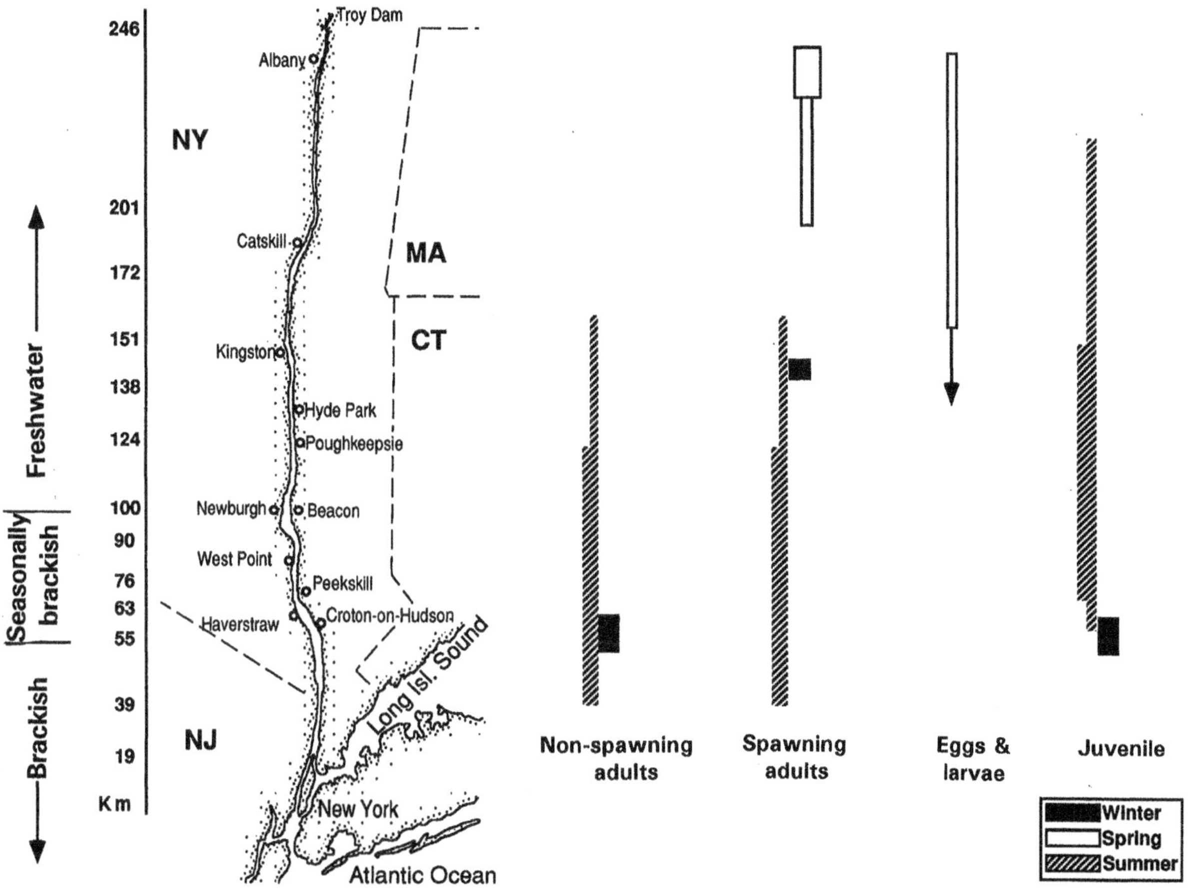

Figure 1. Life intervals and seasonal distribution of shortnose sturgeon in the Hudson River estuary relative to river features, river distances upstream of upper New York City bay, and salinity. Fall and sometimes spring distributions are not shown because these seasons are transitional periods. Width of the distribution lines and symbols indicates relative density of individuals.

tude ranges from 0.82 to 1.43 m causing a tidal volume (mean 5 670 - 8 500 $m^3\ s^{-1}$ depending on location) from 10 to 100 times river discharge (mean 623 $m^3\ s^{-1}$; Limburg et al. 1989). The Hudson River channel is large (mean width 1280 m) and generally deep (mean depth 10 m), but lacking any significant gradient. However, channel morphology varies with sections averaging as much as 5.5 km wide and 34 m deep (maximum depth = 53 m). Much of the river channel is shaped by rock with fine grain (e.g., sand and clay) sediments composing the substrate. The lower 100 km of the Hudson River estuary is saline (> 0.1 ppt salt) during seasons of low freshwater discharge with salinity generally below 10 ppt.

Shortnose sturgeon

Shortnose sturgeon is best described as an amphidromous (defined in McDowall 1987, see also Bemis & Kynard 1997 this volume) species since use of marine waters is limited to the estuaries of their natal rivers (see Kynard 1997 this volume). On one occasion, shortnose sturgeon were reported in waters of coastal New Jersey adjacent to the mouth of the Hudson River (Dovel et al. 1992). Within the Hudson River Estuary, shortnose sturgeon display complex migratory behavior that has been inconsistently described in past investigations. The life history for Hudson River shortnose sturgeon will be reviewed in four intervals that vary in characteristics

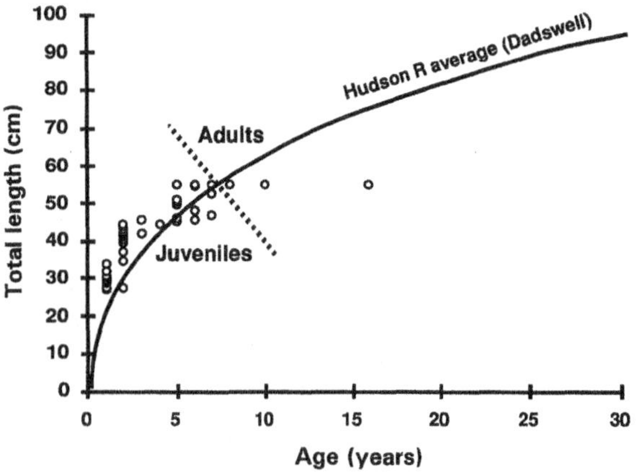

Figure 2. Shortnose sturgeon sizes and ages reported for the Hudson River from Dadswell et al.[5] using their compilation of unpublished data in modified form, and the total length of shortnose sturgeon aged by Dovel et al. (1992; open circles). The dashed line separates juvenile and adult life intervals at 55 cm total length or about 50 cm fork length.

(Figure 1; also see the species review by Kynard 1997 this volume).

Non-spawning adult interval

In many or all populations of shortnose sturgeon, adult fish do not spawn every year. Dadswell (1979) reported that females spawn every third to fifth year, and males every second year in the Saint John River, New Brunswick. This pattern may differ in the Hudson River because Dovel et al. (1992) reported the occurrence of tagged shortnose sturgeon at the spawning grounds in successive years. Non-spawning adults appear to use different habitats and display different migratory behavior than adults within a year of spawning.

The maximum sizes reported (Dadswell et al.[5]) for Hudson River shortnose sturgeon were a female weighing 7.2 kg (94.5 cm fork length [FL], 105 cm total length [TL]) and a male weighing 5.3 kg (89 cm FL, 99 cm TL). However, Dovel et al. (1992) documented an even larger but unsexed shortnose sturgeon from the Hudson River: 107 cm TL and 10.7 kg. The age record for shortnose sturgeon is 67 years with the oldest Hudson River specimen aged at 37 years (Dadswell et al.[5]). Most shortnose sturgeon captured in the Hudson River estuary in research and monitoring programs (1983-1988) were adults ranging in size from about 45 to 80 cm TL (Geoghegan et al. 1992) or about 8 to 20 years of age (Figure 2). Pooled across the sexes, maturity criteria that can be used for the Hudson population of shortnose sturgeon would be 50 cm FL (Table 1) and about 6 years of age (sexes pooled, Figure 2). The 50 cm FL criteria (55 cm TL) is useful for field handled

Table 1. Ages and sizes of the life intervals of shortnose and Atlantic sturgeons in the Hudson River. Data reported are generalized because of minor variations in specific values reported in other studies (see text for discussion of specific data).

Life interval	Age range (yr)	Fork length[a] (cm)	Total length[a] (cm)
Shortnose sturgeon			
Larva	< 0.08		≤ 2
Male juveniles	0.08–≥ 3	~ 2–50	2–55
Female juveniles	0.08–≥ 6	~ 2–50	2–55
Male adults	≥ 3	> 50	> 55
Female adults	≥ 6	> 50	> 55
Atlantic sturgeon			
Larva	< 0.08		≤ 3
Early juveniles	0.08–2	~ 2–44	~ 3–49
Intermediate juveniles	3–6	45–63	50–70
Late juveniles	6–11	> 63–134	> 70–149
Non-spawning adults	≥ 12	≥ 135	≥ 150
Female spawners	≥ 15	≥ 180	≥ 200
Male spawners	12–20	≥ 135–190	≥ 150–210

[a] Fork length and total length sizes were made to fit the conversion formulae reported by Dadswell et al.[5] for shortnose sturgeon: FL = 0.9 × TL; TL = 1.1 × FL.

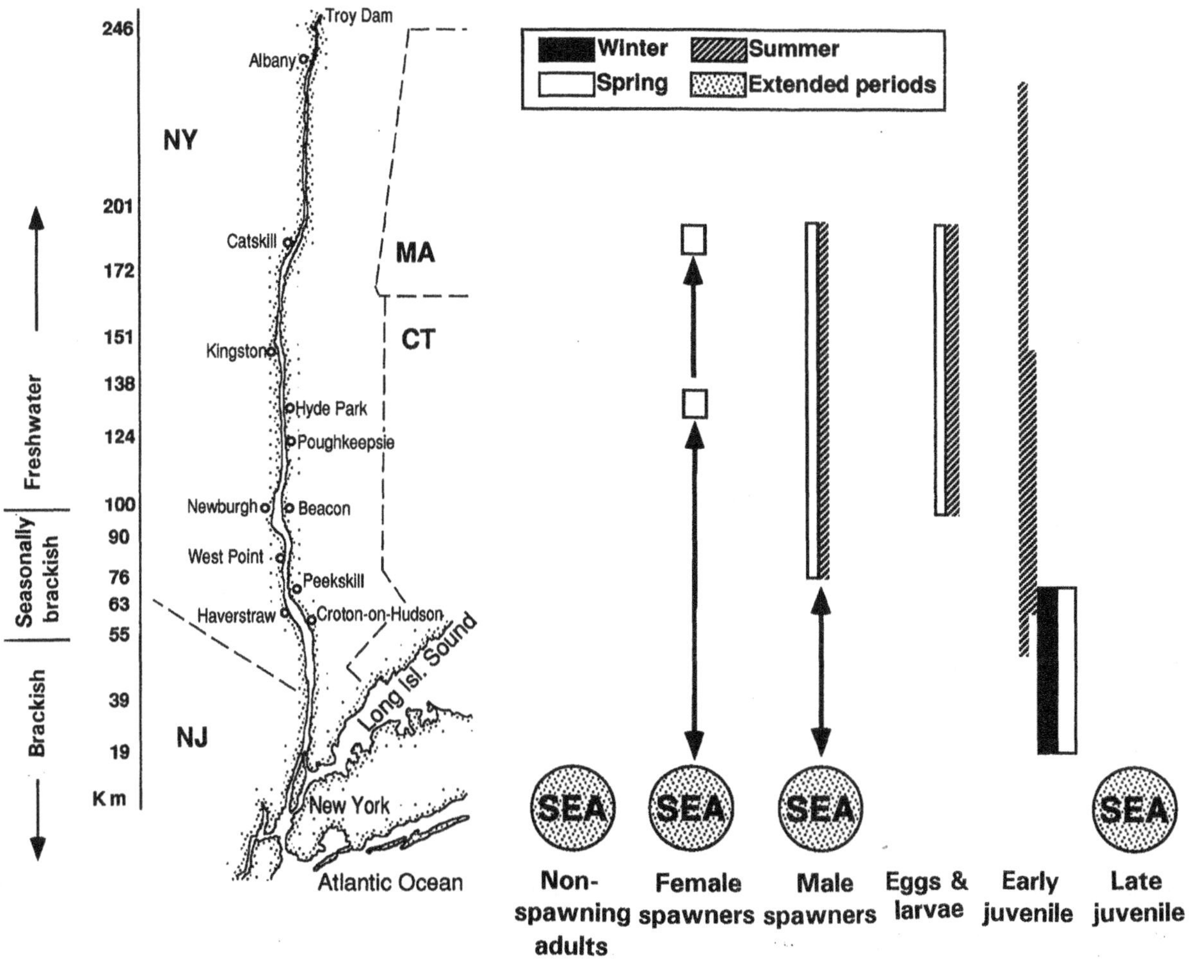

Figure 3. Life intervals and seasonal distribution of Atlantic sturgeon in the Hudson River estuary relative to river features, river distances upstream of upper New York City bay, and salinity. Fall distributions are not shown because this season is transitional. Width of the distribution lines and symbols indicates relative density of individuals. Sea distributions includes long-distance migrations to waters outside the Hudson River estuary.

fish because sex cannot be determined except at the time of spawning by observation of sperm or eggs.

From late spring through early fall, adult shortnose sturgeon are distributed in deep, channel habitats of the freshwater and brackish reaches of the Hudson River estuary. River monitoring (1969–1980) of fish distributions by the Hudson River electric utilities (Hoff et al. 1988) recorded adult shortnose sturgeon from a large portion of the estuary (Figure 1): most captures occurred between km 38 through 122, and no captures upstream of km 166. Later river monitoring (Geoghegan et al. 1992) showed a similar pattern. During this apparent growth and feeding period, the diet of shortnose sturgeon in the Hudson River likely includes insects and crustaceans with molluscs being a major component (25 to 50% of the diet; Curran & Ries[1], Townes[3]).

As water temperature declines in the late fall, adult shortnose sturgeon typically concentrate in a few overwintering areas. Dovel et al. (1992) concluded that most or all adults form an overwinter concentration near Kingston (approximately km 140). However, river monitoring in late fall indicates another concentration near Haverstraw (km 54–61). Life history studies for some shortnose stur-

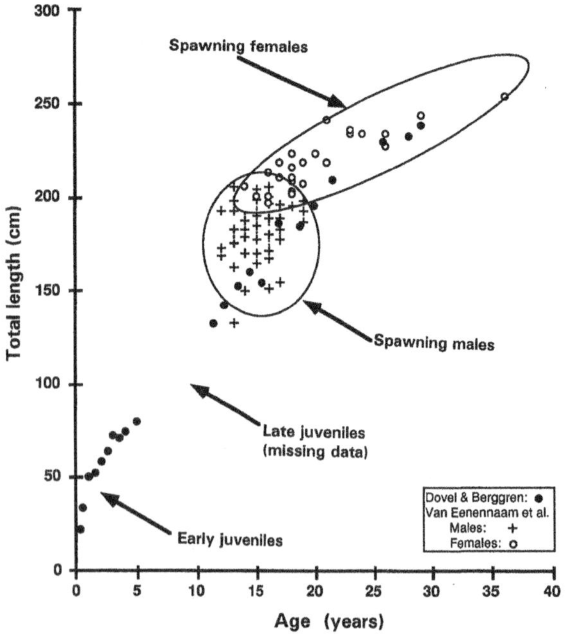

Figure 4. Individual Atlantic sturgeon sizes and ages reported for the Hudson River by Dovel & Berggren (1983; solid dots) and Van Eenennaam et al. (1996; open dots for females, + symbols for males). The circled points indicate individuals determined to be in spawning condition by Van Eenennaam et al. (1996). The arrow for late juveniles indicates a gap in the age and size series corresponding with an absence of fish from the Hudson River.

geon populations (Dadswell et al.[5]) and observations in the Hudson River (Geoghegan et al. 1992) indicate that non-spawning adults behave differently from adults entering reproductive condition. Adults that will not be in reproductive condition the following spring concentrate in brackish waters. In the Hudson, this overwintering area appears to be located between km 54 and 61 (Figure 1). In the spring, these fish migrate upstream and disperse through the tidal portion of the river.

Spawning adult interval

Shortnose sturgeon spawn once in spring, usually at a single location as far upriver as the population ranges. Pre-spawning adults overwinter in one large concentration widely separated from those adults that will not spawn the following spring. Females and males have the same migratory and habitat use behavior so I treat them as one life history interval (Figure 1).

Growth rates for shortnose sturgeon vary by region and sex but all fish mature at approximately the same size throughout their range: 45–55 cm FL (50–60 cm TL) for males and females (Dadswell et al.[5]). For the Hudson River population, Greeley[2] reported that males first spawn at 3 to 4 years of age (average 44.5 cm FL), and females first spawn at 6 to 8 years of age (average 51.5 cm FL, Table 1). However, Dadswell (1979) concluded from fin ray interannular increments that first spawning may follow maturation by 1 to 2 years in males and as much as 5 years in females. Therefore, Greeley[2] may have overestimated the age at maturity.

From late spring through early fall, all adult shortnose sturgeon have a dispersed distribution as described above for non-spawning adults. Adult shortnose sturgeon that will spawn the following spring congregate in an overwintering site near the spawning grounds. In the Hudson, a single large overwintering concentration of pre-spawning adults is well documented to form annually in deep, channel habitats a few kilometers downstream of Sturgeon Point (km 139). Many fish were readily captured at this site by Dovel et al. (1992), and it was known as a productive fishing area prior to protection of the species. From information on other populations (Dadswell 1979), females at the overwintering site may not feed prior to spawning, but males do feed during this period. Food items are probably similar to those reported above for non-spawning adults. In mid-April, adult fish move upstream to the spawning grounds extending from below the Federal Dam at Troy to about Coxsackie (km 239–190; Dovel et al. 1992, Hoff et al. 1988). Spawning occurs from late-April to early May. Afterward, the adults disperse downriver into the summer range.

Egg, embryo and larva interval

Eggs of shortnose sturgeon adhere to solid objects on the river bottom, and newly hatched embryos remain on the bottom (Buckley & Kynard 1981, Taubert 1980). Hatching size ranges from 7 to 11 mm TL

(Buckley & Kynard 1981, Taubert 1980), with Hudson River embryos ranging in size from 15 to 18 mm TL at 10 to 15 days of age (Pekovitch[6]). After hatching, embryos gradually disperse downstream over much of the Hudson River estuary (Hoff et al. 1988). Shortnose sturgeon larvae captured in the Hudson River were associated with deep waters and strong currents (Pekovitch[6], Hoff et al. 1988). At 20 mm TL, shortnose sturgeon in the Hudson River had fully developed external characteristics indicating a transition to the juvenile interval (Pekovitch[6]; Table 1). No further information is available on this interval of the shortnose sturgeon life cycle.

Juvenile interval

Juvenile shortnose sturgeon (2–55 cm TL; Table 1), use a large portion of the tidal reach of the Hudson River. Dovel et al. (1992) indicated that yearling juvenile sturgeon grow rapidly (to 30 cm TL in first year, Figure 2) and disperse downriver to about km 55 by fall. Juveniles have been captured in the same deep channel habitats used by adults. During midsummer, the juvenile distribution centers on the mid-river region (Geoghegan et al. 1992). By late fall and early winter, most juveniles occupy the broad region of the Hudson River near Haverstraw (km 55–63; Dovel et al. 1992, Geoghegan et al. 1992). However, there is no evidence that juveniles move out of the lower river into coastal marine waters.

Juvenile shortnose sturgeon feed on smaller and somewhat different organisms than do adults (Carlson & Simpson 1987). Common prey items are aquatic insects (chironomids), isopods, and amphipods. Unlike adults, molluscs do not appear to be an important part of their diet (Dadswell 1979).

[6] Pekovitch, A.W. 1979. Distribution and some life history aspects of the shortnose sturgeon (*Acipenser brevirostrum*). Hazleton Environmental Sciences Corp., Northbrook. 23 pp.

Atlantic sturgeon

Atlantic sturgeon are anadromous. Spawning occurs in freshwater, but male and female fish reside for many years in marine waters. Atlantic sturgeon undertake long-distance migrations along the Atlantic coast. Atlantic sturgeon marked in the Hudson River by Dovel & Berggren (1983) were recaptured in marine waters and river mouths from just south of Cape Hatteras, North Carolina to just north of Cape Cod, Massachusetts. In addition to these marine movements, Atlantic sturgeon display complex migratory behavior within the Hudson River. Here, I review the life cycle for Atlantic sturgeon in the Hudson River in six intervals that vary by habitat, migratory behavior, and size (Figure 3). Also see Smith & Clugston (1997 this volume) for a general review of Atlantic sturgeon life history and fishery.

Non-spawning adult interval

The inter-spawning period for Atlantic sturgeon is thought to range from 3 to 5 years depending on sex (discussed below). During non-spawning years, adults use marine waters either all year or seasonally. Little is known about their behavior in marine waters except that adult-size fish (≥ 150 cm TL, Table 1) marked in the Hudson River have been recaptured in coastal waters and river mouths from North Carolina to Massachusetts. The largest commercial harvest of adult Atlantic sturgeon from the Hudson River population occurs in marine waters throughout the New York Bight (Waldman et al. 1996). Female Atlantic sturgeon apparently grow in marine waters, whereas males appear to grow little after maturity (Figure 4). In marine habitats, Atlantic sturgeon eat amphipods, isopods, shrimps, molluscs, and fish (Scott & Crossman 1973).

The maximum age for the species is 30 years (Scott & Crossman 1973) with a similar estimate for the Hudson River (T.I.J. Smith 1985). The largest known Atlantic sturgeon was a female 427 cm TL, and 368 kg (Saint John River , New Brunswick; Van

Den Avyle[7]). Large Atlantic sturgeon are likely to be females because of marked sexual dimorphism (Figure 4).

Female spawning interval

Adult female Atlantic sturgeon differ sharply from adult males in size, growth, migratory behavior, and age structure (Figure 3). Spawning female sturgeon are age 15 or older, weigh more than 34 kg, and are greater than 200 cm TL (Van Eenennaam et al. 1996, Table 1). Dovel & Berggren (1983) reported a slightly older age at first spawning (18 years) but the same minimum size. Age and growth data (Van Eenennaam et al. 1996) clearly indicate steady growth in females (Figure 4), and data from Dovel & Berggren (1983) are consistent with this pattern.

Adult females enter the Hudson River Estuary for spawning beginning in mid-May. They migrate directly to the spawning grounds which are deep, channel or off-channel habitats (Dovel & Berggren 1983). The female sturgeon return to marine waters quickly after spawning (C.L. Smith 1985). The spawning period ranges from May through July or possibly August in the Hudson River estuary (Dovel & Berggren 1983, Van Eenennaam et al. 1996). Female sturgeon do not appear to feed on the spawning run in freshwater (T. I. J. Smith 1985).

Dovel & Berggren (1983) report that spawning occurs near the salt wedge (km 55) early in the season (late May), moving upstream to km 136 during June and early July. However, Van Eenennaam et al. (1996) collected spawning Atlantic sturgeon only at two historically important fishing sites known to be spawning areas (Figure 3): near Hyde Park (km 130) and Catskill (km 182). Van Eenennaam et al. (1996) argue that spawning is unlikely to occur near brackish water because sturgeon eggs, embryos and larvae are intolerant of saline conditions, and some significant length of river habitat is needed downstream of a spawning site to accommodate dispersal of embryos and larvae.

[7] Van Den Avyle, M.J. 1984. Species profiles: life histories and environmental requirements of coastal fishes and invertebrates (South Atlantic) – Atlantic sturgeon. U.S. Fish and Wildlife Service FWS/OBS-82/11.25, Washington, D.C. 17 pp.

Male spawning interval

Mature, male Atlantic sturgeon enter the Hudson River starting in April and at least some remain in the Hudson River as late as November (Dovel & Berggren 1983). Spawning males are 12 or more years old and from 150 to 210 cm TL (Van Eenennaam et al. 1996, Table 1). Van Den Avyle[7] reported that the maximum size for males is 213 cm TL which is similar to the sizes recorded in the Hudson River spawning stock (Figure 4). No spawning males over 20 years old have been recorded in the Hudson River. Male Atlantic sturgeon may not spawn annually, and the period between spawnings has been estimated to range from 1 to 5 years (T.I.J. Smith 1985).

From limited sturgeon telemetry by Dovel & Berggren (1983), males appear to move upstream on incoming tides and then remain stationary for several hours. During their upstream migration, male sturgeon meander back and forth across the channel, but stay in water greater than 7.6 m deep. Van Eenennaam et al. (1996) observed that adult male sturgeon appear at spawning sites in association with females, indicating that they search for females while moving about in the river.

Egg, embryo and larva interval

Eggs of Atlantic sturgeon are adhesive and the embryos remain on the bottom in deep channel habitats. Atlantic sturgeon embryos have been recorded in the Hudson River from km 60 through 148 (Dovel & Berggren 1983); a range including some brackish waters. Sturgeon embryos and larvae have limited salt tolerance, so their habitat must be well upstream of the salt front (Van Eenennaam et al. 1996; as illustrated in Figure 3). No further information is available on this interval of the Atlantic sturgeon life cycle.

Atlantic sturgeon embryos are about 7 mm TL at hatching, and in hatcheries, they reached 19.9 mm TL in 20 days (Smith et al. 1980). The transition from larva to juveniles appears to occur at about 30 mm TL (Table 1) based on Hudson River specimens (Bath et al. 1981).

Juvenile riverine interval

The juvenile period of the Atlantic sturgeon life cycle is marked by major ecological changes, and it can be divided into two life history intervals: early and late juvenile (Figure 3). The precise division between these intervals is unclear because changes are gradual, although growth is very rapid (Figure 4). Consequently, I added a third intermediate interval for age and growth statistics shown in Table 1. The first juvenile interval is limited to riverine habitats. Relatively good information is available for this interval due to research in the Hudson River estuary.

Juvenile Atlantic sturgeon are well distributed over much of the Hudson River from July through September, and they use deep channel habitats as in other life intervals (Figure 3). The largest numbers of juveniles appear to be located from km 63 to 140 (Dovel & Berggren 1983). As water temperature drops below 20° C in the fall, juveniles form an overwintering distribution in brackish water between km 19 to 74 (Dovel & Berggren 1983). From October through June, this region of the Hudson River contains many juveniles and they appear to move little during the period. Upstream dispersion of juveniles begins in late spring. Some juvenile Atlantic sturgeon have been recorded in the overwintering area used by pre-spawning, adult shortnose sturgeon (Esopus Meadows, km 134) as early as mid-April which indicates some variation in the general migration pattern.

Juvenile Atlantic sturgeon grow quickly in the first three years of life (70 cm TL at age 3, Figure 3) but growth slows considerably if they remain in the Hudson River estuary (Dovel & Berggren 1983). Riverine juveniles feed on aquatic insects, amphipods, isopods, and small molluscs (Scott & Crossman 1973).

Juvenile marine interval

After 2 to 6 years of residence in the Hudson River, juvenile Atlantic sturgeon migrate to marine waters. Dovel & Berggren (1983) reported that some males leave the river in year 2, while females may stay in the river until year 5 or 6. This migration to marine waters marks a major change in ecology, behavior, and growth for Atlantic sturgeon. Table 1 shows approximate ages and sizes for early (riverine) juveniles, late (sea migrant) juveniles, and intermediate juveniles because the later includes the group that gradually emigrates from the river during a period of rapid growth. After about 10 years at sea, juvenile sturgeon reach adult size (about 150 cm TL, Table 1 for sexes pooled).

Little is known about Atlantic sturgeon in marine waters except that large juveniles are often captured in Long Island Sound and off the Long Island and New Jersey coasts in commercial fishing gear. Reviews of Atlantic sturgeon life history (e.g., Van Den Avyle[7]) and information specific to the Hudson River estuary (Dovel & Berggren 1983, C.L. Smith 1985) describe post-emmigration juveniles as inhabitants of marine waters. However, large juveniles (50–150 cm TL) may reside in riverine habitats along the Atlantic coast during warm months. Atlantic sturgeon sampling in the Hudson River has documented the occurrence of large juveniles (sometimes called pre-adults; Dovel & Berggren 1983, Van Eenennaam et al. 1996). Data of Dovel & Berggren (1983) on tag recaptures show that most fish were reported from river mouths and the lower sections of coastal rivers from Cape Cod to Chesapeake Bay. Murawski & Pacheco[8] described a similar pattern for tagging and recaptures in the St. Lawrence River, Quebec. Late juvenile Atlantic sturgeon often enter and reside in rivers that lack active spawning sites (e.g. Merrimack River, Massachusetts; Kieffer & Kynard 1993). Most Atlantic sturgeon in rivers of the central US Atlantic coast are probably from the Hudson River population

[8] Murawski, S.A. & A.L. Pacheco. 1977. Biological and fisheries data on Atlantic sturgeon, *Acipenser oxyrhynchus* (Mitchell). National Marine Fisheries Service Technical Series 10, Highlands. 69 pp.

(Waldman et al. 1996). Consequently, late juvenile Atlantic sturgeon from the Hudson River may annually use other riverine habitats during warm months before returning to the Hudson for spawning.

Discussion

Sturgeon (family Acipenseridae) are the modern descendants of the original ray-finned fish that achieved greatest abundance and diversity 280 to 345 million years ago. Atlantic, shortnose, and all other sturgeon retain many ancestral body characteristics and ways of living that distinguish them as relict fishes (see Bemis et al. 1997 this volume). Among North American fishes, sturgeons exhibit a unique combination of life history attributes: advanced age and large size at maturity, eggs that are numerous and small in relation to body size, and spawning that is episodic and seasonal (Winemiller & Rose 1992). Beyond being unique, these characteristics make sturgeon especially vulnerable to population collapse due to overfishing (Boreman 1997 this volume). Life history information on the Hudson River sturgeons fits these generalizations and it substantiates the need to carefully conserve these species. In addition, life history details such as seasonal areas of concentration, migration times and routes, and specific spawning locations highlight the vulnerability of both shortnose and Atlantic sturgeon to easy exploitation and habitat disruption. Fortunately, in the case of the Hudson River estuary, key habitats for spawning, rearing, and overwintering are intact and suitable for the species. Also, both species of sturgeon are managed through either endangered species protection (shortnose sturgeon; US Endangered Species Act) or fishery restrictions (Smith & Clugston 1997 this volume), even though the latter may not be adequate to sustain the current population (Young et al. 1988, Boreman 1997 this volume).

The two sturgeons in the Hudson River share many common life history attributes. Both are long-lived and mature at advanced age compared to almost all other fishes in the Hudson River. Both species have rapid and similar growth rates during the first few years of life. In general, sturgeon are characterized as indiscriminate bottom-feeding carnivores, and specific information on diet indicates they feed on generally the same food items in the Hudson River. Both sturgeon have complex migratory patterns in the Hudson River with distinct, seasonal, and predictable concentration areas. Finally, both sturgeons primarily use deep channel habitats for all life intervals.

Despite many similarities in life history, Atlantic and shortnose sturgeons differ in some obvious ways. Adult sizes are greatly different, and the sizes and ages at maturity diverge. The timing and location of spawning is so different that it appears impossible that the two species behaviorally interact during this key life interval. Use of marine habitats and long-distance coastal migrations are restricted to Atlantic sturgeon. With respect to management, one species is heavily exploited while the other is fully protected under the US Endangered Species Act.

Widespread occurrence of Atlantic and shortnose sturgeons in many Atlantic Coast rivers of North America raises questions as to how two species can co-exist with so many shared life history attributes. The prevailing view (e.g., Dadswell et al.[5], Dovel et al. 1992, Kieffer & Kynard 1993) has been that the two species are spatially segregated in rivers in association with salinity; with shortnose sturgeon oriented to freshwater, and Atlantic sturgeon concentrated in brackish water except at spawning and very early life. However, a review of the movements and habitat use of both species in the Hudson River estuary conflicts with these interpretations.

Juvenile shortnose sturgeon and early juvenile Atlantic sturgeon have virtually identical distributions in the Hudson River estuary during all seasons. During this period of co-occurrence, both species are very similar in size, grow at about the same rate, feed on similar foods, and share deep channel habitats. Furthermore, the distribution of adult shortnose sturgeon overlaps with that of juvenile Atlantic sturgeon. Interestingly, the period of river emigration of juvenile Atlantic sturgeon closely corresponds with the age (intermediate juveniles in Table 1) when they reach a size (ca. 55 cm TL) equal to the minimum adult size of shortnose sturgeon.

The protracted period of Atlantic sturgeon emigration (4 years) indicates that the two species overlap considerably in space, food, and habitat. Also, the pattern of emigration in conjunction with comparability in size and habits between the species suggests that co-exploitation of space and food resources may be important in the migratory behavior of juvenile Atlantic sturgeon.

The apparently extensive co-occurrence of the two sturgeons in the Hudson River estuary has not been clearly identified in previous investigations on the Hudson River. Although sturgeon biologists working on the Hudson River undoubtedly captured both species simultaneously in their work, analyses and reports have always been oriented to a single-species. This review is the first to simultaneously report details of the life history of both sturgeons in the Hudson River. The conclusion that the two species are not spatially segregated for large parts of their life histories indicates that the Hudson River estuary may be unique within the joint ranges of the two species.

Acknowledgements

I thank William E. Bemis for his strong encouragement and confidence that motivated the writing of this paper. I also thank Vadim J. Birstein, Robert H. Boyle, and John Waldman for organizing the International Conference on Sturgeon Biodiversity and Conservation, where a first version of this paper was presented. William E. Bemis, Vadim J. Birstein, and John Waldman deserve much credit for working with the authors and papers included in this volume. This work is a contribution from research sponsored by the Hudson River Foundation with supplemental support provided by the U. S. Army Corps of Engineers. Jeremy Knight, John Waldman, Andy Kahnle, Steve Nack, and Nancy Haley reviewed an early version of the manuscript, and provided constructive comments and suggestions for improvement.

References cited

Barnthouse, L.W., J. Boreman, S.W. Christensen, C.P. Goodyear, W. Van Winkle & D.S. Vaughan. 1984. Population biology in the courtroom: the Hudson River controversy. BioScience 34: 14–19.

Bath, D.W., J.M. O'Connor, J.B. Alber & L.G. Arvidson. 1981. Development and identification of larval Atlantic sturgeon (*Acipenser oxyrhynchus*) and shortnose sturgeon (*A. brevirostrum*) from the Hudson River estuary. Copeia 1981: 711–717.

Bemis, W.E. & B. Kynard. 1997. Sturgeon rivers: an introduction to acipenseriform biogeography and life history. Env. Biol. Fish. (this volume).

Bemis, W.E., E.K. Findeis & L. Grande. 1997. An overview of Acipenseriformes. Env. Biol. Fish. (this volume).

Boreman, J. 1997. Sensitivity of North American sturgeons and paddlefish to fishing mortality. Env. Biol. Fish. (this volume).

Buckley, J. & B. Kynard. 1981. Spawning and rearing of shortnose sturgeon from the Connecticut River. Prog. Fish-Cult. 43: 74–76.

Carlson, D.M. & K.W. Simpson. 1987. Gut contents of juvenile shortnose sturgeon in the upper Hudson estuary. Copeia 1987: 796–802.

Coch, N.K. & H.J. Bokuniewicz. 1986. Oceanographic and geologic framework of the Hudson system. Northeastern Geol. 8: 96–108.

Dadswell, M.J. 1979. Biology and population characteristics of the shortnose sturgeon, *Acipenser brevirostrum* LeSueur, 1818 (Osteichthyes: Acipenseridae), in the Saint John River estuary, New Brunswick, Canada. Can. J. Zool. 57: 2186–2210.

Dovel, W.L. & T.J. Berggren. 1983. Atlantic sturgeon of the Hudson Estuary, New York. New York Fish and Game J. 30: 140-172.

Dovel, W.L., A.W. Pekovitch & T.J. Berggren. 1992. Biology of the shortnose sturgeon (*Acipenser brevirostrum* Leseur, 1818) in the Hudson River estuary, New York. pp. 187-216. *In:* C. L. Smith (ed.) Estuarine Research in the 1980s, State Univ. New York Press, Albany.

Geoghegan, P., M.T. Mattson & R.G. Keppel. 1992. Distribution of the shortnose sturgeon in the Hudson River estuary, 1984-1988. pp. 217-277. *In*: C. L. Smith (ed.) Estuarine Research in the 1980s, State Univ. New York Press, Albany.

Hoff, T.B., R.J. Klauda & J.R. Young. 1988. Contribution to the biology of shortnose sturgeon in the Hudson River estuary. pp. 171-189. *In:* C.L. Smith (ed.) Fisheries Research in the Hudson River, State Univ. New York Press, Albany.

Kieffer, M.C. & B. Kynard. 1993. Annual movements of shortnose and Atlantic sturgeons in the Merrimack River, Massachusetts. Trans. Amer. Fish. Soc. 122: 1088–1103.

Kynard, B. 1997. Life history, latitudinal patterns, and status of the shortnose sturgeon, *Acipenser brevirostrum*. Env. Biol. Fish. (this volume).

Limburg, K.E., S.A. Levin & R.E. Brandt. 1989. Perspectives on management of the Hudson River ecosystem. pp. 265-291. *In:*

D.P. Dodge (ed.) Proceedings of the International Large River Symposium, Can. Spec. Publ. Fish. Aquat. Sci. 106.

McDowall, R.M. 1987. The occurrence and distribution of diadromy among fishes. Amer. Fish. Soc. Symp. 1: 1–13.

Scott, W.B. & E.J. Crossman. 1973. Freshwater fishes of Canada. Fish. Res. Board Can. Bull. 184. 966 pp.

Smith, C.L. 1985. The inland fishes of New York State. The New York State Department of Environmental Conservation, Albany. 522 pp.

Smith, C.L. 1992. Estuarine research in the 1980s. State Univ. New York Press, Albany.

Smith, T.I.J. 1985. The fishery, biology, and management of Atlantic sturgeon, *Acipenser oxyrhynchus*, in North America. Env. Biol. Fish. 14: 61–72.

Smith, T.I.J. & J.P. Clugston. 1997. Status and management of Atlantic sturgeon, *Acipenser oxyrinchus*, in North America. Env. Biol. Fish. (this volume).

Smith, T.I.J., E.K. Dingley & D.E. Marchette. 1980. Induced spawning and culture of the Atlantic sturgeon, *Acipenser oxyrhynchus* (Mitchell). Prog. Fish-Cult. 42: 147–151.

Taubert, B., 1980. Reproduction of shortnose sturgeon (*Acipenser brevirostrum*) in Holyoke Pool, Connecticut River, Massachusetts. Copeia 1980: 114–117

Young, J.R., T.B. Hoff, W.P. Dey & J.G. Hoff. 1988. Management recommendations for a Hudson River Atlantic sturgeon fishery based on an age-structured population model. pp. 353–365. *In:* C.L. Smith (ed.) Fisheries Research in the Hudson River, State Univ. New York Press, Albany.

Van Eenennaam, J.P., S.I. Doroshov, G.P. Moberg, J.G. Watson, D.S. Moore & J. Linares. 1996 . Reproductive conditions of the Atlantic sturgeon (*Acipenser oxyrhynchus*) in the Hudson River. Estuaries (in press).

Waldman, J.R., J. Hart & I. Wirgin. 1996. Stock composition of the New York Bight Atlantic sturgeon fishery, based on analysis of mitochondrial DNA. Trans. Amer. Fish. Soc. 125: 364–371.

Winemiller, K.O & K.A. Rose. 1992. Patterns of life-history diversification in North American fishes: implications for population regulation. Can. J. Fish. Aquat. Sci. 49: 2196–2218.

Environmental Biology of Fishes **48**: 359–370, 1997.

Biological characteristics of European Atlantic sturgeon, *Acipenser sturio*, as the basis for a restoration program in France

Patrick Williot, Eric Rochard, Gérard Castelnaud, Thierry Rouault, Rémy Brun, Mario Lepage & Pierre Elie
Cemagref, Division Aquaculture et Pêche, BP 3, 33612 Cestas, France

Received 11.10.1994 Accepted 8.4.1996

Key words: population structure, artificial reproduction, wild juvenile farming, endangered species

Synopsis

The European Atlantic Sturgeon, *Acipenser sturio*, has received increased attention in France because of population declines due to overfishing and deterioration of spawning grounds. Conservation of this species requires many actions, including publicizing the necessity to protect this fish and its habitats, investigations on catches and probable spawning grounds, and on artificial reproduction, which is still in an experimental stage. During its sea life, European Atlantic sturgeon occur from the Bay of Biscay to the Bristol Channel and North Sea. Presently, the number of young fish in the Gironde Estuary during summertime is low and the population has a unimodal age structure. At the time this paper was written, the last recorded reproduction of sturgeon in the Gironde system occurred in 1988 (new evidence of reproduction was discovered in 1995). Growth of young fish from the 1988 cohort was faster than that previously determined for others cohorts. Over our study period (1980–1994), the availability of wild broodfish declined. Successful artificial reproduction of wild-caught females requires an optimal physiological state; any delay decreases their reproductive potential. Acclimatization of wild-caught juveniles to fresh water was most successful when fish were transported and held upon arrival in low salinity water. For such wild-caught juveniles, the first food intake usually occurs several months after capture, but remains irregular. These findings will be used to improve ongoing efforts to restore *A. sturio*.

Introduction

The European Atlantic sturgeon, *Acipenser sturio*, is a species distinct from the North American Atlantic sturgeon, *A. oxyrinchus*, although many biological features are similar (see Birstein & Bemis 1997 this volume). Since the middle of the 19th century, populations of *A. sturio* have been declining, the decline occurring faster in the northern part of their overall distribution area (Roule 1922, Berg 1948, Magnin 1959b, Kinzelbach 1987, Holčík et al. 1989, Maitland & Lyle 1990, Debus 1995). In the middle of the twentieth century, European Atlantic sturgeon existed along the southwest coast of France (Magnin 1962), Portugal and Spain (Classen 1944, Gutierrez Rodriguez 1962), in the Adriatic Sea (Holčík et al. 1989) and in the Black Sea (Antipa 1934, Ninua 1976). Populations have since declined along the Iberian Peninsula (Almaça 1988, Elvira et al. 1991, Elvira & Almodovar 1993), France (Castelnaud et al. 1991), and the Romanian course of the Danube River (Bacalbaça-Dobrovici 1991, 1997 this volume). At present, we know of only two relict populations: one inhabits the Gironde and its tributaries, the Garonne and Dordogne rivers in France; the other occurs in the Rioni River basin (Black Sea) in Georgia (former USSR).

In France, *A. sturio* is considered endangered

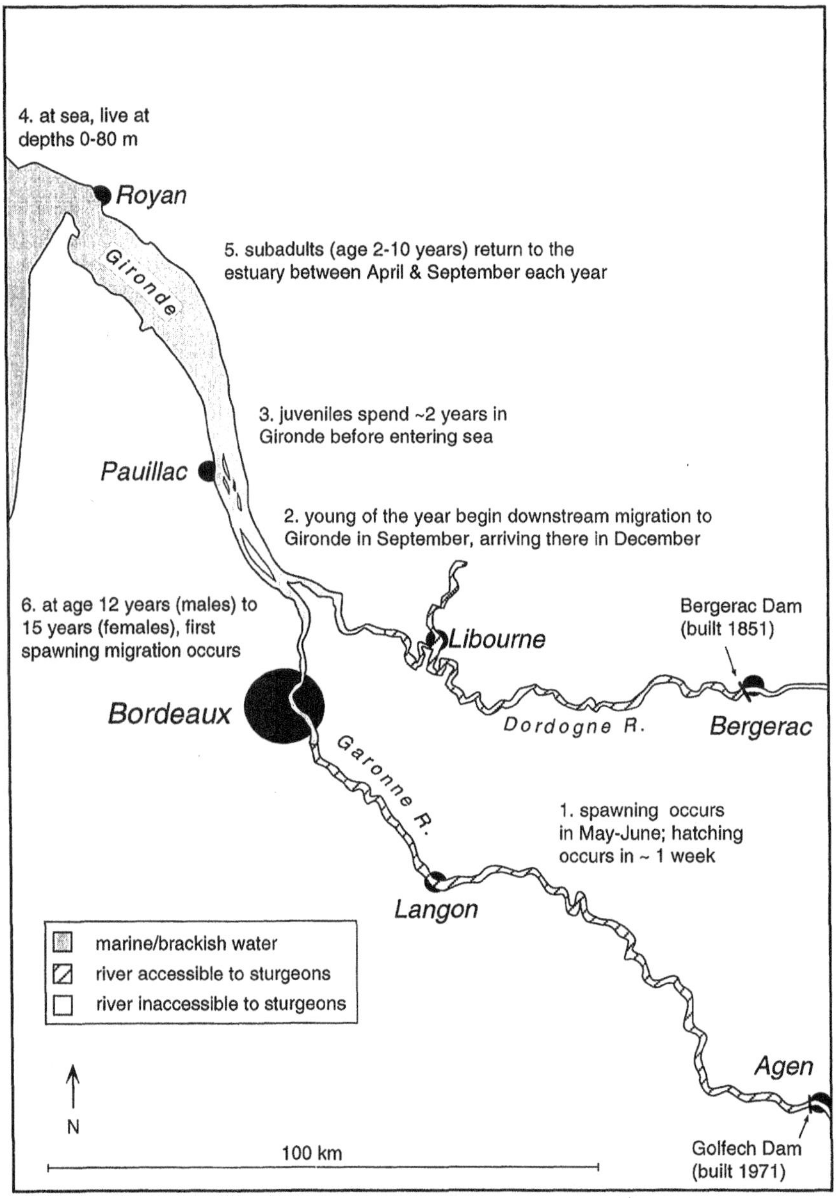

Figure 1. Life history of *A. sturio* in the Gironde Estuary system. Major points in the life history are numbered. Modified from Anonymous (1995).

(Lepage & Rochard 1995). In the past, it was fished for its flesh (Benecke 1986), mainly in the southwest region of the country (Laporte 1853, Roule 1922). Sturgeon were exploited in the Garonne and Dordogne rivers, in the Gironde Estuary, and adjacent continental shelf waters (Letaconnoux 1961). Caviar fishing occurred from 1920 to 1970 (Castelnaud et al. 1991). Demand for caviar increased fishing pressure on adults while the juveniles continued to be exploited for flesh. Increased fishing effort, combined with the impact of dams and gravel extraction which limit the reproduction by decreasing the availability of spawning grounds (Trouvery et al. 1984, Rochard et al. 1990), resulted in decline of the species.

By the early 1970s, *A. sturio* was no longer an economically important resource and it became clear that its survival would require a long term program encompassing protection of both the species and its habitats, improvement of biological and ecological knowledge, and probably, restocking. It was shown in the former USSR (Barannikova 1987, Khodorevskaya et al. 1997 this volume) that restocking not only improved sturgeon catches in the Caspian and Azov seas, but also saved the beluga, *Huso huso*, population in the Caspian Sea from extinction. Most individuals of *H. huso* in the Caspian Sea are now of hatchery origin.

We started our investigations on the wild *A. sturio* in the Gironde system during the late 1970s. Basic life history information on this population is summarized in Figure 1. In the early 1980s, we performed studies of artificial reproduction on Siberian sturgeon, *A. baerii*, using it as a biological model to avoid additional impact on indigenous populations of *A. sturio*. The availability during the 1980s of Siberian sturgeon also allowed us to initiate studies on nutrition and on methods of early sex determination.

This paper presents information on *A. sturio* including the status of wild populations, characteristics of available broodfish, results of experiments on artificial breeding and rearing of larvae, and adaptation of juveniles to farm conditions. We also present the main directions of our proposed restoration program.

Materials and methods

Capture-mark-recapture

Our tagging program focusing on large juveniles, which enter the lower Gironde Estuary for summer feeding (Magnin 1962), started in 1981. Fishing operations are conducted with the collaboration of local fishermen using drifting trammel nets. Fishermen are compensated on the basis of the mean expected loss of income. The fish are measured, weighed, and tagged (Petersen disc) to delineate the distribution and migration pattern of this population (Trouvery et al. 1984). In 1985, we modified tagging experiments in order to assess (using the Jolly-Seber method) the size of the stock which enters the Gironde Estuary in summer (Rochard 1992). We used a specially designed tie-on tag made of a stainless steel wire with a vinyl tube marker (Hallprint Pty. Ltd.) placed through the muscle anterior to the dorsal fin (Castelnaud 1988). Trawling has been performed since 1986 to enhance our sampling effort (Castelnaud et al. 1991).

Age determination

We modified (Rochard & Jatteau 1991) the aging method for sturgeon (Classen 1944, Cuerrier 1951, Magnin 1959a) to avoid additional mortality on this endangered species as already reported by Kohlhorst (1979). Only a small piece of the first ray of a pectoral fin was used (Cochnauer et al. 1985). We used the Walford graph and the Von Bertalanffy growth function (Rochard & Jatteau 1991) to establish seasonal age-growth relationship (Rochard 1992).

Location of spawning grounds

We established an inventory of potential spawning grounds in the lower part of the Garonne and Dordogne rivers (Trouvery 1980) using echo sounding equipment and information obtained from experienced fishermen.

Artificial reproduction

Each year before the expected spawning migration, we asked fishermen (via the local newspaper) to contact Cemagref in case of accidental catches of *A. sturio* during the lamprey and shad fishing season. The only way we can obtain wild broodfish is to pay a high price for them (despite the fact that sturgeon fishing is illegal). Fish were transported carefully to the hatchery for measurement (total length, TL and body weight, W) and immediate determination of sex, as most do not exhibit external characteristics of gender. Abdominal massage may lead to sperm

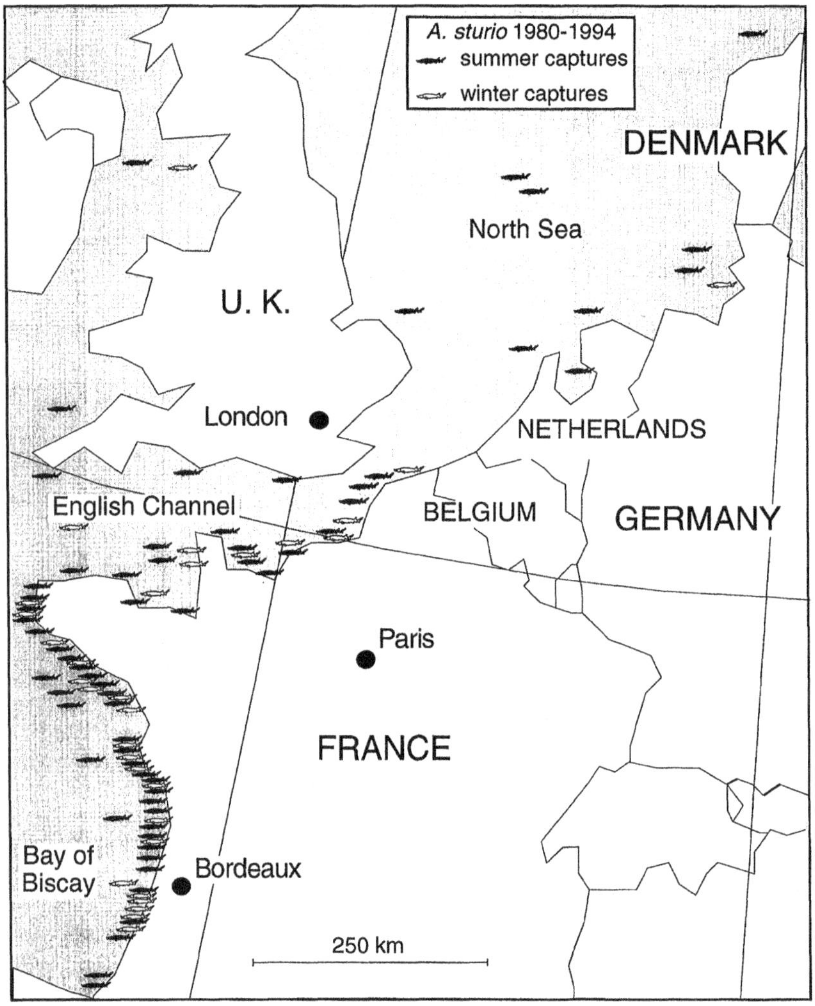

Figure 2. Map of the distribution of captures of *A. sturio* in the eastern Atlantic and North Sea.

emission from males. In the other cases, we checked the gonads by making a small abdominal cut. We determined the reproductive potential of females by examining three characteristics of the eggs: their mean largest size (n = 15); their polarization index (PI), defined as the ratio of the distance separating the germinal vesicle from the animal pole to the largest dimension of the egg (Kazanskii et al. 1978); and the in vitro maturation competence of eggs (observation of germinal vesicle break down: GVBD; n = 30), according to the method described for *A. baerii* (Williot et al. 1991) and also used on white sturgeon *A. transmontanus* (Lutes et al. 1987). We characterized males as immature (do not yield any sperm, even by stripping), running (sperm is running naturally), and non-motile sperm (running but spermatozoa do not move in water).

To induce spawning, we injected fish with carp hypophysis powder at a rate of 5 mg kg^{-1} and 2 mg kg^{-1} of body weight for the females and males, respectively. Sperm was removed from the males by massage and ova were collected by laparotomy (approximately 5 cm long) near and above the genital opening. The abdominal cut was sewn up with four cross stitches. Processes used for insemination, neutralization of the sticky egg envelope, and incubation methods, were the same as those used for *A. baerii* (Williot et al. 1991). During all of the above

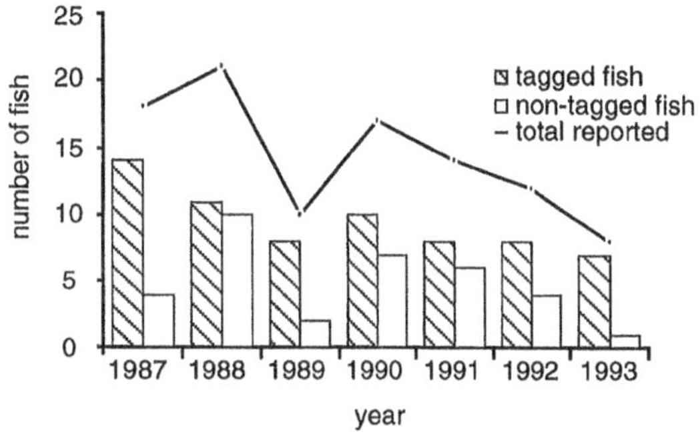

Figure 3. Accidental captures of *A. sturio* between 1987 and 1993, showing numbers of tagged versus non-tagged individuals reported.

stated operations, the water temperature was ~17°C.

Rearing of larvae

Larvae were fed mainly with live nauplii of *Artemia salina* and other zooplankton or frozen natural food (chironomid larvae or *Tubifex* sp.) alone or mixed with chicken egg yolk, beef spleen, and an experimental artificial diet made of beef liver and yeast (Kaushik et al. 1986). Water temperature for rearing was approximately 17°C.

Adaptation of wild immature fish to farm conditions

We first tested direct adaptation of large juveniles fish from brackish (15 to 20‰) to fresh water. The analysis of published data (Magnin 1962) and use of the updated growth information (Rochard & Jatteau 1991) imply a possible successful adaptation of sturgeon ≥ 105 cm TL to freshwater. Our experiments used riverine or well water at a constant temperature of 18°C. Methods for long term acclimatization of wild immature European Atlantic sturgeon to farm conditions were based on previous success with other sturgeon species, *A. naccarii* (Arlati et al. 1988), *A. transmontanus* (Struffenegger 1992), and *H. huso* (Goncharov personal communication).

Communication with the public

We printed and distributed booklets and posters about *A. sturio* to fishermen of the Atlantic coast of France. We mailed more specific materials to all French fisheries administrators, organizations of fishermen, and to French and western European fish research laboratories. We organized local meetings to explain our goals and previous results, and established an information network. To increase public awareness, we collaborated with the media (local, regional, and national newspapers; radio and television). To foster scientific cooperation, we organized the first international symposium on sturgeon, held in Bordeaux in 1989 (Williot 1991).

Results and discussion

Current status of the population

Figure 1 summarizes background information necessary to understand the Gironde population of *Acipenser sturio*, which has currently the only known spawning stock of this species in western Europe. During its sea life, *A. sturio* occurs over a wide range (Figure 2), from the Bay of Biscay to the Bris-

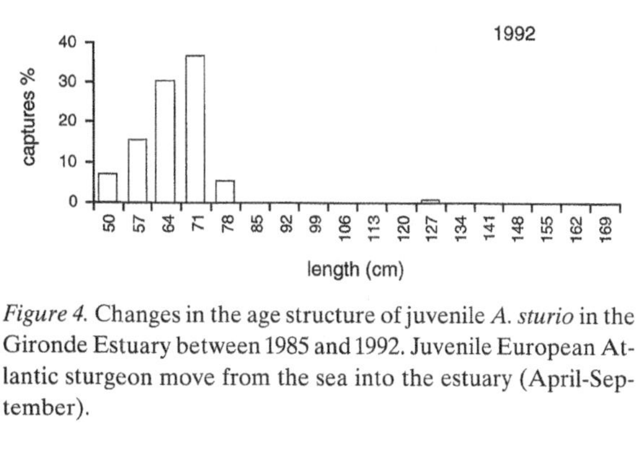

Figure 4. Changes in the age structure of juvenile *A. sturio* in the Gironde Estuary between 1985 and 1992. Juvenile European Atlantic sturgeon move from the sea into the estuary (April-September).

tol Channel and North Sea (Castelnaud et al. 1991). It inhabits shallow littoral areas, with 83% of catches occurring between 10 and 70 m depth. From 1988 to present, the number of accidental catches on the continental shelf area has decreased progressively (Figure 3).

Declarations of accidental captures of *A. sturio* increased after our public awareness campaign, so public awareness is essential for the success of our restoration program. More tagged fish than untagged fish were reported among accidental captures at sea (Figure 3). Perhaps fishermen informed us and released the fish more often when a fish was tagged. Curiously, tags used for scientific purposes, which sometimes increase mortality due to tag injuries or other effects, may in this case actually prevent some mortality.

The population (1984–1988) in the Gironde Estuary had a low number (500–2000 individuals) of young fish (3–8 years) during the summer (Castelnaud et al. 1991, Rochard 1992). During our study period (1987–1994), the last recorded natural reproduction (as evidenced by winter catches of small fish approximately 25.0 cm TL) likely occurred in summer 1988.[1]

The structure of the stock that enters the Gironde during the summer has changed over time (Figure 4). Before 1990, we observed a typically polymodal structure (Castelnaud et al. 1991) as a result of yearly reproduction. After 1990, the structure of the stock presented only one mode, corresponding to the 1988 cohort. No sign of pathology has been detected. The age 0–5 year growth of the 1988 cohort was significantly faster (Rochard & Jatteau 1991, Rochard 1992) than that determined previously by Magnin (1962) when the stock was far larger. The difference is not attributable only to methodological differences between studies (Rochard 1992) but may be a density dependent effect (Therrien et al. 1988).

Thanks to the sex determination methods of Cuisset (1993, developed originally for *A. baerii*) we shall be able to obtain data on the sex ration of *A. sturio* by measuring plasma concentrations of 11-ketotestosterone or vitellogenin.

[1] In the spring of 1995, Cemagref reported the first evidence of natural reproduction in 1994 of European Atlantic sturgeon in the Gironde since 1988 (editors note, March 1996).

Habitats

No new physical obstacles to spawning migration have been erected since the construction of the dams of Bergerac on the Dordogne River in 1851 and of Golfech near Agen on the Garonne River in 1971 (Figure 1). In 1981, gravel extraction stopped in the Dordogne River; only one site is still exploited in the lower part of the Garonne River (Rochard et al. 1990). Cadmium (mainly dissolved) in the Gironde Estuary is at concentrations 10 to 20 times higher than those measured in other French Atlantic estuaries (Maurice 1994). This heavy metal comes from old mines in the upper part of the Garonne River basin, which stopped working in the early 1970s.

Exploitation

Acipenser sturio has been protected in France since 1982; fishing, transport, and commerce are strictly forbidden. Nevertheless, some sturgeon are caught in the near continental shelf and in the mouth of some estuaries as by-catch of sole and other bottom fishes. At present, this is the main direct anthropogenic impact on this stock.

Status of wild spawners

Of 40 catches of wild sturgeon, 75% were males and 25% females. Most of them (90%) were caught before 1989, and the last female was caught in 1987. This sex ratio is exactly the opposite (74%F – 24%M) of that reported by Magnin (1962) for a 5 year sampling period in the Gironde (n = 96) and by Elvira et al. (1991) for captures in the Guadalquivir River (Southwestern Spain) from 1932–1943. The latter study indicated that the skewed sex ratio was an artefact of selectivity of the nets, which could also be true in our study. Ripe broodfish occurred in the Gironde basin from mid-April to the end of

Table 1. Results of attempts to spawn wild caught females of *Acipenser sturio* (1981–1994).

1981	The only female ovulated, but due to over-maturation, the quality of eggs was poor. Nearly 700 embryos hatched, but only ten survived the first month of rearing, after which rearing was discontinued.
1983	The two females did not ovulate, probably because of under-maturation.
1984	The PI of eggs of the only captured female did not decrease by 6 days after stocking; the eggs became softer and more delicate as probable expression of damage. Because male *A. sturio* were not available simultaneously we attempted hybridization of *A. sturio* with Siberian sturgeon *A. baerii*, but ovulation and spermiation were poor, and only a few abnormal embryos hatched and died quickly.
1985	The two females ovulated perfectly, and the fertilization rate was close to 80% in both cases; incubation lasted 4.6 days and mean hatching success was 80%, but all free embryos died.
1986	The first female was weak and lacked the ability to reproduce (GVBD = 0). It died 2 days later and autopsy showed fatty gonads, a sign of under-maturation. A second female, caught very late, had already spawned.
1987	The reproductive potential of the only female was low (GVBD = 9%) without any change in one week. Hormonal stimulation at a rate of 0.6 mg kg^{-1} (12% of the normal dose used for reproduction) increased the sensitivity of eggs (GVBD = 63%), but at too a low level to expect successful reproduction (Williot et al. 1991). Sensitivity (GVBD) decreased two days after to only 23%. New hormonal treatment did not improve the physiological condition of this female.
1988	The only female caught was not mature.

June; the only fish caught later was a female that had already spawned. Most catches (68%) took place from early May to 15 June. Half of the fish were caught in the Garonne River, one third in the Dordogne River (more concentrated in the lower part), and the rest in the Gironde Estuary. One male was caught three times (1981, 1984, and 1987), always in the Dordogne River. Of the fish caught in the Gironde Estuary, none was suitable for reproduction, because the males and one female were immature and the second female had already spawned.

Females weighed almost twice as much as males (41.6 kg and 23.7 kg, respectively) and females averaged 30 cm longer than males (193 cm and 163 cm, respectively). Assuming that the age-size relationship established by Rochard & Jatteau (1991) is valid for these fish, we estimated the minimum mean age (using mean TL) to be 10 and 15 years for males and females, respectively.

The upstream migration period of broodfish in the Gironde Estuary does not seem to have changed since Magnin (1962). He reported that 78% of the spawning fish were caught in May and first half of June; we caught 68% during that period. Our age estimates are very close to those of the previously studied stock in the Guadalquivir River (Spain, see Classen 1944), with some fish being older than those of the Rioni River (Georgia, see Ninua 1976).

Artificial reproduction

From 1981 until 1994, we collected 10 females for artificial collection of eggs (Table 1). Only 20% (n = 2) of the females showed an optimal physiological state, the others being under-mature or over-mature. We rarely had both sexes simultaneously. Also, we collected only five mature males (Table 2).

We induced successful ovulation only when females were caught in an optimal physiological state,

Table 2. Results of attempts to spawn wild caught males of *Acipenser sturio* (1981–1987).

1981	Both males yielded sperm after 9 and 6 days of stocking, respectively (even one that was not running upon arrival).
1985	Two males yielded sperm 3 and 4 weeks after stocking, but only with two hormonal injections at 12 hour intervals, and at rates of 20% and 100% of the normal dose, respectively.
1987	One male yielded sperm after 4 weeks of stocking (one injection) and, another one, in poor condition upon arrival, did not yield any sperm 6 days after stocking. Between 1981–1994, it seems that the quality of spermatozoa became poorer, as measured by its motility in water.

as occurred in 1985. In the case of immature fish, it was difficult or impossible to bring them into reproductive condition. This also has been reported for *A. stellatus*, a species for which Dettlaff & Davydova (1979) show that good reproduction can be achieved when fish are held for fewer than four days. In contrast, Doroshov & Lutes (1984) achieved good results with wild females of *A. transmontanus* held up to 4 months. It is important to ascertain whether these differences are species-specific or due to our procedures.

The physiological condition of males appears to be deteriorating as spermatozoa are increasingly non-motile. The possible influence of cadmium on spermatogenesis should be explored as some pollutants disturb the reproductive cycle (Billard & Gillet 1984). Second samples of sperm may be obtainable from males three to four weeks after capture. That has been achieved with a two hormonal injection procedure (instead of one), consisting respectively of 20% and 100% of the normal dose (2 mg kg^{-1}) of carp hypophysis powder at a 12 hour interval, the fish being held at the same water temperature. Because females and males are seldom caught simultaneously, it would be useful to confirm whether such a process could be improved by appropriate management of water temperature (Kazanskii 1981). Also, it is necessary to evaluate long term preservation of *A. sturio* sperm, as has been done with several other sturgeon species (Cherepanov et al. 1993, Drokin et al. 1993). Using sperm from several males would also increase genetic heterogeneity of the offsprings. Although we found that hybridization between *A. sturio* and *A. baerii* is possible, further investigation is needed to conclude if the non-viable embryos obtained were due to the poor quality of the sexual products, or to genetic incompatibility.

Rearing of larvae

We recovered nearly 700 embryos in 1981, and 230 000 and 70 000 from the two females stripped in 1985. In 1981, about ten larvae were still alive after one month of rearing, after which rearing was discontinued. In 1985, all free embryos died at approximate age of 3 weeks. Some had started to feed, but live food (cladocerans) obstructed the digestive tract, causing death.

In the future, it will be necessary to select digestible prey species available in late spring in large quantities (or easy to produce) with an appropriate size (worms, insect larvae, small crustaceans). In mass rearing, this procedure is time-consuming, costly, and can introduce pathogens. Specific weaning practice with inert food is also needed, even though that is far more difficult than for other sturgeon species (Monaco et al. 1981, Dabrowski et al. 1985, Giovannini et al. 1991). The imposed experimental water temperature (17° C) was slightly lower than in the rivers at the time of capture. Also, it has been shown for teleost species such as *Dicentrarchus labrax*, *Saprus aurata* (Barnabé 1989) and *Coregonus lavaretus* (Champigneulle 1988) that better rearing occurs at water temperatures higher than that of spawning. For both these reasons, the value of higher rearing temperatures (22–24° C) should be tested on *A. sturio*.

Adaptation of wild immature sturgeon to farm conditions

We caught 20 large juveniles fish in early July 1991, of which only one met the size criterion (TL < 105 cm) for successful acclimatization to fresh water. We used 2 m diameter tanks. Upon arrival, most fish had lost their righting response as a result of the stress of both catching and transport. Heavy mortality occurred in the first 24 hours for fish < 105 cm TL (mean TL = 79 cm) after direct transfer to fresh water. Survivors were divided into two groups, one remaining in fresh water and the other in brackish water (salinity = 5–10‰); however, all of these fish eventually died, the last one three weeks after stocking and without having fed over that time. The first fish transferred to brackish water died later than the others (< 0.05, Mann-Whitney U test). The largest fish (TL = 130 cm, W = 10.8 kg) was still alive in fresh water by the end of August, but never fed naturally. We force-fed it pellets from one to three times a day. After having lost about 25% of its weight, this individual regained weight, but never recovered its natural feeding behavior.

These observations show that we must minimize stresses to successfully adapt fish to farm conditions. In 1993, we used larger tanks (3 and 4 m diameter) supplied with low salinity (5‰) thermoregulated (18°C) water in darkness to keep fish calm. Soon after the arrival of fish, salinity was slowly lowered to 0‰. Each tank ran as a closed system in which nitrogen compounds, pH, and temperature were controlled. The bottom of one tank was covered with 10 cm depth of 8–10 mm gravel. Two fish caught in mid-August were stocked into one tank and two others into the second tank in late September. Their sizes (TL = 105.8 ± 1.2 cm, W = 5.4 ± 0.2 kg) were close to the goal of < 105 cm TL. The fish were fed shrimps, mostly frozen *Palaemonetes varians, Crangon crangon*, and *Palaemon longirostris*, and when available, live *Palaemon longirostris*.

All four fish caught in 1993 were still alive in 1994: three began to eat shrimp after 160 days, and after 200 days, the last individual fed. During the period of starvation they lost about 20% of their body weight. They were progressively weaned onto pellets, but shrimp remained the preferred food. Whatever the food, their consumption is irregular, and we did not find any relationship between feeding and the two types of tank bottom.

We suggest that quick adaptation to fresh water of fish ≥ 105 cm is possible but that normal food intake is needed within 5.5 to 6.5 months. In the future, better results may be obtained by reducing all possible causes of stress. We intend to improve catch, transport, and stocking conditions, particularly using brackish water (15‰) until fish begin to feed, and only thereafter begin the transfer to fresh water.[2]

Restoration program for *A. sturio*

Biological features of *A. sturio* in western Europe (unimodal population structure, decline in occasional marine catches, increased growth rates, and non-motile sperm) are symptoms of the dramatic decline of this species. We met with many difficulties (inability for the females to achieve reproductive maturation in captivity, feeding of larvae, and adaptation to farm condition).[2]

Four main problems must be solved for successful restoration. The first is obtaining broodstock and producing *A. sturio* for restocking. This involves determining the best methods of capture and transportation of wild fish to our research station, assessment of good conditions for short and long term acclimatization to farm conditions, maturation of broodstock emphasizing water temperature, improving reproduction methods for females and males, production of juveniles, preparation of juveniles before release (counting, tagging, transport), release of fish, and finally, in case mature wild spawning fish are caught, experiments spanning artificial stripping to release of juveniles. The second problem is increasing natural reproduction of *A. sturio*. We plan to study spawning grounds in order to be able to improve those which are degraded and to create artificial ones, as has already been done elsewhere (Vlasenko 1974, Gendron 1988). The third problem is increasing our basic knowledge of the biology and status of the present population, with special emphasis on its distribution, migration, and structures. Fourth, biologists involved in the restoration program must attract the attention of, and educate, the public about the current condition of the species.

It is often difficult to obtain financial support because the effectiveness of restoration programs can be estimated only on a long term basis. But, from a long term financial point of view it has already been demonstrated that 'the management of endangered species is intrinsically a policy of loss minimization' (Point 1991). Such awareness must be promoted. Because all sturgeon species are more or less endangered, and because they cross many international boundaries, their management needs international cooperation and investment.

[2] Several of these problems were solved during the 1995 season, when Cemagref successfully captured and artificially stripped adults and subsequently stocked juveniles into the Gironde system (editors note, March, 1996).

Acknowledgements

We thank V. J. Birstein, J. R. Waldman, and R. Boyle for the invitation to present this program at the sturgeon symposium. An important part of this work has been supported by a grant Contrat de Plan Etat – Région Aquitaine. We thank M. Trouvery, O. Rooryck, P. Jatteau, D. Mercier, and M. Pelard for their participation in this work and Maria Melin-Castelnaud for her help with English. Special thanks to F. Chapman, B. R. Konkle, P. Lorka, and F. Parauka for their critical review of the manuscript. The constructive comments and time of the reviewers are gratefully acknowledged. John R. Waldman, V. J. Birstein, and W. E. Bemis edited the text. William E. Bemis prepared the maps and figures.

References cited

Almaça, C. 1988. On the sturgeon, *Acipenser sturio*, in the Portuguese rivers and sea. Folia Zool. 37: 183–191.

Anonymous. 1995. L'esturgeon européen ne doit pas disparaître! Fleuves et Rivières 1: 57–64.

Antipa, G. 1934. Les esturgeons de la mer Noire. Leur biologie et les mesures nécessaires pour leur protection. Commission Internationale pour l'Exploration Scientifique de la Méditérranée. Rapports et Procés Verbaux 8: 61–75.

Arlati, G., P. Bronzi, L. Colombo & G. Giovannini. 1988. Induced breeding of the Italian sturgeon (*Acipenser naccarii*) raised in captivity. Riv. Ital. Acquacol. 23: 94–96.

Bacalbaça-Dobrovici, N. 1991. Statut des différentes espèces d'esturgeons dans le Danube Roumain: problèmes liés à leur maintenance. pp. 185–192. *In*: P. Williot (ed.) Acipenser, Cemagref Publ., Bordeaux.

Bacalbaça-Dobrovici, N. 1997. Endangered migratory sturgeons of the lower Danube River and its delta. Env. Biol. Fish. (this volume).

Barannikova, I.A. 1987. Review of sturgeon farming in the Soviet Union. J. Ichthyol. 27: 62–71.

Barnabé, G. 1989. L'élevage du loup et de la daurade. pp. 675–720. *In*: G. Barnabé (coord.) Aquaculture, Technique et Documentation Lavoisier.

Benecke, N. 1986. Some remarks on sturgeon fishing in the southern Baltic region in medieval times. pp. 9–17. *In*: D.C. Brinkhuizen & A.T. Clason (ed.) Fish and Archaeology, Studies in Osteometry, Taphonomy, Seasonality and Fishing Methods, (BAR International Series 294).

Berg, L. S. 1948. The freshwater fishes of the USSR and adjacent countries. 4th ed. Akademia Nauk USSR, Moscow & Leningrad, Part 1. 466 pp. (in Russian, English translation published by Israel Program for Scientific Translations, Jerusalem. 505 pp.)

Billard, R. & C. Gillet. 1984. Influence de quelques facteurs de lenvironnement sur la fonction de reproduction chez les poissons. Cahiers Lab. Montereau 15: 45–54.

Birstein, V.J. & W.E. Bemis. 1997. How many species are there within the genus *Acipenser*? Env. Biol. Fish. (this volume).

Castelnaud, G. 1988. L'opération de marquage de l'esturgeon dans l'estuaire de la Gironde (France) une dimension Européenne. International Council for the Exploration of the Sea, Anadromous and Catadromous Fish Committee, CM 1988/M28. 12 pp.

Castelnaud, G., E. Rochard, P. Jatteau & M. Lepage. 1991. Données actuelles sur la biologie d'*Acipenser sturio* dans l'estuaire de la Gironde. pp. 251–275. *In*: P. Williot (ed.) Acipenser, Cemagref Publ., Bordeaux.

Champigneulle, A. 1988. A first experiment in mass rearing of coregonid larvae in tanks with dry food. Aquaculture 74: 249–261.

Cherepanov, V. V., S. I. Drokin, S. I. Ochkur, B. B. Dzuba, A. S. Chikhachov & E. F. Kopeika. 1993. Freezing of sperm of the Azov-Black sea acipenserids. pp. 63–64. *In*: International Symposium on Sturgeons, Abstract Bulletin, 6–11 September 1993, VNIRO, Moscow.

Classen, T. E. 1944. Estudio bio-estadistico del esturion o sollo del Guadalquivir (*Acipenser sturio* L.). Ministerio de Marina Instituto Español de Oceanografia, Madrid. 112 pp.

Cochnauer, T. G., J. R. Lukens & F. E. Partridge. 1985. Status of white sturgeon, *Acipenser transmontanus*, in Idaho. pp. 127–133. *In*: F.P. Binkowski & S. Doroshov (ed.) North American Sturgeons: Biology and Aquaculture Potential, Dr W. Junk Publishers, Dordrecht.

Cuerrier, J.P. 1951. The use of pectoral fin rays for determining age of sturgeon and other species of fish. Can. Fish Cult. 11: 10–18.

Cuisset, B. 1993. Etude endocrinologique de la fonction de reproduction chez l'esturgeon Sibérien *Acipenser baeri* Brandt: application au sexage des populations sauvages ou élevées en acipensériculture. Thèse 1039, Université Bordeaux I, Bordeaux. 215 pp.

Dabrowski, K., S.J. Kaushik & B. Fauconneau. 1985. Rearing of sturgeon (*Acipenser baeri* Brandt) larvae. I. Feeding trial. Aquaculture 47: 185–192.

Debus, L. 1995. Former and recent distribution of *Acipenser sturio* in the north part and Baltic Sea. pp. 189–203. *In*: A.D. Gershanovich & T.I.J. Smith (ed.) Proceedings of the International Symposium on Sturgeons, 6–11 September 1993, VNIRO, Moscow.

Dettlaff, T.A. & S.I Davydova. 1979. Differential sensitivity of cells of follicular epithelium and oocytes in the stellate sturgeon to unfavorable conditions, and correlating influence of triiodothyronine. Gen. Compar. Endocrin. 39: 236–243.

Doroshov, S.I. & P.B Lutes. 1984. Preliminary data on the induction of ovulation in white sturgeon (*Acipenser transmontanus* Richardson). Aquaculture 38: 221–227.

Drokin, S.I., V.V. Cherepanov, E.F. Kopeika & N.I. Shilin. 1993.

Cryopreservation of the sperm of Sakhalin sturgeon (*Acipenser medirostris micadoi*): problems and prospects for cryopreserved sperm collection from rare and endangered sturgeon species. pp. 64–65. *In*: International Symposium on Sturgeons, Abstract Bulletin, 6–11 September 1993, VNIRO, Moscow.

Ehrenbaum, E. 1923. Die Eider als Störfluss und die Schonung des Störs. Fischerboten, Norddeutsche Fisch 5: 77–83.

Elvira, B. & A. Almodovar. 1993. Notice about the survival of sturgeon (*Acipenser sturio* L., 1758) in the Guadalquivir estuary (S.W. Spain). Arch. Hydrobiol. 129: 253–255.

Elvira, B., A. Almodovar & J. Lobon-Cervia. 1991. Sturgeon (*Acipenser sturio*) in Spain. The population of the river Guadalquivir: a case history and a claim for a restoration program. pp. 337–347. *In*: P. Williot (ed.) Acipenser, Cemagref Publ., Bordeaux.

Gendron, M. 1988. Riviére des praires, suivi de l'aménagement du haut fond, synthese 1982–1988. Le groupe de recherche SEEEQ Ltée, pour le service de recherches en environnement et santé publique, vice présidence environnement, Hydro Québec. 95 pp.

Giovannini, G., L. Colombo, P. Bronzi & G. Arlati. 1991. Growth of hatchery produced juveniles of Italian sturgeon, *Acipenser naccarii* Bonaparte, reared intensively in fresh water. pp. 401–404. *In*: P. Williot (ed.) Acipenser, Cemagref Publ., Bordeaux.

Gutierrez Rodriguez, F. 1962. El esturión del rio Guadalquivir. Ministerio de Agricultura Servicio Nacional de Pesca Fluvial y Caza, Folleto Informativo 5. 58 pp.

Holčík, J., R. Kinzelbach, L.I. Sokolov & V.P. Vasilev. 1989. *Acipenser sturio* Linneaus, 1758. pp. 367–394. *In*: J. Holčík (ed.) The Freshwater Fishes of Europe, Vol. 1, Part II, General Introduction to Fishes, Acipenseriformes, Aula Verlag, Wiesbaden.

Kazanskii, B.N. 1981. The ecological-evolutive principles of the reproduction of fish population. Vestnik Leningradskogo Universiteta, Zoologiya 21: 52–58 (in Russian).

Kazanskii, B.N., Y.A. Feklov, S.B. Poduska & A.N. Molodtsov. 1978. Quick method for assessment of sexual maturity in sturgeon close to spawning. Rybnoe Khozaistvo 2: 24–27 (in Russian).

Kaushik, S., K. Dabrowski & P. Bergot. 1986. Metabolic aspects of dry diet utilization by juvenile coregonids. Arch. Hydrobiol. Beih. Ergebn. Limnol. 22: 161–169.

Khodorevskaya, R.P., G.F. Dovgopol, O.L. Zhuravleva & A.D. Vlasenko. 1997. Present status of commercial stocks of sturgeons in the Caspian Sea basin. Env. Biol. Fish. (this volume).

Kinzelbach, R. 1987. Das Ehemalige Vorkommen des Störs, *Acipenser sturio* (Linnaeus, 1758), im Einzugsgebiet des Rheins (Chondrostei, Acipenseridae). Zeitsch. Ange. Zool. 74: 167–200.

Kohlhorst, D.W. 1979. Effect of first pectoral fin ray removal on survival and estimated harvest rate of white sturgeon in the Sacramento-San Joaquin estuary. Calif. Fish Game 65: 173–177.

Laporte, E. 1853. Faune ichthyologique du département de la Gironde. Act. Soc. Lin. Bordeaux 19: 158–224.

Lepage, M. & E. Rochard. 1995. Threatened fishes of the world: *Acipenser sturio* Linnaeus, 1758 (Acipenseridae). Env. Biol. Fish. 43: 28.

Letaconnoux, R. 1961. Note sur la fréquence de la distribution des captures d'esturgeons (*Acipenser sturio* L.) dans le Golfe de Gascogne. Rev. Trav. Inst. Pêch. Marit. 25: 253–261.

Lutes, P.B., S.I. Doroshov, F. Chapman, J. Harrah, R. Fitzgerald & M. Fitzpatrick. 1987. Morpho-physiological predictors of ovulatory success in white sturgeon, *Acipenser transmontanus* Richardson. Aquaculture 66: 43–52.

Magnin, E. 1959a. Détermination de l'âge et croissance de l'*Acipenser sturio* L. de la Gironde. Bull. Franç. Pisc. 193: 152–159.

Magnin, E. 1959b. Répartition actuelle des Acipenseridés. Rev. Trav. Inst. Pêch. Marit. 23: 277–285.

Magnin, E. 1962. Recherches sur la systématique et la biologie des Acipensérìdés, *Acipenser sturio* L., *Acipenser oxyrhynchus* Mitchill et *Acipenser fulvescens* Raf. Ann. Stat. Cent. Hydrobiol. Appl. 9: 7–242.

Maitland, P.S. & A.A. Lyle. 1990. Practical conservation of British fishes: current action on six declining species. J. Fish Biol. 37 (Supplement A): 255–256.

Maurice, L. 1994. La qualité des eaux de l'estuaire. pp. 32–45 *In*: J.L. Mauvais & J.F. Guillaud (coord.) Livre blanc de l'estuaire de la Gironde. Agence de l'eau Adour Garonne.

Monaco, G., R.K. Buddington & S.I. Doroshov. 1981. Growth of white sturgeon (*Acipenser transmontanus*) under hatchery conditions. J. Maricult. Soc. 12: 113–121.

Ninua, N. S. 1976. The Atlantic sturgeon of Rioni River. Metsniereba, Tbilissi. 121 pp. (in Russian).

Point, P., 1991. Quelques réflexions sur la prise en compte de l'incertitude dans la politique de gestion d'une espèce sauvage: l'exemple de l'esturgeon ouest européen (*Acipenser sturio*). pp. 459–474. *In*: P. Williot (ed.) Acipenser, Cemagref Publ., Bordeaux.

Rochard, E. 1992. Mise au point d'une méthode de suivi de l'abondance des amphihalins dans le système fluvio-estuarien de la Gironde, application à l'étude écobiologique de l'esturgeon *Acipenser sturio*. Thèse no. 841, Université de Rennes I/Cemagref de Bordeaux. 315 pp.

Rochard, E., G. Castelnaud & M. Lepage. 1990. Sturgeons (Pisces: Acipenseridae); threats and prospects. J. Fish Biol. 37 (Suppl. A): 123–132.

Rochard, E. & P. Jatteau. 1991. Amélioration de la méthode de détermination de l'âge de l'esturgeon commun *Acipenser sturio* et premières applications. pp. 193–208. *In*: P. Williot (ed.) Acipenser, Cemagref Publ., Bordeaux.

Roule, L. 1922. Etude sur l'esturgeon du golfe de Gascogne et du bassin Girondin. Office scientifique et technique des pêches maritimes. Notes et Mémoires 20. 12 pp.

Struffenegger, P. 1992. Sturgeon farming in California: a promising new industry. Aquaculture Europe 17: 6–9.

Therrien, J. 1988. Avis scientifique sur le statut de la population d'esturgeon noir dans l'estuaire du Saint-Laurent. Avis scientifique 88/1 MLCP Québec et M.A.P.A.C. 23 pp.

Trouvery, M. 1980. Recueil de données sur l'état des zones de pêche et des frayères potentielles de l'esturgeon (*Acipenser sturio*) en Dordogne (du Bec d'Ambes à Castillon La Bataille)

et en Garonne (du Bec d'Ambes à Cassueil). Rapport préliminaire CTGREF, Division A.L.A., Bordeaux. 12 pp.
Trouvery, M., P. Williot & G. Castelnaud. 1984. Biologie et écologie d'*Acipenser sturio*. Etude de la pêcherie. Etude no. 17, Série esturgeon no. 1. Cemagref de Bordeaux, Div. ALA/Agedra. 79 pp.
Vlasenko, A.D. 1974. Problems concerning artificial spawning grounds of the Kuban River sturgeons. Trudy VNIRO 102: 3–29 (in Russian).
Williot, P. (ed.). 1991. Acipenser. Cemagref Publ., Bordeaux. 519 pp.
Williot, P., R. Brun, T. Rouault & O. Rooryck. 1991. Management of female spawners of the Siberian sturgeon, *Acipenser baeri* Brandt: first results. pp. 365–379. *In*: P. Williot (ed.) Acipenser, Cemagref Publ., Bordeaux.

Part 3: Controversies, conservation and summary

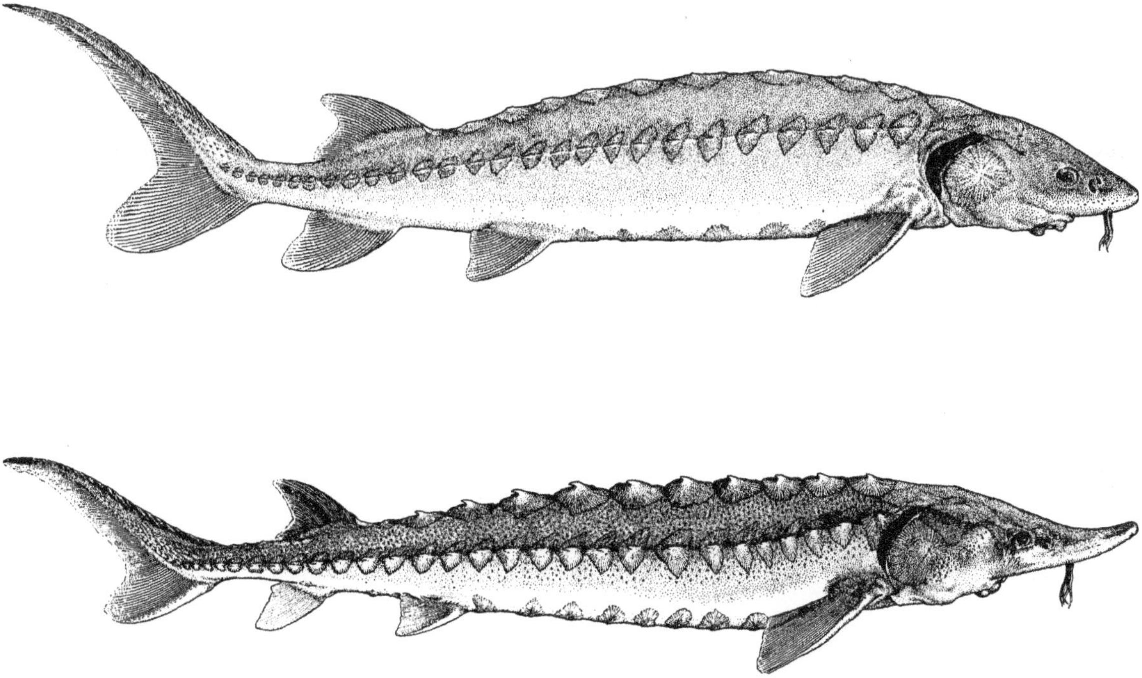

Acipenser brevirostrum female 58 cm long from the Hudson River near Kingston, New York, above *Acipenser oxyrinchus* male 58 cm long from the Lawrence River near St. Vallier, Quebec, by Paul I. Voevodine from Vladykov & Greeley (1963).

Sturgeon landing on the Volga River. From an engraving in Moynet[1], p. 85.

[1] Moynet, M. 1867. La Volga. Le Tour du Monde 15: 81–96.

Environmental Biology of Fishes **48**: 373–380, 1997.

Sturgeons and the Aral Sea ecological catastrophe

Iliya Zholdasova
Institute of Bioecology, Karakalpak Branch of the Academy of Sciences of Uzbekistan, Nukus, Republic Karakalpakstan

Received 5.4.1995 Accepted 16.3.1996

Key words: Amu-Dar River, Syr-Dar River, *Pseudoscaphirhynchus kaufmanni*, *P. fedtschenkoi*, *P. hermanni*, *Acipenser nudiventris*, pollution, acclimatization, *Nitzschia*

Synopsis

A short description of the catastrophic changes in the ecology of the Aral Sea basin during the three last decades is presented. These changes have influenced the status of two acipenserid endemics to the area, the large Amu-Dar shovelnose, *Pseudoscaphirhynchus kaufmanni*, and the ship sturgeon, *Acipenser nudiventris*. The main biological characteristics of both species in the new environmental conditions are given. Previous unsuccessful attempts to introduce other acipenserid species into the area are also described. International cooperation is needed for saving the last surviving species representing the genus *Pseudoscaphirhynchus*. The only two other species of the same genus, *P. fedtschenkoi* and *P. hermanni*, have already become victims of the Aral Sea catastrophe and are apparently extinct.

Introduction

Historically the endemic fauna of the Aral Sea basin included four acipenserid species: the large Amu-Dar shovelnose sturgeon, *Pseudoscaphirhynchus kaufmanni* (Bogdanov, 1874), small Amu-Dar shovelnose sturgeon, *P. hermanni* (Kessler, 1877), Syr-Dar shovelnose sturgeon, *P. fedtschenkoi* (Kessler, 1872), and ship sturgeon, *Acipenser nudiventris* (Lovetsky, 1828) (Nikolskii 1938, 1940, Berg 1948, Tleuov & Sagitov 1973, Tleuov & Tleubergnov 1974). All shovelnose sturgeons are freshwater species, while the ship sturgeon is an anadromous fish.

During the last 30 years, the environmental conditions in this region changed drastically because of the drying of the Aral Sea and extreme pollution (Aladin & Potts 1992); the whole change is known now as the Aral Sea environmental disaster or catastrophe (Feshbach & Friendly 1992, Peterson 1993). Two sturgeon species, the Syr-Dar and small Amu-Dar shovelnose sturgeons, were among the first victims of this disaster and seem to be extinct. The Syr-Dar shovelnose sturgeon was a regional endemic and occurred in the Syr Darya River basin only (Berg 1948). The last report on this sturgeon in Kazakhstan refers to 1952–1953 (Dairbaev 1959, Mitrofanov et al. 1986). The small Amu-Dar shovelnose sturgeon was a regional endemic of the Amu Darya River, and initially it was described as a rare species (Nikolskii 1938, Berg 1948). It has not been found since 1982. It is believed that this species has disappeared completely (Pavlov et al. 1985)[1]. Description of the contemporary status of the two remaining endemic species, *Pseudoscaphirhynchus kaufmanni* and *Acipenser nudiventris*, in the con-

[1] In April 1996 three *P. hermanni* were cought near the town of Kerki (Salnikov et al. 1996).

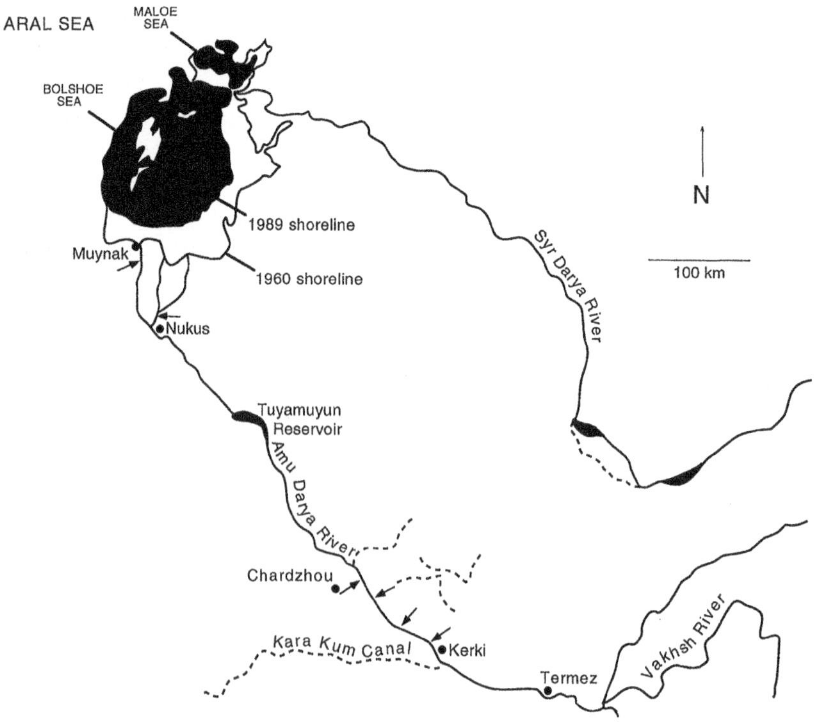

Figure 1. Map of the Aral Sea area. Two shorelines are shown: 1960 and 1989. Locations of catches of the large Amu-Dar shovelnose sturgeon between 1989 and 1993 are marked with arrows.

text of environmental changes in the area, is the main goal of this paper.

The Aral Sea disaster: catastrophic changes in the ecosystem of the Amu Darya River and Aral Sea

Changes in the Amu Darya River

The Amu Darya River is the longest river of Central Asia, being 2325 km long. It starts in the mountain regions of Pamir and Hindu Kush and has no tributaries downstream for the last 1257 km of its length. In the past, it entered the Aral Sea near the town of Muynak which was located in the Amu Darya River delta; at present, the river ends at a distance of 100 km from the shrunken sea. A map of the region is given in Figure 1.

The Aral Sea ecological catastrophe was initially caused by intensive and irrevocable removal of water from the Amu Darya and Syr Darya rivers for irrigation. This resulted in a considerable reduction in the discharge of water from the rivers into the Aral Sea, which, in turn, caused the drying out of the sea, a gradual shrinking of its size, and a gradual increase in its salinity. The most significant impact was in the lower reaches of the Amu Darya River. Supplying irrigation channels, lakes, and artificial reservoirs with water became impossible without special dams, which were constructed in the lower reaches of the river beginning in 1967. These dams are still needed for distribution of water from the river, even after construction of the Takhiatash Dam (220 km from the mouth) and Tuyamuyun Hydrocomplex with its system of reservoirs (450 km from the river mouth).

Dams constructed on the Amu Darya River in the 1960s cut off many anadromous fishes, including the ship sturgeon, from the sea. This caused a sharp decline in abundance of the ship sturgeon by the middle 1970s. Moreover, use of mineral fertilizers and pesticides for cotton agriculture resulted in an incredibly high level of pollution. Water of the rivers and lakes was extremely contaminated by chemicals due to disposal of drainage waste from the area of irrigated land cultivation (Chembarisov 1989). The increased mineralization (from 1000 to 1420 mg l^{-1}) and contamination of the drainage waters with pesticides, mineral fertilizers, and salts changed the hydrochemical regime of the rivers and lakes. In the 1950s, the mean content of ions in the Amu-Darya water was 540 mg l^{-1} (Rogov 1957), while during the last decade it varied between 600 and 1500 mg l^{-1}.

The highest level of pollution is in the lower reaches of the Amu Darya River. In 1989[2], the mean annual mineralization of water was 1525.5 mg l^{-1} near the town of Nukus (215 km from the mouth), and in 1990, 946.8 mg l^{-1}.

Changes in the hydrochemical regime of the Amu Darya River were augmented by decreases in the annual discharge. The mineral content of water in the Amu Darya River depends to a great extent on fluctuations in the volume of water discharge. Until the time when river flow became regulated (the 1960s), the fluctuations were caused mostly by variations in the melting of glacier-snow in the upper reaches. A spring-summer peak of water and a dry fall were quite distinct. Even now these seasonal fluctuations in the water level are expressed in the upper and middle reaches of the river. In the lower reaches these seasonal variations are less pronounced because so much water is removed by irrigation channels and reservoirs. In this area the salinity of water is very high because of a low water flow. Thus, in 1989 and 1990 mineralization during the full-water period in the lower reaches was 437–679 and 336–598 mg l^{-1}, respectively. It increased significantly in the dry season to 922–3999 mg l^{-1} in 1989 and to 973–1868 mg l^{-1} in 1990.

Increases in mineralization were accompanied by the general contamination of water in the river. Pollutants included almost all dangerous chemical substances: oil products, phenols, heavy metals, organic substances, etc. (Table 1). In 1989–1990, a tendency towards a decrease in pesticide content in the Amu-Darya water was noticed. In 1990 and 1991, the concentration of DDT and its metabolites in the Tuyamuyun Reservoir was 0.1–20.0 mg l^{-1}, and the concentration of other organochlorine pesticides, 0.015–0.616 mg l^{-1}. Pollution caused death of silver carp, *Hypophthalmichthys molitrix*, on a massive scale in this reservoir.

[2] Yearbook on Surface Water Quality and Efficiency of Water Protection Measures in Uzbekistan in 1989. 1991. Goskomgidromet USSR, Tashkent, Pt. 1, Vol. 4, Book 1-5 (in Russian).

Changes in the Aral Sea

The Aral Sea ecosystem has undergone considerable change during the last 30 years. It was the fourth biggest lake on the planet. It changed from a brackish-water lake with an average water salinity 10.3 in 1961 to a salt-water one with a salinity of 37–38. Water lev-

Table 1. Water pollution in two areas of the Amu Darya River, near towns of Termez (middle reaches) and Nukus (lower reaches).

Chemicals	Termez				Nukus			
	1989		1990		1989		1990	
	Concentration	MAC factor*	Concentration	MAC factor*	Concentration	MAC factor*	Concentration	MAC factor*
Phenols	0.004 mg l^{-1}	43	0.008 mg l^{-1}	83	0.004 mg l^{-1}	43	0.010 mg l^{-1}	103
Oil products	0.18 mg l^{-1}	3.63	0.04 mg l^{-1}	0.83	0.15 mg l^{-1}	33	–	–
Copper	8.8 µg l^{-1}	8.83	10.0 µg l^{-1}	103	10.0 µg l^{-1}	103	10.7 µg l^{-1}	10.63
Nitrate nitrogen	0.037 mg l^{-1}	1.93	0.027 mg l^{-1}	1.43	0.042 mg l^{-1}	0.23	1.42	2.13
α-benzene hexachloride	0.028 mg l^{-1}	2.83	0.019 mg l^{-1}	1.93	ND	–	ND	–
γ-benzene hexachloride	0.013 mg l^{-1}	1.33	0.015 mg l^{-1}	1.53	ND	–	ND	–

* Number of times of MAC (Maximum Allowable Concentrations, PDK in Russian), are given according to the Soviet standards (see Feshbach & Friendly 1992); ND = not determined.

el had decreased by 16.4 m by the end of 1993. Water volume decreased more than three times, and now it is less than 300 km^3. The size of the sea was reduced from 68.4 to 30.0 thousand km^2, and the shoreline receded in some places by 100 km or more. The southern brackish-water Adjibaiskii, Muynakskii, Sarbas, Abbas, and Zhaltyrbas bays, which different fishes used as spawning areas, are at present dry.

By 1987 the Aral Sea was divided by a sand bar into two basins, the Bolshoe (Large) and Maloe (Small) seas (Figure 1). The Bolshoe Sea receives water from the Amu Darya River; the drainage waste water from the irrigated area to the south from the Aral Sea also goes into this basin. The bed in the mouth part of the Syr Darya River has been changed and it now enters into the Maloe Sea.

At present, both rivers are disconnected from the Aral Sea: the Syr Darya River has not reached the sea since 1975, and the Amu Darya, since 1982. Connection of the Amu Darya River with the sea was first interrupted in the middle 1970s, when the volume of the river water upstream from the Takhiatash Dam decreased to such an extent that water was fully withdrawn for irrigation and no downstream flow took place. After this, the delta of the river considerably changed. Regulated flows of clear water (historically water in the Amu Darya River was extremely muddy) from the Tuyamuyun Reservoir caused deepening of the river bed by 4–5 m or more below the dam. Water goes to the sea through a system of small lakes located in the natural bed of the river and in the former sea bays. The artificially regulated discharge volume of water is extremely small. The sea level continues to decrease and its water salinity increases.

As a consequence of all these events, fishing in the Aral Sea stopped in 1984. At present, practically all endemic fish species have perished. They were replaced by the species introduced in the 1960s, such as *Hypophthalmichthys molitrix*, *Aristichthys nobilis*, *Parabramis perkinensis*, *Ctenopharyngodon idella*, and others (Zholdasova et al. 1991).

Due to the contemporary level of water in the Aral Sea, as well as in the Amu and Syr Darya rivers, there is no hope that populations of anadromous fishes, including the ship sturgeon, could be restored.

Acipenserid endemics of the Aral Sea basin under conditions of the environmental disaster: biology and status

The large shovelnose sturgeon

The large Amu-Darya shovelnose sturgeon, *Pseudoscaphirhynchus kaufmanni*, is an endemic of the Amu Darya River. It is an endangered species on the verge of extinction. It was included in the Uzbek SSR Red Data Book[3] and in the USSR Red Data Book[4]. Until the 1960s, it inhabited the Amu Darya River from its source to its mouth (Nikolskii 1938, Tleuov & Sagitov 1973).

The major part of the large shovelnose sturgeon population was located in the foothill and valley areas of the river. Sturgeon were largely concentrated near the villages of Kerki-Chardzhou-Ildzhik in the upper and middle reaches of the river and were caught there commercially in the 1930s. *Pseudoscaphirhynchus kaufmanni* was also numerous in the lower reaches of the Amu Darya River. In the late 1970s, large shovelnose sturgeon juveniles constituted up to 26% of all young fish in the lower reaches of the Amu Darya River (Tleuov 1981). *Pseudoscaphirhynchus kaufmanni* inhabited small irrigation channels connected with the river as well. Although it is a freshwater species, the shovelnose sturgeon can tolerate some salinity: it was found in the near-mouth area of the Aral Sea in salinity 8.5‰ (Gosteeva 1953).

At present, this species is preserved only in the middle reaches of the Amu Darya River. Studies carried out in 1989 after a 15-year break (Zholdasova et al. 1990) and in 1991 (together with Sergei Gamalei, Moscow Aquarium, Russia) showed that large shovelnose sturgeon are still concentrated in their usual habitats in the middle reaches of the river, within a region between the two towns of Kerki and Chardzhou. This species was not reported downstream from Chardzhou in 1989 and in 1990–1991 (Zholdasova et al. 1991). Sturgeon were also

[3] Uzbek SSR Red Data Book. 1983. Vol. 1, Vertebrates. FAN, Tashkent. 128 pp. (in Russian).

[4] USSR Red Data Book. 1984. Vol. 1. Lesnaya Promyshlennost, Moscow. 390 pp. (in Russian).

not found in the lower reaches of the river during the 1980s.

The large Amu-Dar shovelnose sturgeon lives in turbid, muddy waters. It inhabits shallow-water parts of the river with sandy or stony-pebble bottoms (Nikolskii 1938, Tleuov & Sagitov 1973). In 1989–1991, I also found sturgeon in the main channel of the river with sandy bottom at a depth of 1.0–1.5 m, as well as at the edges of pools with turbid water, and near sand bars.

In the past, the maximum size reported of the large shovelnose sturgeon was 75 cm, and the maximum weight, 2 kg (Berg 1948). In 1965–1966, the average body length was 37 cm and the average weight, 241 g (Tleuov & Sagitov 1973). The size of fish studied by us in 1989 and 1991 varied from 9.3 to 38.0 cm, being 23.6 cm on average, and the weight, from 3.2 to 270 g (100.2 g on average). The age of individuals caught was from 1 to 6 years. The whole length of the largest fish, including the tail filament, was 69.4 cm (33.5 cm without the filament), and the weight was 250 g.

The age of individuals in the population has changed compared to that in the 1960s. In the 1960s, the age of fish varied from 1+ to 14+ with a predominance of 3 to 6 year old fish (Tleuov & Sagitov 1973). In the late 1980s, the population consisted mostly of young (from 1 to 6 years old) fish, with a predominance of 3-year (36.8%) and 4-year old individuals (41.6%) (Zholdasova et al. 1990). The rate of linear growth has also slowed down compared with the 1960s. Apparently, this decrease in growth rate was caused by a change in sturgeon diet: an analysis of the content of stomachs showed a decrease in a proportion of fishes consumed by sturgeon compared to the 1960s.

Adult shovelnose sturgeon are benthophagous with predatory inclinations (Nikolskii 1938, Berg 1948, Tleuov & Sagitov 1973). In the 1960s, in the middle reaches of the Amu Darya River, five fish species constituted up to 64.5% of the diet of large shovelnose sturgeon: juvenile *Barbus brachycephalus*, *Aspius aspius*, *Acipenser nudiventris*, *Capoetobrama kuschakewitschi*, and *Noemacheilus oxianus*. At present, the diet is more diverse; in addition to fishes (36.6% of the food biomass), the larvae of 15 species of midges of the family Chironomidae comprise 30.5% of the food. Fishes are represented by *C. kuschakewitschi* and species of the genus *Rhinogobius*.

Under the current conditions, the large shovelnose sturgeon continues to reproduce in the Amu Darya River. In April–May 1989 we did not find ripe females ready for spawning or already spawned. Females were mostly at stage II or II–III of maturity [according to the scale of Nedoshivin (1928)]. However, there were males at stage IV or IV–V. According to the results obtained by Boris Goncharov (personal communication), in the region near the town of Chardzhou in 21–30 September 1991, individuals at stage IV predominated among males, while females were mostly at stage II. The rate of reproduction of the population is very low. In 1991, only a single late embryo and two one-year old juveniles were collected in the low reaches of the river (Pavlovskaya & Zholdasova 1991). Early development and larvae of *P. kaufmanni* were described recently (Schmalhausen et al. 1991, Dettlaff et al. 1993).

In 1993, I caught three large shovelnose sturgeon during my expedition to the Ordybai Channel of the Amu Darya River delta. They were young individuals of about 20 cm in length (without the tail filament) and about 50 g in weight. A shovelnose sturgeon of a similar size, caught near the town of Nukus in the Kattyagar irrigation channel was also reported in summer 1993. The appearance of sturgeon in their former habitats in the lower reaches of the Amu Darya River is evidently related to a significant volume and stable downstream flow in 1993. It is noteworthy that during the last three years the mean discharge in the Amu Darya River has been 3.7 times higher than during the previous decade. Also, the river flow is turbid and goes apparently directly through the Tuyamuyun Reservoir. Unfortunately, I have no information on the level of chemical pollution of the water in 1992–1993. But a general decrease in usage of pesticides and mineral fertilizers has recently been observed, mostly due to economic problems.

The large shovelnose sturgeon occurs exclusively in the fast running turbid waters of the Amu Darya River. Since these fish were never found in the lakes, Nikolskii (1938) believed that in stagnant water sturgeon die rapidly. I assume that in the low-water period from the middle 1970s until the late 1980s, the decrease in the flow velocity, clearing of water and the decrease in the size of river beds prevented migration of shovelnose sturgeon from the middle to the lower reaches of the Amu Darya River.

In 1989 and 1991, sturgeon were concentrated in the regions of the river where the drainage waste enters into it. Possibly, this was caused by the existence of organisms inhabiting these waste waters because of their enriched salt and organic materials content. This influence of drainage waters possibly resulted in a much greater biodiversity of the benthic fauna: the number of species increased from 55 (10 higher taxa) in 1974 to 83 species (17 taxa) in 1989. Species diversity of chironomids, ephemeropterans, and molluscs increased; new immigrant mysids and shrimps appeared. Fifteen species and forms of chironomids were found in diet of sturgeons in May and October 1989. Studies on the distribution of benthic fauna in the river bed and areas of its concentration may facilitate a search for sturgeons.

Unfortunately, the attempts to breed *P. kaufmanni* in captivity in 1983–1985 were unsuccessful: in all experiments the embryos died at 10–12 days after hatching (Goncharov et al. 1991).

Ship sturgeon

The ship sturgeon, *Acipenser nudiventris*, is a large anadromous fish occurring in the Black, Caspian, and Aral Sea basins (Berg 1948, Sokolov & Vasilev 1989). In the Aral Sea the ship sturgeon was the only representative of the genus *Acipenser* (Nikolskii 1940). Ship sturgeon reached a length of 160 cm and a weight of 45 kg (Tleuov & Sagitov 1973). In 1933–1934, ship sturgeon were introduced from the Aral Sea into Lake Balkhash (Kazakhstan), where they grew faster than in the Aral Sea (Dombrovskii et al. 1972).

Before the flow in the Amu Darya and Syr Darya rivers was regulated, the ship sturgeon thrived everywhere in the Aral Sea. For spawning they migrated mainly into the Syr Darya River in which they moved 1800 km or more upstream (Mitrofanov et al. 1986). In the Amu Darya River, the spawning areas of this species were located in a large region from the Kyzylzhar Cape (103 km upstream from the river mouth) up to Faizabadkal Cape (more than 1500 km from the river mouth).

There were two forms of the ship sturgeon, the winter and summer ones. The Aral Sea population of sturgeon was comprised predominantly of the winter form. Its migration into the Amu Darya River usually started at the beginning of spring flood, in the second half of April. Mass migration continued from the beginning of May until September, sometimes until the end of October. Sturgeon stayed the whole winter in the river and spawned next year in spring. The age of fish migrating for spawning into the rivers was from 7 to 30 years with a predominance (82.1%) of 20–21 year old individuals (Tleuov 1981).

The main spawning grounds of the ship sturgeon were located in the middle reaches of the Amu Darya River, between the towns of Chardzhou and Turtkulem; here about 50–60% of the progeny appeared (Tleuov & Sagitov 1973). Spawning also took place near the town of Nukus. The spawning period of ship sturgeon began in March at a water temperature above 10°C and continued until the end of May at a temperature of 21–23°C. The absolute fecundity of the Aral ship sturgeon varied from 52 259 to 554 700 eggs, being 389 731 on average (Tleuov 1981).

After spawning, ship sturgeon returned to the sea where they fed until the next maturation. They fed mostly on molluscs; among them, *Hipanis minima*, *Dreissena polimorpha*, and *D. caspiana* predominated (88%) (Tleuov 1981). Fish young were rarely found in sturgeon stomachs. After introduction of certain fishes in the Aral Sea, when gobies and sand smelts appeared in great amounts in the sea, a transition of the ship sturgeon to piscivory was noticed. Piscivory started at the age of two years and at the age of three years sturgeon consumed mainly fishes (up to 61% of their food). This transition to more piscivorous feeding occurred in the second half of the 1960s (Tleuov 1981).

From the late 1920s to the 1970s, there were numerous attempts to introduce other sturgeon species (including stellate sturgeon, *Acipenser stellatus*, and Russian sturgeon, *A. gueldenstaedtii*) into the Aral Sea, but none of them was successful (Bykov 1961, 1970, Karpevich 1975, Balymbetov 1981). The first stocking of stellate sturgeon from the Caspian into the Aral Sea in the 1920–1930s greatly impacted the endemic Aral Sea population of the ship sturgeon. A gill trematode, *Nitzschia sturionis*, was introduced together with the stellate sturgeon. It caused an epizooty and the death of ship sturgeon on a massive scale in 1936–1937 (Osmanov 1971). Before this, ship sturgeon was a commercially important species: its catch was 3000–4000 metric tons annually in 1928–1935 (Nikolskii 1940). After the *Nitzschia* epizooty in 1936–1937, commercial fishing was banned on 1 June 1940 (Tleuov & Sagitov 1973, Mitrofanov et al. 1986). In the 1960–1970s, an experimental catch of 700–9300 kg annually was allowed. Later virtually the whole population of ship sturgeon in the Aral Sea was destroyed by illegal fishing during spawning.

From the end of the 1970s, no findings of the ship sturgeon in the lower reaches of the Amu Darya River were reported. Ship sturgeon were not found among young fishes migrating downstream in 1989 (Pavlovskaya & Zholdasova 1991). Ship sturgeon have not been found in the Amu Darya River during the last 7–8 years, except two reports of local fishermen: in December 1990, near the town of Ildzhik (4 kg), and in March 1991, 35 km upstream from the town of Chardzhou (2 kg). Therefore, the Aral Sea form of *A. nudiventris* is practically extinct. Possibly, a small population of this form still exists in Lake Balkhash, where these fish were introduced from the Aral Sea in the 1930s (Pechnikova 1964). If so, this population might be used to reintroduce of the ship sturgeon into the Aral Sea.

Acknowledgements

I am very thankful to Vadim Birstein, John Waldman and Robert Boyle for inviting me to the International Conference on Sturgeon Biodiversity and Conservation. I am also grateful to Vadim Birstein and anonymous reviewers for their editorial notes on the manuscript and to William E. Bemis for drawing the map.

References cited

Aladin, N.V. & W.T. Potts. 1992. Changes in the Aral Sea ecosystem during the period 1960–1990. Hydrobiologia 237: 67–79.

Balymbetov, S.K. 1981. On the salinity resistance of some fish recommended for introduction into the Aral Sea. pp. 26–27. *In*: Biological Foundations of Fishery in the Central Asian and Kazakhstan Water Bodies, Ilim, Frunze (in Russian).

Berg, L.S. 1948. The freshwater fishes of the USSR and adjacent countries, Vol. 1, Part 1. Akademia Nauk USSR, Moscow & Leningrad (in Russian; English translation published by Israel Program for Scientific Translations, Jerusalem. 505 pp.)

Bogdanov, M. 1874. A report on a newly discovered acipenserid fish at the meeting of Zoological Section. Trudy Sankt-Peterburgskogo obshchestva ispytatelei prirody 5: 48 (in Russian).

Bykov, N.E. 1961. Materials on acclimatization of fishes and other vertebrates in the Aral Sea. pp. 45–50. *In*: Biological Foundations of Fishery in Central Asian Republics and Kazakhstan, Frunze (in Russian).

Bykov, N.E. 1970. On acclimatization of stellate sturgeon (*Acipenser stellatus*) in the Aral Sea. Trudy VNIRO 76: 192–195 (in Russian).

Chembarisov, E.I. 1989. Runoff and water mineralization in large collectors in Central Asia. Vodnye Resursy 1: 49–53 (in Russian).

Dairbaev, M.M. 1959. Formation, composition and distribution of ichthyofauna in water bodies of the Syr Darya River irrigation system. pp. 286–299. *In*: Collected Papers on Ichthyology and Hydrobiology 2, Izdatelstvo Akademii Nauk KazakhSSR, Alma-Ata (in Russian).

Dettlaff, T.A., A.S. Ginsburg & O.I. Schmalhausen. 1993. Sturgeon fishes. Developmental biology and aquaculture. Springer Verlag, Berlin. 300 pp.

Dombrovskii, G.V., N.P. Serov & V.N. Dikanskii. 1972. Biology and fishing of the ship sturgeon, *Acipenser nudiventris* Lov., in the Balkhash-Ili basin. Trudy TSNIORKh 4: 146–148 (in Russian).

Feshbach, M. & A. Friendly, Jr. 1992. Ecocide in the USSR. Health and Nature Under Siege. Basic Books, New York. 376 pp.

Goncharov, B.F., O.I Shubravy & V.K. Uteshev. 1991. Reproduction and early development of *Pseudoscaphirhynchus kaufmanni* Bogdanow under artificial environmental conditions. Ontogenez 22: 485–492 (in Russian, English translation Soviet J. Dev. Biol. 22: 296–301).

Gosteeva, M.N. 1953. Finding of the shovelnose sturgeon, *Pseudoscaphirhynchus kaufmanni* (Bogd.), in brackish water. Voprosy Ikhtiologii 1: 115–116 (in Russian).

Karpevich, A.F. 1975. Theory and practice of acclimatization of aquatic organisms. Pishchevaya Promyshlennost, Moscow. 432 pp. (in Russian).

Kessler, K.F. 1872. On a remarkable fish of the family of sturgeons discovered by A. P. Fedtchenko in the Syr Darya River. Izvestiya Obshchestva lyubitelei estestvoznaniya, antropologii i etnografii 10: 70–76 (in Russian).

Kessler, K.F. 1877. Fishes of the Aralo-Caspian-Pontine region. Trudy Aralo-Kaspiiskoi ekspeditsii 4: 190–196 (in Russian).

Lovetsky, A. 1828. On the fishes belonging to the sturgeon genus and inhabiting waters of the Russian Empire. Novyi Magazin Estestvennoi Istorii, Fiziki, Khimii i Svedenii Ekologicheskikh, Izdannyi I. Dvigubskim, Chast 6, Moscow (in Russian, French translation Bull. Sci. Nat. 23: 131–134 (1831)).

Mitrofanov, V.P., G.M. Dukravets, H.E. Peseridi & A.N. Poltorykhina. 1986. Fishes of Kazakhstan, Vol. 1. Izdatelstvo GYLYM, Alma-Ata. 271 pp. (in Russian).

Nedoshivin, A.Ya. 1928. Materials on the Don River fishery. Trudy Azovo-Chernomorskoi Nauchno-Promyslovoi Ekspeditsii 4: 1–175 (in Russian).

Nikolskii, G.V. 1938. Fishes of Tadjikistan. Izdatelstvo Akademii Nauk USSR, Moscow-Leningrad. 228 pp. (in Russian).

Nikolskii, G.V. 1940. Fishes of the Aral Sea. Moscovskoe Obshchestvo Ispytatelei Prirody, Moscow. 216 pp. (in Russian).

Osmanov, S.O. 1971. Parasites of fishes of Uzbekistan. FAN, Tashkent. 532 pp. (in Russian).

Pavlov, D.S., Yu.S. Reshetnikov, M.I. Shatunovsky & N.I. Shilin. 1985. Rare and endangered fish species of the USSR and principles for listing them in the Red Data Book. Voprosy Ikhtiologii 25: 16–25 (in Russian, English translation: J. Ichthyol. 25: 88–99).

Pavlovskaya, L.P. & I.M. Zholdasova. 1991. Anthropogenic changes in the fish fauna of the Amu Darya River (based on data from sampling drift of eggs and larvae). Voprosy Ikhtiologii 31: 585–595 (in Russian, English translation: J. Ichthyol. 31: 106–117).

Pechnikova, N.V. 1964. Results of acclimatization of the Aral Sea ship sturgeon (*Acipenser nudiventris* Lov.) in Lake Balkhash. Voprosy Ikhtiologii 4: 142–152 (in Russian).

Rogov, M.M. 1957. Hydrology of the delta of the Amu Darya River. Gidrometeoizdat, Leningrad. 254 pp. (in Russian).

Salnikov, V.B., V.J. Birstein & R.L. Mayden. 1996. The contemporary status of the two Amu Darya River shovelnose sturgeons, *Pseudoscaphirhynchus kaufmanni* and *P. hermanni*. Sturgeon Quart. 4(3): 8–10.

Schmalhausen, O.I. 1991. Prelarval development of *Pseudoscaphirhynchus kaufmanni*. Ontogenez 22: 493–513 (in Russian, English translation Soviet J. Dev. Biol. 22: 302–315).

Sokolov, L.I. & V.P. Vasilev. 1989a. *Acipenser nudiventris* Lovetsky, 1828. pp. 206–226. *In*: J. Holčík (ed.) The Freshwater Fishes of Europe, Vol. 1, Pt. II, General Introduction to Fishes, Acipenseriformes, AULA-Verlag, Wiesbaden.

Tleuov, R.T. 1981. New regime of the Aral Sea and its influence on ichthyofauna. Izdatelstvo FAN, Tashkent. 190 pp. (In Russian).

Tleuov, R.T. & N.I. Sagitov. 1973. Acipenserid fishes of the Aral Sea. FAN Press, Tashkent. 155 pp. (in Russian).

Tleuov, R.T. & Sh. Tleubergenov. 1974. Fishes of Karakalpakiya. Nukus. 95 pp. (in Russian).

Zholdasova, I.M., L.P. Pavlovskaya, L.N. Guseva & V.T. Utebaeva. 1990. Status of populations of rare and threatened species of the Amu Darya River and measures purposed to protect them. Information No. 483, FAN, Tashkent. 12 pp. (in Russian).

Zholdasova, I.M., L.P. Pavlovskaya & S.K. Lyubimova. 1991. On the death of fishes in the Tuyamuyun Reservoir (Amu Darya River). Bulletin Karakalpakskogo Otdeleniya Akademii Nauk Uzbekskoi SSR 1: 18–24 (in Russian).

Environmental Biology of Fishes **48**: 381–383, 1997.

Threatened fishes of the world: *Pseudoscaphirhynchus* spp. (*Acipenseridae*)

Vadim J. Birstein
The Sturgeon Society, 331 West 57th Street, Suite 159, New York, NY 10019, U.S.A.

Pseudoscaphirhynchus kaufmanni (Bogdanov, 1874)

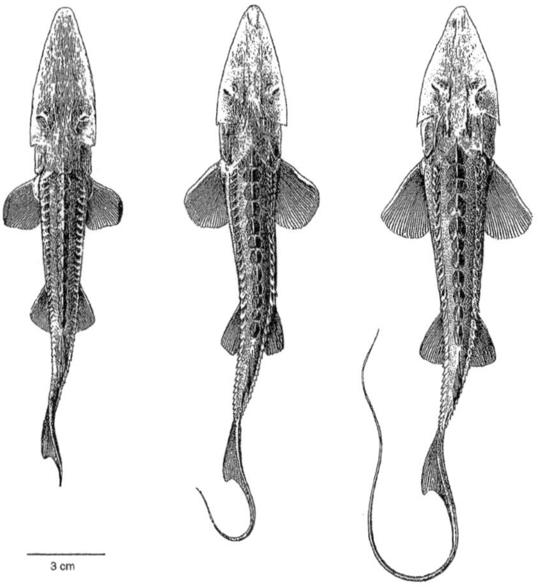

Pseudoscaphirhynchus hermanni — hybrid — *Pseudoscaphirhynchus kaufmanni*

Common names: Large Amu-Dar shovelnose sturgeon (E), Grand nez-pelle de lAmou daria (F), Bolshoi Amudarinskii Lzhelopatonos (R), Sumrai or Beltkumys (Karakalpakian), Elan Luiryk or Tuchkan Kuiryk (Uzbek and Turkmen). **Conservation status:** Endangered (Uzbek SSR Red Data Book 1983, USSR Red Data Book 1984, Turkmen SSR Red Data Book 1985, 1996 IUCN Red List). **Identification:** D 25–37, A 15–24 rays, 10–15 dorsal scutes, 28–40 lateral scutes, and 5–11 ventral scutes. Very unusal appearance. Body is fusiform, the fore part rather thick. Head ends in a broad spade-like snout. There are 2–4 backward pointing spines on the tip of the rostrum. Two similar spines are located on the upper side of the head in front of very small eyes. The snout is flattened on its upper surface and completely flat on the lower surface, where there are two pairs of barbels. Head shields are visible. The upper lip of the transverse mouth is divided in the middle, the lower lip is broader than the upper and is also slightly divided. Regularly distributed granulations lying between the scute rows. Below the dorsal fin base and above the anal fin base there are small flat scutes. All paired fins and the anal fin are rounded. Pectoral fins have a very strong and sharp first ray. The caudal is prolonged on its upper heterocercal lobe in a long filament. Historically, the maximal size of the fish was 75 cm and the 2 kg in weight. In the 1960s, the average size was 37 cm, and 241 g weight. Two forms were described in the 1960s: common and dwarf. Dwarf adults were smaller than the common ones and their dorsal, anal, and ventral fins were located closer to the tail than in common form. In the early 1990s, the dwarf form predominated in the population. Dorsum from grey to almost black, ventrum white. William E. Bemis modified the figures from originals published by Nikolskii (1938). **Distribution:** Endemic to the Amu Darya River and its tributaries (Central Asia). Historically, *P. kaufmanni* was distributed along the river from the upper reaches (Pyandzh River) to the delta. Presently, there are two populations: in the Vakhsh River, and the middle reaches of the Amu Darya River. In the early 1990s, only a few individuals were recorded in the lower reaches of the Amu Darya River. **Abundance:** No exact estimations. **Habitat and ecology:** *P. kaufmanni* live at a depth of 1.0–1.5 m in highly turbulent muddy water. Fish inhabit shallow-water parts of the river with fast current, sandy or stony-pebble grounds. Adults feed mainly on small fish, with insect larvae forming the rest of the diet. **Reproduction:** Takes place in late March-early May at a water temperature of 14–16 C. Males become mature at 5–7 years, and females, at 6–8 years. Intervals between spawning periods possibly last 4–5 years. Fecundity is 3127–36558 (common form) or 996–1910 (dwarfs). **Hybridization:** Historically, easily hybridized with the other species of *Pseudoscaphirhynchus*, *P. hermanni*. **Threats:** Changes in the environment caused by the drying out of the Aral Sea. Presently, the Amu Darya River does not reach the Aral Sea. Dams and channels constructed in the 1970s–1980s affected the water regime of the river. Also, the level of water pollution in the river is very high. **Conservation action:** An international recovery project is planned by scientists of Karakalpakstan (a part of Uzbekistan), Turkmenistan, the United States, and, possibly, Russia.

Pseudoscaphirhynchus hermanni (Kessler, 1877)

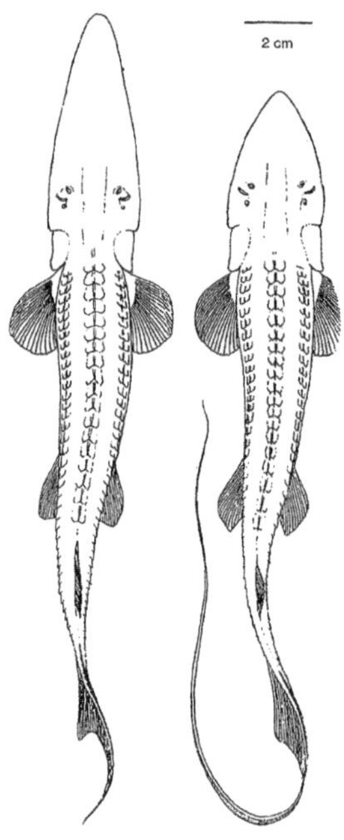

Pseudoscaphirhynchus fedtschenkoi

Common names: Small Amu-Dar shovelnose sturgeon (E), Petit nez-pelle de lAmou daria (F), Malyi Amudarinskii Lzhelopatonos (R).
Conservation status: Critically endangered (1996 IUCN Red List), Endangered (UzbekSSR Red Data Book 1983, USSR Red Data Book 1984, Turkmen SSR Red Data Book 1985).
Identification: D 27–35, A 15–21 rays, 10–13 dorsal scutes, 30–38 lateral scutes, and 6–10 ventral scutes. Morphologically it is similar to *P. kaufmanni*, but smaller (20.7–27.0 cm). Its snout is longer than that of *P. kaufmanni*. This species does not have a long caudal filament. The snout is shovel-shaped and the rostrum is more rounded than in *P. kaufmanni*. There are no spines on the snout. The snout grows longer with age. Pectoral fins have a fold which curls dorsally. Scutes are not armored with spines or the spines are very short. Each dorsal and lateral scute covers almost half of the following one. There are granules between the rows of scutes. As in *P. kaufmanni*, there are 3–4 flat scutes between anal and ventral fins. Eyes are extremely small. Two outer barbels are 2–3 times longer than the inner ones. Dorsum deep brown, ventrum white.
Distribution: Endemic to the Amu Darya River (middle and lower reaches). **Abundance:** Historically rare; in April 1996 three specimens were caught for the first time in the last 15 years. **Habitat and ecology:** Practically unknown. Adults are benthophagous feeding mostly on midge larvae. **Reproduction:** Unknown. **Hybridization:** Historically, easily hybridized with *P. kaufmanni*. **Threats:** Changes in environment caused by the drying out of the Aral Sea, construction of dams and channels which affected the water regime in the Amu Darya River, and a very high level of water pollution in the river. **Conservation action:** None. During the carrying out the recovery plan for *P. kaufmanni*, it will be possible to make the estimation of the status of *P. hermanni*.

Pseudoscaphirhynchus fedtschenkoi (Kessler, 1872)

Common names: Syr-Dar shovelnose sturgeon (E), Nez-pelle du Syr daria (F), Syrdarinskii Lzhelopatonos (R).
Conservation status: Critically endangered (1996 IUCN Red List), Endangered (KazakhSSR Red Data Book 1978, USSR Red Data Book 1983).
Identification: D 30–34, A 19–20 rays, 15–22 dorsal scutes, 38–46 lateral scutes, and 6–11 ventral scutes. It is small: with the caudal filament, up to 36 cm; without the filament, 20.7–27.0 cm. It has more dorsal scutes and a longer snout than the two Amu Darya species, *P. kaufmanni* and *P. hermanni*. The pectoral fins have a fold similar to that in *P. hermanni*. There are no spines on the head. The size and shape of the snout varies considerably. Historically, there were three morphs: (1) a common morph with a long rostrum without a long caudal filament; (2) morpha *brevirostris* Berg with a short snout and a long caudal filament; (3) morpha *intermedia* Berg with a middle-sized snout and a filament.
Distribution: Endemic of the Syr Darya River, middle and lower reaches (Central Asia). **Abundance:** No reports since the 1960s; practically extinct. **Habitat and ecology:** Unknown. Adults were benthophagous feeding mostly on midge larvae. **Reproduction:** Spawning during late April. **Threats:** Changes in the environment caused by the drying out of the Aral Sea. The Syr Darya River, as the Amu Darya River, does not reach the Aral Sea now. **Conservation action:** None.

Berg, L. S. 1905. Fishes of Turkestan. Scientific results of the Aral expedition, No. 6, St. Petersburg. 261 pp. (in Russian).
Berg, L. S. 1948. The freshwater fishes of the USSR and adjacent countries, Vol. 1, Part 1. Akademia Nauk USSR, Moscow & Leningrad (in Russian, English translation published by Israel Program for Scientific Translations, Jerusalem. 505 pp.
Bogdanov, M. 1874. A report on a newly discovered acipenserid fish at the meeting of Zoological Section. Trudy Sankt-Peterburgskogo obshchestva ispytatelei prirody 5: 48 (in Russian).

Kessler, K. F. 1872. On a remarkable fish of the family of sturgeons discovered by A. P. Fedtchenko in the Syr Darya River. Izvestiya Obshchestva lyubitelei estestvoznaniya, antropologii i etnografii 10: 70–76 (in Russian).

Kessler, K. F. 1873. On a remarkable fish of the family of sturgeons discovered by M. A. P. Fedchenko in the River Suir-dar. Ann. & Mag. Nat. Hist. Ser. 4. No. 70: 269–276.

Kessler, K. F. 1877. Fishes of the Aralo-Caspian-Pontine region. Trudy Aralo-Kaspiiskoi ekspeditsii 4: 190–196 (in Russian).

Mitrofanov, V. P., G. M. Dukravets, H. E. Peseridi & A. N. Poltorykhina. 1986. Fishes of Kazakhstan. Vol. 1, GYLYM Press, Alma-Ata. 271 pp. (in Russian).

Nikolskii, G. V. 1938. Fishes of Tadjikistan. Izdatelstvo Academii Nauk SSSR, Moscow-Leningrad. 228 pp. (in Russian).

Salnikov, V.B., V.J. Birstein & R.L. Mayden. 1996. The contemporary status of the two Amu Darya River shovelnose sturgeons, *Pseudoscaphirhynchus kaufmanni* and *P. hermanni*. Sturgeon Quart. 4(3): 8–10.

Tleuov, R. T. & N. I. Sagitov. 1973. Acipenserid fishes of the Aral Sea. FAN Press, Tashkent. 155 pp. (in Russian).

Zholdasova, I. 1997. Sturgeons and the Aral Sea ecological catastrophe. Env. Biol. Fish. (this volume).

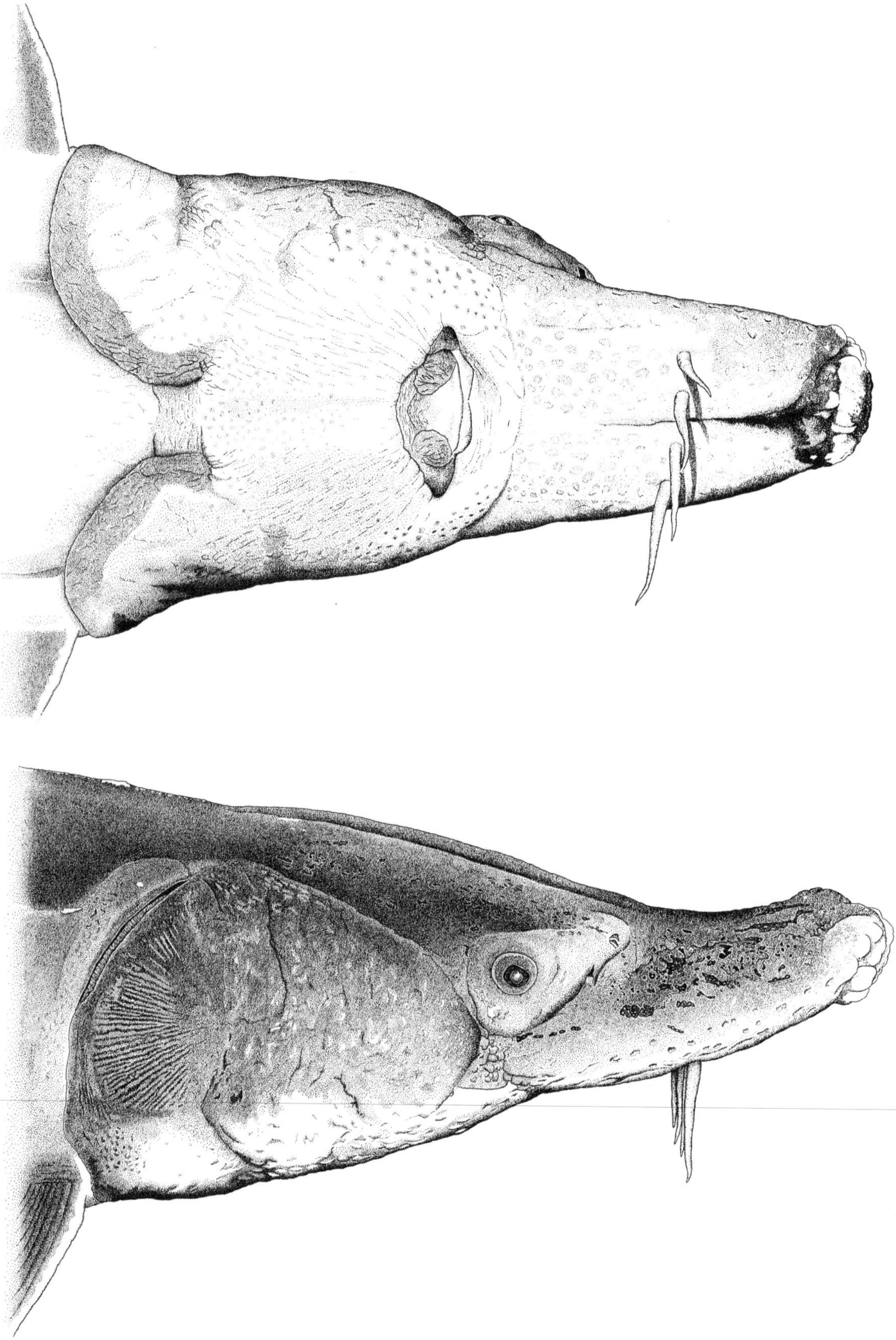

Head from ventral and right side of *Acipenser oxyrinchus* 200 cm TL juvenile, from St. Lawrence River at Kamouraska, Quebec, wh
now resides at the Montreal Biodome (see the typical damage to the tip of the rostrum). Originals by Paul Vecsei. 1996.

Environmental Biology of Fishes **48**: 385–398, 1997.

Molecular analysis in the conservation of sturgeons and paddlefish

Isaac I. Wirgin[1], Joseph E. Stabile[1] & John R. Waldman[2]
[1] *Nelson Institute of Environmental Medicine, New York University Medical Center, Long Meadow Road, Tuxedo, NY 10987, U.S.A.*
[2] *Hudson River Foundation, 40 West 20th Street, Ninth Floor, New York, NY 10011, U.S.A.*

Received 15.4.1995 Accepted 14.5.1996

Key words: mitochondrial DNA, nuclear DNA, *Acipenser, Scaphirhynchus, Polyodon*, stock identification, hybridization

Synopsis

Sturgeon and paddlefish populations worldwide have declined because of anthropogenic influences. The structure and magnitude of genetic diversity of natural populations serves to buffer these fishes against environmental variation and should be maintained. Modern molecular biological techniques provide the ability to sensitively characterize and quantify the extent of genetic variation in natural populations. We provide a summary of those problems in sturgeon population biology that are amenable to investigation with DNA approaches, and their applications to date. These have included genetic identification and discrimination of taxa, identification of hybrids, stock identification, mixed-stock analysis, and estimation of gene flow and homing fidelity. To date, almost all studies have been restricted to North American fauna. Improvements to these technologies, including nondestructive sampling, should permit more widespread application of molecular approaches to problems of acipenseriform conservation. We suggest that the use of more sensitive molecular tools such as analyses of hypervariable repetitive and non-coding single copy nuclear DNA may assist management even in those taxa which exhibit overall low levels of genetic diversity.

Introduction

The worldwide diversity of sturgeons and paddlefishes is currently threatened, and in fact, the extirpation of some forms in Europe and Asia has been reported (Rochard et al. 1990, Birstein 1993, Waldman 1995). At the same time, population abundances of most species are at historically low levels, including many North American taxa (e.g., Kynard 1997 this volume, Smith & Clugston 1997 this volume). Although molecular analysis has rapidly become a primary tool in the management and conservation of fishes (Hallerman & Beckmann 1988, Wirgin & Waldman 1994, Bernatchez 1995), its use for problems involving acipenseriforms has lagged, perhaps due to the scarcity of many of these species and the consequential difficulty of obtaining adequate numbers of tissue samples. However, improvements to DNA analysis technologies, including nonlethal sampling, slackening of tissue preservation requirements, and the ability of PCR to allow analysis of minute amounts of tissue have relaxed many of the constraints that challenged sample collection. Thus, we anticipate an expanding role for molecular analysis in the conservation of these highly threatened fishes.

Wirgin & Waldman (1994) listed 12 areas of investigation in fisheries biology amenable to DNA analyses. Here, we review studies in these areas in which molecular techniques have been applied to problems in the conservation of sturgeons and paddlefishes. We also suggest other areas to which these techniques may be applied. Additionally, we present a genetic comparison between two ambiguously differentiated, putative sturgeon species, *Acipenser oxyrinchus* (American Atlantic sturgeon) and *A. sturio* (European Atlantic sturgeon).

Characteristics of the acipenseriform genome

All acipenseriforms are polyploids and their chromosomes can be arranged as a series 4n–8n–16n (Birstein et al. 1993, Blacklidge & Bidwell 1993, Birstein et al. 1997 this volume). Chromosome numbers are very high: 120, 240, and possibly, 500. Two forms of chromosomes are found: normal-sized 'macrochromosomes' and very small 'microchromosomes'. A number of studies have indicated slow rates of DNA and protein evolution of sturgeons and paddlefishes (Birstein 1993); we believe that extended generation times in comparison with most other fishes may decelerate their molecular evolution. Acipenseriforms hybridize readily in the laboratory and in nature, at even the intergeneric level (Birstein et al. 1997 this volume).

In most respects, the structure and evolutionary change of the acipenseriform mitochondrial DNA (mtDNA) genome is typical of vertebrates. Estimates of the size of mtDNA in *Acipenser transmontanus* (white sturgeon, 16.1–16.7 kilobases, Brown et al. 1992) and *Acipenser fulvescens* (lake sturgeon, 16.6–16.9 kb, Guenette et al. 1993) are within the range of most other vertebrate taxa (16.5 ± 0.5 kb, Meyer 1993). Gene composition and gene order of acipenseriform mtDNA have not been directly investigated; however, the mtDNA control region of *A. transmontanus* is organized similarly to other vertebrates (Buroker et al. 1993).

Differences in the size of the mtDNA molecule have been observed among acipenserids. In *A. transmontanus* (Buroker et al. 1990) and *A. fulvescens* (Ferguson et al. 1993), the overall size of the mtDNA molecule differs among individual specimens (both within and among populations) because of variable numbers of tandem repeat units within the control region. In contrast, no evidence of mtDNA length variation was observed among specimens of *A. oxyrinchus* (Waldman et al. 1996b). Moreover, variation in the size of mtDNA within individuals (heteroplasmy) was seen in *A. transmontanus*, but not *A. oxyrinchus* (Stabile et al. 1996). A thorough review of the acipenseriform genome appears in Birstein et al. (1997 this volume).

Molecular studies of acipenseriforms directed toward management and conservation

Identification and discrimination of taxa

In the U.S., the taxonomic status of threatened or endangered populations of vertebrates is important in determining the degree of protection they may receive under the Endangered Species Act (O'Brien & Mayr 1991). For *Acipenser oxyrinchus*, the Altantic (*A. o. oxyrinchus*) and Gulf forms (*A. o. desotoi*) were designated subspecies based on differentiation of morphological features, although only two Gulf specimens were examined (Vladykov 1955). Wooley (1985) reanalyzed these morphological features of a larger sample of Gulf sturgeon and concluded that only one, relative spleen length, was diagnostic. However, values for relative spleen length of the two forms were not discrete and it may be that differences in these values are ecophenotypic in nature. Ong et al. (1996) used direct sequencing of a hypervariable area (203 base pairs) of the control region of mtDNA to quantify the extent of genetic differentiation between the two putative subspecies. Representatives of each subspecies from populations across their distributions were surveyed and three fixed differences were found among 15 polymorphic sites. An additional two nucleotides were nearly fixed. Although polymorphisms also were detected within populations across the distribution of each subspecies, no fixed differences among populations were found. Ong et al. (1996) concluded that these data strongly supported the designation of subspecies of *A. oxyrinchus*.

The relationship between *A. oxyrinchus* and *A. sturio* has long been debated. These fish are found on opposite sides of the North Atlantic and are phenotypically very similar. *Acipenser oxyrinchus* occurs widely along the Atlantic coast of North Amer-

```
                                1                TAS1                          51
A. sturio                       taagattctacattaaactattctctggccacat..gtctgacccatacca
                                ||||||||||||||||||||||||||||||| ||    ||  || | ||||
A. oxyrinchus (Atlantic)        taagattctacattaaactattctctggccatatcatgctcgcctacacca
                                |||||||||||||||||||||||||||||||||||||||||||||||||||
A. oxyrinchus (Gulf)            taagattctacattaaactattctctggccatatcatgctcgcctacacca

                                52                                              100
A. sturio                       atgtctgcatacattaaattgtacatacataaacatactatgtttaatccc
                                |||| |  ||||||||||||| |||| |||  |||||||||||||||||
A. oxyrinchus (Atlantic)        atgttt..atacattaaattgcttatgcatggacatactatgtttaatccc
                                ||||||  |||||||||||||||||| ||||||||||||||||||||||
A. oxyrinchus (Gulf)            atgttt..atacattaaattgcttatacatggacatactatgtttaatccc
                                         PS1          PS2  *PS3*       PS4

                                101                                             151
A. sturio                       cattaatttctagccaccaatactaatgtttacctatatattaaattatcta
                                ||||||||||||| |||| ||| |||||||||  |||| |||||  || ||
A. oxyrinchus (Atlantic)        cattaatttctagtcacc.atatcaatgtttatatatacattaagccattta
                                |||||||||||||||||| ||| |||||||||||||||||||| ||||||
A. oxyrinchus (Gulf)            cattaatttctagtcacc.ataccaatgtttatatatacattaacccattta
                                                      PS5      PS6   PS7      PS8
                                152                                             203
A. sturio                       agtacatagacatactatgtttaatccccattaatttctagtcaacatatca
                                |||||||  .||||||||||||| |||||||||| ||||||||||||||||
A. oxyrinchus (Atlantic)        agtacatgaacatactatgttttatccccattaacttctagtcaccataaca
                                |||||||| ||||   ||||||||||||| |||||||||||||||| ||||||
A. oxyrinchus (Gulf)            agtacatgaacatgttatgttttatcctcattaacttctagtcactataaca
                                     PS9  *PS10PS11*      PS12    PS13  PS14 PS15
```

Figure 1. Comparison of mtDNA control region sequence from *Acipenser sturio* and the Atlantic and Gulf of Mexico subspecies of *Acipenser oxyrinchus*. Alignment begins nine nucleotides upstream of the terminal association sequence (TAS 1) of tRNA proline. The position of the polymorphic sites observed among all 159 specimens of *A. oxyrinchus* are highlighted in bold. The position of the three fixed sites between the subspecies of *A. oxyrinchus* are noted by asterisks.

ica. *Acipenser sturio*, which once had a similarly broad range in Europe and Asia, now occurs only in very low abundances in both the Gironde River, France, and the Black Sea (Rochard et al. 1990). Some workers have considered the Atlantic sturgeon to be synonymous with or a subspecies of *A. sturio* (Scott & Scott 1988, Birstein 1993), but Vladykov & Greeley (1963) and Magnin (1964) recommended they be given separate specific status, pending additional research. We obtained a tissue sample from one *A. sturio* specimen captured during 1994 in the Gironde River. We then compared a portion (203 bp) of the control region of mtDNA between *A. sturio* and *A. oxyrinchus*.

We found a minimum of 31, and a maximum of 33 nucleotide substitutions between the individual of *A. sturio* and 159 individuals of both subspecies of *A. oxyrinchus* (Figure 1). Three sites also exhibited nucleotide additions or deletions. In comparison, the number of nucleotide substitutions between any pair of specimens of the two subspecies of *A. oxyrinchus* ranged between five and eight, with no additions or deletions. Excluding additions or deletions, nucleotide divergence between *A. oxyrinchus* and *A. sturio* was approximately 15%, much higher than the maximum of 3.5% between two subspecies of *A. oxyrinchus* (Ong et al. 1996). There are no unambiguous criteria for the interpretation of molecular data in determining taxonomic status (e.g., O'Brien & Mayr 1991, Wayne 1992). However, we believe that the level of differentiation observed argues strongly for full species status of each of the western and eastern Atlantic sturgeons.

Scaphirhynchus suttkusi (Alabama sturgeon) has recently been described (Williams & Clemmer 1991); this exceedingly rare species is restricted to the Mobile River basin of Alabama and Mississippi. *Scaphirhynchus suttkusi* are morphologically similar to *S. platorynchus* (shovelnose sturgeon), but differ significantly in six meristic and at least eleven mensural variables. The extent of genetic similarity between *S. suttkusi* (one specimen) and *S. platoryn-*

chus and *S. albus* (pallid sturgeon) was compared using PCR products (amplicons) (Genetic Analyses[1]). *Acipenser fulvescens* and the two subspecies of *A. oxyrinchus* were analyzed to provide additional points of reference. PCR primers were designed based on conserved sequences (among mammals) for eight nDNA genes and these were shown to amplify sturgeon DNA and to produce amplicons. These amplicons were then digested with a battery of different restriction enzymes and DNA band sharing was compared between two putative taxa. The level of genetic differentiation between *S. suttkusi* and the other two species of *Scaphirhynchus* was greater than that between subspecies of *A. oxyrinchus* but less than that between the two species of *Acipenser*. The authors also concluded that distinct species designation for *S. albus* and *S. platorynchus* may not be valid, but instead, that the two forms may represent morphological variants of a single species.

A later, expanded mtDNA study (Campton et al.[2]) also found very low levels of divergence between *S. suttkusi* (N = 3) and the other two *Scaphirhynchus* species. A unique haplotype distinguished the three specimens of *S. suttkusi* from all individuals of the other two species of *Scaphirhynchus* collected from the upper Missouri River, but this haplotype differed from the most common haplotype found among *S. albus* and *S. platorynchus* by only a single nucleotide substitution. Campton et al.[2] concluded that the mtDNA data support the morphological and biogeographic arguments for recognizing *S. suttkusi* as an endangered 'species' of 'distinct population segment' as defined by the U.S. Endangered Species Act. Analysis of additional specimens of *S. albus* and *S. platorynchus* from the lower Mississippi River (nearer the range of *S. suttkusi*) is needed to confirm the genetic discreteness of *S. suttkusi* from its congeners.

Analyses of hybridization and introgression

Scaphirhynchus albus and *S. platorynchus* have been recognized as similar, but distinct species of river sturgeon within the Mississippi River drainage (Forbes & Richardson 1905). Significant differences in meristic, morphometric, and life history characteristics support this taxonomic division. In recent years, a small but increasing number of fish collected from the Missouri and Mississippi rivers exhibited intermediacy in the expression of these discriminatory characters, suggesting the existence of interspecific hybrids (Carlson et al. 1985). Populations of both species, but particularly of *S. albus*, currently are severely depleted, probably due to habitat alterations, and *S. albus* is listed as endangered by the U.S. Fish and Wildlife Service (Dryer & Sandoval[3]). Hybridization between two sturgeon species under these circumstances would not be unexpected; Hubbs (1955) concluded that anthropogenic impacts and sharp imbalances in the abundances of potentially hybridizing species significantly increase the frequency of hybridization events in fishes.

Genetic studies were conducted to quantify the relatedness of *S. albus* and *S. platorynchus* and to identify markers which could be used to identify F_1 and later generation hybrids. Protein electrophoresis studies revealed only low levels of polymorphism within and between these taxa and those loci (3 of 37) which were polymorphic did not display significant interspecific allelic differences (Phelps & Allendorf 1983). Thus, protein electrophoresis proved insensitive in distinguishing these species, perhaps due to its focus on gene products that are essential for survival and that tend to be conservative across taxa.

Recently, direct analyses of nuclear DNA (nDNA) and mtDNA were used to further discrimi-

[1] Genetic Analyses, Inc. 1994. Genetic studies of *Scaphirhynchus* spp. Report to the U.S. Army Corps of Engineers, Omaha District; U.S. Fish and Wildlife Service, Bismarck, North Dakota; U.S. Army Corps of Engineers, Mobile District. 41 pp.

[2] Campton, D.E., A.I. Garcia, B.W. Bowen & F.A. Chapman. 1995. Genetic evaluation of pallid, shovelnose, and Alabama sturgeon (*Scaphirhynchus albus, S. platorynchus*, and *S. suttkusi*) based on control region (D-loop) sequences of mitochondrial DNA. Final Report to the U.S. Fish and Wildlife Service, Bismarck, North Dakota. 35 pp.

[3] Dryer, M.P. & A.J. Sandoval. 1993. Recovery plan for the pallid sturgeon (*Scaphirhynchus albus*). U.S. Fish and Wildlife Service, Denver, 55 pp.

nate between *S. albus* and *S. platorynchus* with the expectation that higher levels of diversity would be detected with these more sensitive approaches. Nuclear DNA sequences were PCR amplified with conserved primers and these PCR products were then digested with a battery of different restriction enzymes and DNA band sharing was compared between the two putative taxa. Variability among individuals in the digestion patterns of amplicons was seen only at a prealbumin-related locus (Genetic Analyses, Inc.[1]). A comparison of allelic frequencies between *S. albus* and *S. platorynchus* at the prealbumin-related locus showed no significant differences.

Campton et al.[2] sequenced more than 400 bp of mtDNA from the control regions of *S. albus* (N = 18) and *S. platorynchus* (N = 20) and found 8 haplotypes from an area of sympatry in the upper Missouri River. However, these haplotypes overlapped between the two species to the extent that maximum parsimony analysis did not reveal two distinct species-congruent branches, but χ^2 analysis did show significant haplotypic frequency differences ($p < 0.001$) between the species. Control region of other acipenserids were similarly sequenced to serve as references; mean nucleotide diversity among specimens of the three scaphirhynchids was 0.58%, compared to 1.20% for *A. transmontanus*. Also, the mean nucleotide diversity was 0.62% between *S. albus* and *S. platorynchus*, compared with 14.1% between *A. transmontanus* and *A. medirostris* (green sturgeon). Campton et al.[2] were unable to dismiss hybridization as a factor in the differentiation of *S. albus* and *S. platorynchus*, but they favored the hypothesis that the two species maintain some degree of reproductive isolation.

Identification of stocks and assessment of genetic variability

Several studies have assessed genetic differentiation among hypothesized stocks of sturgeons and paddlefishes. Identification of genetic stocks requires that individuals from hypothesized stocks be surveyed for genetic variability, usually at rapidly evolving sites. If statistically significant genetic differences exist among two or more spawning populations, they may be considered separate stocks and managed as distinct units (Waldman & Wirgin 1994). Acipenseriforms investigated for stock differences include two freshwater species, *Polyodon spathula* (American paddlefish) and *A. fulvescens*; and three anadromous sturgeons, *S. stellatus* (stellate sturgeon), *A. transmontanus*, and *A. oxyrinchus*.

Genetic studies on populations of *P. spathula* have been conducted on both the protein and DNA levels. Carlson et al. (1982) observed extremely low levels of protein variation in *P. spathula* and did not detect evidence of differentiation among populations. In contrast, using both protein analysis and restriction fragment length polymorphism (RFLP) analysis of mtDNA, Epifanio et al.[4] found greater levels of genetic polymorphisms among *P. spathula* collected from 21 populations, including the Mississippi, Pearl, Alabama, and Neches drainages. However, the majority of protein variation was observed within populations and was of little use in elucidating population structure. Mitochondrial DNA genotypes showed greater geographic partitioning with both northern (mid-Missouri River) and southern haplotypes evident, suggesting the existence of some population structure within their overall distribution.

Acipenser fulvescens also is widely distributed in North America. Porter et al.[5] used protein electrophoresis to compare the genetic status of *A. fulvescens* from Lake Erie with other populations from the Laurentian Great Lakes. Only the lactate dehydrogenase enzyme system (LDH) revealed useful population-level variation. Although the Lake Erie population was the least variable of those surveyed, one of its three LDH phenotypes was not found among other populations. Guenette & Fortin (1993) found low levels of mtDNA variation in *A. fulvescens* from the St. Lawrence River and James

[4] Epifanio, J., M. Nedbal & D.P. Phillipp. 1989. A population genetic analysis of paddlefish (*Polyodon spathula*). Report to Missouri Department of Conservation. 63 pp.

[5] Porter, B., T. Cavender, P. Fuerst & T. Nickell. 1995. The genetic status of lake sturgeon in Lake Erie and other populations from the Laurentian Great Lakes. Report to Ohio Department of Natural Resources. 47 pp.

Bay drainages. No differences in mtDNA haplotypes were detected in sturgeon from different locations in the St. Lawrence River; however, haplotypic frequencies differed from those in the Waswanipi River in the James Bay basin. The lower mtDNA heterogeneity in fish from the St. Lawrence River was interpreted to reflect greater anthropogenic influences on population abundance.

Ferguson et al. (1993, 1997 this volume) used RFLP analysis of mtDNA and direct sequencing of the control region of mtDNA to investigate stock structure of *A. fulvescens* from Canadian systems, with an emphasis on the Moose River basin in Ontario. Little differentiation of mtDNA haplotypes was found among major tributaries of the Moose River basin, suggesting an absence of discrete stocks. However, sturgeon from the Moose River basin exhibited significantly higher levels of mtDNA diversity and were genetically differentiated from fish from the Great Lakes-St. Lawrence drainage and other Hudson Bay-James Bay populations. These workers hypothesized that mtDNA differentiation in sturgeon from Canadian waters resulted from post-Pleistocene colonization from separate refugia in the Mississippi and St. Lawrence River drainages. Further investigation of mtDNA diversity in extant stocks of *A. fulvescens* in eastern and western U.S. drainages should resolve this question.

Acipenser stellatus is an abundant sturgeon species in the Caspian Sea. Stock structure of this species in the southern portion of the Caspian Sea was investigated by restriction enzyme digestion of the PCR amplified ND5/ND6 region of mtDNA. Although polymorphic haplotypes were identified, no significant differences in haplotype frequencies were detected among fish from four regions, which led to the conclusion that *A. stellatus* in the southern Caspian Sea probably represent a single stock (M. Pourkazemi personal communication).

Acipenser transmontanus occurs from California to Alaska with viable fisheries currently centered in the Columbia River, Washington and the Fraser River, British Columbia. Both sequencing of the rapidly evolving control region of mtDNA and RFLP analyses of the entire mtDNA molecule were used to examine the extent of differentiation between populations of *A. transmontanus* from the Fraser and Columbia rivers (Brown et al. 1992, 1993). In the most recent glacial advance, the lower Columbia River was believed to remain an ice-free refugium, whereas the Fraser River was completely glaciated. Sequence variation of the control region was consistent with the hypothesis that the Columbia River was the founder source for the Fraser River population (Brown et al. 1993). However, levels of overall mtDNA diversity were significantly higher among fish from the Fraser River than the Columbia River, despite the greater recency of the Fraser River population (Brown et al. 1992). It was hypothesized that lower genetic diversity in the Columbia River population resulted from reduced access to historic spawning areas due to the construction of dams. Brown et al. (1992) also found significant differences in mtDNA haplotype frequencies between samples from the Fraser River and Columbia River, indicating that these two rivers support genetically distinct populations of sturgeon. Because haplotype differences were not fixed between the two populations, the authors suggested that low levels of gene flow between systems may be occurring. Alternatively, we suggest that given the brief time since divergence of these populations, the accumulation of significant differences in genotype frequencies indicates effective reproductive isolation of these systems.

Populations of *A. oxyrinchus* occur in drainages of the Gulf of Mexico and along the Atlantic coast of North America. *Acipenser oxyrinchus desotoi* (Gulf sturgeon) are considered threatened by the U.S. Fish and Wildlife Service. Historically, *A. o. desotoi* were found in major river systems extending from central Florida to the Mississippi River; many of these drainages still host depleted populations (USFWS and GSMFC[6]). Efforts are being considered to restore depressed populations through hatchery supplementation. However, knowledge of the stock composition of *A. o. desotoi* was federally mandated prior to the initiation of restorative efforts, so that native gene pools are pre-

[6] U.S. Fish and Wildlife Service and Gulf States Marine Fisheries Commission. 1995. Gulf sturgeon recovery plan. Atlanta, Georgia. 170 pp.

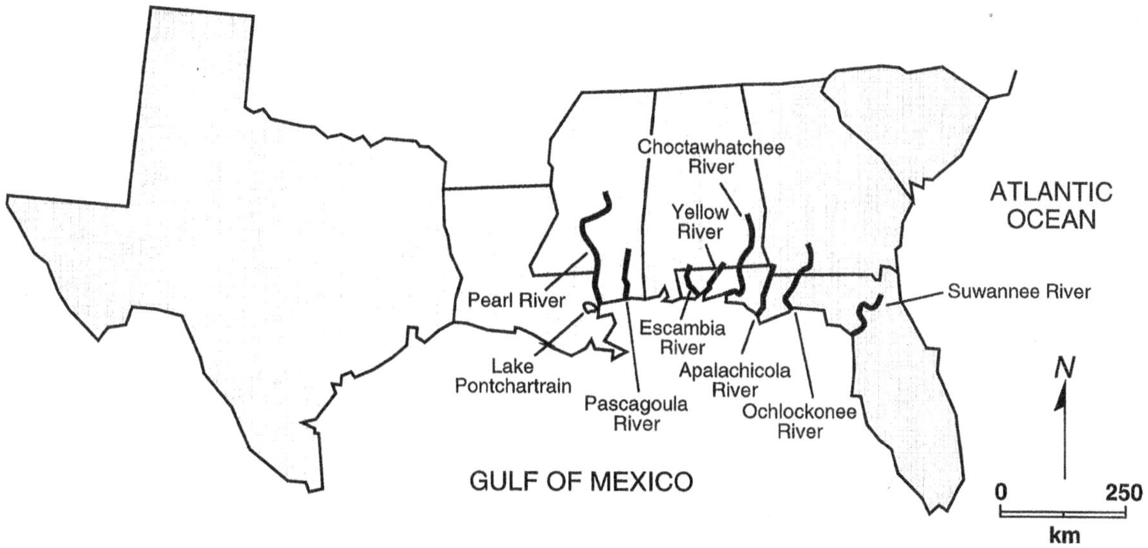

Figure 2. Gulf of Mexico rivers from which specimens of *Acipenser oxyrinchus* were obtained.

served. The Suwannee River population has been extensively studied over the past decade and probably contains the largest population of sturgeon along the Gulf of Mexico. Miracle & Campton (1995) examined the extent of genetic variation in sturgeon from the Suwannee River to determine if its population constituted a single homogeneous unit. Sequence analysis of 268 base pairs of a highly variable area of the mtDNA control region did not reveal significant genetic heterogeneity among sturgeon from different sampling locations.

We used RFLP and sequencing analysis of mtDNA to assess the stock structuring of *A. o. desotoi* populations among eight drainages (Figure 2) extending from the Mississippi River to the Suwannee River (Stabile et al. 1996). RFLP analysis using four diagnostic restriction enzymes yielded eight composite haplotypes; genotypic diversity indices (Nei & Tajima 1981) ranged between 0.173 for the Choctawhatchee River sample to 0.732 for the Yellow River collection. Significant differences ($p < 0.05$) in haplotype frequencies indicated substantial geographic structuring of sturgeon populations; results from the RFLP ($N = 164$) and sequence ($N = 141$) analyses were largely congruent (Figure 3). Five regional or river-specific stocks were identified (from west to east): (1) Lake Ponchartrain and Pearl River, (2) Pascagoula River, (3) Escambia and Yellow rivers, (4) Choctawhatchee River, and (5) Apalachicola, Ochlockonee, and Suwannee rivers. These results suggest strong reproductive isolation of *A. o. desotoi* stocks on at least a regional basis, and point to the inadvisibility of mixing of hatchery-reared progeny of broodstock from different Gulf rivers.

Also, we used RFLP analysis of mtDNA with five diagnostic restriction enzymes to characterize the stock structure of populations of *A. o. oxyrinchus* along the Atlantic coast, including the St. Lawrence River, Quebec; St. John River, New Brunswick; Hudson River, New York; Edisto River, South Carolina; and four rivers in Georgia; the Altamaha, Ogeechee, Savannah, and Satilla (Waldman et al. 1996a, b). Chi-square analysis showed the eight populations could be grouped as three highly differentiated ($p < 0.0001$) stocks: (1) Canadian (St. Lawrence and St. John rivers); (2) Hudson River; and (3) southeastern (Edisto, Savannah, Ogeechee, Altamaha, and Satilla). Composite haplotypes showed a clear cline in genotypic diversity indices (Nei & Tajima 1981) among populations that ranged from complete monomorphism (0.0) of the two Canadian populations to considerable polymorphism among southeastern populations (e.g., Edisto River: 0.646; Ogeechee River: 0.750). A latitudinal cline in genotypic diversity along the Atlantic coast is consistent with founder effects among

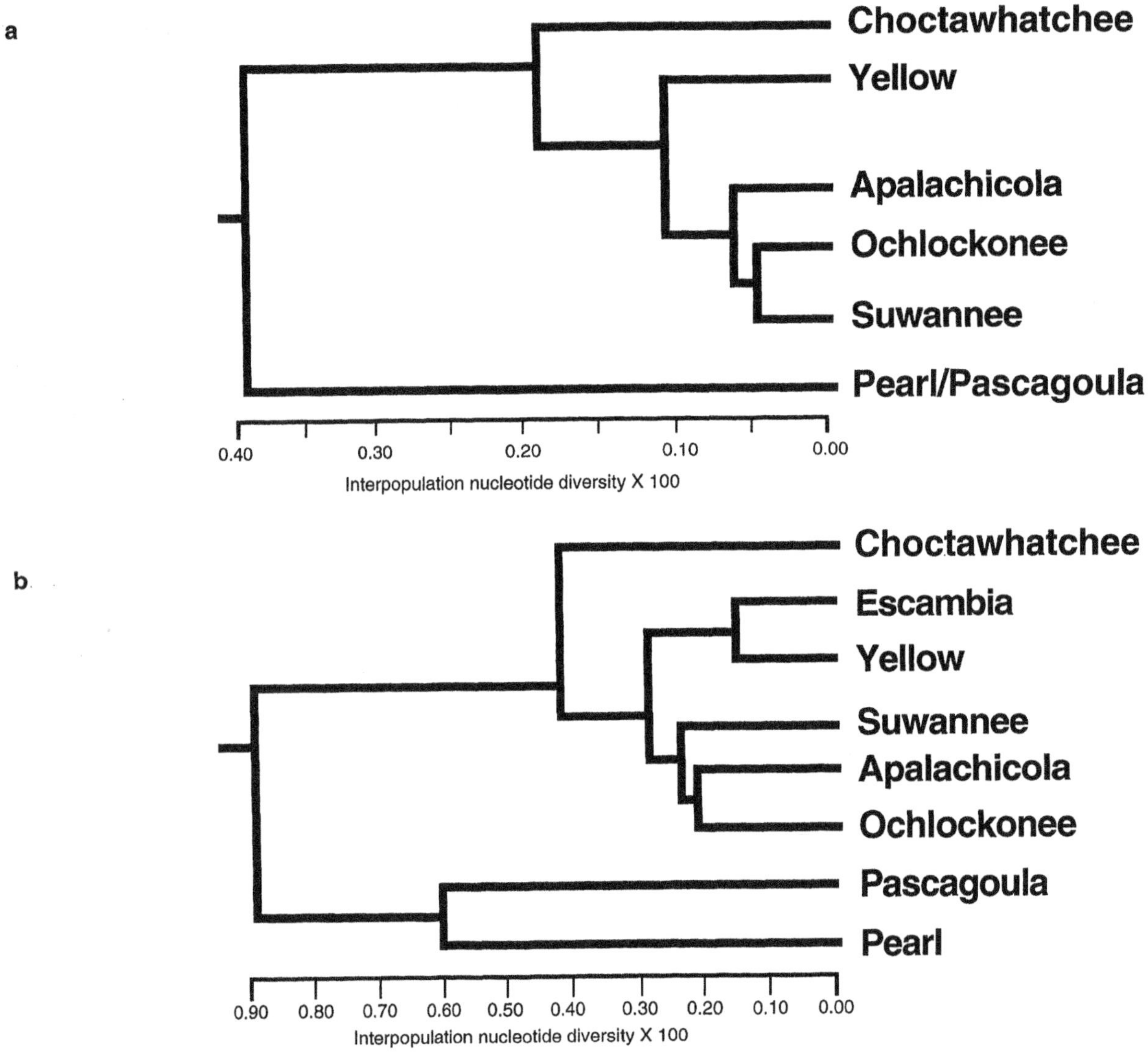

Figure 3. UPGMA phenograms of the interpopulation diversity indices for the (a) RFLP data, and (b) control region sequence data fo *Acipenser oxyrinchus* from the Gulf of Mexico.

northern populations that recolonized glaciated drainages from more genotypically diverse populations in southern, nonglaciated regions.

Mixed-stock analysis

For some wide ranging species, fisheries have developed distant from spawning and nursery areas, and these fisheries may harvest individuals from more than one stock. For management purposes, it is im portant to quantitatively estimate the relative con tributions of individual stocks to mixed fisheries t allow managers to protect threatened stocks at site distant from their natal rivers. Successful applica tion of genetic approaches to mixed-stock analysi is dependent on the existence of significant differ entiation of genetic characters among spawnin stocks which contribute to the mixed fishery (Utte & Ryman 1993, Xu et al. 1994). To conduct mixe stock analysis, frequencies of genotypes must be characterized in reference spawning stocks and in the mixed fishery.

A targeted coastal fishery for *A. o. oxyrinchus* has developed in recent years along the mid-Atlantic coast of New Jersey and New York (New York Bight) and bycatch fisheries have been reported off the southeastern coast of the U.S. (Collins et al. 1996). Waldman et al. (1966a) performed RFLP analysis of specimens of *A. o. oxyrinchus* from eight populations from Canada to Georgia and concluded, based on haplotype frequency differences, that three statistically discrete ($p < 0.0001$) stocks exist: (1) Canadian, (2) Hudson River, and (3) southeastern. Haplotypic frequency data of these stocks were then used in a mixture model (constrained least squares; Xu et al., 1994) to estimate the relative contributions of each of these stocks to a sample of Atlantic sturgeon (N = 112) from the fishery in the New York Bight off New Jersey. This analysis showed a 97% to 99% contribution from the Hudson River stock, with the remainder from the southeastern stock. The overwhelming contribution of the Hudson River stock was attributed both to (1) a hypothesized tendency for marine migrating Atlantic sturgeon to remain within the geographic provinces of their natal rivers (the Hudson River is within the Virginian province), and (2) to the absence of other robust American Atlantic sturgeon populations within the Virginian province.

Gene flow and homing fidelity

Most populations of sturgeons are anadromous or potamadromous and thus, migrate from marine or lake waters to rivers to spawn (Bemis & Kynard 1997 this volume). However, almost nothing is known of the degree of homing fidelity shown by acipenseriforms. Although homing fidelity of fishes may be studied directly by means of capture-mark-recapture (e.g., Melvin et al. 1986), the relative scarcity and high value of sturgeons precludes such an approach. An alternative is to assess homing fidelity indirectly through genetic analysis (Tallman & Healey 1994).

Homing fidelity of sturgeons through genetic analyses would best be assessed among populations in rivers that drain to a common water body and that historically have a stable geographic history to avoid confoundment by founder effects as a consequence of recolonizations. Stabile et al. (1996) used both RFLP and sequencing analysis of mtDNA to estimate gene flow among five stocks of *A. o. desotoi* that occur in eight drainages that feed the Gulf of Mexico between Mississippi and Florida. The five stocks were defined based on χ^2 analyses ($p < 0.05$) of haplotype frequencies; some stocks were equivalent to single populations, whereas others were regional stocks made up to two or more populations. Pairwise gene flow estimates (N_m) between stocks were derived from F_{st} values (Wright, 1943) obtained via AMOVA analysis (Excoffier et al. 1992).

Pairwise estimates of gene flow (Table 1) among the Gulf stocks based on sequencing analysis ranged from 0.15 between the western (Lake Ponchartrain and Pearl River) stock and the Escambia River-Yellow River stock, to 1.2 between the Escambia River-Yellow River stock and the eastern stock

Table 1. Estimates of gene flow among populations of *Acipenser oxyrinchus desotoi*. Values above diagonal are based on data from restriction fragment length polymorphism analysis of mitochondrial DNA (mtDNA); values below diagonal are based on data from sequence analysis of 203 base pairs of mtDNA control region.

	Western	Pascagoula	Escambia-Yellow	Choctawhatchee	Eastern
Western	–	–	0.26	0.11	0.23
Pascagoula	249.75	–	–	–	–
Escambia-Yellow	0.15	0.23	–	0.45	0.66
Choctawhatchee	0.22	0.35	0.79	–	0.09
Eastern	0.17	0.27	1.20	0.97	–

(Apalachicola, Ochlockonee, and Suwannee rivers). Gene flow estimates derived from RFLP analysis were even lower on average, and ranged from 0.09 between the western and Choctawhatchee River stocks to 0.66 between the western and Escambia River-Yellow River stock.

These gene flow values are very low in comparison with estimates for other anadromous fishes. Estimated annual straying rates among populations of Pacific salmons have ranged between about 1% and 27% (reviewed in Adkindson 1996). Laughlin & Turner (1996) used two statistical methods to estimate N_m of *Morone saxatilis* (striped bass) among three Virginia tributaries of Chesapeake Bay; the private allele approach of Barton & Slatkin (1986) yielded an estimate of $N_m = 14.2$, whereas the F_{st} approach yielded an estimate of $N_m = 2.7$. In a capture-mark-recapture study, Melvin et al. (1986) estimated an annual straying rate of 3% among Canadian populations of *Alosa sapidissima* (American shad).

Moreover, the low gene flow estimates for *A. o. desotoi* were obtained across populations that occur in eight rivers, the mouths of which are arrayed across little more than 500 km of coastline. Sturgeon from these rivers have the opportunity to mix in the Gulf of Mexico during winter. These mtDNA data show that despite the geographic proximity of these rivers, stocks of *A. o. oxyrinchus* generally exchange less than one female per generation, a level sufficient to permit differentiation at the stock level (Adkinson 1996). Gene flow estimates also were generally higher among proximal stocks, suggesting that what straying occurs does so in 'stepping stone' fashion (Kimura & Weiss 1964) in which migrants among semi-isolated populations are exchanged chiefly with neighboring populations. If this is true for *A. o. desotoi*, then such spatially restricted straying should have contributed to the geographic structuring observed among these populations (Adkinson 1996). Stabile et al. (1996) hypothesized that the homing imperative of *A. o. desotoi* for spawning purposes is strong, but that it may be reinforced by metabolic constraints. *Acipenser oxyrinchus desotoi* returns to rivers from the Gulf of Mexico to summer near cold water springs; tagging has shown that individuals are recaptured at the same cool water refuges in which they were first tagged (Clugston et al. 1995).

Conclusions and recommendations

The comparatively few molecular analyses directed to conservation of acipenseriforms have yielded unique and important information. Some of these studies have refined notions concerning species and subspecies status; our analysis of mtDNA control region differences between the American and European Atlantic sturgeons shows that should relict populations of the latter become extinct, restocking with *A. oxyrinchus* would constitute introduction of a foreign species. Likewise, the study by Ong et al. (1996) supporting subspecies designations for Atlantic and Gulf forms of *A. oxyrinchus* reinforces the current status of the latter as a threatened subspecies under the U.S. Endangered Species Act.

Unfortunately, to date, conservation directed molecular studies of acipenseriforms at the species level and below have almost exclusively been restricted to North American species. Highly sensitive determination of the genetic relationships among extant species, stocks, and between extinct and closely related extant taxa are possible. Much of the latter analyses will use archived museum samples as sources of DNA. Birstein (1993) provided many examples of Eurasian sturgeons that are phenotypically differentiated (i.e., large and small 'forms') below the species level but that have not received genetic analysis. Even if financial resources are currently not available, samples can be collected and archived for future analyses.

Population surveys also have been limited largely to North America. Genetic analyses of *P. spathula* and *A. fulvescens* – paddlefish and sturgeon species with similar and broad North American distributions – have revealed relatively little genetic variation and only minor substructuring among populations. In comparison, coastal species such as *A. oxyrinchus* and *A. transmontanus* exhibit higher levels of polymorphism and greater geographic population structuring. The reasons for this disparity are not apparent, and are contrary to comparisons among other freshwater and anadromous fishes (Waldman & Wirgin 1994). We cannot be sure if present levels of genetic diversity among these four species reflects prebottleneck levels, given the widespread anthropogenically-imposed bottlenecks on their populations. However, it is likely that the paucity of genetic diversity and shallow genetic differentiation among populations of the two freshwater species is at least partly due to a combination of Pleistocene bottlenecks and recent recolonization of northern waters, prolonged generation times leading to low mutation rates, and continued gene flow among populations that remain largely linked through their inhabitation of few discrete drainages. Although the anadromous sturgeons of North America were not exempt from glacial influences, post-Pleistocene genetic diversity was maintained, except in far northern drainages recolonized by *A. oxyrinchus*.

Genetic analyses to date of the freshwater scaphirhynchids show very limited genetic variation, consistent with *P. spathula* and *A. fulvescens* – species that are largely sympatric with *S. albus* and *S. platorynchus*. Low genetic variation within *Scaphirhynchus* is in accordance with pre-existing biogeographic theories suggesting recent speciation with Wisconsinian glacial events (70 000 to 10 000 years before present). However, the taxonomic boundaries among the three putative scaphirhynchid species remain unclear, but of great importance to their conservation status under the U.S. Endangered Species Act (ESA). Future molecular analyses of *Scaphirhynchus* spp. and other North American acipenseriforms will not only attempt to assess traditional taxonomic divisions such as species and subspecies, but also their Evolutionarily Significant Units as mandated under the Endangered Species Act (Waples 1995).

Molecular analyses to date also are relevant to conservation efforts involving hatchery-based stocking (St. Pierre[7]). Anadromous sturgeons show stock structure, implying that interpopulation transfers will have genetic repercussions; to maximize the likelihood that stocked fish will have high fitness for a particular environment, broodstock from the same environment as wild fish should be used (Krueger et al. 1981). The single study of gene flow in an acipenserid suggests that hatchery-based restocking may be necessary to reestablish sturgeon populations in a reasonable length of time in drainages where they are extinct. That is, if gene flow estimates among populations of *A. o. desotoi* are indicative of general levels of straying for acipenserids, then straying rates less than 1.0 per generation, combined with the long generation times of sturgeons, means that natural restocking may require decades before it even is initiated, notwithstanding additional decades necessary for population growth (Boreman 1997 this volume). Circumstantial evidence for extremely slow rates of natural restocking through straying is provided by the example of the Maryland tributaries of Chesapeake Bay, where there has been no indication of recolonization by *A. oxyrinchus* over many decades in rivers that once supported large populations (David Secor personal communication).

However, a generic problem with the hatchery production of sturgeons is the need to acquire sufficient broodstock to prevent inbreeding (Nelson & Soulé 1987). Commonly accepted guidelines for fish production are for an effective population size of 100 or more individuals (Kincaid 1983, Kapuscinski & Lannan 1986, Allendorf & Ryman 1987). But, given the generally large size and scarcity of sturgeons, annual hatchery reproduction at these effective population levels are unrealistic for most restoration efforts aimed toward single stocks. If lower than recommended numbers of broodstock must be used, one way to reduce inbreeding would be to use genetic screening of individuals. With the data from these analyses, hatchery crosses can be optimized to maximize diversity and yet still maintain stock-specific gene frequencies. We envision that a priori knowledge of the genetic composition of broodstock can help ease the burden of rigorous demands for large numbers of broodstock.

Rapid technical advances in the development of molecular biological approaches will allow for their routine application in the future to problems of the conservation of sturgeons and paddlefishes. It is now possible to obtain DNA sequence information from non-destructively obtained tissues such as barbels, fin clips, or blood, from early life intervals

[7] St. Pierre, R.A. 1996. Breeding and stocking protocol for cultured Atlantic sturgeon. Final Report from the Atlantic Sturgeon Aquaculture and Stocking Committee to the Atlantic States Marine Fisheries Commission Atlantic Sturgeon Management Board. 17 pp.

such as single eggs, embryo or larva, and even from archived museum specimens. Additionally, a variety of DNA level approaches have been developed which allow for investigations which focus on characters whose rate of change varies from extremely slow to exceedingly rapid (Wirgin & Waldman 1994). This permits quantification of genetic relationships extending from the interspecific to interindividual levels. We strongly encourage the use of nDNA-based approaches to resolve management questions concerning taxa such as *Scaphirhynchus* in which low levels of genetic diversity have been reported. Because of the polyploid character of the acipenserid genome, it is highly likely that many duplicated gene loci have been relieved of functional constraints and were free to rapidly evolve. Thus, the nuclear genome of sturgeon species should offer a wealth of rapidly evolving single copy or repetitive DNA sequences for analysis.

Several other avenues for future molecular research of sturgeons and paddlefishes are apparent. Poaching of these fishes across North America and Eurasia is a major threat to their existence; much of this illegal harvest is sold in world markets (Birstein 1993, Bemis & Findeis 1994). A study is now underway (DeSalle & Birstein 1996) to develop forensic molecular markers to help identify illegally procured and mislabeled acipenserid products. Also, most acipenseriforms are difficult to sex except at spawning. Nonlethal molecular gender determination, as has been developed for *Oncorhynchus tshawytscha* (chinook salmon, Devlin et al. 1994), would be useful for many management purposes.

Additionally, contaminant exposure has been proposed as a major factor in the decline of certain sturgeon populations (Birstein 1993). However, quantitative data on comparative exposure histories or possible biological effects are largely lacking. We suggest that a molecular biomarker approach, in which structural alterations at anonymous genetic loci or expression levels of xenobiotically responsive genes are quantified, can help fill this void. The extent of DNA sequence variation at these loci or levels of expression of inducible genes such as cytochrome P4501A or metallothionein can allow for a comparison of the exposure histories of sturgeons from environments with differing degrees of pollution. This approach has been used successfully to quantify the exposure histories of species such as *Parophrys vetulus* (English sole, Stein et al. 1992) and *Microgadus tomcod* (Atlantic tomcod, Wirgin et al. 1994) from North American estuaries. Furthermore, increased levels of gene expression in fish from contaminated environments have been correlated with higher level biological effects at the population level (Wirgin & Garte 1994).

Acknowledgements

We thank Patrick Williot for the tissue sample of European Atlantic sturgeon. Work was funded by the National Oceanic and Atmospheric Administration award # NA46RG0090 to the Research Foundation of the State University of New York for the New York Sea Grant Institute. The U.S. Government is authorized to produce and distribute reprints for governmental purposes notwithstanding any copyright notation that may appear hereon. Views expressed herein are those of the authors and do not necessarily reflect the views of NOAA or any of its subagencies. This work also was supported by the U.S. Fish and Wildlife Service and the Hudson River Foundation and NIEHS center grant ES00260.

References cited

Adkinson, M.D. 1996. Population differentiation in Pacific salmon: local adaptation, genetic drift, or the environment? Can. J. Fish. Aquat. Sci. 52: 2762–2777.

Allendorf, F.W. & N. Ryman. 1987. Genetic management of hatchery stocks. pp. 141–159. *In:* N. Ryman & F. Utter (ed.) Population Genetics and Fishery Management, University of Washington, Seattle.

Barton, N.H. & M. Slatkin. 1986. A quasi-equilibrium theory of the distributions of rare alleles in a subdivided population. Heredity 56: 409–415.

Bemis, W.E. & E.K. Findeis. 1994. The sturgeon's plight. Nature 370: 602.

Bemis, W.E. & B. Kynard. 1997. Sturgeon rivers: an introduction to acipenseriform biogeography and life history. Env. Biol. Fish. (this volume).

Bernatchez, L. 1995. A role for molecular systematics in defining evolutionarily significant units in fishes. Amer. Fish. Soc. Symp. 17: 114–132.

Birstein, V.J. 1993. Sturgeons and paddlefishes: threatened fishes in need of conservation. Conserv. Biol. 7: 773–787.

Birstein, V.J., B.F. Goncharov & A.I. Poletaev. 1993. DNA content in Eurasian sturgeon species determined by flow cytometry. Cytometry 14: 377–383.

Birstein, V.J., R. Hanner & R. DeSalle. 1997. Phylogeny of the Acipenseriformes: cytogenetic and molecular approaches. Env. Biol. Fish. (this volume).

Blacklidge, K.H. & C.A. Bidwell. 1993. Three ploidy levels indicated by genome quantification in acipenseriforms of North America. J. Hered. 84: 427–430.

Boreman, J. 1997. Sensitivity of North American sturgeons and paddlefish to fishing mortality. Env. Biol. Fish. (this volume).

Brown, J.R., K. Beckenbach, A.T. Beckenbach & M.J. Smith. 1996. Length variation, heteroplasmy and sequence divergence in the mitochondrial DNA of four species of sturgeon (*Acipenser*). Genetics 142: 525–535.

Brown, J.R., A.T. Beckenbach & M.J. Smith. 1992. Mitochondrial DNA length variation and heteroplasmy in populations of white sturgeon (*Acipenser transmontanus*). Genetics 132: 221–228.

Brown, J.R., A.T. Beckenbach & M.J. Smith. 1993. Intraspecific DNA sequence variation of the mitochondrial control region of white sturgeon (*Acipenser transmontanus*). Mol. Biol. Evol. 10: 326–341.

Buroker, N.E., J.R. Brown, T.A. Gilbert, P.J. O'Hara, A.T. Beckenbach, W.K. Thomas & M.J. Smith. 1990. Length heteroplasmy of sturgeon mitochondrial DNA: an illegitimate elongation model. Genetics 124: 157–163.

Carlson, D.M., M.K. Kettler, S.E. Fisher & G.S. Whitt. 1982. Low genetic variability in paddlefish populations. Copeia 1982: 721–725.

Carlson, D.M., W.L. Pflieger, L. Trial & P.S. Haverland. 1985. Distribution, biology and hybridization of *Scaphirhynchus albus* and *S. platorynchus* in the Missouri and Mississippi rivers. Env. Biol. Fish. 14: 51–59.

Clugston, J.P., A.M. Foster & S.H. Carr. 1995. Gulf sturgeon, *Acipenser oxyrinchus desotoi*, in the Suwannee River, Florida, USA. pp. 215–224. *In:* A.D. Gershanovich & T.I.J. Smith (ed.) Proceedings International Symposium on Sturgeons, VNIRO Publishing, Moscow.

Collins, M.R., S.G. Rogers & T.I.J. Smith. 1996. Bycatch of sturgeons along the southern Atlantic coast of the USA. N. Amer. J. Fish. Manage. 16: 24–29.

DeSalle, R. & V.J. Birstein. 1996. PCR identification of black caviar. Nature 381: 197–198.

Devlin, R.H., B.K. McNeil, T.D.D. Groves & E.M. Donaldson. 1991. Isolation of a Y-chromosome DNA probe capable of determining genetic sex in chinook salmon (*Oncorhynchus tshawytscha*). Can. J. Fish. Aquat. Sci. 48: 1606–1612.

Excoffier, L., P.E. Smouse & J.M. Quattro. 1992. Analysis of molecular variance inferred from metric distances among DNA haplotypes: application to human mitochondrial DNA restriction data. Genetics 131: 479–491.

Ferguson, M.M., L. Bernatchez, M. Gatt, B.R. Konkle, S. Lee, M.L. Malott & R.S. McKinley. 1993. Distribution of mitochondrial DNA variation in lake sturgeon (*Acipenser fulvescens*) from the Moose River basin, Ontario, Canada. J. Fish Biol. 43 (Suppl. A): 91–101.

Ferguson, M.M. & G.A. Duckworth. 1997. The status and distribution of lake sturgeon, *Acipenser fulvescens*, in the Canadian provinces of Manitoba, Ontario and Quebec: a genetic perspective. Env. Biol. Fish. (this volume).

Forbes, S.A. & R.E. Richardson. 1905. On a new shovelnose sturgeon from the Mississippi River. Bulletin of the Illinois State Laboratory of Natural History 7: 37–44.

Guenette, S. & R. Fortin. 1993. Mitochondrial DNA variation in lake sturgeon (*Acipenser fulvescens*) from the St. Lawrence River and James Bay drainage basins in Quebec, Canada. Can. J. Fish. Aquat. Sci. 50: 659–664.

Hallerman, E.M. & J.S. Beckmann. 1988. DNA-level polymorphism as a tool in fisheries science. Can. J. Fish. Aquat. Sci. 45: 1075–1087.

Hubbs, C.L. 1955. Hybridization between fish species in nature. Syst. Zool. 4: 1–120.

Kapuscinski, A.R.D. & J.E. Lannan. 1986. A conceptual genetic fitness model for fisheries management. Can. J. Fish. Aquat. Sci. 43: 1606–1616.

Kimura, M. & G.H. Weiss. 1964. The stepping stone model of population structure and the decrease of genetic correlation with distance. Genetics 49: 561–576.

Kincaid, H.L. 1983. Inbreeding in fish populations used for aquaculture. Aquaculture 33: 215–227.

Krueger, C.C., A.J. Gharrett, T.R. Dehring & F.W. Allendorf. 1981. Genetic aspects of fisheries rehabilitation programs. Can. J. Fish. Aquat. Sci. 38: 1877–1881.

Kynard, B. 1997. Life history, latitudinal patterns, and status of shortnose sturgeon, *Acipenser brevirostrum*. Env. Biol. Fish. (this volume).

Laughlin, T.F. & B.J. Turner. 1996. Hypervariable DNA markers reveal high genetic variability within striped bass populations of the lower Chesapeake Bay. Trans. Amer. Fish. Soc. 125: 49–55.

Magnin, E. 1964. Validité d'une distinction spécifique entre les deux acipenseridés: *Acipenser sturio* L. d'Europe et *Acipenser oxyrhynchus* d'Amerique du Nord. Natur. Can. 91: 5–20.

Melvin, G.D., M.J. Dadswell & J.D. Martin. 1986. Fidelity of American shad, *Alosa sapidissima*, to its river of previous spawning. Can. J. Fish. Aquat. Sci. 43: 640–646.

Meyer, A. 1993. Evolution of mitochondrial DNA of fishes. pp. 1–38. *In:* P.W. Hochachka & T.P. Mommsen (ed.) Biochemistry and Molecular Biology of Fishes, Vol. 2, Elsevier, Amsterdam.

Miracle, A. & D.E. Campton. 1995. Tandem repeat sequence variation and length heteroplasmy in the mitochondrial DNA D-loop of the threatened Gulf of Mexico sturgeon, *Acipenser oxyrhynchus desotoi*. J. Hered. 86: 22–27.

Nei, M. & F. Tajima. 1981. DNA polymorphism detectable by restriction endonucleases. Genetics 97: 145–163.

Nelson, K. & M. Soulé. 1987. Genetical conservation of exploited fishes. pp. 345–368. *In:* N. Ryman & F. Utter (ed.) Population Genetics and Fishery Management, University of Washington, Seattle.

O'Brien, S.J. & E. Mayr. 1991. Bureaucratic mischief: recognizing endangered species and subspecies. Science 251: 1187–1188.

Ong, T-L., J. Stabile, I. Wirgin & J.R. Waldman. 1996. Genetic divergence between *Acipenser oxyrinchus oxyrinchus* and *A. o. desotoi* as assessed by mitochondrial DNA sequencing analysis. Copeia 1996: 464–469.

Phelps, S.R. & F.W. Allendorf. 1983. Genetic identity of pallid and shovelnose sturgeon (*Scaphirhynchus albus* and *S. platorynchus*). Copeia 1983: 696–700.

Rochard, E., G. Castelnaud & M. Lepage. 1990. Sturgeons (Pisces : Acipenseridae); threats and prospects. J. Fish Biol. (Suppl. A): 123–132.

Scott, W.B. & M.G. Scott. 1988. Atlantic fishes of Canada. Can. Bull. Fish. Aquat. Sci. 219: 1–731.

Smith, T.I.J. & J.P. Clugston. 1997. Status and management of Atlantic sturgeon, *Acipenser oxyrinchus*, in North America. Env. Biol. Fish. (this volume).

Stabile, J., J.R. Waldman, F. Parauka & I. Wirgin. 1996. Stock structure and homing fidelity in Gulf of Mexico sturgeon (*Acipenser oxyrinchus desotoi*) based on RFLP and sequence analyses of mitochondrial DNA. Genetics (in press).

Stein, J.E., T.K. Collier, W.L. Reichert, E. Casillas & U. Varanasi. 1992. Bioindicators of contaminant exposure and sublethal effects: studies with benthic fish in Puget Sound, WA. Env. Toxicol. Chem. 11: 701–714.

Tallman, R.F. & M.C. Healey. 1994. Homing, straying, and gene flow among seasonally separated populations of chum salmon (*Oncorhynchus keta*). Can. J. Fish. Aquat. Sci. 51: 577–588.

Utter, F. & N. Ryman. 1993. Genetic markers and mixed stock fisheries. Fisheries 18: 11–21.

Vladykov, B. 1955. A comparison of Atlantic sea sturgeon with a new subspecies from the Gulf of Mexico (*Acipenser oxyrhynchus desotoi*). J. Fish. Res. Board Can. 12: 754–761.

Vladykov, V.D. & J.R. Greeley. 1963. Family Acipenseridae. Fishes of the Western North Atlantic. Mem. Sears Found. Mar. Res. 1(3): 26–60.

Waldman, J.R. 1995. Sturgeon and paddlefishes: a convergence of biology, politics and greed. Fisheries 20: 20–21, 49.

Waldman, J.R., J.T. Hart & I.I. Wirgin. 1996a. Stock composition of the New York Bight Atlantic sturgeon fishery based on analysis of mitochondrial DNA. Trans. Amer. Fish. Soc. 125: 364–371.

Waldman, J.R., K. Nolan, J. Hart & I.I. Wirgin. 1996b. Genetic differentiation of three key anadromous fish populations of the Hudson River. Estuaries 19: 759–768.

Waldman, J.R. & I.I. Wirgin. 1994. Use of DNA analyses in the management of natural fish populations. pp. 29–64. *In:* S.J. Garte (ed.) Molecular Environmental Biology, Lewis Publishers, Boca Raton.

Waples, R.S. 1995. Evolutionarily significant units and the conservation of biological diversity under the Endangered Species Act. Amer. Fish. Soc. Symp. 17: 8–27.

Wayne, R.K. 1992. On the use of morphologic and molecular genetic characters to investigate species status. Conserv. Biol. 6: 590–592.

Williams, J.D. & G.H. Clemmer. 1991. *Scaphirhynchus suttkusi*, a new sturgeon (Pisces : Acipenseridae) from the Mobile Basin of Alabama and Mississippi. Bull. Alabama Mus. Nat. Hist. 10: 17–31.

Wirgin, I.I. & S.J. Garte. 1994. Assessment of environmental degradation by molecular analysis of a sentinel species: Atlantic tomcod. pp. 117–132. *In:* S.J. Garte (ed.) Molecular Environmental Biology, Lewis Publishers, Boca Raton.

Wirgin, I.I., C. Grunwald, S. Courtenay, G.-L. Kreamer, W.L. Reichert & J.E. Stein. 1994. A biomarker approach to assessing xenobiotic exposure in Atlantic tomcod from the North American coast. Env. Health Perspect. 103: 764–770.

Wirgin, I.I. & J.R. Waldman. 1994. What DNA can do for you. Fisheries 19: 16–27.

Wooley, C.M. 1985. Evaluation of morphometric characters used in taxonomic separation of Gulf of Mexico sturgeon, *Acipenser oxyrhynchus desotoi*. pp. 97–103. *In:* F.P. Binkowski & S.I. Doroshov (ed.) North American Sturgeons: Biology and Aquaculture Potential, Developments in Environmental Biology of Fishes 6, Dr W. Junk Publishers, Dordrecht.

Wright, S. 1943. Isolation by distance. Genetics 28: 114–138.

Xu, S., C.J. Kobak & P.E. Smouse. 1994. Constrained least squares estimation of mixed population stock composition from mtDNA haplotype frequency data. Can. J. Fish. Aquat. Sci. 51: 417–425.

Environmental Biology of Fishes **48**: 399–405, 1997.

Sensitivity of North American sturgeons and paddlefish to fishing mortality

John Boreman
UMass/NOAA Cooperative Marine Education and Research Program, The Environmental Institute, Blaisdell House, University of Massachusetts, P.O. Box 30820, Amherst, MA 01003–0820, U.S.A.

Received 6.10.1994 Accepted 28.3.1996

Key words: reproductive potential, exploitation, endangered species, chondrosteans, eggs per recruit

Synopsis

Sturgeons and paddlefish exhibit unusual combinations of morphology, habits, and life history characteristics, which make them highly vulnerable to impacts from human activities, particularly fisheries. Five North American sturgeons (shortnose, Gulf, pallid, Alabama, and green sturgeon) are listed as endangered or threatened by management authorities. Managers have instituted fishery closures for the three other species of North American sturgeons (Atlantic, white, and shovelnose) and paddlefish because of low stock abundance at some point in this century. Reproductive potential in four species I examined (Atlantic, white, and shortnose sturgeon, and paddlefish) is more sensitive to fishing mortality than it is for three other intensively-fished coastal species in North America: striped bass, winter flounder, and bluefish. The sturgeons and paddlefish are generally longer-lived than the three other coastal species, and also have an older age at full maturity, lower maximum fecundity values, and older ages at which 50% of the lifetime egg production is realized with no fishing mortality.

Introduction

Of the 25 chondrostean species still living in the Northern Hemisphere, the paddlefish, *Polyodon spathula*, and eight species of sturgeon are found in North American waters (Birstein 1993): Atlantic sturgeon, *Acipenser oxyrinchus*, shortnose sturgeon, *Acipenser brevirostrum*, white sturgeon, *Acipenser transmontanus*, lake sturgeon, *Acipenser fulvescens*, green sturgeon, *Acipenser medirostris*, pallid sturgeon *Scaphirhynchus albus*, shovelnose sturgeon *Scaphirhynchus platorynchus*, and Alabama sturgeon, *Scaphirhynchus suttkusi*. They inhabit fresh, brackish, and sea water systems in North America, and exhibit three types of life history patterns (after Rochard et al. 1990): (1) their entire life history is spent in fresh water (paddlefish, lake sturgeon, white sturgeon, pallid sturgeon, Alabama sturgeon, and shovelnose sturgeon); (2) adults move into brackish water (white sturgeon and shortnose sturgeon); or (3) adults move into the ocean (white sturgeon, green sturgeon, and Atlantic sturgeon). All of the species reproduce in fresh water.

Sturgeons and paddlefish exhibit unusual combinations of morphology, habits, and life history characteristics, which make them highly vulnerable to impacts from human activities, particularly fisheries. In North America, human activities known to impact sturgeons and paddlefish are industrial and municipal pollution, blockage of access to habitats by dikes and dams, channelization and elimination of backwater areas, dewatering of streams, physical destruction of spawning grounds, inundation of habitat by reservoirs, and overfishing (Baker 1980, Carlson & Bonislawsky 1981, Trautman 1981, Becker 1983, Kallemeyn 1983, Cochnauer et al. 1985, Rochard et al. 1990, Moyle et al.[1]).

Five of the North American sturgeons are listed as endangered or threatened by management authorities: shortnose sturgeon (Dadswell et al. 1984), green sturgeon (Moyle et al.[1]), pallid sturgeon (Keenlyne & Jenkins 1993), Alabama sturgeon (Williams et al. 1989), and Gulf sturgeon, *Acipenser oxyrinchus desotoi* (Mason & Clugston 1993), which is a subspecies of Atlantic sturgeon. Although not officially listed as endangered or threatened, three other species of North American sturgeons and paddlefish have been reduced to such low densities at some point in this century that fishery closures were instituted. Harvesting of lake sturgeon in the Lake Winnebago system, Wisconsin, was prohibited from 1915 to 1931 due to a concern over the drop in abundance (Folz & Meyers 1985). Harvest of white sturgeon in the Snake River was terminated after 1983 due to a decline in abundance, and a total fishery closure was recommended for sections of the river (Cochnauer et al. 1985). In Wisconsin and Iowa, non-fishing zones below navigation dams were adopted to protect shovelnose sturgeon from overfishing (Becker 1983). A three-year moratorium on commercial harvest of paddlefish was initiated in Louisiana in 1986 and the recreational creel limit was reduced to one fish per day because of declines in stock levels (Reed et al. 1992).

Although draconian fishing restrictions have been instituted for many of the North American chondrosteans, demand for the species has not diminished. Since the middle 1800s, North American sturgeons have been the target of intensive fisheries, primarily for caviar and also for their meat (fresh, smoked, or tinned); angling for sport is also growing more popular, especially for white and lake sturgeons (Rochard et al. 1990). When sturgeons became unavailable for the lucrative caviar market, some fisheries switched to paddlefish (Carlson & Bonislawsky 1981, Reed et al. 1992).

[1] Moyle, P.B., R.M. Yoshiyama, J.E. Williams & E.D. Wikramanayake. 1996. Fish species of special concern in California (second edition). Prepared for the State of California, The Resources Agency, Dept. Fish and Game, Inl. Fish. Div., Rancho Cordova, California (in press).

In this paper, I examine the sensitivity of North American sturgeons and paddlefish to fishing and early life mortality. Specifically, I present the impact of fishing mortality on reproductive potential for several representative sturgeon and paddlefish populations, and examine whether adjustments to fishing mortality could be used to offset reductions in reproductive potential caused by other sources of mortality due to human activities. I also compare the chondrosteans to other fish species currently supporting intensive fisheries in North America (striped bass in the Hudson River, New York, winter flounder in Cape Cod Bay, Massachusetts, and bluefish along the Atlantic coast of North America) to demonstrate how life history characteristics make the chondrosteans more sensitive to fishing mortality.

Methods

Effects of fishing mortality on reproductive potential

For purposes of this paper, I define reproductive potential as the potential lifetime egg production of an age 1 female (eggs-per-recruit, *EPR*). This measure is the sum of the number of eggs she is likely to produce at each age times the probability that she will survive to that age (Boreman et al. 1993):

$$EPR = \sum_{i=2}^{n} \rho_i \varphi_i \prod_{j=1}^{i-1} e^{-(F_j+M_j)}, \qquad (1)$$

where ρ_i is the proportion of females mature at age i, φ_i is the average fecundity of an age-i female, F_j is the instantaneous rate of fishing mortality during period j, M_j is the instantaneous rate of natural mortality during period j, and n is the oldest spawning age. The maximum value for potential lifetime egg production (EPR_{max}) is achieved when no fishing mortality occurs ($F_j = 0$ for all j). As F is increased, *EPR* will decline due to the lessened probability that an age *1* female will survive to the next age, given the increased risk of fishing mortality along with the risk of natural mortality she also must endure. Relative sensitivity of reproductive potential to a specific rate of fishing mortality is the ratio of the *EPR*-value calculated for that rate to EPR_{max}.

Restricting fishing mortality to offset losses from other sources

Fishery managers have the option of regulating fishing mortality to offset impacts of mortality from unknown or largely uncontrollable sources, such as changes in river flows and contaminant toxicity, which usually occur in fish stocks during the first year of life. To maintain stationary population abundance, i.e., population abundance that is neither increasing nor declining over generations, the survival rate of a female egg to age *1* (S_0) must be equal to two times the reciprocal of the *EPR*-value (assuming the sex ratio of deposited eggs is 50:50), so that:

$$EPR \cdot S_0 = 2. \qquad (2)$$

If survival rate during age *0* declines, the population abundance can be maintained if *EPR* is increased to a level that keeps the product of age *0* survival and *EPR* equal to two, assuming that the population has sufficient compensatory capabilities to maintain stationarity under all mortality conditions.

Data sources

Sufficient information exists in available publications to estimate *EPR* values for three sturgeon populations and one paddlefish population in North America: white sturgeon in the Columbia River below Bonneville Dam (Tracy & Wall 1993, DeVore et al. 1996); Atlantic sturgeon in the Hudson River (Kahnle et al. 1992, Kahnle unpublished data); paddlefish in Lake Ponchartrain, Louisiana (Reed et al. 1992); and shortnose sturgeon in the lower Connecticut River (Boreman[2]). The fishery for white sturgeon in the lower Columbia River currently has minimum size limit of 112 cm and a maximum size limit of 168 cm (DeVore personal communication), and the fishery for Atlantic sturgeon in the Hudson River has a minimum size limit of 152 cm (Kahnle et al.[3]). I did not attempt to divide fishing mortality into sport or commercial. For shortnose sturgeon and paddlefish, I assumed that all ages 1 and older would be equally vulnerable to any increase in mortality beyond that already incorporated into the natural mortality rates.

Values for life history parameters of female striped bass in the Hudson River are from Goodyear (1988). The values are for the population prior to closure of the fishery in the river in 1976 due to chemical contamination, and represent a period when coastal landings of striped bass from the Hudson River and elsewhere along the Atlantic coast of North America were at their peak (Boreman & Austin 1985). Boreman et al. (1993) list parameter values for female winter flounder in Cape Cod Bay, and parameter values for female bluefish along the Atlantic coast are given in MAFMC[4].

Results and discussion

Effects of fishing mortality on reproductive potential

To prevent harvesting of spawners below the replacement level of their progeny, Goodyear (1993) recommends maintaining levels of spawning stock biomass per recruit that are at least 20% of the maximum (when $F = 0$), unless evidence exists for exceptionally strong density-dependence in the population. Boreman et al. (1984) used a higher level of 50% of maximum spawning stock biomass per recruit as a target for rebuilding (rather than maintaining) populations of shortnose sturgeon along the Atlantic coast. If spawning stock biomass and egg production are linearly related (fecundity is typically a linear function of female body weight), then the same 20% and 50% target levels should

[2] Boreman, J. 1992. Impact of added mortality on the reproductive success of shortnose sturgeon in the lower Connecticut River. Report prepared for the Northeast Regional Office, National Marine Fisheries Service. 14 pp.

[3] Kahnle, A., K. Hattala & K. McKown. 1992. Proposed New York State Atlantic sturgeon regulations. Prepared for the Atlantic States Marine Fisheries Commission, Atlantic Sturgeon Plan Review Team. 14 pp.

[4] MAFMC (Mid-Atlantic Fishery Management Council). 1990. Fishery management plan for the bluefish fishery. Mid-Atlantic Fishery Management Council, Dover, Delaware. 79 pp.

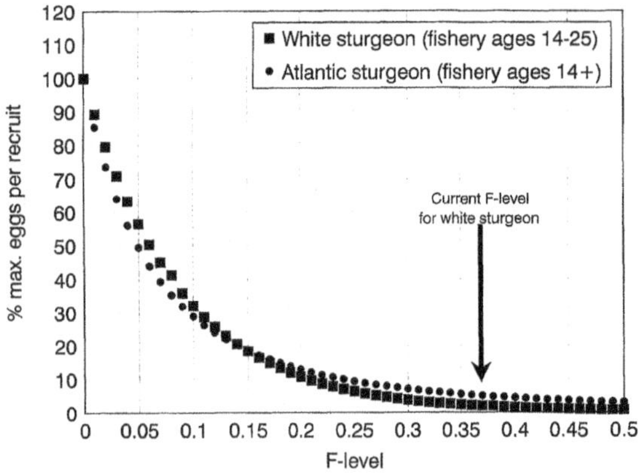

Figure 1. Relationship between fishing mortality rate (F) and corresponding percentage of the maximum lifetime egg production of an age 1 female when F = 0 for white sturgeon in the Columbia River below Bonneville Dam and Atlantic sturgeon in the Hudson River.

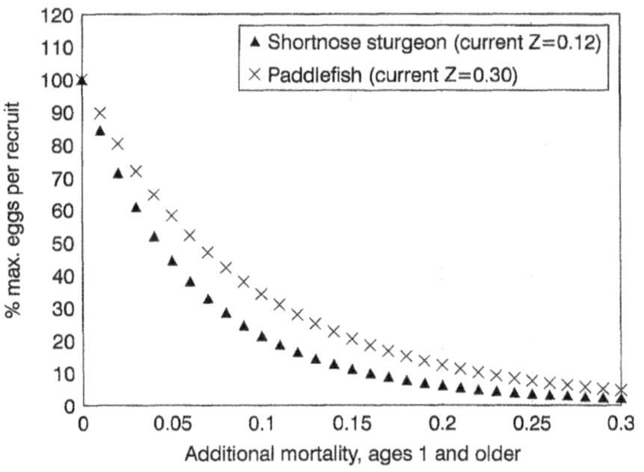

Figure 2. Effects of an increase in the current mortality rate (Z) of age 1 and older females on the corresponding percentage of the maximum lifetime egg production of an age 1 female for shortnose sturgeon in the lower Connecticut River and paddlefish in Lake Ponchartrain.

apply to potential lifetime egg production per recruit (*EPR*).

The fishing mortality rate corresponding to 20% of EPR_{max} for white sturgeon in the lower Columbia River and for Atlantic sturgeon in the Hudson River is $F = 0.14$ (Figure 1). A 50% level of EPR_{max} would be achieved if the fishing mortality is $F = 0.06$ for white sturgeon, and $F = 0.05$ for Atlantic sturgeon. At the 1986–1990 average fishing mortality rate of $F = 0.37$, lifetime egg production of age *1* females in the white sturgeon population is approximately 2% of EPR_{max} – far below the recommended minimum level of 20%. An estimate of the current fishing mortality rate for Atlantic sturgeon in the Hudson River is unavailable.

Mortality due to incidental capture in indirect fisheries is probably incorporated in the current estimates of total mortality (*Z*) for both the shortnose sturgeon population in the lower Connecticut River and the paddlefish population in Lake Ponchartrain. Increasing the total mortality rate on age 1 and older shortnose sturgeon from $Z = 0.12$ (current) to $Z = 0.16$ will reduce potential lifetime egg production of a female recruit to 50% of EPR_{max},

Table 1. Female life history characteristics for white sturgeon in the lower Columbia River, Atlantic sturgeon in the Hudson River, shortnose sturgeon in the lower Connecticut River, paddlefish in Lake Ponchartrain, striped bass in the Hudson River, winter flounder in Cape Cod Bay, and bluefish along the Atlantic coast.

Characteristic	White sturgeon	Atlantic sturgeon	Shortnose sturgeon	Paddlefish	Striped bass	Winter flounder	Bluefish
Maximum age (years)	104	60	30	20	18	12	8
Natural mortality (M)	0.09	0.07	0.12	0.30	0.15	0.35	0.25
Length (cm) at oldest age	309	343	91	120	105	45	89
Fecundity at oldest age	1 500 000	1 800 000	66 000	200 000	3 100 000	2 200 000	5 300 000
Age at first maturity (years)	16	11	5	9	3	2	2
Age at full maturity (years)	35	21	17	10	9	6	3
Years between successive spawnings	3	3	3	2	1	1	1
Age at 50+% EPR_{max} (years)	37	29	17	11	11	6	3
Fishing mortality (F)	0.37	?	0	0	0.39	1.07	0.80
Ages in fishery	14–25	14+	–	–	2+	2+	0+

and increasing total mortality to $Z = 0.23$ will reduce the *EPR* value to 20% of EPR_{max} (Figure 2). For paddlefish, increasing the total mortality rate from $Z = 0.30$ (current) to $Z = 0.36$ will reduce the *EPR* value to 50% of EPR_{max}, and increasing the rate to $Z = 0.45$ will reduce the value to 20% of EPR_{max} (Figure 2).

As a group, chondrostean populations are more sensitive to loss in reproductive potential caused by increases in the mortality rate of age *1* and older females than are striped bass, winter flounder, and bluefish populations (Figure 3). The higher sensitivity of the chondrosteans to mortality in age *1* and older fish is due to a combination of characteristics that determine their population dynamics (Table 1). The chondrosteans are generally longer lived, are later maturing, and have lower natural mortality rates than striped bass, winter flounder, and bluefish. The chondrosteans do not spawn every year once they reach sexual maturity and, except for Atlantic sturgeon and white sturgeon, have substantially lower fecundity than the other three species I examined. A life history characteristic that integrates individual fecundity, natural mortality, age at maturity, and years between successive spawnings is the age at which at least 50% of the maximum lifetime egg production of an age *1* female is achieved when no fishing mortality occurs (EPR_{max}). For white and Atlantic sturgeons, this age is 3–10 times greater than the equivalent age for striped bass, winter flounder, and bluefish (Table 1);

Table 2. Reduction in fishing mortality rate (F) necessary to achieve equivalent lifetime egg production of an age 1 female white sturgeon in the Columbia River below Bonneville Dam when the fraction of females surviving from egg to age 1 (S_0) is reduced.

Reduction in S_0 (%)	F needed to maintain equivalent lifetime egg production	Reduction in F (%)
0	0.370	0
5	0.364	2
10	0.358	3
15	0.352	5
20	0.345	7
25	0.339	8
30	0.331	11

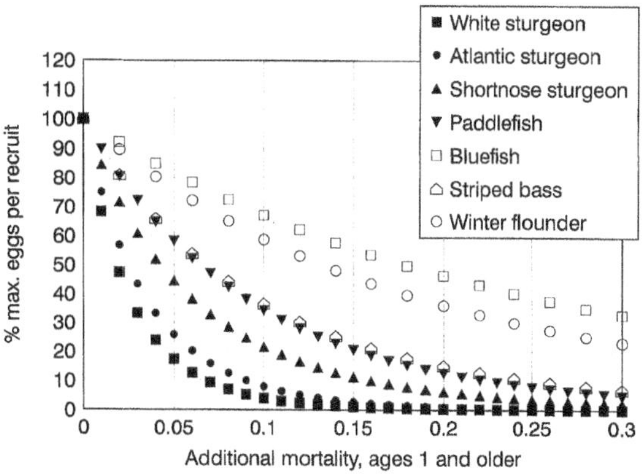

Figure 3. Effects of increasing the total mortality rate (Z) of age 1 and older females above the level when F = 0, on the corresponding percentage of the maximum lifetime egg production of an age 1 female for white sturgeon in the Columbia River below the Bonneville Dam, Atlantic sturgeon in the Hudson River, shortnose sturgeon in the lower Connecticut River, paddlefish in Lake Ponchartrain, striped bass in the Hudson River, winter flounder in Cape Cod Bay, and bluefish along the Atlantic coast.

therefore, the probability of surviving from age *1* to the age of 50% of maximum lifetime egg production is reduced by a power of 3–10 for the sturgeons.

Restricting fishing mortality to offset losses from other sources

For relatively long-lived species such as sturgeons and paddlefish, a small reduction in fishing mortality on the age groups vulnerable to harvest can offset the effects of a relatively large reduction in age *0* survival. This relationship is possible because the age *0* fish are exposed to the risk of reduced survival during only one year in their life; whereas, exposure to the risk of fishing spans many years. As an example, suppose the number of age *0* white sturgeon in the lower Columbia River is reduced by 20% due to contaminant toxicity. A 20% reduction in age *0* survival implies that for every age *1* female that would have survived her first year of life, only 0.8 females are now surviving under the altered conditions. The value for potential lifetime egg production from 0.8 age *1* females with the baseline fishing mortality rate of $F = 0.37$ is equal to the lifetime egg production of one age *1* female and a fishing mortality rate that is reduced by 7%, from $F = 0.37$ to $F = 0.345$ (Table 2).

Even though fishing may not be the reason for an observed decline in abundance of sturgeon and paddlefish populations, reducing fishing mortality is an effective means of offsetting the effects on reproductive potential caused by other, often uncontrollable mortality sources. Restricting fishing mortality may be the only tool available to managers for restoring depleted populations. At a minimum, reducing fishing pressure on long-lived species allows managers time to detect and correct the true causes of population decline. This strategy is currently being employed to rebuild the population of white sturgeon in the lower Columbia River (Columbia River Management Joint Staff[5]). The strategy was also adopted by the Atlantic States Marine Fisheries Commission in the early 1980s to restore the depleted coastal migratory stock of striped bass (ASMFC[6]), which is now producing year classes at record levels (Donald Cosden personal communication).

Acknowledgements

Special thanks go to the people who shared their information during the preparation of the manuscript, especially A. Kahnle, K. Hattala, J. Clugston, V. Birstein, J. DeVore, T. Cochnauer, P. B. Moyle, K. D. Keenlyne, and J. Waldman, to Brad McGowan for providing valuable assistance in locating references, and to John DeVore and an anonymous reviewer for critically reviewing the paper.

[5] Columbia River Management Joint Staff. 1993. Status report: Columbia River fish runs and fisheries 1938–1992. Prepared for Oregon Department of Fish and Wildlife and Washington Department of Fisheries. 257 pp.

[6] ASMFC (Atlantic States Marine Fisheries Commission). 1989. Amendment 4 to the Atlantic States Marine Fisheries Commission interstate striped bass management plan. ASMFC, Washington, DC. 60 pp.

References cited

Baker, J.P. 1980. The distribution, ecology, and management of the lake sturgeon (*Acipenser fulvescens* Rafinesque) in Michigan. Mich. Dept. Nat. Res., Fish. Div., Fish. Res. Rept. No. 1883. 95 pp.

Becker, G.C. 1983. Fishes of Wisconsin. University of Wisconsin Press, Madison. 1052 pp.

Birstein, V.J. 1993. Sturgeons and paddlefishes: threatened fishes in need of conservation. Consv. Biol. 7: 773–787.

Boreman, J. & H.A. Austin. 1985. Production and harvest of anadromous striped bass stocks along the Atlantic coast. Trans. Amer. Fish. Soc. 114: 3–7.

Boreman, J., S. J. Correia & D.B. Witherell. 1993. Effects of changes in age-0 survival and fishing mortality on egg production of winter flounder in Cape Cod Bay. Amer. Fish. Soc. Symp. 14: 39–45.

Boreman, J., W.J. Overholtz & M.P. Sissenwine. 1984. A preliminary analysis of the effects of fishing on shortnose sturgeon. National Marine Fisheries Service, Northeast Fisheries Center, Woods Hole Laboratory Reference Document 84–17. 23 pp.

Carlson, D.M. & P.S. Bonislawsky. 1981. The paddlefish (*Polyodon spathula*) fisheries of the midwestern United States. Fisheries (Bethesda) 6: 17–27.

Cochnauer, T.G., J.R. Lukens & F.E. Partridge. 1985. Status of white sturgeon, *Acipenser transmontanus*, in Idaho. pp. 127–133. *In:* F.P. Binkowski & S.I. Doroshov (ed.) North American Sturgeons. Dr W. Junk Publishers, Dordrecht.

Dadswell, M.J., B.D. Taubert, T.S. Squiers, D. Marchette & J. Buckley. 1984. Synopsis of biological data on shortnose sturgeon, *Acipenser brevirostrum* LeSueur 181. FAO Fish. Synop. No. 140, NOAA Tech. Rept. NMFS 14, US Dept. Commerce, Washington, DC. 45 pp.

DeVore, J.D., B.W. James, C.A. Tracy & D.A. Hale. 1996. Dynamics and potential production of white sturgeon in the unimpounded lower Columbia River. Trans. Amer. Fish. Soc. (in press).

Folz, D.J. & L.S. Meyers. 1985. Management of the lake sturgeon, *Acipenser fulvescens*, population in the Lake Winnebago system, Wisconsin. pp. 135–146. *In:* F.P. Binkowski & S.I. Doroshov (ed.) North American Sturgeons, Dr W. Junk Publishers, Dordrecht.

Goodyear, C.P. 1988. Implications of power plant mortality for management of the Hudson River striped bass fishery. Amer. Fish. Soc. Monogr. 4: 245–254.

Goodyear, C.P. 1993. Spawning stock biomass per recruit in fisheries management: foundation and current use. pp. 67–81. *In:* S.J. Smith, J.J. Hunt & D. Rivard (ed.) Risk Evaluation and Biological Reference Points for Fisheries Management, Can. Spec. Publ. Fish. Aquat. Sci. 120.

Kallemeyn, L. 1983. Status of the pallid sturgeon. Fisheries (Bethesda) 8: 3–9.

Keenlyne, K.D. & L.G. Jenkins. 1993. Age at sexual maturity of the pallid sturgeon. Trans. Amer. Fish. Soc. 122: 393–396.

Mason, W.T., Jr. & J.P. Clugston. 1993. Foods of the Gulf sturgeon in the Suwanee River, Florida. Trans. Amer. Fish. Soc. 122: 378–385.

Reed, B.C., W.E. Kelso & D.A. Rutherford. 1992. Growth, fecundity, and mortality of paddlefish in Louisiana. Trans. Amer. Fish. Soc. 121: 378–384.

Rieman, B.E. & R.C. Beamesdorfer. 1990. White sturgeon in the lower Columbia River: is the stock overexploited? N. Amer. J. Fish. Manag. 10: 388–396.

Rochard, E., G. Castelnaud & M. Lepage. 1990. Sturgeons (Pisces: Acipenseridae); threats and prospects. J. Fish Biol. 37 (Supplement A): 123–132.

Tracy, C.A. & M.F. Wall. 1993. Length at age relationships for white sturgeon, *Acipenser transmontanus*, in the Columbia River downstream from Bonneville Dam. pp. 185–200. *In*: R.C. Beamesderfer & A.A. Nigro (ed.) Status and Habitat Requirements of the White Sturgeon Populations in the Columbia River Downstream From the McNary Dam, Volume 2, Final Report to Bonneville Power Administration, Portland.

Trautman. M.H. 1981. The fishes of Ohio. Ohio State University Press, Columbus. 782 pp.

Williams, J.E., E.J. Johnson, D.A. Hendrickson & S. Contreras-Balderas. 1989. Fishes of North America endangered, threatened, or of special concern. Fisheries (Bethesda) 14: 2–20.

Sturgeons from the western and eastern north Pacific rim: the Sakhalin sturgeon, *Acipenser mikadoi*, 102 cm TL from the Datta (Tumnin) River of the Russian far east, now residing at Propa-Gen International, Komadi, above a white sturgeon, *A. transmontanus*, 86 cm TL from western US stock, now at Szarvas, Hungary. Originals by Paul Vecsei, 1996.

Environmental Biology of Fishes **48**: 407–417, 1997.

Alternatives for the protection and restoration of sturgeons and their habitat

Raymond C.P. Beamesderfer & Ruth A. Farr
Oregon Department of Fish and Wildlife, 2501 Southwest First Avenue, P.O. Box 59, Portland, OR 97207, U.S.A.

Received 1.11.1994 Accepted 18.3.1996

Key words: Acipenseriformes, life history, population dynamics, harvest, culture, Columbia River, hydropower

Synopsis

This paper reviews the life history and habitat requirements of sturgeons, alternatives for their protection and restoration in North America, and a typical protection and enhancement program in the Columbia River. Sturgeon are uniquely adapted to mainstem river systems which are characterized by their large scale, diverse habitats, and dynamic nature. Adaptations include mobility, opportunistic food habits, delayed maturation, longevity, and high individual fecundity. Unfortunately these life history characteristics are now a handicap for sturgeon because of fragmentation and destruction of their habitat. A variety of habitat-related alternatives for the protection and restoration of sturgeon were identified in a review of the literature and a survey of sturgeon biologists and managers throughout North America. However, harvest restrictions and supplementation using aquaculture are much more likely to be implemented than the system-wide measures needed to affect sturgeon habitat. A program for white sturgeon protection and enhancement in the Columbia River is a typical case where harvest management and supplementation measures are being used to optimize production of existing habitat but significant changes in water use and hydropower operation are needed to restore sturgeon to historic levels of production.

Introduction

Given their singular evolutionary, morphological genetic, and physiological traits (Grande & Bemis 1991, Bemis et al. 1997 this volume, Birstein 1993, Birstein et al. 1997 this volume), it is no surprise that sturgeon are also ecologically unique. However, life history traits which have proven adaptive over the last 100 million years are now a disadvantage in the face of drastic habitat changes and overfishing during the last century. Sturgeon are presently depleted, threatened, or extinct almost everywhere they occur (Smith 1990, Birstein 1993). Biologists throughout North America are grappling with the difficulty of developing protection or recovery programs for the nine endemic sturgeon species. In this paper, we discuss key characteristics of sturgeon life history which constrain populations, alternatives for protection and restoration of sturgeon and their habitats which have been identified for North American populations, and protection and enhancement efforts for Columbia River white sturgeon which typify the problems faced in many other populations.

Life history and habitat requirements

Critical habitat requirements and effective protection and restoration measures can be inferred from sturgeon life history. Sturgeon are uniquely adapted to the large mainstem river systems upon which all species rely during all or part of their life cycle (Rochard et al. 1990). Rivers include diverse habitats which are distributed in large scale patterns corresponding to the surrounding topography. Typical transitions include headwaters through tributaries, mainstem, and estuary into an ocean, sea, or large lake. In large basins, rivers may traverse many different regions and climatic zones. Rivers are also extremely dynamic habitats featuring large seasonal and annual variations in physical conditions and resource availability (Sheehan & Rasmussen 1993). Seasonal cycles in weather and runoff drive changes in velocity, morphometry, temperature, substrate, and turbidity. Conditions vary from year to year in unpredictable patterns based on regional weather patterns. Periodic floods and droughts may radically alter the riverine environment. Distribution and abundance of many species of fishes and other organisms vary widely in response to spatial and temporal patterns. For instance, anadromous fishes are seasonally abundant as they move between spawning and feeding areas in portions of many temperate rivers and estuaries.

Sturgeon have evolved life history characteristics which allow them to thrive in these large, diverse, and dynamic river systems. Individuals often range widely to take advantage of scattered and seasonally abundant resources. Regular migrations for spawning and short-term movements for feeding have been observed for many species (Chadwick 1959, Miller 1972a, Haynes et al. 1978, Haynes & Gray 1981, Smith 1985, Wooley & Crateau 1985, Sandilands 1987, Kempinger 1988, Odenkirk 1989, Hall et al. 1991, Mosindy & Rusak[1], O'Herron et al. 1993). Many species are euryhaline and move freely between freshwater, estuaries, and saltwater (Rochard et al. 1990) to further broaden their resource base. Long-distance movements are facilitated by their large size, shape, and swimming ability which allow them to move through heavy current.

Sturgeon are opportunistic predators that eat a variety of prey and switch as prey availability changes. These fish can also withstand long periods of starvation during periods of low food availability or spawning migrations (Dadswell 1979, Mason & Clugston 1993). Sturgeon generally feed on invertebrates in the benthic food chain (Held 1969, Dadswell 1979, Carlson et al. 1985, Sandilands 1987, McCabe et al. 1993) where most production occurs in large river systems (Sheehan & Rasmussen 1993). Fish may also be an important diet component of some sturgeon species (Semakula & Larkin 1968). Large sturgeon can consume large prey. Pursuit and capture of active prey belie an image of sturgeon as sluggish bottom scavengers.

Populations of sturgeon are buffered from annual variation in environmental conditions by delayed maturation, longevity, and high individual fecundity. Delayed maturation (Roussow 1957, Sunde 1961, Dadswell 1979, Conte et al.[2], Chapman 1989, Guenette et al. 1992, Keenlyne & Jenkins 1993) speeds growth to large sizes as energy is devoted to somatic rather than gonadal development. Large size helps reduce predation, lowering natural mortality rate and increasing longevity. A long lifespan (Pycha 1956, Wilson 1987, Rien & Beamesderfer 1994) allows fish numerous opportunities to spawn and reduces the need to spawn in years when conditions are not suitable. Many species have been observed to resorb eggs under these conditions (Artyukhin et al. 1979, Chapman 1989). High fecundity associated with large size improves spawning success in years when suitable conditions are encountered.

Many sturgeon species depend on free-flowing rivers and seasonal floods to provide suitable spawning conditions. Adhesive eggs are typically broadcast over rocky substrates in turbulent, high-velocity areas during high spring runoff (Magnin 1966, Buckley & Kynard 1985, Smith 1985, Kempinger 1988, Hall et al. 1991, Mosindy & Rusak[1], LaHaye et al. 1992, Parsley et al. 1993). Recruitment

[1] Mosindy, T. & J. Rusak. 1991. An assessment of lake sturgeon populations in Lake of the Woods and the Rainy River 1987–90. Ontario Ministry of Natural Resources. 66 pp.

[2] Conte, F.S., S.I. Doroshov & P.B. Lutes. 1988. Hatchery manual for the white sturgeon *Acipenser transmontanus* Richardson with application to other North American Acipenseridae. University of California Cooperative Extension Publication 3322. 104 pp.

has been widely correlated with spring and summer discharge (Stevens & Miller 1970, Khoroshko 1972, Votinov & Kasyanov 1979, Kohlhorst et al. 1991, Veshchev 1991). Flowing water provides oxygen, disperses eggs, and excludes egg predators. Seasonal floods scour substrates free of sand and silt which might suffocate eggs. Seasonal floods and corresponding changes in temperature, velocity, and turbidity may also provide spawning cues (Kempinger 1988, Kohlhorst et al. 1991, LaHaye et al. 1992).

Unfortunately, many of these adaptations to large river systems are now detrimental to sturgeon. Availability of food and critical spawning areas are limited where construction of dams blocks movements among scattered areas and creates homogenous reservoirs which reduce habitat diversity. Dam and reservoir operation for hydropower generation, flood control, irrigation, and navigation reduce seasonal and annual variability in flow which provide suitable spawning and rearing conditions for sturgeon and many of their prey. Altered systems favor development of a new array of prey, predators, and competitors. Benthic feeding and delayed maturation increase vulnerability to bioaccumulation of toxic pollutants (Ruelle & Keenlyne 1993). Longevity and delayed maturation make populations extremely susceptible to overexploitation. Large size and high fecundity increase the value of individual fish and provide incentives for excessive or illegal harvest.

Because of the unique features of their large river habitats and adaptive life history characteristics, sturgeon require a much broader definition of habitat than is typically applied to fishes when alternatives for habitat improvement are considered. Fish habitats are often defined in terms of site-specific conditions like depth, velocity, substrate, and cover. Sturgeon habitat must be defined in terms of system-wide conditions including large areas of diverse habitat; natural variation in flow, velocity, temperature, and turbidity; high water quality; a broad prey base; and free-flowing sections which provide suitable spawning sites (Carlson et al. 1985, Crance[3], Mosindy 1987, Payne 1987, Curtis 1990, Taub[4], Lane 1991, Pitman[5], Beamesderfer 1993, Dryer & Sandoval[6], USFWS & GSMFC[7]).

Alternatives for protection and restoration

To help identify and assess the potential feasibility of alternatives for protecting and restoring sturgeons and their habitat, we recently conducted a mail survey of 268 sturgeon and paddlefish biologists and managers from throughout North America. One page questionnaires including a return address and postage were sent to each person identified in a 'Summary of sturgeon and paddlefish researchers and managers' developed by the United States Fish and Wildlife Service. While survey results from this sample cannot be construed as an unbiased indication of which measures are appropriate, results should be useful in identifying the range of alternatives available.

Survey questions included 'please list measures you believe to be potentially beneficial to the conservation, productivity, or diversity of sturgeon or paddlefish populations with which you are familiar' and 'which of the above alternatives have been implemented and proven beneficial to the targeted sturgeon or paddlefish species?' In addition, each person was asked to (1) identify their experience with sturgeon or paddlefish (basic research, applied research, stock assessment/monitoring, habitat protection, fishery regulation, or aquaculture); (2)

[3] Crance, J.H. 1986. Habitat suitability index models and instream flow suitability curves: shortnose sturgeon. U.S. Fish and Wildlife Service Biological Report 82 (10.129).

[4] Taub, S.H. 1990. Fishery management plan for Atlantic sturgeon (*Acipenser oxyrhynchus oxyrhynchus*). Atlantic States Marine Fisheries Commission Fisheries Management Report 17. 73 pp.

[5] Pitman, V.M. 1992. Texas paddlefish recovery plan. Texas Parks and Wildlife Department, Austin. 30 pp.

[6] Dryer, M.P. & A.J. Sandoval. 1993. Recovery plan for the pallid sturgeon (*Scaphirhynchus albus*). U.S. Fish and Wildlife Service, Denver. 55 pp.

[7] USFWS & GSMFC (United States Fish and Wildlife Service & Gulf States Marine Fisheries Commission). 1995. Gulf sturgeon recovery plan. Atlanta. 170 pp.

Table 1. Specific alternatives identified as potentially beneficial to the conservation, productivity, or diversity of sturgeon or paddlefish populations in a survey of sturgeon and paddlefish biologists throughout North America. Each biologist developed a list of alternatives and ranked them according to potential benefit.

Category Specific response	Number responses	Number implemented	Mean rank
Habitats			
General (e.g. protect or restore critical habitat)	59	18	1.9
Flow (e.g. restore hydrograph)	30	6	1.8
Spawning habitat (e.g. protect)	15	10	2.9
Dredging or channelization	9	4	2.4
Control predators	3	0	5.7
Dams	3	0	3.0
Rearing habitat	2	1	3.0
Construct spawning habitat	1	1	5.0
Harvest			
Partial size specific (e.g. protect broodstock)	37	31	2.2
General (e.g. control harvest)	32	25	2.6
Complete closure	17	11	1.8
Enforcement (e.g. poaching and caviar sales)	14	7	3.0
Commercial closure	10	5	1.8
Bycatch control	2	0	3.5
Research			
Stock assessment	20	6	2.9
Aquaculture	18	5	2.4
Genetics	17	6	3.1
Life history	15	6	2.4
Habitat requirements	14	5	2.6
General	9	4	3.6
Reproduction	9	4	2.2
Monitoring	3	2	2.3
Pollution	3.	0	2.3
Passage	1	0	3.0
Culture stocking			
General	29	16	2.5
To historic ranges	17	7	2.3
Fingerlings	5	5	2.2
Establish cryogenic stock reservoirs	4	3	2.5
Young of the year	1	1	2.0
Use as reservoir for genetic stock	1	1	2.0
Passage			
Improve passage at dams	16	4	3.0
Eliminate dams	12	1	1.7
Run of the river operations	6	6	1.8
General	2	1	2.5
Pollution			
General	19	5	3.1
Contaminants	8	1	4.1
Sediments	5	2	2.8
Nutrient (e.g. feedlot runoff)	2	2	3.5
Planning			
General (e.g. coordinate interstate efforts)	13	8	2.1
Listing/legal protection	9	7	2.0
Recovery plan	4	4	3.0
Management plan	4	3	3.7
Information and education			
Public outreach	15	5	3.1
General	7	2	4.0

name the sturgeon or paddlefish species with which familiar; and (3) provide related articles or reports.

Specific alternatives identified in 151 responses were classified into 8 general categories (Table 1), including general problem areas identified in the 'Framework for the management and conservation of paddlefish and sturgeon species' prepared by a national steering committee of biologists for the U.S. Fish and Wildlife Service. The most frequently identified categories included habitat, harvest, and research (Figure 1). Average ranks based on order of listing were similar for habitat, harvest, passage, culture/stocking, and planning (Table 1).

Habitat-related alternatives most often involved protection of critical habitat, especially for spawning. Effects of the natural hydrograph, dredging or channelization, dams, and predators were also recognized (Table 1). Pollution- and passage-related measures were tabulated separately although they might also be considered as habitat-related measures. Pollution-related alternatives mentioned contaminants, nutrients, and sediments. Passage-related alternatives recognize the widespread construction of dams which are barriers to migration. Harvest-related alternatives involved complete fishery closures, partial restrictions, and more intensive enforcement of restrictions especially with respect to caviar. Specific alternatives to culture-stocking most frequently involved stocking juveniles to supplement or reestablish populations within historic ranges. Research needs on all aspects of biology and management were noted. Frequent references to planning efforts recognize the widespread distribution of sturgeons across several jurisdictional boundaries. Several responses also reiterated a need for public outreach programs.

The most likely measures to be implemented involved planning (73%), harvest restrictions (70%), and aquaculture (58%). Although habitat protection and enhancement measures were the most frequently recognized as potentially beneficial, they appeared least likely to be implemented in cases where identified. Planning efforts were frequently identified as beneficial, perhaps reflecting the relatively low cost of such efforts. A generally poor understanding of sturgeon biology is implied by the frequent mention of a need for additional research

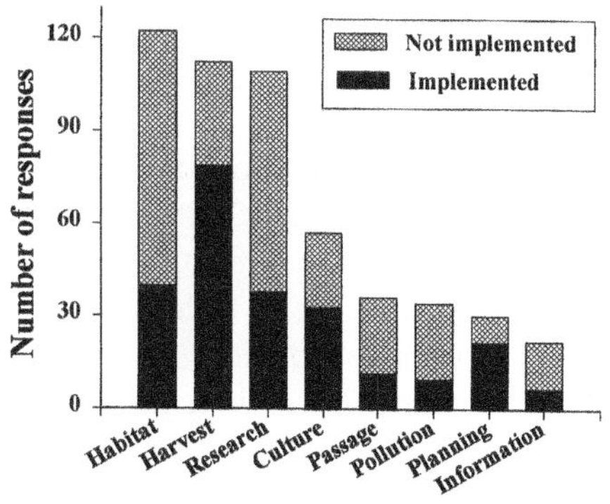

Figure 1. Potentially beneficial alternatives identified and implemented for the conservation, productivity, or diversity of sturgeon or paddlefish populations in a survey of sturgeon and paddlefish managers throughout North America.

but the low incidence where programs had been implemented (35%).

The capture and harvest of sturgeon are restricted almost everywhere they occur in North America. Annual harvest rates greater than 5–10% are almost universally believed to exceed sustainable levels because of resulting low survival to large reproductive sizes (Semakula & Larkin 1968, Miller 1972, Huff[8], Threader & Brousseau 1986, Nowak & Jessup 1987, Young et al. 1988, Rieman & Beamesderfer 1990, Kohlhorst et al. 1991). Closed reasons, protected areas, size limits, bag limits, gear restrictions, and catch-release regulations have all been used for sturgeon and paddlefish (Cochnauer 1983, Cochnauer et al. 1985, Foltz & Meyers 1985, Galbreath 1985, Smith 1985, Hart 1987, Debrot et al.[9], Scarnecchia et al. 1989, Taub[4], PSMFC[10]). Significant fisheries still occur for white sturgeon, *Acipenser transmontanus,* and paddlefish, *Polyodon spathula,*

[8] Huff, J.A. 1975. Life history of gulf of Mexico sturgeon, *Acipenser oxyrhynchus desotoi,* in Suwannee River, Florida. Florida Marine Research Publication No. 16. 32 pp.

[9] Debrot, A.O., H.A. Schaller & M.A. Matylewich. 1989. Estimates of sustainable exploitation rates for Columbia River landlocked white sturgeon: evaluating the importance of a maximum size limit. Columbia River Inter-Tribal fish Commission Technical Report 88-4. 41 pp.

[10] PSMFC (Pacific States Marine Fisheries Commission). 1992. White sturgeon management framework plan. Portland. 201 pp.

(Galbreath 1985, Pitman[11], Graham 1996). Small fisheries remain for lake *Acipenser fulvescens*, green *Acipenser medirostris,* Atlantic *Acipenser oxyrinchus oxyrinchus*, and shovelnose *Scaphirhynchus platorynchus* sturgeons (Smith et al. 1984, Foltz & Meyers 1985, Thuemler 1985, Olver 1987, Smith 1990, Michalenko et al. 1991). Fishing for shortnose *Acipenser brevirostrum,* pallid *Scaphirhynchus albus*, and gulf *Acipenser oxyrinchus desotoi* sturgeons has been curtailed by their federally-recognized status as endangered or threatened species. Alabama sturgeon *Scaphirhynchus suttkusi* are rare and not subject to harvest.

Culture of North America sturgeon currently relies on the capture of wild broodstock which are stimulated to spawn using hormones, although captive broodstock are being developed for several species (Smith 1990). Artificial spawning has been documented for Atlantic, shortnose, pallid, lake, and white sturgeons, and for paddlefish (Conte et al.[2], Smith 1990). Success of several experimental releases of lake sturgeon, shortnose sturgeon, and paddlefish is currently being evaluated (Graham[12], Anderson 1987, Pitman[11], Smith & Jenkins 1991, LaPan et al.[13], Graham 1996). Stocking programs for white sturgeon have been restricted to release of small numbers of juveniles in the Sacramento, Snake, and Willamette rivers as partial mitigation from private hatchery operators for use of wild broodstock. An experimental hatchery program is also being developed to supplement white sturgeon in the Kootenai River which flows through British Columbia, Idaho, and Montana (Apperson & Wakkinen 1992, Kincaid[14]).

Habitat modifications to benefit sturgeon have rarely been implemented. The elimination of daily discharge fluctuation for hydroelectric power generation at a dam in Michigan has increased spawning activity of lake sturgeon (Auer 1996). Effects of experimental releases of water from a Montana reservoir on spawning success of white sturgeon are currently being tested (Marcuson[15]). Dredge and fill operations have been modified or curtailed in spawning areas of lake sturgeon in the St. Lawrence River (Dumont et al. 1987). Successful site-specific habitat alterations have improved spawning by lake sturgeon in several areas where rock substrate was limiting and introduced to stabilize shoreline (Folz & Meyers 1985) or to increase current velocity (Rochard et al. 1990, LaHaye et al. 1992). A fish elevator was operated sporadically from 1938–1969 to lift white sturgeon past Bonneville Dam on the Columbia River (Warren & Beckman[16]).

The broad habitat needs of sturgeon suggest that only large-scale, system-wide habitat protection and improvement programs can be expected to provide significant benefits for those populations that are depleted or threatened by habitat alteration. Except in rare cases, site specific changes can be expected to have little effect. Options for producing system-wide changes to benefit sturgeon are limited because they involve complex issues of water diversion, land use, and hydropower system development or operation whose implementation is constrained by economic and social considerations.

Our survey demonstrated that while several alternatives may be identified, effective options are limited. Managers have had to rely on harvest management and aquaculture because system-wide habitat protection and enhancement measures have been extremely difficult to implement. These measures have effectively maintained populations and provided fishery benefits where habitat degradation is not severe. However, efforts which do not address habitat degradation have generally failed

[11] Pitman, V.M. 1991. Synopsis of paddlefish biology and their utilization and management in Texas. Texas Parks and Wildlife Department, Austin. 70 pp.

[12] Graham, L.K. 1986. Reintroduction of lake sturgeon in Missouri. Final Rep., D.J. Proj. F-1-R-35, Study S-25. Missouri Dep. Conserv., Columbia. 11 pp.

[13] LaPan, S.R., A. Schiavone, R.M. Klindt, W.F. Krise, M.N. DiLauro & K. Fynn-Aikins. 1994. Re-establishment of lake sturgeon in tributaries of the St. Lawrence River, 1993. Report to the Lake Ontario Committee, Great Lakes Fishery Commission. 10 pp.

[14] Kincaid, H.L. 1993. Breeding plan to preserve the genetic variability of the Kootenai River white sturgeon. Bonneville Power Admin., Portland. 18 pp.

[15] Marcuson, P. 1994. Kootenai River white sturgeon investigations annual report. Bonneville Power Admin., Portland. 67 pp.

[16] Warren, J.J. & L.G. Beckman. 1993. Fishway use by white sturgeon to bypass mainstem Columbia River Dams. U.S. Fish Wildl. Sea Grant Extension Proj., Col. R. Series WSG-AG 93-02. 12 pp.

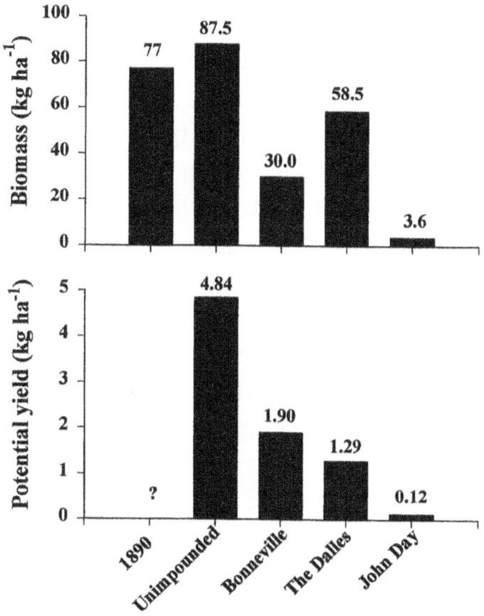

Figure 2. Abundance and productivity of the pristine (1890) and present impounded and unimpounded white sturgeon stocks in the lower Columbia River (Beamesderfer et al. 1995, DeVore et al. 1995).

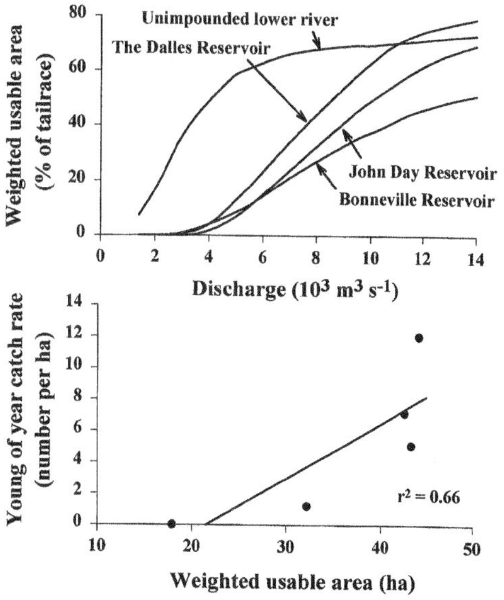

Figure 3. Relations between river discharge, availability of spawning habitat, and annual recruitment (Bonneville Reservoir only) for white sturgeon in the lower Columbia River (adapted from Parsley & Beckman 1994).

to restore sturgeon populations to historic levels of productivity.

A Columbia River example

The Columbia River white sturgeon populations represent a typical situation where habitat changes have drastically affected the stock, harvest has been regulated, supplementation stocking is being considered, but only habitat changes can be expected to restore sturgeon productivity to historic levels. Only an accident of engineering prevented the Columbia River population of white sturgeon from joining the other threatened and endangered sturgeon and paddlefish populations throughout the world. In 1983 Bonneville Dam was completed in the gorge where the river cuts through the Cascade mountain range. If this dam had been built at the bottom of the gorge just 5 miles downstream, it would have flooded or blocked access to critical spawning habitat and destroyed productive commercial and sport sturgeon fisheries in the lower river which annually produce 45 000 fish, yield 350 000 kg, and support 145 000 angler trips (DeVore et al. 1995).

Current sturgeon biomass in the unimpounded 234 km of the lower Columbia River appears similar to levels during pristine conditions prior to significant exploitation in the late 1800s (Figure 2). Upstream from Bonneville Dam, a series of mainstem dams have trapped stocks of white sturgeon in a series of reservoirs. Individual white sturgeon range extensively throughout each reservoir but rarely pass upstream or downstream dams (North et al. 1993). All reservoirs are similar in that hydrologic retention times are short, littoral zone is limited, and current is measurable most of the year. However, reservoirs vary in size, depth, substrate, and length of the free-flowing portion in the tailwater of the upstream dam.

Columbia River reservoirs provide a laboratory for examining limiting factors for white sturgeon. Each stock is presented with a different array of habitat conditions which affect reproduction, growth, and survival and in turn regulate population size and productivity. Fish that historically moved throughout this area to use scattered resources are now trapped in a reservoir which no longer furnishes optimal conditions for different intervals of the life cycle. In some areas recruitment is high but rearing habitat is limited. Elsewhere rearing habitat is abundant but spawning habitat is not. The net result is that biomass and potential yield are less in impounded stocks than in the unimpounded stock (Figure 2).

Productivity of some impounded stocks is especially limited by poor recruitment resulting from lack of suitable spawning habitat. Columbia River white sturgeon spawn in areas of high velocity (> 0.8 ms^{-1}) over large rocky substrate now available only in riverine areas downstream from dams (Parsley et al. 1993). Availability of usable habitat increases with river discharge and recruitment is correlated with habitat availability (Figure 3). Discharge effects vary among areas as a result of differences in channel morphology, and some areas provide little habitat except at very high flows. Discharge regulation in isolated areas of the upstream Kootenai and Snake rivers has resulted in complete reproductive failure (Apperson & Wakkinen 1992, Marcuson[15]).

Harvest levels which could be supported by the productive unimpounded stock cannot be sustained by the impounded stocks (Beamesderfer et al. 1995). As a result, fisheries for several impounded stocks collapsed in the late 1980s after a period of intense exploitation as sport and commercial fisheries switched to sturgeon following declines of salmon fisheries. Hydropower system managers are now cooperating with government agencies and Indian tribes responsible for managing these fish to protect and enhance these impounded stocks.

One element of this program is intensive harvest management. Before 1988 sturgeon stocks throughout the lower Columbia River were managed with similar regulations and only a few key stocks were monitored. Fisheries are now being regulated with stock-specific regulations tailored to the unique attributes of each stock in an attempt to optimize fishery benefits. A more intensive monitoring program has also been undertaken to regulate harvest at optimum levels.

A second program element is evaluating transplants of juveniles from the large and productive unimpounded stock into reservoirs where poor recruitment appears to have understocked the available rearing habitat. We believe that production of sturgeon by the system will ultimately be limited by the carrying capacity of the rearing habitat and that peak production will result from full stocking of all areas. Survival, growth, and condition of transplanted fish will be monitored to determine the costs and benefits of this alternative. Transplants also provide a low-cost means of evaluating the potential for enhancement of reservoir stocks without capital costs, genetic risks, or disease problems of a hatchery operation. Hatchery supplementation will be considered in more detail if transplants are not feasible or effective.

A third program element is developing and supporting recommendations for hydropower system operation to optimize river discharge and velocity during spring periods when water temperature is suitable for spawning. However, the large social and economic costs of modifications in hydropower system operation are likely to preclude changes in water allocation for the sole benefit of sturgeon. Program cooperators are therefore implementing intensive sampling designs for eggs, embryos and larvae in an attempt to identify effects of within-year differences in flow and to develop recommendations for using available water to optimize spawning conditions within each year.

Flow management is the only element in the lower Columbia River program which attempts to enhance sturgeon by directly modifying habitat. Intensive harvest management for each stock and supplementation recognize habitat limitations but maximize productivity of the existing habitat rather than producing habitat improvements.

Conclusions

The flexible and opportunistic life history style of sturgeons may help explain their persistence and success over the last 100 million years. However, system-wide changes in the large river systems they inhabit now pose serious risks to these remarkable creatures. The large scale of detrimental habitat changes make them extremely difficult to control for the sole benefit of sturgeon and so sturgeon managers and biologists have been forced to rely on harvest restrictions and aquaculture programs with limited success.

Sturgeon provide obvious economic and scientific benefits. We believe that sturgeon also serve as very large canaries in the coal mine of riverine ecosystems. These fish are universally threatened because their large riverine habitats are on the verge of ceasing to function at the ecosystem level. Only a combination of alternatives integrating habitat protection and recovery with harvest restrictions and supplementation can be expected to sustain sturgeon populations that are anything more than museum pieces. The challenge of all who recognize these problems will be to push for fundamental changes in how we use these large riverine systems rather than settling solely for alternatives in the constrained sphere of our immediate influence.

Acknowledgements

Many thanks to all the sturgeon researchers and managers who filled out and returned questionnaires and sent material. J. DeVore and M. Parsley provided unpublished information on white sturgeon recruitment in the Columbia River. Research and recovery efforts for Columbia River white sturgeon are funded by the Bonneville Power Administration (Contract DE-A179-86BP63584), the Federal Aid to Fish Restoration Act, and the States of Oregon and Washington.

References cited

Anderson, E.R. 1987. Lake sturgeon (*Acipenser fulvescens*) management and culture in Minnesota and Michigan. pp. 84. *In:* C.H. Olver (ed.) Proceedings of a Workshop on the Lake Sturgeon (*Acipenser fulvescens*), Ont. Fish. Tech. Rep. Ser. No. 23.

Apperson, K.A. & V.D. Wakkinen. 1992. Kootenai River white sturgeon investigations and experimental culture. Bonneville Power Admin., Portland. 37 pp.

Artyukhin, E.N., A.D. Sukhoparova & L.G. Fimukhina. 1979. The gonads of the sturgeon, *Acipenser guldenstadti,* in the zone below the dam of the Volgograd water engineering system. J. Ichthyol. 18: 912–923.

Auer, N.A. 1996. Response of spawning lake sturgeons to change in hydroelectric facility operation. Trans. Amer. Fish. Soc. 125: 66–77.

Beamesderfer, R.C. 1993. A standard weight (W_s) equation for white sturgeon. Calif. Fish Game 79: 63–69.

Beamesderfer, R.C., T.A. Rien & A.A. Nigro. 1995. Differences in the dynamics and potential production of impounded and unimpounded white sturgeon populations in the lower Columbia River. Trans. Amer. Fish. Soc. 124: 857–872.

Bemis, W.E., E. Findeis & L. Grande. 1997. An overview of Acipenseriformes. Env. Biol. Fish. (this volume).

Birstein, V.J. 1993. Sturgeons and paddlefishes: threatened fishes in need of conservation. Cons. Biol. 7: 773–787.

Birstein, V.J., R. Hanner & R.DeSalle. 1997. Phylogeny of the Acipenseriformes: cytogenetic and molecular approaches. Env. Biol. Fish. (this volume).

Buckley, J. & B. Kynard. 1985. Habitat use and behavior of prespawning and spawning shortnose sturgeon, *Acipenser brevirostrum,* in the Connecticut River. pp. 111–117. *In:* F.P. Binkowski & S.I. Doroshov (ed.) North American Sturgeons: Biology and Aquaculture Potential, Dr W. Junk Publishers, Dordrecht.

Carlson, D.M., W.L. Plieger, L. Trial & P.S. Haverland. 1985. Distribution, biology and hybridization of *Scaphirhynchus albus* and *S. platorhynchus* in the Missouri and Mississippi rivers. Env. Biol. Fish. 14: 51–59.

Chadwick, H.K. 1959. California sturgeon tagging studies. Calif. Fish Game 45: 297–301.

Chapman, F.A. 1989. Sexual maturation and reproductive parameters of wild and domestic stocks of white sturgeon, *Acipenser transmontanus.* Ph.D. Dissertation, University of California, Davis. 99 pp.

Cochnauer, T.G. 1983. Abundance, distribution, growth and management of white sturgeon (*Acipenser transmontanus*) in the middle Snake River, Idaho. Ph.D. Dissertation, University of Idaho, Moscow. 52 pp.

Cochnauer, T.G., J.R. Lukens & F.E. Partridge. 1985. Status of white sturgeon, *Acipenser transmontanus*, in Idaho. pp. 127–133. *In:* F.P. Binkowski & S.I. Doroshov (ed.) North American Sturgeons: Biology and Aquaculture Potential, Dr W. Junk Publishers, Dordrecht.

Curtis, G.L. 1990. Habitat use by shovelnose sturgeon in pool 13, upper Mississippi River, Iowa. M.Sc. Thesis, Iowa State University, Ames. 79 pp.

Dadswell, M.J. 1979. Biology and population characteristics of the shortnose sturgeon, *Acipenser brevirostrum* LeSueur 1818 (Osteichthyes: Acipenseridae), in the Saint John River estuary, New Brunswick, Canada. Can. J. Zool. 57: 22186–2210.

DeVore, J.D., B.W. James, C.A. Tracy & D.H. Hale. 1995. Dynamics and potential production of white sturgeon in the unimpounded Columbia River. Trans. Amer. Fish. Soc. 124: 845–856.

Dumont, P., R. Fortin, G. Desjardins & M. Bernard. 1987. Biology and exploitation of lake sturgeon (*Acipenser fulvescens*) in the Quebec waters of the Saint-Laurent River. pp. 57–76. *In:* C.H. Olver (ed.) Proceedings of a Workshop on the Lake Sturgeon (*Acipenser fulvescens*), Ont. Fish. Tech. Rep. Ser. No. 23.

Folz, D.J. & L.S. Meyers. 1985. Management of the lake sturgeon, *Acipenser fulvescens,* population in the Lake Winnebago system, Wisconsin. pp. 135–146. *In:* F.P. Binkowski & S.I. Doroshov (ed.) North American Sturgeons: Biology and Aquaculture Potential, Dr W. Junk Publishers, Dordrecht.

Galbreath, J.L. 1985. Status, life history, and management of Columbia River white sturgeon, *Acipenser transmontanus* pp. 119–125. *In:* F.P. Binkowski & S.I. Doroshov (ed.) North American Sturgeons: Biology and Aquaculture Potential, Dr W. Junk Publishers, Dordrecht.

Graham, K. 1997. Contemporary status of the North American paddlefish, *Polyodon spathula*. Env. Biol. Fish. (this volume).

Grande, L. & W. Bemis. 1991. Osteology and phylogenetic relationships of fossil and recent paddlefishes (Polyodontidae) with comments on the interrelationships of Acipenseriformes. J. Vert. Paleon. 11 (supplement 1): 1–121.

Guenette, S., D. Goyette, R. Fortin, J. Leclerc, N. Fournier, G. Roy & P. Dumon. 1992. La periodicite de la croissance chez la femelle de l'esturgeon jaune (*Acipenser fulvescens*) du fleuve Saint-Laurent est-elle reliee a la periodicite de la reproduction? Can. J. Fish. Aquat. Sci. 49: 1336–1342.

Hall, J.W., T.I.J. Smith & S.D. Lamprecht. 1991. Movements and habitats of shortnose sturgeon, *Acipenser brevirostrum* in the Savannah River. Copeia 1991: 695–702.

Hart, M.L. 1987. Considerations for the management of lake sturgeon (*Acipenser fulvescens*) commercial fisheries in Ontario. pp. 85–90. *In*: C.H. Olver (ed.) Proceedings of a Workshop on the Lake Sturgeon (*Acipenser fulvescens*), Ont. Fish. Tech. Rep. Ser. No. 23.

Haynes, J.M., R.H. Gray & J.C. Montgomery. 1978. Sesaonal movements of white sturgeon (*Acipenser transmontanus*) in the mid-Columbia River. Trans. Amer. Fish. Soc. 107: 275–280.

Haynes, J.M. & R.H. Gray 1981. Diel and seasonal movements of white sturgeon, *Acipenser transmontanus*, in the mid-Columbia River. U.S. Fish. Bull. 79: 367–370.

Held, J.W. 1969. Some early summer foods of the shovelnose sturgeon in the Missouri River. Trans. Amer. Fish. Soc. 98: 514–517.

Keenlyne, K.D. & L.G. Jenkins. 1993. Age at sexual maturity of the pallid sturgeon. Trans. Amer. Fish. Soc. 122: 393–396.

Kempinger, J.J. 1988. Spawning and early life history of lake sturgeon in the Lake Winnebago system, Wisconsin. Amer. Fish. Soc. Symp. 5: 110–122.

Khoroshko, P.N. 1972. The amount of water in the Volga Basin and its effect on the reproduction of sturgeons (*Acipenseridae)* under conditions of normal and regulated discharge. J. Ichthyol. 12: 608–616.

Kohlhorst, D.W., L.W. Botsford, L.S. Brennan & G.M. Cailliet. 1991. Aspects of the structure and dynamics of an exploited central California population of white sturgeon (*Acipenser transmontanus*). pp. 277–293. *In:* P. Williot (ed.) Acts du Premier Colloque International sur l'Esturgeon, Centre National du Machinisme Agricole du Genie Rural des Eaux et des Forets, Bordeaux.

LaHaye, M., A. Branchaud, M. Gendron, R. Verdon & R. Fortin. 1992. Reproduction, early life history, and characteristics of the spawning grounds of the lake sturgeon (*Acipenser fulvescens*) in Des Prairies and L'Assomption rivers, near Montreal, Quebec. Can. J. Zool. 70: 1681–1689.

Lane, E.D. 1991. Status of the white sturgeon, *Acipenser transmontanus*, in Canada. Can. Field Naturalist 105: 161–168.

Magnin, E. 1966. Quelques donnees biologiques sur la reproduction des esturgeons *Acipenser fulvescens* Raf. de la Riviere Nottaway, tributaire de la Baie James. Can. J. Zool. 44: 257–263.

Mason, W.T., Jr. & J.P. Clugston. 1993. Foods of the Gulf sturgeon in the Suwannee River. Trans. Amer. Fish. Soc. 122: 378–385.

McCabe, G.T., Jr. R.L. Emmett & S.A. Hinton. 1993. Feeding ecology of juvenile white sturgeon (*Acipenser transmontanus*) in the lower Columbia River. Northwest Sci. 67: 170–180.

Michalenko, G., L. Marcolgliese & the Muskrat Dam Band. 1991. The subsistence lake sturgeon (*Acipenser fulvescens*) fishery of the Indian village of Muskrat Dam in northern Ontario, Canada. pp. 447–458. *In:* P. Williot (ed.) Actes du Premier Colloque International sur l'Esturgeon, Centre National du Machinisme Agricole du Genie Rural Eaux et des Forets, Bordeau.

Miller, L.W. 1972a. Migrations of sturgeon tagged in the Sacramento-San Joaquin estuary. Calif. Fish Game 58: 102–106.

Miller, L.W. 1972b. White sturgeon population characteristics in the Sacramento-San Joaquin estuary as measured by tagging. Calif. Fish Game 58: 94–101.

Mosindy, T. 1987. The lake sturgeon (*Acipenser fulvescens*) fishery of Lake of the Woods Ontario. pp. 48–56. *In:* C.H. Olver (ed.) Proceedings of a Workshop on the Lake Sturgeon (*Acipenser fulvescens*), Ont. Fish. Tech. Rep. Ser. No. 23.

North, J.A., R.C. Beamesderfer & T.A. Rien. 1993. Distribution and movements of white sturgeon in three lower Columbia River reservoirs. Northwest Sci. 67: 105–111.

Nowak, A.M. & C.S. Jessop. 1987. Biology and management of the lake sturgeon (*Acipenser fulvescens*) in the Groundhog and Mattagami rivers, Ontario. pp. 20–32. *In:* C.H. Olver (ed.) Proceedings of a Workshop on the Lake Sturgeon (*Acipenser fulvescens*), Ont. Fish. Tech. Rep. Ser. No. 23.

Odenkirk, J.S. 1989. Movements of Gulf of Mexico sturgeon in the Apalachicola River, Florida. Proc. Annu. Conf. Southeast. Assoc. Fish Wildl. Agencies 43: 230–238.

O'Herron, J.C., II, K.A. Able & R.W. Hastings. 1993. Movements of shortnose sturgeon (*Acipenser brevirostrum*) in the Delaware River. Estuaries 16: 235–240.

Olver, C.H. (ed.) 1987. Proceedings of a workshop on the lake sturgeon (*Acipenser fulvescens*). Ont. Fish. Tech. Rep. Ser. No 23. 99 pp.

Parsley, M.J. & L.G. Beckman. 1994. White sturgeon spawning and rearing habitat in the lower Columbia River. N. Amer. J. Fish. Man. 14: 812–827.

Parsley, M.J., L.G. Beckman & G.T. McCabe, Jr. 1993. Spawning and rearing habitat use by white sturgeons in the Columbia river downstream from McNary Dam. Trans. Amer. Fish. Soc. 122: 217–228.

Payne, D.A. 1987. Biology and population dynamics of lake sturgeon (*Acipenser fulvescens*) from the Frederick House, Abitibi, and Mattagami rivers, Ontario. pp. 10–19. *In:* C.H. Olver (ed.) Proceedings of a Workshop on the Lake Sturgeon (*Acipenser fulvescens*), Ont. Fish. Tech. Rep. Ser. No. 23.

Pycha, R.L. 1956. Progress report on white sturgeon studies. Calif. Fish. Game 23–35.

Rieman, B.E. & R.C. Beamesderfer. 1990. White sturgeon in the lower Columbia River: is the stock overexploited? N. Amer. J. Fish. Man. 10: 388–396.

Rien, T.A. & R.C. Beamesderfer. 1994. Accuracy and precision of white sturgeon age estimates from pectoral fin rays. Trans. Amer. Fish. Soc. 123: 255–265.

Rochard, E., G. Castelnaud & M. LePage. 1990. Sturgeons (Pisces: Acipenseridae); threats and prospects. J. Fish Biol. 37A: 123–132.

Roussow, G. 1957. Some considerations concerning sturgeon spawning periodicity. J. Fish. Res. Board Can. 14: 553–572.

Ruelle, R. & K.D. Keenlyne. 1993. Contaminants in Missouri River pallid sturgeon. Bull. Env. Contam. Toxicol. 50: 898–906.

Sandilands, A.P. 1987. Biology of the lake sturgeon (*Acipenser fulvescens*) in the Kenogami River, Ontario. pp. 33–46. *In:* C.H. Olver (ed.) Proceedings of a Workshop on the Lake Sturgeon (*Acipenser fulvescens*), Ont. Fish. Tech. Rep. Ser. No. 23.

Scarnecchia, D.L., T.W. Gengerke & C.T. Moen. 1989. Rationale for a harvest slot limit for paddlefish in the upper Mississippi River. N. Amer. J. Fish. Man. 9: 477–487.

Semakula, S.N. & P.A. Larkin. 1968. Age, growth, food, and yield of the white sturgeon (*Acipenser transmontanus*) of the Fraser River, British Columbia. J. Fish. Res. Board Can. 25: 2589–2602.

Sheehan, R.J. & J.L. Rasmussen. 1993. Large rivers. pp. 445–468. *In:* C.C. Kohler & W.A. Hubert (ed.) Inland Fisheries Management in North America, Amer. Fish. Soc. Bethesda.

Smith, T.I.J. 1985. The fishery, biology, and management of Atlantic sturgeon, *Acipenser oxyrhynchus,* in North America. Env. Biol. Fish. 14: 61–72.

Smith, T.I.J. 1990. Culture of North American sturgeons for fishery enhancement. National Marine Fisheries Service Tech. Rep. 85: 19–27.

Smith, T.I.J. & W.E. Jenkins. 1991. Development of a shortnose sturgeon, *Acipenser brevirostrum*, stock enhancement program in North America. pp. 329–336. *In:* P. Williot (ed.) Actes du Premier Colloque International sur l'Esturgeon, Centre National du Machinisme Agricole du Genie Rural des Eaux et des Forets, Bordeau.

Smith, T.I.J., D.E. Marchette & G.F. Ulrich. 1984. The Atlantic sturgeon fishery in South Carolina. N. Amer. J. Fish. Man. 4: 164–176.

Stevens, D.E. & L.W. Miller. 1970. Distribution of sturgeon larvae in the Sacramento-San Joaquin River system. Calif. Fish Game 56: 80–86.

Sunde, L.A. 1961. Growth and reproduction of the lake sturgeon (*Acipenser fulvescens* Rafinesque) of the Nelson River in Manitoba. M. Sc. Thesis, University of British Columbia, Vancouver. 93 pp.

Threader, R.W. & C.S. Brousseau. 1986. Biology and management of the lake sturgeon in the Moose River, Ontario. N. Amer. J. Fish. Man. 6: 383–390.

Thuemler, T.F. 1985. The lake sturgeon, *Acipenser fulvescens*, in the Menominee River, Wisconsin-Michigan. Env. Biol. Fish. 14: 73–78.

Veshchev, P.V. 1991. Effectiveness of natural production in the stellate sturgeon, *Acipenser stellatus*, under conditions of regulated flow in the Volga. J. Ichthyol. 30: 76–83.

Votinov, N.P. & V.P. Kasyanov. 1979. The ecology and reproductive efficiency of the Siberian sturgeon, *Acipenser baeri,* in the Ob as affected by hydraulic engineering works. J. Ichthyol. 18: 20–29.

Wilson, N.C. 1987. Age determination of lake sturgeon (*Acipenser fulvescens*) by use of the marginal pectoral fin ray. pp. 77–83. *In:* C.H. Olver (ed.) Proceedings of a Workshop on the Lake Sturgeon (*Acipenser fulvescens*), Ont. Fish. Tech. Rep. Ser. No. 23.

Wooley, C.M. & E.J. Crateau. 1985. Movement, microhabitat, exploitation, and management of Gulf of Mexico sturgeon, Apalachicola River, Florida. N. Amer. J. Fish. Man. 5: 590–605.

Young, J.R., T.B. Hoff, W.P. Dey & J.G. Hoff. 1988. Management recommendations for a Hudson River Atlantic sturgeon fishery based on an age-structured population model. pp. 353–367. *In:* C.L. Smith (ed.) Fisheries Research in the Hudson River, St. Univ. New York Press, New York.

Environmental Biology of Fishes **48**: 418–419, 1997.

Threatened fishes of the world: *Scaphirhynchus suttkusi* Williams & Clemmer, 1991 (Acipenseridae)

Richard L. Mayden & Bernard R. Kuhajda
Department of Biological Sciences, Box 870344, University of Alabama, Tuscaloosa, AL 35487-0344, U.S.A.

Common names: Alabama shovelnose sturgeon, Alabama sturgeon (E).

Conservation status: Recommended for protection (Ramsey 1976, 1986). Proposed as Federally Endangered 15 June 1993 (Federal Register 1993); listing postponed 21 June 1994 (Federal Register 1994a), listing withdrawn 15 December 1994 (Federal Register 1994b). Currently receives no protection.

Identification: One of three species of *Scaphirhynchus* distinguished by having orangish-yellow, brassy-orange, to brownish-tan head, body, and fin coloration in mature adults; significantly larger orbit diameter; no sharp, retrose spiness on snout tip and anterodorsal to eye; and poorly developed squamation on venter. Other features include significantly different number of dorsal plates, anus to anal fin plates, plates posterior to anal fin, lateral plates anterior to dorsal fin origin, dorsal fin rays, and 11 head, fin, and body proportions, relative to its sister species *S. platorynchus*. Complete description by Williams & Clemmer (1991) and Mayden & Kuhajda (1996). Photograph by John Caruso.

Distribution: Endemic to larger rivers of Mobile Basin. Former distribution included Tombigbee, Alabama, Cahaba, Coosa, Alabama and Mississippi rivers. Recent collections are only from the Alabama River, Alabama below the lowermost two dams. **Abundance**: Species verey rare; only 36 specimens known from museum records or photo documentation. Contrary to decisions by the U.S. Fish and Wildlife Service (Federal Register 1994b), the species is not extinct; two adult specimens were captured in the Alabama River in Spring 1995 and one in Spring 1996. All current data, extending from the general degradation of the Mobile Basin ecosystem (U.S. Fish and Wildlife Service 1994) to the paucity of records and specimens of the Alabama sturgeon, indicate that this is one of the most endangered species in the United States. However, this has not always been the case, as this species once maintained a healthy population size, but has declined in this century due to anthropogenic changes to the ecosystem. In a report to the U.S. Congress in 1898 (U.S. Commission of Fish and Fisheries 1898) the total catch of 'shovelnose sturgeon' from rivers in alabama was 19 500 kg. Of this, captures of the shovelnose sturgeon, *S. platorynchus*, from the Tennessee River numbered only 500 kg, while captures of ***S. suttkusi*** from the Alabama and Black Warrior (Tombigbee) rivers numbered 19 000 kg. Given that an Alabama sturgeon averages 1 kg, this represents a substantial harvest of a species that is currently extremely rare. **Habitat and ecology**: Life history aspects of *S. suttkusi* are poorly known. Appears to prefer relatively stable substrates of gravel and sand in areas of current in large river channels, but will also occur over softer sediments. Its Spring diet is dominated by macroinvertebrates that typically bury in sandy substrates in both riffle and lentic depositional areas. Occasionally, small fish, plant items, and some mollusks and snails associated with relatively stable and silt-free substrates occur in stomachs. Based on food items present in stomachs and lack of items (invertebrate taxa) indicative of certain habitats, Spring-collected Alabama sturgeon appear to feed in sandy depositional areas with very little silt and slow to moderate current (see Burke & Ramsey 1995, Mayden & Kuhajda 1996). **Reproduction**: Reproductive biology not well known; most information is inferred from its closest relative, *S. platorynchus*. Spawning season probably extends from February to July. Species of *Scaphirynchus* spawn in freshwater and are known to migrate upstream to spawning areas as 'spawning runs' that may be triggered by rising water levels in the Spring and early Summer. Probably spawns in larger rivers in swift current and coarse, rock or gravel substrates, but may also spawn over hard bottom substrates in main-channel areas or in tributaries to major rivers. Like other members of the family, individuals of this species probably do not spawn every year. Rather, following sexual maturity at five to seven years of age, spawning may occur every one to three years with a hiatus of even greater number of years possible. Although unknown for the Alabama sturgeon, eggs of shovelnose sturgeon are adhesive and require current for proper development, indicating that both a stable and silt-free substrate is necessary for their successful development. Hatching occurs in five to eight days under proper conditions (see Mayden & Kuhajda 1996). **Threats**: Recruitment of Alabama sturgeon, indicative of spawning success, has declined precipitously during this century. Because this species has not been sought in the 20th century in a commercial fishery, this decline can be correlated directly with habitat degradation, modifications of the rivers for navigational purposes, mining operations in and adjacent to the rivers, and the construction of dams in the Mobile Basin, all of which have flourished in this century. All of these changes result in increased siltation of benthic habitats, reduced overall current velocities, elimination of natural seasonal flooding of river flood plains, irregular flow regimes within the river channels, the loss of and/or change in structure of reverine macroinvertebrate faunas, and obstruction of upstream migratory routes for migratory species. The latter constraint severely limits available spawning areas for the Alabama sturgeon; irregular flow regimes and flood/navigational controls eliminate natural signals to initiate spawning runs; and increased siltation and loss of stable substrate result in the loss of food sources and habitat appropriate for spawned eggs (current and clean surfaces). Macroinvertebrates found in the sturgeon's diet are adversely impacted by the higher silt deposits occurring over stable substrates in low current areas, a consequence that will have an indirect, yet adverse impact on the survival of this sturgeon species. **Conservation action**: The U.S. Fish and Wildlife Service proposed this species for endangered status with critical habitat on 15 June 1993 (Federal Register 1993). On 21 June 1994 (Federal Register 1994a), the Service postponed listing for six months to provide addditional time to assess the conservation status of the species through sampling. During this six month period no specimens of the sturgeon were captured. Consequently, the listing proposal was withdrawn by the Service on 15 December 1994 (Federal Register 1994b) due to 'insufficient information to justify listing a species that may no longer exist'. On 2 December 1993, less than one year before either the postponement or withdrawal of the listing, a mature male specimen of *S. suttkusi* was captured from the Alabama River. In April and May 1995 and April 1996, three additional specimens were discovered in the Alabama River. Currently, the Alabama sturgeon receives no protection and is not under official consideration by the Service. **Conservation recommendation**: There can be little doubt that the imperilment of the many sturgeon species worldwide is a direct result of the gross changes that have occurred in their natural habitats due to modifications of river channels and water flow patterns. The current predicament of the Alabama sturgeon is no exception. The protection and recovery of this sturgeon species, and others, will be a challenge, but one that can be accomplished through concerted and novel efforts. An effective recovery plan for the Alabama sturgeon most minimally includes efforts to (1) increase appropriate spawning habitats, (2) increase access to upstream and downstream river stretches across dams, (3) establish minimum flow regimes, and (4) decrease silt loads. As observed for some populations of shovelnose sturgeon (Cross 1967), the initiation of spawning activity of the Alabama sturgeon may be triggered by high-water conditions in the Spring. Should this be the case, an effective recovery program for the Alabama sturgeon must address the need for both irregular and high water flow in and downstream of spawning areas for this species. Critical habitat must be designated to assist in securing the species' future. To compile additional biological information on this species, additional specimens need to be captured, radio-tagged, released and followed through telemetry studies. Finally, adults should be captured for a propagational program, as has been developed for the endangered pallid sturgeon, *S. albus*, in the Missouri River drainage. **Remarks**: The listing of this species has been charged with abundant political and industrial opposition, actions due to potential environmental regulations that may be placed on existing hydroelectric dams and industrial users of the large rivers inhabited by this species.

Burke, J.S. & J.S. Ramsey. 1995. Present and recent historic habitat of the Alabama sturgeon, *Scaphirhynchus suttkusi* Williams and Clemmer, in the Mobile Basin. Bull. Alabama Mus. Nat. Hist. 17: 17–24.

Cross, F.B. 1967. Handbook of fishes of Kansas. Misc. Publ. Mus. Nat. Hist., University of Kansas, Lawrence. 357 pp.

Federal Register. 1993. Proposed endangered status and designation of critical habitat for the Alabama sturgeon (*Scaphirhynchus suttkusi*). Vol. 58, No. 113 (15 June 1993): 33148–33154.

Federal Register. 1994a. Extension of the final decision to list the Mobile River System population of the Alabama sturgeon as an endangered species with critical habitat. Vol. 59, No. 118 (21 June 1994): 31970–31974.

Federal Register. 1994b. Withdraw of proposed rule for endangered status and critical habitat for the Alabama sturgeon. Vol. 59, No. 240 (15 December 1994): 64784–64809.

Mayden, R.L. & B.R. Kuhajda. 1996. Systematics, taxonomy, and conservation status of the endangered Alabama sturgeon, *Scaphirhynchus suttkusi* Williams and Clemmer (Actinopterygii, Acipenseridae). Copeia 1996: 241–273.

Ramsey, J.S. 1976. Freshwater fishes. pp. 53–65. *In*: H.T. Boschung (ed.) Endangered and Threatened Plants and Animals of Alabama, Bull. Alabama Mus. Nat. Hist. 2.

Ramsey, J.S. 1986. Alabama shovelnose sturgeon. pp. 18–19. *In*: R.H. Mount (ed.) Vertebrate Animals of Alabama in Need of Special Attention, Alabama Agricultural Experimental Station, Auburn University, Auburn.

U.S. Commission of Fish & Fisheries. 1898. Statistics of the fisheries of the interior waters of the United States. A report to the 55th Congress, House of Representatives: 489–497, 531–533.

U.S. Fish & Wildlife Service. 1994. Mobile River Basin ecosystem recovery plan. Unpubl. technical/agency draft, Jackson, Mississippi. 128 pp.

Williams, J.D. & G.H. Clemmer. 1991. *Scaphirhynchus suttkusi*, a new sturgeon from the Mobile Basin of Alabama and Mississippi. Bull. Alabama Mus. Nat. Hist. 10: 17–31.

Environmental Biology of Fishes **48**: 420–421, 1997.

Threatened fishes of the world: *Scaphirhynchus albus* (Forbes & Richardson, 1905) (Acipenseridae)

Richard L. Mayden & Bernard R. Kuhajda
Department of Biological Sciences, Box 870344, University of Alabama, Tuscaloosa, AL 35487-0344, U.S.A.

Common name: Pallid sturgeon (E).
Conservation status: Listed under Endangered Species Act as Federally Endangered (Federal Register 1990). Currently receives federal protection; detailed recovery plan prepared by Dryer & Sandvol (1993).

Identification: One of three species of *Scaphirhynchus* distinguished by having pallid cream, gray, or whitish head, body, and fin coloration in mature adults; very small orbit diameter; elongate and pointed snout; inner barbels positioned anterior to outer barbels; few to no sharp, retrose spines on snout tip and anterodorsal to eye; and poorly developed squamation on venter. Other features include different number of dorsal plates, anus to anal fin plates, plates posterior to anal fin, lateral plates anterior to dorsal fin origin, anal and dorsal fin rays, and head, fin, and body proportions, relative to its close relatives *S. platorynchus* and *S. suttkusi*. One of the most detailed morphological analyses currently available was presented by Bailey & Cross (1954); recent morphological comparisons and analyses presented by Keenlyne et al. (1994a) and Mayden & Kuhajda (1996). Illustration from Forbes & Richardson (1920).
Distribution: Endemic to Mississippi River Basin but naturally limited to the Missouri and lower Mississippi river drainages (Bailey & Cross 1954). Historic range includes the Atchafalaya River, lower Mississippi River upstream to confluence with Missouri River, and Missouri River (Keenlyne 1995). The species has never been found to occur in either the Ohio or upper Mississippi rivers where the shovelnose sturgeon, *S. platorynchus*, typically abounds. **Abundance**: Species rare, abundance declined markedly following channelization and dam construction in the lower Mississippi and Missouri rivers. These activities not only limit migratory routes of the species but have largely curtailed natural Spring flooding periods that are thought to trigger spawning. Habitat alterations have also impacted the naturally turbid characteristics of the Missouri and lower Mississippi rivers to the extent that forage species are declining and the typical turbid, large-river habitat of the pallid sturgeon has declined. Riverine ecosystems historically occurring in these major waterways are being replaced with lentic habitats that are less turbid and with aquatic species adapted to these clear, lentic environments (Pflieger & Grace 1987). Dryer & Sandvol (1993) provide detailed account of the distribution and abundance of this species. **Habitat and ecology**: Life history aspects of *S. albus* are relatively poorly known. Species is found in large river channels with considerable diversity in microhabitats. They are usually associated with rapid current over sand, gravel or rocky substrates (Kallemeyn 1983, Carlson et al. 1985, Erickson 1992). Known to prefer turbid water conditions that historically characterized the Missouri and lower Mississippi rivers. Diet of adults is dominated by fishes, typically large-river minnows and shiners (Cyprinidae) (Erickson 1992). Based on observed food habits, the pallid sturgeon depends on the historical, naturally occurring turbid water conditions to conceal itself from prey items (Keenlyne 1995). The documented decline of many cyprinid species known to serve as regular food items for the pallid sturgeon is likely involved in its imperilment. Relative to the shovelnose sturgeon, growth is much more rapid throughout various age groups (Carlander 1969, Ruelle & Keenlyne 1993). Pallid sturgeon may weigh up to 45 kg (Brown 1971); males reach 39 years of age, while females may live as long as 41 years (Ruelle & Keenlyne 1993). **Reproduction**: Reproductive biology poorly known. Spawning believed to occur between April and mid-June, depending upon latitude (Keenlyne & Jenkins 1993). Males mature at 53 to 58 cm or 5 to 7 or 9 years, with 2- or 3-year intervals between spawning; size of females at maturity unknown, but estimated to occur by 9 to 12 or 15 to 20 years of age, with 3- to 10-year intervals between spawning (Kallemeyn 1983, Dryer & Sandvol 1993, Keenlyne & Jenkins 1993). Like shovelnose sturgeon, pallid sturgeon have been observed to possess mature gametes during periods coinciding with high river flow levels, possibly indicating that onset of spawning is initiated by typical Spring flooding of rivers. Spawning habitat not known. Gross habitat modifications made to the large river habitats in the Mississippi and Missouri river drainages through channelization and dams for navigational purposes preclude an accurate appraisal of the natural spawning habitats of this rare species. Under natural conditions it likely spawns in fast-flowing sections of the main-stem portions of the rivers. Because very few records exist for the species outside of main rivers, this species may not ascend smaller tributaries to spawn as does the shovelnose sturgeon. Although unknown for pallid sturgeon, eggs of shovelnose sturgeon are adhesive and require current for proper development, indicating that both a stable and silt-free substrate is necessary for their successful development. Hatching probably occurs in five to eight days under natural conditions (see Mayden & Kuhajda 1996). **Threats**: The gross human-induced habitat modifications in the Missouri and lower Mississippi rivers are the primary factors involved in the decline of the pallid sturgeon. These alterations, made primarily under the guise of navigation and flood control, have resulted in regulated flow patterns of these major rivers and have created habitats more lentic than lotic. Both of these conditions differ radically from the natural habitats to which the pallid sturgeon is adapted (e.g., braided channels, irregular flow patterns, flooding of terrestrial habitats, extensive

microhabitat diversity, and turbid waters). These changes have also reduced the natural forage base of the pallid sturgeon, another likely reason for its decline. Purported cases of hybridization with the shovelnose sturgeon (incidentally or intentionally occurring) may also be detrimental to the pallid sturgeon populations. **Conservation action**: This species was listed as endangered by the U.S. Fish and Wildlife Service on 6 September 1990 (Federal Register 1990). A panel of scientists has been assembled to serve as an advisory group for the recovery of this species; they have developed a recovery plan that may eventually lead to downlisting the pallid sturgeon (Dryer & Sandvol 1993). The major elements in the recovery of the species include establishing three wild-caught broodstock populations in different hatcheries; captive breeding, propagation, and stocking; protection of wild individuals; and habitat restoration in designated areas of the Missouri and lower Mississippi rivers. **Conservation recommendation**: Restoration of natural habitat and migratory patterns are essential. The historic habitats in the Missouri and lower Mississippi rivers must be restored in sections of these systems to provide appropriate microhabitats for pallid sturgeon foraging, spawning, and migration. Natural migratory patterns may be reestablished for the pallid sturgeon with the development of novel structures associated with dams that assist this speciess and others with overcoming these barriers. **Remarks**: Considerable interest exists as to whether the pallid and shovelnose sturgeon are diferent species. Much of this concern stems from the genetic study by Phelps & Allendorf (1983) wherein 'hybrid' and parental sturgeon were examined and no genetic differences were detected for these species at 37 loci. This study is technically and theoretically flawed and should not be used as either a basis for the existence of hybridization between these sturgeon species or for determining genetic variation either within or between these species. Their study employed buffer media standard for salmonid fishes, did not provide an adequate examination of differing environmental conditions for electrophoretic examination of protein variation, and did not demonstrate any empirical evidence for the existence of purported hybrids between these species. Unfortunately, because of the above study and that by Carlson et al. (1985), some biologists and laypersons have preconceived notions that hybridization between the pallid and shovelnose sturgeon is common in the wild. However, there is no empirical evidence to support this premise. To the contrary, in addition to these species possessing different geographic distributions, there are abundant morphological, behavioral and ecological attributes that may be used to distinguish these species.

Bailey, R.M. & F.B. Cross. 1954. River sturgeons of the American genus *Scaphirhynchus*: characters, distribution and syhonymy. Michigan Acad. Sci. Arts and Letters 39: 169–208.

Brown, C.J.D. 1971. fishes of Montana. Montana State University, Bozeman. 207 pp.

Carlander, K.D. 1969. Handbook of freshwater fishery biology. Iowa State University Press, Ames. 752 pp.

Carlson, D.M., W.L. Pflieger, L. Trial & P.S. Haverland. 1985. Distribution, biology and hybridization of *Scaphirhynchus albus* and *S. platorynchus* in the Missouri and Mississippi rivers. Env. Biol. Fish. 14: 51–59.

Dryer, M.P. & A.J. Sandvol. 1993. Pallid sturgeon (*Scaphirhynchus albus*) recovery plan. U.S. Fish & Wildlife Service, Bismarck. 55 pp.

Erickson, J.D. 1992. Habitat selection and movement of pallid sturgeon in Lake Sharpe, South Dakota. M.S. Thesis, South Dakota State University, Brookings. 70 pp.

Federal Register. 1990. Determination of endangered status for the pallid sturgeon. Vol. 55, No. 173 (6 September 1990): 36641–36647.

Forbes, S.A. & R.E. Richardson. 1920. The fishes of Illinois, 2nd ed. Illinois Nat. Hist. Surv., Champaign. 357 pp.

Kallemeyn, L.W. 1983. Status of pallid sturgeon (*Scaphirhynchus albus*). Fisheries 8: 3–9.

Keenlyne, K.D. 1995. Recent North American studies on pallid sturgeon, *Scaphirhynchus albus* (Forbes and Richardson). pp. 225–234. *In*: A.D. Gershanovich & T.I.J. Smith (ed.) Proceedings of the September 6–11 1993 International Symposium on Sturgeons, Moscow-Kostroma-Moscow, VNIRO Publications, Moscow.

Keenlyne, K.D. & L.G. Jenkins. 1993. Age at sexual maturity of the pallid sturgeon. Trans. Amer. Fish. Soc. 122: 393–396.

Keenlyne, K.D., C.J. Henry, A. Tews & P. Clancy. 1994a. Morphometric comparisons of upper Missouri River sturgeons. Trans. Amer. Fish. Soc. 123: 779–785.

Mayden, R.L. & B.R. Kuhajda. 1996. Systematics, taxonomy, and conservation status of the endangered Alabama sturgeon, *Scaphirhynchus suttkusi* Williams and Clemmer (Actinopterygii, Acipenseridae). Copeia 1996: 241–273.

Pflieger, W.L. & T.B. Grace. 1987. Changes in the fish fauna of the lower Missouri River, 1940–1983. pp. 166–177. *In*: W.J. Matthews & D.C. Heins (ed.) Community and Evolutionary Ecology of North America Stream Fishes, University of Oklahoma Press, Norman.

Phelps, S.R. & F.W. Allendorf. 1983. Genetic identity of pallid and shovelnose sturgeon (*Scaphirhynchus albus* and *S. platorynchus*). Copeia 1983: 696–700.

Ruelle, R. & K.D. Keenlyne. 1993. Contaminants in Missouri River pallid sturgeon. Bull. Environ. Contam. Toxicol. 50: 898–906.

Labels from caviar jars bought in New York. The survey of the New York City area shops using the molecular method of species identification (DeSalle & Birstein[1]) showed that currently the number of illegal replacements of commercial sturgeon species caviar (beluga, sevruga, and Russian sturgeon caviar or osetra) by the caviar of American or non-commercial Eurasian species is very high: in December 1995 and April 1996, it was 17% and in December 1996, it was already 32% (Birstein et al.[2]). Such companies as Petrossian Inc. (the upper row) continue to sell excellent caviar without misrepresenting.

[1] DeSalle, R. & V.J. Birstein. 1996. PCR identification of black caviar. Nature 381: 197–198.

[2] Birstein, V.J., P. Doukakis, B. Sorkin & R. DeSalle. 1997. Population aggregation analysis of caviar producing species of sturgeons and implications for diagnosis of black caviar. Cons. Biol. (submitted).

Environmental Biology of Fishes **48**: 423–426, 1997.

Sturgeon poaching and black market caviar: a case study

Andrew Cohen
U.S. Department of Commerce, National Oceanic and Atmospheric Administration, National Marine Fisheries Service, Office of Enforcement, Seatte, WA 98115, U.S.A.

Received 20.3.1996 Accepted 7.4.1996

Key words: Acipenser transmontanus, Columbia River, Lacey Act, beluga caviar, illegal harvesting, environmental crimes

Synopsis

This paper documents a recent United States Federal prosecution of members of a poaching ring that sold caviar derived from illegally taken Columbia River white sturgeon, *Acipenser transmontanus*. Experts estimated that over 2000 adult sturgeon were killed in the process of illegally harvesting the more than 1500 kg of caviar involved in the case. Case studies of illegal activities related to exploitation of natural resources are rare. These crimes are difficult to discover and prosecute, for secrecy is essential, and by the time the facts are publicly available, irreparable environmental damage may have already been done. Sturgeons and paddlefishes have long life spans but take many years to reach reproductive maturity; they reproduce infrequently and rely upon large, often urban rivers for their spawning migrations. These basic biological characteristics render these fishes especially susceptible to illegal exploitation, particularly when stocks have already been damaged by overfishing, dam construction or pollution (as has the Columbia River population of white sturgeon). Given the often exorbitant prices for sturgeon and paddlefish caviar, and the relative ease of capturing these fishes during their spawning migrations, persons may be tempted to circumvent state and federal regulations designed to protect acipenseriforms. Additionally, those involved in the distribution and sale of caviar can be motivated to fraudulently mislabel the product; for instance, in this case, white sturgeon caviar was marked as beluga caviar and sold at approximately five times the normal price of white sturgeon caviar. Despite the clear evidence of an environmental crime, the scale of the abuse, and the convictions, sentencing was light, a discouraging sign for those who hope to limit such destructive crimes in the future.

Between 1985 and 1990 poachers in the Pacific Northwest shipped more than 3000 lbs (> 1352 kg) of high quality caviar made from the roe of white sturgeon, *Acipenser transmontanus*, to a caviar company in New Jersey. The poachers sent containers of caviar via the Federal Express shipping company, using a fictitious business name, and a variety of nonexistent return addresses. The president/owner of the caviar company paid the poachers, whom he had never met in person, by mailing packages of cash to various post office boxes in Washington. The owner of the company, a well known fifth generation caviar merchant, has been profiled in both Playboy and People magazines, and is often quoted as an industry expert.

How was this poaching ring finally cracked? The story began with a bank robbery in the rural southern Washington State town of Dollars Corner, not far from the Columbia River, near Portland, Oregon.

During November 1990, a poacher and his confederate were producing high quality caviar in a cheap motel room in Vancouver, Washington. Much of the caviar came from sturgeon caught by the poachers, and some of it was made from roe illegally purchased from sport fishermen on the banks of the Columbia River. The poachers had paid the month's rent, approximately $ 900, in cash, and then told the motel manager that they did not want any maid service. The manager thought that the request for privacy was unusual, and became suspicious.

About the time the poachers were checking into the motel a bank robbery occurred in nearby Dollars Corner, Washington. The bank robbers escaped with a package of cash that contained an exploding dye pack. The pack detonated, spewing red ink and staining both the cash and the thieves.

The day after the robbery another local bank received a cash deposit from the motel where the poachers had rented the room. This deposit included red dye stained money. The bank teller recognized the stains on the bills as stains from the dye pack and notified the FBI. The FBI went to the motel and asked the manager where she had obtained the cash. There were only two rooms paid for in cash, one of which belonged to the poachers. The manager then told the FBI about her suspicions. The FBI immediately began to conduct surveillance of the poacher's activities, suspecting that they were the bank robbers.

For the next few days the FBI agents hid in a nearby motel room and watched the suspected robbers through the window blinds. They looked like a pair of fishing buddies, with their ragged clothes and small boat trailered behind an old red pickup truck. At one point, the agents observed one of the poachers deposit a bag of trash in the motel dumpster. An agent casually took a look at the garbage, and found an empty salt box. Although the agent did not know it at the time, he had discovered an important piece of evidence; salt, of course, is a major ingredient in caviar.

The next day the fisherman left the motel room with two large boxes. They loaded the boxes into the back of the red pickup truck and drove to a Federal Express office in Portland, Oregon. The agents followed the truck as it crossed the state line, and watched the driver mail the boxes at the Federal Express office. The poachers then left, and the FBI agents soon lost the truck in traffic. The agents returned to the Federal Express Office and spoke to a clerk who showed them the airbills and the boxes. Both boxes were addressed to a caviar company in New Jersey. The Federal Express clerk told the agents that the man who shipped the boxes looked like a fisherman, and smelled like fish. The FBI agents thought that the case was looking less like a bank robbery with every new bit of information. Without a search warrant, they could not look inside the boxes, so after recording the information from the shipping labels, the boxes were loaded on a plane bound for New Jersey.

In the meantime, the motel manager was concerned about the poachers' unusual request for no room service and the FBI's interest in both the cash and the men. She took it upon herself to enter the room with a passkey. She had read that drug dealers sometime manufactured methamphetamines in motels, and suspected that perhaps that was occurring in her motel. To her surprise, she found no drug manufacturing paraphernalia, but lots of fishing gear, an outboard engine, and buckets of brine. She later testified in court that the whole place smelled like fish.

The FBI agents soon realized that the suspects were not involved in the bank robbery. Nevertheless, their actions were strange for a couple of buddies who were on a fishing holiday. They contacted the Washington Department of Fisheries and the National Marine Fisheries Service, two agencies who specialize in resource related crimes. Thus began a two year investigation of an organized poaching ring that spread from the Columbia River to New Jersey.

Evidence began to trickle in as a federal Grand Jury in Seattle issued more than 30 subpoenas for records from telephone companies, banks, the caviar company, accountants, auditors and others. No stone was left unturned; after obtaining one of the poachers' one-gallon sized shipping containers, the agents contacted the plastic company that manufactured the jars to obtain a list of everyone who had ever purchased that type of container in quantity. The caviar merchant provided a few documents after they were demanded by subpoena, however, he denied any contact with the poachers to occur before 1990. Agents later learned that the records of over 50 earlier shipments had been destroyed. Fortunately, copies of the early records had been secreted away by a company employee and were discovered during the execution of a federal search warrant. When all of the evidence was compiled, the agents determined that at least 67 shipments had taken place. The poachers had been paid an estimated $ 247 176 tax free dollars for 1462 kg of American sturgeon caviar. The poaching ring had been active for over five years.

Seized records and court testimony indicated that when the caviar company repacked the illegally harvested product for resale, much of it was labeled as imported beluga or osetra caviar. Although the poachers' caviar was of high quality, it was still merely American sturgeon caviar, which sells for about $ 89 wholesale and about $ 130 per pound (0.454 kg) retail. Imported beluga caviar can sell for as much as $ 600 per pound. During trial in federal court, the president of the caviar company testified that he sold the American caviar as *imported* caviar to customers such as the Rainbow Room, the Waldorf Astoria and Pan American Airlines. He stated, in substance, that he would not sell the mislabeled caviar to those customers whom he thought would recognize the subtle differences in American, beluga and osetra caviar, however he thought that the mislabeled caviar was acceptable for some restaurants and other commercial or institutional customers. If all of the poachers' caviar was resold as beluga, then it was worth almost two million dollars.

The caviar company's records indicated that the poachers were paid up to $ 100 a pound for the illicit caviar. The poachers did not pay income tax on his profits from the unlawful sales. It is impossible to distinguish a male from a female sturgeon in the field, without first killing the fish. Therefore, it must be assumed that the poachers probably killed as many adult male sturgeon as female in their search for the valuable eggs. Of course, only a fraction of adult female sturgeon will be ripe at any particular time, and those fish would also have to be killed and eviscerated to determine if they contained roe. Caviar experts estimated that during the process of transforming sturgeon roe into caviar, about forty percent of the original weight of the eggs is lost. The Washington Department of Fisheries estimated that to obtain 3220 pounds of finished caviar, approximately two thousand male, female, ripe, and unripe sturgeon were killed in a five year period. That figure represents a significant part of the sturgeon population in the lower Columbia River, where the poachers were operating.

After a two and a half week trial, a federal jury found the owner of the caviar company guilty of conspiracy to violate the Lacey Act. That statute prohibits the interstate transportation, purchase, sale or possession of fisheries products, like caviar, if the product was taken in violation of any state law. He was also found guilty of four misdemeanor counts of the Lacey Act itself, and one felony count of Obstruction of Justice. The Obstruction charge stemmed from the destruction of documents that were under grand jury subpoena, and then lying to the grand jury to hide his involvement. He was sentenced to eighteen months in federal prison, $ 4175 in fines and penalties, three years probation, plus the costs of his imprisonment and probation.

The caviar company itself was charged as a separate defendant, and was also found guilty of criminal Conspiracy and four misdemeanor counts of the Lacey Act. Fines and penalties ($ 20 625) were assessed, along with three years' probation, although $ 10 000 of the fine was suspended by the judge.

One of the poachers plead guilty to a Conspiracy, four felony Lacey Act counts and one felony income tax count. His plea-bargain arrangement included testimony against the other defendants at trial, in return for a lesser sentence. He was sentenced to eight months in federal prison and $ 2675 in fines and penalties.

The resale of the American caviar mislabeled as imported beluga and osetra caviar was not charged in the indictment. Nevertheless, when the caviar company president took the witness stand, he testified that the mislabeling took place.

When a consumer pays several hundred dollars a pound for a product, there is a presumption that he or she is buying credibility. After all, 'you get what you pay for'. At the same time, most consumers can not afford to eat top of the line caviar often enough to truly develop an educated palate, and rely on the caviar tin's label and the brand name to ensure quality. Had it not been an unrelated bank robbery, this cycle of wholesale commercial poaching and product mislabeling might still be going on.

The jury returned the guilty verdicts on 22 October 1993, exactly eight years to the day after the first known shipment between the poachers and the caviar company.[1]

[1] Convicted were: Hansen Caviar Co. Inc.; Mr. Arnold Hansen-Sturm, the president of Hansen Caviar Co. Inc.; and Mr. Stephen Gale Darnell, the lead poacher (the second poacher served as a witness for the prosecution). This case was documented by the commercial news media at the time of the indictments, trial and convictions (e.g. Anonymous 1993, 1994, Boss 1994, Houtz 1994a, b, O'Neill 1993) but this and other poaching cases are little known to the scientific community. In part, we think that this may stem from general unfamiliarity with (or unwillingness to recognize) the catastrophic biological impact that can be made by small numbers of environmental criminals. If we are to have any impact in reducing environmental crime, better awareness and swifter condemnation by biologists seem essential (editors' note, March 1996).

References cited

Anonymous. 1993. New Jersey firm guilty in roe case. The New York Times, 24 October.

Anonymous. 1994. Caviar cons. Seafood Leader 14.

Boss, K. 1994. Caught fishing. Pacific Magazine, 13 March: 12–20 (prepared by Seattle Times).

Houtz, J. 1994a. The courts. U.S. District Court, Seattle. The Seattle Times, 14 February: B2.

Houtz, J. 1994b. Man gets 8 months for selling caviar from protected sturgeon. The Seattle Times, 15 January.

O'Neill, M. 1993. Caviar distributor and 2 fishermen charged with evading limits on scarce U.S. roe. The New York Times, 4 April: 40.

Environmental Biology of Fishes **48**: 427–435, 1997.

The threatened status of acipenseriform species: a summary

Vadim J. Birstein[1], William E. Bemis[2] & John R. Waldman[3]
[1] *The Sturgeon Society, 331 W 57th Street, Suite 159, New York, NY 10019, U.S.A.*
[2] *Department of Biology and Graduate Program in Organismic and Evolutionary Biology, University of Massachusetts, Amherst, MA 01003, U.S.A.*
[3] *Hudson River Foundation, 40 West 20th Street, Ninth Floor, New York, NY 10011, U.S.A.*

Received 29.5.1996 Accepted 3.6.1996

Key words: IUCN, CITES, listing, overfishing, pollution, dam construction, sturgeons, paddlefishes

> ... increased demand has recently driven the price of black market smoked sturgeon as high as $ 26 a kilogram. With poachers standing to gain roughly a third of this price [besides the much higher price of caviar], a large fish could be worth thousands of dollars.
>
> Gary Hamilton
> in *Canadian Geographic*, July/August 1996, p. 62

Papers in this volume describe many factors that expose acipenseriform fishes to particular risks of population decline and extinction. These range from such basic problems as how to define species boundaries and recognize different species of sturgeons, factors that necessarily impact all regulatory and law enforcement efforts. Other factors concern the extreme sensitivity of sturgeons to overfishing, their dependence on large, often polluted urban river systems for spawning, and migration routes blocked by hydroelectric dams. The value of the traditional reaction to such problems – stocking hatchery reared fish – is increasingly debated, particularly if the stocking occurs 'on top of' a remnant population of sturgeons. The prognosis for most species is extremely bleak, and has worsened during the few years that we have been recording information. Technical developments – such as the use of genetic markers to recognize different species of sturgeons (and their caviar) – may offer some new regulatory tools. Improved basic knowledge – especially behavioral and ecological data obtained by telemetry of wild fish – may help governments to protect sensitive and important sites, particularly spawning areas.

Stocking can be a mixed blessing, though it certainly helped historically to sustain some species, such as *Huso huso* in the Volga River and Caspian Sea. Reintroduction to portions of ranges from which a species has been extirpated may seem to be a laudable goal, although this has not yet been achieved for any acipenseriform species. As in cases of attempted restorations of salmoniform fishes, serious questions surround the choice of stocks that might be used for reintroduction. Given the short time during which we can hope to act to preserve the global biodiversity of sturgeons and paddlefishes, greater international awareness, better regulation and stricter enforcement of existing laws are essential. It seems especially important that all interested persons act to assemble the best possible data on the current status of their local species of sturgeons and paddlefishes.

Most species and many populations of sturgeons

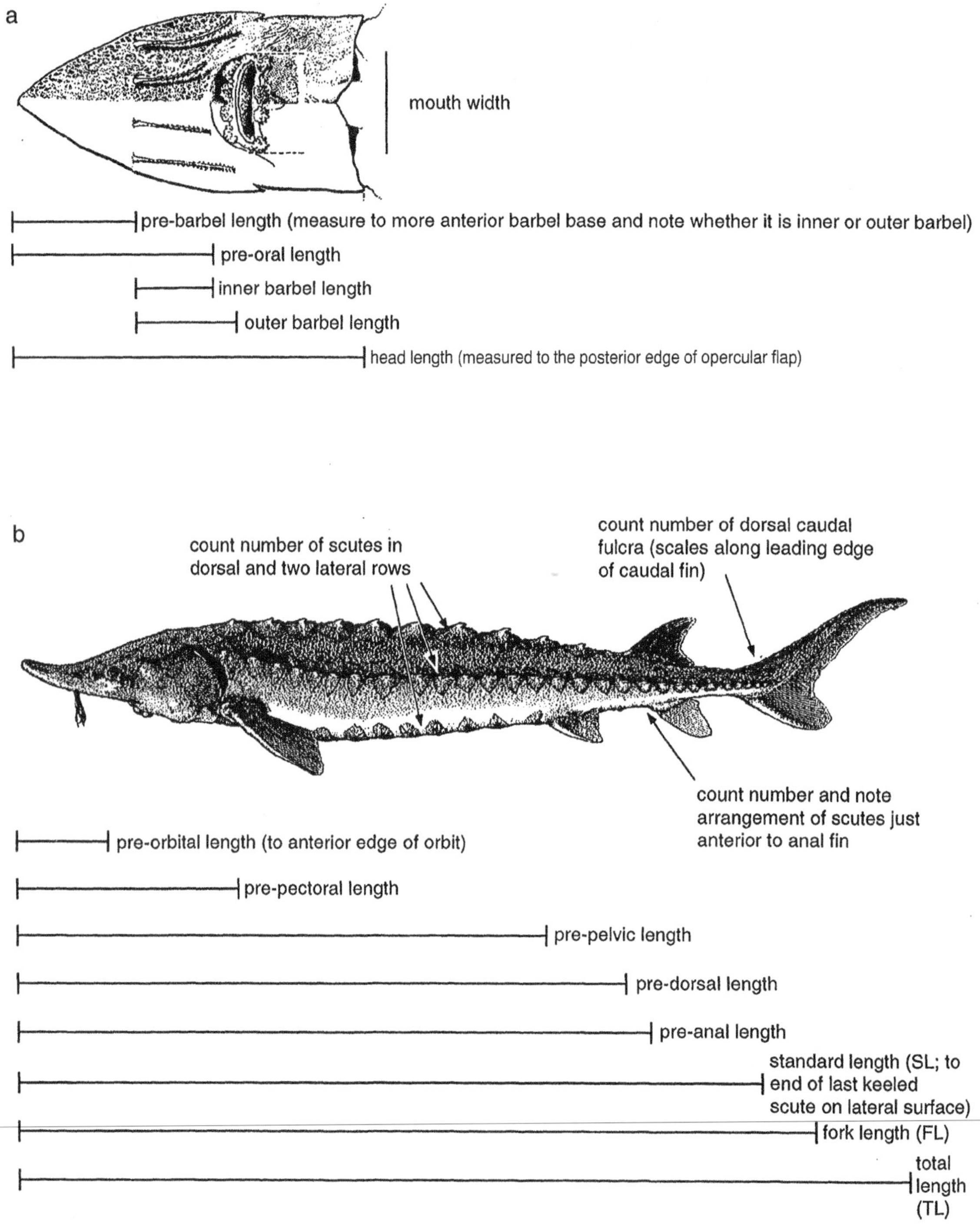

Figure 1. Standard measurements for sturgeons: a – Ventral view of mouth based on an illustration of *Scaphirhynchus platorynchus* modified from Bailey & Cross (1954). b – Lateral view of body based on an illustration of *Acipenser oxyrinchus* (518 mm TL) modified from Vladykov & Greeley (1963). A color photograph including a metric scale bar should be made showing the ventral view of the head (as in a) and lateral view of the body (as in b). All measurements should be recorded from each specimen, together with date and exact locality data, water conditions, observers name and institution, and fate of the specimen.

and paddlefishes, particularly anadromous forms, are in trouble (Birstein 1993, Bemis & Findeis 1994, Waldman 1995). Declines of sturgeon and paddlefish populations are described in many papers in this volume (e.g., Bacalbaşa-Dobrovici 1997, Graham 1997, Hensel & Holčík 1997, Khodoreskaya et al. 1997, Krykhtin & Svirskii 1997, Ruban 1997, Wei et al. 1997, Zholdasova 1997). Like other anadromous fishes, such as salmonids, sturgeons are extremely sensitive to overfishing (Boreman 1997, this volume). Overfishing, including unprecedented levels of poaching, is the main threat to sturgeon survival in Europe (especially in Russia), Siberia and China (Birstein 1993, 1996, Dumont 1995, Anonymous 1995a, De Meulenaer & Raymakers 1996, Ruban 1996). Poaching also plagues certain populations in the United States, such as Columbia River white sturgeon (Cohen 1997 this volume). Other factors, including pollution, at present play a less important role in the decline of populations (for instance, Khodorevskaya et al. 1997, Ruban 1997). Even species that are not fished for either caviar or meat, such as all three species of *Pseudoscaphirhynchus*, have declined, in this case primarily in response to the drying of the Aral Sea (Zholdasova 1997).

Persistent problems in identifying species of *Acipenser* outside of their supposedly native ranges cause us to make two practical suggestions. First, document by photography and measurement external features of live wild sturgeons. Intraspecific variation, particularly in wide ranging species such as *Acipenser ruthenus*, is extremely confusing and no single researcher is ever likely to have access to all river systems in which such species occur. Figure 1 proposes a series of measurements to be recorded, together with locality data including water flows, associated fauna, etc. These data will be most valuable when coupled with color photographs of the live specimen showing its natural coloration. Second, systematics cannot be stronger than the specimens and collections upon which it is based. Voucher specimens, especially large fish and ontogenetic series – even partial or salvaged specimens – with good locality data are needed for many species of acipenseriforms from many areas of the world, so we must take active roles in the growth and maintenance of sturgeon materials in permanent natural history collections. (As an aside, Grande & Bemis 1991, found it easier to find museum specimens of the Eocene Green River paddlefish, *Crossopholis* than the extant Chinese paddlefish, *Psephurus gladius*, which is quite the reverse of most paleoichthyological experience.)

For many years, stocking of artificially reared young has been used to maintain populations of acipenseriforms in the former Soviet Union (e.g., the three main commercial species of Caspian Sea sturgeons, *Huso huso, A. stellatus*, and *A. gueldenstaedtii*, Khodorevskaya et al. 1997) and in the United States (e.g., *Polyodon*, Graham 1997). The fragility of this approach is well-illustrated by events in Russia since the late 1980s, when stocking of Caspian Sea sturgeons began to decrease (Khodorevskaya et al. 1997). Not only do fewer hatcheries now operate on the Volga River, but also they are unable to catch enough brood stock, so that the beluga, *H. huso*, has become extremely threatened. Because of dam construction, beluga lost access to practically all of the spawning grounds in the Volga River. In 1995, the number of females caught in the Volga River delta was insufficient for artificial breeding. Therefore, in 1995 there was no natural or artificial reproduction of *H. huso* in the Volga River. The situation on the Danube River is similarly discouraging, for the two dams at the Iron Gates now prevent the historic migration of beluga between the Black Sea and the middle reaches of the Danube River. Artificial breeding of *H. huso* in the Danube River (in the Serbian part of the river) has also been unsuccessful, and there is no indication that the situation will improve in the near future. We must make the case throughout the world that even the very best stocking programs can only provide short-term solutions unless they are coupled to plans for protecting and increasing levels of natural reproduction.

Technology may aid in enforcing existing regulations and learning what to protect in nature, but our efforts as scientists must be focused not only on what we can learn about sturgeons and paddlefishes but also how to translate that knowledge into practical measures (Wirgin et al. 1997, this volume). For example, it is now possible to identify the caviar of certain species of sturgeons using species – specific

Table 1. Threatened status of acipenseriforms.

Species	English name	Distribution	Status (national listing or latest studies)		IUCN listing		CITES 1996
			Status[1]	References	1994	1996[2]	
Family Acipenseridae							
Acipenser baerii	Siberian sturgeon	Main Siberian rivers				VU	
A. baerii baerii	Siberian sturgeon	Ob River basin	EN	Ruban 1996, 1997		EN	
A. baerii stenorrhynchus	Lena River sturgeon	Basins of the East Siberian rivers: Yenisey, Lena, Indigirka, Kolyma, and Anadyr	VU	Ruban 1997		VU	
A. baerii baicalensis	Baikal sturgeon	Lake Baikal (Siberia)	VU	RSFSR Red Data Book 1983, Pavlov et al. 1985, 1994		EN	
A. brevirostrum[3]	Shortnose sturgeon	Rivers, estuaries and ocean along east coast of North America from Indian River (Florida) to Saint John River (New Brunswick)	T (Canada & USA)	USFWS 1967, Williams et al. 1989, Manci 1993	VU (Canada & USA)	VU	Appendix I
			V (Canada)	Campbell 1991			
A. dabryanus	Yangtze or Dabry's sturgeon	Yangtze River system	EN (category I of state protection)[4]	Wei et al. 1997		CR	
A. fulvescens	lake sturgeon	Great Lakes and lakes of southern Canada	T (Canada & USA)	Williams et al. 1989, Manci 1993	VU (Canada & USA)	VU	
A. gueldenstaedtii	Russian sturgeon	Black, Azov, Caspian seas and rivers entering into them	VU	Lelek 1987		EN	
		Caspian Sea population	H	Khodorevskaya et al. 1997		EN	
		Black Sea population				EN	
		Danube River population, Hungary	EN	Guti 1995			
		Danube River population, Romania	VU	Bănărescu 1995			
		Dnepr River population (Black Sea)	EN	Gringevsky 1994			
		Sea of Azov population	VU, H	Volovik et al. 1993		EN	
A. medirostris	Green sturgeon	Pacific coast of North America from Aleutian Islands and Gulf of Alaska to Ensenada, Mexico	V (Canada)	Campbell 1991		VU	
			T (USA)	Moyle et al. 1994			
A. mikadoi	Sakhalin sturgeon	Pacific Ocean from Amur River to northern Japan, Korea, and Bering Sea, Tumnin (Datta) River	EN	USSR Red Data Book 1984, Pavlov et al. 1994		EN	
A. naccarii	Adriatic sturgeon	Adriatic Sea, Po and Adige Rivers	VU	Lelek 1987		VU	
A. nudiventris	Ship sturgeon	Aral, Caspian, Black seas and rivers entering into them				EN	
		Caspian Sea and rivers entering into it	EN	Pavlov et al. 1985, Lelek 1987		EN	
			H	Sokolov & Vasilev 1989			
		Black Sea and rivers entering into them (Russia, Ukraine)	EN	Pavlov et al. 1985, 1994		EN	
			H	Sokolov & Vasilev 1989			
		Danube River population				CR	
		Danube River population, Hungary	EN	Guti 1995			
		Danube River population, Romania	EX	Bănărescu 1995			
		Aral Sea (Central Asia)	EX	Zholdasova 1997		EX	

Table 1. Continued.

A. oxyrinchus	Atlantic sturgeon	Atlantic Ocean (Canada and USA, east coast)				Lr (nt)	
A. oxyrinchus desotoi[5]	Gulf sturgeon	Gulf of Mexico and northern coast of South America	T	Williams et al. 1989, USFWS 1990b, Manci 1993		VU	
A. o. oxyrinchus	Atlantic sturgeon	Rivers, estuaries and ocean along east coast of North America from the St. Johns River (Florida) to Hamilton Inlet (Labrador)	SC (USA)	Williams et al. 1989	VU (Canada & USA)	LR (nt)	Appendix II
A. persicus	Persian sturgeon	Caspian and Black seas and rivers entering into them	EN	Lelek 1987		EN	
		Caspian Sea population				VU	
		Black Sea population	R	Pavlov et al. 1994		EN	
A. ruthenus	Sterlet	Drainages of main rivers entering the Caspian and Black seas (Volga, Danube)	EN	Lelek 1987		VU	
		Volga River population				LR (lc)	
		Danube River population				VU	
		Danube River population, Hungary	VU	Guti 1995			
		Danube River population, Romania	VU	Bănărescu 1995			
		Dnepr River population	EN	Gringevsky 1994			
		Ob, Irtysh, Yenisey rivers (Siberia)				VU	
A. schrenckii	Amur River sturgeon	Amur River system (Siberia)	T	Manci 1993	VU (China & Russia)	EN	
			EN	Krykhtin & Svirskii 1997			
A. sinensis	Chinese sturgeon	Yangtze River system (China)	EN, H (category I of state protection)[4]	Wei et al. 1997		EN	
A. stellatus	Stellate sturgeon or sevruga	Caspian, Azov, Black and Aegean seas and rivers entering into them				EN	
		Caspian Sea population				VU	
			H	Khodorevskaya et al. 1997			
		Black Sea population				EN	
		Dnepr River population (Black Sea)	EN	Gringevsky 1994			
		Danube River population, Hungary	EN	Guti 1995			
		Danube River population, Romania	EN	Bănărescu 1995			
		Sea of Azov	VU, H	Volovik et al. 1993		EN	
A. sturio	Atlantic (Baltic) sturgeon	Baltic, Eastern North Atlantic, Mediterranean, and Black Sea	EN[6]	USSR Red Data Book 1984, Lelek 1987, Holcik et al. 1989	EN	CR	Appendix I
A. transmontanus	White sturgeon	Rivers and Pacific coast of North America from the Gulf of Alaska to Baja California	V (Canada)	Campbell 1991		LR (nt)	
			V (California, USA)	Moyle 1994			
		Kootenai River and Kootenai Lake in Idaho, Montana, and British Columbia downstream of Libby Dam in Montana	EN	USFWS 1994[7]		EN	

Table 1. Continued.

Huso dauricus	Kaluga sturgeon	Amur River system	EN	Krykhtin & Svirskii 1997, Wei et al. 1997	EN	EN	
H. huso	Giant sturgeon or beluga	Caspian, Black, and Adriatic seas and rivers entering into them	VU	Lelek 1987		EN	
		Caspian Sea population				EN	
			H	Khodorevskaya et al. 1997			
		Black Sea population				EN	
		Dnepr River population (Black Sea)	EN	Gringevsky 1994			
		Danube River population, Hungary	EN	Guti 1995			
		Danube River population, Romania	VU	Bănărescu 1995			
		Azov Sea population	EN	Pavlov et al. 1994		CR	
		Adriatic Sea population				EX	
Pseudoscaphirhynchus fedtschenkoi	Syr-Dar shovelnose sturgeon	Syr-Darya River (Kazakhstan, Central Asian)	EN	USSR Red Data Book 1984		CR	
			EX	Pavlov et al. 1985, 1994			
P. hermanni	Small Amu-Dar shovelnose sturgeon	Amu-Darya River (Uzbekistan, Central Asia)	EN	USSR Red Data Book 1984		CR	
			EX	Pavlov et al. 1985			
P. kaufmanni	Large Amu-Dar shovelnose sturgeon	Amu-Darya River (Turkmenistan, Uzbekistan & Tadjikistan, Central Asia)	EN	USSR Red Data Book 1984		EN	
			CR	Zholdasova 1997			
Scaphirhynchus albus	Pallid sturgeon	Missouri and Mississippi River basins	EN[8]	Williams et al. 1989, USFWS 1990b, Manci 1993	EN	EN	
S. platorynchus	Shovelnose sturgeon	Missouri and Mississippi River basins	E	Williams et al. 1989		VU	
S. suttkusi	Alabama sturgeon	Mobil basin in Alabama and Mississippi	EN[9]	Williams et al. 1989, Williams & Clemmer 1991, Manci 1993, 1994	EN	CR	
Family Polyodontidae							
Polyodon spathula	North American paddlefish	Mississippi River system, particularly Missouri and its tributaries	SC (USA & Canada)	Williams et al. 1989	VU (USA)	VU	Appendix II
Psephurus gladius	Chinese paddlefish	Yangtze River system	EN (category I of state protection)[4]	Wei et al. 1997	VU	CR	

[1] Categories are given in the new IUCN system (IUCN Red List Categories 1994): EX = extinct; CR = critically endangered; EN = endangered; VU = vulnerable; LR = low risk; LR(nt) = near threatened; LC(lc) = least concern; or in the U.S. Office of Endangered Species system: E = endangered; T = threatened; SC = special concern. H (Hatcheries) designates species whose natural reproduction is limited; such species are artificially bred and juveniles obtained are released into their natural habitat.

[2] Proposals of the Sturgeon Specialist Group, IUCN

[3] All populations of *A. brevirostrum* along the east coast of the USA and Canada are listed as endangered by the USFWS, Title 50, parts 17.11, 17.12 (USFWS, 1967; DOI, 1973).

[4] A list of wild animals by the states special protection in category I and II. 14 pp. (in Chinese).

[5] Populations of *A. oxyrinchus desotoi* are listed as endangered in AL, FL, GA, LA, and MS by the USFWS, Title 50, parts 17.11, 17.12; federally threatened status from September 30, 1990 (USFWS 1990b).

[6] According to the IUCN Red List (1994), the status of *A. sturio* is different in different countries: Albania (EN), Algeria (EN), Belgium (EX?), Finland (EX?), France (EN), Germany (EX?), Greece (EN), Iceland (EX?), Ireland (EX?), Italy (EN), Morocco (E), Netherlands (EX?), Norway (EX?), Poland (EX?), Portugal (EN), Romania (EN), Russia (EN), Spain (EX?), Switzerland (?), Turkey (EN), Ukraine (EN), United Kingdom (EN), Yugoslavia (EX?). The status of *A. sturio* for Spain and Netherlands should be considered as E since in 1992 sturgeons were caught in both countries (Volz & DeGroot 1992, Elvira & Almodovar 1993). In 1995, a few live *A. sturio* were caught in Albania (Tamas Gulyas personal communication).

[7] Kootenai River population of *A. transmontanus* is listed as federally endangered from 6 September, 1994 (USFWS 1994).

[8] Populations of *S. albus* are listed as endangered in AR, IA, IL, KS, KY, LA, MO, MS, MT, ND, NE, SD and TN by the USFWS, Title 50, parts 17.11, 17.12; federally endangered status from September 6, 1990 (USFWS 1990a).

[9] Proposed listing of *S. suttkusi* as endangered has been withdrawn (Federal Register 59, No. 240: 64794–6409 (1994)). For the present, USFWS has placed this species in Category 2 (those species for which insufficient information is available to determine whether to proceed with a proposed rule to list or to consider the species extinct). At its meeting in Edmonton (Canada), 15–19 June 1995, the American Society of Ichthyologists and Herpetologists urged USFWS to list *S. suttkusi* as an endangered species (Anonymous 1995b, Mayden & Kuhajda 1996).

DNA primers (DeSalle & Birstein 1996). This may help law enforcement agencies to detect violations of CITES and other regulations, and coupled to the development of this technology must be increased willingness to speak publicly on matters concerning enforcement of environmental laws. As another example of technology and efforts to conserve sturgeons, radio telemetric studies revealed the spawning sites of shortnose sturgeon in the Connecticut River (e.g., Buckley & Kynard 1985). One of these sites lies just below Holyoke Dam, Holyoke, Massachusetts, where the river passes through a highly disturbed urban environment. Knowledge of the existence of this spawning site enables public utilities and state highway officials to limit their further impact on this portion of the river, particularly during the spring spawning season. In the future, it may become necessary to seek specific regulations protecting individual spawning sites from dredging or other destructive impacts.

The highly threatened status of all extant acipenseriform species is summarized in Table 1, which updates information given by Birstein (1993). Data for the main basins inhabited by sturgeons are given separately. For the Danube River, evaluations are shown for both the middle (Slovak-Hungarian) and lower (Romanian) reaches. Restocking efforts are also mentioned in Table 1. The international evaluation of status is given for 1994 (IUCN Red List) and 1996. Data for 1996 were collected by the Sturgeon Specialist Group created within the Species Survival Commission of IUCN in 1994 (Birstein 1995). The last column of Table 1 shows the present listings of species of sturgeons on the Appendices I or II of CITES (Convention on International Trade in Endangered Species of Wild Fauna and Flora).

It is evident from Table 1 that all European and Asian sturgeon species are in trouble. For various reasons, however, only the European Atlantic sturgeon, *A. sturio*, attracts serious international attention to its conservation (Elvira & Gessner 1996, Williot et al. 1997, this volume). Meanwhile, the situation for many other species worsens. The construction of Three Gorges Dam on the Yangtze River continues. When it is completed, the spawning grounds of all three endemic Chinese species, of which two (*Acipenser dabryanus* and *Psephurus gladius*) are already critically endangered, will be destroyed. Plans to develop oil fields in the northern part of the Caspian Sea in Kazakhstan and Turkmenistan (Sager 1994, Dumont 1995) threaten the future of all sturgeons in the Caspian Sea. But even without these fundamental environmental changes it is evident that we may soon lose at least some of the Eurasian sturgeon species. 'Like the Californian condor, the sturgeons only chance of survival may be in captivity' (Dumont 1995).

The status of the American species of Acipenseriformes is comparatively much better than that of the Eurasian species (Table 1). Considering the many ongoing recovery programs for almost all American species (partly described in this volume by Bain 1997, Beamesderfer & Farr 1997, Graham 1997, Kynard 1997, Smith & Clugston 1997), the future of American species seems to be much better than that of the Eurasian ones, especially those with many extirpated populations. But as suggested above, we must be careful not to become so reliant on artificial stocking of certain species that we neglect to develop ways to encourage – or at least to permit – natural spawning to play as large a role as possible in maintaining populations.

Clearly, Table 1 presents an initial step of the evaluation of the status of acipenseriforms; a complete picture of the status of the group revealed river by river in each basin will take much effort and time. But it seems that this effort is necessary if we are to understand what is really left of the former range of the extant acipenseriform species. Time is short, and we will be grateful for any forthcoming improvements to our data base.

Acknowledgements

We thank the many readers and contributors to *The Sturgeon Quarterly*, which since its inception in 1993 has helped to keep all of us aware of the global circumstances of sturgeons and paddlefishes.

References cited

Anonymous. 1995a. From the editors. Sturgeon Quart. 3(4): 1.

Anonymous. 1995b. Annual Business Meeting, 19 June 1995. Copeia 1995: 1042–1945.

Bacalbaşa-Dobrovici, N. 1997. Endangered migratory sturgeons of the lower Danube River and its delta. Env. Biol. Fish. (this volume).

Bailey, R.M. & F.B. Cross. 1954. River sturgeons of the American genus *Scaphirhynchus*: characters, distribution, and synonymy. Pap. Mich. Acad. Sci. 39: 169–208.

Bain, M. 1997. Atlantic and shortnose sturgeons of the Hudson River: common and divergent life history attributes. Env. Biol. Fish. (this volume).

Banarescu, P.M. 1994. The present-day conservation status of the fresh water fish fauna of Romania. Ocrot. nat. med. inconj. 38: 5–20.

Beamesderfer, R.C.P. & R.A. Farr. 1997. Alternatives for the protection and restoration of sturgeons and their habitat. Env. Biol. Fish. (this volume).

Bemis, W.E. & E.K. Findeis. 1994. The sturgeons plight. Nature 370: 602.

Birstein, V.J. 1993. Sturgeons and paddlefishes: threatened fishes in need of conservation. Cons. Biol. 7: 773–787.

Birstein, V.J. 1995. Sturgeon specialist group. Species 24: 63.

Birstein, V.J. 1996. Sturgeons may soon disappear from the Caspian Sea. Russian Cons. News 7: 15–16.

Boreman, J. 1997. Sensitivity of North American sturgeons and paddlefish to fishing mortality. Env. Biol. Fish. (this volume).

Buckley, J. & B. Kynard. 1985. Habitat use and behavior of prespawning and spawning shortnose sturgeon, *Acipenser brevirostrum*, in the Connecticut River. pp. 111–117. *In*: F.P. Binkowski & S.I. Doroshov (ed.) North American Sturgeons, Dr W. Junk Publishers, Dordrecht.

Campbell, R.P. 1991. Rare and endangered fishes and marine mammals of Canada. COSEWIC Fish and Marine Mammal Subcommittee Status Reports: VII. Canad. Field-Natur. 105: 151–156.

CITES (Convention on International Trade in Endangered Species of Wild Fauna and Flora). 1992. Appendices I and II. Adopted by the Conference of the Parties, valid as of 11 June 1992.

Cohen A. 1997. Sturgeon poaching and black market caviar: a case study. Env. Biol. Fish. (this volume).

De Meulenaer, T. & C. Raymakers. 1996. Sturgeons of the Caspian Sea and the international trade in caviar. TRAFFIC International, Cambridge. 71 pp.

DeSalle, R. & V.J. Birstein. 1996. PCR identification of black caviar. Nature 381: 197–198.

Dumont, H. 1995. Ecocide in the Caspian Sea. Nature 377: 673–674.

Elvira, B. & A. Almodovar. 1993. Notice about the survival of sturgeon (*Acipenser sturio* L., 1758) in the Guadalquivir estuary (S. W. Spain). Arch. Hydrobiol. 129: 253–255.

Elvira, B. & J. Gessner. 1996. The society to save the sturgeon *Acipenser sturio*. Sturgeon Quart. 4(1/2): 7.

Graham, K. 1997. Contemporary status of the North American paddlefish, *Polyodon spathula*. Env. Biol. Fish. (this volume).

Grande, L. & W.E. Bemis. 1991. Osteology and phylogenetic relationships of fossil and recent paddlefishes (Polyodontidae) with comments on the interrelationships of Acipenseriformes. J. Vert. Paleo. 11, supplement 1: 1–121.

Gringevsky, N.V. 1995. Evaluation of the contemporary status of sturgeons in the Dnieper River basin. Sturgeon Quart. 3(1): 9.

Guti, G. 1995. Conservation status of fishes in Hungary. Opusc. Zool. Budapest 27/28: 153–158.

Hensel, K. & J. Holčík. 1997. Past and current status of sturgeons in the upper and middle Danube River. Env. Biol. Fish. (this volume).

Holčík, J., R. Kinzelbach, L.I. Sokolov & V.P. Vasilev. 1989. *Acipenser sturio* Linnaeus, 1758. pp. 366-394. *In:* J. Holčík (ed.) The Freshwater Fishes of Europe, Vol. 1, Pt. II. General Introduction to Fishes, Acipenseriformes, AULA-Verlag, Wiesbaden.

IUCN Red List Categories. 1994. IUCN, Gland. 21 pp.

Khodorevskaya, R.P., G.F. Dovgopol, O.L. Zhuravleva & A.D. Vlasenko. 1997. Present status of commercial stocks of sturgeons in the Caspian Sea basin. Env. Biol. Fish. (this volume).

Krykhtin, M.L. & V.G. Svirskii. 1997. Endemic sturgeons of the Amur River: kaluga, *Huso dauricus*, and Amur sturgeon, *Acipenser schrenckii*. Env. Biol. Fish. (this volume).

Kynard, B. 1997. Life history, latitudinal patterns, and status of the shortnose sturgeon, *Acipenser brevirostrum*. Env. Biol. Fish. (this volume).

Lelek, A. 1987. The freshwater fishes of Europe, Vol. 9, Threatened fishes of Europe. AULA-Verlag, Wiesbaden.

Manci, W. E. 1993. Sturgeons. pp. 1160–1168. *In:* G. Lee (ed.) Endangered Wildlife of the World, Vol. 9, Marshall Cavendish Co., New York.

Mayden, R.L. & B.R. Kuhajda. 1996. Systematics, taxonomy, and conservation status of the endangered Alabama sturgeon, *Scaphirhynchus suttkusi* Williams and Clemmer (Actinopterygii, Acipenseridae). Copeia 1996: 241–273.

Moyle, P.B. 1994. The decline of anadromous fishes in California. Cons. Biol. 8: 869–870.

Moyle, P.B., P.J. Foley & R.M. Yoshiyama. 1994. Status and biology of the green sturgeon, *Acipenser medirostris*. Sturgeon Quart. 2(1): 7.

Pavlov, D.S., Yu. S. Reshetnikov, M.I. Shatunovsky & N.I. Shilin. 1985. Rare and endangered fish species of the USSR and principles of their inclusion in the Red Data Book. Voprosy Ikhtiologii 25: 16–25 (in Russian, English translation: J. Ichthyol. 25: 88–99).

Pavlov, D.S, K.A. Savvaitova, L.I. Sokolov & S.S. Alekseev. 1994. Rare and endangered animals. Fishes. Vysshaya Shkola, Moscow. 334 pp. (in Russian).

RSFSR [Russian Federation] Red Data Book. 1983. Animals. Rosselkhozizdat, Moscow (in Russian).

Ruban, G. I. 1996. The Siberian sturgeon, *Acipenser baerii baerii*, population status in the Ob River. Sturgeon Quart. 4(1–2): 8–10.

Ruban, G.I. 1997. Species structure, contemporary distribution and status of the Siberian sturgeon, *Acipenser baerii*. Env. Biol. Fish. (this volume).

Sagers, M.J. 1994. The oil industry in the southern-tier former Soviet republics. Post-Soviet Geography 35: 267–298.

Smith, T.I.J. & J.P. Clugston. 1997. Status and management of Atlantic sturgeon, *Acipenser oxyrinchus*, in North America. Env. Biol. Fish. (this volume).

Sokolov, L.I. & V.P. Vasilev. 1989. *Acipenser nudiventris* Lovetsky, 1828. pp. 206-226. *In:* J. Holčík (ed.) The Freshwater Fishes of Europe, Vol. 1, Pt. II. General Introduction to Fishes, Acipenseriformes, AULA-Verlag, Wiesbaden.

U.S. Department of the Interior. 1973. Threatened wildlife of the United States. U.S. Bureau of Sport Fisheries and Wildlife. Resource Publication # 114, Washington, D.C.

U.S. Fish and Wildlife Service. 1967. Endangered and threatened wildlife and plants; determination of endangered status for the pallid sturgeon. Federal Register (March 11) 32: 4001.

U.S. Fish and Wildlife Service. 1990a. Endangered and threatened wildlife and plants; determination of endangered status for the pallid sturgeon. Federal Register (September 6) 55: 36641.

U.S. Fish and Wildlife Service. 1990b. Endangered and threatened wildlife and plants; determination of threatened status for the Gulf sturgeon. Federal Register (September 30) 56 (189): 49653–49658.

U.S. Fish and Wildlife Service. 1994. Endangered and threatened wildlife and plants; determination of threatened status for the Kootenai River population of the white sturgeon. Federal Register (September 6) 59 (171): 45989–46002.

USSR Red Data Book. 1984. 2nd ed. Vol. 1. Lesnaya Promyshlennost, Moscow. 390 pp. (in Russian).

Vladykov, V. & J.R. Greeley. 1963. Order Acipenseroidei. pp. 24-60. *In:* H.B. Bigelow & W.C. Schroeder (ed.) Fishes of the Western North Atlantic, Sears Foundation for Marine Research, Yale University, New Haven.

Volovik, S.P., V.G. Dubinina & A.D. Semenov. 1993. Hydrobiology and dynamics of fisheries in the Sea of Azov. pp. 1–58. *In*: Studies and Reviews, Federal Council for the Mediterranean, no. 64, FAO, Rome.

Volz, J. & S.J. DeGroot. 1992. Erster Nachweis des Störs (*Acipenser sturio*) im niederlandischen Rhein seit 40 Jahren. Fischkologie 6: 3–6.

Waldman, J.R. 1995. Sturgeon and paddlefishes: a convergence of biology, politics and greed. Fisheries 20: 20–21, 49.

Wei, Q., F. Ke, J. Zhang, P. Zhuang, J. Luo, R. Zhou & W. Yang. 1997. Biology, fisheries, and conservation of sturgeons and paddlefish in China. Env. Biol. Fish. (this volume).

Williams, J.E. & G.H. Clemmer. 1991. *Scaphirhynchus suttkusi*, a new sturgeon (Pisces: Acipenseridae) from the Mobile Basin of Alabama and Mississippi. Bull. Alabama Mus. Nat. Hist. 10: 17–31.

Williams, J.E., E.J. Johnson, D.A. Hendrickson & S. Contreras-Balderas. 1989. Fishes of North America endangered, threatened, or of special concern: 1989. Fisheries 14: 2–20.

Williot, P., E. Rochard, G. Castelnaud, T. Rouault, R. Brun, M. Lepage & P. Elie. 1997. Biological characteristics of the European Atlantic sturgeon, *Acipenser sturio*, as the basis for a restoration program in France. Env. Biol. Fish. (this volume).

Wirgin, I.I., J.E. Stabile & J.R. Waldman. 1997. Molecular analysis in the conservation of sturgeons and paddlefish. Env. Biol. Fish. (this volume).

Zholdasova, I. 1997. Sturgeons and the Aral Sea ecological catastrophe. Env. Biol. Fish. (this volume).

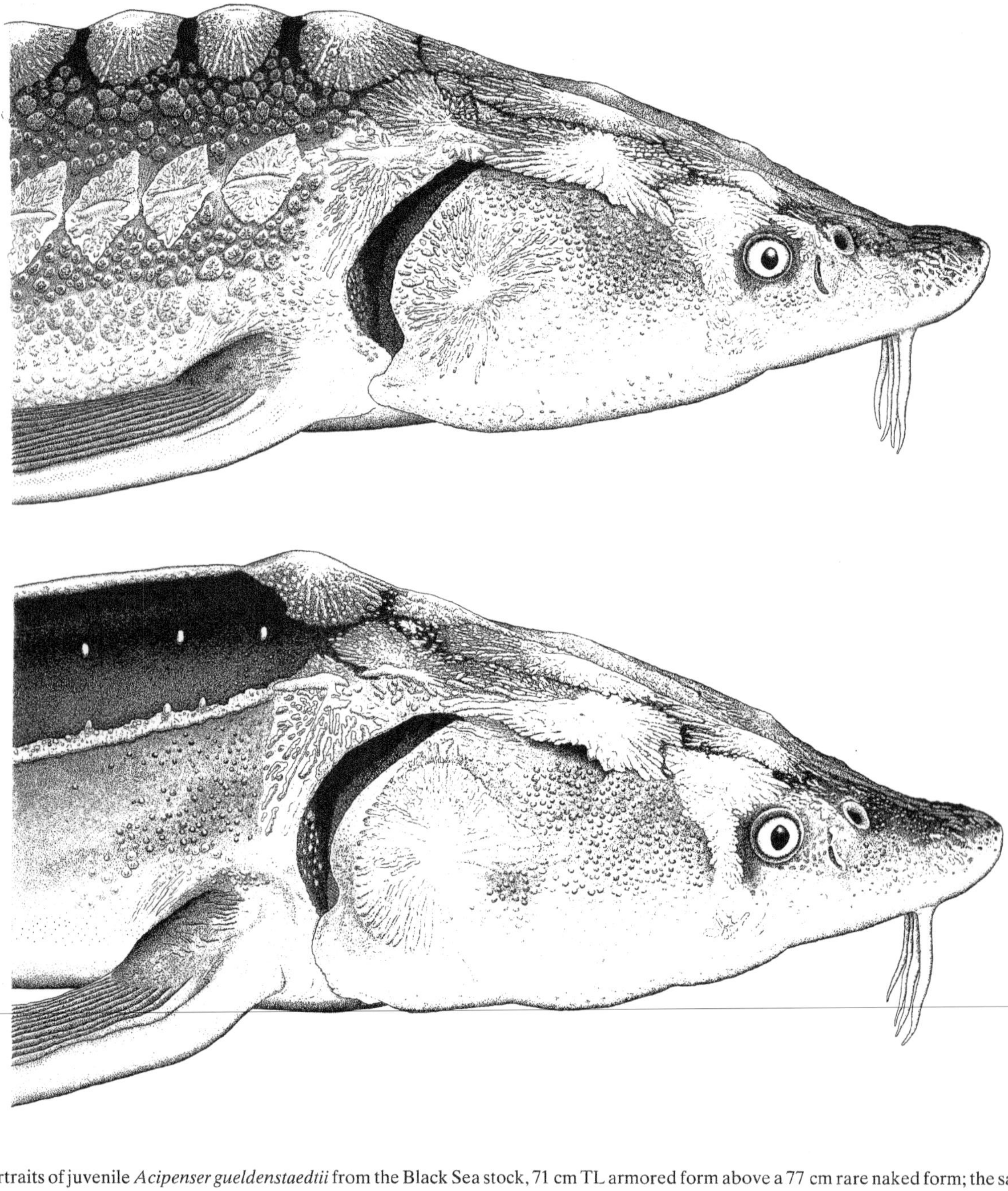

Portraits of juvenile *Acipenser gueldenstaedtii* from the Black Sea stock, 71 cm TL armored form above a 77 cm rare naked form; the same individuals as on the frontispiece photographs. Originals by Paul Vecsei, 1996.

Species and subject index*

* Prepared by Alice G. Klingener, University of Massachusetts, Amherst, MA 01003, U.S.A.

GPSR Compliance

The European Union's (EU) General Product Safety Regulation (GPSR) is a set of rules that requires consumer products to be safe and our obligations to ensure this.

If you have any concerns about our products, you can contact us on ProductSafety@springernature.com

In case Publisher is established outside the EU, the EU authorized representative is:

Springer Nature Customer Service Center GmbH
Europaplatz 3
69115 Heidelberg, Germany

Batch number: 10800072

Printed by Printforce, the Netherlands